AF675619

PETERSON'S

JOB OPPORTUNITIES

FOR

ENGINEERING and COMPUTER SCIENCE MAJORS

1999

PETERSON'S
Princeton, New Jersey

About Peterson's

Peterson's is the country's largest educational information/communications company, providing the academic, consumer, and professional communities with books, software, and online services in support of lifelong education access and career choice. Well-known references include Peterson's annual guides to private schools, summer programs, colleges and universities, graduate and professional programs, financial aid, international study, adult learning, and career guidance. Peterson's Web site at petersons.com is the only comprehensive—and most heavily traveled—education resource on the Internet. The site carries all of Peterson's fully searchable major databases and includes financial aid sources, test-prep help, job postings, direct inquiry and application features, and specially created Virtual Campuses for every accredited academic institution and summer program in the U.S. and Canada that offers in-depth narratives, announcements, and multimedia features.

Visit Peterson's Education Center on the Internet (World Wide Web) at www.petersons.com

To order additional copies of these books, call Peterson's Customer Service at 800-338-3282.

Editorial inquiries concerning this book should be addressed to the editor at Peterson's, P.O. Box 2123, Princeton, New Jersey 08543-2123.

ISSN 1094-2300
ISBN 0-7689-0027-1

Printed in the United States of America

10 9 8 7 6 5 4 3 2 1

CONTENTS

HOW TO USE THIS BOOK

Peterson's Job Opportunities for Engineering and Computer Science Majors 1999 provides job hunters with vital information on more than 2,000 of America's major employers. With this information, readers can identify and research companies that are seeking their skills and, by surveying the opportunities presented, build a picture of the various career alternatives possible at major businesses.

About Employer Profiles

This guide is not a definitive listing of America's largest organizations. If you do not find a particular company within the pages of this guide, it doesn't necessarily mean that the company has no employment opportunities. Similarly, a listing in the guide does not mean that the organization has immediate openings in the field of expertise mentioned. It simply means that the organization, by virtue of its business, typically seeks employees with the background cited.

Companies and other organizations profiled in this guide were selected based on the number of people they employ and were then contacted for specific information on the type of employees sought and appropriate contacts for employment inquiries. Employers were also asked to provide a contact for recent college graduates, if college recruiting is handled separately from overall employment recruiting within the organization. In addition to the above, profiles contain a business description; annual sales and number of employees (last reported publicly available figures as of December 1997); and in most instances a founding date. Founding dates generally refer to the year in which the organization was first established. However, they may in some cases indicate when the organization experienced the most recent major change resulting in its current corporate structure. Profiles also contain the number of professional and managerial hires, if the company surveyed was able to provide this information.

About the Data

The information for the company profiles in *Peterson's Job Opportunities for Engineering and Computer Science Majors 1999* was collected from February through May 1998 through telephone interviews with representatives of each company. Although every company's information has been updated, the reader should be aware that the information provided by these companies may have changed since the book was published. Because of the extensive system of checks performed on the data that appear in the company profiles, we believe that the information presented in this guide is accurate. Nonetheless, errors and omissions are possible in a data collection and processing endeavor of this scope. It is always a good idea to contact the company directly, using the names and telephone numbers provided in the company profile, to obtain the company's most current information.

Peterson's does not make any claim concerning the hiring policies or practices of the companies profiled in this book.

About Employer Descriptions

Some organizations have chosen to augment their profile with narrative information about their activities and job opportunities. In these cases, a short description follows the organizational profile. Longer descriptions, in which organizations provide detailed information about their businesses, future employment opportunities, quality of work life, and application information, appear in a separate section following the Employer Profiles.

In addition to the information about specific organizations, this book also includes the following valuable features:

Workplace 2000: Engineering

Read how changes in corporate America are affecting today's employment picture. The skills that companies are looking for in employees today and what makes an employee valuable have changed. As a job seeker, are you prepared to take the initiative in your career? Do you thrive on constant change and ambiguity? Can you adjust your sights toward a mobile career path in a global marketplace? Are you willing to keep expanding your knowledge and skills, both technical and interpersonal, to keep yourself employable? Find out what to expect in today's and tomorrow's workplace.

The Internet and the World Wide Web—Promises and Pitfalls

It's a new technology—surfing the "Net" to find a job. This article is packed with helpful hints on where to look, what to beware of, and how to get your resume in a Web site that employers will see.

Researching, Getting, and Keeping Your Ideal Job

Reading the want-ads, keeping your resume neat and updated, and hoping an employer will call you isn't the way to a top job. Today's job market is tough and highly competitive. Read this section for tips on how to tackle the job of landing a job and keeping it.

The Most Desired Job Skill in the World: Communication

What is the number one skill that employers look for when hiring new employees? The answer may surprise you. This article tells you what that skill is and how to develop it and use it to your advantage in getting and keeping your desired job.

WORKPLACE 2000:

ENGINEERING

by Tom Ferrell

Most time travelers would consider a ten-year jaunt to be hardly worth the effort. But consider the changes that engineers and the engineering disciplines have experienced in the past decade. Industries that once dominated the engineering rolls—the military-driven aerospace industry, for example—have given way to smaller, more streamlined organizations that bring high technology into every aspect of our daily lives. While budget cutbacks and changing personnel philosophies within government organizations and large companies have created an environment in which traditional stability is difficult to find, these same factors have spawned unprecedented entrepreneurial opportunity.

What does this mean for you? It means a new way of looking at things. Engineers are no longer asked to be specialty players in the corporate organization. Today, engineers not only design new products, they often manage the development process and, in many cases, market and sell the finished product. Today's successful engineer is the corporate renaissance person, moving easily across the same lines that once hindered career advancement and diversification.

"We're riding a boom wave," proclaims Manny Contomonolis, director of career services at Rochester Institute of Technology in Rochester, New York. "The outlook is very solid for all disciplines, and I expect the trend to continue. We've reached the point of applying and integrating technology on a broad scale. It's not just theoretical anymore."

The high-tech society is upon us, and engineers are at the forefront. In this decade they've already delivered one of the most dramatic evolutions in communication that society has ever seen: the Internet. But the Internet is just a starting point. Entire industries are reinventing themselves. Television is poised on the brink of a major change as the industry moves toward High Definition Television (HDTV). Telecommunications advances are tearing down the traditional barriers of geography and redefining our concept of the workplace. Of course, engineering figures into our lives in many less noticeable (though no less important) ways. Engineers have made our cars safer. They've lessened the environmental impact of manufacturing. Virtually every vision of the future requires engineers to make it real. And those who can re-engineer their own ideas of employment and career success have a bright road to follow.

THE NEW ENGINEER

The new engineer has to know more than numbers. Employers are looking beyond traditional skill sets, and in return they offer more than traditional opportunities. These employers may be smaller companies, rather than the sprawling corporations that once formed the bedrock of the engineering career economy. Regardless of size, however, they are almost always companies dealing in fast-paced markets and developing technologies that are in constant evolution.

Compatible Systems Corporation of Boulder, Colorado, thrives on change. With slightly more than 50 employees, Compatible Systems is a small company by most standards, but it competes in a big market—the network communications industry—and it relies on engineering innovation in order to fuel 80 percent annual growth rates. Matt McConnell, president of Compatible Systems Corporation, encourages engineering students to "think outside the box" and find the ways that they can contribute not only to technical advances but also to the business mission of their employers.

"The lines between engineering development and execution of the corporate business plan are more blurred than ever before,"

McConnell observes. "Engineers can't just work out of view. They need to have a concept of their industry, their competitors, and the way their technology is perceived in the marketplace. Particularly in areas where standards-based products are the norm, one element of insight or added value can make a huge difference in the success or failure of a technology."

Jack Wilson has seen it all in his work with Career Sciences, a placement firm that helps both new and experienced engineers. He believes that the heads-up engineer is always looking for that "something extra" that will separate him or her during the job search process.

"There is an overwhelming belief that if a student can just get a good grade point average, or if a career engineer simply has experience with well-known companies, then it proves by definition an adequate level of technical skill. The mistake they make is that they think technical skills are the only thing they need to get a good job. That's just not true anymore."

Companies use engineering personnel on product development teams, in customer support positions, and often for sales training and education. And successful businesses today aim to stretch their hiring dollars, which means finding candidates with broad skill sets and particularly strong interpersonal skills—traits that were not nearly as important to engineering positions in earlier days.

Colleges are responding to the demand for more well-rounded engineers. RIT's Contomonolis preaches to his students that technical competency is only the starting point.

"Your technical skill set is necessary but no longer sufficient," he says. "Today you can't win based only on knowledge and past achievements. You have to market your potential and your ability to be flexible, to learn, and to work in an unstructured and changing environment. It's impossible to predict even two to three years into the engineering future. The successful engineer today is the one who can get into the position of having choices."

For those students who can make the transition and demonstrate added value and skills, the engineering job market continues to be strong, with the promise of sustained growth in most areas for the next ten years.

A LOOK AT THE NUMBERS

According to the National Bureau of Labor Statistics (BLS) *Occupational Outlook Handbook,* the number of people holding engineering jobs continues to rise. The total number of private engineering jobs is split fairly evenly between manufacturing and non-manufacturing industries, with most of the recent growth occurring on the non-manufacturing side. An additional 190,000 engineers clock in for federal, state, and local government jobs.

The BLS expects employment opportunities in engineering to expand through the year 2005. Though projections for some engineering disciplines—aerospace, for example—are not as strong as others, these setbacks do not affect the overall picture. Higher-than-average job-growth figures, combined with a slow but steady decline in the number of bachelor's degrees being awarded in engineering, ensure that job seekers will continue to find opportunity in the marketplace. Because the engineering job market is in constant evolution, specialties in demand within the various engineering disciplines are often relatively new. The engineers who are working in these areas must rely on skills gained through on-the-job training, particularly for "hybrid" engineering jobs that straddle disciplines.

Another factor to consider—and a good reason for maintaining a broad scope during your job search—is that a company or industry that is hiring large numbers of one type of engineer will often need smaller numbers of another. The personal computer industry, for example, is most often viewed as an area for electrical engineers, but the industry also offers significant opportunities for mechanical engineers to work on product packaging and the design of peripheral devices such as tape and disk drives.

For the most part, those seeking work in all of the traditional disciplines—except for the aerospace industry and nuclear engineering, which rely heavily on the federal government—have cause to feel optimistic. Some of the hottest areas are described below.

Civil Engineering

Few fields in any employment area have more jobs than qualified candidates. Yet the BLS predicts this will be the case for civil engineers at least through the turn of the century. During that time, civil engineers will enjoy the highest demand for jobs of all the best-known engineering disciplines. A key element in this clamor for civil engineers is the fact that America's infrastructure—highways, transportation systems, and urban areas—largely came into being in the boom times following World Wars I and II. Age and population growth have taken their tolls, and virtually every city is pouring resources into its infrastructure.

In addition, the effects of fifty years of industrial-age growth and expansion have created a need for a new type of civil engineer—the environmental engineer. Required compliance with EPA regulations, combined with a rise in construction during this period of economic recovery, should create continuing opportunities in the environmental fields of civil engineering for many years to come. The size of the market opportunity is attracting more traditional civil engineering firms and leading to increased hiring.

Finally, the easing of trade barriers between the U.S. and its neighbors has created a unified front in terms of environmental regulations. Both the General Agreement on Trade and Tariffs (GATT) and the North American Free Trade Agreement (NAFTA) mandate that treaty participants observe certain environmental standards. As a result, foreign companies are now required to adhere to the same environmental regulations as U.S. companies, leveling the playing field and making U.S.-based engineering firms more competitive in a number of areas.

Electrical Engineering

Another field with strong growth potential—here again characterized by some nontraditional tendencies in terms of what the jobs entail—is electrical engineering. Once the realm of "geeks" and "nerds," electrical engineering is moving more into the mainstream because of communications technologies such as the Internet and also because of the proliferation of smaller, contract-based firms. This means that today's electrical engineering candidate is more likely to be well-versed in real-world skills, including writing and team-building, than any previous generation of electrical engineers. Both college engineering programs and corporate training programs reflect this trend. The new electrical engineer has come out of the wiring closet, so to speak.

While electrical engineering has been especially vulnerable to aerospace and other defense industry cutbacks, many electrical engineers are finding themselves in demand for more general, or mainstream, positions such as technical sales, education, and product management.

"It's critical to diversify," says Chuck Jordan, an electrical engineer and partner in Atlanta-based Brewer and Scala Engineering. "It's especially important for career advancement. The people who do well have become strong technically. The ones who really separate themselves from the pack are the ones who can get out and sell. We're juggling hundreds of clients. Project management and customer relations are huge parts of what we do. You have to be able to communicate."

Software Engineering

In the corporate world, programmers often work with systems analysts, writing software that allows the computer systems to achieve the goals of management. Programming can be an excellent vehicle for later movement into systems analyst or other management positions.

With advanced, graphical-based languages like C++ at their disposal, programmers are capable of dazzling achievements without much of the torturous coding their predecessors had to endure. These advances lead many industry experts to pick programming as the brightest spot of all in the computing industry.

In addition, new server-based programming languages, such as Java, are opening new doors for software authors. Skilled programmers with degrees and relevant experience will have little trouble finding suitable employment. Though job growth is not expected to match the rate of other computing professions, a good programmer will always be worth his or her weight in gold.

Electronics Engineers

The electronics engineering disciplines almost always benefit from the rapid pace of change in technology. Automation of factory and office processes creates demand for new systems development.

"Most chip vendors are shifting people from desktop PC development," notes Loring Wirbel of *Electronics Engineering Times* magazine. "Some areas of the field have been gutted, but for the most part we're defining slow growth as low double-digits. Wireless development is huge, so RF (radio frequency) is a good area to pursue. And keep an eye on vertical markets such as health care and transportation. There are so many good opportunities."

Many of those opportunities have more to do with information signaling than with the more traditional movement of electricity. The availability of new information processing technology has allowed for new automation applications that simply were not possible just a few years ago.

"Traditional electronics engineering is giving way to information technology," says Dudley Kay of the IEEE (Institute of Electrical and Electronics Engineers). "Internet, wireless, satellite—that's the industry right now and that's where the research money is going. MEMs (microelectromechanical systems) technology is in high demand. These devices gather information and effect an action. You've simply got to think about what you can do with the information."

Industrial/Mechanical Engineering

Technology continues to drive opportunity, even in the traditionally "low-tech" disciplines of industrial and manufacturing technology. When the recession of the early 1990s caused corporations to adopt a "do more with less" philosophy, industrial and mechanical engineering became a touchstone for increasing overall efficiency within the corporation.

Emerging quality standards, such as Europe's ISO 9000 specifications (ISO stands for the International Organization for Standardization) and employee participation programs like Total Quality Management (TQM), have broadened horizons for industrial and mechanical engineers. In addition, new applications of communications and software technology are requiring new housings, new designs, and new ways of thinking. The mechanical design of a product has become a crucial element of the overall positioning and marketing of the technology. And as the trend of integrating high technology into more mainstream products continues, the future for mechanical and industrial engineers brightens even more.

NEW OPPORTUNITIES

One piece of advice when you're starting your search: Don't look only to the traditional sources of engineering jobs. Just as job seekers today must demonstrate a broad skill set, they must also apply a broad vision to the job search process in terms of the types of employers to consider and approach. The industrial and manufacturing machines that gobbled up engineers throughout the past 30 years have found their appetite sated.

The BLS predicts very low job growth prospects in a number of engineering fields once considered standard. For instance, nuclear engineering and petroleum engineering—disciplines in high demand for the past 30 years—are showing flat and sometimes even negative job growth. It's a trend that shows up again and again in the Fortune 1000.

What is picking up the slack? According to a number of career counselors, it's the software industry. Take the example of Microsoft Corporation. For the position of software engineer, Microsoft will accept an application from anyone with a degree or work experience in any engineering discipline. With new software markets constantly opening up as a result of advances in conventional engineering areas, such as aerospace and civil engineering, and newer areas, such as environmental engineering and communications, job candidates who combine a knowledge of the discipline with an ability to write application software find themselves in great demand.

Another area of rapidly growing opportunity for engineers is large-scale consulting. Andersen Consulting is the largest information technology contracting firm in the world. A spokesperson for Andersen notes that the company is interested in applicants who have demonstrated creativity, discipline, and the application of technical principles during their college and work careers. There is room for both new graduates and more experienced engineers re-entering the workforce.

Alan Waksman, a principle at the engineering consulting firm Applied Concepts and a spokesman for the National Association of Computer Consulting Businesses, paints a bright future for engineers interested in consulting careers. "What's in demand? Well, what isn't?" Waksman asks. "There's obviously a place for newcomers hungry to gain experience, but there's plenty of room for middle-aged candidates as well. Legacy systems, for example. Today's students often don't know much about computer systems in the pre-network days. A more experienced candidate who can help companies mine that data from the old mainframe system is valuable for sure."

"The trend toward consulting positions is a mixed blessing for many students considering engineering careers," notes RIT's Contomonolis. But it also contributes to a broader scale of opportunity. "We're still working with students who envision a single-employer paradigm. While some of those jobs may still exist, we're seeing a move toward smaller companies and more transient jobs. The real opportunities are for the young people willing to take risks. And, those risks often come in the form of working with consulting companies."

Consulting opportunities are now cropping up even at established "name" engineering players. At IBM, for instance, 40 percent of revenue now comes from consulting. The service sector is a leading high-tech force, and their need for engineering firepower should continue to grow as advanced communication, design, and manufacturing technologies seep into the commercial mainstream.

THE "BACKYARD" APPROACH

If you were to take a quick glance at the engineering job marketplace and rank it based on Fortune 100 recruiting, you might find some surprising results. Fortune 100 recruiting is down nationwide, sometimes as much as 75 percent. Does this mean that companies aren't hiring? Not necessarily. But after the economic upheaval of the past several years, many companies are opting to stay in their own back yard.

What this means to the job seeker is that the market is much more regional than it used to be. As a result, the engineering job opportunity landscape varies dramatically according to geographic location.

"Geographical pockets of activity are everywhere," says Brewer and Skala's Jordan. "In the Southeast, for instance, you can't find enough talented people. The growth is mind-boggling, and the salaries that talented young engineers can command is almost crazy."

Job seekers in any discipline should consider geographical economies. Mechanical engineers, for instance, will likely find slow going in the job search process in areas such as southern California, where there is little heavy manufacturing industry. However, mechanical engineering students at Purdue, located in the manufacturing-rich Midwest, may find a number of excellent opportunities.

ENGINEERING EARNINGS

According to the Bureau of Labor Statistics, engineering employees as a whole enjoy high earnings compared to other vocational disciplines. Below are current annual salaries for engineers as reported by the BLS.

Median Salaries, Individual Fields	
Aerospace	$50,200
Chemical	$53,100
Civil	$44,700
Electrical	$48,000
Industrial	$40,900
Mechanical	$46,400
Average Salaries, All Fields	
Entry-level positions, starting salary	$33,900
Mid-level engineer, no supervisory responsibility	$54,400
Senior engineer, managerial responsibility	$87,048

In the end, if you are set on following a certain engineering discipline, you should be aware of where the demand is greatest for your skills. At the very least you may need to plan your own trip to interview with specific employers, and it may even be necessary to relocate prior to landing a job.

PUTTING TECHNOLOGY TO WORK

Today's engineering job seekers have the chance to enhance their prospects, contacts, and overall knowledge of their field as members of a worldwide online community—the Internet. This granddaddy of all computer networks is accessible from virtually every personal computer using simple and inexpensive communications hardware and software. Used properly, the Internet can be a valuable and vital part of a job-seeker's search. As an engineering or computer science major, there may be no greater way to promote yourself as a job candidate than on the World Wide Web.

Using e-mail, candidates can maintain instantaneous contact with colleagues, potential employers, and engineering organizations and societies. Through any number of UsENET bulletin board discussion groups, engineers can share experiences and stay abreast of openings and other developments within their specific field of interest.

You can add online job banks and employment agencies to the list of recent network phenomena. Some online employment services

ADDITIONAL RESOURCES

It's a good idea to contact (and to join) professional organizations in the field, many of which offer career assistance and sponsor conferences and training programs. The following list includes two organizations that represent various engineering disciplines (ASEE and NSPE) as well as several that represent specific disciplines.

American Society for Engineering Education (ASEE)
1818 N Street NW, Suite 600
Washington, D.C. 20036
202-331-3500
http://www.ns.asee.org

National Society of Professional Engineers (NSPE)
1420 King Street
Alexandria, Virginia 22314–2794
703-684-2800
http://www.nspe.org

American Institute of Chemical Engineers
345 East 47th Street
New York, New York 10017-2395
212-705-7338 or
800-242-4363 (toll-free)
http://www.aiche.org

American Chemical Society: Career Services
1155 16th Street NW
Washington, D.C. 20036
202-872-4600
http://www.acs.org

American Society of Civil Engineers
1801 Alexander Bell Drive
Reston, Virginia 20191-4400
703-295-6300 or
800-548-2723 (toll-free)
http://www.asce.org

Institute of Electrical and Electronics Engineers
Operations Center
445 Hoes Lane
Piscataway, New Jersey 08855-1331
732-981-0060
http://www.ieee.org

Institute of Industrial Engineers, Inc.
25 Technology Park/Atlanta
Norcross, Georgia 30092
770-449-0461
http://www.iienet.org

The American Society of Mechanical Engineers
345 East 47th Street
New York, New York 10017
212-705-7158
http://www.asme.org

provide a database of jobs and allow prospective employees to search the databases for free. Others operate on a traditional fee-based employment agency model. (For more information on how to use the Web to your advantage, see page 10).

Networking technology is also playing a greater role in new jobs themselves. Many consultants use high-speed ISDN (Integrated Switched Digital Network) to deliver products to customers located hundreds or thousands of miles away. Companies use video conferencing to save on travel costs and to create "virtual teams," in which team members can be located in various spots around the world.

Communications technology is one of the most influential developments in recent years, and engineers are once more at the heart of packaging the technology for public use. Any professional who has a command over the capabilities and applications of networks immediately adds value to an organization.

WHERE DOES IT GO FROM HERE?

The near-term employment picture for the U.S. is rosy. Unemployment is at an all-time low, and small businesses born of large-industry downsizing are beginning to create jobs in a number of technological fields. Engineering job growth may not fully reflect the robust gains in overall employment simply because the recovery cycle dictates that employers first raise production capacity to handle increased demand, and then gradually increase spending on new product research and development.

Regardless of short-term outlooks, however, forecasts from the BLS indicate that growth for engineers in nearly all disciplines should continue to grow. Civil engineering holds the most promise, with its pleasant anomaly of more openings than candidates. Electrical and electronics areas also appear to be strong, with a solid equilibrium between jobs and candidates. Mechanical engineers will still find stiff competition in the job market, though the BLS notes improvement over the painfully low ratio of jobs to new hires that prevailed a decade ago.

And don't forget the "emerging" engineering disciplines. The BLS projects that some of the best opportunities for engineering jobs will come in less traditional disciplines, where competition is lower, the demand higher, and corporate cultures still focus on innovation. These areas include software engineering and industrial and manufacturing engineering.

A PARTING WORD

The best advice for surviving—and thriving—in the new engineering workplace is to keep your ear to the ground. Watch for developing technologies and be innovative in their applications. Pay attention to business conditions and keep an eye on companies that have a knack for satisfying the demands of their technical customers. It's a fact now that engineers will change jobs more often than in past generations, so make job awareness a part of your overall skill set. When you get the opportunity to participate in training or educational programs, grab it.

"What we're going through right now is a change that is second only to the industrial revolution," says Career Sciences' Jack Wilson. "The notion of going to school for four years, graduating, and getting a lifetime job with one company is going to be very, very rare."

Today's engineering professional is not just a role player anymore. Stay in the middle of the action, and your career will take care of itself.

Tom Ferrell is a business communications consultant and writer based in Boulder, Colorado.

THE INTERNET AND THE WORLD WIDE WEB—

PROMISES AND PITFALLS

by Cristopher J. Maloney

Just when you thought you were doing everything you needed to do to find a wonderful position with the perfect company, someone asks you if you've put your resume on the World Wide Web. Whether you answered "yes," "not yet," or "world-wide-what?," you may be surprised that there is a lot to learn even for seasoned Web wizards. Complete novices will be happy to know that understanding what a job hunter needs to know about the Web, and its big brother the Internet, is easier than many buzzword-using, keyboard-tapping, computer technocrats might have inadvertently led you to believe. In this article, I'll be discussing the opportunities and risks provided by the Web for job hunters, including what to do and what not to do on the Web with, for example, your resume.

It is impossible to describe in detail exactly what each reader of this book will see on his or her computer when using the Internet. Big and small companies provide consumers with dozens of different ways to get connected to the Internet and the Web (America Online, AT&T, and EarthLink come to mind as just a few of the popular players), and many of these access providers allow their subscribers to use different brands of software within their systems. This article will provide you with insight about the Web and give the wary an understanding of how it all works. I will assume that your access provider and the software you select will make the various steps mentioned obvious once you become a "Web user." In fact, if you are using the Web and you can't figure it out, you have good reason to switch providers or switch software. Using the Internet should be EASY!

If you don't know about the Internet, newsgroups, or what a World Wide Web page is, here is a quick tutorial.

The Internet

Are you one of those people who feels that there must be a rule about knowing how the Internet works before you're allowed to use it?

The only thing most people need to know about the Internet is that it is similar to the worldwide network that connects everyone's telephone. If you pay to have access, you can call anyone in the world. In the United States that means you need to buy or lease a phone and pay a local access company a monthly fee to connect you to their regional phone network, which in turn is connected to the network owned by the long-distance provider of your choice. Long-distance companies connect you to the worldwide network of telephone systems. Of course, you pay the local access company for what they consider to be long-distance calls within their region, and you pay your long-distance service provider for calls that must be carried on their network to phones that are far away. In the end this means that you can pick up your phone and call someone on the other side of the planet—even if you don't have a clue about how the telephone system works.

On the Internet side of the fence, it is actually a little simpler—you need to buy or lease a computer, install some easy-to-use software, plug your computer's modem into your phone jack, pay to have calls you make with your computer connected to the Internet (typically a monthly fee like basic phone service) and, voilá , you can pick up your mouse and point-n-click your way to another computer connected to the Internet that might be down the street or 5,000 miles away. Alternately, you could purchase a product like WebTV or WebTV PLUS. It has integrated Web browsing software, a modem, and a hand-held pointer, like a remote control for a TV set, for making and submitting selections on Web pages (more about that later). However, it has some significant limitations. For example, you'll need a keyboard if you want to type your resume into a company's Web-based resume service. But you can't store your resume on it like you can with a computer, so you have to retype your resume every time you find someplace you want to submit it. For the purposes of this article, I'll assume you are using a computer and not a product like WebTV.

Skip the next paragraph unless you want to read a lay explanation of how the Internet works.

Hook a few computers together (you don't have to know how) so that all the computers can send and receive messages (i.e., e-mail), share documents, and/or print to a single printer and you've got a small, personal computer network. Establish a common set of diplomatic protocols that all the connected computer networks can agree to use when one network is trying to talk to another network. Now, take all those small computer networks from all over the globe and connect them to each other. The computer networks, the millions of computer users that all the networks now connect, and the services (like newsgroups and World Wide Web pages that computers on these smaller networks provide) are collectively called the Internet.

World Wide Web

The World Wide Web is the portion of the Internet that has caused all the hoopla over the past couple of years. In fact, the World Wide Web has become so popular that it is generally just called the Web and many people incorrectly use "Internet" and "Web" interchangeably.

The Web allows novices and experts alike to more easily and enjoyably navigate the Internet. It also gives more companies and individuals the opportunity to electronically publish (and distribute more widely) their resource and outreach materials. Millions and millions of users now have access to millions of pages on the Web.

A page on the World Wide Web can contain graphics as well as text. Well-designed Web pages are easy for computer novices to navigate and can be interesting and fun to explore.

The person or company that publishes a Web page can build links on the page that allow you, with the click of your computer's mouse button, to connect to any other page on the World Wide Web. For example, you might use your computer to go to a Web page filled with jobs available at many different companies. This aggregator's page typically allows you to do some type of search by job type and geographic region.

Let's say you want to find a job in St. Louis, Missouri. In the results from a search you may read about a job you really like. Clicking on the name of the company in the job description might take you to more details about the employer and from there you might be invited to link to a Web page operated not by the aggregator but by the headquarters of the hiring company. Once you're on the hiring company's Web site, you might find a link that takes you to a Web site operated by the hiring division.

The aggregator's Web site may be operating out of Princeton, New Jersey, and the hiring company's headquarters could be in Tokyo. Yet, the hiring division (and the computer publishing its Web site) might be in St. Louis. In a matter of seconds, you crossed several mountain ranges, an ocean, and the International Date Line (twice) and all you did was click a button a few times. The Web transcends the boundaries that normally restrict us—time, access to huge sums of travel money, and political borders. Of course, as long as the job isn't too far from where you live, you probably don't care how far or how fast your search took you around the globe.

Get Your Mouse in the Door

Beyond the task of finding a company with a job that matches your interests, it is now easy to regularly visit companies in a way not possible before.

Let's face it, if down the road from your house is a company you have always wanted to work for and you showed up every morning asking if maybe today there was a job opening for you—well, at best you wouldn't be arrested for acting crazy but you certainly wouldn't get hired. But if they have a presence on the World Wide Web, you can stop by the company virtually (no pun intended) every day, every hour if you like. And guess what? Companies love it when people visit their Web sites!

It is probably bad form to show up for an interview if you haven't visited and extensively browsed a would-be employer's Web site. Hiring managers and human resource professionals are pleased to learn that you have followed the company in recent newspaper coverage and can demonstrate knowledge that you might have learned by reading about the company in a book such as this one. Just imagine how pleased they would be if you told them how much you learned from their Web site! Company personnel are typically proud of their Web sites and enjoy discussing them.

You might also feel more comfortable in an interview if you've seen a picture of the hiring manager and had a chance to learn a bit about his or her history via an online biography. Depending on what a company has decided to publish, you might be able to find these things at the company's URL.

Web-Speak: What Is a URL?

Here comes another telephone analogy. URL, or Uniform Resource Locator, is the address (or phone number) of a Web page. You type a URL into the Web browser software, hit the "Enter" key and your software silently calls another computer on the Internet. If the other computer is there (if there aren't any busy signals) then the answering computer "looks" at the URL you entered into your software and determines if you can "speak" to that specific page. The computer you are trying to reach has to determine two things: it has see if the page exists and, assuming it does, the computer has to check its programming to see if you are allowed to access the page (Web pages can be password protected).

This is no more complicated than a high school boy calling his new girlfriend only to have her father answer the telephone. The boy asks, "Is Victoria there, please?" The father asks, "Who is calling, please?" The phone at the girlfriend's house is like the computer with the Web pages, Victoria is like the page, and the father is checking his programming to determine whether the caller is granted access.

Of course, instead of answering by saying "Hello," the computer answers by sending instructions to your computer telling it how to display the Web page you requested. On the other hand, Web pages can include audio files that are automatically played by the receiving computer—so don't be surprised if the computer *does* answer by saying "Hello."

Is the Internet or the Web a Job Hunter's Promised Land?

Not long ago, use of the Internet to find a job was pretty much limited to reading postings in newsgroups—arcane, if not extremely popular and valuable, text-based areas of the Internet. Reading through postings on newsgroups was something that, more often than not, only technically oriented people would consider and, not surprisingly, the jobs that could most often be found were typically very technical in nature. In fact, even to this day, the jobs posted in newsgroups tend to be for computer programmers, network administrators, and systems analysts.

So, is job hunting on the Internet a waste of time for sales and marketing professionals, administrative assistants, teachers, and executives? Not at all!

Job postings on the World Wide Web regularly list just about every kind of position that might otherwise be advertised in a local, regional, or national newspaper. So those of you who don't stay up all night learning the latest in programming algorithms will be happy to learn that the jobs you'll find listed on Web pages are not strictly computer related. You are just as likely to find job listings for executive secretary, freelance writer, summer camp counselor, or vice president of sales as you are for computer programmer.

Be Prepared

You are likely to find a place on the Web page where you are given the opportunity to fill in a job application or enter your resume. If you can have your word processing software up and running while you're looking for jobs via the Internet, it may save you time to cut and paste between the resume on your computer and the Web page you're visiting. This can also help protect you against entering embarrassing spelling errors that would have been corrected in your word processor software.

Letting the Internet Work for You

Have you ever thought, "I bet there is a company out there that would just love to have somebody with my skills. If only they could find me." Well, believe it or not, now they can.

A company that you've never heard of can search Web-based collections of resumes and find *you.*

Where You Shouldn't Put Your Resume On Line

More and more people have been adding something like this at the end of their cover letters and e-mails:

My resume is available on the World Wide Web at http://www.hostingservice.com/~mywebpage/.

Putting your resume on the Web in a resume aggregator's system is probably a very good idea. Intermingling it with a personal Web page, however, can be a very bad idea. Too many personal Web pages are much too personal for the tastes of most employers. Not that there is anything wrong with the Web pages themselves or expressing and supporting your beliefs in a strong and forthright fashion, but when was the last time you started an interview by evangelizing your political, sexual, and religious preferences?

I can't tell you how many times I've heard or read that you can never dress too conservatively during the interview process. My point is, if you're going to invite a potential employer to visit your Web page, dress it conservatively.

Is There Just One Place to Remember?

The list of job and career resources on the Web grows every day. Peterson's, the publisher of this and other career books, also publishes The Education & Career Center on the World Wide Web. Its URL is www.petersons.com. You can use this site free of charge to search for a job, get tips for preparing your resume, read the latest articles from career magazines (including *Graduating Engineer, Minority M.B.A.*, and *Computer and Engineering Horizons,*) learn about over 4,000 employers, post your resume (let them find you), e-mail employers, and look for educational opportunities that could improve your readiness for a job or promotion.

It is probably best to think of the Internet and the Web as a library that gives you access to material and services from a wide range of publishers. Evaluate the publishers. Do they have a history of producing accurate and trusted information? Evaluate the services. Does the service help you? You may find more than one that suits your needs. In the end, the one thing that everyone will agree on is that, clearly, the Web is an important tool for people in the job market.

Not every company has a Web page/URL address. (That's why you still need this book to find the companies with the jobs!) This edition of *Peterson's Job Opportunities for Engineering and Computer Science Majors* has the company URL listed as the last element in the profile. Use it as an opportunity to learn even more about a company you may be interested in as a potential employer and see if the company has any openings posted at petersons.com.

RESEARCHING, GETTING, AND KEEPING YOUR IDEAL JOB

How to Track an Industry

Once you have used this book to identify employers you may be interested in, take the time to investigate their industry group. The current financial performance and future prospects of an industry, as a whole, in which you take a job will directly affect your job security, along with your chance for raises and promotions. It is easy to become better informed. Consult with general business magazines such as *Business Week*, *Forbes*, or *Fortune*. They do most of the work for you.

For instance, each year these magazines publish industry-by-industry analyses of the top corporations in the United States. Use these analyses to judge the overall prospects of a company you are interested in, regardless of whether it is actually listed. Look for industries that are rated above average in sales growth, profits as a percentage of sales, and growth in the number of employees. Keep in mind that one or two years of bad results for an industry does not necessarily mean the industry is in permanent decline; some industries are more sensitive to short-term economic conditions than others.

How to Track a Company

Once you have checked the overall performance of a company's industry, you can use the library or your computer to gather more information about the company itself. Perform database searches to look for mention in the press of new products or new contracts, along with comments from industry analysts. You invariably will impress the people who are conducting an interview by informing yourself about their company before you get there. Most interviewees have little or no knowledge of the companies to which they apply. An hour's worth of research time will make you look sharp and prepared and put you in a position to interview *them* while they are interviewing you, *and* you will stand head and shoulders above anyone else applying for the job.

If a company is publicly traded on a stock exchange, you can also take advantage of a good indicator of a company's future performance. Look at the P/E (price to earnings) ratio of companies you are interested in over the last year or so. Compare this ratio with other companies in the same industry. You can find up-to-date figures for most publicly traded companies in the *Wall Street Journal* or *Barron's*. If the P/E of a company in a healthy industry is significantly higher than other similar companies, the stock market's judgment is that the near- to mid-term prospects for profitability of that company are good.

Companies that are achieving profitability through sales growth and high profit margins typically will be hiring. The stock market is full of smart people doing their best to judge the prospects of companies year in and year out. They are not always right, of course, but they have more sources of information at their disposal than the average job seeker. Take advantage of them.

The point is that tracking an industry takes many different forms. You should study the raw data—the numbers—but look at the intangibles and, especially, at the future.

The Search

In searching for a job, remember that the ideas and techniques that have worked for years still are the most valid and reliable techniques today. Never forget how important a resume is in the job search process. There are plenty of resources you can use to help you create the best resume possible. Peterson's *New 90-Minute Resume* is an excellent source for everything you need to know to create a great resume. The idea isn't to dazzle the prospective employer. A good, professional impression will go a long way toward getting you through the interview process and, eventually, to the job offer. Be aware of your appearance and your presentation manner, and play up what you consider your strongest points. A thank-you note after the interview never hurts.

1) Experience counts. Over and over again you will hear that employers are looking for engineers who already have specific job experience in a given field. If you haven't graduated yet, look into co-op programs, suggests Gordon Gray of the University of Colorado. When you hit the job market, you will have a running start.

2) Be wary of growth potential. Despite what you might hear during the interview, the job you first get hired for is often what you are doing ten years later. Do your best to make sure that what you are learning on the job will be of interest to potential future employers. If not, your next job change may be difficult.

3) Use a standard resume format that highlights your applicable experience. The employers who are hiring are receiving large numbers of resumes. Give yourself a better chance of being noticed by using a format that employers are used to reading. They want to see who you worked for, in chronological order, and what you did while you were there.

4) But take advantage of new communications technology. If you have access to the Internet, try posting your resume in an applicable news group or mailing list. Peruse industry-specific computer bulletin boards. Many recent graduates have found jobs this way. The fact that you know how to use today's communications technology highlights one of your skills before you even get an interview.

5) There's no substitute for connections. Begin any job search by talking to ex-classmates and former co-workers. A good reference from someone who knows you and is trusted in the company you are interviewing with counts as much, and sometimes more, than a great interview. And remember, you can network electronically as well (see tip 4 above).

6) Personalize your resume and cover letter to fit the job you are looking for. There's nothing wrong with highlighting your applicable experience as much as possible. But, don't go overboard (i.e., don't lie), or you are likely to find yourself living through the "interview from hell."

7) Check your references. Contact your references and let them know a potential employer may be calling them about you. Ask them if they are comfortable with serving as a reference for you. *You* may be almost sure that someone thought the world of you and your work at your last job. Be absolutely certain. There's nothing worse than giving someone as a reference who turns out to be halfhearted about your talents.

8) If it's a choice between money and learning, take learning. Especially when it's early in your career, learning the right set of skills can keep you in good stead for many years to come. Taking a job offer just because of a higher salary might put you into a dead-end specialty within your field.

9) Don't forget sales jobs. People with engineering backgrounds and good interpersonal skills can make excellent technical salespeople. If you got into engineering as much for the starting salaries as you did for the technical challenges, consider sales. If you're good at it, you might end up making a better living than your peers who stay in the lab.

10) It can be tough out there. Don't give up. If you feel you are up against people with much more experience, sell youth and enthusiasm. If you're up against youth, sell experience. Be active, never lackadaisical, even when you feel down.

When approaching the job market, you'll want to determine what is unique or special about any given company. And you'll want to be aware of some of the major trends driving growth in the small-business sector: outsourcing activities of larger companies; the ever-increasing reliance on technology; the expanding service sector of the economy; the waves of growth stemming from trade liberalization south of the border; and the job gains being yielded by women-owned businesses.

Small companies cover a broad range of industries and geographic locations. As in big business, most job creation will occur in the services sector. High-technology companies in particular have long graced the rosters of *Inc.* magazine's annual Inc. 100 list of fastest-growing small public companies. Indeed, the sluggish Silicon Valley has made a comeback, with small biotechnology and software companies redressing the defense industry fallout. Small companies are also finding success in servicing health-related industries, as evidenced by increasing numbers showing up in the Inc. 100 and other barometers of high-growth businesses.

In any industry, the area in which small companies tend to shine is in supplying goods and services to their larger counterparts. The best of them capitalize on an uncommon business strategy or a unique market niche, offering a highly innovative product or specialized service.

Keeping up with trends—in business, among consumers, in society at large—pays off in a job search focused on rapid-growth companies every bit as much as with corporate giants. It's a useful means of determining whether a company's "unique proposition" makes good business sense and has staying power. If your instincts tell you "that's a smart idea," you just might want to get in touch!

Of course, it isn't always—or only—the brand-spanking-new product or service that dictates success. The "unique proposition" for some small companies is simply that they do what they do very well, perhaps in a better or different way than it's been done by competitors for awhile. Papa John's International Inc. (Louisville, KY) and Outback Steakhouse (Tampa, FL), for example, each took traditional fare—pizza and steaks—to new heights with smart management practices and meals that appeal to value-conscious consumers.

The battle between fad and fashion rages wildly in the small-business sector, however. The overall failure rate is high; today's hot commodity becomes tomorrow's fond memory. In particular, small companies with a fresh idea are delectable prey for large company copycats. Typically, big business has more resources at its disposal to churn out, and often improve, products introduced by smaller competitors. Yet small companies have speed and flexibility on their side. And those that effectively manage change and can recreate their businesses, even in the midst of success, are the ones that stick around.

So how can you assess the elusive factor of "staying power"? One key task, when interviewing with small rapid-growth companies, is to ask about their plans for the future. While a company's not apt to pull out the blueprints or divulge trade secrets, try to get a sense of how it's planning to build on its current success and whether it's already thinking of ways to diversify its product lines or tap into new markets. And if you can throw out some ideas of your own for them to consider (meaning you did a little homework and "imagineering" before you got to the interview), you're likely to shine.

Rapid-growth companies have a penchant for being in the right place at the right time. And the right places in the next few years could well be Mexico and Latin America. Several regions or individual countries around the globe—including Eastern Europe, India, and the Pacific Rim countries—are currently ripe for U.S. business expansion, but none more so than those south of the border. Over the last few years, small businesses provided a wellspring of support for the passage of the North American Free Trade Agreement (NAFTA), opening up trade opportunities in Mexico, and for GATT (General Agreement on Tariffs and Trade), the global trade compact under which trade liberalization in Latin America is covered. Large U.S. corporations already have sparked a surge of business activity in both regions. However, the wealth of import and export opportunities in these regions is equally significant for small businesses that are geared up and ready to take the plunge.

Mexico is currently the world's second-largest importer of U.S. manufactured goods and the third-largest importer of U.S. agricultural products. As local companies in Mexico and its neighboring countries in Latin America emerge into a free-market, global economy, U.S. businesses can help pave the way. High on the list of priorities in these regions is state-of-the-art technology. But that list goes on, and it encompasses a wide variety of needs: office equipment, food-processing equipment, gas and oil-field equipment, electricity and gas distribution equipment, construction machinery, lumber, computers, telecommunications gear, iron and steel tubing, cars and car parts, and medical supplies. These needs translate to small-business opportunities in the manufacture, distribution, and/or direct sales of products to a burgeoning marketplace. Additional opportunities are surfacing in the pursuit of joint ventures with local companies and certain government entities.

Small companies will be major players in these emerging economies. To get in on the action, bone up on your knowledge of Mexican and Latin American markets. Learn about the ways and means of doing business there. Learn about the culture. Learn Spanish. If you're a bilingual marketing pro, market yourself to ad agencies or companies in any industry as being able to prepare brochures and sales materials in Spanish. If account management is your trade, dazzle potential employers with your cross-cultural negotiating skills and knowledge of the intricate document and contract preparation required in these regions. There's room for everyone from human resources specialists to computer whizzes in small companies that aim to be in Mexico and Latin America at the right time.

Women Mean Business

Among the rapid-growth companies not to be overlooked—perhaps especially if you're a woman—are women-owned businesses.

There are 6.5 million women-owned businesses in the U.S. today, virtually all of them with fewer than 250 employees. In each of the last five years, more than half of the 3 million businesses formed in the U.S. were started by women. Moreover, the gender profile of the personnel in a woman-owned company is typically two thirds female. By simple arithmetic, one can see that women are being recruited by growth companies more frequently than men.

While women entrepreneurs are starting new businesses at a faster clip than their male counterparts, the facts of business life are that they typically cannot attract capital as easily. Consequently, rather than asset-based or capital equipment manufacturing companies, women tend to start trust-based businesses—those in which the customer pays in advance. They also substitute labor for capital, and the labor most in demand is skilled in affinity-marketing, communicating with suppliers to extend credit, desktop publishing, information gathering, and customer service. Have a skill in these areas? It's for hire by a woman-owned small business.

Women entrepreneurs have proven themselves heroines in the current U.S. economic recovery. With the best and most expansive of them growing from an average initial capital of $38,000 to an average valuation of $90 million in fifteen years, they also are providing training grounds for their employees to leave and launch their own businesses, which creates an ever-widening circle of women hiring mostly women. A shared work ethic is key, but male job seekers can rest assured that EEO regulations still apply: Connections to the old-boy network may not carry much weight, but inimitable, in-demand business skills count.

Acclimating to the Small-Business Environment

Just as certain types of people tend to start and grow successful small businesses, employees of a certain temperament tend to thrive in a small-business setting. If you'd like to put rapid-growth companies into your mix of employment prospects, keep in mind these hints and caveats:

- Small companies could give lessons to the giants on doing more with less. Dealing with resource constraints and, presuming you're financially conscientious, treating every dollar of the company's money as if it were your own are typical of small company life.
- Be prepared to wear many hats. Resource constraints often extend to personnel, with employees playing various roles, filling in for each other, and doing without an extensive support staff. The good news is that you get to try your hand at different functions, find out what you like best and are best at, and can develop multiple skills to bring to your next employer.
- Starting salaries may be lower than big company offerings. Frequently, though, other forms of compensation come into play, not only equity opportunities (such as stock options) but the "psychic rewards" of having a more immediate, direct, and visible impact on the business.
- On the other hand, some rapid-growth companies offer big bucks to bring new employees aboard—especially those in urgent need of specialized or technical skills.

(But remember the trade-off: small businesses tend to fall into the high-risk category, considering their overall failure rate.)

- It's no different than in big business: you've got to sell yourself, your skills, and the way in which you, unlike anybody else, can add value to the company. Small companies simply can't afford to hire nonproductive, noncontributing employees; the impact on the business is too immediate. They also have access now to experienced folks "downsized" by big companies, so you've got to communicate your edge on the competition.
- Some job seekers are attached to the prestige of working for a "brand-name" large corporation. However, proponents of working for lesser-known, rapid-growth companies cite the personal satisfaction of getting in on, or slightly above, the ground floor and helping the company make a name for itself.
- Enthusiasm goes far in a small company—so do hard work, long hours, bright ideas, self-motivated work habits, calculated risk-taking, an ability to pitch in wherever needed and to shift gears quickly, and a healthy respect for authority (the CEO) tempered with the ability to question authority (the CEO)—diplomatically, of course.

The working environment in rapid-growth companies is usually exciting, chaotic, and ever-changing. Just keep in mind that not every rapid-growth company keeps growing rapidly. Again, examine the company's "unique proposition." Ask about future growth projections and product/service lines. Ask about employee turnover rates. Talk to a few different employees to get their take on the business and day-to-day office life. Discreetly explore sources and adequacy of financing. And definitely get a sense of a company's management style and philosophy. If you pick a winner in the rapid-growth ranks, and the climate suits you, the opportunity is there for you to grow with the business and make your mark.

Top Skills Employers Look For

No matter what job you hold and what industry you work in, you'll need to be proficient in a number of important skills if you want to prosper in the ever-more-mobile and ever-more-global business world of the next century.

Communication skills: Speaking and writing clearly and persuasively.

Computer literacy: Being adept at accessing, analyzing, and storing data and communicating with other employees via computer.

Flexibility: Having a "go with the flow" mentality that allows you to do different tasks—and even switch jobs—in response to changing business conditions.

Team player: Demonstrating the ability to cooperate and work with others productively to achieve a common goal, putting the needs of the group above your own.

Self-motivation: Getting things done on your own, without having someone else give you guidance and direction on what to do and how to do it.

Cross-Cultural Skills: Being able to communicate in other languages and understand different cultures.

Quick Tips for Job Search Success

In the four lists that follow, we've provided some proven tips, techniques, and reminders that we're sure will improve even an experienced job seeker's chances of success. To make these lists even more useful, we've left space on each page for you to personalize them by adding items based on your own hard-learned lessons from interviews that were less than perfect. Experience is a great teacher and, since you pay dearly for the lessons you learn, you might as well benefit by improving your checklist. It will be invaluable in your future job searches.

We recommend taking the "During the Interview" checklist with you to your job interview. A last-minute glance at it will remind you of the essential things to do to make a terrific first impression.

Good luck!

CHECKLIST #1: APPLYING FOR A JOB

- Thoroughly research the companies you apply to—a good reference librarian can guide you to the best sources for your research, including annual reports, Web sites, trade press articles, and *Who's Who* for information on founders or top executives.
- Avoid form letters. Customize your letter to fit each company that you apply to.
- Send a query letter without a resume; your intent at this point is to gather information. You can follow up with a tailored resume once you learn more about the qualifications the company is seeking. You can elaborate on the points of your resume during the interview.
- Make your approach to the hiring manager. Research the person's name, title, and address and make sure you spell everything correctly in any correspondence.
- Don't apply to a company unless you can think of good reasons why you would be a valuable asset to their company. If you can't, they probably won't be able to either.
- Give your letter a reasonable amount of time to be delivered to the recipient, then make a call. Be prepared to cite your accomplishments and potential value to the company and the reasons you want to work there.
- Be sure your resume is current and that it has no typographical errors. Contact your references now so that you will be ready to provide their names if asked during the interview.

NOTES:

CHECKLIST #2: PREPARING FOR THE INTERVIEW

- Ask if you will be interviewing with more than one person. Find out the exact names and titles of all the people who will interview you. Take extra copies of your resume with you.
- Find out how long the interview is expected to last. This will help you to determine how much time you have to ask your questions and provide information about yourself.
- If you've never been to the company before, make a trial run the day before to see how long it will take to get there. The day of the interview, give yourself plenty of time and get there early.
- Make sure your hands, hair, and skin are clean and well-groomed and that your clothes are pressed.
- Dress professionally. Minimize accessories and excess baggage. Carry only a small portfolio with copies of your resume and any other materials that demonstrate your skills and accomplishments.
- Bring a book or magazine to read, in case you end up waiting for your interviewer. Be judicious in your selection of reading material; comic books would be a poor choice. Remember—everything you say or do makes a statement about who you are.
- Prepare a list of questions that you would like to have answered during the interview. If you don't have any questions, it may appear that you are not interested in the job.
- Be prepared for the many questions that might be asked of you, even inappropriate ones, and rehearse your responses. If possible, ask a friend to conduct a mock interview with you and focus on difficult questions.

NOTES:

CHECKLIST #3: DURING THE INTERVIEW

- Smile. It will make the interviewer think you are happy and enthusiastic, even if in reality you are nervous.
- Shake hands firmly and make eye contact.
- Exhibit poise, enthusiasm, and good manners.
- Think before you speak. It is perfectly OK to gather your thoughts before you respond to a question.
- Speak clearly and confidently and in complete sentences. Avoid using slang expressions—they may make you appear unintelligent and immature.
- Listen attentively—sit forward in your chair. Don't interrupt, argue, tell jokes, fidget, chew your nails, or play with your hair or jewelry.
- Take notes and check your list of questions to make sure that they have been answered.
- It is all right to ask about the salary range if your interviewer has not mentioned it. Don't ask specifics and don't discuss it any further at this stage of the interview process.
- Never criticize or complain about your former employer, professors, or anyone for that matter. Nobody else can understand what it was really like at your last job or in the classroom, and you will come across as a disgruntled complainer.
- Never lie about your credentials or abilities.

NOTES:

CHECKLIST #4: AFTER THE INTERVIEW

- Thank your interviewer (or interviewers) for their time and the opportunity to learn something about their company and the job.
- Reiterate your interest in the job if, indeed, you are still interested.
- Ask what the time frame is for filling the position and ask if it is OK for you to call your interviewer to check on the status of hiring.
- Send a thank you note the same day or the next day at the latest.
- If you haven't heard from the interviewer within a week, make a follow-up call to check on your status.

NOTES:

Making Yourself A Stand-Out Candidate

Getting up to speed on current business trends and deciding what kind of job opportunities you want to tap are important steps in the job-hunting process. Your next task is to communicate that knowledge and your particular assets to potential employers. In other words, you need to build an airtight case for hiring you by letting a company know precisely how you can contribute to its success. This is the time to express your passion and to strut out some new search strategies. And here are some guidelines that could help.

- ***Adopt an actor's mentality.*** As the old joke goes, how do you get to Carnegie Hall . . . ? Practice. The same theory applies to honing your job seeking "craft." Be ready for the audition. Don't take rejection personally. Rehearse your lines. Pay attention to the casting director. And use feedback to improve your performance skills.
- ***Tailor your tools to your targets.*** Sending out hundreds of resumes may feel like an industrious activity, but generic mailings typically get minimal to no results. Of 1,500 successful job seekers surveyed by the *National Business Employment Weekly*, most got in through contacts, only 2 percent through unsolicited mailings.
- ***Create your own job security.*** Finding a job doesn't mean you will never have to network, trend-track, or update your resume again. Research indicates that most people will work for 10 or more employers over a lifetime. Job security, then, comes not from your current employer but from taking charge of your career and keeping yourself employable.
- ***Make change a way of life.*** The old order changeth . . . what about you? In a survey of senior managers conducted by EquiPro International, their most frequently cited complaints were about employees accepting the status quo, not taking risks, reacting instead of initiating—in other words, resisting change. As companies live or die based on their ability to manage change, it's fairly obvious that they want to hire people who view change as a challenge and an opportunity.
- ***Learn.*** The American Society for Training and Development (ASTD) projects that by the year 2000 more than 65 percent of all jobs will require some education beyond high school; 23 million people will be employed in professional and technical jobs—the largest single occupational category—that require ongoing training. The key is to determine the kind of education and training that is valued by your market niche (or one you'd like to enter) at different stages of your career.
- ***Upgrade your computer-ease.*** The need to be technologically literate is no longer limited to programmers, number crunchers, and data processors. Technology will continue to transform virtually every working environment, affecting how work gets done and, thus, the types of jobs available.
- ***Join the team.*** Scrap the phrase "it's not in my job description" from your vocabulary. Potential employers want to hear that you're willing to roll up your sleeves and do whatever it takes to get the job done. Any evidence you can offer of your ability to both lead teams and be a good team player will win you points.
- ***Get personal.*** Whenever possible, present yourself to potential employers either in person or through a person. Resumes and cover letters have their place. But paper is often best used—especially by job seekers who have several years' experience or want to switch fields—to support your image rather than establish it.
- ***Get noticed.*** Seize every chance to become known or better known (note: this is another activity that shouldn't stop once you land a job). Take an active role in an industry association or a civic group. Publish an article in a trade journal and send "FYI" copies to key contacts and potential employers. Send an item of interest to employers with whom you had a good interview but who weren't hiring at the time. Teach. Give speeches. And always send thank-you letters.
- ***Take the scenic route.*** One nontraditional way to approach the job market is to become part of the growing contingent workforce—"temporary professionals" or "interim executives" who are matched with needy employers by employment firms that specialize in this area. Assignments are usually specific and

short-term (3 to 18 months), which lets you check out different employers and career options and sometimes leads to full-time employment. It tends to work out best if your skills are easily transferable and/or highly specialized and you can handle the uncertainty of when and where you'll be assigned next.

Notes on Resume Design

Your resume is an important tool in the job-hunt process. In it, you are trying to give the employer an overview of your work life, your job objective, and your accomplishments. Not an easy thing to consolidate onto one piece of paper and a task which could take hours and hours of your time. Find a good resource for advice about writing and designing a resume. There are lots of books on this subject out there, but we, of course, recommend the *New 90-Minute Resume.*

Three things make for a winning resume. First it must be visually inviting. Use a design that is clean, readable, and scannable. Second, it should highlight what you have to offer a particular employer. This may mean modifying your resume for each and every job you apply for. This is easy to do if you have a computer. If you don't, you may want to put your job objective in your cover letter rather than on your resume page. And third, it should highlight your accomplishments, the things that identify you as an outstanding person and employee.

The following is advice on the major points you should consider when creating your resume.

DO strive for consistency in your use of graphic elements. If you capitalized your first job title, subsequent job titles should also be capitalized.

DO use point sizes consistently. For example, using 12-point text and 14-point headings makes good design sense and ensures that your resume won't be misread by a scanner.

DON'T mix typefaces. Choose a typeface with a clean and simple design, e.g., Helvetica, Ariel, Times Roman.

DO make a sophisticated choice when it comes to paper. The color should be white or beige and the finish linen or laid. Keep in mind that using high-quality paper is only important if a person is reviewing your resume. If it is being scanned, plain white paper is the best choice.

DO use the same stock for your cover letter.

DO use black ink. Don't show off by using colors or graphics. Unless you are an expert designer, it will come off as amateurish.

DO print out each resume so that it looks like an original. If you are using a typewriter, make sure the copier you are using is one of high-quality, such as can be found at a copy shop.

DO keep the design simple and straightforward. If you attempt to be clever or humorous, it may be misinterpreted.

DO proofread the copy several times, and if you are using a computer, use the spell check option every time you make changes or corrections.

DO give it to someone else to read before you print the final version.

DO put your name at the top of each page of your resume.

DON'T include your photograph.

DON'T fold or staple your resume. Send it in an 8½ x 11 envelope and use paper clips.

SAMPLE COVER LETTER

Michael DeAngelis
202 Balmoray Drive
Durham, NC 77101
717-555-7812

June 16, 1998

Mr. Andrew Uster
Perot Systems Corporation
12377 Merit Drive
Dallas, Texas 75251-3233

Dear Mr. Uster:

Having recently earned my Master of Business Administration/Master of Arts in Engineering Administration and with fourteen years of engineering experience, I am ready to move on to the next level of my career. My goal is to gain a managerial position in a consulting firm that specializes in public sector clients.

The job-matching sector on your Web site helped me identify a Project Manager's position that is available in Dallas, Texas, that matches my qualifications perfectly. It also lists a 60 percent travel time requirement. I would have no problem relocating to Dallas, as I have family there, and I would look forward to the travel aspect of this position.

My research has shown that Perot Systems Corporation is the largest management consulting firm in the country. Your list of clients is impressive and your company's reputation in the consulting field is unmatched. My management and organizational skills, as well as my reputation for efficiency, are the assets that I could contribute to your company.

I will call you next week to see if we can arrange a time to talk.

Sincerely,

Michael DeAngelis

SAMPLE RESUME

Michael DeAngelis

202 Balmoray Drive
Durham, North Carolina 77107
717-555-7812

JOB OBJECTIVE To move from the public to the private sector to a managerial position with a management consulting firm specializing in public sector clients

SUMMARY OF QUALIFICATIONS Fourteen years of experience as a civil engineer; eight years of management and fiscal administration; M.B.A., and a Master of Arts in Engineering Management; expertise in project management, budget projects, and clear report writing; reputation for efficiency and working within budget; received citations for superlative performance from management in all jobs held

AREAS OF EXPERTISE

Management

- Managed Johnson Parkway Project, a $2.5M federally funded project; received national Infrastructure Award by *City and County Magazine*, 1995
- Coordinated preparation of the city's five-year capital improvement plan, which was cited by mayor as being principal catalyst behind $10M new business investments by private companies, 1992
- Managed engineering services division; enacted employee suggestions on reducing water meter installation and reduced average installation time by three days, 1992
- Administered 200-employee Department of Public Works; reduced staff by 15 percent through reallocation of job responsibilities and early retirement options, 1990

Fiscal Administration

- Prepared, allocated, and managed an operating and capital budget of $10M, 1996
- Coordinated budget and capital improvement plan preparation of $7M, 1990
- Initiated two major utility rate studies; recommended change that resulted in savings of $1M annually to city, 1990

Employee Supervision

- Developed self-directed teams to create improved policies and procedures for servicing public and internal customer needs; was cited as one of top twenty innovative employee plans by American Management Association, 1995
- Increased organizational efficiency and accountability by reorganizing the engineering department into four divisions—Design, Transportation, Utilities, and Development, 1994

Civil Engineering

- Devised design criteria explaining the City's design standards for public improvements for use by private developers and engineers; reduced processing time by 25 percent, 1993
- Supervised coordination of technical review for water, sewer, storm drain, and street elements aspects of a $5M development
- Served as project manager for a $1M highway department road upgrade program; finished the project two months ahead of schedule, 1986
- Supervised the planning, design, consultant coordination, and administration of public facilities, including water systems, wastewater systems, and streets; cited three times for superior performance, 1985–1989

JOB HISTORY

1992–1997 *City Engineer, City of Durham, North Carolina*

1989–1992 *Assistant to the Director of Public Works, City of Charleston, West Virginia*

1983–1989 *Engineer II: Civil, City of Reading, Pennsylvania*
Started as an engineering aide and held five positions prior to being made Engineer II in 1987

EDUCATION

Master of Business Administration and Master of Arts in Engineering Administration, Duke University, 1997

Bachelor of Science, Civil Engineering, Carnegie Mellon University, 1983

PROFESSIONAL CREDENTIALS AND INVOLVEMENTS

Civil Engineer, Pennsylvania, 678143; West Virginia, 087657; North Carolina, C-576439
Vice President, American Public Works Association, Durham Chapter, 1994–97
Fellow, American Society of Civil Engineers
Member, American Water Works Association

INTERESTS

Bicycling (have competed in over thirty long-distance events)
Medieval jousting (perform at a half dozen Renaissance fairs annually)
Reading mysteries

THE MOST DESIRED JOB SKILL IN THE WORLD

COMMUNICATION

by Paul Dyer

What is the number one skill that employers look for when hiring new employees? The answer may surprise you. Certainly, employers evaluate a job candidate's educational background, his or her work-related experience, and the candidate's technical and computer skills. Employees who possess these attributes and skills help an organization achieve its strategic goals. Of course, every organization wants to achieve its goals, so employers try very hard to hire employees who possess these skills. But guess what? None of these things serves as the number one qualification that employers most desire when hiring new employees.

Employers assume that many, if not most, applicants possess enough technical knowledge to do the job. If not, employers are often willing to train employees for those aspects of the work. So what job skills do employers consider most valuable? Numerous recent studies and surveys show that the skills employers most strongly desire relate to *communication.* They want people who can transfer meaning from one human being to another; they want people who can communicate effectively. Not only do communication skills predict a new hire's likelihood of achieving initial success, but employers also promote employees with top-notch communication skills quickly.

As any employer knows, however, finding employees with good communication skills is extraordinarily difficult. One recent survey of personnel interviewers reported that only about half of those interviewing for jobs demonstrated effective communication skills. Employers are looking for skilled communicators at least as hard as you're looking for a job.

Communication Skills and the New Realities

Communication skills were not always at the top of employment recruiters' lists. In a survey conducted in 1975, employers didn't even rate communication skills as one of the top five desired job qualifications. However, just eight years later, a similar survey found that employers rated oral and written communication as the most important of all job qualifications. What changed between 1975 and 1983? A lot. In fact, by the early 1980s a whole new set of rules began to govern the world's economy. These rules continue to intensify through today.

Actions taken by Middle East oil producers on October 16, 1973, signaled profound economic changes that created these new rules. On that date, these producers decided unilaterally to raise the price of oil. The next day, OPEC agreed to cut back oil production voluntarily. Consumers felt the impact of these two actions immediately. The price of oil skyrocketed and long lines formed at gas stations around the world. Although oil prices have greatly moderated since that volatile period, the October 16, 1973, date now holds a historic place in economics. As Harvard's John Kotter has pointed out, this date serves as a point of demarcation: the day the world moved from a U.S.-dominated economy to a global economy. And, oh my, how the rules have changed.

Any organization wishing to find success in the new global economy *must* have few levels of hierarchy and empower employees to work in teams. These teams must make decisions and take action quickly in order to be very responsive to customers' needs. In the modern organization, loose boundaries exist between functions, departments, and even other organizations. More and more companies find success through partnering and joint ventures with other organizations—coming together to meet customer needs, disbanding, and then reforming into some new business arrangement.

These new realities radically changed the way businesses operate: teams, teams, and more teams; very intense global competition; greater focus on quality and customer service; more employee involvement in decision making; and a business pace approaching the speed of light. The world of work is a very dynamic place at the turn of the century! Above all, the global economy's new realities demand employees who communicate very well. As we move into the new millennium, effectively conveying ideas, information, and feelings is much more important than in the old days—even though the old days were only fifteen to twenty years ago.

Communication Defined

Three primary concepts capture the most important practical aspects of communication. First, a skilled communicator must be able to understand clearly (or decode) the messages sent by others. Second, a good communicator must be able to send (or project) messages effectively. Third, people send and receive these messages through one or more of the five senses: seeing, hearing, touching, smelling, and tasting.

How important is it that people accurately and effectively send and receive messages? Sometimes, communication errors present only a minor inconvenience, perhaps with even humorous outcomes. For example, almost all of us have played the "telephone" game. In this game, a group of people sits in a circle. One person begins by whispering a brief story in the ear of the person next to him or her. The person receiving the message then tells the same story to the next person. This communication chain continues until the last person relates the story to the individual who originated the story. The outcome is often hilarious. Participants almost always greatly distort characters' names, event times, and other significant parts of the story's content. While the game is fun, in the real world, this sort of distortion can have grave consequences. Sending the wrong messages or misunderstanding messages sent, can drive customers away, create major—sometimes even life-threatening—errors, and destroy teamwork. How important are good communication skills? Once again, in today's business climate they are simply the most important skills of all. And because you're in the midst of a job search, you not only need to possess these skills, you also must be able to demonstrate to a potential employer that you have them.

Speaking Effectively

To speak effectively, you need to speak clearly and concisely both in one-on-one settings and in groups. The goal when you speak in either setting is simple enough: to get your point across. Simple, yes, but ask yourself these three important questions: How do you know when you've achieved this goal? Do you have any speaking habits that interfere with your ability to effectively convey your intended meaning? What can you do to ensure that the message you want to send is indeed the message that is received?

Before you begin speaking, create a brief outline in your mind. Determine the two or three main points you wish to make, and then make them. Throughout your conversation, or during a presentation, you should check for meaning. In other words, ask listeners to summarize what you've said. This will let you know if the listener caught your intended meaning. Confirm the points that are on target and clarify the ones that aren't.

Even though you may feel that you already speak effectively, you may unknowingly possess one or more speaking habits that annoy others. Perhaps you speak too quickly or too slowly. Maybe your speech is too loud or too soft. You may speak too formally or use too much slang. Ask a trusted friend, your spouse, or a respected colleague to provide honest feedback about your speech habits. Regardless of the feedback's content, do not become defensive. Listen carefully and receive it for what it is: a gift that will improve your communication skills. If you identify areas that need improvement, ask a friend or coworker to provide ongoing feedback on those particular areas. It's best if you receive this feedback immediately after you engage in one of the bad habits or after you show noticeable improvement. Here are several other important tips for speaking effectively:

- Make sure you balance your speaking with listening. Nothing is more annoying than participating in a conversation in which one party dominates the discussion by rambling on and on. In casual conversations, do the majority of your speaking in no more than 15- to 45-second chunks. *Very* rarely should your point exceed 1 minute in a one-on-one dialogue.

- Be sure that you understand others' meanings before you respond. Paraphrase their statements, and ask if you are correct. This way you'll be sure you understand, and you'll have a few more seconds to formulate your ideas.
- Build your vocabulary daily. On a sheet of paper, or on index cards, create a *New Words List.* Carry the list with you at all times. Each time you hear or read a word whose meaning is unfamiliar, jot it down on the *New Words List.* Find the word's definition in the dictionary, and then use the word several times the next day. Periodically review the list until the words become a part of your everyday vocabulary.
- Tape record one of your conversations or a presentation. Listen to the tape a few days later. Then identify at least one thing you can do to improve your speaking skills. Share this area with a friend, your spouse, or a colleague, and ask for ongoing feedback as you go about improving.
- To improve your presentation and public speaking skills, consider joining a Toastmasters Club. You can contact Toastmasters on the World Wide Web at www.Toastmasters.org.
- Learn to read nonverbal cues as discussed in the section below.

Listening to Others

This is perhaps the most neglected, and the most important, of all communication competencies. Listening is not sitting quietly as you wait your turn to speak! It is a process that requires active participation and analysis. Listening expert Lyman Steil's four-step listening formula (SIER) serves as a great tool for quickly improving your listening performance.

Sensing. This first step forms the bedrock, the foundation, for listening. When sensing a message, you take in information from others through all of your five senses. Unfortunately, many people rush through this first step (Sensing) and on to steps three and four (Evaluating and Responding). If you fail to sense all of the relative verbal, nonverbal, and other cues, you will destroy the rest of the listening process. Your job during this step is simply to gather information.

Interpreting. Now it's time for you to decode the message. At this point you should not determine whether you agree or disagree with the speaker. You should not yet begin to formulate what you can add to the dialogue. Your essential question during the Interpreting step is, "Do we have shared meaning?" When the speaker finishes talking, you may wish to paraphrase his or her words. Move on in the listening process to the Evaluating step only after the speaker acknowledges, verbally or nonverbally, "Yes, you've got it."

Evaluating. As previously mentioned, poor listeners move much too quickly to this step and then rush on to Responding. By doing so, they very often misinterpret the content of the message. (In addition, poor listening habits can break down relationships. One of the best ways to show people that they matter, that they're important, is to listen. By not taking the time to gather information and to understand clearly the speaker's intended message, poor listeners send another, often unintended message: "You're not valued by me." Obviously, this message undermines relationship building and will lead to further deterioration of the communication process.)Again, to be a great listener, begin your evaluation of the message only after you complete Sensing and Interpreting. You can then determine whether you agree or disagree, what you can add to the dialogue, or if you need more information.

Responding. Notice the word for this step is not "reacting." Responding's roots are in the learning process, whereas reacting comes from emotions. After you take in information through your senses, interpret the message, and evaluate the content, you can decide how best to respond. Here are several other important tips for building effective listening skills:

- Eliminate as many distractions as possible. Find quiet places to conduct dialogues. Don't type, arrange papers, or anything else when on the telephone.

- Encourage others to say more through body language (e.g., smiling, sitting forward in your chair, etc.). Or use a simple statement such as, "Can you say more?"
- Don't interrupt. Listen to *all* of the speaker's words.
- Use open-ended questions or phrases to gather more information, rather than questions that can be answered with a yes or no. For example: "Tell me about," or "how did this occur," or "please explain."
- Use the reflecting technique. Also known as mirroring, this technique allows the listener to reflect the speaker's emotions without agreeing or disagreeing with the content. This technique is especially valuable in situations of conflict. Simply make a short statement that identifies the speaker's emotions and mentions something about the situation. Begin the statement with "you." For example, "You seem upset about not getting the promotion. How can I help?" Or "You appear very excited about winning the award."
- Stay in the *now*. While listening, do not think about the past and do not think about the future. Discipline yourself to actively listen and to focus on the speaker as you gather information. Departures from the *now* are devastating to the listening process.
- Use silence to foster effective communication. During presentations, or in one-on-one conversations, do *not* answer your own questions. After you ask a question, wait; count seconds in your head. Virtually no one can endure the silence for more than 8 to 10 seconds. This is one of the most useful of all communication tips. It takes discipline, but incorporating silence into your communication skills will immediately yield valuable results.
- Use self-disclosure. Share an experience similar to the one the speaker is expressing. Be open, honest, and authentic. Be careful not to engage in one-upmanship.

Nonverbal Communication

Much of the skill in sending and receiving information is unrelated to the content of the message. In other words, it's not just what you say, it's how you say it. Body language such as eye contact, facial expression, movement of the head, and posture all convey meaning. Hand gestures such as a "thumbs up," as well as voice tone, rate of speech, and volume also send powerful messages. Those with good communication skills understand these nonverbal signals and respond to them appropriately.

Good communicators also use appropriate nonverbals to help convey the intended message. For example, if you lean forward in your chair, maintain eye contact, and move your head up and down as someone is speaking, you make it clear that the speaker's message is one that you value.

To improve your ability to send and receive nonverbal communication, try this three-week activity. As a first step, keep a record of nonverbal behaviors you observe. On a piece of paper, create three columns with the three headings shown in the example at the bottom of this page.

During the first week, observe others' communication. During meetings, at restaurants, or in any other setting, make notes in the first column of any nonverbal behaviors you observe. Look for nonverbals such as facial expression, gestures, eye contact, and body posture. In the second column, record the message you think these nonverbal behaviors send. Finally, in the last column, write why you believe the person engaged in the nonverbal behavior.

After the first week, begin to focus on your own communication for one week. Create a second chart with the same three headings. Again, during formal meetings, group or team activities, or one-on-one dialogue, observe and record nonverbals. Note both the nonverbals others display and those you find yourself exhibiting. Again, complete the second and third columns.

At the beginning of the third week, determine the types of messages you'd like to send using

Nonverbals Observed	Message Sent	Most Likely Reason for the Message
Bill looking around the room during Jane's presentation	This is boring	He was not engaged in active listening

nonverbal communication. Create a third chart with the same headings. Each time you *purposely* use a nonverbal, record it in the first column and then complete the last two columns. This exercise will improve your ability to both understand others' nonverbals and effectively manage your own.

In many situations, nonverbal communication skills are more important than verbal or written skills in transferring meaning. As Carol Rudman said in her engaging book *Frames of Reference,* "If your nonverbals contradict your words rather than underline them, people tend to believe the nonverbals as more trustworthy indicators of what you really mean."

Written Communication

Poor writing confuses and annoys the reader. All of us have read documents containing words, words, words . . . and little else. Poorly written e-mails, memos, letters, and reports undermine communication because they fail to transfer meaning. Effective writing is concise, clear, and to the point.

Write from a *you* perspective. Write for your audience, not for yourself. Before you begin writing, ask yourself these two questions: "What does the reader already know? How does the reader feel about me and this subject?" Throughout your writing, maintain a "you" rather than a "me" focus.

Be yourself. Don't try to copy someone else's style. "Write as you speak" serves as a good rule of thumb in business writing. Just be cautious not to become too informal.

Revise and rewrite. The great writer Robert Louis Stevenson once said, "When I say writing, oh believe me, it is rewriting I have chiefly in mind." Of course it takes time and discipline, but *rewriting* dramatically improves writing quality. Here are four good rewriting tips. First, don't procrastinate; begin writing projects early enough to allow time for revisions. Second, after you complete the work, proofread it. Make sure the writing contains no typos, misspellings, or grammatical errors. If writing on a computer, always use the spelling and grammar check. Third, ask someone whose writing you respect and enjoy reading to proof your document. If you're a technical person writing for a nontechnical audience, you should have a nontechnical person proof your materials. Fourth, as soon as possible, seek feedback about what you've written from the readers to whom you sent the information. Find out if you conveyed the message you intended to convey. Then ask, "What could I have done to make this clearer?"

Write clearly. "The potential for adequate coverage of all process factors and elements is somewhat feasible, but unlikely given our capacity limitations and our current functional alignments." Huh? As the writer David Lambuth wrote, "If you have a nail to hit, hit it on the head."

Complex sentences and flowery language will not impress readers. They will, however, stop you from communicating effectively. Put simply, if your readers don't understand your writing, you did not communicate. Along with writing from a "you" perspective, being yourself, and rewriting, here are several ideas to improve your writing clarity.

- Begin writing by outlining the key points and subpoints you wish to communicate.
- Keep your sentence length fairly short: between fifteen and eighteen words.
- Avoid jargon and acronyms. Readers unfamiliar with the terms find this particularly maddening.
- Avoid euphemisms. For example, there are no potholes in Tucson, only "pavement deficiencies."
- Only include as much detail as the reader needs.
- Finally, practice writing in the active voice to add interest and clarity to your writing. Take a look at these two sentences:

  The hiring manager is interested in Jim's job qualifications. (passive voice)

  Jim's job qualifications interest the hiring manager. (active voice)

  The second sentence is much easier to read and understand. To write in the active voice, show the action in the verb rather than in the noun. You can rewrite any passive sentence in the active voice. Here is another example:

  I am excited by this job opportunity.

  This job opportunity excites me.

Writing in the active voice will enhance your ability to communicate with clarity, interest, and power.

Demonstrating Your Communication Skills to Prospective Employers

Of course, if employers want people with good communication skills, you will want to show them that you have these skills. Employers evaluate job applicants' communication skills through three primary avenues: cover letters, resumes, and interview performance. In all three cases, employers evaluate both *content,* what is said or written, and *presentation,* how it is said or written.

In reviewing content, employers look for past achievements that demonstrate applicants' communication skills. Because employers know that past behavior is the best predictor of future behavior, applicants' achievements play an essential role in organizations' hiring decisions. Past leadership positions, speech classes or clubs, writing or speaking awards, or any other achievement requiring communication skills will gain employers' interest. Your cover letters, resume, and interview performance should reflect these past achievements.

My book, *The Ultimate Job Search Survival Guide,* provides detailed "how to" information on identifying and marketing past achievements, preparing professional cover letters and resumes, and achieving peak performance during the interview.

Practice the tips and skills presented here and you'll improve your written and oral communication skills. Improve your communication skills and you will possess the qualities most sought by employers.

Paul L. Dyer is a teacher, consultant, and author of Peterson's The Ultimate Job Search Survival Guide. *You can reach him on the World Wide Web at www.pauldyer.com.*

EMPLOYER PROFILES

A listing of major U.S. companies and selected government organizations, organized in geographical order

ALABAMA

ALTEC INDUSTRIES, INC.
210 INVERNESS CENTER DRIVE
BIRMINGHAM, ALABAMA 35242

OVERVIEW: Manufactures utility trucks. Established: 1929. Ownership: Private.

KEY STATISTICS: Annual Sales: $384 million. Number of Employees: 1,500.

HIRING HISTORY: Number of professional employees hired in 1997: 70. 1996: 50.

EXPERTISE/EDUCATION SOUGHT: Research, accounting, sales, information systems, computer programming, program analysis, engineering, human resources

ALTERNATIVES TO FULL-TIME EMPLOYMENT: Part-time jobs, paid internships

COMPANY-PAID TRAINING is available.

INTERNATIONAL ASSIGNMENTS are available to employees.

CONTACT: For individuals with previous experience: Mr. Bill Riley, Director of Human Resources Administration; Phone: 205-991-7733, ext. 6983; Fax: 205-995-0670. E-mail: billo.riley@altec.com.

COMPANY URL: http://www.altec.com

AMERICAN FINE WIRE CORPORATION
907 RAVENWOOD DRIVE
SELMA, ALABAMA 36701-6724

OVERVIEW: Manufactures bonding wire. Established: 1995. Parent Company: Kulicke and Saffa Industries. Ownership: Public.

KEY STATISTICS: Annual Sales: $67 million. Number of Employees: 313.

HIRING HISTORY: Number of professional employees hired in 1997: 20.

EXPERTISE/EDUCATION SOUGHT: Mechanical engineering, accounting, data processing, finance

ALTERNATIVES TO FULL-TIME EMPLOYMENT: Paid internships

COMPANY-PAID TRAINING is available.

CONTACT: For individuals with previous experience: Manager of Personnel; Phone: 334-875-4040; Fax: 334-874-1351.

AMERICAN HOMESTAR CORPORATION
HIGHWAY 78 WEST
GUIN, ALABAMA 35563

MAILING ADDRESS: PO Box 69, Guin, AL 35563

OVERVIEW: Builds mobile homes. Established: 1994. Ownership: Public.

KEY STATISTICS: Annual Sales: $60 million. Number of Employees: 600.

EXPERTISE/EDUCATION SOUGHT: Manufacturing

CONTACT: Mr. Jeff Skaggs, Human Resource Director; Phone: 205-468-3076; Fax: 205-468-2828.

AVEX ELECTRONICS, INC.
4807 BRADFORD DRIVE
HUNTSVILLE, ALABAMA 35805-1948

OVERVIEW: Manufactures circuit boards. Established: 1987. Parent Company: J. M. Huber Corporation. Ownership: Private.

KEY STATISTICS: Annual Sales: $717.1 million. Number of Employees: 2,500.

EXPERTISE/EDUCATION SOUGHT: Electrical engineering, mechanical engineering, chemical engineering

COMPANY-PAID TRAINING is available.

INTERNATIONAL ASSIGNMENTS are available to employees.

CONTACT: Director of Human Resources; Phone: 205-722-6000; Fax: 205-722-6232.

COMPANY URL: http://www.avex.com

EBSCO INDUSTRIES, INC.
5724 HIGHWAY 280
BIRMINGHAM, ALABAMA 35242-6818

OVERVIEW: Provides publication subscription, printing, and telemarketing services; manufactures metal and wire products. Established: 1943. Ownership: Private.

KEY STATISTICS: Annual Sales: $950 million. Number of Employees: 4,100.

EXPERTISE/EDUCATION SOUGHT: Sales, management

COMPANY-PAID TRAINING is available.

INTERNATIONAL ASSIGNMENTS are available to employees.

CONTACT: For individuals with previous experience: Mr. Leon Sizemore, Corporate Recruiter; Phone: 205-991-1474; Fax: 205-991-1479.

GULF STATES PAPER CORPORATION
1400 RIVER ROAD, NE
TUSCALOOSA, ALABAMA 35404

OVERVIEW: Manufactures paper, wood, and forest products. Established: 1884. Ownership: Private.

KEY STATISTICS: Annual Sales: $300 million. Number of Employees: 2,100.

EXPERTISE/EDUCATION SOUGHT: Chemical engineering, pulp and paper science, industrial engineering

ALTERNATIVES TO FULL-TIME EMPLOYMENT: Paid internships

CONTACT: Mr. Jim Merrell, Vice President of Human Resources; Phone: 205-553-6200; Fax: 205-553-6095.

INTERGRAPH CORPORATION
289 DUNLOP ROAD
HUNTSVILLE, ALABAMA 35824

OVERVIEW: Manufactures computers and software for industrial use. Established: 1969. Ownership: Public.

KEY STATISTICS: Annual Sales: $1.1 billion. Number of Employees: 8,200.

EXPERTISE/EDUCATION SOUGHT: Electrical engineering, computer engineering, design engineering, computer-aided design

ALTERNATIVES TO FULL-TIME EMPLOYMENT: Part-time jobs

COMPANY-PAID TRAINING is available.

INTERNATIONAL ASSIGNMENTS are available to employees.

CONTACT: For individuals with previous experience: Mr. Dave Cummings, Executive Manager of Professional Development; Phone: 256-730-7965; Fax: 256-730-7252. E-mail: intergraph@ingr.com.

COMPANY URL: http://www.ingr.com

INTERGRAPH PUBLIC SAFETY
PO BOX 64158
HUNTSVILLE, ALABAMA 35824

OVERVIEW: Manufactures electronic parts and equipment. Established: 1996. Parent Company: Intergraph Corporation. Ownership: Public.

KEY STATISTICS: Annual Sales: $70 million. Number of Employees: 500.

HIRING HISTORY: Number of professional employees hired in 1997: 25.

EXPERTISE/EDUCATION SOUGHT: Software engineering/development

INTERNATIONAL ASSIGNMENTS are available to employees.

CONTACT: For individuals with previous experience: Mr. Derek Johns, Senior Human Resources Representative; Phone: 205-730-7884; Fax: 205-730-8006. E-mail: dmjohns@ingr.com.

COMPANY URL: http://www.intergraph.com

INTER-NATIONAL CHILDRENSWEAR LLC
3500 6TH AVENUE SOUTH
BIRMINGHAM, ALABAMA 35222-2406

OVERVIEW: Manufactures men's and boys' trousers and slacks. Established: 1997. Ownership: Private.

KEY STATISTICS: Annual Sales: $72 million. Number of Employees: 400.

Hiring History: Number of professional employees hired in 1997: 20. 1996: 10.

Expertise/Education Sought: Engineering, design, production management

Alternatives to Full-Time Employment: Internships

International Assignments are available to employees.

Contact: For individuals with previous experience: Human Resources Department; Phone: 205-251-9143; Fax: 205-251-9124. E-mail: wgrill@liberty-kids.com.

Company URL: http://www.libertykids.com

JVC AMERICA, INC.
2 JVC ROAD
TUSCALOOSA, ALABAMA 35405-3548

Overview: Manufactures floppy disks, VCR tapes, and compact disks. Established: 1986. Ownership: Private.

Key Statistics: Annual Sales: $160 million. Number of Employees: 650.

Expertise/Education Sought: Marketing, sales, engineering, electrical engineering, chemical engineering, mechanical engineering

Alternatives to Full-Time Employment: Paid internships

Contact: For individuals with previous experience: Human Resources Department; Phone: 205-556-7111.

QMS, INC.
1 MAGNUM PASS
MOBILE, ALABAMA 36618-3412

Overview: Manufactures thermal and laser printers. Established: 1977. Ownership: Public.

Key Statistics: Annual Sales: $124.6 million. Number of Employees: 725.

Hiring History: Number of professional employees hired in 1997: 75.

Expertise/Education Sought: Software engineering/development, manufacturing engineering, quality engineering, customer service and support, technical engineering

Alternatives to Full-Time Employment: Paid internships

Company-Paid Training is available.

Contact: Mr. Todd St. Mary, Manager of Staffing; Phone: 334-633-4300; Fax: 334-633-7253. E-mail: staffing@qms.com.

Company URL: http://www.qms.com

RUSSELL CORPORATION
755 LEE STREET
ALEXANDER CITY, ALABAMA 35010-0272

Mailing Address: PO Box 272, Alexander City, AL 35011-0272

Overview: Manufactures textiles, knitwear, and clothing. Established: 1902. Ownership: Public.

Key Statistics: Annual Sales: $1.2 billion. Number of Employees: 17,843.

Hiring History: Number of professional employees hired in 1996: 100.

Expertise/Education Sought: Sales, marketing, manufacturing engineering, mechanical engineering, electrical engineering, accounting, textiles, computer programming

Alternatives to Full-Time Employment: Part-time jobs, paid internships

Company-Paid Training is available.

International Assignments are available to employees.

Contact: Mr. Melvin Ervin, Corporate Staff Representative; Phone: 205-329-5587; Fax: 205-329-4709. E-mail: employment@russellcorp.com.

Company URL: http://www.russellcorp.com

SCI SYSTEMS, INC.
2101 CLINTON AVENUE WEST
HUNTSVILLE, ALABAMA 35805-3007

Mailing Address: PO Box 1000, Huntsville, AL 35807

Overview: Manufactures instrumentation and computer equipment. Established: 1961. Ownership: Public.

Key Statistics: Annual Sales: $5.8 billion. Number of Employees: 18,470.

Expertise/Education Sought: Computer engineering, software engineering/development, electrical engineering, mechanical engineering, radio frequency engineering

International Assignments are available to employees.

Contact: Mr. Kirk Scruggs, Professional Staffing; Phone: 205-882-4800; Fax: 205-882-4804.

TCI CABLEVISION OF ALABAMA
3443 LORNA LANE
BIRMINGHAM, ALABAMA 35216

Overview: Provides cable television services. Established: 1986. Ownership: Public.

Key Statistics: Annual Sales: $134.3 million. Number of Employees: 1,000.

Hiring History: Number of professional employees hired in 1996: 20.

Expertise/Education Sought: Electrical engineering, customer service and support

Alternatives to Full-Time Employment: Part-time jobs

Contact: For individuals with previous experience: Mr. Philip Lusco, Manager of Business Relations; Phone: 205-822-8731; Fax: 205-978-9397.

Company URL: http://www.tci.com

VULCAN MATERIALS COMPANY
1 METROPLEX DRIVE
BIRMINGHAM, ALABAMA 35209-6893

Mailing Address: PO Box 530187, Birmingham, AL 35253-0187

Overview: Manufactures construction materials and industrial chemicals. Established: 1916. Ownership: Public.

Key Statistics: Annual Sales: $1.6 billion. Number of Employees: 6,926.

Hiring History: Number of professional employees hired in 1997: 100. 1996: 90.

Expertise/Education Sought: Finance, auditing, accounting, management information systems, computer programming, data processing, law, civil engineering, electrical engineering, mining engineering

Alternatives to Full-Time Employment: Paid internships

International Assignments are available to employees.

Contact: Mr. Jack Riley, Manager of Human Resources, PO Box 530187, Birmingham, Alabama, 35253-0187; Phone: 205-877-3000; Fax: 205-877-3048. E-mail: jobs@vulcanmat.com. Job Hotline: 205-877-3986.

Company URL: http://www.vul.com

ARIZONA

ARIZONA ELECTRIC POWER COOPERATIVE
1000 SOUTH HIGHWAY 80
BENSON, ARIZONA 85602-0670

Overview: Electric utility. Established: 1961. Ownership: Private.

Key Statistics: Annual Sales: $169.1 million. Number of Employees: 292.

Hiring History: Number of professional employees hired in 1996: 48.

Expertise/Education Sought: Business administration, accounting, human resources, engineering

Alternatives to Full-Time Employment: Paid internships

Arizona Electric Power Cooperative (continued)

Company-Paid Training is available.

Contact: Ms. Yvette Kroeze, Employment and Benefits Administrator; Phone: 520-586-5342; Fax: 520-586-5249. E-mail: employment@aepnet.com. Job Hotline: 800-562-5355.

Company URL: http://www.aepnet.com

CITY MEAT AND PROVISIONS COMPANY, INC.

1418 NORTH 27TH AVENUE
PHOENIX, ARIZONA 85009

Overview: Wholesales fresh meats. Established: 1996. Parent Company: Alliant Food Services. Ownership: Private.

Key Statistics: Annual Sales: $125 million. Number of Employees: 500.

Hiring History: Number of professional employees hired in 1997: 177. 1996: 120.

Expertise/Education Sought: Sales, transportation, production

Contact: For individuals with previous experience: Human Resources Department; Phone: 602-269-7717; Fax: 602-269-1084. E-mail: davidrogers@alliantff.com.

CSK AUTO, INC.

645 EAST MISSOURI AVENUE
PHOENIX, ARIZONA 85012

Overview: Operates chain of automotive parts stores. Established: 1996. Parent Company: Northern Retail Corporation. Ownership: Private.

Key Statistics: Annual Sales: $793.1 million. Number of Employees: 8,200.

Expertise/Education Sought: Accounting, marketing, finance, law, industrial engineering

Alternatives to Full-Time Employment: Part-time jobs, paid internships

Company-Paid Training is available.

Contact: Ms. Catherine Medina, Employment Manager; Phone: 602-631-7117; Fax: 602-266-4144. E-mail: cmedina@cskauto.com. Job Hotline: 800-548-8471.

Company URL: http://www.checkerauto.com

FNF CONSTRUCTION, INC.

115 SOUTH 48TH STREET
TEMPE, ARIZONA 85281

Overview: Provides heavy road construction services. Established: 1987. Ownership: Private.

Key Statistics: Annual Sales: $87.7 million. Number of Employees: 300.

Expertise/Education Sought: Civil engineering, business administration

Alternatives to Full-Time Employment: Paid internships

Contact: For individuals with previous experience: Ms. Debbie Remos, Human Resources Manager; Phone: 602-784-2910; Fax: 602-829-8607.

INTESYS TECHNOLOGIES, INC.

1300 NORTH FIESTA BOULEVARD
GILBERT, ARIZONA 85233-1604

Overview: Manufactures plastic moldings for computers and cars. Established: 1992. Ownership: Private.

Key Statistics: Annual Sales: $145 million. Number of Employees: 1,550.

Expertise/Education Sought: Data processing, accounting, finance

Alternatives to Full-Time Employment: Paid internships

Company-Paid Training is available.

Contact: Mr. Nick Smeed, Vice President of Administration; Phone: 602-497-1401, ext. 1520; Fax: 602-497-7000. Job Hotline: 602-497-1401.

KVOA COMMUNICATIONS, INC.

209 WEST ELM STREET
TUCSON, ARIZONA 85705

Mailing Address: PO Box 5188, Tucson, AZ 85703

Overview: Operates television broadcasting facilities. Parent Company: Evening Post Publishing, Inc. Ownership: Private.

Key Statistics: Annual Sales: $10 million. Number of Employees: 112.

Expertise/Education Sought: Engineering

Alternatives to Full-Time Employment: Internships

Contact: For individuals with previous experience: Ms. Sandra Anuhdson, Business Manager; Phone: 520-792-2270; Fax: 520-620-1309.

Company URL: http://www.kvoa.com

PHELPS DODGE CORPORATION

2600 NORTH CENTRAL AVENUE
PHOENIX, ARIZONA 85004-3014

Overview: Mines copper ore. Established: 1885. Ownership: Public.

Key Statistics: Annual Sales: $3.8 billion. Number of Employees: 16,033.

Expertise/Education Sought: Accounting, benefits administration

Contact: For individuals with previous experience: Human Resources.

PIMALCO, INC.

6833 WEST WILLIS ROAD, BOX 5050
CHANDLER, ARIZONA 85226-5107

Overview: Manufactures aluminum tubing, ingots, and other products. Established: 1985. Parent Company: Aluminum Company of America. Ownership: Public.

Key Statistics: Annual Sales: $64 million. Number of Employees: 700.

Hiring History: Number of professional employees hired in 1996: 100.

Expertise/Education Sought: Metallurgical engineering, mechanical engineering, accounting, finance, electrical engineering, data processing, quality control

International Assignments are available to employees.

Contact: For individuals with previous experience: Human Resources Department; Phone: 602-598-2232; Fax: 602-598-2266.

SIMPLIFIED BUSINESS SOLUTIONS

7776 SOUTH POINTE PARKWAY WEST, SUITE 138
PHOENIX, ARIZONA 85044-5424

Overview: Employee leasing service. Established: 1996. Ownership: Private.

Key Statistics: Annual Sales: $10 million. Number of Employees: 1,413.

Expertise/Education Sought: Engineering

Alternatives to Full-Time Employment: Internships

Contact: For individuals with previous experience: Human Resources Department; Phone: 602-414-0062; Fax: 602-414-0020.

TECH GROUP, INC.

7975 NORTH HAYDEN ROAD
SCOTTSDALE, ARIZONA 85258-3241

Overview: Manufactures injection molded plastics. Established: 1967. Ownership: Private.

Key Statistics: Annual Sales: $104.5 million. Number of Employees: 1,000.

Hiring History: Number of professional employees hired in 1997: 20. 1996: 15.

Expertise/Education Sought: Quality engineering, plastics engineering, process engineering

Alternatives to Full-Time Employment: Part-time jobs

Contact: Mr. David McWatters, Human Resources; Phone: 602-951-3000. E-mail: dave.mcwatters@techgrp.com.

Company URL: http://www.techgrp.com

TESSENDERLO KERLEY, INC.
2801 WEST OSBORN ROAD
PHOENIX, ARIZONA 85017

OVERVIEW: Produces nitrogenous fertilizers. Established: 1995. Parent Company: Tessenderlo Chemie, Inc. Ownership: Private.

KEY STATISTICS: Annual Sales: $170.4 million. Number of Employees: 550.

EXPERTISE/EDUCATION SOUGHT: Manufacturing, engineering, engineering management, engineering science, engineering technology

COMPANY-PAID TRAINING is available.

CONTACT: Mr. Fernando Solano, Director of Human Resources; Phone: 602-528-0600; Fax: 602-528-0683.

TOOH DINEH INDUSTRIES, INC.
LUEPP SCHOOL, BUILDING 4
LEUPP, ARIZONA 86035

MAILING ADDRESS: HC 61 Box E, Winslow, AZ 86047

OVERVIEW: Manufactures semiconductors. Established: 1983. Parent Company: Dineh Cooperatives, Inc. Ownership: Private.

KEY STATISTICS: Annual Sales: $42 million. Number of Employees: 80.

HIRING HISTORY: Number of professional employees hired in 1997: 20. 1996: 18.

EXPERTISE/EDUCATION SOUGHT: Quality engineering, electrical engineering, quality control, technology, data processing, process engineering

ALTERNATIVES TO FULL-TIME EMPLOYMENT: Paid internships

COMPANY-PAID TRAINING is available.

CONTACT: Ms. Mary Rose Begay, Human Resources Supervisor, Luepp School Building 4, Leupp, Arizona, 86035; Phone: 520-686-6411; Fax: 520-686-6516. E-mail: toohdinehindustries@worldnet.att.net.

VENTANA MEDICAL SYSTEMS, INC.
3865 NORTH BUSINESS CENTER DRIVE
TUCSON, ARIZONA 85705-2945

OVERVIEW: Manufactures medical instruments and diagnostic systems. Established: 1985. Ownership: Public.

KEY STATISTICS: Annual Sales: $24.1 million. Number of Employees: 100.

EXPERTISE/EDUCATION SOUGHT: Software engineering/development, electrical engineering

ALTERNATIVES TO FULL-TIME EMPLOYMENT: Internships

COMPANY-PAID TRAINING is available.

INTERNATIONAL ASSIGNMENTS are available to employees.

CONTACT: For individuals with previous experience: Human Resources.

WEISER LOCK CORPORATION
6700 SOUTH WEISER LOCK DRIVE
TUCSON, ARIZONA 85746-6007

OVERVIEW: Manufactures locks, bolts, and other miscellaneous metal products. Established: 1992. Ownership: Public.

KEY STATISTICS: Annual Sales: $60 million. Number of Employees: 730.

EXPERTISE/EDUCATION SOUGHT: Engineering

INTERNATIONAL ASSIGNMENTS are available to employees.

CONTACT: For individuals with previous experience: Human Resources Department; Phone: 520-741-6200; Fax: 520-741-6236.

WINCUP HOLDINGS, INC.
7980 WEST BUCKEYE ROAD
PHOENIX, ARIZONA 85043

OVERVIEW: Manufactures plastic foam cups. Established: 1995. Ownership: Private.

KEY STATISTICS: Annual Sales: $108.7 million. Number of Employees: 1,058.

EXPERTISE/EDUCATION SOUGHT: Marketing, accounting, engineering

CONTACT: For individuals with previous experience: Human Resources Department; Phone: 602-936-1791.

ARKANSAS

ABF FREIGHT SYSTEM, INC.
3801 OLD GREENWOOD ROAD
FORT SMITH, ARKANSAS 72903

OVERVIEW: Provides trucking and transport services. Established: 1988. Parent Company: Arkansas Best Corporation. Ownership: Public.

KEY STATISTICS: Annual Sales: $1.1 billion. Number of Employees: 13,022.

HIRING HISTORY: Number of professional employees hired in 1997: 20.

EXPERTISE/EDUCATION SOUGHT: Sales, marketing, transportation, business, accounting, auditing, finance

INTERNATIONAL ASSIGNMENTS are available to employees.

CONTACT: For individuals with previous experience: Mr. Ron Orick, Manager of Personnel; Phone: 501-785-8700; Fax: 501-785-8912. E-mail: rorick@abfs.com.

COMPANY URL: http://www.abfs.com

ACXIOM MAILING SERVICES
301 INDUSTRIAL BOULEVARD
CONWAY, ARKANSAS 72032-7168

OVERVIEW: Provides direct mail advertising services. Established: 1987. Ownership: Public.

KEY STATISTICS: Annual Sales: $11.1 million. Number of Employees: 300.

EXPERTISE/EDUCATION SOUGHT: Administration, customer service and support, software engineering/development

ALTERNATIVES TO FULL-TIME EMPLOYMENT: Paid internships

COMPANY-PAID TRAINING is available.

INTERNATIONAL ASSIGNMENTS are available to employees.

CONTACT: For college students and recent graduates: Mr. Tim McKeena, College Recruiter; Phone: 501-336-1522. E-mail: jobs@acxiom.com. For individuals with previous experience: Human Resources Department; Phone: 501-336-1000; Fax: 501-336-3724. E-mail: jobs@acxiom.com.

COMPANY URL: http://www.acxiom.com

ALLTEL CORPORATION
11025 ANDERSON DRIVE, SUITE 325
LITTLE ROCK, ARKANSAS 72212

OVERVIEW: Supplies telecommunications equipment and services. Established: 1960. Ownership: Public.

KEY STATISTICS: Annual Sales: $3.2 billion. Number of Employees: 16,307.

HIRING HISTORY: Number of professional employees hired in 1997: 1,500.

EXPERTISE/EDUCATION SOUGHT: Information systems, telecommunications, electronics/electronics engineering, computer programming, database management, network analysis

ALTERNATIVES TO FULL-TIME EMPLOYMENT: Part-time jobs, paid internships

COMPANY-PAID TRAINING is available.

INTERNATIONAL ASSIGNMENTS are available to employees.

CONTACT: For college students and recent graduates: Ms. Allison Nicholas, Manager of College Recruiting; Phone: 501-220-7620. For individuals with previous experience: Ms. Anne Kerr, Manager for Professional Recruitment, PO Box 2177, Little Rock, Arkansas, 72203; Phone: 501-220-7620; Fax: 501-220-7606. Job Hotline: 501-220-4104.

COMPANY URL: http://www.alltel.com

ARKANSAS BEST CORPORATION

3801 OLD GREENWOOD ROAD
FORT SMITH, ARKANSAS 72903

OVERVIEW: Provides worldwide freight transportation and distribution, truck tire remanufacturing, and information services. Established: 1935. Ownership: Public.

KEY STATISTICS: Annual Sales: $1.6 billion.

HIRING HISTORY: Number of professional employees hired in 1997: 16. 1996: 14.

EXPERTISE/EDUCATION SOUGHT: Management information systems, computer programming, systems analysis, computer science, industrial engineering

ALTERNATIVES TO FULL-TIME EMPLOYMENT: Paid internships

CONTACT: Mr. Timothy R. Harmon, Account Analyst, PO Box 10048, Fort Smith, Arkansas, 72917-0048; Phone: 501-784-8573; Fax: 501-784-8599. E-mail: tim.harmon@data-tronics.com.

COMPANY URL: http://www.arkbest.com

CORPORATE STATEMENT

Several exciting opportunities exist within Arkansas Best Corporation subsidiaries. At **ABF Freight Systems, Inc.**, *pricing analysts evaluate and analyze special pricing requests for existing and potential ABF customers. At* **Data-Tronics Corp.**, *systems analyst/programmers analyze, design, program, and implement information systems. Positions are located in the scenic and relaxed environment of northwest Arkansas.*

ARROW AUTOMOTIVE INDUSTRIES

23 NABCO AVENUE
CONWAY, ARKANSAS 72032

OVERVIEW: Manufactures generators, alternators, and distributors for aircraft and automobiles. Established: 1929. Ownership: Public.

KEY STATISTICS: Annual Sales: $87.5 million. Number of Employees: 1,282.

HIRING HISTORY: Number of professional employees hired in 1997: 400. 1996: 200.

EXPERTISE/EDUCATION SOUGHT: Manufacturing

COMPANY-PAID TRAINING is available.

CONTACT: Mr. Terry Renalds, Vice President of Human Resources; Phone: 501-327-6904; Fax: 501-327-0332.

ENTERGY ARKANSAS

425 WEST CAPITOL AVENUE
LITTLE ROCK, ARKANSAS 72201

MAILING ADDRESS: PO Box 551, Little Rock, AR 72203

OVERVIEW: Electric utility. Established: 1993. Parent Company: Entergy Corporation. Ownership: Public.

KEY STATISTICS: Annual Sales: $1.7 billion. Number of Employees: 1,455.

EXPERTISE/EDUCATION SOUGHT: Mechanical engineering, electrical engineering, industrial engineering, chemical engineering, marketing, sales, finance, accounting

ALTERNATIVES TO FULL-TIME EMPLOYMENT: Paid internships

COMPANY-PAID TRAINING is available.

INTERNATIONAL ASSIGNMENTS are available to employees.

CONTACT: Mr. Philip Conn, Human Resource Operator; Phone: 501-377-4111; Fax: 501-377-3636. E-mail: pconn@entergey.com. Job Hotline: 501-377-3618.

COMPANY URL: http://www.entergy.com

ESI GROUP, INC.

2100 BROOKWOOD DRIVE
LITTLE ROCK, ARKANSAS 72202

OVERVIEW: Designs computer systems and offers computer integration services. Established: 1996. Ownership: Private.

KEY STATISTICS: Number of Employees: 135.

HIRING HISTORY: Number of professional employees hired in 1997: 15. 1996: 7.

EXPERTISE/EDUCATION SOUGHT: Computer science, computer programming, engineering technology

ALTERNATIVES TO FULL-TIME EMPLOYMENT: Paid internships

COMPANY-PAID TRAINING is available.

CONTACT: For individuals with previous experience: Personnel; Phone: 501-666-4233, ext. 260; Fax: 501-666-4169.

COMPANY URL: http://www.esigroup.com

GEMALA TRAILER HOLDING CORPORATION

2800 BP WAY
JONESBORO, ARKANSAS 72401

OVERVIEW: Manufactures truck trailers. Established: 1992. Parent Company: Gemala Trailer Company of North America. Ownership: Private.

KEY STATISTICS: Annual Sales: $35.3 million. Number of Employees: 330.

HIRING HISTORY: Number of professional employees hired in 1997: 350. 1996: 200.

EXPERTISE/EDUCATION SOUGHT: Accounting, engineering, management

COMPANY-PAID TRAINING is available.

CONTACT: Mr. Rick Sparks, Human Resources Manager; Phone: 870-931-4529; Fax: 870-972-9843.

J.B. HUNT TRANSPORT, INC.

615 J. B. HUNT CORPORATE DRIVE
LOWELL, ARKANSAS 72745

OVERVIEW: Provides truck and intermodal shipping services. Established: 1961. Ownership: Public.

KEY STATISTICS: Annual Sales: $1.5 billion. Number of Employees: 11,575.

EXPERTISE/EDUCATION SOUGHT: Data processing, management, logistics

ALTERNATIVES TO FULL-TIME EMPLOYMENT: Paid internships

COMPANY-PAID TRAINING is available.

CONTACT: Ms. Sherry Moncrief, Personnel Manager; Phone: 501-820-0000; Fax: 501-820-8249.

TERRA NITROGEN CORPORATION

HIGHWAY 18 EAST
ARMOREL, ARKANSAS 72310

OVERVIEW: Manufactures fertilizer. Established: 1965. Parent Company: Peria Industries. Ownership: Private.

KEY STATISTICS: Annual Sales: $20.6 million. Number of Employees: 100.

HIRING HISTORY: Number of professional employees hired in 1997: 4.

EXPERTISE/EDUCATION SOUGHT: Engineering, operations, maintenance

COMPANY-PAID TRAINING is available.

INTERNATIONAL ASSIGNMENTS are available to employees.

CONTACT: For individuals with previous experience: Human Resource Department; Phone: 870-763-0800; Fax: 870-762-3723.

TOWNES TELE-COMMUNICATIONS

JUNCTION HIGHWAY 82 & 29
LEWISVILLE, ARKANSAS 71845

OVERVIEW: Local telephone communications. Established: 1981. Ownership: Private.

KEY STATISTICS: Annual Sales: $10.7 million. Number of Employees: 93.

EXPERTISE/EDUCATION SOUGHT: Engineering

INTERNATIONAL ASSIGNMENTS are available to employees.

CONTACT: For individuals with previous experience: Human Resources; Phone: 870-921-4224; Fax: 870-921-4400.

TYSON FOODS, INC.
2210 WEST OAKLAWN ROAD
SPRINGDALE, ARKANSAS 72762-6999

OVERVIEW: Produces and processes poultry products. Established: 1935. Ownership: Public.

KEY STATISTICS: Annual Sales: $6.4 billion. Number of Employees: 58,300.

EXPERTISE/EDUCATION SOUGHT: Sales, finance, business management, marketing, production, human resources, research and development, accounting

ALTERNATIVES TO FULL-TIME EMPLOYMENT: Paid internships

COMPANY-PAID TRAINING is available.

INTERNATIONAL ASSIGNMENTS are available to employees.

CONTACT: Mr. Russel Tooley, Manager of Personnel; Phone: 501-290-4000, ext. 7338; Fax: 501-290-4217.

COMPANY URL: http://www.tyson.com

CALIFORNIA

ABM INDUSTRIES, INC.
160 PACIFIC AVENUE, SUITE 222
SAN FRANCISCO, CALIFORNIA 94111

MAILING ADDRESS: PO Box 193224, San Francisco, CA 94119-3224

OVERVIEW: Provides janitorial, security, and building maintenance services. Established: 1909. Ownership: Public.

KEY STATISTICS: Annual Sales: $1.1 billion. Number of Employees: 47,000.

HIRING HISTORY: Number of professional employees hired in 1997: 10,000. 1996: 10,000.

EXPERTISE/EDUCATION SOUGHT: Computer programming, building engineering

ALTERNATIVES TO FULL-TIME EMPLOYMENT: Paid internships

COMPANY-PAID TRAINING is available.

INTERNATIONAL ASSIGNMENTS are available to employees.

CONTACT: For individuals with previous experience: Ms. Joan Lyon, Employee Relations Manager, 50 Fremont Street, San Francisco, California, 94105-2230; Phone: 415-733-4000; Fax: 415-733-7333.

COMPANY URL: http://www.abm.com

ACE TECHNOLOGIES, INC.
2860 ZANKER ROAD, SUITE 204
SAN JOSE, CALIFORNIA 95134-2120

OVERVIEW: Provides custom computer programming services. Established: 1993. Ownership: Private.

KEY STATISTICS: Annual Sales: $5 million. Number of Employees: 100.

EXPERTISE/EDUCATION SOUGHT: Computer programming, software engineering/development

CONTACT: For individuals with previous experience: Human Resources Department; Phone: 408-577-1680; Fax: 408-577-1761. E-mail: acetech@ccnet.com.

ACTEL CORPORATION
955 EAST ARQUES AVENUE
SUNNYVALE, CALIFORNIA 94086-4533

OVERVIEW: Manufactures integrated circuits and develops software. Established: 1985. Ownership: Public.

KEY STATISTICS: Annual Sales: $148.8 million. Number of Employees: 340.

HIRING HISTORY: Number of professional employees hired in 1997: 126. 1996: 119.

EXPERTISE/EDUCATION SOUGHT: Design engineering, manufacturing engineering, software engineering/development, applied engineering, systems analysis, test engineering, marketing, product engineering

ALTERNATIVES TO FULL-TIME EMPLOYMENT: Part-time jobs, paid internships

COMPANY-PAID TRAINING is available.

CONTACT: Mr. John Velcamp, Staffing Specialist; Phone: 408-739-1010; Fax: 408-522-4485. E-mail: staffing@actel.com.

COMPANY URL: http://www.actel.com

ADC MOBILE SYSTEMS, INC.
9645 SCRANTON ROAD
SAN DIEGO, CALIFORNIA 92121-1761

OVERVIEW: Manufactures radio and television equipment. Established: 1996. Parent Company: ADC Telecommunications. Ownership: Public.

KEY STATISTICS: Annual Sales: $28.3 million. Number of Employees: 333.

HIRING HISTORY: Number of professional employees hired in 1997: 50.

EXPERTISE/EDUCATION SOUGHT: Electrical engineering, hardware design, hardware engineering, software engineering/development

CONTACT: Human Resources; Phone: 619-642-7481; Fax: 619-550-0759.

COMPANY URL: http://www.adc.com

ADOBE SYSTEMS, INC.
345 PARK AVENUE
SAN JOSE, CALIFORNIA 95110

OVERVIEW: Provides graphics, file management, and page description software services. Established: 1982. Ownership: Public.

KEY STATISTICS: Annual Sales: $786.6 million. Number of Employees: 2,266.

HIRING HISTORY: Number of professional employees hired in 1997: 200.

EXPERTISE/EDUCATION SOUGHT: Graphic arts, software engineering/development, electrical engineering, computer science, business management, information systems, management information systems, software engineering/development

ALTERNATIVES TO FULL-TIME EMPLOYMENT: Part-time jobs, paid internships

INTERNATIONAL ASSIGNMENTS are available to employees.

CONTACT: Mrs. Revatchi Iyer, Staffing Manager; Phone: 408-536-6000; Fax: 408-537-6313.

COMPANY URL: http://www.adobe.com

ADVANTEST AMERICA RESEARCH DEVELOPMENT CENTER
3201 SCOTT BOULEVARD
SANTA CLARA, CALIFORNIA 95054-3008

OVERVIEW: Provides computer hardware research and development services. Established: 1994. Parent Company: Advantest. Ownership: Private.

KEY STATISTICS: Annual Sales: $10 million. Number of Employees: 42.

HIRING HISTORY: Number of professional employees hired in 1997: 8. 1996: 30.

EXPERTISE/EDUCATION SOUGHT: Hardware engineering, software engineering/development, applications engineering, administration

ALTERNATIVES TO FULL-TIME EMPLOYMENT: Part-time jobs, paid internships

COMPANY-PAID TRAINING is available.

INTERNATIONAL ASSIGNMENTS are available to employees.

CONTACT: Mr. John Catania, Human Resources Manager; Phone: 408-727-2222, ext. 299; Fax: 408-727-0740. E-mail: j.catania@advantest.com.

COMPANY URL: http://www.advantest.com

ALL DATA CORPORATION
9412 BIG HORN BOULEVARD
ELK GROVE, CALIFORNIA 95758-1101

OVERVIEW: Provides information retrieval and repair services. Established: 1996. Parent Company: AutoZone. Ownership: Public.
KEY STATISTICS: Annual Sales: $28 million. Number of Employees: 275.
HIRING HISTORY: Number of professional employees hired in 1997: 60. 1996: 60.
EXPERTISE/EDUCATION SOUGHT: Software engineering/development, sales
CONTACT: For individuals with previous experience: Personnel Department; Phone: 916-684-5200; Fax: 916-684-5225.
COMPANY URL: http://www.alldata.com

ALLIED TELESYN INTERNATIONAL CORPORATION
950 KIFER ROAD
SUNNYVALE, CALIFORNIA 94086-5206

OVERVIEW: Manufactures connectors and repeaters for network systems. Established: 1992. Ownership: Private.
KEY STATISTICS: Annual Sales: $181.3 million. Number of Employees: 450.
HIRING HISTORY: Number of professional employees hired in 1997: 30. 1996: 150.
EXPERTISE/EDUCATION SOUGHT: Hardware engineering, marketing, sales, software engineering/development, ASCI/MMIC design, finance, administration, information technology
ALTERNATIVES TO FULL-TIME EMPLOYMENT: Part-time jobs, paid internships
COMPANY-PAID TRAINING is available.
INTERNATIONAL ASSIGNMENTS are available to employees.
CONTACT: For individuals with previous experience: Mr. Ben Wong, Professional Recruiter; Phone: 408-730-0950; Fax: 408-736-0343.
COMPANY URL: http://www.alliedtelesyn.com

ALTERA CORPORATION
101 INNNOVATION DRIVE
SAN JOSE, CALIFORNIA 95134

OVERVIEW: Semiconductor company specializing in program logic devices and software to design. Established: 1983. Ownership: Public.
KEY STATISTICS: Annual Sales: $497.3 million. Number of Employees: 918.
HIRING HISTORY: Number of professional employees hired in 1997: 350. 1996: 300.
EXPERTISE/EDUCATION SOUGHT: Software engineering/development, product engineering, design engineering, marketing, design
ALTERNATIVES TO FULL-TIME EMPLOYMENT: Part-time jobs, paid internships
INTERNATIONAL ASSIGNMENTS are available to employees.
CONTACT: For college students and recent graduates: Ms. Michelle Bladow, College Relations Manager; Phone: 408-544-8079; Fax: 408-544-6409. E-mail: hr@altera.com. For individuals with previous experience: Ms. Katherine Fairbanks, Senior Manager of Staffing; Phone: 408-544-7725; Fax: 408-544-6409. E-mail: hr@altera.com.
COMPANY URL: http://www.altera.com

ALTOCOM, INC.
257 CASTRO STREET, SUITE 223
MOUNTAIN VIEW, CALIFORNIA 94041-1286

OVERVIEW: Provides computer programming services and software development. Established: 1997. Ownership: Private.
KEY STATISTICS: Annual Sales: $16 million.
EXPERTISE/EDUCATION SOUGHT: Engineering
CONTACT: For individuals with previous experience: Human Resource Department; Phone: 650-691-4502; Fax: 650-691-4503. E-mail: sales@altocom.com.
COMPANY URL: http://www.altocom.com

AMAZING CONTROLS, INC.
1615 WYATT DRIVE
SANTA CLARA, CALIFORNIA 95054-1587

OVERVIEW: Manufactures industrial control and process control instruments. Established: 1996. Ownership: Private.
KEY STATISTICS: Annual Sales: $10 million. Number of Employees: 25.
EXPERTISE/EDUCATION SOUGHT: Control engineering
COMPANY-PAID TRAINING is available.
CONTACT: Human Resources; Phone: 408-566-0300; Fax: 408-748-7724.
COMPANY URL: http://www.amazingcontrols.com

AMD
1 AMD PLACE, PO BOX 3453, MAIL STOP 935
SUNNYVALE, CALIFORNIA 94088-3453

OVERVIEW: Manufactures microprocessors and computer memory products. Established: 1969. Ownership: Public.
KEY STATISTICS: Annual Sales: $2 billion. Number of Employees: 12,200.
EXPERTISE/EDUCATION SOUGHT: Electrical engineering, computer engineering, materials science
ALTERNATIVES TO FULL-TIME EMPLOYMENT: Part-time jobs
COMPANY-PAID TRAINING is available.
INTERNATIONAL ASSIGNMENTS are available to employees.
CONTACT: Mr. Robert Lake, Manager of Employment; Phone: 408-749-4197; Fax: 408-774-7023.
COMPANY URL: http://www.amd.com

AMDAHL CORPORATION
1250 EAST ARQUES AVENUE
SUNNYVALE, CALIFORNIA 94088-3470

MAILING ADDRESS: PO Box 3470, Sunnyvale, CA 94088-3470
OVERVIEW: Manufactures mainframe computers, communications equipment, and software. Established: 1970. Ownership: Private.
KEY STATISTICS: Annual Sales: $1.6 billion. Number of Employees: 9,900.
HIRING HISTORY: Number of professional employees hired in 1997: 124. 1996: 500.
EXPERTISE/EDUCATION SOUGHT: Finance, marketing, human resources, customer service and support, software engineering/development, systems integration, systems analysis, computer programming
ALTERNATIVES TO FULL-TIME EMPLOYMENT: Part-time jobs, paid internships
COMPANY-PAID TRAINING is available.
INTERNATIONAL ASSIGNMENTS are available to employees.
CONTACT: For college students and recent graduates: Ms. Theresa Lawhead, College Recruiter; Phone: 408-746-6000; Fax: 408-746-8207. For individuals with previous experience: Ms. Regine Von Aspe, Staffing Specialist; Phone: 408-746-6000; Fax: 408-746-8207.
COMPANY URL: http://www.amdahl.com

AMERICAN RACING EQUIPMENT, INC.
19200 SOUTH REYES AVENUE
RANCHO DOMINGUEZ, CALIFORNIA 90221-5813

OVERVIEW: Manufactures aluminum wheels for automobiles. Established: 1988. Parent Company: Noranda Aluminum, Inc. Ownership: Private.
KEY STATISTICS: Annual Sales: $233.1 million. Number of Employees: 2,500.
HIRING HISTORY: Number of professional employees hired in 1997: 25.
EXPERTISE/EDUCATION SOUGHT: Sales, accounting, management information systems, marketing, engineering
COMPANY-PAID TRAINING is available.
CONTACT: For individuals with previous experience: Mr. Joaquin Correa, Corporate Employee Relations Manager; Phone: 310-635-7806; Fax: 310-537-6762. E-mail: california corporate joaquin@amracing.attmail.com.
COMPANY URL: http://www.americanracing.com

APPLE COMPUTER, INC.
1 INFINITE LOOP
CUPERTINO, CALIFORNIA 95014-2083

Overview: Develops, manufactures, and markets personal computer products. Established: 1976. Ownership: Public.

Key Statistics: Annual Sales: $7.1 billion. Number of Employees: 10,896.

Expertise/Education Sought: Business administration, software engineering/development, finance, computer programming, marketing, computer science, computer engineering

Alternatives to Full-Time Employment: Part-time jobs, paid internships

Company-Paid Training is available.

International Assignments are available to employees.

Contact: Human Resources; Phone: 800-473-7411; Fax: 408-974-6694. Job Hotline: 800-331-4496.

Company URL: http://www.apple.com

APPLIED MEDICAL RESOURCES
26051 MERIT CIRCLE, SUITE 103
LAGUNA HILLS, CALIFORNIA 92653-7023

Overview: Manufactures medical and surgical instruments. Established: 1987. Ownership: Private.

Key Statistics: Annual Sales: $21.6 million. Number of Employees: 240.

Expertise/Education Sought: Manufacturing engineering

Alternatives to Full-Time Employment: Paid internships

Contact: For individuals with previous experience: Human Resources Department; Phone: 714-582-6120.

APPLIED TECHNOLOGY ASSOCIATES, INC.
4151 MIDDLEFIELD ROAD, SUITE 101
PALO ALTO, CALIFORNIA 94303

Overview: Researches and developments technology for use in satellites and space exploration. Established: 1977. Ownership: Private.

Key Statistics: Annual Sales: $49.6 million. Number of Employees: 820.

Hiring History: Number of professional employees hired in 1997: 150.

Expertise/Education Sought: Computer science, information systems, human resources, computer programming, management, engineering, software engineering/development

Alternatives to Full-Time Employment: Paid internships

Contact: For individuals with previous experience: Human Resources Department; Phone: 650-496-2400; Fax: 650-496-2418. E-mail: hr_corp@atainc-de.com.

ARAMARK UNIFORM SERVICES, INC.
115 NORTH 1ST STREET
BURBANK, CALIFORNIA 91502

Overview: Provides industrial laundry services. Established: 1977. Parent Company: Aramark Corporation. Ownership: Private.

Key Statistics: Annual Sales: $623.4 million. Number of Employees: 13,500.

Expertise/Education Sought: Accounting, finance, management information systems, sales, operations

Company-Paid Training is available.

Contact: Human Resources Department; Phone: 818-973-3700.

Company URL: http://www.aramark-uniform.com

ARTECON, INC.
6305 EL CAMINO REAL
CARLSBAD, CALIFORNIA 92009

Overview: Provides computer systems and integration design services. Established: 1984. Parent Company: Nordic Systems. Ownership: Private.

Key Statistics: Annual Sales: $83 million. Number of Employees: 250.

Expertise/Education Sought: Manufacturing engineering, mechanical engineering, electrical engineering

Alternatives to Full-Time Employment: Part-time jobs

Contact: Human Resources Department; Phone: 619-931-5500. E-mail: apply@artecon.com.

Company URL: http://www.artecon.com

ASANTE TECHNOLOGIES, INC.
821 FOX LANE
SAN JOSE, CALIFORNIA 95131-1601

Overview: Manufactures data networking products. Established: 1988. Ownership: Public.

Key Statistics: Annual Sales: $83.3 million. Number of Employees: 197.

Hiring History: Number of professional employees hired in 1997: 69. 1996: 50.

Expertise/Education Sought: Manufacturing engineering, software engineering/development, design engineering, marketing, ASCI/MMIC design, product management, information systems, technical support

Alternatives to Full-Time Employment: Paid internships

Company-Paid Training is available.

International Assignments are available to employees.

Contact: For individuals with previous experience: Ms. Bette Young, Human Resources Director; Phone: 408-435-8401; Fax: 408-894-9150. E-mail: hr@asante.com.

Company URL: http://www.asante.com

ASTEC AMERICA, INC.
6339 PASEO DEL LAGO
CARLSBAD, CALIFORNIA 92009-1309

Overview: Manufactures power supply equipment for computers. Established: 1975. Parent Company: Emerson Electric, Inc. Ownership: Private.

Key Statistics: Annual Sales: $550 million. Number of Employees: 2,000.

Expertise/Education Sought: Electrical engineering, sales, administration, technical

International Assignments are available to employees.

Contact: Ms. Lois Dehne, Human Resources Manager; Phone: 760-930-4755; Fax: 760-930-0698.

Company URL: http://www.astec.com

ATMEL CORPORATION
2325 ORCHARD PARKWAY
SAN JOSE, CALIFORNIA 95131-2032

Overview: Manufactures integrated microcircuits. Established: 1984. Ownership: Public.

Key Statistics: Annual Sales: $1.1 billion. Number of Employees: 3,914.

Hiring History: Number of professional employees hired in 1997: 202. 1996: 200.

Expertise/Education Sought: Design engineering, product engineering, test engineering, applications engineering, equipment operation, quality assurance, ASCI/MMIC design

Alternatives to Full-Time Employment: Part-time jobs, paid internships

International Assignments are available to employees.

Contact: For college students and recent graduates: Ms. Donna Gilmour, College Recruiter; Phone: 408-441-0311; Fax: 408-451-4828. E-mail: resume@atmel.com. For individuals with previous experience: Mr. Chris Shols, Human Resources Representative; Phone: 408-436-4234; Fax: 408-451-4828. E-mail: resume@atmel.com.

Company URL: http://www.atmel.com

AURORA ELECTRONICS
9477 WAPLES STREET
SAN DIEGO, CALIFORNIA 92121

OVERVIEW: Manufactures semiconductors. Established: 1992. Ownership: Public.

KEY STATISTICS: Annual Sales: $98 million. Number of Employees: 350.

EXPERTISE/EDUCATION SOUGHT: Manufacturing engineering, test engineering, sales, applied engineering, purchasing

ALTERNATIVES TO FULL-TIME EMPLOYMENT: Part-time jobs

CONTACT: For individuals with previous experience: Human Resources; Phone: 619-552-1213; Fax: 619-552-1402.

COMPANY URL: http://www.aur.com

AUSPEX SYSTEMS, INC.
2300 CENTRAL EXPRESSWAY
SANTA CLARA, CALIFORNIA 95050-2516

OVERVIEW: Manufactures computer file servers. Established: 1987. Ownership: Public.

KEY STATISTICS: Annual Sales: $202.5 million. Number of Employees: 619.

HIRING HISTORY: Number of professional employees hired in 1996: 220.

EXPERTISE/EDUCATION SOUGHT: Manufacturing engineering, mechanical engineering, software engineering/development, sales, marketing

ALTERNATIVES TO FULL-TIME EMPLOYMENT: Part-time jobs, paid internships

COMPANY-PAID TRAINING is available.

INTERNATIONAL ASSIGNMENTS are available to employees.

CONTACT: For individuals with previous experience: Mr. Tom Perry, Senior Technical Recruiter; Phone: 408-566-2329; Fax: 408-566-2366. E-mail: tperry@corp.auspex.com.

COMPANY URL: http://www.auspex.com

BARCLAYS GLOBAL INVESTORS
45 FREMONT STREET, FLOOR 17
SAN FRANCISCO, CALIFORNIA 94105-2204

OVERVIEW: Provides investment advisory services. Established: 1995. Parent Company: Barclay's Bank. Ownership: Private.

KEY STATISTICS: Annual Sales: $108.3 million. Number of Employees: 801.

EXPERTISE/EDUCATION SOUGHT: Business administration, software engineering/development, accounting

ALTERNATIVES TO FULL-TIME EMPLOYMENT: Part-time jobs

CONTACT: Professional Staffing; Fax: 415-597-2492. E-mail: staffing@bglobal.com.

COMPANY URL: http://www.bglobal.com

BEA SYSTEMS, INC.
385 MOFFETT PARK DRIVE
SUNNYVALE, CALIFORNIA 94089-1208

OVERVIEW: Provides computer systems design and application services. Established: 1995. Ownership: Private.

KEY STATISTICS: Annual Sales: $61.6 million. Number of Employees: 459.

HIRING HISTORY: Number of professional employees hired in 1997: 450. 1996: 400.

EXPERTISE/EDUCATION SOUGHT: Software engineering/development

ALTERNATIVES TO FULL-TIME EMPLOYMENT: Paid internships

INTERNATIONAL ASSIGNMENTS are available to employees.

CONTACT: Human Resources Department; Phone: 408-743-4000; Fax: 408-734-9234. E-mail: resumes@beasys.com.

COMPANY URL: http://www.beasys.com

BECHTEL SYSTEMS INFRASTRUCTURE
50 BEALE STREET
SAN FRANCISCO, CALIFORNIA 94105-1813

OVERVIEW: Provides civil engineering services. Established: 1996. Ownership: Private.

KEY STATISTICS: Annual Sales: $850 million. Number of Employees: 4,200.

HIRING HISTORY: Number of professional employees hired in 1997: 9,000.

EXPERTISE/EDUCATION SOUGHT: Engineering, computer programming

ALTERNATIVES TO FULL-TIME EMPLOYMENT: Paid internships

INTERNATIONAL ASSIGNMENTS are available to employees.

CONTACT: For individuals with previous experience: Human Resource Department; Phone: 415-768-5222; Fax: 415-768-1104.

COMPANY URL: http://www.bechtel.com

BIRD PRODUCTS CORPORATION
1100 BIRD CENTER DRIVE
PALM SPRINGS, CALIFORNIA 92262-8000

OVERVIEW: Manufactures medical equipment and supplies. Established: 1995. Parent Company: Thermo Electron Corporation. Ownership: Public.

KEY STATISTICS: Annual Sales: $53.8 million. Number of Employees: 400.

HIRING HISTORY: Number of professional employees hired in 1997: 20. 1996: 20.

EXPERTISE/EDUCATION SOUGHT: Electrical engineering, mechanical engineering, software engineering/development, mechanical engineering

COMPANY-PAID TRAINING is available.

INTERNATIONAL ASSIGNMENTS are available to employees.

CONTACT: Ms. Della Adams, Personnel Supervisor; Phone: 760-778-7253; Fax: 760-778-7338.

CADENCE DESIGN SYSTEMS, INC.
555 RIVER OAKS PARKWAY
SAN JOSE, CALIFORNIA 95134

OVERVIEW: Develops electronic design automation software. Established: 1982. Ownership: Public.

KEY STATISTICS: Annual Sales: $741.5 million. Number of Employees: 3,190.

HIRING HISTORY: Number of professional employees hired in 1996: 1,300.

EXPERTISE/EDUCATION SOUGHT: Electrical engineering, design engineering

ALTERNATIVES TO FULL-TIME EMPLOYMENT: Part-time jobs, paid internships

INTERNATIONAL ASSIGNMENTS are available to employees.

CONTACT: Staffing Department; Phone: 408-944-7428; Fax: 408-955-0820. Job Hotline: 408-894-2600.

COMPANY URL: http://www.cadence.com

CAERE CORPORATION
100 COOPER COURT
LOS GATOS, CALIFORNIA 95030-3321

OVERVIEW: Develops information recognition software products. Established: 1973. Ownership: Public.

KEY STATISTICS: Annual Sales: $54.5 million. Number of Employees: 272.

HIRING HISTORY: Number of professional employees hired in 1997: 10. 1996: 10.

EXPERTISE/EDUCATION SOUGHT: Computer programming, software engineering/development, program analysis, marketing, sales, software engineering/development, software engineering/development, technical support

ALTERNATIVES TO FULL-TIME EMPLOYMENT: Part-time jobs, paid internships

COMPANY-PAID TRAINING is available.

International Assignments are available to employees.

Contact: For individuals with previous experience: Human Resources Department; Phone: 408-395-7000; Fax: 408-354-5166. Job Hotline: 408-354-5166.

Company URL: http://www.caere.com

CALIFORNIA AMPLIFIER, INC.

460 CALLE SAN PABLO
CAMARILLO, CALIFORNIA 93012

Overview: Manufactures systems for microwave distribution of television signals. Established: 1981. Ownership: Public.

Key Statistics: Annual Sales: $49.3 million. Number of Employees: 348.

Expertise/Education Sought: Microwave engineering, radio frequency engineering, electrical engineering, manufacturing

Alternatives to Full-Time Employment: Paid internships

International Assignments are available to employees.

Contact: Ms. Jackie Sheehan, Human Resources Manager; Phone: 805-987-9000, ext. 8245; Fax: 805-987-2655. E-mail: jackies@calamp.com.

CALIFORNIA PUBLIC UTILITIES COMMISSION

505 VAN NESS AVENUE
SAN FRANCISCO, CALIFORNIA 94102-3298

Overview: Regulates state utilities. Established: 1911. Ownership: Private.

Key Statistics: Number of Employees: 820.

Expertise/Education Sought: Analysis, accounting, engineering

Alternatives to Full-Time Employment: Internships

Company-Paid Training is available.

Contact: Mrs. Michelle Diamonon, Recruitment Coordinator; Phone: 415-703-2160; Fax: 415-703-1758. E-mail: diamonon@razorlogic.com.

Company URL: http://www.cpuc.ca.gov

CANDLE CORPORATION

2425 OLYMPIC BOULEVARD
SANTA MONICA, CALIFORNIA 90404-4030

Overview: Develops systems software. Established: 1977. Ownership: Private.

Key Statistics: Annual Sales: $300 million. Number of Employees: 1,400.

Hiring History: Number of professional employees hired in 1997: 500. 1996: 200.

Expertise/Education Sought: Software engineering/development, computer programming, computer science

Alternatives to Full-Time Employment: Part-time jobs, paid internships

International Assignments are available to employees.

Contact: For individuals with previous experience: Ms. Judy Wortsman, Human Resources Manager; Phone: 310-582-4174; Fax: 310-582-4262. E-mail: judy_wordsmith@candle.com.

Company URL: http://www.candle.com

CANON BUSINESS MACHINES, INC.

3191 RED HILL AVENUE
COSTA MESA, CALIFORNIA 92626-3401

Overview: Manufactures typewriters, copying machines, and other office equipment. Established: 1971. Parent Company: Canon, Inc. Ownership: Private.

Key Statistics: Annual Sales: $286.2 million. Number of Employees: 850.

Expertise/Education Sought: Software engineering/development, electrical engineering, mechanical engineering, design engineering, accounting, computer programming, quality engineering, supply engineering, network management systems

Alternatives to Full-Time Employment: Paid internships

Company-Paid Training is available.

Contact: Ms. Bonnie Robinson, Manager of Human Resources; Phone: 714-556-4700; Fax: 714-433-8103. E-mail: resumes@cbm.canon.com.

Company URL: http://www.careermosaic.com/cm/cbm

CAST ALLOYS, INC.

20409 PRAIRIE STREET
CHATSWORTH, CALIFORNIA 91311

Overview: Manufactures and supplies castings for golf clubs. Established: 1984. Parent Company: Nieman Porter and Company. Ownership: Private.

Key Statistics: Annual Sales: $50 million. Number of Employees: 1,500.

Expertise/Education Sought: Metallurgical engineering, mechanical engineering

International Assignments are available to employees.

Contact: Director of Human Resources; Phone: 818-718-3131; Fax: 818-407-8981.

CCI/TRIAD

3055 TRIAD DRIVE
LIVERMORE, CALIFORNIA 94550-9559

Overview: Provides turnkey computer systems for the automotive after-market and retail hardware industries. Established: 1972. Parent Company: Cooperative Computing, Inc. Ownership: Private.

Key Statistics: Annual Sales: $175.7 million. Number of Employees: 1,503.

Hiring History: Number of professional employees hired in 1996: 150.

Expertise/Education Sought: Management information systems, systems analysis

Alternatives to Full-Time Employment: Internships

Company-Paid Training is available.

International Assignments are available to employees.

Contact: For individuals with previous experience: Ms. Ann Patterson, Corporate Hiring Manager; Phone: 510-449-0606; Fax: 510-606-2105. Job Hotline: 800-289-2677.

Company URL: http://www.cci/triad.com

CENDANT SOFTWARE

19840 PIONEER AVENUE
TORRANCE, CALIFORNIA 90503-1660

Overview: Develops prepackaged software. Established: 1996. Parent Company: Cendant Corp. Ownership: Public.

Key Statistics: Annual Sales: $500 million. Number of Employees: 1,806.

Expertise/Education Sought: Software engineering/development

Alternatives to Full-Time Employment: Paid internships

Contact: Mr. Michael Jaimes, Recruiter; Phone: 310-793-0600, ext. 1871; Fax: 310-793-0601. E-mail: mjaimes@cendantsoft.com. Job Hotline: 310-793-0599.

Company URL: http://www.davd.com

CENTRAL CORPORATION

811 HANSEN WAY
PALO ALTO, CALIFORNIA 94303-0750

Overview: Provides custom computer programming services. Established: 1996. Ownership: Private.

Key Statistics: Annual Sales: $18 million.

Hiring History: Number of professional employees hired in 1997: 3.

Expertise/Education Sought: Software engineering/development, sales, marketing

Contact: Human Resources; Phone: 650-846-3612; Fax: 650-858-0454.

Company URL: http://www.centraal.com

CENTURA SOFTWARE CORPORATION
975 ISLAND DRIVE
REDWOOD CITY, CALIFORNIA 94065

Overview: Develops client/server software. Established: 1984. Ownership: Public.
Key Statistics: Annual Sales: $63.2 million. Number of Employees: 270.
Expertise/Education Sought: Software engineering/development
Alternatives to Full-Time Employment: Part-time jobs, paid internships
International Assignments are available to employees.
Contact: For individuals with previous experience: Employment Services Human Resources; Phone: 650-596-3400; Fax: 650-596-4980. E-mail: jobs@centurasoft.com.
Company URL: http://www.centurasoft.com

CHEVRON CORPORATION
575 MARKET STREET, ROOM 1246
SAN FRANCISCO, CALIFORNIA 94105

Overview: Produces and refines oil and gas. Established: 1879. Ownership: Public.
Key Statistics: Annual Sales: $42.8 billion. Number of Employees: 40,820.
Hiring History: Number of professional employees hired in 1997: 200. 1996: 150.
Expertise/Education Sought: Chemical engineering, petroleum/petrochemical engineering, mechanical engineering
Alternatives to Full-Time Employment: Part-time jobs, paid internships
Company-Paid Training is available.
International Assignments are available to employees.
Contact: Mr. Mark Witzke, Team Leader of Recruiting; Phone: 415-894-4322; Fax: 415-894-5233.
Company URL: http://www.chevron.com

CIRRUS LOGIC, INC.
3100 WEST WARREN AVENUE
FREMONT, CALIFORNIA 94538-6423

Overview: Manufactures semiconductors and integrated circuits. Established: 1981. Ownership: Public.
Key Statistics: Annual Sales: $917.2 million. Number of Employees: 2,135.
Hiring History: Number of professional employees hired in 1997: 200. 1996: 200.
Expertise/Education Sought: Accounting, finance, software engineering/development, design engineering, product engineering, test engineering, applications engineering
Alternatives to Full-Time Employment: Internships
Company-Paid Training is available.
International Assignments are available to employees.
Contact: For individuals with previous experience: Human Resources; Phone: 510-623-8300; Fax: 510-624-7140. Job Hotline: 888-224-7787.
Company URL: http://www.cirrus.com

CISCO SYSTEMS, INC.
170 WEST TASMIN DRIVE
SAN JOSE, CALIFORNIA 95134

Overview: Manufactures and markets computer networking systems.
Expertise/Education Sought: Hardware engineering, test engineering, manufacturing engineering, finance, information services, accounting, customer service and support
Alternatives to Full-Time Employment: Part-time jobs, paid internships
International Assignments are available to employees.
Contact: Human Resources.
Company URL: http://www.cisco.com.

CLARK PACIFIC
1980 SOUTH RIVER ROAD
WEST SACRAMENTO, CALIFORNIA 95691-2817

Overview: Manufactures architectural slabs. Established: 1978. Ownership: Private.
Key Statistics: Annual Sales: $30 million. Number of Employees: 250.
Hiring History: Number of professional employees hired in 1997: 20. 1996: 25.
Expertise/Education Sought: Civil engineering, construction engineering, structural engineering
Contact: For individuals with previous experience: Mr. Robert E. Clark, Chief Engineer; Phone: 916-371-0305.
Company URL: http://www.clarkpacific.com

CMOS SOFT, INC.
3040 CORONADO DRIVE
SANTA CLARA, CALIFORNIA 95054-3203

Overview: Provides computer services. Established: 1996. Parent Company: CMOS Technologies. Ownership: Private.
Key Statistics: Number of Employees: 200.
Hiring History: Number of professional employees hired in 1997: 50. 1996: 20.
Expertise/Education Sought: Design, software engineering/development
International Assignments are available to employees.
Contact: For individuals with previous experience: Human Resources Department; Phone: 408-982-9997; Fax: 408-982-0818.
Company URL: http://www.cmos.com

CNET TECHNOLOGY TOTAL NETWORK SOLUTIONS
1455 MCCANDLESS DRIVE
MILPITAS, CALIFORNIA 95035

Overview: Manufactures local area networking products. Established: 1987. Ownership: Private.
Key Statistics: Annual Sales: $65 million. Number of Employees: 250.
Expertise/Education Sought: Sales, technical engineering, marketing
Alternatives to Full-Time Employment: Part-time jobs
Company-Paid Training is available.
International Assignments are available to employees.
Contact: For individuals with previous experience: Human Resources; Phone: 408-934-0800; Fax: 408-934-0900.
Company URL: http://www.cnet.com

COMAIR ROTRON, INC.
2675 CUSTOMHOUSE COURT
SAN YSINDRO, CALIFORNIA 92173-3618

Overview: Designs and manufactures commercial cooling fans. Established: 1996. Ownership: Private.
Key Statistics: Annual Sales: $65 million. Number of Employees: 1,000.
Hiring History: Number of professional employees hired in 1997: 4. 1996: 3.
Expertise/Education Sought: Design engineering, administration, electrical engineering, computer programming, data processing, design engineering
Alternatives to Full-Time Employment: Part-time jobs, paid internships
International Assignments are available to employees.
Contact: For individuals with previous experience: Mr. Kenneth Brarrera, Treasurer; Phone: 619-661-6688; Fax: 619-661-6057.
Company URL: http://www.comairrotron.com

CONTOUR AEROSPACE
428 BERRY WAY
BREA, CALIFORNIA 92821-3114

Overview: Manufactures aircraft parts and equipment. Established: 1996. Ownership: Private.

Key Statistics: Annual Sales: $63 million. Number of Employees: 375.
Expertise/Education Sought: Industrial engineering, mechanical engineering
Company-Paid Training is available.
International Assignments are available to employees.
Contact: For individuals with previous experience: Human Resources; Phone: 714-990-9281; Fax: 714-672-6585.

CON-WAY TRANSPORTATION SERVICES, INC.
3340 HILLVIEW AVENUE
PALO ALTO, CALIFORNIA 94304

Overview: Provides truck and transport services. Established: 1988. Parent Company: CNF Transportation, Inc. Ownership: Public.
Key Statistics: Annual Sales: $1 billion. Number of Employees: 14,100.
Expertise/Education Sought: Logistics, sales, distribution, transportation, information systems
Alternatives to Full-Time Employment: Part-time jobs, paid internships
International Assignments are available to employees.
Contact: Mr. Robert Coon, Director of Human Resources; Phone: 650-424-1400; Fax: 650-854-8219. E-mail: coon.robert@con-way.com.
Company URL: http://www.con-way.com

CORE-VENT BIOENGINEERING
27030 MALIBU HILLS ROAD
CALABASAS HILLS, CALIFORNIA 91301-5332

Overview: Manufactures surgical appliances and supplies surgical implants. Established: 1989. Ownership: Private.
Key Statistics: Annual Sales: $12.1 million. Number of Employees: 123.
Expertise/Education Sought: Precision machining
Contact: For individuals with previous experience: Human Resources Department; Phone: 818-878-7818; Fax: 818-878-7828.

CPI, TRAVELING WAVE TUBE PRODUCTS DIVISION
607 HANSEN WAY, PO BOX 51110
PALO ALTO, CALIFORNIA 94303-1110

Overview: Manufactures vacuum tubes. Established: 1995. Ownership: Private.
Key Statistics: Annual Sales: $257.4 million. Number of Employees: 1,800.
Expertise/Education Sought: Engineering, manufacturing engineering, engineering, engineering
Alternatives to Full-Time Employment: Part-time jobs
Contact: Mr. Mario Bertucci, Human Resources Director; Phone: 650-846-3143; Fax: 650-856-8047. E-mail: careers@twtp.cpii.com.
Company URL: http://www.twtc.com

CRANE HYDRO-AIRE
3000 WINONA AVENUE
BURBANK, CALIFORNIA 91504-2540

Overview: Manufactures aircraft parts and equipment. Established: 1997. Parent Company: CRANE, Inc. Ownership: Public.
Key Statistics: Annual Sales: $41.8 million. Number of Employees: 500.
Hiring History: Number of professional employees hired in 1997: 140. 1996: 70.
Expertise/Education Sought: Systems engineering, manufacturing engineering, manufacturing management, tool and die engineering, software engineering/development, technical writing, packaging engineering
Alternatives to Full-Time Employment: Paid internships
International Assignments are available to employees.
Contact: Mr. Michael Baral, Senior Technical Recruiter; Phone: 818-526-2600, ext. 2662; Fax: 818-526-2235. E-mail: hr@hydroaire.com.
Company URL: http://www.hydroaire.com

CROCKETT COGENERATION CA, LP
550 LORING AVENUE
CROCKETT, CALIFORNIA 94525-1232

Overview: Generates electric service. Established: 1993. Ownership: Private.
Key Statistics: Annual Sales: $46 million.
Hiring History: Number of professional employees hired in 1997: 5. 1996: 4.
Expertise/Education Sought: Instrumentation engineering
Alternatives to Full-Time Employment: Paid internships
Company-Paid Training is available.
Contact: For individuals with previous experience: Human Resources Department; Phone: 510-787-4100; Fax: 510-787-4150.
Company URL: http://www.crockettcogen.com

CROWLEY MARITIME CORPORATION
155 GRAND AVENUE
OAKLAND, CALIFORNIA 94612-3758

Overview: Provides marine shipping services. Established: 1892. Ownership: Private.
Key Statistics: Annual Sales: $1.1 billion. Number of Employees: 5,000.
Expertise/Education Sought: Management, management information systems, systems analysis, computer programming, personal computers
Alternatives to Full-Time Employment: Paid internships
Company-Paid Training is available.
International Assignments are available to employees.
Contact: Mrs. Cindy Oylar, Human Resources Coordinator; Phone: 510-251-7500; Fax: 510-251-7610.
Company URL: http://www.crowley.com

CUBIC CORPORATION
9333 BALBOA AVENUE
SAN DIEGO, CALIFORNIA 92123-1305

Overview: Manufactures measurement and control devices. Established: 1972. Ownership: Public.
Key Statistics: Annual Sales: $119.4 million. Number of Employees: 230.
Hiring History: Number of professional employees hired in 1997: 280. 1996: 374.
Expertise/Education Sought: Electronics/electronics engineering, computer science, systems engineering, hardware engineering, project management, quality engineering, manufacturing engineering, hardware design, software engineering/development, management
Alternatives to Full-Time Employment: Part-time jobs, paid internships
International Assignments are available to employees.
Contact: For college students and recent graduates: Mr. Joe Castiglione, College Recruiter; Phone: 619-505-2854; Fax: 619-505-1524. E-mail: joe.castiglione@cubic.com. For individuals with previous experience: Mr. Bob Stamp, Staffing Manager, PO Box 85587, San Diego, California, 92186-5587; Phone: 619-505-2275; Fax: 619-505-1524. E-mail: bob.stamp@cubic.com. Job Hotline: 619-505-1540.
Company URL: http://www.cubic.com

CYPRESS SEMICONDUCTOR CORPORATION
3901 NORTH 1ST STREET
SAN JOSE, CALIFORNIA 95134-1599

Overview: Manufactures semiconductors. Established: 1982. Ownership: Public.
Key Statistics: Annual Sales: $528.4 million. Number of Employees: 2,171.
Expertise/Education Sought: Design engineering, electrical engineering, computer programming, program analysis, process engineering
Alternatives to Full-Time Employment: Part-time jobs, paid internships

Cypress Semiconductor Corporation (continued)

Company-Paid Training is available.
International Assignments are available to employees.
Contact: For individuals with previous experience: Ms. Brie Silberia, Staffing Supervisor. E-mail: bks@cypress.com.
Company URL: http://www.cypress.com

DATA EXCHANGE CORPORATION
3600 VIA PESCADOR
CAMARILLO, CALIFORNIA 93012-5051

Overview: Provides computer repair services. Established: 1980. Ownership: Private.
Key Statistics: Annual Sales: $40 million. Number of Employees: 400.
Hiring History: Number of professional employees hired in 1997: 25.
Expertise/Education Sought: Electronics/electronics engineering, computer programming, mechanical engineering, sales, logistics, microwave engineering, radio frequency engineering, electrical engineering, manufacturing
Company-Paid Training is available.
International Assignments are available to employees.
Contact: Human Resources Department; Phone: 805-388-1711, ext. 332; Fax: 805-445-2941. E-mail: hr@dex.com.
Company URL: http://www.dex.com

DEN-MAT CORPORATION
2727 SKYWAY DRIVE
SANTA MARIA, CALIFORNIA 93455-1413

Mailing Address: PO Box 1729, Santa Maria, CA 93456
Overview: Manufactures dental products. Established: 1972. Ownership: Private.
Key Statistics: Annual Sales: $82 million. Number of Employees: 450.
Expertise/Education Sought: Dentistry/dental hygiene, accounting, finance, data processing, chemistry, pharmacy, marketing
Contact: Mr. Al Petit, Manager of Personnel; Phone: 805-922-8491, ext. 3403; Fax: 805-739-0578.

DIAGNOSTIC PRODUCTS CORPORATION
5700 WEST 96TH STREET
LOS ANGELES, CALIFORNIA 90045

Overview: Manufactures diagnostic test kits. Established: 1971. Ownership: Public.
Key Statistics: Annual Sales: $176.8 million. Number of Employees: 1,325.
Hiring History: Number of professional employees hired in 1996: 75.
Expertise/Education Sought: Chemistry, research, biochemistry, biology, molecular biology, quality control
Alternatives to Full-Time Employment: Paid internships
Contact: For individuals with previous experience: Ms. Casandra Davis, Human Resources Recruiter; Phone: 213-776-0180, ext. 2509; Fax: 213-776-0204. E-mail: cdavis@dpconline.com. Job Hotline: 213-776-0180, ext. 260.

DINWIDDIE CONSTRUCTION COMPANY
275 BATTERY STREET, SUITE 300
SAN FRANCISCO, CALIFORNIA 94111

Overview: Builds and renovates commercial office buildings. Established: 1996. Parent Company: Hathaway Dinwiddie Construction Group. Ownership: Private.
Key Statistics: Annual Sales: $48.5 million. Number of Employees: 200.
Hiring History: Number of professional employees hired in 1997: 20. 1996: 25.
Expertise/Education Sought: Accounting, sales, manufacturing engineering, management, property management
Alternatives to Full-Time Employment: Paid internships
Company-Paid Training is available.
Contact: For individuals with previous experience: Human Resources Department; Phone: 415-986-2718; Fax: 415-956-5669. E-mail: milnea@hdcco.com.
Company URL: http://www.hdcco.com

DYNAMIS, INC.
1125 MUSCAT AVENUE
SANGER, CALIFORNIA 93657-4000

Overview: Provides civil engineering services. Established: 1988. Ownership: Private.
Hiring History: Number of professional employees hired in 1997: 6.
Expertise/Education Sought: Electrical engineering
Alternatives to Full-Time Employment: Internships
Company-Paid Training is available.
International Assignments are available to employees.
Contact: Mr. George Carroll, General Manager; Phone: 209-875-0800; Fax: 209-875-0667.

ELECTRONIC ARTS
1450 FASHION ISLAND BOULEVARD
SAN MATEO, CALIFORNIA 94404-2063

Overview: Develops interactive entertainment, education, and information software. Established: 1982. Ownership: Public.
Key Statistics: Annual Sales: $624.8 million. Number of Employees: 1,700.
Hiring History: Number of professional employees hired in 1996: 200.
Expertise/Education Sought: Sales, software engineering/development, computer programming, accounting, information systems
Alternatives to Full-Time Employment: Part-time jobs
Company-Paid Training is available.
International Assignments are available to employees.
Contact: Human Resource Department; Phone: 650-571-7171; Fax: 650-513-7160.
Company URL: http://www.ea.com

ENDOSONICS CORPORATION
2870 KILGORE ROAD
RANCHO CORDOVA, CALIFORNIA 95670

Overview: Manufactures medical and surgical equipment. Established: 1984. Ownership: Public.
Key Statistics: Annual Sales: $24.4 million. Number of Employees: 300.
Hiring History: Number of professional employees hired in 1996: 60.
Expertise/Education Sought: Engineering, sales, marketing, medical technology
Alternatives to Full-Time Employment: Internships
International Assignments are available to employees.
Contact: Ms. Norma Munoz, Human Resource Manager; Phone: 916-638-8008. E-mail: hr@endosonics.com.
Company URL: http://www.endosonics.com

ENVIORMENTAL ADVANCED ENGINEERING, INC.
300 NORTH LAKE AVENUE, SUITE 920
PASADENA, CALIFORNIA 91101

Overview: Provides management services. Established: 1987. Ownership: Private.
Key Statistics: Annual Sales: $13.9 million. Number of Employees: 394.
Expertise/Education Sought: Environmental engineering
Contact: For individuals with previous experience: Human Resources; Phone: 818-577-9936; Fax: 818-304-9327.

ETAK, INC.

1430 O'BRIEN DRIVE
MENLO PARK, CALIFORNIA 94025

OVERVIEW: Provides mapping services for cassette tapes and compact disks. Established: 1996. Parent Company: Sony Corporation. Ownership: Private.

KEY STATISTICS: Annual Sales: $39.8 million. Number of Employees: 600.

HIRING HISTORY: Number of professional employees hired in 1997: 50.

EXPERTISE/EDUCATION SOUGHT: Finance, marketing, sales, engineering

ALTERNATIVES TO FULL-TIME EMPLOYMENT: Part-time jobs

COMPANY-PAID TRAINING is available.

CONTACT: For individuals with previous experience: Mr. Steve Brieger, Staffing Consultant; Phone: 415-328-3825, ext. 1403.

COMPANY URL: http://www.etak.com

EXSIL, INC.

6541 VIA DEL ORO
SAN JOSE, CALIFORNIA 95119-1207

OVERVIEW: Recycles semiconductor reject wafers. Established: 1984. Ownership: Private.

KEY STATISTICS: Annual Sales: $18.4 million. Number of Employees: 150.

HIRING HISTORY: Number of professional employees hired in 1997: 45. 1996: 25.

EXPERTISE/EDUCATION SOUGHT: Production, process engineering

ALTERNATIVES TO FULL-TIME EMPLOYMENT: Paid internships

CONTACT: Ms. Sherry Bunnell, Human Resources Manager; Phone: 408-629-3142; Fax: 408-629-3168. E-mail: exsilsj@aol.com.

FAILURE ANALYSIS ASSOCIATES

PO BOX 3015
MENLO PARK, CALIFORNIA 94025-1133

OVERVIEW: Engineering consulting firm. Established: 1967. Parent Company: The Failure Group. Ownership: Public.

KEY STATISTICS: Annual Sales: $52.1 million. Number of Employees: 320.

HIRING HISTORY: Number of professional employees hired in 1997: 80. 1996: 80.

EXPERTISE/EDUCATION SOUGHT: Mechanical engineering, automotive engineering, materials science, polymers, petroleum/petrochemical engineering, civil engineering, structural engineering, geotechnology, biomedical engineering, thermal engineering

ALTERNATIVES TO FULL-TIME EMPLOYMENT: Paid internships

CONTACT: Ms. Jannine Davis, Recruiter; Phone: 650-326-9400; Fax: 650-328-3049. E-mail: hr@fail.com.

COMPANY URL: http://www.fail.com

FAIR, ISAAC AND COMPANY, INC.

120 NORTH REDWOOD DRIVE
SAN RAFAEL, CALIFORNIA 94903

OVERVIEW: Develops software. Established: 1955. Ownership: Public.

KEY STATISTICS: Annual Sales: $148.7 million. Number of Employees: 1,037.

HIRING HISTORY: Number of professional employees hired in 1997: 200. 1996: 150.

EXPERTISE/EDUCATION SOUGHT: Credit/credit analysis, computer programming, program analysis, statistics

ALTERNATIVES TO FULL-TIME EMPLOYMENT: Paid internships

INTERNATIONAL ASSIGNMENTS are available to employees.

CONTACT: For college students and recent graduates: Ms. Katharine Filak, College Recruiter; Phone: 415-491-5284; Fax: 415-491-5100. E-mail: kfilak@fairisaac.com. For individuals with previous experience: Ms. Mick Wets, Staffing Manager; Phone: 415-472-2211; Fax: 415-491-5100. E-mail: mickwets@fairisaac.com. Job Hotline: 415-491-7077.

COMPANY URL: http://www.fairisaac.com

FILENET CORPORATION

3565 HARBOR BOULEVARD
COSTA MESA, CALIFORNIA 92626-1405

OVERVIEW: Manufactures and markets document imaging and management systems. Established: 1982. Ownership: Public.

KEY STATISTICS: Annual Sales: $268.9 million. Number of Employees: 1,463.

HIRING HISTORY: Number of professional employees hired in 1997: 364. 1996: 390.

EXPERTISE/EDUCATION SOUGHT: Computer science, finance, systems engineering, accounting, management information systems, administration, computer programming, sales, marketing

ALTERNATIVES TO FULL-TIME EMPLOYMENT: Part-time jobs, paid internships

COMPANY-PAID TRAINING is available.

INTERNATIONAL ASSIGNMENTS are available to employees.

CONTACT: For college students and recent graduates: Ms. Julia Miller, College Recruiter; Fax: 714-966-3440. E-mail: resumes@filenet.com. For individuals with previous experience: Mr. Kenneth Ross, Manager of Global Employment; Phone: 714-966-3400; Fax: 714-966-3440. E-mail: resumes@filenet.com.

COMPANY URL: http://www.filenet.com

FLEETWOOD ENTERPRISES, INC.

3125 MYERS STREET
RIVERSIDE, CALIFORNIA 92503-5527

OVERVIEW: Manufactures recreational vehicles and houses. Established: 1950. Ownership: Public.

KEY STATISTICS: Annual Sales: $2.9 billion. Number of Employees: 17,132.

EXPERTISE/EDUCATION SOUGHT: Accounting, finance, marketing, computer programming, purchasing, sales

ALTERNATIVES TO FULL-TIME EMPLOYMENT: Part-time jobs

COMPANY-PAID TRAINING is available.

CONTACT: For individuals with previous experience: Human Resources Department; Phone: 909-351-3903; Fax: 909-351-3312. Job Hotline: 909-788-5627.

COMPANY URL: http://www.fleetwood.com

FLEXTRONICS INTERNATIONAL USA, INC.

2090 FORTUNE DRIVE
SAN JOSE, CALIFORNIA 95131

OVERVIEW: Produces solid state modules. Established: 1995. Ownership: Private.

KEY STATISTICS: Annual Sales: $77 million. Number of Employees: 1,000.

HIRING HISTORY: Number of professional employees hired in 1997: 800.

EXPERTISE/EDUCATION SOUGHT: Electrical engineering, manufacturing engineering, mechanical engineering, design engineering, technical engineering, process engineering

ALTERNATIVES TO FULL-TIME EMPLOYMENT: Part-time jobs

COMPANY-PAID TRAINING is available.

INTERNATIONAL ASSIGNMENTS are available to employees.

CONTACT: For individuals with previous experience: Human Resources; Phone: 408-428-1300; Fax: 408-428-0237.

COMPANY URL: http://www.flextronics.com

FLUOR DANIEL, INC.

3353 MICHELSON DRIVE
IRVINE, CALIFORNIA 92698

OVERVIEW: Engineering and construction firm serving the petroleum and other process based industries. Established: 1912. Ownership: Public.

KEY STATISTICS: Annual Sales: $11 billion. Number of Employees: 52,461.

EXPERTISE/EDUCATION SOUGHT: Financial analysis, accounting, design engineering

Fluor Daniel, Inc. (continued)

Alternatives to Full-Time Employment: Paid internships
Contact: For college students and recent graduates: College Recruiter; Phone: 714-975-2134; Fax: 714-975-5639. For individuals with previous experience: Human Resources; Phone: 714-975-2134; Fax: 714-975-5639.
Company URL: http://www.fluor.com

FUJITSU COMPUTER PRODUCTS OF AMERICA, INC.

2904 ORCHARD PARKWAY
SAN JOSE, CALIFORNIA 95134-2009

Overview: Manufactures computer peripheral devices. Established: 1968. Parent Company: Fujitsu America. Ownership: Private.
Key Statistics: Annual Sales: $500 million. Number of Employees: 750.
Hiring History: Number of professional employees hired in 1997: 120.
Expertise/Education Sought: Marketing, accounting, sales
Alternatives to Full-Time Employment: Part-time jobs, paid internships
Company-Paid Training is available.
Contact: For college students and recent graduates: Ms. Kathy Jensen, Senior Recruiter. E-mail: kjensen@fcpa.fujitsu.com. For individuals with previous experience: Ms. Claudia Trujillo, Human Resources Manager; Phone: 408-894-3631. E-mail: ctrujillo@fcpa.fujitsu.com.
Company URL: http://www.fcpa.com

FUJITSU MICROELECTRONICS, INC.

3545 NORTH 1ST STREET
SAN JOSE, CALIFORNIA 95134-1804

Overview: Manufactures semiconductors. Established: 1979. Parent Company: Fujitsu, Ltd. Ownership: Private.
Key Statistics: Annual Sales: $100.1 million. Number of Employees: 1,300.
Expertise/Education Sought: Design engineering, electrical engineering, software engineering/development, hardware engineering
Alternatives to Full-Time Employment: Part-time jobs
Contact: For individuals with previous experience: Human Resources Department; Phone: 408-922-9000; Fax: 408-943-0407.
Company URL: http://www.fujitsumicro.com

FUJITSU SOFTWARE CORPORATION

3055 ORCHARD DRIVE
SAN JOSE, CALIFORNIA 95134

Overview: Manufactures computer equipment and software. Established: 1968. Parent Company: Fujitsu, Ltd. Ownership: Private.
Key Statistics: Annual Sales: $1.7 billion. Number of Employees: 4,000.
Hiring History: Number of professional employees hired in 1997: 40. 1996: 85.
Expertise/Education Sought: Software engineering/development, marketing, sales
Alternatives to Full-Time Employment: Paid internships
Company-Paid Training is available.
Contact: Human Resources; Phone: 408-432-1300; Fax: 408-456-7888. E-mail: jobs@fsc.fujitsu.com.
Company URL: http://www.fsc.fujitsu.com

GENENTECH, INC.

1 DNA WAY
SOUTH SAN FRANCISCO, CALIFORNIA 94080-4918

Overview: Develops and manufactures genetically engineered pharmaceuticals. Established: 1990. Ownership: Private.
Key Statistics: Annual Sales: $582.8 million. Number of Employees: 2,842.
Hiring History: Number of professional employees hired in 1997: 500.
Expertise/Education Sought: Biology, chemistry, physics, biotechnology, pharmaceutical chemistry, pharmaceutical technology, chemical engineering
Alternatives to Full-Time Employment: Part-time jobs, paid internships
Company-Paid Training is available.
Contact: Human Resources Department; Phone: 650-225-1000. E-mail: jobs@gene.com. Job Hotline: 650-225-2580.
Company URL: http://www.gene.com

GEOMATRIX CONSULTANTS, INC.

100 PINE STREET, 10TH FLOOR
SAN FRANCISCO, CALIFORNIA 94111-5102

Overview: Provides hazardous waste consulting services. Established: 1984. Ownership: Private.
Key Statistics: Annual Sales: $30.7 million. Number of Employees: 285.
Hiring History: Number of professional employees hired in 1997: 75. 1996: 50.
Expertise/Education Sought: Environmental engineering, civil engineering, chemical engineering, geology, geoscience, risk management
Contact: For individuals with previous experience: Human Resources; Phone: 415-434-9400. Job Hotline: 415-434-9400, ext. 7555.

GRANITE CONSTRUCTION, INC.

585 WEST BEACH STREET
WATSONVILLE, CALIFORNIA 95076-5125

Overview: Provides roadway construction services. Established: 1990. Ownership: Private.
Key Statistics: Annual Sales: $901.8 million. Number of Employees: 1,800.
Hiring History: Number of professional employees hired in 1997: 60. 1996: 50.
Expertise/Education Sought: Civil engineering, accounting, construction management, information systems
Alternatives to Full-Time Employment: Part-time jobs, paid internships
Company-Paid Training is available.
Contact: Mr. Brian Fox, Director of Recruiting; Phone: 408-724-1011; Fax: 408-761-7871. E-mail: clyoung@granite-net.com.
Company URL: http://www.graniteconstruction.com

GTE CALIFORNIA, INC.

ONE GTE PLACE
THOUSAND OAKS, CALIFORNIA 91362-3811

Overview: Provides sales and dial-tone services. Established: 1939. Parent Company: GTE Corporation. Ownership: Public.
Key Statistics: Annual Sales: $3.1 billion. Number of Employees: 11,937.
Hiring History: Number of professional employees hired in 1997: 200.
Expertise/Education Sought: Electrical engineering, telecommunications engineering
Alternatives to Full-Time Employment: Part-time jobs, paid internships
Company-Paid Training is available.
Contact: For individuals with previous experience: Management Staffing; Phone: 805-372-6000.
Company URL: http://www.gte.com

HARDING LAWSON ASSOCIATES, INC.

7655 REDWOOD BOULEVARD
NOVATO, CALIFORNIA 94945-1400

Overview: Provides environmental engineering services. Established: 1959. Parent Company: Harding Lawson Associates Group, Inc. Ownership: Public.
Key Statistics: Annual Sales: $123.4 million. Number of Employees: 835.

Hiring History: Number of professional employees hired in 1997: 175. 1996: 195.
Expertise/Education Sought: Environmental engineering, chemical engineering, civil engineering, biological engineering, land management, accounting
Alternatives to Full-Time Employment: Paid internships
Company-Paid Training is available.
Contact: Ms. Debbie Sharp, Staffing Coordinator, PO Box 578, Novato, California, 94948-0578. E-mail: dsharp@harding.com.
Company URL: http://www.harding.com

HATHAWAY DINWIDDIE CONSTRUCTION GROUP
565 LAURELWOOD ROAD
SANTA CLARA, CALIFORNIA 95054-2419

Overview: Builds office buildings. Established: 1996. Ownership: Private.
Key Statistics: Annual Sales: $382 million. Number of Employees: 650.
Expertise/Education Sought: Administration, engineering, management
Alternatives to Full-Time Employment: Paid internships
Contact: For individuals with previous experience: Human Resources; Fax: 408-988-1958. E-mail: sandyh@eahathaway.com.
Company URL: http://www.hdcco.com

HEALTH VALLEY COMPANY, INC.
16100 FOOTHILL BOULEVARD
IRWINDALE, CALIFORNIA 91706

Overview: Wholesales health foods. Established: 1996. Ownership: Private.
Key Statistics: Annual Sales: $53.9 million. Number of Employees: 282.
Expertise/Education Sought: Production, administration
Contact: For individuals with previous experience: Director of Human Resources; Phone: 818-334-3241; Fax: 818-334-8911.

HUNT-WESSON, INC.
1645 WEST VALENCIA DRIVE
FULLERTON, CALIFORNIA 92833-3899

Overview: Produces canned fruits and vegetables and cooking oils. Established: 1990. Parent Company: ConAgra, Inc. Ownership: Public.
Key Statistics: Annual Sales: $1.5 billion. Number of Employees: 7,000.
Hiring History: Number of professional employees hired in 1997: 220. 1996: 115.
Expertise/Education Sought: Marketing, data processing, finance, research, accounting, sales, business administration, business management
Alternatives to Full-Time Employment: Paid internships
Company-Paid Training is available.
International Assignments are available to employees.
Contact: Ms. Deborah J. Key, Manager of Employment; Phone: 714-680-1000; Fax: 714-449-5191. E-mail: dkey@hwfoods.com.

ICF KAISER ENGINEERS, INC.
2101 WEBSTER STREET, SUITE 1000
OAKLAND, CALIFORNIA 94612-3429

Overview: Provides engineering services. Established: 1959. Ownership: Private.
Key Statistics: Annual Sales: $10.9 million. Number of Employees: 180.
Expertise/Education Sought: Engineering
Contact: For individuals with previous experience: Human Resources; Phone: 510-419-5115; Fax: 510-419-5122.
Company URL: http://www.icfkaiser.com

IMPACT WITH QUALITY, INC.
7751 EDGEWATER DRIVE
OAKLAND, CALIFORNIA 94621-3013

Overview: Designs computer systems and offers computer integration services. Established: 1997. Ownership: Private.
Key Statistics: Annual Sales: $10 million. Number of Employees: 25.
Hiring History: Number of professional employees hired in 1997: 40.
Expertise/Education Sought: Database management, software engineering/development, computer operations
Alternatives to Full-Time Employment: Paid internships
International Assignments are available to employees.
Contact: Mr. David Loy, Administration; Phone: 510-563-3907; Fax: 510-563-3999. E-mail: dloy@iqlabs.com.

INTELINET COMMUNICATIONS
1900 MCCARTHY BOULEVARD, SUITE 204
MILPITAS, CALIFORNIA 95035

Overview: Provides telephone broker services. Established: 1996. Ownership: Private.
Expertise/Education Sought: Telecommunications
Alternatives to Full-Time Employment: Paid internships
Company-Paid Training is available.
Contact: For individuals with previous experience: Mr. John Diatte, President of Operations; Phone: 408-383-9200; Fax: 408-383-0909.

INTERMEDIA PARTNERS
235 MONTGOMERY STREET, SUITE 420
SAN FRANCISCO, CALIFORNIA 94104

Overview: Owns and operates multiple cable television systems. Established: 1988. Ownership: Private.
Key Statistics: Annual Sales: $107.3 million. Number of Employees: 800.
Expertise/Education Sought: Sales, telecommunications, operations, accounting, marketing
Company-Paid Training is available.
Contact: Ms. Donna Dearman, Executive Director of Human Resources; Phone: 415-397-4121; Fax: 415-397-3920.

INTERNATIONAL COMPONENTS TECHNOLOGY COMPANY
2360 ZANKER ROAD
SAN JOSE, CALIFORNIA 95131-1115

Overview: Manufactures automotive air bags and aluminum substrates for memory systems. Established: 1994. Parent Company: Furukawa Electric Company. Ownership: Private.
Key Statistics: Annual Sales: $65 million. Number of Employees: 500.
Hiring History: Number of professional employees hired in 1997: 110.
Expertise/Education Sought: Mechanical engineering
Contact: For individuals with previous experience: Human Resources Department; Phone: 408-435-8780; Fax: 408-922-0544.

INTUIT
2550 GARCIA AVENUE
MOUNTAIN VIEW, CALIFORNIA 94039-7850

Overview: Develops personal finance and small business software. Established: 1983. Ownership: Public.
Key Statistics: Annual Sales: $538.6 million. Number of Employees: 3,184.
Hiring History: Number of professional employees hired in 1997: 60.
Expertise/Education Sought: Software engineering/development, mechanical engineering, marketing, finance, program analysis
Alternatives to Full-Time Employment: Paid internships
Company-Paid Training is available.

Intuit (continued)

INTERNATIONAL ASSIGNMENTS are available to employees.

CONTACT: For college students and recent graduates: Miss Erin Olson, Human Resources Representative; Phone: 650-944-2960; Fax: 650-944-2700. For individuals with previous experience: Mrs. Traci Wicks, Human Resources Administrator; Phone: 650-944-6943; Fax: 650-944-2700.

COMPANY URL: http://www.intuit.com

ISADRA, INC.

1000 ELWELL COURT, SUITE 230
PALO ALTO, CALIFORNIA 94303-4306

OVERVIEW: Provides computer programming services and software development and applications. Established: 1993. Ownership: Private.

KEY STATISTICS: Annual Sales: $10 million.

HIRING HISTORY: Number of professional employees hired in 1997: 20.

EXPERTISE/EDUCATION SOUGHT: Software engineering/development

CONTACT: For individuals with previous experience: Human Resources; Phone: 650-691-1100; Fax: 650-962-5390.

JACOBS ENGINEERING GROUP, INC.

1111 SOUTH ARROYO PARKWAY
PASADENA, CALIFORNIA 91105

MAILING ADDRESS: PO Box 7084, Pasadena, CA 91109-7084

OVERVIEW: Provides engineering, consulting, and construction management services. Established: 1947. Ownership: Public.

KEY STATISTICS: Annual Sales: $1.8 billion. Number of Employees: 7,600.

EXPERTISE/EDUCATION SOUGHT: Mechanical engineering, civil engineering, computer graphics, design engineering, environmental engineering, electrical engineering, petroleum/petrochemical engineering, computer-aided design, industrial engineering

ALTERNATIVES TO FULL-TIME EMPLOYMENT: Paid internships

INTERNATIONAL ASSIGNMENTS are available to employees.

CONTACT: Ms. Maryanne Willetts, Manager of Personnel; Phone: 626-578-3500; Fax: 626-578-6914.

JACUZZI, INC.

2121 NORTH CALIFORNIA BOULEVARD, SUITE 475
WALNUT CREEK, CALIFORNIA 94596

OVERVIEW: Manufactures Jacuzzis and whirlpool baths. Established: 1989. Parent Company: US Industries. Ownership: Public.

KEY STATISTICS: Annual Sales: $10 million. Number of Employees: 1,200.

EXPERTISE/EDUCATION SOUGHT: Mechanical engineering, accounting, customer service and support, data processing, marketing, sales

CONTACT: For individuals with previous experience: Ms. Jenelle Balonon, Manager of Human Resources; Phone: 510-988-6212.

KENPAK, INC.

5416 UNION PACIFIC STREET
LOS ANGELES, CALIFORNIA 90022-5117

OVERVIEW: Manufactures sterile packaging for medical equipment. Established: 1975. Ownership: Private.

KEY STATISTICS: Annual Sales: $35 million. Number of Employees: 315.

HIRING HISTORY: Number of professional employees hired in 1997: 50. 1996: 33.

EXPERTISE/EDUCATION SOUGHT: Mechanical engineering, electrical engineering, drafting, sales, data processing, machinery

ALTERNATIVES TO FULL-TIME EMPLOYMENT: Paid internships

CONTACT: Mr. Jaime Benavides, Human Resources Director; Phone: 213-721-6777; Fax: 213-722-1631. E-mail: jaime@kenpak.com.

COMPANY URL: http://www.kenpak.com

KINETIC SYSTEMS, INC.

3080 RAYMOND STREET
SANTA CLARA, CALIFORNIA 95054-3430

OVERVIEW: Plumbing, heating, and air conditioning. Established: 1988. Parent Company: US Filter. Ownership: Private.

KEY STATISTICS: Annual Sales: $212.2 million. Number of Employees: 2,246.

HIRING HISTORY: Number of professional employees hired in 1997: 400.

EXPERTISE/EDUCATION SOUGHT: Electronics/electronics engineering

ALTERNATIVES TO FULL-TIME EMPLOYMENT: Paid internships

COMPANY-PAID TRAINING is available.

INTERNATIONAL ASSIGNMENTS are available to employees.

CONTACT: For college students and recent graduates: Mr. Bill Young, Director of Staffing; Phone: 408-588-4475; Fax: 408-496-6162. E-mail: hr@kineticsys.com. For individuals with previous experience: Ms. Lisa Kusaba, Employment Representative; Phone: 408-588-4418; Fax: 408-496-6162.

KOMAG, INC.

1704 AUTOMATION PARKWAY
SAN JOSE, CALIFORNIA 95131-5417

OVERVIEW: Manufactures thin film and leads for use in hard disk drives. Established: 1983. Ownership: Public.

KEY STATISTICS: Annual Sales: $577.8 million. Number of Employees: 4,101.

EXPERTISE/EDUCATION SOUGHT: Chemical engineering, industrial engineering, mechanical engineering, electrical engineering

ALTERNATIVES TO FULL-TIME EMPLOYMENT: Paid internships

COMPANY-PAID TRAINING is available.

INTERNATIONAL ASSIGNMENTS are available to employees.

CONTACT: Human Resources Manager; Phone: 408-576-2000.

COMPANY URL: http://www.komag.com

KYOCERA AMERICA, INC.

8611 BALBOA AVENUE
SAN DIEGO, CALIFORNIA 92123-1501

OVERVIEW: Manufactures semiconductors, photographic equipment, communications equipment, and printers.. Established: 1969. Parent Company: Kyocera International, Inc. Ownership: Private.

KEY STATISTICS: Annual Sales: $138.6 million. Number of Employees: 2,100.

EXPERTISE/EDUCATION SOUGHT: Chemical engineering, industrial engineering, mechanical engineering, process engineering, electrical engineering, production engineering, quality engineering

ALTERNATIVES TO FULL-TIME EMPLOYMENT: Part-time jobs, paid internships

COMPANY-PAID TRAINING is available.

CONTACT: Recruiters; Phone: 800-226-6828; Fax: 619-268-3035.

COMPANY URL: http://www.kyocera.com

L.A. CELLULAR

17785 CENTER COURT DRIVE NORTH
CERRITOS, CALIFORNIA 90703-3960

MAILING ADDRESS: PO Box 3960, Cerritos, CA 90703-3960

OVERVIEW: Provides cellular telephone services. Established: 1984. Parent Company: AT&T and BellSouth. Ownership: Private.

KEY STATISTICS: Annual Sales: $320.7 million. Number of Employees: 2,500.

HIRING HISTORY: Number of professional employees hired in 1997: 50. 1996: 75.

EXPERTISE/EDUCATION SOUGHT: Telecommunications, sales, computer science

COMPANY-PAID TRAINING is available.

CONTACT: For individuals with previous experience: Human Resources; Phone: 562-924-0000, ext. 6023; Fax: 562-403-1906. Job Hotline: 562-403-8519.

LOGICON, INC.
3701 SKYPARK DRIVE, SUITE 200
TORRANCE, CALIFORNIA 90505

OVERVIEW: Provides computer software and engineering services. Established: 1961. Parent Company: Northrup Grumman Corporation. Ownership: Public.

KEY STATISTICS: Annual Sales: $562.9 million. Number of Employees: 5,000.

HIRING HISTORY: Number of professional employees hired in 1997: 200. 1996: 200.

EXPERTISE/EDUCATION SOUGHT: Software engineering/development, computer science, nuclear engineering, information systems, operations research, electrical engineering, electronics/electronics engineering, aerospace engineering

ALTERNATIVES TO FULL-TIME EMPLOYMENT: Paid internships

COMPANY-PAID TRAINING is available.

CONTACT: Ms. Judith Keller, Director of Human Resources; Phone: 310-373-0220, ext. 3276; Fax: 310-791-0411. E-mail: jkeller@logicon.com.

COMPANY URL: http://www.logicon.com

LOOK SMART, LTD.
600 TOWNSEND STREET, SUITE 49E
SAN FRANCISCO, CALIFORNIA 94103-4945

OVERVIEW: Provides telecommunications services and Internet host service. Established: 1996. Ownership: Private.

KEY STATISTICS: Annual Sales: $13.1 million. Number of Employees: 83.

HIRING HISTORY: Number of professional employees hired in 1997: 32. 1996: 4.

EXPERTISE/EDUCATION SOUGHT: Computer programming, software engineering/development, administration, sales, marketing

ALTERNATIVES TO FULL-TIME EMPLOYMENT: Paid internships

COMPANY-PAID TRAINING is available.

INTERNATIONAL ASSIGNMENTS are available to employees.

CONTACT: Mr. Grant Van Ulbrich, Human Resources Manager; Phone: 415-437-3820; Fax: 415-437-3829. E-mail: jobs@looksmart.com.

COMPANY URL: http://www.looksmart.com

LUCENT TECHNOLOGIES-OCTEL MESSAGING DIVISION
1001 MURPHY RANCH ROAD
MILPITAS, CALIFORNIA 95035-7912

OVERVIEW: Manufactures voice processing systems. Established: 1982. Ownership: Public.

KEY STATISTICS: Annual Sales: $277.7 million. Number of Employees: 2,900.

HIRING HISTORY: Number of professional employees hired in 1997: 600.

EXPERTISE/EDUCATION SOUGHT: Software engineering/development, design engineering, hardware engineering, computer science, marketing, business administration, software engineering/development, electrical engineering, test engineering

ALTERNATIVES TO FULL-TIME EMPLOYMENT: Part-time jobs, paid internships

COMPANY-PAID TRAINING is available.

INTERNATIONAL ASSIGNMENTS are available to employees.

CONTACT: For college students and recent graduates: Miss Erica Booth, College Relations Coordinator; Phone: 408-324-2699; Fax: 408-324-0613. For individuals with previous experience: Mr. Jim Baughman, Staffing Manager; Phone: 408-324-6380; Fax: 408-321-9823. E-mail: jbaughma@lucent.com.

COMPANY URL: http://www.octel.com

LUCKY STORES, INC.
1701 MARINA BOULEVARD
SAN LEANDRO, CALIFORNIA 94577

OVERVIEW: Operates chain of food stores. Established: 1988. Parent Company: American Stores Company. Ownership: Public.

KEY STATISTICS: Annual Sales: $4.9 billion. Number of Employees: 45,000.

EXPERTISE/EDUCATION SOUGHT: Sales, marketing, business management, accounting, law

ALTERNATIVES TO FULL-TIME EMPLOYMENT: Paid internships

COMPANY-PAID TRAINING is available.

CONTACT: Ms. Jeanine Schreiber, Vice President of Human Resources; Phone: 510-678-4607; Fax: 510-678-4838.

MAGEPOWER SEMICONDUCTOR CORPORATION
2150 COMMERCE DRIVE
SAN JOSE, CALIFORNIA 95131-1804

OVERVIEW: Manufactures semiconductors. Established: 1995. Ownership: Private.

KEY STATISTICS: Annual Sales: $21 million. Number of Employees: 170.

HIRING HISTORY: Number of professional employees hired in 1997: 70.

EXPERTISE/EDUCATION SOUGHT: Hardware engineering

COMPANY-PAID TRAINING is available.

INTERNATIONAL ASSIGNMENTS are available to employees.

CONTACT: For individuals with previous experience: Human Resources; Phone: 408-526-9800; Fax: 408-526-9600. E-mail: resume@magepower.com.

COMPANY URL: http://www.magepower.com

MAGNECOMP CORPORATION
26201 YNEZ ROAD
TEMECULA, CALIFORNIA 92591

OVERVIEW: Manufactures spring assemblies for computer disk drives. Established: 1991. Parent Company: Magnecomp Group. Ownership: Private.

KEY STATISTICS: Annual Sales: $30 million. Number of Employees: 220.

HIRING HISTORY: Number of professional employees hired in 1997: 100. 1996: 150.

EXPERTISE/EDUCATION SOUGHT: Mechanical engineering, industrial engineering, accounting, manufacturing engineering, quality engineering

ALTERNATIVES TO FULL-TIME EMPLOYMENT: Internships

CONTACT: For individuals with previous experience: Mr. Michael Schnell, Director of Human Resources; Phone: 909-693-9130; Fax: 909-693-4150.

MAXIM INTEGRATED PRODUCTS
120 SAN GABRIEL DRIVE
SUNNYVALE, CALIFORNIA 94086-5125

OVERVIEW: Manufactures semiconductors. Established: 1983. Ownership: Public.

KEY STATISTICS: Annual Sales: $433.7 million. Number of Employees: 2,444.

HIRING HISTORY: Number of professional employees hired in 1996: 300.

EXPERTISE/EDUCATION SOUGHT: Electronics/electronics engineering, electrical engineering, mechanical engineering, computer science

CONTACT: Department of Personnel; Phone: 408-737-7600.

MAXTOR CORPORATION
510 COTTONWOOD DRIVE
MILPITAS, CALIFORNIA 95035

OVERVIEW: Manufactures magnetic disk drives. Established: 1996. Parent Company: Hyundai Electronics America. Ownership: Private.

KEY STATISTICS: Annual Sales: $483.3 million. Number of Employees: 4,600.

EXPERTISE/EDUCATION SOUGHT: Accounting, manufacturing engineering, finance, marketing, electrical engineering

ALTERNATIVES TO FULL-TIME EMPLOYMENT: Part-time jobs, paid internships

Maxtor Corporation (continued)

International Assignments are available to employees.
Contact: For individuals with previous experience: Director of Human Resources; Phone: 408-432-4343; Fax: 408-432-4185.
Company URL: http://www.maxtor.com

MAXWELL TECHNOLOGY
9275 SKY PARK COURT, SUITE 400
SAN DIEGO, CALIFORNIA 92123

Overview: Produces electronic capacitors. Established: 1996. Ownership: Public.
Key Statistics: Annual Sales: $8.3 million. Number of Employees: 130.
Expertise/Education Sought: Engineering, manufacturing engineering, mechanical engineering, systems analysis
Alternatives to Full-Time Employment: Part-time jobs
Company-Paid Training is available.
Contact: Human Resource Department; Phone: 619-279-5100; Fax: 619-576-5100.
Company URL: http://www.maxwell.com

MCGHAN MEDICAL CORPORATION
700 WARD DRIVE
SANTA BARBARA, CALIFORNIA 93111

Overview: Manufactures tissue extenders and implants. Established: 1985. Parent Company: Inamed Corporation. Ownership: Public.
Key Statistics: Annual Sales: $25.8 million. Number of Employees: 340.
Expertise/Education Sought: Mechanical engineering, chemical engineering
Alternatives to Full-Time Employment: Paid internships
Contact: For individuals with previous experience: Employee Services; Phone: 805-683-6761; Fax: 805-967-6249. Job Hotline: 805-683-6761, ext. 1300.

MCM CONSTRUCTION, INC.
6413 32ND STREET
NORTH HIGHLANDS, CALIFORNIA 95660-3001

Overview: Provides construction services. Established: 1973. Ownership: Private.
Key Statistics: Annual Sales: $57.4 million. Number of Employees: 370.
Hiring History: Number of professional employees hired in 1997: 5.
Expertise/Education Sought: Civil engineering
Contact: Mr. Jim Carter, President; Phone: 916-334-1221; Fax: 916-334-0562.

MICROSEMI CORPORATION
2830 SOUTH FAIRVIEW STREET
SANTA ANA, CALIFORNIA 92704-5948

Overview: Designs and manufactures semiconductor assemblies and other power conditioning products. Established: 1960. Ownership: Public.
Key Statistics: Annual Sales: $157.4 million. Number of Employees: 2,305.
Expertise/Education Sought: Accounting, finance, sales, marketing, computer science, data processing, business management, engineering
Alternatives to Full-Time Employment: Paid internships
International Assignments are available to employees.
Contact: Mrs. Maritza Casas, Human Resources Assistant; Phone: 714-979-8220, ext. 244; Fax: 714-979-0608.
Company URL: http://www.microsemi.com

MISSION RESEARCH CORPORATION
735 STATE STREET
SANTA BARBARA, CALIFORNIA 93101

Overview: Provides commercial physical research services. Established: 1970. Ownership: Private.
Key Statistics: Annual Sales: $60.7 million. Number of Employees: 320.
Hiring History: Number of professional employees hired in 1997: 25. 1996: 30.
Expertise/Education Sought: Research and development
Alternatives to Full-Time Employment: Part-time jobs
Contact: For individuals with previous experience: Director of Human Resources; Phone: 805-963-8761, ext. 363; Fax: 805-962-8530. E-mail: hresources@mrcsb.com.
Company URL: http://www.mrcsb.com

MITSUBISHI ELECTRONICS AMERICA, INC.
5665 PLAZA DRIVE
CYPRESS, CALIFORNIA 90630-5023

Overview: Manufactures and distributes electrical and electronics products. Established: 1973. Parent Company: Mitsubishi Electric America, Inc. Ownership: Private.
Key Statistics: Annual Sales: $4 billion. Number of Employees: 3,900.
Hiring History: Number of professional employees hired in 1997: 100. 1996: 150.
Expertise/Education Sought: Sales, marketing
Alternatives to Full-Time Employment: Paid internships
Company-Paid Training is available.
International Assignments are available to employees.
Contact: For individuals with previous experience: Ms. Debbie Fudge, Human Resources Manager, 5757 Plaza Drive, Cypress, California, 90630-5029; Phone: 714-220-4613; Fax: 714-229-3898.

MONIERLIFETILE LLC
1 PARK PLAZA, SUITE 900
IRVINE, CALIFORNIA 92614

Overview: Manufactures concrete and clay roofing materials. Established: 1987. Parent Company: Boral Industries/LeFarge. Ownership: Private.
Key Statistics: Annual Sales: $84.6 million. Number of Employees: 840.
Hiring History: Number of professional employees hired in 1997: 85. 1996: 50.
Expertise/Education Sought: Accounting, finance, data processing, product engineering, mechanical engineering, electrical engineering, computer analysis
Company-Paid Training is available.
Contact: Ms. Maureen Acton, Director of Human Resources, PO Box 19792, Irvine, California, 92713-5999; Phone: 714-756-1605; Fax: 714-756-2401.

MONTEREY BAY AERO SPORTSPLEX
3261 IMJIN ROAD
MARINA, CALIFORNIA 93933-5103

Overview: Operates amusements and recreation services, offers parachute training and hot air balloon rides. Established: 1996. Ownership: Private.
Key Statistics: Number of Employees: 1,015.
Expertise/Education Sought: Engineering
Alternatives to Full-Time Employment: Paid internships
Contact: For individuals with previous experience: Human Resources; Phone: 408-384-3483. E-mail: skydive@mbay.net.

MYCOGEN CORPORATION

5501 OBERLIN DRIVE
SAN DIEGO, CALIFORNIA 92121-1718

OVERVIEW: Develops and markets pesticides and other agricultural products. Established: 1982. Parent Company: Dow AgroSciences. Ownership: Public.

KEY STATISTICS: Annual Sales: $155.6 million. Number of Employees: 887.

EXPERTISE/EDUCATION SOUGHT: Biology, chemistry, biotechnology

CONTACT: Human Resources Department; Phone: 619-453-8030; Fax: 619-453-0142. Job Hotline: 619-453-7812.

COMPANY URL: http://www.mycogen.com

NATIONAL O-RING

11634 PATTON ROAD
DOWNEY, CALIFORNIA 90241

OVERVIEW: Manufactures gaskets and sealing devices. Established: 1996. Parent Company: Hutchinson Seal Corporation. Ownership: Private.

KEY STATISTICS: Annual Sales: $36 million. Number of Employees: 700.

EXPERTISE/EDUCATION SOUGHT: Manufacturing engineering, sales, industrial engineering, marketing, accounting

CONTACT: Mr. Jack Saigan, Director of Human Resources; Phone: 562-862-8163; Fax: 562-861-4955.

NATIONAL SEMICONDUCTOR CORPORATION

2900 SEMICONDUCTOR WAY
SANTA CLARA, CALIFORNIA 95051-0606

OVERVIEW: Manufactures semiconductors and integrated circuits. Established: 1959. Ownership: Public.

KEY STATISTICS: Annual Sales: $2.5 billion. Number of Employees: 12,400.

EXPERTISE/EDUCATION SOUGHT: Electrical engineering, computer engineering, computer science, finance, analog engineering

ALTERNATIVES TO FULL-TIME EMPLOYMENT: Part-time jobs

INTERNATIONAL ASSIGNMENTS are available to employees.

CONTACT: For individuals with previous experience: Human Resource.

COMPANY URL: http://www.national.com

NEON TECHNOLOGY, INC.

89 WEST MONTAGUE EXPRESSWAY
MILPITAS, CALIFORNIA 95035-6264

OVERVIEW: Develops computer software systems. Established: 1995. Ownership: Private.

KEY STATISTICS: Annual Sales: $10 million.

EXPERTISE/EDUCATION SOUGHT: Engineering

CONTACT: For individuals with previous experience: Human Resources; Phone: 408-934-5816; Fax: 408-263-2161.

COMPANY URL: http://www.neontech.com

NETWORK DESIGNS AND INTEGRATION SERVICES

43220 CHRISTY STREET
FREMONT, CALIFORNIA 94538

OVERVIEW: Provides computer integrated systems services. Established: 1994. Ownership: Private.

EXPERTISE/EDUCATION SOUGHT: Engineering, sales

ALTERNATIVES TO FULL-TIME EMPLOYMENT: Part-time jobs

COMPANY-PAID TRAINING is available.

CONTACT: Ms. Angela Hunter, Human Resources Manager; Phone: 510-249-9549; Fax: 510-249-9545.

COMPANY URL: http://www.networkdesigns.com

NIELSEN DILLINGHAM BUILDERS

3127 JEFFERSON STREET
SAN DIEGO, CALIFORNIA 92110-4422

OVERVIEW: Builds public buildings. Established: 1994. Parent Company: Dillingham Holdings. Ownership: Private.

KEY STATISTICS: Annual Sales: $330 million. Number of Employees: 210.

HIRING HISTORY: Number of professional employees hired in 1997: 50. 1996: 50.

EXPERTISE/EDUCATION SOUGHT: Engineering, management, estimating, construction management

ALTERNATIVES TO FULL-TIME EMPLOYMENT: Paid internships

CONTACT: For individuals with previous experience: Ms. Denise Howe, Recruiter; Phone: 619-291-6330, ext. 243; Fax: 619-291-5485. E-mail: dhowe@ndbi.com.

COMPANY URL: http://www.nielsendillingham.com

NOVELLUS SYSTEMS, INC.

3970 NORTH FIRST STREET
SAN JOSE, CALIFORNIA 95134-1510

OVERVIEW: Manufactures equipment for fabricating integrated circuits. Established: 1984. Ownership: Public.

KEY STATISTICS: Annual Sales: $461.7 million. Number of Employees: 1,143.

HIRING HISTORY: Number of professional employees hired in 1997: 500. 1996: 400.

EXPERTISE/EDUCATION SOUGHT: Process engineering, electrical engineering, mechanical engineering, chemistry, software engineering/development, computer science

ALTERNATIVES TO FULL-TIME EMPLOYMENT: Paid internships

COMPANY-PAID TRAINING is available.

INTERNATIONAL ASSIGNMENTS are available to employees.

CONTACT: Professional Staffing; Phone: 408-943-9700; Fax: 408-943-3419. E-mail: careers@novellus.com.

COMPANY URL: http://www.novellus.com

OAK TECHNOLOGY, INC.

139 KIFER COURT
SUNNYVALE, CALIFORNIA 94086

OVERVIEW: Manufactures computer integrated circuits. Established: 1987. Ownership: Public.

KEY STATISTICS: Annual Sales: $167.4 million. Number of Employees: 430.

HIRING HISTORY: Number of professional employees hired in 1997: 80. 1996: 60.

EXPERTISE/EDUCATION SOUGHT: Design engineering, software engineering/development

ALTERNATIVES TO FULL-TIME EMPLOYMENT: Part-time jobs

INTERNATIONAL ASSIGNMENTS are available to employees.

CONTACT: Mr. Tom Weed, Internal Recruiter; Phone: 408-737-0888; Fax: 408-737-3838. E-mail: hr@oaktech.com.

COMPANY URL: http://www.oaktech.com

ODETICS TELECOM, INC.

1585 SOUTH MANCHESTER AVENUE
ANAHEIM, CALIFORNIA 92802

OVERVIEW: Manufactures industrial controls and relays. Established: 1997. Ownership: Private.

KEY STATISTICS: Annual Sales: $36.8 million. Number of Employees: 500.

HIRING HISTORY: Number of professional employees hired in 1997: 110.

EXPERTISE/EDUCATION SOUGHT: Engineering, manufacturing

ALTERNATIVES TO FULL-TIME EMPLOYMENT: Paid internships

INTERNATIONAL ASSIGNMENTS are available to employees.

CONTACT: Human Resources; Phone: 714-758-0400; Fax: 714-780-7999. E-mail: jobs@odetics.com. Job Hotline: 714-774-5000.

COMPANY URL: http://www.odetics.com

OPTICSCAN, LTD. CORPORATION

1355 BOBOLINK PLACE
LOS ANGELES, CALIFORNIA 90069-1716

Overview: Manufactures optical instruments. Established: 1995. Ownership: Private.

Key Statistics: Annual Sales: $10 million.

Expertise/Education Sought: Sales, engineering, plastics

Company-Paid Training is available.

Contact: For individuals with previous experience: Ms. Rita Sherman, Vice President; Phone: 310-273-1132; Fax: 310-550-6363.

OPTI, INC.

888 TASMAN DRIVE
MILPITAS, CALIFORNIA 95035

Overview: Manufactures microcomputer chips. Established: 1989. Ownership: Public.

Key Statistics: Annual Sales: $118.7 million. Number of Employees: 210.

Expertise/Education Sought: Applied engineering, design engineering

Alternatives to Full-Time Employment: Part-time jobs

Contact: Human Resource Manager; Phone: 408-486-8000.

Company URL: http://www.opti.com

ORC TECHNOLOGY

1300 OPTICAL DRIVE
AZUSA, CALIFORNIA 91702-3251

Overview: Manufactures electronics, plastic eyeglass lenses, optical lenses, interocular lenses, and high intensity lamps. Established: 1996. Parent Company: BEC Group, Inc. Ownership: Public.

Key Statistics: Annual Sales: $45 million. Number of Employees: 300.

Hiring History: Number of professional employees hired in 1997: 40. 1996: 50.

Expertise/Education Sought: Electrical engineering, mechanical engineering, accounting, software engineering/development

International Assignments are available to employees.

Contact: Ms. Hilda Geese, Manager of Human Resources; Phone: 626-815-3000, ext. 3032; Fax: 626-815-3031.

ORINCON CORPORATION

9363 TOWNE CENTRE DRIVE
SAN DIEGO, CALIFORNIA 92121-3016

Overview: Provides computer software services. Established: 1973. Ownership: Private.

Key Statistics: Annual Sales: $13.5 million. Number of Employees: 155.

Expertise/Education Sought: Information engineering, software engineering/development, computer engineering, systems engineering, process engineering

Alternatives to Full-Time Employment: Part-time jobs, paid internships

Contact: Ms. Beth Mitchell, Human Resources Administrator; Phone: 619-455-5530; Fax: 619-453-9274. E-mail: blmitchell@orincon.com.

Company URL: http://www.orincon.com

PACIFIC GAS & ELECTRIC COMPANY

77 BEALE STREET
SAN FRANCISCO, CALIFORNIA 94105-1814

Overview: Provides electric service. Established: 1995. Parent Company: PG&E Corporation. Ownership: Private.

Key Statistics: Annual Sales: $9.6 billion. Number of Employees: 22,000.

Hiring History: Number of professional employees hired in 1997: 1,000. 1996: 900.

Expertise/Education Sought: Engineering, technology, accounting, customer service and support, technical engineering, engineering technology, engineering management, technical

Alternatives to Full-Time Employment: Paid internships

Company-Paid Training is available.

International Assignments are available to employees.

Contact: For individuals with previous experience: Human Resources Department; Phone: 415-973-7793. Job Hotline: 800-967-5743.

Company URL: http://www.pge.com

PACIFIC PRIDE BAKING COMPANY

2069 ALDERGROVE AVENUE
ESCONDIDO, CALIFORNIA 92029-1902

Overview: Manufactures bakery products. Established: 1996. Parent Company: Grupo Industrial Bimbo. Ownership: Private.

Key Statistics: Annual Sales: $40 million. Number of Employees: 400.

Expertise/Education Sought: Sales, production, sales, production

International Assignments are available to employees.

Contact: For individuals with previous experience: Human Relations Department; Phone: 947-603-3833.

PACIFIC TELESIS GROUP

2410 CAMINO RAMON
SAN RAMON, CALIFORNIA 94583

Overview: Provides telecommunications and related services. Established: 1906. Ownership: Public.

Key Statistics: Annual Sales: $6.2 billion. Number of Employees: 48,330.

Expertise/Education Sought: Computer programming, accounting, computer science, telecommunications, systems analysis

Contact: For individuals with previous experience: Ms. Karol Ann Meier, Human Resources Manager.

PACKARD BELL NEC, INC.

1 PACKARD BELL WAY
SACRAMENTO, CALIFORNIA 95828-0903

Overview: Manufactures personal computers, software, and peripherals. Established: 1986. Ownership: Private.

Key Statistics: Annual Sales: $8 billion. Number of Employees: 8,000.

Expertise/Education Sought: Purchasing, accounting, finance, computer science, technical, quality control, production, engineering, electrical engineering

Alternatives to Full-Time Employment: Part-time jobs, paid internships

Company-Paid Training is available.

International Assignments are available to employees.

Contact: Mr. Dough Divelbiss, Recruiter, 8350 Fruitridge Road, Sacramento, California, 95826-0903; Phone: 916-388-6260; Fax: 916-388-5459. Job Hotline: 916-379-6100.

Company URL: http://www.packardbell.com

PEPSI COLA COMPANY

1150 EAST NORTH AVENUE
FRESNO, CALIFORNIA 93725-1929

Overview: Produces soft drinks. Established: 1988. Ownership: Public.

Key Statistics: Annual Sales: $94.2 million. Number of Employees: 625.

Hiring History: Number of professional employees hired in 1997: 50.

Expertise/Education Sought: Warehousing, sales, production

Company-Paid Training is available.

International Assignments are available to employees.

Contact: Ms. Keri Roth, Human Relations Manager; Phone: 209-485-5050; Fax: 209-486-0203.

PLATINUM SOFTWARE CORPORATION

195 TECHNOLOGY DRIVE
IRVINE, CALIFORNIA 92618-2402

Overview: Develops financial and management information software. Established: 1984. Ownership: Public.

Key Statistics: Annual Sales: $58.1 million. Number of Employees: 350.
Expertise/Education Sought: Computer programming, program analysis, software engineering/development, sales, marketing
Alternatives to Full-Time Employment: Part-time jobs, paid internships
Company-Paid Training is available.
International Assignments are available to employees.
Contact: Ms. Carol Pearson, Employment Specialist. E-mail: resumes@platsoft.com.
Company URL: http://www.platsoft.com

PLAXICON COMPANY
10660 ACACIA STREET
RANCHO CUCAMONGA, CALIFORNIA 91730-5409

Overview: Manufactures plastics. Established: 1996. Parent Company: Liquid Container Company. Ownership: Private.
Key Statistics: Annual Sales: $40 million. Number of Employees: 260.
Hiring History: Number of professional employees hired in 1996: 24.
Expertise/Education Sought: Chemical engineering, plastics, plastics engineering, mechanical engineering, electrical engineering, hydrology, accounting, finance, marketing
Contact: For individuals with previous experience: Human Resources; Phone: 909-944-6868; Fax: 909-944-0110.

READ-RITE CORPORATION
345 LOS COCHES STREET
MILPITAS, CALIFORNIA 95035-5428

Overview: Manufactures thin film magnetic disk drive heads. Established: 1983. Ownership: Public.
Key Statistics: Annual Sales: $991.1 million. Number of Employees: 19,507.
Expertise/Education Sought: Engineering, research and development, process engineering, manufacturing engineering
Alternatives to Full-Time Employment: Paid internships
Company-Paid Training is available.
International Assignments are available to employees.
Contact: For individuals with previous experience: Mr. Jack Warren, Director of Staffing; Phone: 408-956-2480; Fax: 408-956-3212. E-mail: jack.warren@readrite.com.
Company URL: http://www.readrite.com

REER
18034 VENTURA BOULEVARD
ENCINO, CALIFORNIA 91316-3516

Overview: Manufactures electronic and microwave components. Established: 1990. Parent Company: AES Technologies. Ownership: Private.
Key Statistics: Annual Sales: $8.2 million. Number of Employees: 122.
Hiring History: Number of professional employees hired in 1997: 12. 1996: 10.
Expertise/Education Sought: Project engineering, civil engineering, communication engineering, systems engineering
International Assignments are available to employees.
Contact: For individuals with previous experience: Mr. Dave Lancer, Senior Vice President; Phone: 818-757-7347; Fax: 818-344-4892.

REXHALL INDUSTRIES, INC.
46147 7TH STREET WEST
LANCASTER, CALIFORNIA 93534

Overview: Manufactures motor homes. Established: 1986. Ownership: Public.
Key Statistics: Annual Sales: $65 million. Number of Employees: 310.
Hiring History: Number of professional employees hired in 1997: 80.
Expertise/Education Sought: Accounting, finance, data processing, electrical engineering, mechanical engineering
Contact: For individuals with previous experience: Ms. Cindy Bowen, Human Resource Payroll Administration. E-mail: marcom@rexall.com.

ROTONICS MANUFACTURING, INC.
17038 SOUTH FIGUEROA STREET
GARDENA, CALIFORNIA 90248-3019

Overview: Manufactures molded plastic products. Established: 1943. Ownership: Public.
Key Statistics: Annual Sales: $39.4 million. Number of Employees: 425.
Hiring History: Number of professional employees hired in 1997: 6. 1996: 4.
Expertise/Education Sought: Mechanical engineering, chemical engineering, plastics engineering, accounting, finance, plastics
Alternatives to Full-Time Employment: Paid internships
Contact: For individuals with previous experience: Human Resources Department; Phone: 310-538-4932; Fax: 310-538-5579.

SAFESKIN CORPORATION
12671 HIGH BLUFF DRIVE
SAN DIEGO, CALIFORNIA 92130

Overview: Manufactures latex examination and surgical gloves. Established: 1985. Ownership: Public.
Key Statistics: Annual Sales: $146.1 million. Number of Employees: 4,500.
Hiring History: Number of professional employees hired in 1997: 73.
Expertise/Education Sought: Chemical engineering, sales, marketing, accounting, finance
Alternatives to Full-Time Employment: Internships
Company-Paid Training is available.
International Assignments are available to employees.
Contact: Mrs. Cathy Jain, Professional Recruiter; Phone: 800-462-9993; Fax: 619-350-2378.
Company URL: http://www.safeskin.com

SALEM COMMUNICATION CORPORATION
4880 SANTA ROSA ROAD, SUITE 300
CAMARILLO, CALIFORNIA 93012-5190

Overview: Provides broadcasting services. Established: 1979. Ownership: Private.
Key Statistics: Annual Sales: $48.2 million. Number of Employees: 600.
Expertise/Education Sought: Radio broadcasting, engineering, sales
Alternatives to Full-Time Employment: Paid internships
Contact: For individuals with previous experience: Mr. Russ Hauth, Vice President of Resource Development; Phone: 805-987-0400; Fax: 805-482-8570.

SASCO ELECTRIC
12900 ALONDRA BOULEVARD
CERRITOS, CALIFORNIA 90703

Overview: Provides electrical products and services. Established: 1968. Ownership: Private.
Key Statistics: Annual Sales: $297.1 million. Number of Employees: 2,500.
Hiring History: Number of professional employees hired in 1997: 40. 1996: 40.
Expertise/Education Sought: Electrical engineering
Alternatives to Full-Time Employment: Paid internships
Company-Paid Training is available.
Contact: For individuals with previous experience: Mrs. Jan Harrigan, Personnel Manager; Phone: 562-926-0900; Fax: 562-926-1399.

SEGASOFT, INC.
150 SHORELINE DRIVE
REDWOOD CITY, CALIFORNIA 94065

Overview: Wholesales computers, peripherals, and software. Established: 1996. Ownership: Private.

Key Statistics: Annual Sales: $51.9 million. Number of Employees: 250.

Hiring History: Number of professional employees hired in 1997: 75. 1996: 100.

Expertise/Education Sought: Computer programming, accounting, marketing, software engineering/development

Alternatives to Full-Time Employment: Part-time jobs

Contact: Human Resources; Phone: 650-654-3400; Fax: 650-654-2590. E-mail: segasoft.jobs@segasoft.com.

Company URL: http://www.segasoft.com

SIEMENS SOLAR INDUSTRIES
4650 ADOHR LANE
CAMARILLO, CALIFORNIA 93012

Overview: Manufactures semiconductors. Established: 1990. Parent Company: Siemens Corporation. Ownership: Private.

Key Statistics: Annual Sales: $30.6 million. Number of Employees: 398.

Hiring History: Number of professional employees hired in 1997: 20. 1996: 25.

Expertise/Education Sought: Chemical engineering, electronics/electronics engineering, mechanical engineering, information systems

Alternatives to Full-Time Employment: Paid internships

Company-Paid Training is available.

International Assignments are available to employees.

Contact: Ms. Sandra Hall, Human Resources Representative; Phone: 805-482-6800; Fax: 805-388-6394. E-mail: shall@solarpv.com.

Company URL: http://www.solarpv.com

SILICON SPICE, INC.
1001 NORTH RENGSTORFF AVENUE
MOUNTAIN VIEW, CALIFORNIA 94043-1715

Overview: Manufactures semiconductors. Established: 1996. Ownership: Private.

Key Statistics: Annual Sales: $10 million.

Expertise/Education Sought: Computer engineering

Contact: Mrs. Michelle Nelson, Human Resource Recruiter; Phone: 650-940-7777; Fax: 650-940-7770. E-mail: meksilicon-spice.com.

Company URL: http://www.silicon.com

SILICON SYSTEMS, INC.
14351 MYFORD ROAD
TUSTIN, CALIFORNIA 92680-7039

Overview: Manufactures semiconductors and related components. Established: 1989. Parent Company: Texas Instruments. Ownership: Public.

Key Statistics: Annual Sales: $560 million. Number of Employees: 2,039.

Hiring History: Number of professional employees hired in 1997: 172. 1996: 170.

Expertise/Education Sought: Accounting, finance, product design and development, data processing, software engineering/development, design engineering, marketing, quality assurance

Alternatives to Full-Time Employment: Part-time jobs, paid internships

Contact: Ms. Jennifer Hughes, Staffing Program Manager; Phone: 714-731-7110, ext. 6252; Fax: 714-573-6901. E-mail: jennifer.hughes@tus.ssi1.com.

Company URL: http://www.ssi1.com

SILICON VALLEY GROUP, INC.
2240 RINGWOOD AVENUE
SAN JOSE, CALIFORNIA 95131

Overview: Manufactures semiconductor production equipment. Established: 1977. Ownership: Public.

Key Statistics: Annual Sales: $595 million. Number of Employees: 2,875.

Expertise/Education Sought: Software engineering/development, electrical engineering, mechanical engineering, manufacturing engineering

Alternatives to Full-Time Employment: Part-time jobs, paid internships

Company-Paid Training is available.

International Assignments are available to employees.

Contact: For individuals with previous experience: Human Resources; Phone: 408-434-0500.

Company URL: http://www.svg.com

SIMMTRON TECHNOLOGY
5893 OBERLIN DRIVE, SUITE 108
SAN DIEGO, CALIFORNIA 92121-3760

Overview: Operates retail personal computer stores. Established: 1984. Parent Company: Simmtron Technology, LLC. Ownership: Private.

Key Statistics: Annual Sales: $40 million.

Hiring History: Number of professional employees hired in 1996: 5.

Expertise/Education Sought: Engineering, sales, electrical engineering, computer programming

Alternatives to Full-Time Employment: Part-time jobs

Contact: For individuals with previous experience: Operation Manager. E-mail: swilliams@simmtron.com.

Company URL: http://www.simmtron.com

SIRA TECHNOLOGY SOLUTIONS, INC.
3155 KEARNEY STREET, SUITE 120
FREMONT, CALIFORNIA 94538

Overview: Provides computer related consulting services. Established: 1994. Ownership: Private.

Key Statistics: Annual Sales: $6 million. Number of Employees: 35.

Expertise/Education Sought: Software engineering/development, consulting

Contact: For individuals with previous experience: Mr. Sunit Sachdev, Marketing Manager; Phone: 510-445-0600; Fax: 510-445-0610.

Company URL: http://www.sira.com

SMARTFLEX SYSTEMS, INC.
14312 FRANKLIN AVENUE
TUSTIN, CALIFORNIA 92781-2085

Overview: Subcontract assembler, specializing in flexible printed circuits. Established: 1985. Ownership: Public.

Key Statistics: Annual Sales: $146.1 million. Number of Employees: 1,006.

Hiring History: Number of professional employees hired in 1997: 35. 1996: 30.

Expertise/Education Sought: Electrical engineering, manufacturing engineering, quality control

International Assignments are available to employees.

Contact: Ms. Corrie Owens, Human Resources Representative; Phone: 714-838-8737; Fax: 714-838-8787. E-mail: owenc@sflex.com.

Company URL: http://www.smartflex.com

SOLAR TURBINES, INC.
2200 PACIFIC HIGHWAY
SAN DIEGO, CALIFORNIA 92101-1745

Mailing Address: PO Box 85376, San Diego, CA 92186

Overview: Supplies turbine machinery products to the oil, gas, and power generation industries. Established: 1981. Parent Company: Caterpillar, Inc. Ownership: Public.

Key Statistics: Annual Sales: $1 billion. Number of Employees: 4,800.

Expertise/Education Sought: Manufacturing engineering, industrial engineering, design engineering, program analysis, electrical engineering, facilities management, mechanical engineering, computer programming

Alternatives to Full-Time Employment: Internships

International Assignments are available to employees.

Contact: Mr. Antonio Torres, Manager of Equal Employment Opportunity Office; Phone: 619-544-5674; Fax: 619-544-2876. Job Hotline: 619-694-6046.

SOLECTRON CORPORATION

847 GIBRALTER DRIVE
MILPITAS, CALIFORNIA 95035-6332

Overview: Manufactures semiconductors and electronic components. Established: 1977. Ownership: Public.

Key Statistics: Annual Sales: $2.8 billion. Number of Employees: 12,999.

Expertise/Education Sought: Electronics/electronics engineering, computer science

Alternatives to Full-Time Employment: Paid internships

Company-Paid Training is available.

International Assignments are available to employees.

Contact: Human Resources Department; Phone: 408-957-8500. Job Hotline: 408-956-6200.

Company URL: http://www.solectron.com

SORIN BIOMEDICAL, INC.

17600 GILLETTE AVENUE
IRVINE, CALIFORNIA 92614

Overview: Manufactures cardiopulmonary and respiratory products. Established: 1994. Parent Company: Sorin Biomedica. Ownership: Private.

Key Statistics: Annual Sales: $50 million. Number of Employees: 280.

Hiring History: Number of professional employees hired in 1997: 25.

Expertise/Education Sought: Mechanical engineering, manufacturing engineering, plastics engineering, accounting, marketing

Company-Paid Training is available.

Contact: For individuals with previous experience: Mrs. Sue Brosnan, Human Resources Representative; Phone: 714-250-8394; Fax: 714-250-8202. E-mail: sbrosnan@sorinbio.com.

SOUTHERN CALIFORNIA EDISON COMPANY

2244 WALNUT GROVE AVENUE
ROSEMEAD, CALIFORNIA 91770-3714

Mailing Address: PO Box 800, Rosemead, CA 91770

Overview: Electric utility. Established: 1988. Parent Company: Edison International. Ownership: Public.

Key Statistics: Annual Sales: $7.6 billion. Number of Employees: 12,057.

Expertise/Education Sought: Customer service and support, engineering

Alternatives to Full-Time Employment: Part-time jobs, paid internships

Contact: Human Resources Department; Phone: 818-302-9850.

Company URL: http://www.edisonx.com

SPACE SYSTEMS/LORAL, INC.

3825 FABIAN WAY
PALO ALTO, CALIFORNIA 94303-4604

Overview: Builds telecommunication and weather satellites. Established: 1992. Parent Company: Loral Space And Communications Corporation. Ownership: Public.

Key Statistics: Annual Sales: $1.4 billion. Number of Employees: 3,800.

Hiring History: Number of professional employees hired in 1997: 600. 1996: 900.

Expertise/Education Sought: Computer science, electrical engineering, mechanical engineering, manufacturing engineering, telecommunications

Alternatives to Full-Time Employment: Part-time jobs

Company-Paid Training is available.

International Assignments are available to employees.

Contact: For college students and recent graduates: Mr. Tim Wyatt, Employment Representative; Phone: 650-852-6269; Fax: 650-852-4825. E-mail: wyatt.tim@ssd.loral.com. For individuals with previous experience: Mr. Dan Thompson, Manager of Employment; Phone: 650-852-6269; Fax: 650-852-4825. E-mail: thompson.dan@ssd.loral.com.

Company URL: http://www.ssloral.com

SPLASH TECHNOLOGY, INC.

555 DEL REY AVENUE
SUNNYVALE, CALIFORNIA 94086-3514

Overview: Develops, produces, and markets color servers. Established: 1995. Ownership: Public.

Key Statistics: Annual Sales: $47 million.

Hiring History: Number of professional employees hired in 1997: 60. 1996: 10.

Expertise/Education Sought: Software engineering/development, nursing, human resources, health care

Alternatives to Full-Time Employment: Paid internships

Company-Paid Training is available.

International Assignments are available to employees.

Contact: For individuals with previous experience: Human Resources; Phone: 408-328-6451; Fax: 408-328-6400. E-mail: employment@splashtech.com.

Company URL: http://www.splashtech.com

SPM/ANAHEIM

3125 EAST CORONADO STREET
ANAHEIM, CALIFORNIA 92806

Overview: Manufactures plastic components for computer parts. Established: 1996. Parent Company: Dynacast. Ownership: Private.

Key Statistics: Annual Sales: $82.7 million. Number of Employees: 1,100.

Hiring History: Number of professional employees hired in 1996: 100.

Expertise/Education Sought: Plastics

Alternatives to Full-Time Employment: Paid internships

International Assignments are available to employees.

Contact: For individuals with previous experience: Human Resources Department; Fax: 714-630-3759.

SSE TELECOM, INC.

47823 WESTINGHOUSE DRIVE
FREEMONT, CALIFORNIA 94539

Overview: Manufactures satellite to earth ground stations. Established: 1989. Ownership: Public.

Key Statistics: Annual Sales: $46.2 million. Number of Employees: 460.

Hiring History: Number of professional employees hired in 1997: 50. 1996: 50.

Expertise/Education Sought: Research and development, electrical engineering, electronics/electronics engineering, manufacturing engineering

Alternatives to Full-Time Employment: Part-time jobs

International Assignments are available to employees.

Contact: Ms. Claire Mendonca, Human Resources Director; Phone: 510-624-8353; Fax: 510-657-1047. E-mail: claire.mendonca@sse.com.

Company URL: http://www.sset.com

SUNRISE MEDICAL, INC.

2382 FARADAY AVENUE
CARLSBAD, CALIFORNIA 92008

Overview: Manufactures medical rehabilitation and recovery equipment. Established: 1983. Ownership: Public.

Key Statistics: Annual Sales: $656.7 million. Number of Employees: 4,254.

Expertise/Education Sought: Administration, finance, accounting, communications

Alternatives to Full-Time Employment: Paid internships

Company-Paid Training is available.

International Assignments are available to employees.

Contact: For college students and recent graduates: Dr. Roberta Baade, Vice President Human Resources; Phone: 760-930-1500. For individuals with previous experience: Ms. Debi Ward, Human Resources Administrator; Phone: 760-930-1500.

Company URL: http://www.sunrisemedical.com

SUPERIOR INDUSTRIES INTERNATIONAL

7800 WOODLEY AVENUE
VAN NUYS, CALIFORNIA 91406-1722

Overview: Designs, manufactures, and markets automobile accessories and related products. Established: 1957. Ownership: Public.

Key Statistics: Annual Sales: $504.2 million. Number of Employees: 4,500.

Hiring History: Number of professional employees hired in 1997: 395. 1996: 275.

Expertise/Education Sought: Quality control, accounting, mechanical engineering, chemical engineering, automotive engineering, maintenance, research and development

Alternatives to Full-Time Employment: Paid internships

Company-Paid Training is available.

International Assignments are available to employees.

Contact: Ms. Griselda Padilla, Personnel Administrator; Phone: 818-902-2530; Fax: 818-902-2531.

SUTMYN STORAGE CORPORATION

2251 LAWSON LANE, BUILDING 7
SANTA CLARA, CALIFORNIA 95056-5687

Overview: Provides computers, peripherals, and software. Established: 1996. Ownership: Private.

Key Statistics: Annual Sales: $50 million.

Hiring History: Number of professional employees hired in 1997: 30. 1996: 50.

Expertise/Education Sought: Electronic design and development, electronics/electronics engineering, software engineering/development, sales, marketing

Alternatives to Full-Time Employment: Paid internships

Company-Paid Training is available.

Contact: Human Resources Department; Phone: 408-992-3461; Fax: 408-992-3622. E-mail: hr@sutmyn.com.

Company URL: http://www.sutmyn.com

SYMANTEC CORPORATION

10201 TORRE AVENUE
CUPERTINO, CALIFORNIA 95014

Overview: Develops information management and productivity enhancement software. Established: 1982. Ownership: Public.

Key Statistics: Annual Sales: $472.2 million. Number of Employees: 2,000.

Hiring History: Number of professional employees hired in 1997: 700. 1996: 500.

Expertise/Education Sought: Software engineering/development, customer service and support, quality assurance, computer programming, marketing, finance, management information systems, technical support

Alternatives to Full-Time Employment: Part-time jobs, paid internships

Company-Paid Training is available.

Contact: Professional Staffing; Phone: 408-253-9600; Fax: 408-366-5972. E-mail: jobs@symantec.com.

Company URL: http://www.symantec.com

SYNOPSYS, INC.

700 EAST MIDDLEFIELD ROAD
MOUNTAIN VIEW, CALIFORNIA 94043-4033

Overview: Designs and develops software. Established: 1986. Ownership: Public.

Key Statistics: Annual Sales: $353.5 million. Number of Employees: 1,716.

Expertise/Education Sought: Software engineering/development

Alternatives to Full-Time Employment: Paid internships

Contact: Corporate Employment; Phone: 650-962-5000; Fax: 650-694-4133.

Company URL: http://www.synopsys.com

TCG CERFNET, INC.

PO BOX 919014
SAN DIEGO, CALIFORNIA 92191-0194

Overview: Provides Internet services. Established: 1988. Parent Company: Teleport Communications Group, Inc. Ownership: Private.

Key Statistics: Annual Sales: $20 million. Number of Employees: 50.

Expertise/Education Sought: Systems administration, marketing, technical support, network management systems, information systems, planning, customer service and support, operations, sales

Alternatives to Full-Time Employment: Part-time jobs, paid internships

Company-Paid Training is available.

Contact: Recruitment; Phone: 619-812-4600; Fax: 619-812-4662. E-mail: jobs@cerf.net. Job Hotline: 619-812-4600.

Company URL: http://www.cerf.net

TEKELEC

26580 AGOURA ROAD
CALABASAS, CALIFORNIA 91302-1921

Overview: Manufactures communications and diagnostic equipment. Established: 1971. Ownership: Public.

Key Statistics: Annual Sales: $72.1 million. Number of Employees: 367.

Hiring History: Number of professional employees hired in 1996: 40.

Expertise/Education Sought: Electrical engineering, mechanical engineering, software engineering/development

Alternatives to Full-Time Employment: Part-time jobs

Contact: Ms. Donna Jamerson, Human Resources Administrator; Phone: 818-880-5656; Fax: 818-880-8894.

Company URL: http://www.tekelec.com

TOSHIBA AMERICA MRI, INC.

280 UTAH AVENUE
SOUTH SAN FRANCISCO, CALIFORNIA 94080

Overview: Manufactures magnetic resonance imaging equipment. Established: 1989. Parent Company: Toshiba America Medical Systems. Ownership: Private.

Key Statistics: Annual Sales: $40 million. Number of Employees: 185.

Hiring History: Number of professional employees hired in 1996: 40.

Expertise/Education Sought: Research and development, manufacturing engineering, electrical engineering, software engineering/development, hardware engineering

Company-Paid Training is available.

Contact: For individuals with previous experience: Ms. Julie Kersting, Manager of Human Resources.

Company URL: http://www.toshiba.com

TOYOTA MOTOR, NORTH AMERICA
19001 SOUTH WESTERN AVENUE
TORRANCE, CALIFORNIA 90501

Overview: Manufactures cars and trucks. Established: 1996. Parent Company: Toyota Motor Corporation. Ownership: Private.

Key Statistics: Number of Employees: 11,000.

Expertise/Education Sought: Engineering, information systems, sales, administration

Alternatives to Full-Time Employment: Paid internships

Company-Paid Training is available.

Contact: For college students and recent graduates: Ms. Gretchen Thompson, College Recruiter; Phone: 310-618-1000; Fax: 310-618-7816. For individuals with previous experience: Human Resources; Phone: 310-618-1000; Fax: 310-618-7816. Job Hotline: 800-544-0815.

TRANSAMERICA INTELLITECH
1860 HOWE AVENUE, SUITE 455
SACRAMENTO, CALIFORNIA 95825

Overview: Provides information retrieval services. Established: 1996. Parent Company: Transamerica Corporation. Ownership: Public.

Key Statistics: Annual Sales: $35.9 million. Number of Employees: 505.

Hiring History: Number of professional employees hired in 1996: 60.

Expertise/Education Sought: Software engineering/development

Alternatives to Full-Time Employment: Paid internships

Company-Paid Training is available.

Contact: For individuals with previous experience: Ms. Joy Pado, Director of Human Resources. E-mail: joy.atkari-pado@transamerica.com.

Company URL: http://www.transamerica.com

TRIDENT MICROSYSTEMS, INC.
189 NORTH BERNARDO DRIVE
MOUNTAIN VIEW, CALIFORNIA 94043

Overview: Manufactures graphics and flat panel controllers. Established: 1987. Ownership: Public.

Key Statistics: Annual Sales: $177.9 million. Number of Employees: 387.

Hiring History: Number of professional employees hired in 1996: 60.

Expertise/Education Sought: Design engineering, electronics/electronics engineering, marketing, sales

Alternatives to Full-Time Employment: Paid internships

Company-Paid Training is available.

International Assignments are available to employees.

Contact: Ms. Rachel Lester, Recruiter; Phone: 650-944-7067; Fax: 650-968-4827. E-mail: staffing@tridentmicro.com.

Company URL: http://www.tridentmicro.com

UMAX COMPUTER CORPORATION
47470 SEABRIDGE DRIVE
FREMONT, CALIFORNIA 94538-6548

Overview: Provides computer integrated systems services. Established: 1996. Parent Company: Umax Data Systems. Ownership: Private.

Key Statistics: Annual Sales: $35.8 million. Number of Employees: 160.

Expertise/Education Sought: Software engineering/development, hardware engineering

Alternatives to Full-Time Employment: Part-time jobs

Contact: Ms. Pam Taylor, Human Resource Manager; Fax: 510-623-7350. E-mail: ptaylor@umaxcc.com.

Company URL: http://www.supermac.com

USCS INTERNATIONAL
2969 PROSPECT PARK DRIVE
RANCHO CORDOVA, CALIFORNIA 95670-6006

Mailing Address: 4511 Golden Foothill Park, El Dorado Hills, CA 95762

Overview: Computer systems integration and software developer and billing services for the cable television industry. Established: 1969. Ownership: Public.

Key Statistics: Annual Sales: $263.2 million. Number of Employees: 2,038.

Hiring History: Number of professional employees hired in 1997: 220. 1996: 200.

Expertise/Education Sought: Computer programming, electrical engineering, mechanical engineering, software engineering/development, computer programming, software engineering/development

Alternatives to Full-Time Employment: Part-time jobs, paid internships

International Assignments are available to employees.

Contact: For college students and recent graduates: Mr. Mark Zoeller, Corporate Staffing Manager; Phone: 916-939-5451; Fax: 916-939-5417. For individuals with previous experience: Ms. Fran Yates, Corporate Staffing Manager, 4511 Golden Foothill Parkway, El Dorado Hills, California, 95762-6006; Phone: 916-939-5451. E-mail: jobs@uscs.com.

Company URL: http://www.uscs.com

VALOR ELECTRONICS
9715 BUSINESS PARK AVENUE
SAN DIEGO, CALIFORNIA 92131-1642

Overview: Manufactures electronic components. Established: 1956. Parent Company: GTI. Ownership: Public.

Key Statistics: Annual Sales: $92.5 million. Number of Employees: 8,075.

Hiring History: Number of professional employees hired in 1997: 50.

Expertise/Education Sought: Engineering

Company-Paid Training is available.

International Assignments are available to employees.

Contact: For individuals with previous experience: Human Resources Department; Phone: 619-537-2500; Fax: 619-537-2525.

VALTRON TECHNOLOGIES, INC.
28309 AVENUE CROCKER
SANTA CLARITA, CALIFORNIA 91355-1251

Overview: Provides computer and data processing equipment maintenance and repair services. Established: 1996. Ownership: Private.

Key Statistics: Annual Sales: $6.1 million. Number of Employees: 110.

Hiring History: Number of professional employees hired in 1996: 20.

Expertise/Education Sought: Administration, technology

Alternatives to Full-Time Employment: Part-time jobs

Contact: For individuals with previous experience: Human Resources Department; Phone: 805-257-0333; Fax: 805-257-0113.

Company URL: http://www.valtron.com

VERTEL CORPORATION
21300 VICTORY BOULEVARD
WOODLAND HILLS, CALIFORNIA 91367-2525

Overview: Provides telecommunications management services. Established: 1995. Ownership: Public.

Key Statistics: Annual Sales: $11.5 million. Number of Employees: 100.

Hiring History: Number of professional employees hired in 1997: 40. 1996: 42.

Expertise/Education Sought: Engineering

Alternatives to Full-Time Employment: Part-time jobs, paid internships

International Assignments are available to employees.

Contact: For individuals with previous experience: Human Resources; Phone: 818-227-5729; Fax: 818-598-0049. E-mail: human-resources@vertel.com.

Company URL: http://www.vertel.com

WALKER INTERACTIVE SYSTEMS

303 SECOND STREET
SAN FRANCISCO, CALIFORNIA 94107

Overview: Develops financial and control application software. Established: 1969. Ownership: Public.

Key Statistics: Annual Sales: $62.8 million. Number of Employees: 388.

Expertise/Education Sought: Software engineering/development, marketing, accounting, sales

Alternatives to Full-Time Employment: Part-time jobs

Company-Paid Training is available.

Contact: Mr. Gregory Allen, Recruiter; Phone: 415-243-2843; Fax: 415-281-1573. E-mail: gja@walker.com.

Company URL: http://www.walker.com

WATKINS MANUFACTURING CORPORATION

1280 PARK CENTER DRIVE
VISTA, CALIFORNIA 92083-8314

Overview: Manufactures portable spas. Established: 1986. Parent Company: Masco Corporation. Ownership: Public.

Key Statistics: Annual Sales: $25 million. Number of Employees: 600.

Hiring History: Number of professional employees hired in 1997: 90. 1996: 60.

Expertise/Education Sought: Sales, finance, mechanical engineering, electrical engineering, customer service and support, purchasing, manufacturing

Alternatives to Full-Time Employment: Paid internships

International Assignments are available to employees.

Contact: Ms. Sandra Shuda, Human Resources Director; Phone: 760-598-6464, ext. 148; Fax: 760-598-5534.

Company URL: http://www.hotspring.com

WESTEC SECURITY GROUP

100 BAYVIEW CIRCLE, SUITE 1000
NEWPORT BEACH, CALIFORNIA 92660-2985

Overview: Owns and operates Westech and other companies. Established: 1987. Ownership: Private.

Key Statistics: Annual Sales: $219.5 million. Number of Employees: 2,500.

Hiring History: Number of professional employees hired in 1997: 65. 1996: 20.

Expertise/Education Sought: Accounting, finance, information technology, telecommunications

Alternatives to Full-Time Employment: Paid internships

Company-Paid Training is available.

Contact: For individuals with previous experience: Mr. Dennis Buster, Vice President of Human Resources; Phone: 714-725-6200; Fax: 714-725-6293. E-mail: aranier-westec@att.net.

WESTERN DIGITAL CORPORATION

8105 IRVINE CENTER DRIVE
IRVINE, CALIFORNIA 92718-2902

Overview: Manufactures magnetic disk drives and computer components. Established: 1970. Ownership: Public.

Key Statistics: Annual Sales: $4.2 billion. Number of Employees: 13,507.

Hiring History: Number of professional employees hired in 1997: 880. 1996: 450.

Expertise/Education Sought: Electronics/electronics engineering, computer science, finance, marketing, mechanical engineering

Alternatives to Full-Time Employment: Part-time jobs, paid internships

International Assignments are available to employees.

Contact: For college students and recent graduates: Ms. Karen Shaw, College Recruiter, 2119 Tasman Drive, Irvine, California, 95054-2902; Phone: 714-932-5087; Fax: 714-932-6629. E-mail: karen.shaw@corp.wdc.com. For individuals with previous experience: Mr. Larry Turchick, Staffing Manager; Phone: 714-932-5138; Fax: 714-932-6629. E-mail: lawrence.s.turchick@corp.wdc.com.

Company URL: http://www.wbc.com

WESTERN FARM SERVICE, INC.

2787 WEST BULLARD AVENUE
FRESNO, CALIFORNIA 93711-2273

Overview: Wholesales fertilizers and agricultural chemicals. Established: 1988. Parent Company: Agrium US, Inc. Ownership: Private.

Key Statistics: Annual Sales: $393.6 million. Number of Employees: 1,400.

Expertise/Education Sought: Agriculture, agricultural engineering, agricultural science

Alternatives to Full-Time Employment: Part-time jobs, paid internships

International Assignments are available to employees.

Contact: Mr. Walter Ormonde, Human Resources; Phone: 209-436-0450; Fax: 209-438-9362.

Company URL: http://www.agrium.com

WHEREHOUSE ENTERTAINMENT, INC.

19701 HAMILTON AVENUE, SUITE 200
TORRANCE, CALIFORNIA 90502-1311

Overview: Distributes and markets audio tapes, CDs, and video tapes. Established: 1992. Ownership: Private.

Key Statistics: Annual Sales: $365.5 million. Number of Employees: 4,450.

Expertise/Education Sought: Operations, field service, finance, analysis, training and development, advertising, project management, merchandising, accounting

Alternatives to Full-Time Employment: Paid internships

Company-Paid Training is available.

International Assignments are available to employees.

Contact: Recruiters; Phone: 310-538-2314; Fax: 310-538-8798. Job Hotline: 310-538-2314, ext. 2565.

COLORADO

ARCADIS GERAGHTY AND MILLER, INC.

1099 18TH STREET, SUITE 2100
DENVER, COLORADO 80202

Overview: Engineering and architectural consulting firm. Established: 1982. Parent Company: Arcadis, NV. Ownership: Private.

Key Statistics: Annual Sales: $28.1 million. Number of Employees: 390.

Hiring History: Number of professional employees hired in 1997: 200.

Expertise/Education Sought: Mechanical engineering, structural engineering, electronics/electronics engineering, electrical engineering, transportation, environmental engineering, civil engineering

Company-Paid Training is available.

Contact: For individuals with previous experience: Ms. Jacqueline Maidel, Vice President of Human Resources; Phone: 303-294-1200; Fax: 303-294-1219.

BALL CORPORATION
10 LONGS PEAK DRIVE
BROOMFIELD, COLORADO 80021-2510

Overview: Manufactures packaging, consumer products, and aerospace systems. Established: 1880. Ownership: Public.

Key Statistics: Annual Sales: $2.2 billion. Number of Employees: 7,500.

Expertise/Education Sought: Accounting, finance, data processing

International Assignments are available to employees.

Contact: Human Resources, 345 South High Street, Muncie, Indiana, 47305; Phone: 303-469-5511; Fax: 303-460-2315.

CODAR TECHNOLOGY, INC.
2405 TRADE CENTER AVENUE
LONGMONT, COLORADO 80503-7601

Overview: Manufactures electronic computers. Established: 1990. Parent Company: NAI Technologies, Inc. Ownership: Public.

Key Statistics: Annual Sales: $10.7 million. Number of Employees: 80.

Expertise/Education Sought: Electronics/electronics engineering, design, drafting

Contact: Human Resources Department; Phone: 303-776-0472; Fax: 303-776-8532.

Company URL: http://www.codar.com

COMMNET CELLULAR, INC.
8350 EAST CRESCENT PARKWAY, SUITE 400
ENGLEWOOD, COLORADO 80111

Overview: Manufactures cellular telephones. Established: 1983. Ownership: Public.

Key Statistics: Annual Sales: $115.2 million. Number of Employees: 510.

Hiring History: Number of professional employees hired in 1997: 4.

Expertise/Education Sought: Radio frequency engineering, customer service and support, telecommunications, network management systems, sales, collection

Contact: Mr. John Closs, Director of Employment Compensation; Phone: 303-694-3234; Fax: 303-721-3580.

Company URL: http://www.cels.com

COORS BREWING COMPANY
311 10TH STREET
GOLDEN, COLORADO 80401-0030

Overview: Manufactures and distributes beer and other malt beverages. Established: 1873. Ownership: Public.

Key Statistics: Annual Sales: $1.7 billion. Number of Employees: 5,800.

Hiring History: Number of professional employees hired in 1997: 500. 1996: 500.

Expertise/Education Sought: Accounting, marketing, sales, electrical engineering, information technology, manufacturing, mechanical engineering, packaging engineering, chemical engineering, industrial engineering

Alternatives to Full-Time Employment: Paid internships

Company-Paid Training is available.

International Assignments are available to employees.

Contact: For college students and recent graduates: Mrs. Debbie Helburg, College Recruiter. For individuals with previous experience: Human Resources Department. Job Hotline: 303-277-3450.

CRYSTAL SPECIALTIES, INC.
2853 JANITELL ROAD
COLORADO SPRINGS, COLORADO 80906-4104

Overview: Manufactures industrial inorganic chemicals. Established: 1994. Ownership: Private.

Key Statistics: Annual Sales: $7 million. Number of Employees: 55.

Expertise/Education Sought: Manufacturing

Company-Paid Training is available.

Contact: For individuals with previous experience: Human Resources; Phone: 719-540-0990; Fax: 719-540-0994.

EATON METAL PRODUCTS COMPANY
4800 YORK STREET
DENVER, COLORADO 80216

Overview: Manufactures steel tanks, bins, hoppers, and silos. Established: 1919. Ownership: Private.

Key Statistics: Annual Sales: $127 million. Number of Employees: 700.

Expertise/Education Sought: Mechanical engineering

Contact: For individuals with previous experience: Ms. Karen Mayner, Manager of Personnel Payroll; Phone: 303-296-5725; Fax: 303-296-5736.

ECHOSTAR COMMUNICATIONS CORPORATION
5701 SOUTH SANTE FE
LITTLETON, COLORADO 80210-5317

Overview: Provides direct broadcasting satellite programming services. Established: 1993. Ownership: Public.

Key Statistics: Annual Sales: $209.7 million.

Expertise/Education Sought: Engineering

International Assignments are available to employees.

Contact: For individuals with previous experience: Human Resource Department; Phone: 303-799-8222; Fax: 303-799-0353.

THE GATES RUBBER COMPANY
990 SOUTH BROADWAY
DENVER, COLORADO 80209

Overview: Manufactures belts and hoses. Established: 1996. Parent Company: Tompkins, PLC. Ownership: Private.

Key Statistics: Annual Sales: $1.2 billion. Number of Employees: 13,360.

Hiring History: Number of professional employees hired in 1997: 150. 1996: 110.

Expertise/Education Sought: Mechanical engineering, quality control, public relations, industrial engineering, business, manufacturing engineering, marketing, sales, finance

Alternatives to Full-Time Employment: Paid internships

International Assignments are available to employees.

Contact: For individuals with previous experience: Ms. Sandra Skuball, Manager of Human Resources; Phone: 303-744-5527; Fax: 303-744-4905. E-mail: sskuball@gates.com. Job Hotline: 303-744-5900.

Company URL: http://www.gates.com

HENSEL PHELPS CONSTRUCTION COMPANY
420 6TH AVENUE
GREELEY, COLORADO 80631

Overview: Provides industrial building, highway, and street construction services. Established: 1960. Ownership: Private.

Key Statistics: Annual Sales: $876.2 million. Number of Employees: 1,500.

Expertise/Education Sought: Construction engineering, field engineering

Alternatives to Full-Time Employment: Paid internships

Contact: For individuals with previous experience: Human Resources Department; Phone: 970-352-6565; Fax: 970-352-9311.

KAISER-HILL COMPANY LLC
PO BOX 464
GOLDEN, COLORADO 80402-0464

Overview: Provides environmental research and development services. Parent Company: ICF Kaiser and CHEM Hill.

Expertise/Education Sought: Environmental engineering

Kaiser-Hill Company LLC (continued)

Company-Paid Training is available.
Contact: For individuals with previous experience: Mr. Dana Dorr, Employee Relations Manager; Phone: 303-966-7978; Fax: 303-966-4991. Job Hotline: 800-835-7298.

MEDIA ONE
9785 MAROON CIRCLE, SUITE 400
ENGLEWOOD, COLORADO 80112

Overview: Provides cable television services. Established: 1963. Parent Company: U S West Media Group. Ownership: Public.
Key Statistics: Annual Sales: $1.2 billion. Number of Employees: 9,200.
Expertise/Education Sought: Computer engineering, electrical engineering
Alternatives to Full-Time Employment: Paid internships
Company-Paid Training is available.
Contact: For individuals with previous experience: Ms. Susan Ellis, Director of Human Resources; Fax: 303-754-4366.
Company URL: http://www.mediaone.com

MONFORT, INC.
1918 AA STREET
GREELEY, COLORADO 80631

Overview: Produces meat products and prepared foods. Established: 1987. Parent Company: ConAgra, Inc. Ownership: Public.
Key Statistics: Annual Sales: $3.6 billion. Number of Employees: 14,000.
Expertise/Education Sought: Laboratory technology, microbiology, chemistry, electrical engineering, accounting, mechanical engineering, finance, manufacturing
Alternatives to Full-Time Employment: Paid internships
Company-Paid Training is available.
Contact: For individuals with previous experience: Ms. Sue Toner, Recruiting Manager, PO Box 2600, Greeley, Colorado, 80631; Phone: 970-395-8399; Fax: 970-351-0467. E-mail: hropport@crfc.com. Job Hotline: 800-258-9596.

NEW CENTURY ENERGIES, INC.
1730 BLAKE STREET, SUITE 100
DENVER, COLORADO 80202-5534

Overview: Electric and gas utility. Established: 1906. Ownership: Private.
Key Statistics: Annual Sales: $1.1 billion. Number of Employees: 4,000.
Hiring History: Number of professional employees hired in 1997: 300.
Expertise/Education Sought: Finance, marketing, accounting, electrical engineering, mechanical engineering, human resources
Alternatives to Full-Time Employment: Part-time jobs, paid internships
Company-Paid Training is available.
International Assignments are available to employees.
Contact: Mr. Richard Paul, Manager of Corporate Staffing; Phone: 303-294-8041; Fax: 303-294-8533. E-mail: thomasboy187@msp.psco.com. Job Hotline: 303-571-7563.
Company URL: http://www.psco.com

NICOLET VASCULAR, INC.
6355 JOYCE DRIVE
GOLDEN, COLORADO 80403

Overview: Manufactures Doppler system products for physicians, clinics, and hospitals. Established: 1976. Parent Company: Nicolet Biomedical, Inc. Ownership: Public.
Key Statistics: Annual Sales: $7 million. Number of Employees: 75.
Hiring History: Number of professional employees hired in 1997: 9. 1996: 8.
Expertise/Education Sought: Engineering, marketing
Alternatives to Full-Time Employment: Paid internships
Company-Paid Training is available.
International Assignments are available to employees.
Contact: For individuals with previous experience: Human Resources; Phone: 303-431-9400, ext. 251; Fax: 303-431-0429. E-mail: imxkelly@aol.com.

RESORT COMPUTER CORPORATION
12596 WEST BAYAUD AVENUE
LAKEWOOD, COLORADO 80228-2000

Overview: Develops computer software and applications. Established: 1981. Parent Company: Cendant. Ownership: Private.
Key Statistics: Annual Sales: $8.2 million. Number of Employees: 90.
Expertise/Education Sought: Software engineering/development
Alternatives to Full-Time Employment: Internships
Company-Paid Training is available.
Contact: Human Resources Department; Phone: 303-232-9417; Fax: 303-232-9556. E-mail: recruiter@rccden.com.
Company URL: http://www.rccden.com

STORAGE TECHNOLOGY CORPORATION
2270 SOUTH 88TH STREET
LOUISVILLE, COLORADO 80028-1170

Overview: Manufactures and markets magnetic disk drives, tape drives, communications equipment and software. Established: 1969. Ownership: Public.
Key Statistics: Annual Sales: $2 billion. Number of Employees: 8,300.
Hiring History: Number of professional employees hired in 1997: 500. 1996: 350.
Expertise/Education Sought: Computer engineering, sales, marketing, electrical engineering
Alternatives to Full-Time Employment: Part-time jobs, paid internships
Company-Paid Training is available.
International Assignments are available to employees.
Contact: For college students and recent graduates: Ms. Debra Coyle, College Recruiter; Phone: 303-673-5151; Fax: 303-661-7637. For individuals with previous experience: Mr. David Armstead, Manager of Recruiting Services; Phone: 303-673-5151; Fax: 303-661-7637.
Company URL: http://www.storagetek.com

SUNDSTRAND FLUID HANDLING CORPORATION
14845 WEST 64TH AVENUE
ARVADA, COLORADO 80007-7523

Overview: Manufactures pumps and pumping equipment. Established: 1996. Parent Company: Sundstrand Corporation. Ownership: Public.
Key Statistics: Annual Sales: $50 million. Number of Employees: 386.
Expertise/Education Sought: Engineering, marketing, graphic arts, purchasing
Alternatives to Full-Time Employment: Part-time jobs, paid internships
Company-Paid Training is available.
International Assignments are available to employees.
Contact: Ms. Patricia Butler, Human Resources Administrator; Phone: 303-425-0800; Fax: 303-425-0896. Job Hotline: 303-940-3112.
Company URL: http://www.sfh.com

TRI-STATE GENERATION AND TRANSMISSION
100 WEST 116TH AVENUE
WESTMINSTER, COLORADO 80030

Mailing Address: PO Box 33695, Denver, CO 80233
Overview: Electric utility. Established: 1952. Ownership: Private.
Key Statistics: Annual Sales: $429.5 million. Number of Employees: 835.
Hiring History: Number of professional employees hired in 1996: 10.

Expertise/Education Sought: Electrical engineering, mechanical engineering

Company-Paid Training is available.

Contact: For individuals with previous experience: Mr. Placido Herrera, Recruitment Specialist; Phone: 303-452-6111; Fax: 303-254-6018.

UNITED ARTIST THEATER CIRCUIT, INC.

9110 EAST NICHOLS AVENUE, SUITE 200
ENGLEWOOD, COLORADO 80112

Overview: Owns and operates movie theaters, produces motion pictures, and provides cable television services. Established: 1992. Ownership: Private.

Key Statistics: Annual Sales: $677.5 million. Number of Employees: 11,000.

Hiring History: Number of professional employees hired in 1997: 20.

Expertise/Education Sought: Accounting, finance, law, marketing, human resources, administration, film/film production

Contact: For individuals with previous experience: Human Resources Department; Phone: 303-792-8613; Fax: 303-792-8221.

Company URL: http://www.uatc.com

VERIBEST, INC.

6101 LOOKOUT ROAD
BOULDER, COLORADO 80301-3359

Overview: Provides computer software development services. Established: 1995. Parent Company: Intergraph. Ownership: Public.

Key Statistics: Annual Sales: $24.3 million. Number of Employees: 310.

Expertise/Education Sought: Software engineering/development

Alternatives to Full-Time Employment: Internships

Company-Paid Training is available.

International Assignments are available to employees.

Contact: For individuals with previous experience: Mr. Dan Peters, Director of Human Resources; Phone: 303-581-2303; Fax: 303-581-9972. E-mail: djpeters@veribest.com.

Company URL: http://www.veribest.com

WESTERN GAS RESOURCES, INC.

12200 NORTH PECOS STREET
DENVER, COLORADO 80234-3439

Overview: Extracts petroleum and natural gas. Established: 1989. Ownership: Public.

Key Statistics: Annual Sales: $2.1 billion. Number of Employees: 920.

Hiring History: Number of professional employees hired in 1997: 125. 1996: 125.

Expertise/Education Sought: Petroleum/petrochemical engineering, geology, mechanical engineering, marketing, accounting, information technology

Alternatives to Full-Time Employment: Paid internships

Company-Paid Training is available.

Contact: Mr. David DeBell, Director of Human Resources; Phone: 800-677-5603; Fax: 303-457-8482.

Company URL: http://www.westerngas.com

CONNECTICUT

ACME UNITED CORPORATION

75 KINGS HIGHWAY CUTOFF
FAIRFIELD, CONNECTICUT 06430-5340

Overview: Manufactures scissors and hand shears. Established: 1867. Ownership: Public.

Key Statistics: Annual Sales: $47.5 million. Number of Employees: 384.

Hiring History: Number of professional employees hired in 1997: 70. 1996: 15.

Expertise/Education Sought: Manufacturing engineering, engineering, finance, information systems, business management

Alternatives to Full-Time Employment: Paid internships

Company-Paid Training is available.

International Assignments are available to employees.

Contact: For individuals with previous experience: Human Resources Department.

ADVANCED TECHNOLOGY MATERIALS, INC.

7 COMMERCE DRIVE
DANBURY, CONNECTICUT 06810

Overview: Develops and manufactures high performance materials and equipment for the semiconductor industry.

Hiring History: Number of professional employees hired in 1997: 173. 1996: 40.

Expertise/Education Sought: Electrical engineering, chemical engineering, physics, materials science, analytical chemistry, inorganic chemistry, mechanical engineering

Alternatives to Full-Time Employment: Part-time jobs, paid internships

Company-Paid Training is available.

International Assignments are available to employees.

Contact: Manager of Staffing and University Relations; Phone: 203-794-1100; Fax: 203-792-8040.

Company URL: http://www.atmi.com

AMERICAN RADIO RELAY LEAGUE, INC.

225 MAIN STREET
NEWINGTON, CONNECTICUT 06111-1400

Overview: Provides emergency radio service to agencies and people affected by disaster. Established: 1914. Ownership: Private.

Key Statistics: Annual Sales: $12 million. Number of Employees: 120.

Expertise/Education Sought: Telecommunications

Contact: For individuals with previous experience: Mr. Robert Boucher, Personnel Manager; Phone: 860-594-0200; Fax: 860-594-0298. E-mail: rboucher@arrl.org.

Company URL: http://www.arrl.org

AMPHENOL CORPORATION

358 HALL AVENUE
WALLINGFORD, CONNECTICUT 06492-3574

Overview: Manufactures electrical connectors, cables, and telecommunications equipment. Established: 1932. Ownership: Private.

Key Statistics: Annual Sales: $436.8 million. Number of Employees: 6,200.

Expertise/Education Sought: Marketing, accounting, sales, electronics/electronics engineering

Company-Paid Training is available.

International Assignments are available to employees.

Contact: Mr. Cary Martin, Manager of Human Resources; Phone: 203-265-8765; Fax: 203-265-8746.

AMS SERVICES, INC.

3 WATERSIDE CROSSING
WINDSOR, CONNECTICUT 06095

Overview: Manufactures software for insurance agents. Established: 1983. Ownership: Public.

Key Statistics: Annual Sales: $158.9 million. Number of Employees: 1,187.

Hiring History: Number of professional employees hired in 1997: 3,000. 1996: 50.

Expertise/Education Sought: Software engineering/development, computer programming

Alternatives to Full-Time Employment: Paid internships

Contact: For individuals with previous experience: Ms. Beth Bernardo, Human Resources Representative; Phone: 860-602-6000; Fax: 860-602-6006.

ANALYSIS AND TECHNOLOGY, INC.

ROUTE 2
NORTH STONINGTON, CONNECTICUT 06359-9801

Mailing Address: PO Box 220, North Stonington, CT 06359

Overview: Develops software and computer systems. Established: 1969. Ownership: Public.

Key Statistics: Annual Sales: $142.5 million. Number of Employees: 1,371.

Expertise/Education Sought: Accounting, software engineering/development, systems analysis, electrical engineering, operations engineering, systems engineering

Alternatives to Full-Time Employment: Paid internships

Company-Paid Training is available.

International Assignments are available to employees.

Contact: Ms. Dawn Vernieri, Human Resources Administrator; Phone: 860-599-3910, ext. 2415; Fax: 860-599-6516. E-mail: dvernieri@atinc.com.

Company URL: http://www.aati.com

ANAMET, INC.

698 SOUTH MAIN STREET
WATERBURY, CONNECTICUT 06706

Overview: Manufactures wiring devices, industrial machinery, and motor vehicle parts and accessories. Established: 1984. Ownership: Private.

Key Statistics: Annual Sales: $75 million. Number of Employees: 1,000.

Hiring History: Number of professional employees hired in 1996: 5.

Expertise/Education Sought: Finance

Contact: Mr. Eric Nelson, Human Resources Representative; Phone: 203-574-8500; Fax: 203-574-8623.

APEX MACHINE TOOL COMPANY, INC.

21 SPRING LANE
FARMINGTON, CONNECTICUT 06032-3128

Overview: Provides engineering services. Established: 1944. Ownership: Private.

Key Statistics: Annual Sales: $15.5 million. Number of Employees: 165.

Hiring History: Number of professional employees hired in 1997: 30. 1996: 25.

Expertise/Education Sought: Manufacturing, engineering

Alternatives to Full-Time Employment: Paid internships

Company-Paid Training is available.

Contact: For individuals with previous experience: Human Resource Department; Phone: 860-677-2884, ext. 304; Fax: 860-676-4347.

Company URL: http://www.apexmachinetool.com

AUTOMOTIVE CONTROLS CORPORATION

1 ECHLIN ROAD
BRANFORD, CONNECTICUT 06405-4254

Overview: Manufactures engine electrical apparatus. Established: 1964. Parent Company: Eichlin, Inc. Ownership: Public.

Key Statistics: Annual Sales: $243.9 million. Number of Employees: 1,500.

Hiring History: Number of professional employees hired in 1997: 9. 1996: 18.

Expertise/Education Sought: Design engineering, mechanical engineering, manufacturing engineering, industrial engineering, computer programming, automotive engineering, network management systems, quality engineering, electrical engineering, systems analysis

Alternatives to Full-Time Employment: Paid internships

Company-Paid Training is available.

International Assignments are available to employees.

Contact: For individuals with previous experience: Mr. John Donovan, Manager of Human Resources; Phone: 203-481-5771; Fax: 203-488-8797.

BALDWIN GRAPHIC PRODUCTS

12 COMMERCE DRIVE
SHELTON, CONNECTICUT 06484

Overview: Manufactures equipment for the printing industry. Established: 1918. Parent Company: Baldwin Technology Company, Inc. Ownership: Public.

Key Statistics: Annual Sales: $244.1 million. Number of Employees: 1,047.

Hiring History: Number of professional employees hired in 1996: 10.

Expertise/Education Sought: Technical engineering, sales, marketing, engineering

Alternatives to Full-Time Employment: Paid internships

Contact: For individuals with previous experience: Manager of Human Resources; Phone: 203-925-1100; Fax: 203-929-7856.

THE BARDEN CORPORATION

200 PARK AVENUE
DANBURY, CONNECTICUT 06810-7553

Overview: Manufactures ball bearings. Established: 1990. Parent Company: FAG Corporation. Ownership: Private.

Key Statistics: Annual Sales: $70.6 million. Number of Employees: 580.

Hiring History: Number of professional employees hired in 1996: 15.

Expertise/Education Sought: Manufacturing

Company-Paid Training is available.

Contact: Mr. Phillip Brockelman, Human Resources Director; Phone: 203-744-2211.

BICRON ELECTRONICS COMPANY

50 BARLOW STREET
CANAAN, CONNECTICUT 06018-2035

Overview: Manufactures transformers. Established: 1964. Ownership: Private.

Key Statistics: Annual Sales: $9.5 million. Number of Employees: 136.

Expertise/Education Sought: Engineering

Company-Paid Training is available.

Contact: For individuals with previous experience: Human Resources; Phone: 860-824-5125, ext. 112.

BRANSON ULTRASONICS CORPORATION

41 EAGLE ROAD
DANBURY, CONNECTICUT 06813

Overview: Manufactures ultrasonic welders and cleaning equipment. Established: 1984. Parent Company: Emerson Electric Company. Ownership: Public.

Key Statistics: Annual Sales: $129 million. Number of Employees: 1,600.

Hiring History: Number of professional employees hired in 1997: 50.

Expertise/Education Sought: Mechanical engineering, software engineering/development, control engineering, technology research

Alternatives to Full-Time Employment: Paid internships

International Assignments are available to employees.

Contact: Mrs. Laura Sheehy, Manager of Employee Relations; Phone: 203-796-0347; Fax: 203-796-9802.

Company URL: http://www.branson.com

CANBERRA INDUSTRIES

800 RESEARCH PARKWAY
MERIDEN, CONNECTICUT 06450-7127

Overview: Manufactures measuring and controlling devices. Established: 1965. Parent Company: Packard BioScience. Ownership: Private.

Key Statistics: Annual Sales: $87.9 million. Number of Employees: 1,000.

Expertise/Education Sought: Engineering, manufacturing engineering

Contact: For individuals with previous experience: Human Resource Manager; Phone: 203-639-2106.

Company URL: http://www.canberra.com

CARLINGSWITCH, INC.

60 JOHNSON AVENUE
PLAINVILLE, CONNECTICUT 06062-1177

Overview: Manufactures electric switches. Established: 1920. Ownership: Private.

Key Statistics: Annual Sales: $50 million. Number of Employees: 1,800.

Hiring History: Number of professional employees hired in 1997: 6. 1996: 10.

Expertise/Education Sought: Manufacturing engineering, design engineering, electronics/electronics engineering, sales, marketing

Contact: For individuals with previous experience: Ms. Karen Olsen, Human Resource Manager; Phone: 860-793-9266; Fax: 860-747-0296.

Company URL: http://www.carlingswitch.com

CARRIER CORPORATION

1 CARRIER PLACE
FARMINGTON, CONNECTICUT 06032

Overview: Manufactures heating and air conditioning equipment. Established: 1979. Parent Company: United Technologies Corporation. Ownership: Public.

Key Statistics: Annual Sales: $5.5 billion. Number of Employees: 33,000.

Expertise/Education Sought: Management, business administration

Alternatives to Full-Time Employment: Part-time jobs, paid internships

Company-Paid Training is available.

International Assignments are available to employees.

Contact: Ms. Heather Beering, Manager of Professional Recruiting, PO Box 4015, Farmington, Connecticut, 06034-4015; Phone: 860-674-3117; Fax: 860-674-3125.

Company URL: http://www.carrier.com

CHAMPION INTERNATIONAL CORPORATION

1 CHAMPION PLAZA
STAMFORD, CONNECTICUT 06921

Overview: Produces wood and paper products. Established: 1937. Ownership: Public.

Key Statistics: Annual Sales: $5.9 billion. Number of Employees: 24,400.

Expertise/Education Sought: Management information systems, accounting, management

Alternatives to Full-Time Employment: Paid internships

Company-Paid Training is available.

International Assignments are available to employees.

Contact: Staffing Manager; Phone: 203-358-7006; Fax: 203-358-6483.

Company URL: http://www.championpaper.com

CONAIR CORPORATION

1 CUMMINGS POINT ROAD
STAMFORD, CONNECTICUT 06904

Overview: Manufactures personal care products and small appliances. Established: 1959. Ownership: Public.

Key Statistics: Annual Sales: $654.7 million. Number of Employees: 3,200.

Expertise/Education Sought: Electrical engineering, sales, marketing, accounting

Alternatives to Full-Time Employment: Internships

Company-Paid Training is available.

International Assignments are available to employees.

Contact: For individuals with previous experience: Ms. Barbara Melbourne, Director of Human Resources; Phone: 203-351-9048; Fax: 203-351-9234.

Company URL: http://www.conair.com

CONTINENTAL PLASTIC CONTAINER

301 MERRITT 7 CORPORATE PARK
NORWALK, CONNECTICUT 06856-5410

Overview: Manufactures plastic film, plastic bottles, and food cans. Established: 1970. Parent Company: Continental Can Company, Inc. Ownership: Public.

Key Statistics: Annual Sales: $585 million. Number of Employees: 3,700.

Expertise/Education Sought: Civil engineering, structural engineering, design engineering, electrical engineering

Alternatives to Full-Time Employment: Paid internships

Company-Paid Training is available.

Contact: For individuals with previous experience: Human Resources Department.

Company URL: http://www.cpchome.com

COROMETRICS MEDICAL SYSTEMS

61 BARNES PARK ROAD NORTH
WALLINGFORD, CONNECTICUT 06492-0333

Overview: Manufactures electronic instrumentation, diagnostic ultrasound systems, and related accessories. Established: 1994. Parent Company: Marquette Medical Systems. Ownership: Public.

Key Statistics: Annual Sales: $55 million. Number of Employees: 525.

Hiring History: Number of professional employees hired in 1997: 50. 1996: 50.

Expertise/Education Sought: Software engineering/development, electrical engineering, biomedical engineering, manufacturing engineering, finance, accounting, marketing, clinical research

Alternatives to Full-Time Employment: Part-time jobs, paid internships

Company-Paid Training is available.

International Assignments are available to employees.

Contact: Mr. John Kubilus, Manager of Human Resources; Phone: 800-243-3952; Fax: 203-949-2460.

Company URL: http://www.mei.com

DAVIS-STANDARD CORPORATION

1 EXTRUSION DRIVE
PAWCATUCK, CONNECTICUT 06379-2313

Overview: Manufactures pumps and pumping equipment. Established: 1995. Parent Company: Crompton And Knowles Corporation. Ownership: Public.

Davis-Standard Corporation (continued)

Key Statistics: Annual Sales: $310 million. Number of Employees: 1,100.
Expertise/Education Sought: Extrusion
Company-Paid Training is available.
International Assignments are available to employees.
Contact: Mr. Russell Chinni, Director of Human Resources.

DEVTEC CORPORATION
812 BLOOMFIELD AVENUE
WINDSOR, CONNECTICUT 06095-2321

Overview: Manufactures aircraft parts and equipment. Established: 1991. Parent Company: Teleflex Lionel-Dupont. Ownership: Private.
Key Statistics: Annual Sales: $55.6 million. Number of Employees: 211.
Hiring History: Number of professional employees hired in 1996: 5.
Expertise/Education Sought: Sales, manufacturing
Company-Paid Training is available.
Contact: Ms. Debbie Dorman, Human Resources Manager; Phone: 860-688-4986; Fax: 860-688-7773.

DEXTER NONWOVENS
2 ELM STREET
WINDSOR LOCKS, CONNECTICUT 06096-2334

Overview: Manufactures specialty nonwoven materials. Established: 1767. Parent Company: Dexter Corporation. Ownership: Public.
Key Statistics: Annual Sales: $1.1 billion. Number of Employees: 4,600.
Hiring History: Number of professional employees hired in 1996: 44.
Expertise/Education Sought: Sales, marketing, management information systems, engineering, finance, research and development
Alternatives to Full-Time Employment: Paid internships
Company-Paid Training is available.
International Assignments are available to employees.
Contact: Ms. Michelle Virkler, Executive Recruiter; Phone: 860-654-8473; Fax: 860-654-8596.

DICTAPHONE CORPORATION
3191 BROADBRIDGE AVENUE
STRATFORD, CONNECTICUT 06497-2566

Overview: Manufactures and markets dictation, voice processing, and communications equipment. Established: 1995. Ownership: Private.
Key Statistics: Annual Sales: $332.5 million. Number of Employees: 2,600.
Hiring History: Number of professional employees hired in 1997: 300. 1996: 200.
Expertise/Education Sought: Accounting, marketing, sales, engineering, information systems
Alternatives to Full-Time Employment: Part-time jobs, paid internships
Company-Paid Training is available.
International Assignments are available to employees.
Contact: Human Resources Department; Phone: 203-381-7136; Fax: 203-386-8566. E-mail: hr@dictaphone.com.
Company URL: http://www.dictaphone.com

DNE SYSTEMS, INC.
50 BARNES INDUSTRIAL PARK
WALLINGFORD, CONNECTICUT 06492-1846

Overview: Manufactures telephone and telegraph equipment. Established: 1995. Parent Company: Superior Telecom, Inc. Ownership: Public.
Key Statistics: Annual Sales: $25 million. Number of Employees: 175.
Hiring History: Number of professional employees hired in 1997: 20. 1996: 20.
Expertise/Education Sought: Electrical engineering, software engineering/development, management, sales, manufacturing
Company-Paid Training is available.
Contact: Mr. Ronald Hanley, Director of Human Resources; Phone: 203-265-7151; Fax: 203-265-9101. E-mail: djs@dne.com.

DURACELL NORTH ATLANTIC GROUP
BERKSHIRE CORPORATE PARK
BETHEL, CONNECTICUT 06801

Overview: Manufactures batteries. Established: 1916. Parent Company: Gilette Company. Ownership: Public.
Key Statistics: Annual Sales: $2.3 billion. Number of Employees: 8,000.
Expertise/Education Sought: Accounting, finance, computer programming, sales, marketing, design engineering, project engineering, mechanical engineering, electrical engineering
Alternatives to Full-Time Employment: Paid internships
Company-Paid Training is available.
Contact: Ms. Deborah Adames, Assistant Manager of Corporate Staffing; Phone: 203-796-4000; Fax: 203-791-3039.
Company URL: http://www.duracellusa.com

EDWARDS SYSTEMS TECHNOLOGIES
90 FIELDSTONE COURT
CHESHIRE, CONNECTICUT 06410-1212

Overview: Manufactures lighting equipment. Established: 1989. Parent Company: General Signal. Ownership: Public.
Key Statistics: Annual Sales: $182 million. Number of Employees: 2,107.
Hiring History: Number of professional employees hired in 1997: 10.
Expertise/Education Sought: Administration, finance, information systems, human resources, accounting, engineering
Alternatives to Full-Time Employment: Paid internships
Company-Paid Training is available.
International Assignments are available to employees.
Contact: For individuals with previous experience: Human Resources; Phone: 203-699-3000; Fax: 203-699-3321.

ELECTROCAL DESIGNS, INC.
78 EDWIN ROAD
SOUTH WINDSOR, CONNECTICUT 06074-2414

Overview: Provides commercial printing services. Established: 1985. Ownership: Private.
Key Statistics: Annual Sales: $14.1 million. Number of Employees: 155.
Hiring History: Number of professional employees hired in 1997: 37. 1996: 39.
Expertise/Education Sought: Manufacturing engineering, marketing
Alternatives to Full-Time Employment: Paid internships
Company-Paid Training is available.
International Assignments are available to employees.
Contact: Mr. Kyle Boyle, Vice President; Phone: 860-290-3916; Fax: 860-289-7847.

EMHART INDUSTRIES, INC.
510 RIVER ROAD
SHELTON, CONNECTICUT 06484

Overview: Manufactures hardware, locks, bolts, and other miscellaneous metal products. Established: 1989. Parent Company: Black And Decker Corporation. Ownership: Public.
Key Statistics: Annual Sales: $1.7 billion. Number of Employees: 20,000.
Hiring History: Number of professional employees hired in 1997: 50.
Expertise/Education Sought: Engineering, research and development, manufacturing engineering, mechanical engineering
Alternatives to Full-Time Employment: Paid internships

Company-Paid Training is available.
Contact: For individuals with previous experience: Human Resource Department; Phone: 203-925-3112; Fax: 203-925-3199.
Company URL: http://www.emhart.com

EMSON, INC.
1100 BOSTON AVENUE
BRIDGEPORT, CONNECTICUT 06610-2654

Overview: Manufactures metal products, including spray nozzles aerosol valves. Established: 1956. Ownership: Private.
Key Statistics: Annual Sales: $86 million. Number of Employees: 385.
Expertise/Education Sought: Administration, engineering, computer science, sales, purchasing
International Assignments are available to employees.
Contact: Human Resource Department; Phone: 203-366-4501; Fax: 203-331-4444.

EXECUTONE INFORMATION SYSTEMS
478 WHEELERS FARMS ROAD
MILFORD, CONNECTICUT 06460-1847

Overview: Wholesales telephone and telegraphic equipment. Established: 1996. Ownership: Private.
Key Statistics: Annual Sales: $280.2 million. Number of Employees: 1,400.
Hiring History: Number of professional employees hired in 1997: 50.
Expertise/Education Sought: Telecommunications
Contact: For individuals with previous experience: Human Resources Department; Phone: 203-882-6364; Fax: 203-882-0398.
Company URL: http://www.executone.com

FAG BEARINGS CORPORATION
200 PARK AVENUE
DANBURY, CONNECTICUT 06813

Overview: Manufactures ball and roller bearings. Established: 1968. Parent Company: FAG Holding Corporation. Ownership: Private.
Key Statistics: Annual Sales: $349.9 million. Number of Employees: 1,300.
Expertise/Education Sought: Mechanical engineering, design engineering, mathematics, customer service and support, sales
Company-Paid Training is available.
Contact: For individuals with previous experience: Corporate Resource Manager; Phone: 203-830-8111.

THOMAS G. FARIA CORPORATION
385 NORWICH NEW LONDON TURNPIKE
UNCASVILLE, CONNECTICUT 06382-2432

Mailing Address: PO Box 983, Uncasville, CT 06382
Overview: Manufactures fluid meters and counting devices. Established: 1956. Ownership: Private.
Key Statistics: Annual Sales: $24 million. Number of Employees: 300.
Expertise/Education Sought: Manufacturing engineering
Alternatives to Full-Time Employment: Part-time jobs
International Assignments are available to employees.
Contact: Human Relations Department.
Company URL: http://www.Faria-instruments.com

FIRE-LITE ALARMS, INC.
12 CLINTONVILLE ROAD
NORTHFORD, CONNECTICUT 06472-1653

Overview: Manufactures fire alarms. Established: 1984. Parent Company: Pittway. Ownership: Public.
Key Statistics: Annual Sales: $43.8 million. Number of Employees: 500.
Expertise/Education Sought: Accounting, engineering, computer programming, computer science
Alternatives to Full-Time Employment: Paid internships
Company-Paid Training is available.
International Assignments are available to employees.
Contact: Ms. Susanna Schroeder, Recruitment Manager; Phone: 203-484-7161; Fax: 203-484-1037.

FIRST BRANDS CORPORATION
83 WOOSTER HEIGHTS ROAD
DANBURY, CONNECTICUT 06813

Overview: Manufactures and markets consumer, household, and automotive products. Established: 1986. Ownership: Public.
Key Statistics: Annual Sales: $1.1 billion. Number of Employees: 4,800.
Expertise/Education Sought: Chemical engineering, computer programming, mechanical engineering, electrical engineering, sales, industrial engineering, accounting, human resources
Alternatives to Full-Time Employment: Part-time jobs, paid internships
International Assignments are available to employees.
Contact: Mr. Vic Munger, Manager of Recruiting; Phone: 203-731-2300; Fax: 203-731-2087.
Company URL: http://www.firstbrands.com

FORTUNE BRANDS, INC.
1700 EAST PUTNAM AVENUE
OLD GREENWICH, CONNECTICUT 06870-1321

Overview: Manufactures and markets consumer products. Established: 1904. Ownership: Public.
Key Statistics: Annual Sales: $11.6 billion. Number of Employees: 28,000.
Hiring History: Number of professional employees hired in 1997: 10. 1996: 10.
Expertise/Education Sought: Finance, accounting
Company-Paid Training is available.
Contact: For individuals with previous experience: Ms. Alean N. Timm, Assistant Manager of Corporate Human Resources; Phone: 203-698-5000; Fax: 203-637-5695.
Company URL: http://www.fortunebrands.com

FUSCO CORPORATION
555 LONG WHARF DRIVE
NEW HAVEN, CONNECTICUT 06511-6107

Overview: Builds office buildings and shopping centers. Established: 1990. Ownership: Private.
Key Statistics: Annual Sales: $54.1 million. Number of Employees: 182.
Hiring History: Number of professional employees hired in 1996: 3.
Expertise/Education Sought: Construction, construction engineering, construction management, accounting
Alternatives to Full-Time Employment: Internships
Company-Paid Training is available.
Contact: For individuals with previous experience: Human Resources; Phone: 203-777-7451; Fax: 203-777-8325. E-mail: bscaramozza@fusco.com.
Company URL: http://www.fusco.com

GE
3135 EASTON TURNPIKE
FAIRFIELD, CONNECTICUT 06431

Overview: Diversified company that includes high technology, service, and manufacturing businesses. Established: 1892. Ownership: Public.
Key Statistics: Annual Sales: $79.2 billion. Number of Employees: 239,000.
Expertise/Education Sought: Business administration, computer science, accounting, finance, engineering, human resources, management, manufacturing, marketing, information technology

GE (continued)

ALTERNATIVES TO FULL-TIME EMPLOYMENT: Part-time jobs, paid internships
COMPANY-PAID TRAINING is available.
INTERNATIONAL ASSIGNMENTS are available to employees.
CONTACT: For individuals with previous experience: Manager of Recruiting; Fax: 513-583-7340. E-mail: resume@opportunities.ge.com.
COMPANY URL: http://www.ge.com

GENERAL SIGNAL CORPORATION

1 HIGH RIDGE PARK
STAMFORD, CONNECTICUT 06904

OVERVIEW: Manufactures process control, electrical, electronic, and communications equipment. Established: 1904. Ownership: Public.
KEY STATISTICS: Annual Sales: $2.1 billion. Number of Employees: 13,000.
EXPERTISE/EDUCATION SOUGHT: Marketing, finance, manufacturing engineering
ALTERNATIVES TO FULL-TIME EMPLOYMENT: Part-time jobs
COMPANY-PAID TRAINING is available.
INTERNATIONAL ASSIGNMENTS are available to employees.
CONTACT: Mrs. Anita Wheeler, Director of Staffing, PO Box 10010, Stamford, Connecticut, 06904; Phone: 203-329-4100; Fax: 203-329-4223.
COMPANY URL: http://www.generalsignal.com

GERBER SCIENTIFIC, INC.

83 GERBER ROAD WEST
SOUTH WINDSOR, CONNECTICUT 06074-3230

OVERVIEW: Provides services for computer-aided applications. Established: 1948. Ownership: Public.
KEY STATISTICS: Annual Sales: $380.9 million. Number of Employees: 1,900.
EXPERTISE/EDUCATION SOUGHT: Electrical engineering, computer science, mechanical engineering
COMPANY-PAID TRAINING is available.
INTERNATIONAL ASSIGNMENTS are available to employees.
CONTACT: For individuals with previous experience: Human Relations Department; Phone: 860-644-1551; Fax: 860-648-8291.
COMPANY URL: http://www.gerberscientific.com

THE C.R. GIBSON COMPANY

32 KNIGHT STREET
NORWALK, CONNECTICUT 06856

OVERVIEW: Manufactures notebooks, albums, and binders. Established: 1995. Parent Company: Thomas Nelson Gifts. Ownership: Public.
KEY STATISTICS: Annual Sales: $34.6 million. Number of Employees: 625.
HIRING HISTORY: Number of professional employees hired in 1997: 5.
EXPERTISE/EDUCATION SOUGHT: Manufacturing
CONTACT: For individuals with previous experience: Mr. Richard Graham, Human Resources Director; Phone: 203-847-4543; Fax: 203-847-7613.

HANDY AND HARMAN REFINING GROUP

300 RYE STREET
SOUTH WINDSOR, CONNECTICUT 06074-1220

OVERVIEW: Produces secondary nonferrous metals. Established: 1996. Ownership: Private.
KEY STATISTICS: Annual Sales: $350 million. Number of Employees: 210.
HIRING HISTORY: Number of professional employees hired in 1997: 5.
EXPERTISE/EDUCATION SOUGHT: Accounting, auditing, chemistry, computer programming, engineering, finance, marketing, manufacturing engineering
CONTACT: Mr. Doug Gillette, Human Resource Manager; Phone: 860-282-0871, ext. 431; Fax: 860-528-4748.

HARCO LABORATORIES, INC.

186 CEDAR STREET
BRANFORD, CONNECTICUT 06405-3700

OVERVIEW: Manufactures temperature sensors and thermocouples. Established: 1993. Ownership: Private.
KEY STATISTICS: Annual Sales: $12.8 million. Number of Employees: 110.
EXPERTISE/EDUCATION SOUGHT: Manufacturing
CONTACT: For individuals with previous experience: Human Resource Manager; Phone: 203-483-3733; Fax: 203-483-0391.

HEATING OIL PARTNERS, LP

1120 POST ROAD
DARIEN, CONNECTICUT 06820

OVERVIEW: Provides fuel oil services. Established: 1995. Ownership: Private.
KEY STATISTICS: Annual Sales: $78 million. Number of Employees: 600.
HIRING HISTORY: Number of professional employees hired in 1997: 40. 1996: 100.
EXPERTISE/EDUCATION SOUGHT: Transportation, accounting, customer service and support
ALTERNATIVES TO FULL-TIME EMPLOYMENT: Paid internships
COMPANY-PAID TRAINING is available.
CONTACT: For individuals with previous experience: Ms. Eileen Pivar, Manager of Personnel and Benefits Administration; Phone: 203-655-8290, ext. 117; Fax: 203-655-9383.

HUBBELL, INC.

584 DERBY MILFORD ROAD
ORANGE, CONNECTICUT 06477-4024

OVERVIEW: Manufactures wiring, cable, lighting, electronics, and industrial control equipment. Established: 1888. Ownership: Public.
KEY STATISTICS: Annual Sales: $1.3 billion. Number of Employees: 8,178.
EXPERTISE/EDUCATION SOUGHT: Finance, engineering
COMPANY-PAID TRAINING is available.
INTERNATIONAL ASSIGNMENTS are available to employees.
CONTACT: For individuals with previous experience: Mr. George Zurman, Vice President of Human Resources, PO Box 549, Orange, Connecticut, 06477; Phone: 203-799-4255; Fax: 203-799-4254. E-mail: gzurman@corp.hubbell.com.
COMPANY URL: http://www.hubbell.com

HUNT IMAGING, LLC

15 VALLEY DRIVE
GREENWICH, CONNECTICUT 06831-5205

OVERVIEW: Produces chemical preparations. Established: 1996. Ownership: Private.
KEY STATISTICS: Annual Sales: $15 million. Number of Employees: 40.
HIRING HISTORY: Number of professional employees hired in 1997: 5.
EXPERTISE/EDUCATION SOUGHT: Business management, mechanical engineering, chemical engineering, computer science, accounting, finance
CONTACT: Mr. Michael Stanek, Vice President and Chief Financial Officer; Phone: 440-826-0433, ext. 224; Fax: 440-234-2474.

ITT CANNON RF PRODUCTS

585 EAST MAIN STREET
NEW BRITAIN, CONNECTICUT 06051-2025

OVERVIEW: Produces electronic connectors. Established: 1988. Parent Company: ITT Industries. Ownership: Public.

Key Statistics: Annual Sales: $18.4 million. Number of Employees: 196.
Hiring History: Number of professional employees hired in 1997: 4.
Expertise/Education Sought: Manufacturing engineering, management, administration
Company-Paid Training is available.
Contact: For individuals with previous experience: Human Resources Department; Phone: 860-223-2700.

THE LEE COMPANY
2 PETTIPAUG ROAD
WESTBROOK, CONNECTICUT 06498-1591

Overview: Manufactures fluid circuit devices. Established: 1949. Ownership: Private.
Key Statistics: Annual Sales: $63.2 million. Number of Employees: 537.
Hiring History: Number of professional employees hired in 1997: 67. 1996: 20.
Expertise/Education Sought: Manufacturing engineering
Alternatives to Full-Time Employment: Paid internships
International Assignments are available to employees.
Contact: Mr. Richard Broderick, Personnel Manager, PO Box 424, Westbrook, Connecticut, 06498-0424; Phone: 860-399-6281; Fax: 860-399-8025.

LEXA INTERNATIONAL CORPORATION
300 ATLANTIC STREET
STAMFORD, CONNECTICUT 06901-3530

Overview: Wholesales fuel oil and gasoline. Established: 1920. Ownership: Private.
Key Statistics: Annual Sales: $1.3 billion. Number of Employees: 1,640.
Expertise/Education Sought: Telecommunications, engineering, energy/energy management
Alternatives to Full-Time Employment: Internships
Company-Paid Training is available.
International Assignments are available to employees.
Contact: For individuals with previous experience: Human Resources; Phone: 203-326-5200.

LOCTITE CORPORATION
1001 TROUT BROOK CROSSING
ROCKY HILL, CONNECTICUT 06067

Overview: Manufactures sealants and adhesives. Established: 1953. Parent Company: Henkel. Ownership: Private.
Key Statistics: Annual Sales: $679 million. Number of Employees: 4,300.
Expertise/Education Sought: Marketing, finance, sales, accounting, chemistry
Alternatives to Full-Time Employment: Internships
Company-Paid Training is available.
Contact: Ms. Pam Brunelle, Recruiter; Phone: 860-571-5140; Fax: 860-571-5294. E-mail: pam.brunelle@loctite.com.

LORAD-DIVISION OF TREX MEDICAL
36 APPLE RIDGE ROAD
DANBURY, CONNECTICUT 06810-7301

Overview: Manufactures radiographic x-ray apparatus. Established: 1995. Parent Company: Trex Medical Corporation. Ownership: Public.
Key Statistics: Annual Sales: $150.2 million. Number of Employees: 500.
Expertise/Education Sought: Manufacturing, manufacturing, marketing, sales
Alternatives to Full-Time Employment: Internships
International Assignments are available to employees.
Contact: For individuals with previous experience: Human Resources Department; Phone: 203-790-1188; Fax: 203-731-8398.

MACDERMID, INC.
245 FREIGHT STREET
WATERBURY, CONNECTICUT 06702-1802

Overview: Produces compounds for plating metal and resisting rust. Established: 1922. Ownership: Public.
Key Statistics: Annual Sales: $293.7 million. Number of Employees: 1,083.
Hiring History: Number of professional employees hired in 1996: 100.
Expertise/Education Sought: Manufacturing engineering, chemistry, chemical engineering, electrical engineering, manufacturing
Alternatives to Full-Time Employment: Paid internships
Company-Paid Training is available.
International Assignments are available to employees.
Contact: Miss Merna Hill, Personnel Manager; Phone: 203-575-5700; Fax: 203-575-7969. E-mail: mhill@macdermid.com.
Company URL: http://www.macdermid.com

MARK EYELET, INC.
63 WAKELEE ROAD
WOLCOTT, CONNECTICUT 06716-2609

Overview: Produces electronic connectors. Established: 1996. Ownership: Private.
Key Statistics: Annual Sales: $5.2 million. Number of Employees: 45.
Expertise/Education Sought: Manufacturing
Company-Paid Training is available.
International Assignments are available to employees.
Contact: For individuals with previous experience: Human Resources Department; Phone: 203-756-8847; Fax: 203-755-9410.

MECA SOFTWARE, LLC
115 CORPORATE DRIVE
TRUMBULL, CONNECTICUT 06611

Overview: Provides computer software development services. Established: 1995. Ownership: Private.
Key Statistics: Annual Sales: $11.8 million. Number of Employees: 350.
Expertise/Education Sought: Computer programming, design engineering, information systems, software engineering/development
Alternatives to Full-Time Employment: Paid internships
Company-Paid Training is available.
Contact: For individuals with previous experience: Ms. Lynn Woodhall, Director of Human Resources; Phone: 203-452-2634; Fax: 203-452-8700.

METHODE NEW ENGLAND COMPANY, INC.
50 MCDERMOTT ROAD
NORTH HAVEN, CONNECTICUT 06473-3612

Overview: Manufactures computer interface equipment. Established: 1994. Parent Company: Methode Electronics. Ownership: Public.
Key Statistics: Annual Sales: $6 million. Number of Employees: 89.
Expertise/Education Sought: Manufacturing engineering
Company-Paid Training is available.
International Assignments are available to employees.
Contact: For individuals with previous experience: Human Resources Department. E-mail: methode@attmail.com.

MICROPHASE CORPORATION
587 CONNECTICUT AVENUE
NORWALK, CONNECTICUT 06854-1711

Overview: Manufactures microwave communications equipment. Established: 1955. Ownership: Private.
Key Statistics: Annual Sales: $8 million. Number of Employees: 93.

Microphase Corporation (continued)

Hiring History: Number of professional employees hired in 1997: 15. 1996: 10.
Expertise/Education Sought: Manufacturing, production, design
Alternatives to Full-Time Employment: Part-time jobs, paid internships
Contact: For individuals with previous experience: Ms. Angela Coombs, Director of Human Resources; Phone: 203-866-8000; Fax: 203-857-0439. E-mail: acoombs@microphase.com.
Company URL: http://www.microphase.com

O.F. MOSSBERG AND SONS, INC.
7 GRASSO AVENUE
NORTH HAVEN, CONNECTICUT 06473-3237

Overview: Manufactures shotgun parts. Established: 1993. Ownership: Private.
Key Statistics: Annual Sales: $78 million. Number of Employees: 300.
Hiring History: Number of professional employees hired in 1997: 60.
Expertise/Education Sought: Manufacturing engineering, design engineering
Alternatives to Full-Time Employment: Paid internships
Company-Paid Training is available.
Contact: Mr. Larry Lacoste, Director of Human Resources; Phone: 203-230-5473; Fax: 203-230-5475.
Company URL: http://www.mossberg.com

THE NASH ENGINEERING COMPANY
9 TREFOIL DRIVE
TRUMBULL, CONNECTICUT 06611-1330

Overview: Manufactures vacuum pumps for the chemical, petrochemical, health-care, and pulp and paper industries. Established: 1905. Ownership: Private.
Key Statistics: Annual Sales: $170 million. Number of Employees: 922.
Expertise/Education Sought: Sales, accounting, engineering
Contact: Human Resources Department, PO Box 4000, Trumbull, Connecticut, 06151-0095; Phone: 203-459-3900; Fax: 203-459-3880.
Company URL: http://www.nasheng.com

THE NAUGATUCK GLASS COMPANY
THE CORNER OF CHURCH AND BRIDGE STREET
NAUGATUCK, CONNECTICUT 06770

Overview: Produces glass products. Established: 1928. Parent Company: Erie Scientific. Ownership: Private.
Key Statistics: Annual Sales: $18 million. Number of Employees: 200.
Expertise/Education Sought: Manufacturing
Contact: For individuals with previous experience: Ms. Phyllis Tarsi, Human Resource Manager.

THE JM NEY COMPANY
NEY INDUSTRIAL PARK
BLOOMFIELD, CONNECTICUT 06002

Overview: Produces electronic components. Established: 1991. Parent Company: Andersen Group, Inc. Ownership: Public.
Key Statistics: Annual Sales: $24 million. Number of Employees: 406.
Hiring History: Number of professional employees hired in 1997: 47. 1996: 30.
Expertise/Education Sought: Manufacturing engineering
Alternatives to Full-Time Employment: Paid internships
Contact: Ms. Karen Ruhlemann, Employment Recruiter; Phone: 860-242-2281, ext. 148; Fax: 860-286-6113.

NORTHEAST ELECTRONICS CORPORATION
455 BIC DRIVE
MILFORD, CONNECTICUT 06460-1702

Overview: Produces hermetic seals for electronic equipment. Established: 1961. Ownership: Private.
Key Statistics: Annual Sales: $5.7 million. Number of Employees: 100.
Hiring History: Number of professional employees hired in 1996: 12.
Expertise/Education Sought: Engineering
Contact: For individuals with previous experience: Mr. Frank Gaudiano, Vice President and General Manager; Phone: 203-878-3511; Fax: 203-877-5694.

NORTHROP GRUMMAN NORDEN SYSTEMS
10 NORDEN PLACE
NORWALK, CONNECTICUT 06855-1436

Overview: Manufactures search and navigational equipment. Established: 1996. Ownership: Public.
Key Statistics: Annual Sales: $122.1 million. Number of Employees: 1,500.
Expertise/Education Sought: Engineering
Contact: For individuals with previous experience: Mr. Mike Magaletta, Human Resource Department.

OMEGA ENGINEERING, INC.
RIVER BEND EXECUTIVE CENTER
STAMFORD, CONNECTICUT 06907

Overview: Manufactures process control instruments and computer interface equipment. Established: 1962. Ownership: Private.
Key Statistics: Annual Sales: $86.1 million. Number of Employees: 1,000.
Expertise/Education Sought: Engineering
Alternatives to Full-Time Employment: Part-time jobs
Company-Paid Training is available.
International Assignments are available to employees.
Contact: Mr. Ronald Hemphill, Employee Relations Coordinator; Phone: 203-359-7623; Fax: 203-359-7979.
Company URL: http://www.omega.com

OTIS ELEVATOR COMPANY
10 FARM SPRINGS
FARMINGTON, CONNECTICUT 06032-2526

Overview: Manufactures elevators and escalators. Established: 1976. Parent Company: United Technologies Corporation. Ownership: Public.
Key Statistics: Annual Sales: $5.6 billion. Number of Employees: 51,000.
Expertise/Education Sought: Design engineering, electrical engineering, mechanical engineering, software engineering/development, management, sales, field service
Alternatives to Full-Time Employment: Paid internships
Company-Paid Training is available.
International Assignments are available to employees.
Contact: Ms. Andrea Lanese, Director Human Resources, Staffing and Recruiting; Phone: 860-676-6000, ext. 5897; Fax: 860-676-6549.

PHOTRONICS, INC.
PO BOX 5226
BROOKFIELD, CONNECTICUT 06804-3937

Overview: Manufactures photomasks for semiconductor industry. Established: 1969. Ownership: Public.
Key Statistics: Annual Sales: $160.1 million. Number of Employees: 900.
Expertise/Education Sought: Accounting, mechanical engineering, electrical engineering

Alternatives to Full-Time Employment: Paid internships
Company-Paid Training is available.
International Assignments are available to employees.
Contact: For individuals with previous experience: Ms. Valerie Moss, Human Resources Administrator; Phone: 203-775-9000; Fax: 203-740-5630.
Company URL: http://www.Photronics.com

PH PREFERRED HOLDING CORPORATION

PO BOX S
SOMERSVILLE, CONNECTICUT 06072

Overview: Produces electronic components. Established: 1987. Ownership: Private.
Key Statistics: Annual Sales: $14 million. Number of Employees: 220.
Hiring History: Number of professional employees hired in 1996: 10.
Expertise/Education Sought: Electrical engineering, sales, production control
Contact: For individuals with previous experience: Human Resources Department; Phone: 860-763-7059; Fax: 860-763-7066.

PITNEY BOWES, INC.

1 ELMCROFT ROAD
STAMFORD, CONNECTICUT 06926-0700

Overview: Provides business equipment and business and financial services. Established: 1920. Ownership: Public.
Key Statistics: Annual Sales: $3.9 billion. Number of Employees: 24,054.
Expertise/Education Sought: Software engineering/development, mechanical engineering, electrical engineering, computer programming, systems engineering
Alternatives to Full-Time Employment: Part-time jobs, internships
Company-Paid Training is available.
International Assignments are available to employees.
Contact: For college students and recent graduates: Ms. Carolyn Davis, University Relations Consultant; Phone: 203-351-7289; Fax: 203-351-6332. E-mail: staffing@pb.com. For individuals with previous experience: Ms. Mary Tetro, Director of Staffing; Phone: 203-351-6765; Fax: 203-351-6332. E-mail: staffing@pb.com.
Company URL: http://www.pb.com

PRECISION METAL PRODUCTS, INC.

307 PEPES FARM ROAD
MILFORD, CONNECTICUT 06460-3671

Overview: Manufactures surgical and medical instruments. Established: 1977. Ownership: Private.
Key Statistics: Annual Sales: $13.8 million. Number of Employees: 150.
Hiring History: Number of professional employees hired in 1996: 12.
Expertise/Education Sought: Manufacturing, customer service and support, accounting
Alternatives to Full-Time Employment: Paid internships
Company-Paid Training is available.
Contact: For individuals with previous experience: Personnel; Phone: 203-877-4258; Fax: 203-878-8353.

PREFERRED UTILITIES MANUFACTURING CORPORATION

31-35 SOUTH STREET
DANBURY, CONNECTICUT 06810-8013

Overview: Manufactures measurement and control devices. Established: 1986. Ownership: Private.
Key Statistics: Annual Sales: $10 million. Number of Employees: 82.
Hiring History: Number of professional employees hired in 1997: 4.
Expertise/Education Sought: Engineering, technology
Alternatives to Full-Time Employment: Paid internships
Company-Paid Training is available.
Contact: For individuals with previous experience: Human Resource Department; Phone: 203-743-6741; Fax: 203-798-7313.
Company URL: http://www.preferred-mfg.com

THE PRODUCTO MACHINE COMPANY

990 HOUSATONIC AVENUE
BRIDGEPORT, CONNECTICUT 06606-5716

Overview: Manufactures metal cutting machine tools. Established: 1928. Ownership: Private.
Key Statistics: Annual Sales: $53.1 million. Number of Employees: 475.
Hiring History: Number of professional employees hired in 1996: 12.
Expertise/Education Sought: Manufacturing engineering
International Assignments are available to employees.
Contact: For individuals with previous experience: Mr. John Stawarky, Vice President of Human Resources; Phone: 203-367-8675, ext. 333; Fax: 203-368-2597.
Company URL: http://www.productomachine.com

RAPID POWER TECHNOLOGIES, INC.

18 GRAYS BRIDGE ROAD
BROOKFIELD, CONNECTICUT 06804-2607

Overview: Produces electronic components. Established: 1980. Ownership: Private.
Key Statistics: Annual Sales: $23.6 million. Number of Employees: 175.
Hiring History: Number of professional employees hired in 1996: 30.
Expertise/Education Sought: Accounting, electrical engineering, engineering, marketing, management, field engineering, research and development, drafting, purchasing
Company-Paid Training is available.
International Assignments are available to employees.
Contact: For individuals with previous experience: Human Relations Department; Phone: 203-775-0411; Fax: 203-775-0666.

M.H. RHODES, INC.

99 THOMPSON ROAD
AVON, CONNECTICUT 06001-3000

Overview: Manufactures process control instruments. Established: 1930. Ownership: Public.
Key Statistics: Annual Sales: $8 million. Number of Employees: 90.
Hiring History: Number of professional employees hired in 1996: 3.
Expertise/Education Sought: Sales, marketing
Contact: For individuals with previous experience: Personnel; Phone: 860-673-3281; Fax: 860-673-8633.

RISK CAPITAL REINSURANCE COMPANY

20 HORSENECK LANE
GREENWICH, CONNECTICUT 06830

Overview: Provides insurance services. Established: 1995. Parent Company: Risk Capital Holding, Inc. Ownership: Public.
Key Statistics: Annual Sales: $70 million.
Hiring History: Number of professional employees hired in 1997: 4. 1996: 20.
Expertise/Education Sought: Law, underwriting, finance, operations
Contact: For individuals with previous experience: Human Resources; Phone: 203-861-4300; Fax: 203-861-7240.

ROGERS CORPORATION

1 TECHNOLOGY DRIVE
ROGERS, CONNECTICUT 06263

Overview: Manufactures polymer materials and components. Established: 1832. Ownership: Public.

Rogers Corporation (continued)

Key Statistics: Annual Sales: $141.5 million. Number of Employees: 854.
Expertise/Education Sought: Chemical engineering, engineering
Alternatives to Full-Time Employment: Paid internships
Company-Paid Training is available.
Contact: Human Resources Department, PO Box 188, Rogers, Connecticut, 06263-0188; Phone: 860-774-9605.
Company URL: http://www.rogers-corp.com

SARGENT MANUFACTURING
100 SARGENT DRIVE
NEW HAVEN, CONNECTICUT 06511

Overview: Manufactures power-driven hand tools. Established: 1989. Parent Company: Assa Abloy, Inc. Ownership: Private.
Key Statistics: Annual Sales: $181.7 million. Number of Employees: 2,150.
Expertise/Education Sought: Engineering, data processing
International Assignments are available to employees.
Contact: For individuals with previous experience: Personnel Department; Fax: 203-821-5738.

SERVICE AMERICA CORPORATION
300 FIRST STAMFORD PLACE
STAMFORD, CONNECTICUT 06904

Overview: Provides food service and vending machines. Established: 1994. Parent Company: Servam Corporation. Ownership: Private.
Key Statistics: Annual Sales: $158 million. Number of Employees: 3,000.
Hiring History: Number of professional employees hired in 1997: 20. 1996: 75.
Expertise/Education Sought: Management information systems
Alternatives to Full-Time Employment: Paid internships
Company-Paid Training is available.
International Assignments are available to employees.
Contact: Mr. Angelo Fortuna, Corporate Vice President of Human Resources; Phone: 203-975-5914; Fax: 203-975-5912.

SIMKINS INDUSTRIES, INC.
260 EAST STREET
NEW HAVEN, CONNECTICUT 06511-5839

Overview: Manufactures boxboard. Established: 1901. Ownership: Public.
Key Statistics: Annual Sales: $197.5 million. Number of Employees: 1,900.
Hiring History: Number of professional employees hired in 1997: 1,500. 1996: 1,700.
Expertise/Education Sought: Manufacturing
Contact: For individuals with previous experience: Human Resource Department.

SOUTHERN NEW ENGLAND TELECOM CORPORATION
227 CHURCH STREET
NEW HAVEN, CONNECTICUT 06510

Overview: Provides telecommunications services. Established: 1986. Ownership: Private.
Key Statistics: Annual Sales: $1.5 billion. Number of Employees: 8,702.
Expertise/Education Sought: Telecommunications, electronics/electronics engineering, accounting
Alternatives to Full-Time Employment: Paid internships
Company-Paid Training is available.
Contact: Ms. Claudia Blanchette, Manager of Recruitment; Phone: 203-771-5200; Fax: 203-773-3609.

STOLT PARCEL TANKERS
8 SOUND SHORE DRIVE
GREENWICH, CONNECTICUT 06830-7279

Overview: Operates truck terminals and provides freight services. Established: 1959. Parent Company: Stolt-Nielsen SA. Ownership: Private.
Key Statistics: Annual Sales: $1.4 billion. Number of Employees: 6,700.
Expertise/Education Sought: Logistics, transportation, sales
Alternatives to Full-Time Employment: Paid internships
Company-Paid Training is available.
International Assignments are available to employees.
Contact: Ms. Marion Pace, Supervisor of Employee Relations; Phone: 203-625-9400; Fax: 203-625-3917.

THE TORRINGTON COMPANY
59 FIELD STREET
TORRINGTON, CONNECTICUT 06790

Overview: Manufactures bearing and textile machine needles. Established: 1968. Parent Company: Ingersoll-Rand Company. Ownership: Public.
Key Statistics: Annual Sales: $1 billion. Number of Employees: 11,400.
Hiring History: Number of professional employees hired in 1997: 40. 1996: 35.
Expertise/Education Sought: Mechanical engineering, sales, product design and development, manufacturing engineering, technical sales
Alternatives to Full-Time Employment: Part-time jobs, paid internships
Company-Paid Training is available.
Contact: Mr. Ken Keane, Manager of Professional Recruitment; Phone: 860-626-2287; Fax: 860-496-3603. E-mail: keanek@torrington.com.
Company URL: http://www.torrington.com

TRANS-LUX CORPORATION
110 RICHARDS AVENUE
NORWALK, CONNECTICUT 06854-1622

Overview: Manufactures electric signs. Established: 1920. Ownership: Public.
Key Statistics: Annual Sales: $45.3 million. Number of Employees: 636.
Hiring History: Number of professional employees hired in 1997: 75. 1996: 75.
Expertise/Education Sought: Electrical engineering
Alternatives to Full-Time Employment: Paid internships
Contact: For individuals with previous experience: Mrs. Jane Bauer, Human Resources Specialist; Phone: 203-853-4321; Fax: 203-854-6891. E-mail: employment@trans-lux.com.
Company URL: http://www.trans-lux.com

TRANSPRO, INC.
100 GANDO DRIVE
NEW HAVEN, CONNECTICUT 06513-1049

Overview: Manufactures motor vehicle parts. Established: 1995. Ownership: Public.
Key Statistics: Annual Sales: $264.1 million. Number of Employees: 2,139.
Hiring History: Number of professional employees hired in 1997: 100.
Expertise/Education Sought: Manufacturing engineering, product engineering
Company-Paid Training is available.
Contact: For individuals with previous experience: Mr. Jeff Jackson, Vice President of Human Resources; Phone: 203-562-5121; Fax: 203-865-3723. E-mail: jjackson@transpro.com.

ULBRICH STAINLESS STEELS AND SPECIALS METALS, INC.

57 DODGE AVENUE
NORTH HAVEN, CONNECTICUT 06473-1119

Overview: Produces cold-rolled steel strip. Established: 1924. Ownership: Private.

Key Statistics: Annual Sales: $180 million. Number of Employees: 480.

Expertise/Education Sought: Machinery, allied health

International Assignments are available to employees.

Contact: Mr. Tom O'Leary, Director of Human Resources; Phone: 203-239-4481; Fax: 203-269-2507.

UNION CARBIDE CORPORATION

39 OLD RIDGEBURY ROAD
DANBURY, CONNECTICUT 06817-0001

Overview: Produces industrial organic chemicals. Established: 1917. Ownership: Public.

Key Statistics: Annual Sales: $6.1 billion. Number of Employees: 11,745.

Expertise/Education Sought: Chemical engineering, mechanical engineering, computer science, manufacturing, purchasing, research and development, design engineering, technical sales

Alternatives to Full-Time Employment: Part-time jobs, paid internships

Company-Paid Training is available.

International Assignments are available to employees.

Contact: Mr. Shawn Coker, Director of Corporate Staffing; Fax: 203-794-5333.

Company URL: http://www.jobweb.org/employer/unionc.htm

UNITED STATES SURGICAL CORPORATION

150 GLOVER AVENUE
NORWALK, CONNECTICUT 06850-1308

Overview: Manufactures surgical equipment. Established: 1964. Ownership: Public.

Key Statistics: Annual Sales: $1.1 billion. Number of Employees: 6,133.

Expertise/Education Sought: Nutrition, physical therapy

Contact: Mr. Paul O'Brien, Director of Employment; Phone: 203-845-5000; Fax: 203-492-8226.

UNITED TECHNOLOGIES CORPORATION

1 FINANCIAL PLAZA
HARTFORD, CONNECTICUT 06101

Overview: Manufactures advanced technology products for the aerospace, building, and automotive industries. Established: 1928. Ownership: Public.

Key Statistics: Annual Sales: $23.3 billion. Number of Employees: 173,800.

Hiring History: Number of professional employees hired in 1996: 200.

Expertise/Education Sought: Accounting, finance, information systems, law, human resources

Alternatives to Full-Time Employment: Paid internships

Company-Paid Training is available.

International Assignments are available to employees.

Contact: Ms. Heidi Sandling, Manager of Leadership Development; Phone: 860-728-7827; Fax: 860-728-6568. E-mail: sandlihj@corphq.utc.com.

Company URL: http://www.utc.com

VITRAMON, INC.

10 MAIN STREET
MONROE, CONNECTICUT 06468

Overview: Manufactures electronic capacitors. Established: 1994. Parent Company: Visay. Ownership: Public.

Key Statistics: Annual Sales: $82 million. Number of Employees: 1,700.

Hiring History: Number of professional employees hired in 1996: 50.

Expertise/Education Sought: Ceramic engineering, design engineering, mechanical engineering, accounting

International Assignments are available to employees.

Contact: For individuals with previous experience: Mr. Gerald Sargent, Jr., Human Resources Director, PO Box 544, Monroe, Connecticut, 06621-0544; Phone: 203-268-6261, ext. 217; Fax: 203-452-5613.

WELBILT CORPORATION

225 HIGH RIDGE ROAD
STAMFORD, CONNECTICUT 06905-3000

Overview: Manufactures commercial kitchen equipment. Established: 1995. Parent Company: Berisford, PLC. Ownership: Private.

Key Statistics: Annual Sales: $249.5 million. Number of Employees: 2,000.

Expertise/Education Sought: Accounting

Alternatives to Full-Time Employment: Part-time jobs

Contact: For individuals with previous experience: Human Resources Department.

Company URL: http://www.welbilt.com

WITCO CORPORATION

1 AMERICAN LANE
GREENWHICH, CONNECTICUT 06831-2560

Overview: Global specialty chemical manufacturer. Established: 1920. Ownership: Public.

Key Statistics: Annual Sales: $2.3 billion. Number of Employees: 6,724.

Expertise/Education Sought: Engineering

Alternatives to Full-Time Employment: Paid internships

Company-Paid Training is available.

International Assignments are available to employees.

Contact: For individuals with previous experience: Human Resources; Phone: 203-552-6000.

Company URL: http://www.witco.com

WYMAN-GORDON INVESTMENT CASTINGS, INC.

839 POQUONNOCK ROAD
GROTON, CONNECTICUT 06340-4537

Overview: Operates steel investment foundry. Established: 1990. Ownership: Public.

Key Statistics: Annual Sales: $143 million. Number of Employees: 1,500.

Expertise/Education Sought: Manufacturing, engineering

Alternatives to Full-Time Employment: Paid internships

Company-Paid Training is available.

Contact: Mr. Ken Buttinger, Human Resource Manager; Phone: 860-445-3125; Fax: 860-449-8052.

DISTRICT OF COLUMBIA

BUREAU OF OCEANS AND INTERNATIONAL ENVIRONMENTAL AND SCIENTIFIC AFFAIRS

2201 C STREET, NW
WASHINGTON, D.C. 20520

Overview: Monitors U.S. and international ocean and environmental issues.

Bureau of Oceans and International Environmental and Scientific Affairs (continued)

Expertise/Education Sought: Electrical engineering, accounting, management, business management, finance
Alternatives to Full-Time Employment: Paid internships
Company-Paid Training is available.
International Assignments are available to employees.
Contact: For individuals with previous experience: Ms. Stephanie Kinney, Executive Director; Phone: 202-647-3622; Fax: 202-647-0774.

DEPARTMENT OF COMMERCE
14TH STREET AND CONSTITUTION AVENUE
WASHINGTON, D.C. 20230

Overview: U.S. government agency that promotes fair trade and economic growth.
Expertise/Education Sought: Law, finance, data processing, accounting, economics, statistics
Alternatives to Full-Time Employment: Paid internships
Company-Paid Training is available.
Contact: For individuals with previous experience: Ms. Ausma Karlsons, Director of Staffing and Recruitment; Phone: 202-482-3827; Fax: 202-236-0227. Job Hotline: 202-482-5138.

DEPARTMENT OF HOUSING AND URBAN DEVELOPMENT
451 7TH STREET, SW
WASHINGTON, D.C. 20410

Overview: U.S. agency that administers programs to help families become home owners or obtain housing.
Expertise/Education Sought: Planning, law, accounting, business administration, finance, printing
Alternatives to Full-Time Employment: Paid internships
Company-Paid Training is available.
Contact: For individuals with previous experience: Human Relations Department; Phone: 202-708-4310; Fax: 202-401-2565. Job Hotline: 999-999-999.
Company URL: http://www.hud.gov

DEPARTMENT OF JUSTICE
1331 PENNSYLVANIA AVENUE, NW, SUITE 1100
WASHINGTON, D.C. 20530

Overview: U.S. government agency that supervises and directs the administration of justice through the courts of law.
Expertise/Education Sought: Criminal justice, law, treasury management, accounting, finance, computer science
Alternatives to Full-Time Employment: Paid internships
Contact: For individuals with previous experience: JMD Personnel; Phone: 202-514-6788; Fax: 202-514-6827.
Company URL: http://www.usdoj.gov

DEPARTMENT OF LABOR
200 CONSTITUTION AVENUE, NW
WASHINGTON, D.C. 20210

Overview: U.S. agency that administers and directs labor policies and programs. Established: 1913.
Expertise/Education Sought: Computer science, electrical engineering, industrial hygiene, mathematics, statistics
Alternatives to Full-Time Employment: Part-time jobs, paid internships
Company-Paid Training is available.
International Assignments are available to employees.
Contact: Career Management and Planning; Phone: 202-219-0118; Fax: 202-219-8127. Job Hotline: 202-606-2700.
Company URL: http://www.usajobs.opm.gov

DEPARTMENT OF THE ARMY
6800 ARMY PENTAGON
WASHINGTON, D.C. 20310

Overview: U.S. armed forces organization providing land based military protection for the United States.
Expertise/Education Sought: Computer science
Alternatives to Full-Time Employment: Paid internships
Company-Paid Training is available.
Contact: For individuals with previous experience: Ms. Barbara Hagle, Chief of Personnel Services; Phone: 703-695-3881; Fax: 703-693-6777.

DEPARTMENT OF THE INTERIOR
1849 C STREET, NW
WASHINGTON, D.C. 20240

Overview: U.S. agency that directs the use and conservation of public lands and natural resources.
Expertise/Education Sought: Accounting, finance, law, data processing, economics, business management, environmental science, land management, computer science
Company-Paid Training is available.
International Assignments are available to employees.
Contact: For individuals with previous experience: Mr. Ronald Brown, Recruiter; Phone: 202-208-6459; Fax: 202-208-4522.

ENVIRONMENTAL PROTECTION AGENCY
401 M STREET, SW
WASHINGTON, D.C. 20460

Overview: Provides environmental protection and improvement services.
Expertise/Education Sought: Environmental engineering, computer analysis, accounting, budgeting, data processing, chemical engineering
Alternatives to Full-Time Employment: Paid internships
International Assignments are available to employees.
Contact: Mr. Rick Garman, Employee Services, 401 M Street, SW, Mail Stop 3606, Washington, D.C., 20460; Phone: 202-260-3267; Fax: 202-260-3263. Job Hotline: 202-260-6000.
Company URL: http://www.epa.gov

FEDERAL AVIATION ADMINISTRATION
800 INDEPENDENCE AVENUE, SW
WASHINGTON, D.C. 20591

Overview: Ensures safe and efficient national airspace. Division of the United States Department of Transportation.
Expertise/Education Sought: Aviation, electrical engineering, security, accounting, finance, safety, business management, operations research, computer science
Alternatives to Full-Time Employment: Part-time jobs, paid internships
International Assignments are available to employees.
Contact: For individuals with previous experience: Human Resources Division; Phone: 202-267-3456; Fax: 202-267-3977. Job Hotline: 202-267-8008.
Company URL: http://www.faa.gov

FEDERAL EMERGENCY MANAGEMENT AGENCY
500 C STREET, SW, FEDERAL CENTER PLAZA
WASHINGTON, D.C. 20472

Overview: U.S. agency that provides emergency and disaster relief services.
Expertise/Education Sought: Accounting, business administration, computer science, criminal justice, civil engineering, electrical engineering, finance, law
Alternatives to Full-Time Employment: Paid internships

Company-Paid Training is available.

Contact: For college students and recent graduates: Ms. Cheri Allen, College Recruiter; Phone: 202-646-3072; Fax: 202-646-3350. For individuals with previous experience: Mr. Barry Socks, Director Headquarters Personnel, Room 816, Washington, D.C., 20472; Phone: 202-646-4040; Fax: 202-646-3350. Job Hotline: 202-646-3244.

Company URL: http://www.fema.gov

FEDERAL MARITIME COMMISSION

800 NORTH CAPITOL STREET
WASHINGTON, D.C. 20573

Overview: U.S. agency that administers policies and activities regarding maritime transportation.

Hiring History: Number of professional employees hired in 1996: 11.

Expertise/Education Sought: Law, transportation, international finance, economics, business administration

Alternatives to Full-Time Employment: Part-time jobs

Company-Paid Training is available.

Contact: For individuals with previous experience: Mrs. Hatsie Charbonneau, Director of Personnel; Phone: 202-523-5773; Fax: 202-523-7842.

Company URL: http://www.fmc.gov

FEDERAL TRADE COMMISSION

6 PENNSYLVANIA AVENUE
WASHINGTON, D.C. 20580

Overview: U.S. agency that enforces fair trade practices.

Hiring History: Number of professional employees hired in 1997: 160. 1996: 120.

Expertise/Education Sought: Economics

Contact: Mr. Elliot Davis, Director of Personnel; Phone: 202-326-2022. E-mail: edavis@aftc.gov. Job Hotline: 703-326-2020.

Company URL: http://www.ftc.gov

GEICO CORPORATION

1 GEICO PLAZA
WASHINGTON, D.C. 20076

Overview: Provides property and casualty insurance. Established: 1996. Ownership: Public.

Key Statistics: Assets: $3.1 billion. Number of Employees: 8,800.

Hiring History: Number of professional employees hired in 1997: 300.

Expertise/Education Sought: Facilities management, mathematics, building engineering, computer programming

Alternatives to Full-Time Employment: Paid internships

Company-Paid Training is available.

International Assignments are available to employees.

Contact: Ms. Deborah Lipsey, Professional Recruiter; Phone: 301-986-2954; Fax: 301-986-3092. E-mail: jobs@geico.com.

Company URL: http://www.geico.com

GEMINI AIR CARGO

PO BOX 16254, WASHINGTON/DULLES INTERNATIONAL AIR
WASHINGTON, D.C. 20041-6254

Overview: Provides freight transportation services. Established: 1995. Ownership: Private.

Key Statistics: Annual Sales: $36 million. Number of Employees: 200.

Hiring History: Number of professional employees hired in 1997: 125. 1996: 200.

Expertise/Education Sought: Maintenance

Alternatives to Full-Time Employment: Paid internships

Company-Paid Training is available.

Contact: For individuals with previous experience: Human Resources; Phone: 703-260-8100; Fax: 703-260-8164.

Company URL: http://www.gac-cargo.com

GENERAL SERVICES ADMINISTRATION

1800 F STREET, NW
WASHINGTON, D.C. 20405

Overview: Monitors and evaluates government procurement policies and practices.

Expertise/Education Sought: Finance, accounting, construction engineering, telecommunications, computer science, distribution, purchasing

Alternatives to Full-Time Employment: Paid internships

Company-Paid Training is available.

International Assignments are available to employees.

Contact: For individuals with previous experience: Ms. Lil Stewart, Human Resource Division; Phone: 202-501-0040. Job Hotline: 888-562-7472.

INTERNATIONAL TELECOMMUNICATIONS SATELLITE ORGANIZATION (INTELSAT)

3400 INTERNATIONAL DRIVE, NW
WASHINGTON, D.C. 20008

Overview: International consortium for communications satellite operations. Established: 1964. Ownership: Private.

Key Statistics: Annual Sales: $911.4 million. Number of Employees: 700.

Hiring History: Number of professional employees hired in 1996: 70.

Expertise/Education Sought: Sales, marketing, accounting, finance, human resources, engineering

Alternatives to Full-Time Employment: Part-time jobs, paid internships

Company-Paid Training is available.

International Assignments are available to employees.

Contact: For individuals with previous experience: Human Resources; Phone: 202-944-7895; Fax: 202-944-7150. E-mail: jobs@career.intelsat.int.

Company URL: http://www.intelsat.int

MARINE CORPS RECRUITING COMMAND

#2 NAVY ANNEX BUILDING
WASHINGTON, D.C. 20380

Overview: Administers officer commissioning programs. Division of the All Marine Corps Are Parent Companies.

Hiring History: Number of professional employees hired in 1997: 45,000.

Expertise/Education Sought: Computer science, data processing

Alternatives to Full-Time Employment: Paid internships

Company-Paid Training is available.

International Assignments are available to employees.

Contact: Lt. Dennis Sabal, Head Officer of Procurement Branch, Commanding General/Headquarters #, 2 Navy Annex Building, Washington, D.C., 20380-1775; Phone: 703-614-1983; Fax: 703-614-1801.

Company URL: http://www.marines.com

NATIONAL PARK SERVICE

1849 C STREET, NW
WASHINGTON, D.C. 20240

Overview: U.S. agency that administers and maintains national parks, historic areas, and national recreation facilities.

Expertise/Education Sought: Automation, accounting, finance, editing, architecture, history

Alternatives to Full-Time Employment: Part-time jobs, paid internships

Company-Paid Training is available.

Contact: For individuals with previous experience: Human Resources Department; Phone: 202-208-5574.

NAVAL RESEARCH LABORATORY

4555 OVERLOOK AVENUE, SW
WASHINGTON, D.C. 20375-5324

OVERVIEW: The Navy's corporate laboratory for scientific research and advanced technological development. Established: 1923.

KEY STATISTICS: Number of Employees: 3,283.

EXPERTISE/EDUCATION SOUGHT: Ceramic engineering, materials engineering, electronics/electronics engineering, computer science, mechanical engineering, chemistry, physics, meteorology, oceanography, geophysics

ALTERNATIVES TO FULL-TIME EMPLOYMENT: Part-time jobs, paid internships

COMPANY-PAID TRAINING is available.

CONTACT: Human Resources Office, Code 98 1810PG; Phone: 202-767-3030.

For more information, see full organization profile, pg. 288.

OCCUPATIONAL SAFETY AND HEALTH ADMINISTRATION

200 CONSTITUTION AVENUE, NW, ROOM 3308
WASHINGTON, D.C. 20210

OVERVIEW: U.S. agency that ensures the provision and maintenance of safe work places. Division of the U.S. Department Of Labor.

HIRING HISTORY: Number of professional employees hired in 1997: 50. 1996: 30.

EXPERTISE/EDUCATION SOUGHT: Industrial hygiene, safety engineering, health and safety

COMPANY-PAID TRAINING is available.

CONTACT: Ms. Floria Jones, Supervisor of Staffing; Phone: 202-219-8013; Fax: 202-219-6296.

PEACE CORPS

OFFICE OF HUMAN RESOURCES MANAGEMENT, 1990 K STREET, SUITE 4100
WASHINGTON, D.C. 20526

OVERVIEW: Provides volunteer services to foreign nations in need.

EXPERTISE/EDUCATION SOUGHT: Medical technology, occupational therapy, health science, education, training and development, computer programming, business development, administration, agriculture

ALTERNATIVES TO FULL-TIME EMPLOYMENT: Part-time jobs, paid internships

COMPANY-PAID TRAINING is available.

INTERNATIONAL ASSIGNMENTS are available to employees.

CONTACT: For individuals with previous experience: Director of Human Resources; Phone: 202-606-3950; Fax: 202-606-3006. Job Hotline: 800-818-9579.

COMPANY URL: http://www.peacecorps.gov

SMALL BUSINESS ADMINISTRATION

409 THIRD STREET, SW, SUITE 4200
WASHINGTON, D.C. 20416

OVERVIEW: Provides assistance and loans to small businesses.

EXPERTISE/EDUCATION SOUGHT: Computer analysis, accounting, computer analysis, project management, operations

ALTERNATIVES TO FULL-TIME EMPLOYMENT: Part-time jobs, internships

COMPANY-PAID TRAINING is available.

CONTACT: Mr. Jose Mendez, Personnel Staffing Specialist; Phone: 202-205-6178; Fax: 202-205-6172. E-mail: jose.mendez@sba.gov.

COMPANY URL: http://www.sba.gov

SURFACE TRANSPORTATION BOARD

1925 K STREET, NW, SUITE 880
WASHINGTON, D.C. 20423

OVERVIEW: Monitors and regulates interstate trade and business policies and practices.

HIRING HISTORY: Number of professional employees hired in 1996: 3.

EXPERTISE/EDUCATION SOUGHT: Transportation, law

ALTERNATIVES TO FULL-TIME EMPLOYMENT: Paid internships

COMPANY-PAID TRAINING is available.

CONTACT: Ms. Mary Sylvester, Director of Personnel; Phone: 202-565-1691; Fax: 202-565-9019. E-mail: sylvesterm@stb.dot.gov.

U.S. POSTAL SERVICE

475 L'ENFANTE PLAZA
WASHINGTON, D.C. 20260-4261

OVERVIEW: Provides mail delivery service.

EXPERTISE/EDUCATION SOUGHT: Accounting, finance, law, electrical engineering, mechanical engineering, computer science, economics, psychology, communications

ALTERNATIVES TO FULL-TIME EMPLOYMENT: Paid internships

COMPANY-PAID TRAINING is available.

INTERNATIONAL ASSIGNMENTS are available to employees.

CONTACT: For individuals with previous experience: Mr. Stephen A. Leavey, Manager of Corporate Personnel; Phone: 202-268-3643; Fax: 202-268-6195. Job Hotline: 800-562-8777.

UNITED STATES SERVICE INDUSTRIES

1424 K STREET, NW
WASHINGTON, D.C. 20005-2410

OVERVIEW: Provides building maintenance services. Established: 1969. Ownership: Private.

KEY STATISTICS: Annual Sales: $30 million. Number of Employees: 1,500.

EXPERTISE/EDUCATION SOUGHT: Accounting

ALTERNATIVES TO FULL-TIME EMPLOYMENT: Paid internships

CONTACT: Mr. Juan Lopez, Personnel Director; Phone: 301-587-7443; Fax: 301-588-7563.

U.S. ARMY CORPS OF ENGINEERS

20 MASSACHUSETTS AVENUE, NW
WASHINGTON, D.C. 20314-1000

OVERVIEW: Largest public engineering organization. Established: 1775.

KEY STATISTICS: Number of Employees: 38,000.

EXPERTISE/EDUCATION SOUGHT: Engineering, science, computer science, planning, design, construction management, water resources/quality

CONTACT: Human Resources Department.

COMPANY URL: http://www.usace.army.mil

For more information, see full organization profile, pg. 306.

USDA FOREST SERVICE

1621 NORTH KENT STREET, ROOM 1010 RPE
WASHINGTON, D.C. 20090

OVERVIEW: Conserves and manages the USDA national forests and grasslands.

EXPERTISE/EDUCATION SOUGHT: Research, administration, engineering, physical science, biological sciences

ALTERNATIVES TO FULL-TIME EMPLOYMENT: Paid internships

COMPANY-PAID TRAINING is available.

CONTACT: Human Resources Department, PO Box 96090, Washington, D.C., 20090; Phone: 703-235-8102; Fax: 703-235-1034.

U.S. DEPARTMENT OF TRANSPORTATION, FEDERAL HIGHWAY ADMINISTRATION

400 7TH STREET, SW, ROOM 4323
WASHINGTON, D.C. 20590

OVERVIEW: Organization responsible for design, construction, and maintenance of highway systems. Established: 1967.

HIRING HISTORY: Number of professional employees hired in 1997: 40.

EXPERTISE/EDUCATION SOUGHT: Civil engineering, transportation engineering, transportation planning, environmental planning

ALTERNATIVES TO FULL-TIME EMPLOYMENT: Paid internships

CONTACT: For individuals with previous experience: Mr. Martin Cohen, Personnel Management Specialist; Phone: 202-366-1161. Job Hotline: 202-366-1196.

COMPANY URL: http://www.fhwa.dot.gov

U.S. GENERAL ACCOUNTING OFFICE

441 G STREET, NW
WASHINGTON, D.C. 20548

OVERVIEW: Provides financial investigation services.

EXPERTISE/EDUCATION SOUGHT: Economics, accounting, law, finance, auditing

ALTERNATIVES TO FULL-TIME EMPLOYMENT: Paid internships

CONTACT: Director of Recruiting; Phone: 202-512-4900; Fax: 202-512-2539. E-mail: recruit@gao.gov. Job Hotline: 800-967-5426.

COMPANY URL: http://www.gao.gov

U.S. GOVERNMENT PRINTING OFFICE

EMPLOYMENT BRANCH STOP PSE
WASHINGTON, D.C. 20401

OVERVIEW: Prints and distributes government documents and publications.

HIRING HISTORY: Number of professional employees hired in 1996: 20.

EXPERTISE/EDUCATION SOUGHT: Purchasing, computer science, information systems, printing, publishing, accounting, auditing

COMPANY-PAID TRAINING is available.

CONTACT: For college students and recent graduates: Mr. Charles Kirkpatrick, Personnel Staffing Specialist; Phone: 202-512-1137; Fax: 202-512-1292. For individuals with previous experience: Ms. Margrette Nanney, Chief of Employment Branch; Phone: 202-512-1200; Fax: 202-512-1292. E-mail: bclower@gpo.gov. Job Hotline: 202-512-1200.

COMPANY URL: http://www.admin@www.access.gpo.gov

U.S. NUCLEAR REGULATORY COMMISSION

OFFICE OF HUMAN RESOURCES
WASHINGTON, D.C. 20555

OVERVIEW: Regulates the uses of commercial nuclear energy and radioactive materials to ensure protection of the public health, safety, and the environment. Established: 1973.

KEY STATISTICS: Number of Employees: 3,100.

HIRING HISTORY: Number of professional employees hired in 1997: 12.

EXPERTISE/EDUCATION SOUGHT: Nuclear engineering, mechanical engineering, chemical engineering, environmental engineering, health physics

ALTERNATIVES TO FULL-TIME EMPLOYMENT: Paid internships

COMPANY-PAID TRAINING is available.

CONTACT: Mr. Henry Rubin, Human Resources Specialist; Phone: 301-415-1374; Fax: 301-415-3818. E-mail: har1@nrc.gov.

COMPANY URL: http://www.nrc.gov

For more information, see full organization profile, pg. 308.

THE WASHINGTON POST

1150 15TH STREET, NW
WASHINGTON, D.C. 20071

OVERVIEW: Publishes newspapers. Established: 1887. Parent Company: The Washington Post Company. Ownership: Public.

KEY STATISTICS: Annual Sales: $1.9 billion. Number of Employees: 7,300.

HIRING HISTORY: Number of professional employees hired in 1997: 100.

EXPERTISE/EDUCATION SOUGHT: Accounting, journalism, graphic arts, data processing, business management

ALTERNATIVES TO FULL-TIME EMPLOYMENT: Internships

CONTACT: For individuals with previous experience: Mr. Carl Williams, Director of Personnel; Phone: 202-334-6000, ext. 7174; Fax: 202-334-5658. Job Hotline: 202-334-5350.

COMPANY URL: http://www.washingtonpost.com

DELAWARE

HERCULES, INC.

1313 NORTH MARKET STREET
WILMINGTON, DELAWARE 19801

OVERVIEW: Manufactures specialty chemicals. Established: 1912. Ownership: Public.

KEY STATISTICS: Annual Sales: $2.1 billion. Number of Employees: 7,114.

HIRING HISTORY: Number of professional employees hired in 1996: 25.

EXPERTISE/EDUCATION SOUGHT: Accounting, finance, data processing, chemistry, environmental engineering, chemical engineering, mechanical engineering

ALTERNATIVES TO FULL-TIME EMPLOYMENT: Internships

INTERNATIONAL ASSIGNMENTS are available to employees.

CONTACT: Dr. Diane Dixon, Director of Organizational And People Development; Phone: 302-594-5000, ext. 6114; Fax: 302-594-7097. E-mail: ddixon@herc.com.

MACDERMID, INC.

701 INDUSTRIAL DRIVE
MIDDLETOWN, DELAWARE 19709-1085

OVERVIEW: Produces electronic circuits. Established: 1995. Ownership: Public.

KEY STATISTICS: Annual Sales: $80.5 million. Number of Employees: 152.

EXPERTISE/EDUCATION SOUGHT: Chemical engineering, sales

COMPANY-PAID TRAINING is available.

INTERNATIONAL ASSIGNMENTS are available to employees.

CONTACT: Human Resources Department; Phone: 302-378-3100; Fax: 302-378-3123.

FLORIDA

ADDISON STEEL, INC.

7351 OVERLAND ROAD
ORLANDO, FLORIDA 32810

OVERVIEW: Manufactures fabricated structural metal. Established: 1995. Ownership: Private.

Addison Steel, Inc. (continued)

Key Statistics: Annual Sales: $63.7 million. Number of Employees: 253.

Expertise/Education Sought: Engineering

Contact: Mr. Glen Davis, President; Phone: 407-295-6434, ext. 135; Fax: 407-295-6434. E-mail: 110241.1556@compuserve.com.

ADT SECURITY SERVICES, INC.

1750 CLINT MOORE ROAD
BOCA RATON, FLORIDA 33431

Overview: Provides security systems services. Established: 1982. Parent Company: Tyco International, Ltd. Ownership: Private.

Key Statistics: Annual Sales: $167.2 million. Number of Employees: 12,000.

Expertise/Education Sought: Technical, sales, administration

Company-Paid Training is available.

International Assignments are available to employees.

Contact: Human Resources; Phone: 561-988-3600; Fax: 561-988-3631.

Company URL: http://www.adt.com

AIRTRAN AIRLINES

9955 AIRTRAN BOULEVARD
ORLANDO, FLORIDA 32827

Overview: Operates airline. Established: 1992. Ownership: Private.

Key Statistics: Annual Sales: $219.6 million. Number of Employees: 2,000.

Hiring History: Number of professional employees hired in 1996: 1,500.

Expertise/Education Sought: Engineering, accounting, management, flight attendant, aviation

Contact: For individuals with previous experience: Recruiting Office; Phone: 407-251-5600; Fax: 407-251-5890.

Company URL: http://www.airtran.com

AMERICAN AUTOMOBILE ASSOCIATION

1000 AAA DRIVE
HEATHROW, FLORIDA 32746-5063

Overview: Provides insurance, travel, and related automotive services to members. Established: 1910. Ownership: Private.

Key Statistics: Annual Sales: $474 million. Number of Employees: 1,500.

Hiring History: Number of professional employees hired in 1997: 380. 1996: 280.

Expertise/Education Sought: Accounting, publishing, health insurance, public relations, human resources, market research, publishing, management, finance, information services

Alternatives to Full-Time Employment: Paid internships

Company-Paid Training is available.

Contact: Ms. Barbara Amansec, Human Resources Generalist; Phone: 407-444-7526; Fax: 407-444-7504. E-mail: bamansec@national.AAA.com. Job Hotline: 407-444-7500.

Company URL: http://www.AAA.com

ANCHOR GLASS CONTAINER CORPORATION

4343 ANCHOR PLAZA PARKWAY
TAMPA, FLORIDA 33634-7537

Overview: Manufactures and distributes glassware and glass containers. Established: 1989. Parent Company: Consumers Packaging, Inc. Ownership: Private.

Expertise/Education Sought: Accounting, marketing, finance, ceramic engineering, electrical engineering, mechanical engineering, process engineering, quality engineering

Company-Paid Training is available.

International Assignments are available to employees.

Contact: Mr. Mark Karrenbauer, Vice President Human Resources; Phone: 813-884-0000; Fax: 813-882-7826.

AUTOMOBILI LAMBORGHINI U S A

7601 CENTURION PARKWAY
JACKSONVILLE, FLORIDA 32256-0520

Overview: Manufactures automobiles and related motor vehicles. Established: 1990. Ownership: Private.

Key Statistics: Annual Sales: $42.6 million. Number of Employees: 150.

Hiring History: Number of professional employees hired in 1997: 5.

Expertise/Education Sought: Business administration, mechanical engineering

Alternatives to Full-Time Employment: Internships

Company-Paid Training is available.

International Assignments are available to employees.

Contact: For individuals with previous experience: Supervisor; Phone: 904-569-9100; Fax: 904-569-9108.

Company URL: http://www.lamborghini.com

CARNIVAL CRUISE LINES

3655 NORTHWEST 87TH AVENUE
MIAMI, FLORIDA 33178-2418

Overview: Operates cruise ships and hotels. Established: 1974. Ownership: Public.

Key Statistics: Annual Sales: $2.2 billion. Number of Employees: 18,110.

Expertise/Education Sought: Food services, hospitality/hotel and restaurant management, sales, data processing, law, accounting, finance

International Assignments are available to employees.

Contact: Mr. Herb Schleier, Recruiting Specialist; Phone: 305-599-2600, ext. 3054; Fax: 305-599-8619. Job Hotline: 305-406-4780.

Company URL: http://www.carnival.com

CHS LATIN AMERICA

2010 NORTHWEST 84TH AVENUE
MIAMI, FLORIDA 33122

Overview: Wholesales computers, peripherals, and software. Established: 1995. Parent Company: CHS Electronics. Ownership: Public.

Key Statistics: Annual Sales: $115 million. Number of Employees: 37.

Hiring History: Number of professional employees hired in 1997: 40. 1996: 8.

Expertise/Education Sought: Computer programming, computer science, accounting, sales, software engineering/development, engineering, finance, human resources, marketing, management

International Assignments are available to employees.

Contact: Ms. Ledia Martinez, Human Resources Manager; Phone: 305-908-6800; Fax: 305-599-7008.

Company URL: http://www.chs.com

CMS, INC.

4904 EISENHOWER BOULEVARD, SUITE 310
TAMPA, FLORIDA 33634-6328

Overview: Provides environmental restoration services. Established: 1995. Ownership: Private.

Key Statistics: Annual Sales: $37.8 million. Number of Employees: 550.

Expertise/Education Sought: Data processing, electrical engineering, electronics/electronics engineering, manufacturing engineering, industrial engineering, chemical engineering

Contact: For individuals with previous experience: Ms. Robin Miller, Human Resources Manager; Phone: 813-882-4477; Fax: 813-884-1876. E-mail: cmshr@msn.com. Job Hotline: 888-267-3630.

CORDIS

14201 NORTHWEST 60TH AVENUE
MIAMI LAKES, FLORIDA 33014

Overview: Manufactures medical disposables. Established: 1996. Parent Company: Johnson & Johnson Corporation. Ownership: Public.

Key Statistics: Annual Sales: $304.8 million. Number of Employees: 3,620.

Expertise/Education Sought: Production engineering, product engineering, industrial engineering, quality engineering, manufacturing engineering, mechanical engineering, biomedical engineering

Alternatives to Full-Time Employment: Paid internships

Company-Paid Training is available.

International Assignments are available to employees.

Contact: Ms. Judy Beck, Senior Manager; Phone: 305-824-5289; Fax: 305-824-2020.

CROWLEY AMERICAN TRANSPORT, INC.

9487 REGENCY SQUARE BOULEVARD
JACKSONVILLE, FLORIDA 32225-8126

Mailing Address: PO Box 2110, Jacksonville, FL 32203

Overview: Provides deep sea transportation services. Established: 1954. Parent Company: Crowley Maritime Corporation. Ownership: Private.

Key Statistics: Annual Sales: $814.7 million. Number of Employees: 2,568.

Expertise/Education Sought: Management information systems, accounting

Company-Paid Training is available.

International Assignments are available to employees.

Contact: For college students and recent graduates: Mr. Gregg Bardea, Director of Organization Development; Phone: 904-727-2200; Fax: 904-727-2501. For individuals with previous experience: Mr. Lester Williams, Marine Personnel; Phone: 904-727-2200; Fax: 904-727-2501. Job Hotline: 904-727-4287.

CSX TRANSPORTATION

500 WATER STREET
JACKSONVILLE, FLORIDA 32202

Overview: Provides railroad freight services. Established: 1987. Parent Company: CSX Corporation. Ownership: Public.

Key Statistics: Annual Sales: $205 million. Number of Employees: 780.

Hiring History: Number of professional employees hired in 1997: 2,500. 1996: 1,500.

Expertise/Education Sought: Management information systems, transportation, administration, data processing

Alternatives to Full-Time Employment: Part-time jobs, paid internships

Company-Paid Training is available.

International Assignments are available to employees.

Contact: For college students and recent graduates: Ms. Lisa Kemph, Manager of College Relations; Phone: 904-359-5405. For individuals with previous experience: Mr. Michael Buckner, Manager of Work Force Solution; Phone: 904-366-5880; Fax: 904-359-3932. Job Hotline: 800-521-1658.

Company URL: http://www.csx.com

DATA MANAGEMENT SYSTEMS, INC.

1280 COURT STREET
CLEARWATER, FLORIDA 33756-5806

Overview: Designs computer systems. Established: 1995. Ownership: Private.

Key Statistics: Annual Sales: $12 million.

Expertise/Education Sought: Systems engineering

Contact: For individuals with previous experience: Human Resources; Phone: 813-443-4700.

Company URL: http://www.dmsus.com

DISTRIBUTED PROCESSING TECHNOLOGY

140 CANDACE DRIVE
MAITLAND, FLORIDA 32751

Overview: Manufactures adapters for personal computers. Established: 1977. Ownership: Private.

Key Statistics: Annual Sales: $50 million. Number of Employees: 230.

Expertise/Education Sought: Plastics engineering, software engineering/development, computer engineering, manufacturing engineering, mechanical engineering, electrical engineering, sales

Alternatives to Full-Time Employment: Part-time jobs

International Assignments are available to employees.

Contact: For individuals with previous experience: Personnel; Phone: 407-830-5522, ext. 241; Fax: 407-260-5366.

Company URL: http://www.dpt.com

EXIGENT INTERNATIONAL, INC.

1225 EVANS ROAD
MELBOURNE, FLORIDA 32904-2314

Overview: Provides systems and software engineering services. Established: 1996. Ownership: Public.

Key Statistics: Annual Sales: $29.9 million. Number of Employees: 270.

Expertise/Education Sought: Software engineering/development

Alternatives to Full-Time Employment: Paid internships

International Assignments are available to employees.

Contact: Mr. Mark Phillips, Corporate Recruiter; Phone: 407-952-7550; Fax: 407-952-7555. E-mail: markphillips@xgnt.com.

Company URL: http://www.xgnt.com

W.R. GRACE AND COMPANY, INC.

1 TOWN CENTER ROAD
BOCA RATON, FLORIDA 33486-1010

Overview: Manufactures chemicals, building materials, and medical supplies. Established: 1854. Ownership: Public.

Key Statistics: Annual Sales: $3.5 billion. Number of Employees: 17,400.

Hiring History: Number of professional employees hired in 1996: 75.

Expertise/Education Sought: Accounting, law, chemical engineering, manufacturing engineering, technical engineering

Alternatives to Full-Time Employment: Paid internships

International Assignments are available to employees.

Contact: Ms. Irene Kuenzel, Manager of Employee Relations; Phone: 561-362-2000; Fax: 561-362-2306.

Company URL: http://www.grace.com

GREATER ORLANDO AVIATION AUTHORITY

1 AIRPORT BOULEVARD
ORLANDO, FLORIDA 32827-4399

Overview: Operates airport. Established: 1957. Ownership: Private.

Key Statistics: Annual Sales: $184.8 million. Number of Employees: 600.

Expertise/Education Sought: Engineering, accounting, finance, human resources, property management, operations, information systems, program analysis

Alternatives to Full-Time Employment: Paid internships

Company-Paid Training is available.

Contact: For individuals with previous experience: Human Resources; Phone: 407-825-2253; Fax: 407-825-2099. Job Hotline: 407-825-2253.

GTE DATA SERVICES, INC.

1 EAST TELECOM PARKWAY
TEMPLE TERRACE, FLORIDA 33637

Mailing Address: PO Box 290152, Temple Terrace, FL 33687-0152

GTE Data Services, Inc. (continued)

Overview: Provides software development and information processing services. Established: 1967. Parent Company: GTE Corporation. Ownership: Public.

Key Statistics: Annual Sales: $830.4 million. Number of Employees: 3,800.

Hiring History: Number of professional employees hired in 1997: 500. 1996: 200.

Expertise/Education Sought: Systems analysis, computer programming, software engineering/development, database management, network management systems, business information analysts, networking, systems architecture

Alternatives to Full-Time Employment: Part-time jobs, paid internships

Company-Paid Training is available.

Contact: For individuals with previous experience: Ms. Kathy Costenbader, Staff Administrator; Phone: 813-978-4925; Fax: 813-987-1290. E-mail: kathy.costenbader@telops.gte.com.

Company URL: http://www.gteds.com

HARRIS

HARRIS CORPORATION

1025 WEST NASA BOULEVARD
MELBOURNE, FLORIDA 32919

Overview: Manufactures electronic systems, semiconductors, communications systems, and office equipment. Established: 1895. Ownership: Public.

Key Statistics: Annual Sales: $4 billion. Number of Employees: 29,000.

Expertise/Education Sought: Electrical engineering, computer engineering, computer science, software engineering/development, business administration

Alternatives to Full-Time Employment: Paid internships

Contact: College Relations.

Company URL: http://www.Harris.com

For more information, see full corporate profile, pg. 278.

HI-STAT MANUFACTURING COMPANY, INC.

7290 26TH COURT EAST
SARASOTA, FLORIDA 34243

Overview: Manufactures sensory automotive components. Established: 1967. Ownership: Private.

Key Statistics: Annual Sales: $100 million. Number of Employees: 2,100.

Hiring History: Number of professional employees hired in 1997: 75.

Expertise/Education Sought: Electrical engineering, design engineering, mechanical engineering, automotive engineering, project engineering

Alternatives to Full-Time Employment: Paid internships

Company-Paid Training is available.

Contact: For individuals with previous experience: Mr. Merv Marshall, Corporate Human Resource Manager. E-mail: lombardc@histat.com.

Company URL: http://www.histat.com

HYDRO ALUMINUM ROCKLEDGE

100 GUS HIPP BOULEVARD
ROCKLEDGE, FLORIDA 32955-4701

Overview: Manufactures aluminum extrusions and tubing. Established: 1990. Parent Company: USA Naco, Inc. Ownership: Private.

Key Statistics: Annual Sales: $29 million. Number of Employees: 220.

Expertise/Education Sought: Sales, marketing, engineering

Alternatives to Full-Time Employment: Internships

Company-Paid Training is available.

International Assignments are available to employees.

Contact: For individuals with previous experience: Human Resources Department; Phone: 407-636-8147, ext. 2600; Fax: 407-636-8288.

INJECTION FOOTWEAR CORPORATION

8730 NORTHWEST 36TH AVENUE
MIAMI, FLORIDA 33147-3934

Overview: Manufactures shoes. Established: 1978. Ownership: Private.

Key Statistics: Annual Sales: $31.9 million. Number of Employees: 500.

Expertise/Education Sought: Marketing, finance, data processing, electrical engineering, mechanical engineering

Contact: For individuals with previous experience: Mr. Julio Magrisso, Personnel Manager; Phone: 305-696-4611; Fax: 305-696-0018.

INTERACTIVE MEDIA SYSTEMS

2200 PARK CENTRAL BOULEVARD NORTH, SUITE 600
POMPANO BEACH, FLORIDA 33064-2232

Overview: Provides computer software development services. Established: 1994. Ownership: Private.

Key Statistics: Annual Sales: $25 million.

Hiring History: Number of professional employees hired in 1996: 8.

Expertise/Education Sought: Engineering, marketing, sales

Contact: For individuals with previous experience: Human Resource Department; Phone: 954-974-4441, ext. 14; Fax: 954-974-5551.

KAYDON CORPORATION

19345 US HIGHWAY 19 NORTH
CLEARWATER, FLORIDA 33764-3148

Overview: Manufactures ball bearings, rings, and seals. Established: 1941. Ownership: Public.

Key Statistics: Annual Sales: $290.7 million. Number of Employees: 2,265.

Expertise/Education Sought: Electrical engineering, mechanical engineering, chemical engineering, manufacturing engineering, accounting

Alternatives to Full-Time Employment: Part-time jobs

International Assignments are available to employees.

Contact: Mr. John Brocci, Vice President of Administration; Phone: 813-531-1101, ext. 308.

Company URL: http://www.kaydoncorp.com

KNIGHT RIDDER

1 HERALD PLAZA
MIAMI, FLORIDA 33132-1609

Overview: Publishes newspapers and provides electronic information services. Established: 1892. Ownership: Public.

Key Statistics: Annual Sales: $2.8 billion. Number of Employees: 10,263.

Expertise/Education Sought: Production, auditing, journalism, data processing, marketing, finance

Alternatives to Full-Time Employment: Paid internships

Contact: For individuals with previous experience: Assistant Vice President for Human Resources; Phone: 305-376-3800; Fax: 305-995-8164.

Company URL: http://www.knight-ridder.com

LINVATEC CORPORATION

11311 CONCEPT BOULEVARD
LARGO, FLORIDA 33773

Overview: Manufactures surgical, arthroscopic, endoscopic, orthopedic, and hospital equipment. Established: 1990. Parent Company: ConMed. Ownership: Public.

Key Statistics: Annual Sales: $126.2 million. Number of Employees: 1,500.

Hiring History: Number of professional employees hired in 1996: 20.

Expertise/Education Sought: Mechanical engineering, electrical engineering, accounting, marketing, finance, computer programming

Alternatives to Full-Time Employment: Paid internships

International Assignments are available to employees.

Contact: For individuals with previous experience: Mr. Steve Brodbeck, Human Resources Manger; Phone: 813-399-5129; Fax: 813-399-9900.

MARK III INDUSTRIES/TIARA

5401 NORTHWEST 44TH AVENUE
OCALA, FLORIDA 34482

Mailing Address: PO Box 2525, Ocala, FL 34478

Overview: Manufactures conversion vans. Established: 1978. Ownership: Private.

Key Statistics: Annual Sales: $93.8 million. Number of Employees: 750.

Expertise/Education Sought: Design engineering, mechanical engineering, sales

Alternatives to Full-Time Employment: Paid internships

Company-Paid Training is available.

International Assignments are available to employees.

Contact: For individuals with previous experience: Ms. Emily Carter, Human Resources Director; Phone: 352-732-5878; Fax: 352-351-1017. E-mail: hr@markIII.com.

MCDONALD CONSTRUCTION CORPORATION

5610 SOUTH FLORIDA AVENUE
LAKELAND, FLORIDA 33813

Overview: Provides earth moving services. Established: 1955. Ownership: Private.

Key Statistics: Annual Sales: $48.5 million. Number of Employees: 425.

Expertise/Education Sought: Construction

Company-Paid Training is available.

Contact: Mr. Wiley Johnson, Vice President of Operations; Phone: 941-646-5763; Fax: 941-646-4228.

METAL INDUSTRIES, INC.

1310 NORTH HERCULES AVENUE
CLEARWATER, FLORIDA 34625-1922

Overview: Manufactures heating and air conditioning registers. Established: 1949. Ownership: Private.

Key Statistics: Annual Sales: $94.6 million. Number of Employees: 1,066.

Expertise/Education Sought: Accounting, computer programming, data processing, engineering

Company-Paid Training is available.

Contact: Ms. Hayme Nader, Production Manager; Phone: 813-441-2651; Fax: 813-441-9626.

MULBERRY CORPORATION

4000 STATE ROAD 60 EAST
MULBERRY, FLORIDA 33860

Mailing Address: PO Drawer 797, Mulberry, FL 33860-0797

Overview: Manufactures phosphate fertilizers. Established: 1994. Ownership: Private.

Key Statistics: Annual Sales: $140 million. Number of Employees: 280.

Expertise/Education Sought: Safety, chemical engineering, mechanical engineering, accounting, process engineering, environmental engineering

Contact: For individuals with previous experience: Human Resources; Phone: 941-425-1176; Fax: 941-425-5446.

NORWEGIAN CRUISE LINE, LTD.

7665 CORPORATE CENTER DRIVE
MIAMI, FLORIDA 33126

Overview: Operates cruise ships. Established: 1986. Ownership: Private.

Key Statistics: Annual Sales: $657.5 million. Number of Employees: 5,000.

Expertise/Education Sought: Accounting, marketing, data processing, computer programming

Alternatives to Full-Time Employment: Internships

Company-Paid Training is available.

International Assignments are available to employees.

Contact: For individuals with previous experience: Human Resources; Phone: 305-436-4000; Fax: 305-436-4101.

ORLANDO UTILITIES COMMISSION

500 SOUTH ORANGE AVENUE
ORLANDO, FLORIDA 32802-3193

Overview: Distributes electric power. Established: 1923. Ownership: Private.

Key Statistics: Annual Sales: $340.7 million. Number of Employees: 1,027.

Hiring History: Number of professional employees hired in 1997: 60. 1996: 10.

Expertise/Education Sought: Accounting, finance, computer programming

Alternatives to Full-Time Employment: Paid internships

Contact: Ms. Carol Sciarro, Staffing Administrator; Phone: 407-423-9100. E-mail: bcollier@ouc.com. Job Hotline: 407-423-9191.

Company URL: http://www.ouc.com

PCS PHOSPHATE-WHITE SPRINGS

HIGHWAY 137
WHITE SPRINGS, FLORIDA 32096

Mailing Address: PO Box 300, White Springs, FL 32096

Overview: Mines phosphate rock. Established: 1995. Parent Company: Potash Corporation Of Saskatchewan. Ownership: Private.

Key Statistics: Annual Sales: $305.9 million. Number of Employees: 1,311.

Hiring History: Number of professional employees hired in 1997: 20.

Expertise/Education Sought: Administration, production, technical

Company-Paid Training is available.

International Assignments are available to employees.

Contact: Mr. Paul H. Barrett, Human Resources Manager; Phone: 904-397-8308; Fax: 904-397-8371.

PROFESSIONAL ENGINEERING AND INSPECTION COMPANY

4350 WEST SUNRISE BOULEVARD, SUITE 103
FORT LAUDERDALE, FLORIDA 33313-6709

Overview: Provides engineering services. Established: 1993. Parent Company: US Laboratories, Inc. Ownership: Private.

Key Statistics: Annual Sales: $8.1 million. Number of Employees: 112.

Hiring History: Number of professional employees hired in 1997: 8. 1996: 6.

Expertise/Education Sought: Structural engineering, structural design, material testing, structural inspection

Alternatives to Full-Time Employment: Paid internships

Company-Paid Training is available.

Contact: Mr. Gary Elzweig, President; Phone: 954-581-2008; Fax: 954-581-0433.

PTT TELEKOM, INC.

1000 ORLANDO AVENUE
WINTER PARK, FLORIDA 32789-3359

Overview: Provides local and long distance telephone services. Established: 1995. Ownership: Private.

Key Statistics: Annual Sales: $12 million. Number of Employees: 28.

Hiring History: Number of professional employees hired in 1997: 20. 1996: 20.

Expertise/Education Sought: Manufacturing

Company-Paid Training is available.

International Assignments are available to employees.

Contact: For individuals with previous experience: Human Resources; Phone: 407-648-5300; Fax: 407-648-0900.

Company URL: http://www.ptt-telekom.com

RANGER CONSTRUCTION INDUSTRIES, MARTIN DIVISION

1801 SOUTH NOVA ROAD
DAYTONA BEACH, FLORIDA 32119-1733

OVERVIEW: Provides road construction services. Established: 1971. Parent Company: Ranger Construction Industries. Ownership: Private.

KEY STATISTICS: Annual Sales: $45 million. Number of Employees: 270.

HIRING HISTORY: Number of professional employees hired in 1997: 120.

EXPERTISE/EDUCATION SOUGHT: Civil engineering

COMPANY-PAID TRAINING is available.

CONTACT: For individuals with previous experience: Ms. Betty Muztafago, Director of Human Resources; Phone: 904-761-8383, ext. 340; Fax: 904-756-4326.

RECOTON CORPORATION

2950 LAKE EMMA ROAD
LAKE MARY, FLORIDA 32746-3705

OVERVIEW: Manufactures household audio and video equipment. Established: 1936. Ownership: Public.

KEY STATISTICS: Annual Sales: $331.7 million. Number of Employees: 2,500.

EXPERTISE/EDUCATION SOUGHT: Warehousing, radio frequency engineering, administration supervision, engineering, accounting, customer service and support

CONTACT: For individuals with previous experience: Human Resources; Phone: 407-333-0900, ext. 1206; Fax: 407-333-1964. Job Hotline: 407-333-0900, ext. 1599.

REGENCY ELECTRIC COMPANY, INC.

6601 SOUTHPOINT DRIVE NORTH, SUITE 300
JACKSONVILLE, FLORIDA 32216

OVERVIEW: Electrical work. Established: 1996. Ownership: Private.

KEY STATISTICS: Annual Sales: $64.4 million. Number of Employees: 700.

HIRING HISTORY: Number of professional employees hired in 1997: 200.

EXPERTISE/EDUCATION SOUGHT: Electrical engineering

COMPANY-PAID TRAINING is available.

CONTACT: For individuals with previous experience: Human Resources Department; Phone: 904-281-0600; Fax: 904-281-0599.

ROYAL CARIBBEAN CRUISES LIMITED

1050 CARIBBEAN WAY
MIAMI, FLORIDA 33132-2028

OVERVIEW: Operates cruise ships. Established: 1988. Ownership: Private.

KEY STATISTICS: Annual Sales: $1.4 billion. Number of Employees: 12,500.

HIRING HISTORY: Number of professional employees hired in 1997: 400.

EXPERTISE/EDUCATION SOUGHT: Accounting, finance, sales, marketing, data processing

ALTERNATIVES TO FULL-TIME EMPLOYMENT: Internships

COMPANY-PAID TRAINING is available.

INTERNATIONAL ASSIGNMENTS are available to employees.

CONTACT: Mr. Marc Grad, Human Resources Generalist; Phone: 305-539-6916. Job Hotline: 305-539-6096.

SOLITRON DEVICES, INC.

3301 ELECTRONICS WAY
WEST PALM BEACH, FLORIDA 33407-4620

OVERVIEW: Manufactures transistors and hybrid integrated circuits. Established: 1959. Ownership: Public.

KEY STATISTICS: Annual Sales: $7.2 million. Number of Employees: 100.

HIRING HISTORY: Number of professional employees hired in 1996: 108.

EXPERTISE/EDUCATION SOUGHT: Engineering

CONTACT: Ms. Alicia Egerborn, Human Resources Executive Assistant; Phone: 561-848-4311, ext. 315; Fax: 561-881-5652. E-mail: humanresources@solitrondevices.com.

COMPANY URL: http://www.solitrondevices.com

SPRINT

555 LAKE BORDER DRIVE
APOPKA, FLORIDA 32703-5815

OVERVIEW: Provides telecommunications services. Established: 1967. Parent Company: Sprint Corporation of Kansas City. Ownership: Public.

KEY STATISTICS: Annual Sales: $1.2 billion. Number of Employees: 5,562.

EXPERTISE/EDUCATION SOUGHT: Electrical engineering, computer science, marketing, management information systems, industrial engineering

COMPANY-PAID TRAINING is available.

CONTACT: Ms. Anne Marconi, Manager of Recruiting, PO Box 165000, Apopka, Florida, 32716-5000; Phone: 407-889-6208; Fax: 407-884-8482. Job Hotline: 407-263-5405.

SUNJET AIRLINES

4700 140TH AVENUE NORTH
CLEARWATER, FLORIDA 33762-3846

OVERVIEW: Operates charter airline. Established: 1995. Ownership: Private.

KEY STATISTICS: Annual Sales: $30.9 million. Number of Employees: 220.

HIRING HISTORY: Number of professional employees hired in 1996: 120.

EXPERTISE/EDUCATION SOUGHT: Aviation, flight attendant

ALTERNATIVES TO FULL-TIME EMPLOYMENT: Paid internships

COMPANY-PAID TRAINING is available.

CONTACT: For individuals with previous experience: Ms. Pam Stiers, Director of Human Resources; Phone: 813-530-1515; Fax: 813-530-1615.

SYKES ENTERPRISES, INC.

100 NORTH TAMPA STREET
TAMPA, FLORIDA 33602-5809

OVERVIEW: Provides technical writing, telecommunications, and software development services. Established: 1981. Ownership: Public.

KEY STATISTICS: Annual Sales: $117 million. Number of Employees: 4,000.

HIRING HISTORY: Number of professional employees hired in 1997: 17. 1996: 20.

EXPERTISE/EDUCATION SOUGHT: Software engineering/development, technical writing, program analysis

COMPANY-PAID TRAINING is available.

INTERNATIONAL ASSIGNMENTS are available to employees.

CONTACT: For individuals with previous experience: Ms. Susan Felix, Office Administrator; Phone: 813-274-1000, ext. 441; Fax: 813-273-6570. E-mail: quality@corp.sykes.com.

COMPANY URL: http://www.sykes.com

TECO ENERGY

702 NORTH FRANKLIN STREET
TAMPA, FLORIDA 33602-4418

OVERVIEW: Electric utility. Established: 1981. Parent Company: TECO Energy, Inc. Ownership: Public.

KEY STATISTICS: Assets: $1.1 billion. Number of Employees: 2,798.

HIRING HISTORY: Number of professional employees hired in 1997: 150. 1996: 112.

EXPERTISE/EDUCATION SOUGHT: Chemical engineering, electrical engineering, industrial engineering, civil engineering, engineering, data processing, mechanical engineering, accounting

ALTERNATIVES TO FULL-TIME EMPLOYMENT: Paid internships

COMPANY-PAID TRAINING is available.

INTERNATIONAL ASSIGNMENTS are available to employees.

CONTACT: For college students and recent graduates: Ms. Karen Hobart, Coordinator of Recruitment; Phone: 813-228-1725; Fax: 813-228-4644. E-mail: karenhobart@tecoenergy.com. For individuals with previous experience: Ms. Teri Sturgill, Manager of Recruitment and Staffing; Phone: 813-228-4350; Fax: 813-228-4644. E-mail: teristurgill@tecoenergy.com.

COMPANY URL: http://www.tecoenergy.com

TELEPHONE COMPANY OF CENTRAL FLORIDA
3575 WEST LAKE MARY BOULEVARD, SUITE 107
LAKE MARY, FLORIDA 32746

Overview: Provides local and long distance phone services. Established: 1996. Ownership: Private.

Key Statistics: Annual Sales: $35 million. Number of Employees: 29.

Hiring History: Number of professional employees hired in 1997: 75. 1996: 12.

Expertise/Education Sought: Accounting, data processing, engineering, finance, human resources, management, sales, marketing, customer service and support, administration

Contact: For individuals with previous experience: Mrs. Andrea Welch, Vice President of Administration.

TURBOCOMBUSTOR TECHNOLOGY, INC.
3651 SOUTHEAST COMMERCE AVENUE
STUART, FLORIDA 34997-4967

Overview: Manufactures aircraft engines and engine parts. Established: 1987. Parent Company: Chromalloy Gas Turbine Corporation. Ownership: Public.

Key Statistics: Annual Sales: $20 million. Number of Employees: 210.

Hiring History: Number of professional employees hired in 1996: 30.

Expertise/Education Sought: Process engineering, product design and development, aerospace engineering, quality engineering, estimating, project engineering

Contact: Mr. Dean Stapleton, Direcrtor of Human Resources; Phone: 561-287-7770; Fax: 561-287-0476.

VISTA-UNITED TELECOMMUNICATIONS
3100 BONNET CREEK ROAD
LAKE BUENA VISTA, FLORIDA 32830

Mailing Address: PO Box 10180, Buena Vista, FL 32830

Overview: Provides telephone services. Established: 1971. Parent Company: Walt Disney Company, Inc. Ownership: Private.

Key Statistics: Annual Sales: $61.3 million. Number of Employees: 550.

Expertise/Education Sought: Industrial engineering, marketing, engineering

Alternatives to Full-Time Employment: Paid internships

Company-Paid Training is available.

International Assignments are available to employees.

Contact: Professional Staffing; Phone: 407-827-2000; Fax: 407-827-2096.

WALT DISNEY WORLD COMPANY
1515 BUENA VISTA DRIVE
ORLANDO, FLORIDA 32830-1000

Mailing Address: PO Box 10000, Lake Buena Vista, FL 32830-1000

Overview: Owns and operates amusement parks, hotels, and resorts. Established: 1964. Ownership: Public.

Key Statistics: Annual Sales: $3.2 billion. Number of Employees: 34,000.

Expertise/Education Sought: Accounting, finance, marketing, electrical engineering, mechanical engineering, industrial engineering, sales, information systems, food services

Alternatives to Full-Time Employment: Part-time jobs, paid internships

Company-Paid Training is available.

International Assignments are available to employees.

Contact: For individuals with previous experience: Ms. Kelly Frank, Human Resources; Phone: 407-828-1000; Fax: 407-828-1571. Job Hotline: 407-828-1000.

Company URL: http://www.disney.com

WHITE CONSTRUCTION COMPANY
PO DRAWER 790
CHIEFLAND, FLORIDA 32644

Overview: Provides highway construction services. Established: 1945. Ownership: Private.

Key Statistics: Annual Sales: $69.6 million. Number of Employees: 600.

Hiring History: Number of professional employees hired in 1997: 200.

Expertise/Education Sought: Accounting, civil engineering

Contact: For individuals with previous experience: Mr. Steve Pomeroy, Office Manager, Highway 19 North, Chiefland, Florida, 32626; Phone: 352-493-1444; Fax: 352-493-9943.

WPBT PUBLIC TV, CHANNEL 2
14901 NORTHEAST 20TH AVENUE
MIAMI, FLORIDA 33181

Overview: Operates TV broadcasting station. Established: 1956. Ownership: Private.

Key Statistics: Annual Sales: $17.8 million. Number of Employees: 200.

Hiring History: Number of professional employees hired in 1997: 30. 1996: 20.

Expertise/Education Sought: Production, computer programming, human resources, accounting, marketing, operations engineering, finance, administration

Alternatives to Full-Time Employment: Part-time jobs, internships

Contact: For individuals with previous experience: Mrs. Diane Herrons, Manager of Human Resources; Phone: 305-949-8321, ext. 406; Fax: 305-944-4211.

Company URL: http://www.channel2.org

XL COMPUTING CORPORATION
10305 102ND TERRACE
SEBASTIAN, FLORIDA 32958-7823

Overview: Provides computer systems selling services. Established: 1996. Parent Company: Cycomm International. Ownership: Public.

Key Statistics: Annual Sales: $7 million. Number of Employees: 95.

Expertise/Education Sought: Electrical engineering, computer science, management information systems, mechanical engineering, accounting

Alternatives to Full-Time Employment: Part-time jobs, paid internships

Company-Paid Training is available.

International Assignments are available to employees.

Contact: Ms. Daphne Cumberland, Director of Human Resources; Phone: 561-589-7344; Fax: 561-388-9274. E-mail: dcumberland@xlvision.com.

Company URL: http://www.xlcomputing.com

XOMED
6743 SOUTHPOINT DRIVE NORTH
JACKSONVILLE, FLORIDA 32216-6218

Overview: Manufactures medical equipment for eyes, ears, nose, and throat. Established: 1994. Ownership: Private.

Key Statistics: Annual Sales: $65.7 million. Number of Employees: 310.

Hiring History: Number of professional employees hired in 1996: 25.

Expertise/Education Sought: Electrical engineering, mechanical engineering, industrial engineering, accounting, data processing, marketing, sales, research and development, information services

Alternatives to Full-Time Employment: Paid internships

International Assignments are available to employees.

Contact: Mr. Bud Stevens, Vice President Human Resources; Phone: 904-279-7509; Fax: 904-279-7593. E-mail: bstevens@xomed.com. Job Hotline: 904-296-9600, ext. 7041.

Company URL: http://www.xomed.com

GEORGIA

ADVANCED CIRCUITRY, INC.

3464 HOWELL STREET
DULUTH, GEORGIA 30096-3326

Overview: Manufactures printed circuit boards. Established: 1992. Ownership: Private.

Hiring History: Number of professional employees hired in 1997: 6. 1996: 4.

Expertise/Education Sought: Electrical technology, circuit design, communications, wireless

Company-Paid Training is available.

International Assignments are available to employees.

Contact: For individuals with previous experience: Mr. Roger Dhanani, Company President; Phone: 770-622-2200.

ALLIED AUTOMOTIVE GROUP, INC.

160 CLAIREMONT AVENUE
DECATUR, GEORGIA 30030-2529

Overview: Operates trucking business. Established: 1995. Parent Company: Allied Holding. Ownership: Public.

Key Statistics: Annual Sales: $128.6 million. Number of Employees: 1,801.

Expertise/Education Sought: Transportation

International Assignments are available to employees.

Contact: Human Resources; Phone: 404-371-0379.

W.C. BRADLEY COMPANY

1005 FRONT AVENUE
COLUMBUS, GEORGIA 31901-0140

Mailing Address: PO Box 140, Columbus, GA 31902-0140

Overview: Manufactures household cooking equipment. Established: 1897. Ownership: Private.

Key Statistics: Annual Sales: $103.7 million. Number of Employees: 1,400.

Hiring History: Number of professional employees hired in 1997: 25.

Expertise/Education Sought: Electrical engineering, design engineering, marketing, computer programming, merchandising, purchasing, management, graphic arts

Alternatives to Full-Time Employment: Paid internships

Company-Paid Training is available.

Contact: For individuals with previous experience: Mr. Rick Woodham, Vice President of Compensation and Benefits; Phone: 706-571-6053, ext. 3400; Fax: 706-571-6081. E-mail: rick_woodham@wcbradley.com.

CALLAWAY CHEMICALS

6003 VETERANS PARKWAY
COLUMBUS, GEORGIA 31909-4663

Overview: Manufactures industrial inorganic chemicals. Established: 1994. Parent Company: Vulcan. Ownership: Public.

Key Statistics: Annual Sales: $175 million. Number of Employees: 450.

Expertise/Education Sought: Manufacturing engineering, chemical engineering

International Assignments are available to employees.

Contact: For individuals with previous experience: Human Resources; Phone: 706-576-2000; Fax: 706-576-6455.

CASTLEBERRY/SNOWS BRANDS

1621 15TH STREET
AUGUSTA, GEORGIA 30901-3929

Mailing Address: PO Box 1010, Augusta, GA 30903-1010

Overview: Produces sausage and other prepared meats. Established: 1994. Ownership: Private.

Key Statistics: Annual Sales: $83.2 million. Number of Employees: 500.

Hiring History: Number of professional employees hired in 1997: 30.

Expertise/Education Sought: Manufacturing

Alternatives to Full-Time Employment: Paid internships

Company-Paid Training is available.

Contact: For individuals with previous experience: Human Resources; Phone: 706-733-7765; Fax: 706-667-5312.

CENTERS FOR DISEASE CONTROL AND PREVENTION

4770 BUFORD HIGHWAY
ATLANTA, GEORGIA 30341

Overview: Monitors and controls physical and psychological disease. Division of the Department Of Health And Human Services.

Hiring History: Number of professional employees hired in 1997: 200. 1996: 500.

Expertise/Education Sought: Behavioral sciences, sociology, public health, statistics, biology, mathematics, pharmacology, epidemiology

Alternatives to Full-Time Employment: Part-time jobs, paid internships

Contact: Human Resources Department; Phone: 770-488-1725; Fax: 770-488-1939. Job Hotline: 888-232-4473.

Company URL: http://www.cdc.gov

CHEMOL COMPANY, INC.

4200 NORTHSIDE PARKWAY, NW
ATLANTA, GEORGIA 30327-3054

Mailing Address: PO Box 16286, Greensboro, NC 27416

Overview: Supplies industrial inorganic chemicals. Established: 1996. Parent Company: Siga. Ownership: Private.

Key Statistics: Annual Sales: $12 million. Number of Employees: 40.

Hiring History: Number of professional employees hired in 1997: 4. 1996: 4.

Expertise/Education Sought: Chemical engineering, laboratory technology, maintenance

Contact: For individuals with previous experience: Mr. Fred Wellons, Vice President; Phone: 336-330-3050; Fax: 336-273-4645.

CIBA VISION CORPORATION

11460 JOHNS CREEK PARKWAY
DULUTH, GEORGIA 30097

Overview: Manufactures vision care products. Established: 1985. Parent Company: Novartis Corporation. Ownership: Private.

Key Statistics: Annual Sales: $400 million. Number of Employees: 3,000.

Hiring History: Number of professional employees hired in 1997: 50.

Expertise/Education Sought: Mechanical engineering, electrical engineering, design engineering, project engineering, manufacturing engineering

Alternatives to Full-Time Employment: Paid internships

International Assignments are available to employees.

Contact: For individuals with previous experience: Mrs. Amanda Kijek, Staffing Specialist; Phone: 770-418-3572; Fax: 770-418-3036. Job Hotline: 770-734-5959.

Company URL: http://www.cvworld.com

CORPORATE EXPRESS DELIVERY SYSTEMS

1080 HOLCOMB BRIDGE ROAD
ROSWELL, GEORGIA 30076

Overview: Provides contract delivery service. Established: 1996. Ownership: Public.

Key Statistics: Annual Sales: $571.4 million. Number of Employees: 8,000.

Expertise/Education Sought: Marketing, engineering

Company-Paid Training is available.

Contact: Human Resources Department; Phone: 770-518-1180; Fax: 770-518-2587.

FIRST DATA CORPORATION
5660 NEW NORTHSIDE DRIVE, SUITE 1300
ATLANTA, GEORGIA 30328-5800

Overview: Provides credit card authorization and check processing services. Established: 1992. Ownership: Public.

Key Statistics: Annual Sales: $484.5 million. Number of Employees: 15,500.

Expertise/Education Sought: Systems analysis, accounting, finance

Alternatives to Full-Time Employment: Part-time jobs

Company-Paid Training is available.

Contact: For individuals with previous experience: Director Employee Relations; Phone: 770-857-0001; Fax: 770-857-0404.

Company URL: http://www.firstdata.com

FLEXIBLE PRODUCTS COMPANY
1007 INDUSTRIAL PARK DRIVE
MARIETTA, GEORGIA 30062

Overview: Manufactures ink used for logos and T-shirts. Established: 1989. Ownership: Private.

Key Statistics: Annual Sales: $38.1 million. Number of Employees: 375.

Hiring History: Number of professional employees hired in 1997: 50. 1996: 50.

Expertise/Education Sought: Business, finance, accounting, data processing, chemistry, sales

Alternatives to Full-Time Employment: Paid internships

Company-Paid Training is available.

International Assignments are available to employees.

Contact: For individuals with previous experience: Mr. Stanley Dean, Tax and Benefits Director, PO Box 3190, Marietta, Georgia, 30061; Phone: 770-428-2684; Fax: 770-428-9431.

Company URL: http://www.flexibleproducts.com

GENERAL TIME CORPORATION
520 GUTHRIDGE COURT
NORCROSS, GEORGIA 30092-3503

Mailing Address: PO Box 4125, Norcross, GA 30091-4125

Overview: Manufactures watches and parts for watches and clocks. Established: 1988. Ownership: Private.

Key Statistics: Annual Sales: $45 million. Number of Employees: 900.

Expertise/Education Sought: Manufacturing, sales, administration

Contact: For individuals with previous experience: Human Resource Department; Phone: 770-447-5300; Fax: 770-242-4009.

GENUINE PARTS COMPANY
2999 CIRCLE 75 PARKWAY, NW
ATLANTA, GEORGIA 30339-3073

Overview: Distributes automotive replacement parts, industrial replacement parts, and office supplies. Established: 1928. Ownership: Public.

Key Statistics: Annual Sales: $5.7 billion. Number of Employees: 24,200.

Expertise/Education Sought: Management

Alternatives to Full-Time Employment: Paid internships

Company-Paid Training is available.

Contact: For individuals with previous experience: Human Resources Department; Phone: 770-953-1700; Fax: 770-956-2205.

GEORGIA GULF CORPORATION
400 PERIMETER CENTER TERRACE
ATLANTA, GEORGIA 30346-1227

Overview: Manufactures chemicals and petroleum products. Established: 1971. Ownership: Public.

Key Statistics: Annual Sales: $896.2 million. Number of Employees: 1,030.

Hiring History: Number of professional employees hired in 1997: 58.

Expertise/Education Sought: Chemical engineering, chemistry, accounting

Alternatives to Full-Time Employment: Paid internships

Company-Paid Training is available.

International Assignments are available to employees.

Contact: For individuals with previous experience: Mr. John Hager, Corporate Personnel Manager; Phone: 770-395-4500; Fax: 770-395-4905.

GOODY PRODUCTS, INC.
600 WESTPARK DRIVE
PEACHTREE CITY, GEORGIA 30269

Overview: Manufactures barrettes and hair combs. Established: 1993. Parent Company: Newell Company. Ownership: Public.

Key Statistics: Annual Sales: $62.5 million. Number of Employees: 1,100.

Expertise/Education Sought: Manufacturing engineering

Alternatives to Full-Time Employment: Internships

International Assignments are available to employees.

Contact: For individuals with previous experience: Human Relations Department.

GREENFIELD INDUSTRIES, INC.
470 OLD EVANS ROAD
EVANS, GEORGIA 30809-3655

Mailing Address: PO Box 2587, Augusta, GA 30903

Overview: Manufactures blades for hand, power, and chain saws. Established: 1996. Parent Company: Kennametal, Inc. Ownership: Public.

Key Statistics: Annual Sales: $37.7 million. Number of Employees: 589.

Hiring History: Number of professional employees hired in 1997: 35.

Expertise/Education Sought: Engineering, sales

Alternatives to Full-Time Employment: Paid internships

Company-Paid Training is available.

International Assignments are available to employees.

Contact: Mr. Wayne Thompson, Manager of Human Resources; Phone: 706-650-4247; Fax: 706-650-4110.

GTE WIRELESS
245 PERIMETER CENTER PARKWAY
ATLANTA, GEORGIA 30346-2304

Overview: Provides cellular telephone services. Established: 1995. Parent Company: GTE Corporation. Ownership: Public.

Key Statistics: Annual Sales: $165.9 million. Number of Employees: 2,387.

Expertise/Education Sought: Electrical engineering, electronics/electronics engineering, design engineering, radio frequency engineering

Alternatives to Full-Time Employment: Part-time jobs, paid internships

Company-Paid Training is available.

International Assignments are available to employees.

Contact: Mrs. Suzanne Jones, Staffing Manager, PO Box 105194, Atlanta, Georgia, 30348; Phone: 770-391-8216; Fax: 770-395-8575.

Company URL: http://www.gte.com

THE HARLAND COMPANY
2939 MILLER ROAD
DECATUR, GEORGIA 30035

Overview: Prints checks and business forms. Established: 1923. Ownership: Public.

Key Statistics: Annual Sales: $609.4 million. Number of Employees: 5,485.

Expertise/Education Sought: Production management, sales, management information systems, network analysis

Alternatives to Full-Time Employment: Paid internships

Company-Paid Training is available.

Contact: Ms. Cecilia Palmer, Employment Coordinator; Phone: 770-981-9460.

HOLDER CONSTRUCTION COMPANY
3333 CUMBERLAND CIRCLE, SE
ATLANTA, GEORGIA 30339-3304

Overview: Commercial and office building construction. Established: 1960. Ownership: Private.

Key Statistics: Annual Sales: $49.2 million. Number of Employees: 230.

Expertise/Education Sought: Business administration, engineering, construction engineering, accounting

Alternatives to Full-Time Employment: Paid internships

Contact: Ms. Lee Johnston, Senior Vice President Human Resources; Phone: 770-988-3000; Fax: 770-988-3205. E-mail: ljohnston@holder.com.

HOLIDAY HOSPITALITY
3 RAVINIA DRIVE, SUITE 2900
ATLANTA, GEORGIA 30346-2118

Overview: Owns and operates hotels. Established: 1989. Parent Company: Bass, PLC. Ownership: Private.

Key Statistics: Annual Sales: $505.1 million.

Expertise/Education Sought: Marketing, finance, management information systems, accounting, law, business management

Alternatives to Full-Time Employment: Part-time jobs

Company-Paid Training is available.

International Assignments are available to employees.

Contact: Manager of Employment; Phone: 770-604-2000; Fax: 770-604-2371.

Company URL: http://www.holiday-inn.com

THE HOME DEPOT
2455 PACES FERRY ROAD
ATLANTA, GEORGIA 30339-4024

Overview: Operates chain of home improvement stores. Established: 1978. Ownership: Public.

Key Statistics: Annual Sales: $19.5 billion. Number of Employees: 98,100.

Hiring History: Number of professional employees hired in 1997: 5,000. 1996: 2,000.

Expertise/Education Sought: Sales, marketing, accounting, computer programming

Alternatives to Full-Time Employment: Paid internships

International Assignments are available to employees.

Contact: For individuals with previous experience: Human Resources Office; Phone: 770-433-8211, ext. 87677. Job Hotline: 770-433-8211, ext. 87677.

INTELLISOURCE SERVICES SOLUTIONS
2100 EAST EXCHANGE PLACE
TUCKER, GEORGIA 30084-5336

Overview: Generates and distributes electric power. Established: 1996. Parent Company: Oglethorpe Power Corporation. Ownership: Private.

Key Statistics: Annual Sales: $41.3 million. Number of Employees: 150.

Hiring History: Number of professional employees hired in 1997: 80. 1996: 20.

Expertise/Education Sought: Engineering, mechanical engineering, electrical engineering, environmental engineering, accounting, auditing, human resources, information systems

Alternatives to Full-Time Employment: Paid internships

Company-Paid Training is available.

Contact: Mrs. Sheila Garza, Human Resource Generalist; Phone: 770-270-7651; Fax: 770-270-7676. E-mail: sheila.garza@opc.com. Job Hotline: 770-270-7939.

Company URL: http://www.Intellisource.com

JOHNSTON INDUSTRIES, INC.
105 13TH STREET
COLUMBUS, GEORGIA 31901-2101

Overview: Produces broadwoven synthetic fabric. Established: 1996. Ownership: Public.

Key Statistics: Annual Sales: $8.6 million. Number of Employees: 3,000.

Expertise/Education Sought: Production, administration

International Assignments are available to employees.

Contact: For individuals with previous experience: Human Resources; Phone: 706-641-3140; Fax: 706-641-3158.

KAWNEER COMPANY, INC.
555 GUTHRIDGE COURT
NORCROSS, GEORGIA 30092-3503

Overview: Manufactures architectural metalwork. Established: 1962. Parent Company: Alumax. Ownership: Public.

Key Statistics: Annual Sales: $278 million. Number of Employees: 2,900.

Hiring History: Number of professional employees hired in 1997: 14. 1996: 17.

Expertise/Education Sought: Manufacturing engineering

Alternatives to Full-Time Employment: Paid internships

Company-Paid Training is available.

International Assignments are available to employees.

Contact: For individuals with previous experience: Human Resources Department; Phone: 770-449-5555; Fax: 770-734-1560.

LANIER WORLDWIDE, INC.
2300 PARKLAKE DRIVE, NE
ATLANTA, GEORGIA 30345-2905

Overview: Distributes copiers, facsimile machines, dictation equipment, and other office equipment. Established: 1989. Parent Company: Harris Corporation. Ownership: Public.

Key Statistics: Annual Sales: $1.1 billion. Number of Employees: 7,100.

Expertise/Education Sought: Marketing, accounting, finance, electrical engineering, computer science

Alternatives to Full-Time Employment: Part-time jobs, paid internships

Company-Paid Training is available.

International Assignments are available to employees.

Contact: Ms. Andrea Roberts, Human Resource Specialist; Phone: 770-621-1409; Fax: 770-621-1403.

Company URL: http://www.lanier.com

LAW ENGINEERING AND ENVIRONMENTAL SERVICES
114 TOWNPARK DRIVE, SUITE 500
KENNESAW, GEORGIA 30144-5508

Overview: Provides environmental consulting services. Established: 1995. Ownership: Private.

Key Statistics: Annual Sales: $150.9 million. Number of Employees: 2,398.

Expertise/Education Sought: Civil engineering, engineering consulting

Alternatives to Full-Time Employment: Paid internships
International Assignments are available to employees.
Contact: Human Relations Department; Phone: 770-421-3657; Fax: 770-421-3666.

MICROBILT CORPORATION
1640 AIRPORT ROAD, SUITE 115
KENNESAW, GEORGIA 30144

Overview: Provides data processing services. Established: 1989. Parent Company: First Data Corporation. Ownership: Public.
Key Statistics: Annual Sales: $35 million. Number of Employees: 515.
Hiring History: Number of professional employees hired in 1997: 100. 1996: 100.
Expertise/Education Sought: Sales, accounting, marketing, finance, electrical engineering, mechanical engineering, customer service and support
Alternatives to Full-Time Employment: Part-time jobs, paid internships
Company-Paid Training is available.
Contact: Human Resource Department; Phone: 770-218-5000; Fax: 770-218-4998. Job Hotline: 770-218-4989.
Company URL: http://www.microbilt.com

MORRISON'S RESTAURANTS, INC.
4893 RIVERDALE ROAD, SUITE 260
ATLANTA, GEORGIA 30337-6087

Overview: Operates chain of family restaurants. Established: 1995. Ownership: Public.
Key Statistics: Annual Sales: $249.6 million. Number of Employees: 7,800.
Expertise/Education Sought: Management, marketing, law, construction, research and development, finance, human resources
Contact: For individuals with previous experience: Human Resources; Phone: 770-991-0351; Fax: 770-991-2702.

NATIONAL LINEN SERVICE
1420 PEACHTREE STREET, NE
ATLANTA, GEORGIA 30309

Overview: Provides linen rental, lighting, and industrial chemicals services. Established: 1928. Parent Company: National Service Industries. Ownership: Public.
Key Statistics: Annual Sales: $2 billion. Number of Employees: 20,600.
Hiring History: Number of professional employees hired in 1997: 300. 1996: 1,000.
Expertise/Education Sought: Sales, customer service and support, human resources, operations, marketing, production management, engineering, management
Alternatives to Full-Time Employment: Paid internships
Company-Paid Training is available.
Contact: Mr. Craig Lloyd, Director of Staffing; Phone: 404-853-6259; Fax: 404-853-6031.

PONY EXPRESS DELIVERY SERVICES, INC.
6165 BARFIELD ROAD SUITE 200
ATLANTA, GEORGIA 30328

Overview: Provides courier services. Established: 1987. Parent Company: Borg-Warner Security Corporation. Ownership: Public.
Key Statistics: Annual Sales: $140 million. Number of Employees: 3,800.
Hiring History: Number of professional employees hired in 1997: 2,500. 1996: 3,000.
Expertise/Education Sought: Transportation, sales
Company-Paid Training is available.
Contact: For individuals with previous experience: Human Resources; Phone: 404-847-3120; Fax: 404-847-3175. E-mail: steve_forman@ponyexp.com.

POWERTEL, INC.
1233 O G SKINNER DRIVE
WEST POINT, GEORGIA 31833-1789

Overview: Provides cellular phone services. Established: 1996. Ownership: Public.
Key Statistics: Annual Sales: $21.2 million. Number of Employees: 300.
Hiring History: Number of professional employees hired in 1996: 400.
Expertise/Education Sought: Marketing, sales, engineering, information systems, customer service and support
Company-Paid Training is available.
Contact: For individuals with previous experience: Human Resources; Phone: 706-645-2000; Fax: 706-645-6476.
Company URL: http://www.powertel.com

RACETRAC PETROLEUM, INC.
PO BOX 105035
ATLANTA, GEORGIA 30348

Overview: Gasoline/convenience store retailer. Established: 1934.
Expertise/Education Sought: Information systems, civil engineering, architectural engineering, construction engineering, risk management
Alternatives to Full-Time Employment: Part-time jobs, paid internships
Contact: For college students and recent graduates: Ms. Sue Jackson, Human Resources Manager; Phone: 770-431-7600, ext. 1205; Fax: 770-319-7944. For individuals with previous experience: Ms. Sue Jackson, Human Resources Manager; Phone: 770-431-7600, ext. 1205; Fax: 770-319-7944.
Company URL: http://www.racetrac.com

RIVERWOOD INTERNATIONAL CORPORATION
3350 CUMBERLAND CIRCLE, SE, SUITE 1400
ATLANTA, GEORGIA 30339-3340

Overview: Manufactures boxboard. Established: 1995. Ownership: Private.
Key Statistics: Annual Sales: $837.8 million. Number of Employees: 6,400.
Expertise/Education Sought: Accounting, engineering, management, finance, sales
Alternatives to Full-Time Employment: Paid internships
International Assignments are available to employees.
Contact: For individuals with previous experience: Mr. Richard Banks, Human Resources Assistant; Phone: 770-644-3000; Fax: 770-644-2962.

ROSSER INTERNATIONAL, INC.
524 WEST PEACHTREE STREET, NW
ATLANTA, GEORGIA 30308-3512

Overview: Provides architectural services. Established: 1947. Ownership: Private.
Key Statistics: Annual Sales: $23.1 million. Number of Employees: 250.
Hiring History: Number of professional employees hired in 1996: 30.
Expertise/Education Sought: Electrical engineering, mechanical engineering, architecture, civil engineering
Alternatives to Full-Time Employment: Paid internships
Company-Paid Training is available.
International Assignments are available to employees.
Contact: Mr. Michael Fields, Recruiting Coordinator; Phone: 404-876-3800; Fax: 404-888-6863. E-mail: mfield@rosser.com.
Company URL: http://www.rosser.com

ROSS SYSTEMS, INC.
2 CONCOURSE PARKWAY 2800
ATLANTA, GEORGIA 30328

Overview: Supplies midrange computer software to the financial, human resources, manufacturing, and distribution industries. Established: 1972. Ownership: Public.

Ross Systems, Inc. (continued)

Key Statistics: Annual Sales: $78.8 million. Number of Employees: 440.

Expertise/Education Sought: Accounting, finance, design engineering, business administration, computer programming, consulting, software engineering/development

Alternatives to Full-Time Employment: Part-time jobs

Contact: Mr. Gary Brown, Director of Recruiting; Phone: 770-351-9600, ext. 3036; Fax: 770-351-0036. E-mail: gary_braun@rossinc.com.

Company URL: http://www.rossinc.com

SANDWELL, INC.

2690 CUMBERLAND PARKWAY, NW SUITE 300
ATLANTA, GEORGIA 30339-3913

Overview: Provides engineering consulting services for pulp and paper products. Established: 1986. Parent Company: Sandwell, Inc. Ownership: Private.

Key Statistics: Annual Sales: $25 million. Number of Employees: 220.

Hiring History: Number of professional employees hired in 1997: 7. 1996: 10.

Expertise/Education Sought: Process engineering, structural engineering, mechanical engineering, electrical engineering

Alternatives to Full-Time Employment: Part-time jobs

Company-Paid Training is available.

International Assignments are available to employees.

Contact: Mr. Richard Low, Manager of Human Resources; Phone: 770-433-9336; Fax: 770-434-4995. E-mail: dlow@sandwell.com.

Company URL: http://www.sandwell.com

SCHWEITZER-MAUDUIT INTERNATIONAL, INC.

100 NORTH POINT CENTER EAST, SUITE 600
ALPHARETTA, GEORGIA 30022

Overview: Manufactures cigarette papers and wrapping paper. Established: 1995. Ownership: Public.

Key Statistics: Annual Sales: $471.3 million. Number of Employees: 2,400.

Hiring History: Number of professional employees hired in 1997: 15. 1996: 35.

Expertise/Education Sought: Manufacturing

Alternatives to Full-Time Employment: Paid internships

Company-Paid Training is available.

International Assignments are available to employees.

Contact: For individuals with previous experience: Human Resources; Phone: 770-569-4207.

SIEMENS ENERGY AND AUTOMATION, INC.

3333 OLD MILTON PARKWAY
ALPHARETTA, GEORGIA 30005

Mailing Address: PO Box 89000, Atlanta, GA 30356-9000

Overview: Manufactures switchgears, electric motors, generators, control equipment, and transformers. Established: 1978. Parent Company: Siemens Corporation. Ownership: Private.

Key Statistics: Annual Sales: $561.4 million. Number of Employees: 8,000.

Expertise/Education Sought: Manufacturing engineering

Alternatives to Full-Time Employment: Paid internships

Company-Paid Training is available.

Contact: For individuals with previous experience: Ms. Tina Scott, Personnel Assistant; Phone: 770-751-2000; Fax: 770-751-4333. E-mail: tina.scott@sea.siemens.com.

Company URL: http://www.sea.siemens.com

SKC, INC.

1147 WEST AVENUE, SW
CONYERS, GEORGIA 30012

Overview: Manufactures plastics film and sheet. Established: 1996. Parent Company: SUNKYONG. Ownership: Private.

Key Statistics: Annual Sales: $250 million. Number of Employees: 30.

Hiring History: Number of professional employees hired in 1997: 40. 1996: 15.

Expertise/Education Sought: Manufacturing, administration

International Assignments are available to employees.

Contact: For individuals with previous experience: Miss Kathy Dobbs, Human Resources Recruiter; Phone: 770-922-9997; Fax: 770-922-8889.

SNAPPER

535 MACON ROAD
MCDONOUGH, GEORGIA 30253-3529

Overview: Manufactures lawn and garden equipment. Established: 1982. Parent Company: Metro Media International Group, Inc. Ownership: Private.

Key Statistics: Annual Sales: $120 million. Number of Employees: 1,200.

Hiring History: Number of professional employees hired in 1996: 460.

Expertise/Education Sought: Engineering, industrial engineering, mechanical engineering

Alternatives to Full-Time Employment: Paid internships

Contact: For individuals with previous experience: Ms. Emily James, Director of Human Resources; Phone: 770-954-9000; Fax: 770-954-2583.

SOUTHERN COMPANY

270 PEACHTREE STREET, SUITE 1700
ATLANTA, GEORGIA 30303

Overview: Electric utility. Established: 1945. Ownership: Public.

Key Statistics: Assets: $4.4 billion. Number of Employees: 13,346.

Expertise/Education Sought: Electrical engineering, mechanical engineering, industrial engineering, civil engineering

Alternatives to Full-Time Employment: Paid internships

Company-Paid Training is available.

International Assignments are available to employees.

Contact: For college students and recent graduates: Ms. Mary Gordon, Recruiting Coordinator; Phone: 404-506-0449; Fax: 404-506-0455. For individuals with previous experience: Human Resources Department; Phone: 404-506-0449; Fax: 404-506-0455.

Company URL: http://www.southernco.com

SOUTHERN COMPANY ENERGY MARKETING, LP

900 ASHWOOD PARKWAY, SUITE 500
ATLANTA, GEORGIA 30338-4780

Overview: Natural gas distributor. Established: 1997. Parent Company: Southern Company. Ownership: Private.

Key Statistics: Annual Sales: $79.7 million. Number of Employees: 300.

Expertise/Education Sought: Finance, marketing, engineering

Alternatives to Full-Time Employment: Paid internships

Company-Paid Training is available.

International Assignments are available to employees.

Contact: Human Resources; Phone: 770-379-6905; Fax: 770-379-7758.

SOUTHERN ENGINEERING COMPANY

1800 PEACHTREE STREET
ATLANTA, GEORGIA 30367-8301

Overview: Provides architectural and technical engineering products and services, specializing in electric power. Established: 1945. Ownership: Private.

Key Statistics: Number of Employees: 180.

Hiring History: Number of professional employees hired in 1997: 15.

Expertise/Education Sought: Electrical engineering, civil engineering, design engineering, architecture, technical engineering

Contact: Mr. Ray Drayton, Human Resources Director; Fax: 404-351-1196. E-mail: rdrayton@soeng.com.

Company URL: http://www.soeng.com

CORPORATE STATEMENT

Southern Engineering Company provides services in a variety of electric utility specialties, including transmission and distribution system planning and design, substation design, system mapping, and financial and rate studies. The firm also provides electric utility applications software for distribution engineering, mapping, and outage management and restoration. Other services include civil engineering and architectural design.

SUNDS DEFIBRATOR, INC.

2900 COURTYARDS DRIVE
NORCROSS, GEORGIA 30071-1554

Overview: Manufactures equipment for paper mills. Established: 1994. Parent Company: Sunds Defibrator Industries, AB. Ownership: Private.

Key Statistics: Annual Sales: $125 million. Number of Employees: 105.

Hiring History: Number of professional employees hired in 1996: 10.

Expertise/Education Sought: Accounting, mechanical engineering, finance, pulp and paper science, design engineering, electrical engineering, chemical engineering

Company-Paid Training is available.

International Assignments are available to employees.

Contact: Mr. Stewart Lacy, Vice President of Human Resources; Phone: 770-263-7863; Fax: 770-441-9652.

TEXTILE RUBBER AND CHEMICAL COMPANY

1300 TIARCO DRIVE, SW
DALTON, GEORGIA 30720-9209

Overview: Produces fabricated rubber products. Established: 1951. Parent Company: Tillotson Rubber. Ownership: Private.

Key Statistics: Annual Sales: $187.5 million. Number of Employees: 385.

Expertise/Education Sought: Chemical engineering, industrial engineering, mechanical engineering

Alternatives to Full-Time Employment: Paid internships

International Assignments are available to employees.

Contact: For individuals with previous experience: Mr. Eddie Fox, Assistant Plant Manager; Phone: 706-277-1300, ext. 2298; Fax: 706-277-3738.

TOTAL SYSTEM SERVICES, INC.

1300 6TH AVENUE
COLUMBUS, GEORGIA 31901

Overview: Provides credit card processing services. Established: 1983. Parent Company: Synovus. Ownership: Public.

Key Statistics: Annual Sales: $159.1 million. Number of Employees: 2,500.

Hiring History: Number of professional employees hired in 1996: 1,200.

Expertise/Education Sought: Program analysis, software engineering/development, systems analysis, computer programming, production, technical writing, training and development

Alternatives to Full-Time Employment: Paid internships

Company-Paid Training is available.

International Assignments are available to employees.

Contact: For college students and recent graduates: Ms. Dora Ditchfield, Employment Coordinator; Phone: 706-641-3984; Fax: 706-649-5797. For individuals with previous experience: Mrs. Helen Johnson, Director of Employment Service; Phone: 706-649-4708. Job Hotline: 706-649-4758.

Company URL: http://www.synovus.com

TURNER BROADCASTING SYSTEM, INC.

1 CNN CENTER, NW
ATLANTA, GEORGIA 30303

Overview: Operates cable television networks. Established: 1996. Parent Company: Superstation, Inc. Ownership: Private.

Key Statistics: Annual Sales: $803.8 million. Number of Employees: 7,000.

Expertise/Education Sought: Television broadcasting, journalism, electronics/electronics engineering, film/film production, journalism, sales, accounting, marketing

Alternatives to Full-Time Employment: Internships

International Assignments are available to employees.

Contact: For college students and recent graduates: Mr. Chris Kerr, Recruiting Assistant. For individuals with previous experience: Human Resources Department; Phone: 404-827-1500.

VENCARE ANCILLARY SERVICES

1105 SANCTUARY PARKWAY, SUITE 100
ALPHARETTA, GEORGIA 30004-4738

Overview: Provides management services for personal care facilities. Established: 1997. Ownership: Public.

Key Statistics: Annual Sales: $245.5 million. Number of Employees: 7,297.

Expertise/Education Sought: Therapy, accounting, marketing, sales

Contact: Human Resource Department, 3300 Aegon Center, 400 West Market, Street, Louisville, Kentucky, 40202; Phone: 502-596-7300; Fax: 502-596-4726.

YOKOGAWA CORPORATION OF AMERICA

2 DART ROAD
NEWNAN, GEORGIA 30265

Overview: Manufactures process control instruments. Established: 1989. Parent Company: Yokogawa Electric Corporation. Ownership: Private.

Key Statistics: Annual Sales: $167.2 million. Number of Employees: 753.

Hiring History: Number of professional employees hired in 1997: 5.

Expertise/Education Sought: Manufacturing engineering, electrical engineering, design engineering

Alternatives to Full-Time Employment: Part-time jobs

Company-Paid Training is available.

Contact: For individuals with previous experience: Ms. Dawn Walburg, Human Resources Manager; Phone: 770-253-7000; Fax: 770-253-8174.

Company URL: http://www.yca.com

IDAHO

ALBERTSON'S, INC.

250 PARK CENTER BOULEVARD
BOISE, IDAHO 83706-3940

Overview: Operates chain of food and drug stores. Established: 1939. Ownership: Public.

Key Statistics: Annual Sales: $13.8 billion. Number of Employees: 88,000.

Expertise/Education Sought: Accounting, architecture, civil engineering, law, computer programming, mechanical engineering

Alternatives to Full-Time Employment: Paid internships

Company-Paid Training is available.

Contact: Ms. Natalie Bridges, Manager of Recruiting; Phone: 208-395-6397; Fax: 208-395-6024. Job Hotline: 208-395-6422.

MICRON TECHNOLOGY, INC.

8000 SOUTH FEDERAL WAY
BOISE, IDAHO 83716

Overview: Manufactures computer chips. Established: 1978. Ownership: Public.

Key Statistics: Annual Sales: $3.5 billion. Number of Employees: 12,200.

Hiring History: Number of professional employees hired in 1997: 1,000. 1996: 1,000.

Expertise/Education Sought: Electrical engineering, chemical engineering, design engineering, chemistry, physics, materials science

Alternatives to Full-Time Employment: Part-time jobs, paid internships

Company-Paid Training is available.

International Assignments are available to employees.

Contact: For college students and recent graduates: Ms. Michelle Rumold, College Relations Recruiter; Phone: 208-368-4000; Fax: 208-368-4641. E-mail: mrumold@micron.com. For individuals with previous experience: Mr. Brad Mabe, Human Resources Manager; Phone: 801-767-5013; Fax: 801-767-5003. E-mail: mabe@micron.com. Job Hotline: 800-932-4991.

Company URL: http://www.micron.com

ORE-IDA FOODS, INC.

220 WEST PARKCENTER BOULEVARD
BOISE, IDAHO 83706-3968

Overview: Manufactures frozen potato, vegetable, and dinner products. Established: 1965. Parent Company: H. J. Heinz Company. Ownership: Public.

Key Statistics: Annual Sales: $1.2 billion. Number of Employees: 3,250.

Hiring History: Number of professional employees hired in 1997: 25. 1996: 100.

Expertise/Education Sought: Marketing, engineering, quality control, computer programming

Alternatives to Full-Time Employment: Part-time jobs, paid internships

Company-Paid Training is available.

Contact: Ms. Karey Bertrand, Manager of Human Resources; Phone: 208-383-6330; Fax: 208-383-6902. E-mail: employment@oreida.com. Job Hotline: 208-383-6167.

Company URL: http://www.oreida.com

ILLINOIS

ABBOTT LABORATORIES

1 ABBOTT PARK ROAD
ABBOTT PARK, ILLINOIS 60064-3500

Overview: Develops, manufactures, and markets pharmaceutical and nutritional products. Established: 1888. Ownership: Public.

Key Statistics: Annual Sales: $11 billion. Number of Employees: 52,817.

Hiring History: Number of professional employees hired in 1996: 1,639.

Expertise/Education Sought: Biological sciences, chemistry, pharmacology, chemical engineering, mechanical engineering, electrical engineering, finance, accounting, human resources

Alternatives to Full-Time Employment: Part-time jobs, paid internships

International Assignments are available to employees.

Contact: For individuals with previous experience: College Relations; Phone: 708-937-9016.

Company URL: http://www.abbott.com

ADVANCE DIAL COMPANY

940 INDUSTRIAL DRIVE
ELMHURST, ILLINOIS 60126

Overview: Manufactures custom injection molded plastics. Established: 1988. Parent Company: Anderson Industries, Inc. Ownership: Private.

Key Statistics: Annual Sales: $40 million. Number of Employees: 493.

Expertise/Education Sought: Injection molding, mechanical engineering, accounting, data processing, finance

Alternatives to Full-Time Employment: Internships

Company-Paid Training is available.

Contact: Mr. Bud Feigenbaum, Human Resources Director; Phone: 660-993-1700; Fax: 660-993-0372.

ALLIANT FOOD SERVICE, INC.

1 PARKWAY NORTH
DEERFIELD, ILLINOIS 60015-2532

Overview: Provides food service products to hospitals and restaurants. Established: 1995. Ownership: Private.

Key Statistics: Annual Sales: $4.5 billion. Number of Employees: 9,700.

Expertise/Education Sought: Finance, marketing

Alternatives to Full-Time Employment: Paid internships

Company-Paid Training is available.

Contact: Mr. Brian Glancy, Director of Human Resources; Phone: 847-405-8500; Fax: 847-405-8813.

Company URL: http://www.alliant.com

AMERITECH CELLULAR SERVICES

2000 WEST AMERITECH CENTER
HOFFMAN ESTATES, ILLINOIS 60195-5000

Overview: Provides cellular communications services. Established: 1983. Parent Company: Ameritech Corporation. Ownership: Public.

Key Statistics: Annual Sales: $200 million. Number of Employees: 4,400.

Hiring History: Number of professional employees hired in 1997: 400. 1996: 300.

Expertise/Education Sought: Accounting, finance, data processing, electronics/electronics engineering, electrical engineering, customer service and support

Alternatives to Full-Time Employment: Paid internships

Contact: For individuals with previous experience: Human Resources Department; Phone: 847-765-5781; Fax: 847-765-5333. Job Hotline: 800-HOT-JOBS.

AMERITECH NETWORK SERVICES
2000 AMERITECH CENTER DRIVE
HOFFMAN ESTATES, ILLINOIS 60196

Overview: Provides telecommunications services. Established: 1984. Parent Company: Ameritech Informations Technologies. Ownership: Public.

Key Statistics: Annual Sales: $2.3 billion. Number of Employees: 8,579.

Hiring History: Number of professional employees hired in 1997: 203.

Expertise/Education Sought: Technical, engineering, operations, management

Alternatives to Full-Time Employment: Part-time jobs, paid internships

Company-Paid Training is available.

Contact: Mr. Fred Mertes, Staffing Manager, 45 Erieview Plaza, Cleveland, Ohio, 44114-1814; Phone: 216-822-6038; Fax: 216-822-6752. E-mail: fred.j.mertes@ameritech.com.

Company URL: http://www.ameritech.com

AMERITECH SERVICE, INC.
2000 WEST AMERITECH CENTER DRIVE
HOFFMAN, ILLINOIS 60196

Overview: Provides telecommunications services. Established: 1984. Ownership: Public.

Key Statistics: Annual Sales: $3.2 billion. Number of Employees: 12,026.

Hiring History: Number of professional employees hired in 1996: 30.

Expertise/Education Sought: Engineering science, telecommunications, computer science

Alternatives to Full-Time Employment: Paid internships

Company-Paid Training is available.

International Assignments are available to employees.

Contact: Human Resources Department; Fax: 419-422-6637.

Company URL: http://www.ameritech.com

ANDREW CORPORATION
10500 153RD STREET
ORLAND PARK, ILLINOIS 60462-3099

Overview: Manufactures telecommunications equipment. Established: 1937. Ownership: Public.

Key Statistics: Annual Sales: $793.6 million. Number of Employees: 4,622.

Expertise/Education Sought: Electrical engineering, mechanical engineering, accounting, sales, computer programming, law, human resources, structural engineering

Alternatives to Full-Time Employment: Part-time jobs, paid internships

International Assignments are available to employees.

Contact: Mr. Roger Blaylock, Director of Human Resources; Phone: 708-349-3300; Fax: 708-873-2166. Job Hotline: 708-349-5277.

Company URL: http://www.andrew.com

APPLIED SYSTEMS, INC.
200 APPLIED PARKWAY
UNIVERSITY PARK, ILLINOIS 60466

Overview: Provides computer systems and integration design services. Established: 1976. Ownership: Private.

Key Statistics: Annual Sales: $65 million. Number of Employees: 800.

Expertise/Education Sought: Software engineering/development, computer programming, program analysis

Alternatives to Full-Time Employment: Part-time jobs, paid internships

Company-Paid Training is available.

International Assignments are available to employees.

Contact: Ms. Janet Van Haren, Director of Human Resources; Phone: 708-534-5575; Fax: 708-534-5943.

Company URL: http://www.appliedsystems.com

ARCHER DANIELS MIDLAND COMPANY
4666 EAST FARIES PARKWAY
DECATUR, ILLINOIS 62526

Overview: Processes, markets, and transports agricultural goods. Established: 1902. Ownership: Public.

Key Statistics: Annual Sales: $13.9 billion. Number of Employees: 17,160.

Hiring History: Number of professional employees hired in 1996: 90.

Expertise/Education Sought: Business administration, marketing, finance, business management, engineering

Alternatives to Full-Time Employment: Part-time jobs, paid internships

Company-Paid Training is available.

International Assignments are available to employees.

Contact: Ms. Sheila Witts-Mannweiler, Director of Personnel; Phone: 217-424-5200.

Company URL: http://www.admworld.com

BELL & HOWELL OPERATING COMPANY
5215 OLD ORCHARD ROAD
SKOKIE, ILLINOIS 60077-1035

Overview: Manufactures mail-handling equipment and provides microfilm and CD-ROM publishing systems and services. Established: 1988. Ownership: Public.

Key Statistics: Annual Sales: $902.8 million. Number of Employees: 6,110.

Hiring History: Number of professional employees hired in 1996: 200.

Expertise/Education Sought: Finance, systems integration, law, computer science, accounting

International Assignments are available to employees.

Contact: For individuals with previous experience: Ms. Lori Wallach, Recruiter; Phone: 847-470-7100. E-mail: lwallach@corp.bellhowell.com. Job Hotline: 847-982-6653.

Company URL: http://www.bellhowell.com

BERGSTROM, INC.
2390 BLACKHAWK ROAD
ROCKFORD, ILLINOIS 61109-3602

Overview: Manufactures heaters, ventilators, and air conditioners for motor vehicles. Established: 1960. Ownership: Private.

Key Statistics: Annual Sales: $25 million. Number of Employees: 400.

Hiring History: Number of professional employees hired in 1996: 15.

Expertise/Education Sought: HVAC engineering, manufacturing engineering, mechanical engineering, electrical engineering

Alternatives to Full-Time Employment: Paid internships

Company-Paid Training is available.

International Assignments are available to employees.

Contact: Ms. Elaine Markou, Director of Human Resources; Phone: 815-874-7821, ext. 352; Fax: 815-874-0932.

Company URL: http://www.bergstrominc.com

BINKS SAMES CORPORATION
9201 WEST BELMONT AVENUE
FRANKLIN PARK, ILLINOIS 60131

Overview: Manufactures finishing and coating equipment. Established: 1890. Ownership: Public.

Key Statistics: Annual Sales: $296.7 million. Number of Employees: 1,293.

Expertise/Education Sought: Manufacturing engineering, design engineering

Alternatives to Full-Time Employment: Part-time jobs

International Assignments are available to employees.

Contact: For individuals with previous experience: Mr. James Lindquist, Director of Personnel; Phone: 847-671-3000, ext. 201; Fax: 847-671-6489.

Company URL: http://www.binks.com

BULK LIFT INTERNATIONAL, INC.

1013 TAMARAC
CARPENTERSVILLE, ILLINOIS 60110

Overview: Manufactures woven polypropylene storage bags. Established: 1978. Ownership: Private.

Key Statistics: Annual Sales: $22 million. Number of Employees: 220.

Hiring History: Number of professional employees hired in 1997: 20.

Expertise/Education Sought: Administration, operations, sales

International Assignments are available to employees.

Contact: Ms. Brenda Kardis, Human Resource Manager; Phone: 847-428-6059, ext. 210; Fax: 847-428-0180.

CARA CORPORATION

1900 SPRING ROAD, SUITE 450
OAK BROOK, ILLINOIS 60523-1481

Overview: Provides computer programming services. Established: 1996. Parent Company: Affiliated Computer Services. Ownership: Private.

Key Statistics: Annual Sales: $54 million. Number of Employees: 520.

Hiring History: Number of professional employees hired in 1997: 400. 1996: 300.

Expertise/Education Sought: Business management, computer programming, computer science, software engineering/development, data processing, design engineering

Alternatives to Full-Time Employment: Paid internships

Company-Paid Training is available.

Contact: Human Resources Department; Phone: 630-990-2272; Fax: 630-368-2800.

Company URL: http://www.caracorp.com

CATERPILLAR, INC.

100 NORTHEAST ADAMS STREET
PEORIA, ILLINOIS 61629-1440

Overview: Manufactures farm and construction equipment. Established: 1925. Ownership: Public.

Key Statistics: Annual Sales: $16.5 billion. Number of Employees: 57,412.

Hiring History: Number of professional employees hired in 1997: 1,000. 1996: 525.

Expertise/Education Sought: Accounting, computer science, information systems, mechanical engineering, electrical engineering

Alternatives to Full-Time Employment: Paid internships

Company-Paid Training is available.

Contact: Mr. Phil Jones, Corporate Employment Manager; Phone: 309-675-5923; Fax: 309-675-6476.

Company URL: http://www.CAT.com

CENTRAL ILLINOIS LIGHT COMPANY

300 LIBERTY STREET
PEORIA, ILLINOIS 61602

Overview: Electric utility. Established: 1985. Parent Company: Cilcorp. Ownership: Public.

Key Statistics: Annual Sales: $628.4 million. Number of Employees: 2,121.

Hiring History: Number of professional employees hired in 1997: 10. 1996: 10.

Expertise/Education Sought: Program analysis, customer service and support, computer programming, electrical engineering, marketing

Alternatives to Full-Time Employment: Paid internships

Company-Paid Training is available.

Contact: Senior Employment Administrator; Phone: 309-677-5448; Fax: 309-677-5432. Job Hotline: 309-672-1050.

Company URL: http://www.cilco.com

CERION TECHNOLOGIES

1401 INTERSTATE DRIVE
CHAMPAIGN, ILLINOIS 61821

Overview: Manufactures precision machined products. Established: 1996. Ownership: Public.

Key Statistics: Annual Sales: $36.5 million. Number of Employees: 400.

Hiring History: Number of professional employees hired in 1997: 197. 1996: 150.

Expertise/Education Sought: Accounting, finance, engineering, customer service and support, quality control, production

Alternatives to Full-Time Employment: Paid internships

Company-Paid Training is available.

Contact: Human Resources.

Company URL: http://www.cerion.com

CHICAGO BRIDGE & IRON COMPANY NV

1501 NORTH DIVISION STREET
PLAINFIELD, ILLINOIS 60544-8984

Overview: Fabricates vessels and tanks. Established: 1997. Ownership: Public.

Key Statistics: Annual Sales: $559.9 million. Number of Employees: 5,950.

Expertise/Education Sought: Civil engineering, mechanical engineering

Alternatives to Full-Time Employment: Internships

Company-Paid Training is available.

International Assignments are available to employees.

Contact: For individuals with previous experience: Human Resources Department; Phone: 815-439-6000; Fax: 815-439-6010.

Company URL: http://www.chicago-bridge.com

CHICAGO TRANSIT AUTHORITY

MERCHANDISE MART PLAZA, PO BOX 3555
CHICAGO, ILLINOIS 60654

Overview: Provides local and suburban transit services. Established: 1945. Ownership: Private.

Key Statistics: Annual Sales: $404.3 million. Number of Employees: 12,500.

Hiring History: Number of professional employees hired in 1997: 1,000.

Expertise/Education Sought: Planning, urban and regional planning, accounting, budgeting, law, engineering, transportation

Alternatives to Full-Time Employment: Part-time jobs, paid internships

Company-Paid Training is available.

Contact: For individuals with previous experience: Mr. Thomas Czech, Manager of Personnel, PO Box 3555, Chicago, Illinois, 60654-0555; Phone: 312-664-7200, ext. 3470; Fax: 312-661-1297.

Company URL: http://www.transitchicago.com

COMED, UNICOM COMPANY

1 FIRST NATIONAL PLAZA
CHICAGO, ILLINOIS 60603

Mailing Address: PO Box 767, Chicago, IL 60690-0767

Overview: Electric utility. Established: 1994. Parent Company: Unicom Corporation. Ownership: Public.

Key Statistics: Annual Sales: $6.9 billion. Number of Employees: 16,845.

Hiring History: Number of professional employees hired in 1997: 750.

Expertise/Education Sought: Electrical engineering, mechanical engineering, nuclear engineering, computer science

Alternatives to Full-Time Employment: Paid internships

Company-Paid Training is available.

Contact: Mr. Howard Nelson, Professional Recruitment Director; Phone: 312-394-4489; Fax: 312-394-2367. E-mail: resume@ucm.com.

Company URL: http://www.ucm.com

CYBORG SYSTEMS, INC.
2 NORTH RIVERSIDE PLAZA, SUITE 1200
CHICAGO, ILLINOIS 60606

Overview: Develops software. Established: 1974. Ownership: Private.
Key Statistics: Annual Sales: $49.9 million. Number of Employees: 391.
Hiring History: Number of professional employees hired in 1997: 100.
Expertise/Education Sought: Computer programming, program analysis, computer science, project management, information systems, consulting, sales
Alternatives to Full-Time Employment: Part-time jobs
Company-Paid Training is available.
International Assignments are available to employees.
Contact: For individuals with previous experience: Human Resources; Phone: 312-454-1865; Fax: 312-454-9182.
Company URL: http://www.cyborg.com

DUKANE CORPORATION
2900 DUKANE DRIVE
SAINT CHARLES, ILLINOIS 60174-3348

Overview: Provides diversified communications, audiovisual, and ultrasonic welding services. Established: 1922. Ownership: Private.
Key Statistics: Annual Sales: $78.1 million. Number of Employees: 664.
Expertise/Education Sought: Audio engineering, electrical engineering, hardware engineering, electronics/electronics engineering, computer engineering, information systems, mechanical engineering
Alternatives to Full-Time Employment: Paid internships
Company-Paid Training is available.
Contact: For college students and recent graduates: Mr. William McNew, Professional Staffing and Training; Phone: 708-584-2300; Fax: 708-584-2370. For individuals with previous experience: Mr. Robert Scarlet, Vice President of Human Resources; Phone: 708-584-2300; Fax: 708-584-2370.

EXHIBITGROUP GILTSPUR, INC.
200 GARY AVENUE
ROSELLE, ILLINOIS 60172-1681

Overview: Provides exhibit construction services. Established: 1996. Parent Company: VIAD. Ownership: Private.
Key Statistics: Annual Sales: $22.2 million. Number of Employees: 1,700.
Expertise/Education Sought: Design, project management, operations, account management
Alternatives to Full-Time Employment: Paid internships
Contact: For individuals with previous experience: Mr. Gary Chapman, Human Resource Generalist; Phone: 630-351-7723; Fax: 630-351-7861. E-mail: gchapman@e-g.com.

FSC CORPORATION
13101 SOUTH PULASKI ROAD
CHICAGO, ILLINOIS 60803

Overview: Owns and operates paper mills; manufactures newsprint. Established: 1993. Parent Company: Dillon Read. Ownership: Private.
Key Statistics: Annual Sales: $60 million. Number of Employees: 150.
Hiring History: Number of professional employees hired in 1997: 5. 1996: 5.
Expertise/Education Sought: Process engineering, engineering
Alternatives to Full-Time Employment: Paid internships
Contact: Mr. Ken Zuncic, Director of Human Resources; Phone: 708-389-8520; Fax: 708-389-8237.

FUJISAWA USA, INC.
3 PARKWAY NORTH
DEERFIELD, ILLINOIS 60015

Overview: Manufactures pharmaceuticals. Established: 1989. Parent Company: Fujisawa Pharmaceutical Company, Ltd. Ownership: Private.
Key Statistics: Annual Sales: $288 million. Number of Employees: 1,000.
Expertise/Education Sought: Packaging engineering, chemical engineering, mechanical engineering, computer programming, marketing, distribution, research and development, quality control, purchasing
Alternatives to Full-Time Employment: Internships
Company-Paid Training is available.
Contact: For college students and recent graduates: Ms. Kathy Thissen, Human Resources Representative; Phone: 847-317-8800; Fax: 847-317-1245. For individuals with previous experience: Mr. Lance Williams, Director Employee Relations; Phone: 847-317-8800; Fax: 847-317-1245.
Company URL: http://www.fujisawausa.com

GALILEO INTERNATIONAL
9700 WEST HIGGINS ROAD, SUITE 400
ROSEMONT, ILLINOIS 60018

Overview: Provides electronic global distribution services for the travel industry. Established: 1997. Ownership: Public.
Key Statistics: Annual Sales: $1.1 billion. Number of Employees: 1,909.
Hiring History: Number of professional employees hired in 1997: 125.
Expertise/Education Sought: Computer operations
Alternatives to Full-Time Employment: Paid internships
Company-Paid Training is available.
International Assignments are available to employees.
Contact: Mr. Paul Maceika, Human Resources Generalist; Phone: 847-518-4048; Fax: 847-518-4917.

GAS RESEARCH INSTITUTE
8600 WEST BRYN MAWR AVENUE
CHICAGO, ILLINOIS 60631

Overview: Nonprofit natural gas research institute.
Expertise/Education Sought: Contracts administration/management, auditing, technology, environmental, chemical engineering
International Assignments are available to employees.
Contact: For individuals with previous experience: Human Resources; Phone: 312-399-4614.

GBC
1 GBC PLAZA
NORTHBROOK, ILLINOIS 60062-4195

Overview: Manufactures and distributes binding, shredding, and printing equipment. Established: 1968. Parent Company: Lane Industries, Inc. Ownership: Private.
Key Statistics: Annual Sales: $536.8 million. Number of Employees: 3,543.
Hiring History: Number of professional employees hired in 1996: 250.
Expertise/Education Sought: Management information systems, accounting, sales, marketing, finance, engineering
Company-Paid Training is available.
International Assignments are available to employees.
Contact: For individuals with previous experience: Mr. Gary Smith, Human Resources Representative; Phone: 847-291-6037; Fax: 847-498-3345. E-mail: dbailey@gbc.com. Job Hotline: 847-291-5627.
Company URL: http://www.gbc.com

HELENE CURTIS, INC.
325 NORTH WELLS STREET
CHICAGO, ILLINOIS 60610-4705

OVERVIEW: Manufactures and distributes cosmetics and fragrances. Established: 1996. Parent Company: Unilever Corporation. Ownership: Private.

KEY STATISTICS: Annual Sales: $1.3 billion. Number of Employees: 2,500.

EXPERTISE/EDUCATION SOUGHT: Manufacturing engineering, mechanical engineering, electrical engineering, chemical engineering, industrial engineering, engineering

ALTERNATIVES TO FULL-TIME EMPLOYMENT: Paid internships

COMPANY-PAID TRAINING is available.

CONTACT: Human Resources Department; Phone: 312-661-0222; Fax: 312-661-2394.

HIRDES FREIGHT, LIMITED
812 THORNDALE AVENUE
BENSENVILLE, ILLINOIS 60106

OVERVIEW: Provides freight transportation services. Established: 1992. Ownership: Private.

HIRING HISTORY: Number of professional employees hired in 1997: 5.

ALTERNATIVES TO FULL-TIME EMPLOYMENT: Paid internships

COMPANY-PAID TRAINING is available.

INTERNATIONAL ASSIGNMENTS are available to employees.

CONTACT: For individuals with previous experience: Human Resources; Phone: 630-860-0040; Fax: 630-860-8480.

J.W. HOBBS CORPORATION
YALE BOULEVARD AND ASH STREET
SPRINGFIELD, ILLINOIS 62705

OVERVIEW: Manufactures lighting fixtures and related components. Established: 1987. Parent Company: British Tire and Rubber Company. Ownership: Private.

KEY STATISTICS: Annual Sales: $45.7 million. Number of Employees: 350.

HIRING HISTORY: Number of professional employees hired in 1996: 15.

EXPERTISE/EDUCATION SOUGHT: Electrical engineering, design engineering

CONTACT: For individuals with previous experience: Human Resources Department; Phone: 217-753-7642; Fax: 217-753-7646.

HOLLISTER, INC.
2000 HOLLISTER DRIVE
LIBERTYVILLE, ILLINOIS 60048

OVERVIEW: Manufactures disposable plastic health products and car products. Established: 1956. Ownership: Private.

KEY STATISTICS: Annual Sales: $250 million. Number of Employees: 2,200.

EXPERTISE/EDUCATION SOUGHT: Mechanical engineering, design engineering, manufacturing engineering, accounting, sales, computer programming

CONTACT: For individuals with previous experience: Human Resources Department.

HU-FRIEDY MANUFACTURING COMPANY
3232 NORTH ROCKWELL STREET
CHICAGO, ILLINOIS 60618

OVERVIEW: Manufactures dental instruments. Established: 1958. Ownership: Private.

KEY STATISTICS: Annual Sales: $21.2 million. Number of Employees: 300.

EXPERTISE/EDUCATION SOUGHT: Design engineering, mechanical engineering, accounting, production

ALTERNATIVES TO FULL-TIME EMPLOYMENT: Part-time jobs, paid internships

COMPANY-PAID TRAINING is available.

INTERNATIONAL ASSIGNMENTS are available to employees.

CONTACT: Human Resources; Phone: 773-975-6100; Fax: 773-975-1683.

COMPANY URL: http://www.hufriedy.com

IBM GLOBAL SERVICES-NETWORK SERVICES
231 NORTH MARTINGALE ROAD
SCHAUMBURG, ILLINOIS 60173

OVERVIEW: Provides voice and data network services. Established: 1992. Parent Company: IBM. Ownership: Private.

KEY STATISTICS: Annual Sales: $342.7 million. Number of Employees: 4,000.

HIRING HISTORY: Number of professional employees hired in 1997: 400. 1996: 300.

EXPERTISE/EDUCATION SOUGHT: Electrical engineering, computer science, accounting, marketing, telecommunications

ALTERNATIVES TO FULL-TIME EMPLOYMENT: Part-time jobs, paid internships

COMPANY-PAID TRAINING is available.

INTERNATIONAL ASSIGNMENTS are available to employees.

CONTACT: Staffing; Phone: 847-240-3000; Fax: 847-240-6800.

COMPANY URL: http://www.ibm.com/globalnetworks/usfacts

IDEX CORPORATION
630 DUNDEE ROAD, SUITE 400
NORTHBROOK, ILLINOIS 60062

OVERVIEW: Manufactures fluid handling and industrial products. Established: 1987. Ownership: Public.

KEY STATISTICS: Annual Sales: $562.6 million. Number of Employees: 3,598.

HIRING HISTORY: Number of professional employees hired in 1997: 5.

EXPERTISE/EDUCATION SOUGHT: Accounting, administration, human resources, manufacturing engineering, mechanical engineering, industrial engineering, design engineering

COMPANY-PAID TRAINING is available.

INTERNATIONAL ASSIGNMENTS are available to employees.

CONTACT: For individuals with previous experience: Mr. Jerry Derck, Vice President of Human Resources; Phone: 847-498-7070.

ILLINOIS POWER COMPANY
500 SOUTH 27TH STREET
DECATUR, ILLINOIS 62525

OVERVIEW: Electric utility. Established: 1994. Parent Company: Illinova Corporation. Ownership: Public.

KEY STATISTICS: Assets: $1.7 billion. Number of Employees: 3,635.

EXPERTISE/EDUCATION SOUGHT: Nuclear engineering, electrical engineering, mechanical engineering

ALTERNATIVES TO FULL-TIME EMPLOYMENT: Paid internships

COMPANY-PAID TRAINING is available.

CONTACT: Company Recruiters; Phone: 217-424-7171; Fax: 217-424-8395.

COMPANY URL: http://www.illinova.com

IMPREGILO INTERNATIONAL USA
9600 WEST 47TH STREET
MCCOOK, ILLINOIS 60525-3201

OVERVIEW: Builds bridges, tunnels, and elevated highways. Established: 1980. Ownership: Private.

KEY STATISTICS: Annual Sales: $78.1 million. Number of Employees: 500.

HIRING HISTORY: Number of professional employees hired in 1996: 5.

EXPERTISE/EDUCATION SOUGHT: Engineering

INTERNATIONAL ASSIGNMENTS are available to employees.

CONTACT: For individuals with previous experience: Human Relations Department.

IGT

INSTITUTE OF GAS TECHNOLOGY

1700 SOUTH MOUNT PROSPECT ROAD
DES PLAINES, ILLINOIS 60018-1804

OVERVIEW: Provides energy and environment-related contract research, education programs, and information services. Established: 1941.

KEY STATISTICS: Annual Sales: $21.6 million. Number of Employees: 200.

EXPERTISE/EDUCATION SOUGHT: Chemical engineering, environmental engineering, research and development, mechanical engineering, design engineering, process engineering

ALTERNATIVES TO FULL-TIME EMPLOYMENT: Part-time jobs, paid internships

COMPANY-PAID TRAINING is available.

CONTACT: Ms. Nancy Sinkovec, Human Resources Administrator; Phone: 847-768-0799; Fax: 847-768-0798. E-mail: sinkovec@igt.org.

COMPANY URL: http://www.igt.org

CORPORATE STATEMENT

IGT's scientists and engineers are on the forefront of developments in catalysis, electrochemistry, energy and environmental biotechnology, environmental engineering, fluidization, kinetics, materials research, and waste destruction and detoxification. Programs range from fundamental investigations in the laboratory through pilot plant and full-scale applications. Technologies developed by IGT are transferred to industry for commercialization.

JUNO LIGHTING, INC.

1300 SOUTH WOLF ROAD
DES PLAINES, ILLINOIS 60018-1808

OVERVIEW: Manufactures and markets lighting fixtures for residential, commercial, and institutional use. Established: 1976. Ownership: Public.

KEY STATISTICS: Annual Sales: $131.5 million. Number of Employees: 1,025.

EXPERTISE/EDUCATION SOUGHT: Manufacturing engineering, mechanical engineering, electrical engineering, design engineering, accounting, sales, marketing, management information systems, purchasing

ALTERNATIVES TO FULL-TIME EMPLOYMENT: Paid internships

COMPANY-PAID TRAINING is available.

CONTACT: Ms. Patricia Perez, Manager of Personnel; Phone: 847-827-5894, ext. 500; Fax: 847-827-3666.

KEEBLER COMPANY

1 HOLLOW TREE LANE
ELMHURST, ILLINOIS 60126-1581

OVERVIEW: Manufactures cookies, crackers, and other snack foods. Established: 1996. Ownership: Private.

KEY STATISTICS: Annual Sales: $1.5 billion. Number of Employees: 8,000.

HIRING HISTORY: Number of professional employees hired in 1997: 100. 1996: 100.

EXPERTISE/EDUCATION SOUGHT: Management information systems, food science

CONTACT: For individuals with previous experience: Ms. Jan Dooley, Manager of Corporate Staffing; Fax: 630-782-2301.

COMPANY URL: http://www.keebler.com

KOMATSU AMERICA INTERNATIONAL COMPANY

440 NORTH FAIRWAY DRIVE
VERNON HILLS, ILLINOIS 60061-1836

OVERVIEW: Manufactures construction machinery. Established: 1997. Parent Company: Komatsu, Ltd. Ownership: Private.

KEY STATISTICS: Annual Sales: $57.4 million. Number of Employees: 500.

EXPERTISE/EDUCATION SOUGHT: Marketing, customer service and support, engineering, accounting

INTERNATIONAL ASSIGNMENTS are available to employees.

CONTACT: For individuals with previous experience: Mr. Eugene Eckert, Director of Labor Relations; Phone: 847-970-4100; Fax: 847-970-4183.

COMPANY URL: http://www.komatsuamerica.com

LIFE FITNESS

10601 WEST BELMONT AVENUE
FRANKLIN PARK, ILLINOIS 60131-1545

OVERVIEW: Manufactures stair-climbing and treadmill fitness equipment. Established: 1991. Parent Company: Brunswick Corporation. Ownership: Private.

KEY STATISTICS: Number of Employees: 620.

HIRING HISTORY: Number of professional employees hired in 1996: 10.

EXPERTISE/EDUCATION SOUGHT: Mechanical engineering, electrical engineering, computer programming, computer science, design engineering, software engineering/development, administration, marketing

ALTERNATIVES TO FULL-TIME EMPLOYMENT: Part-time jobs, paid internships

CONTACT: Human Resources Department; Phone: 847-288-3300; Fax: 847-288-3705. E-mail: rdavenport@lifefitness.com.

COMPANY URL: http://www.lifefitness.com

LYON METAL PRODUCTS, LLC

PO BOX 671
AURORA, ILLINOIS 60507

MAILING ADDRESS: 420 North Maine Street, Montgomery, IL 60538

OVERVIEW: Manufactures metal partitions and commercial fixtures. Established: 1996. Parent Company: The L & D Group. Ownership: Private.

KEY STATISTICS: Annual Sales: $132.9 million. Number of Employees: 1,600.

HIRING HISTORY: Number of professional employees hired in 1996: 34.

EXPERTISE/EDUCATION SOUGHT: Production control, accounting, engineering

ALTERNATIVES TO FULL-TIME EMPLOYMENT: Paid internships

COMPANY-PAID TRAINING is available.

CONTACT: Human Resources; Phone: 630-892-8941; Fax: 630-892-8966.

THE MARMON GROUP, INC.

225 WEST WASHINGTON STREET
CHICAGO, ILLINOIS 60606-3418

OVERVIEW: Builds railway cars. Established: 1995. Ownership: Private.

KEY STATISTICS: Annual Sales: $2.1 billion. Number of Employees: 9,050.

EXPERTISE/EDUCATION SOUGHT: Manufacturing engineering

ALTERNATIVES TO FULL-TIME EMPLOYMENT: Part-time jobs, internships

COMPANY-PAID TRAINING is available.

INTERNATIONAL ASSIGNMENTS are available to employees.

CONTACT: Mr. Larry List, Director of Personnel; Phone: 312-372-9500; Fax: 312-845-5305. E-mail: recruit@marmon.com.

COMPANY URL: http://www.marmon.com

MASONITE CORPORATION

1 SOUTH WACKER DRIVE
CHICAGO, ILLINOIS 60606-4617

OVERVIEW: Manufactures building materials and home improvement products. Established: 1988. Parent Company: International Paper Company. Ownership: Public.

KEY STATISTICS: Annual Sales: $264 million. Number of Employees: 3,100.

EXPERTISE/EDUCATION SOUGHT: Mechanical engineering, electrical engineering, environmental engineering, research and development, chemical engineering, accounting

ALTERNATIVES TO FULL-TIME EMPLOYMENT: Paid internships

COMPANY-PAID TRAINING is available.

Masonite Corporation (continued)

International Assignments are available to employees.
Contact: Human Resources Department; Phone: 312-750-0900; Fax: 312-750-9052.

MCCAIN FOODS, INC.
2905 BUTTERFIELD ROAD
OAK BROOK, ILLINOIS 60521-1102

Overview: Manufactures food products. Established: 1985. Parent Company: McCain Foods, Ltd. Ownership: Private.
Key Statistics: Annual Sales: $300 million. Number of Employees: 3,550.
Expertise/Education Sought: Manufacturing engineering
Alternatives to Full-Time Employment: Part-time jobs
Company-Paid Training is available.
Contact: For individuals with previous experience: Mr. Robert H. Thomas, Vice President of Human Resources; Fax: 630-472-0490.
Company URL: http://www.mccain.com

MIDCON CORPORATION
701 EAST 22ND STREET
LOMBARD, ILLINOIS 60148-5072

Overview: Distributes natural gas. Established: 1985. Parent Company: KN Energy. Ownership: Public.
Key Statistics: Annual Sales: $2.6 billion. Number of Employees: 1,747.
Expertise/Education Sought: Management information systems, computer programming, accounting, mechanical engineering, petroleum/petrochemical engineering, civil engineering
Alternatives to Full-Time Employment: Paid internships
Company-Paid Training is available.
Contact: Personnel Services; Phone: 630-691-3421; Fax: 630-282-2116. E-mail: carolyn_l._thomas@oxy.com.

MI-JACK PRODUCTS, INC.
3111 WEST 167TH STREET
HAZEL CREST, ILLINOIS 60429

Overview: Manufactures lifts and cranes. Established: 1955. Parent Company: Lanco. Ownership: Private.
Key Statistics: Annual Sales: $144.3 million. Number of Employees: 1,250.
Expertise/Education Sought: Hydraulic engineering, mechanical engineering, industrial engineering, accounting
Company-Paid Training is available.
Contact: Mr. Rone Bujan, Director of Human Resources; Phone: 708-596-5200, ext. 201; Fax: 708-225-2316.

MITSUBISHI MOTOR MANUFACTURING OF AMERICA, INC.
100 NORTH MITSUBISHI MOTOR WAY
NORMAL, ILLINOIS 61761-8000

Overview: Manufactures automobiles. Established: 1991. Parent Company: Mitsubishi Motor Corporation. Ownership: Private.
Key Statistics: Annual Sales: $553.4 million. Number of Employees: 3,962.
Expertise/Education Sought: Engineering, industrial technician, computer science, purchasing
Alternatives to Full-Time Employment: Paid internships
Company-Paid Training is available.
Contact: Ms. Jeana Leu, Personnel Representative; Phone: 309-888-8380; Fax: 309-888-7922.

MORTON INTERNATIONAL, INC.
100 NORTH RIVERSIDE PLAZA
CHICAGO, ILLINOIS 60606

Overview: Manufactures salt, specialty chemicals, and automotive air bags. Established: 1848. Ownership: Public.
Key Statistics: Annual Sales: $1.5 billion. Number of Employees: 6,300.
Hiring History: Number of professional employees hired in 1997: 600.
Expertise/Education Sought: Chemical engineering, chemistry, accounting, marketing, sales
Contact: Human Resources Department; Phone: 312-807-2000.
Company URL: http://www.morton.com

MOTOROLA
1301 EAST ALGONQUIN ROAD
SCHAUMBURG, ILLINOIS 60196

Overview: Manufactures and markets wireless communications, semiconductors, and advanced electronics. Established: 1928. Ownership: Public.
Key Statistics: Annual Sales: $28 billion. Number of Employees: 139,000.
Expertise/Education Sought: Electrical engineering, manufacturing engineering, mechanical engineering, software engineering/development, chemical engineering, physics, mathematics, computer science, management information systems
Alternatives to Full-Time Employment: Paid internships
Company-Paid Training is available.
International Assignments are available to employees.
Contact: For individuals with previous experience: Staffing Department; Phone: 847-576-5000; Fax: 847-576-1961.

NALCO CHEMICAL COMPANY
1 NALCO CENTER
NAPERVILLE, ILLINOIS 60563-1198

Overview: Manufactures specialty chemicals. Established: 1928. Ownership: Public.
Key Statistics: Annual Sales: $1.3 billion. Number of Employees: 6,502.
Hiring History: Number of professional employees hired in 1997: 400. 1996: 400.
Expertise/Education Sought: Accounting, finance, sales, customer service and support, law, chemistry, chemical engineering, environmental engineering, mechanical engineering
Alternatives to Full-Time Employment: Part-time jobs, paid internships
Company-Paid Training is available.
International Assignments are available to employees.
Contact: Ms. Karen Nordquist, Manager of Technical Employment; Phone: 630-305-1709; Fax: 630-305-2924. E-mail: knordquist@nalco.com.
Company URL: http://www.nalco.com

NORTH AMERICAN LIGHTING, INC.
20 INDUSTRIAL PARK
FLORA, ILLINOIS 62839-9700

Overview: Manufactures headlight assemblies. Established: 1983. Parent Company: Hella, Inc. Ownership: Private.
Key Statistics: Annual Sales: $269.6 million. Number of Employees: 1,600.
Expertise/Education Sought: Marketing, accounting, engineering, electrical engineering
Alternatives to Full-Time Employment: Paid internships
Company-Paid Training is available.
International Assignments are available to employees.
Contact: Human Resource Department, PO Box 499, Flora, Illinois, 62839-0499; Phone: 618-662-4483; Fax: 618-662-8143.

NORTHROP GRUMMAN

NORTHROP GRUMMAN CORPORATION-ELECTRONICS SYSTEMS

600 HICKS ROAD
ROLLING MEADOWS, ILLINOIS 60008

Overview: Researches, designs, develops, and manufactures radar jamming countermeasure systems. Established: 1965. Ownership: Public.

Key Statistics: Annual Sales: $500 million. Number of Employees: 3,000.

Hiring History: Number of professional employees hired in 1997: 100.

Expertise/Education Sought: Computer science, electrical engineering, electronics/electronics engineering, physics, mathematics

Alternatives to Full-Time Employment: Paid internships

Company-Paid Training is available.

International Assignments are available to employees.

Contact: Ms. Ronda Navidad, Technical Recruiter; Phone: 847-259-9600, ext. 5294; Fax: 847-590-3189. E-mail: navidro@mail.northgrum.com.

Company URL: http://www.northgrum.com/recruit

For more information, see full corporate profile, pg. 292.

OUTBOARD MARINE CORPORATION

100 SEA HORSE DRIVE
WAUKEGAN, ILLINOIS 60085-2141

Overview: Manufactures outboard motors and boats. Established: 1936. Parent Company: GreenMarine, Inc. Ownership: Private.

Key Statistics: Annual Sales: $1.1 billion. Number of Employees: 8,250.

Hiring History: Number of professional employees hired in 1996: 89.

Expertise/Education Sought: Mechanical engineering, electrical engineering, manufacturing engineering, auditing, supply chain management

Contact: Mr. Dick Salzman, Contract Recruiter; Phone: 847-689-6200; Fax: 847-689-5220. E-mail: bsalzman@omc-online.com.

Company URL: http://www.omc-online.com

PANDUIT CORPORATION

17301 RIDGELAND AVENUE
TINLEY PARK, ILLINOIS 60477-3093

Overview: Manufactures wiring and communication products. Established: 1953. Ownership: Private.

Key Statistics: Annual Sales: $290 million. Number of Employees: 2,500.

Hiring History: Number of professional employees hired in 1997: 500.

Expertise/Education Sought: Computer science, production, sales, finance, manufacturing engineering, mechanical engineering, electrical engineering

Alternatives to Full-Time Employment: Paid internships

Company-Paid Training is available.

International Assignments are available to employees.

Contact: Ms. Cheryl Lewis, Recruiter; Phone: 708-532-1800, ext. 1246; Fax: 708-614-8357.

Company URL: http://www.panduit.com

PEPSI-COLA GENERAL BOTTLERS

3501 ALGONQUIN ROAD
ROLLING MEADOWS, ILLINOIS 60008-3103

Overview: Manufactures and bottles soft drinks. Established: 1969. Parent Company: Whitman Corporation. Ownership: Public.

Key Statistics: Annual Sales: $1.5 billion. Number of Employees: 6,100.

Expertise/Education Sought: Accounting, manufacturing engineering, industrial engineering, finance, human resources, sales, marketing

Alternatives to Full-Time Employment: Paid internships

International Assignments are available to employees.

Contact: Mr. William S. Kowalski, Corporate Human Resource Manager; Phone: 847-394-7678; Fax: 847-590-3450. E-mail: WSKowalski@PCGB.com.

PHILIP SERVICES CORPORATION

210 WEST SAND BANK ROAD
COLUMBIA, ILLINOIS 62236

Mailing Address: PO Box 230, Columbia, IL 62236

Overview: Provides environmental consulting, remediation, process design, demolition/decommissioning, and analytical services. Established: 1975.

Expertise/Education Sought: BS/MS in chemical or environmental engineering, consulting engineering, remediation, process design, demolition, geology, industrial hygiene, chemical engineering, civil engineering, environmental engineering

Company-Paid Training is available.

International Assignments are available to employees.

Contact: Ms. Anita McClain, Human Resource Manager; Phone: 618-281-7173; Fax: 618-281-8077.

PLAINFIELD STAMPING AND MOLDING

1351 NORTH DIVISION STREET
PLAINFIELD, ILLINOIS 60544-8926

Overview: Provides custom metal stamping services. Established: 1959. Ownership: Private.

Key Statistics: Annual Sales: $43.2 million. Number of Employees: 350.

Hiring History: Number of professional employees hired in 1997: 60. 1996: 65.

Expertise/Education Sought: Design engineering, process engineering, project engineering

Alternatives to Full-Time Employment: Paid internships

Company-Paid Training is available.

Contact: For individuals with previous experience: Ms. Sandy Franczyk, Director of Human Resources; Phone: 815-436-5671; Fax: 815-439-2970.

PLANO MOLDING COMPANY

431 EAST SOUTH STREET
PLANO, ILLINOIS 60545-1676

Overview: Manufactures and distributes plastic products. Established: 1932. Ownership: Private.

Key Statistics: Annual Sales: $51.4 million. Number of Employees: 686.

Hiring History: Number of professional employees hired in 1997: 4.

Expertise/Education Sought: Computer-aided design, industrial engineering, machinery, process engineering, injection molding

Alternatives to Full-Time Employment: Paid internships

Company-Paid Training is available.

International Assignments are available to employees.

Contact: Ms. Lisa Siroky, Director of Human Resources; Phone: 630-552-3111, ext. 9447; Fax: 630-552-9898.

REVERE WARE CORPORATION

1000 SOUTH SHERMAN STREET
CLINTON, ILLINOIS 61727

Overview: Manufactures stainless steel pots and pans. Established: 1992. Parent Company: Corning, Inc. Ownership: Public.

Key Statistics: Annual Sales: $110 million. Number of Employees: 400.

Hiring History: Number of professional employees hired in 1997: 5. 1996: 7.

Expertise/Education Sought: Accounting, facilities management, data processing, finance, management information systems, engineering

Alternatives to Full-Time Employment: Paid internships

Company-Paid Training is available.

Contact: Mr. Elester Williams, Human Resources Manager; Phone: 217-935-7204; Fax: 217-935-7241. E-mail: williae@ccpci.com.

RICHARDS-WILCOX, INC.
600 SOUTH LAKE STREET
AURORA, ILLINOIS 60506-5582

OVERVIEW: Manufactures metal partitions and fixtures. Established: 1996. Parent Company: Constructor Group. Ownership: Private.
KEY STATISTICS: Annual Sales: $43.9 million. Number of Employees: 340.
HIRING HISTORY: Number of professional employees hired in 1996: 50.
EXPERTISE/EDUCATION SOUGHT: Electrical engineering, engineering, mechanical engineering
ALTERNATIVES TO FULL-TIME EMPLOYMENT: Part-time jobs
COMPANY-PAID TRAINING is available.
CONTACT: Mr. James Sirovatka, Director of Human Resources.
COMPANY URL: http://www.richards-wilcox.com

SAFETY-KLEEN CORPORATION
1 BRINCKMAN WAY
ELGIN, ILLINOIS 60123-7857

OVERVIEW: Recycles hazardous waste. Established: 1961. Ownership: Public.
KEY STATISTICS: Annual Sales: $923.1 million. Number of Employees: 7,300.
HIRING HISTORY: Number of professional employees hired in 1997: 130. 1996: 130.
EXPERTISE/EDUCATION SOUGHT: Accounting, finance, data processing, environmental engineering, computer programming
ALTERNATIVES TO FULL-TIME EMPLOYMENT: Paid internships
COMPANY-PAID TRAINING is available.
CONTACT: Ms. Barbara Schuls, Manager of Corporate Employment; Phone: 847-697-8460, ext. 2102. E-mail: bschuls@safety-kleen.com.

SARGENT AND LUNDY
55 EAST MONROE STREET
CHICAGO, ILLINOIS 60603-5709

OVERVIEW: Provides consulting services for nuclear and fossil fuel plants. Established: 1976. Ownership: Private.
KEY STATISTICS: Annual Sales: $206 million. Number of Employees: 1,800.
HIRING HISTORY: Number of professional employees hired in 1996: 200.
EXPERTISE/EDUCATION SOUGHT: Civil engineering, structural engineering, electrical engineering, mechanical engineering
ALTERNATIVES TO FULL-TIME EMPLOYMENT: Paid internships
INTERNATIONAL ASSIGNMENTS are available to employees.
CONTACT: For college students and recent graduates: Ms. Carol Talaronek, Supervisor of Employment; Phone: 312-269-2000. E-mail: carol.talaronek@slchicago.infonet.com. For individuals with previous experience: Mr. Pete Meehan, Division Manager Human Resources; Phone: 312-269-3573; Fax: 312-269-3596. E-mail: peter.j.meehan@slchicago.infonet.com.

SCHOLLE CORPORATION
200 WEST NORTH AVENUE
NORTHLAKE, ILLINOIS 60164

OVERVIEW: Manufactures plastic bags. Established: 1945. Ownership: Private.
KEY STATISTICS: Annual Sales: $200 million. Number of Employees: 1,500.
EXPERTISE/EDUCATION SOUGHT: Process engineering, accounting, management information systems, industrial engineering
ALTERNATIVES TO FULL-TIME EMPLOYMENT: Paid internships
COMPANY-PAID TRAINING is available.
INTERNATIONAL ASSIGNMENTS are available to employees.
CONTACT: For individuals with previous experience: Human Resources Department; Phone: 708-562-7290; Fax: 708-409-4479. E-mail: schpcusa@ix.netcom.com.

SEI INFORMATION TECHNOLOGY
212 EAST OHIO STREET
CHICAGO, ILLINOIS 60611-3203

OVERVIEW: Provides information systems consulting services. Established: 1969. Ownership: Private.
KEY STATISTICS: Annual Sales: $247.8 million. Number of Employees: 1,133.
HIRING HISTORY: Number of professional employees hired in 1997: 50. 1996: 78.
EXPERTISE/EDUCATION SOUGHT: Information systems, computer programming, systems analysis, project management, software engineering/development, Internet/electronic commerce, object oriented design
ALTERNATIVES TO FULL-TIME EMPLOYMENT: Part-time jobs
COMPANY-PAID TRAINING is available.
CONTACT: For individuals with previous experience: Recruitment; Phone: 312-440-8300; Fax: 312-440-8373. E-mail: recruitment@sei-it.com. Job Hotline: 888-734-8326.
COMPANY URL: http://www.sei-it.com

SERVICEMASTER COMPANY
ONE SERVICEMASTER WAY
DOWNERS GROVE, ILLINOIS 60515

OVERVIEW: Provides building maintenance, sanitation, and industrial services. Established: 1986. Ownership: Public.
KEY STATISTICS: Annual Sales: $3.5 billion. Number of Employees: 39,900.
EXPERTISE/EDUCATION SOUGHT: Clinical engineering, chemical engineering, business management, finance, marketing, logistics, accounting
ALTERNATIVES TO FULL-TIME EMPLOYMENT: Paid internships
CONTACT: For individuals with previous experience: Ms. Terri Welch, Director of Personnel; Phone: 708-271-1300.

STONE CONTAINER CORPORATION
150 NORTH MICHIGAN AVENUE
CHICAGO, ILLINOIS 60601

OVERVIEW: Manufactures packaging, pulp, and wood products. Established: 1926. Ownership: Public.
KEY STATISTICS: Annual Sales: $5.1 billion. Number of Employees: 24,200.
EXPERTISE/EDUCATION SOUGHT: Transportation, computer programming
CONTACT: Ms. Sharon Williams, Recruiting Specialist; Phone: 312-580-4649; Fax: 312-649-4294. E-mail: sharonwilliams@sto.com.
COMPANY URL: http://www.stonecontainer.com

STS CONSULTANTS, LTD.
750 CORPORATE WOODS PARKWAY
VERNON HILLS, ILLINOIS 60061

OVERVIEW: Provides consulting engineering services. Established: 1995. Ownership: Private.
KEY STATISTICS: Annual Sales: $27.9 million. Number of Employees: 380.
HIRING HISTORY: Number of professional employees hired in 1997: 10. 1996: 10.
EXPERTISE/EDUCATION SOUGHT: Environmental engineering, structural engineering, geotechnical engineering
ALTERNATIVES TO FULL-TIME EMPLOYMENT: Paid internships
CONTACT: Ms. Sheila Tomaszewski, Director of Human Resources; Phone: 847-267-8010; Fax: 847-267-8040. E-mail: tomaszewski@stsltd.com.

SUNDSTRAND AEROSPACE
4747 HARRISON AVENUE
ROCKFORD, ILLINOIS 61108-7987

OVERVIEW: Manufactures aerospace and industrial products. Established: 1910. Parent Company: Sundstrand Corporation. Ownership: Public.
KEY STATISTICS: Annual Sales: $1.5 billion. Number of Employees: 9,400.

Expertise/Education Sought: Business administration, mechanical engineering, electrical engineering, manufacturing engineering, materials science, industrial engineering

Alternatives to Full-Time Employment: Paid internships

Company-Paid Training is available.

International Assignments are available to employees.

Contact: For college students and recent graduates: Mr. Wayne Orrison, Human Resources; Phone: 815-226-7966; Fax: 815-226-3887. For individuals with previous experience: Mr. Brian Hayes, Director of Human Resources; Phone: 815-226-6294; Fax: 815-226-3887. Job Hotline: 815-226-6269.

Company URL: http://www.snds.com

SYSTEMS AND SOFTWARE SERVICE, INC.

522 SOUTH NORTHWEST HIGHWAY
BARRINGTON, ILLINOIS 60010-4616

Overview: Provides engineering services. Established: 1993. Ownership: Private.

Key Statistics: Annual Sales: $20 million. Number of Employees: 30.

Hiring History: Number of professional employees hired in 1997: 200. 1996: 150.

Expertise/Education Sought: Engineering, information systems

International Assignments are available to employees.

Contact: Mr. Ernie Sostarics, Recruiter.

TC MANUFACTURING COMPANY, INC.

1527 LYONS STREET
EVANSTON, ILLINOIS 60201-3551

Overview: Manufactures polyethylene film. Established: 1996. Ownership: Private.

Key Statistics: Annual Sales: $32 million. Number of Employees: 250.

Hiring History: Number of professional employees hired in 1997: 3.

Expertise/Education Sought: Accounting, administration, engineering

Contact: For individuals with previous experience: Mrs. Jill Blakemore, Accounting Manager; Phone: 847-866-8500; Fax: 847-866-7411.

TDI BATTERIES

2852 HITCHCOCK AVENUE
DOWNERS GROVE, ILLINOIS 60515-4016

Overview: Manufactures storage batteries. Established: 1982. Ownership: Private.

Key Statistics: Annual Sales: $146.8 million. Number of Employees: 795.

Hiring History: Number of professional employees hired in 1997: 75. 1996: 50.

Expertise/Education Sought: Manufacturing engineering, production engineering

Contact: For individuals with previous experience: Department of Human Resources; Phone: 630-971-3460; Fax: 630-971-3903.

TELLABS, INC.

1000 REMINGTON BOULEVARD, MS 147
BOLINGBROOK, ILLINOIS 60440

Overview: Designs and manufactures voice and data transport and network access systems. Established: 1975. Ownership: Public.

Key Statistics: Annual Sales: $2 billion. Number of Employees: 5,800.

Hiring History: Number of professional employees hired in 1997: 700. 1996: 660.

Expertise/Education Sought: Hardware engineering, software engineering/development, electrical engineering, computer science, computer programming

Alternatives to Full-Time Employment: Paid internships

Contact: For college students and recent graduates: Ms. Julie Cunningham, Manager of Corporate College Relations Department P99; Phone: 630-679-3281; Fax: 630-378-5620. E-mail: julie.cunningham@tellabs.com. For individuals with previous experience: Tellabs Staffing Department; Fax: 630-378-5620.

Company URL: http://www.tellabs.com

CORPORATE PROFILE

Tellabs, a telecommunications company with 5,800 employees worldwide, is experiencing dynamic success and outstanding growth and is proud of its strong tradition of research and development, world-class manufacturing, and superior quality and customer service. By designing, producing, marketing, and servicing sophisticated voice and data transport and network access systems, Tellabs is recognized as an industry innovator and major player on the global telecommunications scene. Tellabs' customers include public telephone companies, local and long-distance service providers, and organizations with private telephone systems. Tellabs is also involved with cutting-edge technology in the cellular, wireless, and telephony-over-cable arenas.

Over the past 27 quarters, Tellabs has posted sales gains over prior-year levels. In 1997, the company earned record sales of $1.2 billion—a figure expected to more than double by the year 2000. Given these current performance levels and excellent future prospects, Tellabs is on course to achieve that mark.

While financial stability is just a part of the company's success formula, its real strength can be traced to the people who make it all happen. That's why the best and brightest graduates are recruited for challenging summer internships and full-time employment opportunities. New employees are paired with experienced mentors to tackle real projects—projects that not only enable each individual to demonstrate and develop the skills they've learned in the classroom, but that also can literally change the way the people in the world communicate.

Tellabs' headquarters can be found in Chicago's west suburban cities of Lisle and Bolingbrook, the heart of the Illinois research and development corridor. It is also where dedicated staff members meet an increasingly global demand for improved manufacturing methods while finding new ways to drive technology forward.

For more information, recent grads and students are invited to visit Tellabs' home page on the World Wide Web at http://www.tellabs.com

TOOTSIE ROLL INDUSTRIES, INC.

7401 SOUTH CICERO AVENUE
CHICAGO, ILLINOIS 60629-5818

Overview: Manufactures candy. Established: 1896. Ownership: Public.

Key Statistics: Annual Sales: $340.9 million. Number of Employees: 1,750.

Hiring History: Number of professional employees hired in 1997: 5. 1996: 10.

Expertise/Education Sought: Manufacturing

Company-Paid Training is available.

Contact: Mr. Michael Hale, Director of Human Resources; Phone: 773-838-3400; Fax: 773-838-3533.

TRANSILWRAP COMPANY, INC.

2828 NORTH PAULINA STREET
CHICAGO, ILLINOIS 60657-4012

Overview: Manufactures plastics. Established: 1931. Ownership: Private.

Key Statistics: Annual Sales: $151.8 million. Number of Employees: 721.

Expertise/Education Sought: Business management, accounting, finance, marketing, information systems

Contact: Human Relations Department; Phone: 773-296-1000; Fax: 773-296-2007.

TRANS UNION CORPORATION

555 WEST ADAMS STREET
CHICAGO, ILLINOIS 60661

Overview: Provides credit reporting services. Established: 1982. Parent Company: The Marmon Group. Ownership: Private.

Trans Union Corporation (continued)

Key Statistics: Annual Sales: $402.9 million. Number of Employees: 3,200.

Hiring History: Number of professional employees hired in 1996: 400.

Expertise/Education Sought: Computer programming, program analysis, analysis, accounting, personal computers, land management, database management, computer programming, computer science, management information systems

Alternatives to Full-Time Employment: Part-time jobs, paid internships

Contact: For individuals with previous experience: Mr. Michael Khadivar, Project Manager Recruiting Operator; Phone: 312-258-1717, ext. 3728; Fax: 312-466-8385.

Company URL: http://www.tuc.com

TRIBUNE COMPANY

435 NORTH MICHIGAN AVENUE
CHICAGO, ILLINOIS 60611

Overview: Publishes newspapers and electronic information. Established: 1991. Ownership: Public.

Key Statistics: Number of Employees: 7,300.

Expertise/Education Sought: Accounting, finance, marketing, publishing

Alternatives to Full-Time Employment: Part-time jobs, paid internships

Company-Paid Training is available.

Contact: For individuals with previous experience: Human Resources Director; Phone: 312-222-4571; Fax: 312-222-2595. E-mail: tribune@isearch.com.

Company URL: http://www.tribune.com

TRIM-RITE FOOD CORPORATION

50 WEST COMMERCIAL AVENUE
ADDISON, ILLINOIS 60101-4502

Overview: Wholesales fresh meats and meat products. Established: 1991. Ownership: Private.

Key Statistics: Annual Sales: $60 million. Number of Employees: 140.

Hiring History: Number of professional employees hired in 1997: 150. 1996: 10.

Expertise/Education Sought: Manufacturing

Contact: For individuals with previous experience: Mr. Dave Pioprowski, Controller; Phone: 630-543-2240; Fax: 630-543-2494.

USG CORPORATION

125 SOUTH FRANKLIN STREET
CHICAGO, ILLINOIS 60606-4605

Overview: Manufactures building and construction materials. Established: 1985. Ownership: Public.

Key Statistics: Annual Sales: $1.4 billion. Number of Employees: 6,100.

Expertise/Education Sought: Engineering, computer programming, economics, accounting, data processing, finance, human resources

Alternatives to Full-Time Employment: Internships

Company-Paid Training is available.

International Assignments are available to employees.

Contact: Mr. Fendley Douge, Director Strategic Staffing; Phone: 312-606-5230; Fax: 312-606-4138.

US TSUBAKI, INC.

301 EAST MARQUARDT DRIVE
WHEELING, ILLINOIS 60090-6431

Overview: Manufactures power transmission equipment. Established: 1972. Parent Company: Tsubakimoto Chain Company. Ownership: Private.

Key Statistics: Annual Sales: $160 million. Number of Employees: 1,025.

Hiring History: Number of professional employees hired in 1997: 30. 1996: 30.

Expertise/Education Sought: Marketing, sales, customer service and support

Alternatives to Full-Time Employment: Paid internships

Company-Paid Training is available.

Contact: For individuals with previous experience: Human Resources Department; Phone: 847-459-9500, ext. 324; Fax: 847-459-8508. E-mail: kmartin@ustsubhai.com.

Company URL: http://www.ustsubhai.com

VAPOR CORPORATION

6420 WEST HOWARD STREET
NILES, ILLINOIS 60714-3302

Overview: Manufactures busses. Established: 1996. Parent Company: WABCO. Ownership: Public.

Key Statistics: Annual Sales: $60 million. Number of Employees: 450.

Expertise/Education Sought: Engineering, electrical engineering

Contact: For individuals with previous experience: Ms. Anne Morales, Secretary; Phone: 847-967-8300. E-mail: vaporhr@mcs.com.

VIDEO JET SYSTEMS INTERNATIONAL

1500 MITTEL BOULEVARD
WOOD DALE, ILLINOIS 60191-1072

Overview: Manufactures ink jet printers and coding equipment. Established: 1985. Parent Company: GEC. Ownership: Private.

Key Statistics: Annual Sales: $200 million. Number of Employees: 1,500.

Hiring History: Number of professional employees hired in 1996: 200.

Expertise/Education Sought: Software engineering/development, mechanical engineering, electrical engineering, physics, sales, chemistry

International Assignments are available to employees.

Contact: Mr. Randy Brown, Manager of Staffing; Phone: 630-860-7300; Fax: 630-616-3678.

Company URL: http://www.videojet.com

VISKASE CORPORATION

6855 WEST 65TH STREET
CHICAGO, ILLINOIS 60638-4803

Overview: Manufactures sausage casings. Established: 1989. Parent Company: Envirodyne Industries, Inc. Ownership: Public.

Key Statistics: Annual Sales: $500 million. Number of Employees: 4,000.

Hiring History: Number of professional employees hired in 1996: 100.

Expertise/Education Sought: Mechanical engineering, electrical engineering, food science, chemical engineering, marketing, business administration, accounting

Alternatives to Full-Time Employment: Paid internships

Company-Paid Training is available.

International Assignments are available to employees.

Contact: For individuals with previous experience: Mr. Lowery Stallings, Manager of Training and Development; Phone: 708-496-4200, ext. 4280; Fax: 708-496-4412.

ZENITH ELECTRONICS CORPORATION

1000 MILWAUKEE AVENUE
GLENVIEW, ILLINOIS 60025

Overview: Manufactures televisions, VCRs, radios, electronic components, and related products. Established: 1995. Ownership: Private.

Key Statistics: Annual Sales: $1.3 billion. Number of Employees: 15,900.

Expertise/Education Sought: Purchasing, human resources, finance, research and development, information systems, materials management

Alternatives to Full-Time Employment: Paid internships

Company-Paid Training is available.

International Assignments are available to employees.

Contact: For individuals with previous experience: Human Resources Department; Phone: 847-391-7000.

Company URL: http://www.zenith.com

INDIANA

AMERICAN ELECTRIC POWER COMPANY
110 EAST WAYNE STREET
FORT WAYNE, INDIANA 46802

Mailing Address: PO Box 60, Fort Wayne, IN 46801
Overview: Electric and gas utility. Established: 1925. Ownership: Public.
Key Statistics: Assets: $1.3 billion. Number of Employees: 3,500.
Expertise/Education Sought: Physics, accounting, purchasing, finance, marketing, sales, computer programming, law, engineering
Company-Paid Training is available.
International Assignments are available to employees.
Contact: Ms. Wendy Anstach, Employment Center Manager; Phone: 219-425-2139.
Company URL: http://www.aep.com

ARVIN INDUSTRIES
1 NOBLITT PLAZA
COLUMBUS, INDIANA 47202-3000

Overview: Manufactures automotive exhaust systems, ride control metal coatings and stampings. Established: 1996. Ownership: Public.
Key Statistics: Annual Sales: $22.5 million.
Hiring History: Number of professional employees hired in 1997: 150.
Expertise/Education Sought: Engineering, finance, manufacturing
Company-Paid Training is available.
International Assignments are available to employees.
Contact: For college students and recent graduates: Mr. Tom Clerkin, Director of Human Resources; Phone: 812-379-3081; Fax: 812-379-3688. For individuals with previous experience: Mr. Raymond Mack, Vice President of Human Resources; Phone: 812-379-3603; Fax: 812-379-3688. E-mail: rpmack@arvin.com.

ARVIN SANGO, INC.
2905 WILSON AVENUE
MADISON, INDIANA 47250

Overview: Manufactures automotive exhaust systems. Established: 1986. Parent Company: Arvin Industries And Sango Company Limited. Ownership: Private.
Key Statistics: Annual Sales: $36.8 million. Number of Employees: 400.
Hiring History: Number of professional employees hired in 1997: 6. 1996: 45.
Expertise/Education Sought: Mechanical engineering, industrial engineering, production engineering, electrical engineering, packaging engineering, quality engineering, electrical technology
Company-Paid Training is available.
Contact: Mr. Jeff Schroer, Human Resources Manager; Phone: 812-265-2888, ext. 111; Fax: 812-265-2894. E-mail: jschroer@arvinsango.com.

BTR AUTOMOTIVE
6920 POINTE INVERNESS WAY
FORT WAYNE, INDIANA 46804-7926

Overview: Manufactures mechanical rubber automotive products. Established: 1996. Parent Company: BTR, Inc. Ownership: Private.
Key Statistics: Annual Sales: $46.3 million. Number of Employees: 738.
Hiring History: Number of professional employees hired in 1997: 43. 1996: 56.
Expertise/Education Sought: Mechanical engineering, industrial engineering, quality engineering
Alternatives to Full-Time Employment: Paid internships
Company-Paid Training is available.
International Assignments are available to employees.
Contact: For individuals with previous experience: Human Resources Department; Phone: 219-434-9800; Fax: 219-434-9811.

CERESTAR USA, INC.
1100 INDIANAPOLIS BOULEVARD
HAMMOND, INDIANA 46320

Overview: Processes corn for food products. Established: 1995. Parent Company: Eridania Beghin-Say. Ownership: Private.
Key Statistics: Annual Sales: $173 million. Number of Employees: 865.
Expertise/Education Sought: Accounting, administration, manufacturing engineering, management, marketing, sales, finance, human resources, customer service and support
Alternatives to Full-Time Employment: Paid internships
Company-Paid Training is available.
International Assignments are available to employees.
Contact: Mr. Patrick Somers, Manager of Recruiting; Phone: 219-659-2000, ext. 575; Fax: 219-473-6604.

CINERGY CORPORATION
1000 EAST MAIN STREET
PLAINFIELD, INDIANA 46168-1765

Overview: Electric utility. Established: 1994. Ownership: Public.
Key Statistics: Assets: $1.3 billion. Number of Employees: 2,211.
Hiring History: Number of professional employees hired in 1997: 23. 1996: 15.
Expertise/Education Sought: Electrical engineering, mechanical engineering, accounting, law, computer programming, sales, administration, customer service and support, construction
Alternatives to Full-Time Employment: Paid internships
Company-Paid Training is available.
Contact: Mr. Jerry Depasse, Lead Staffing Consultant; Phone: 317-838-1755; Fax: 317-838-2427. E-mail: gdepasse@cinergy.com. Job Hotline: 800-428-4337, ext. 2838.
Company URL: http://www.cinergy.com

CREATIVE EXPRESSIONS GROUP, INC.
7240 SHADELAND STATION
INDIANAPOLIS, INDIANA 46256-3928

Overview: Manufactures paper plates, cups, and dishes. Established: 1996. Ownership: Private.
Key Statistics: Annual Sales: $45 million. Number of Employees: 450.
Hiring History: Number of professional employees hired in 1997: 100. 1996: 50.
Expertise/Education Sought: Marketing, sales, operations, design
Alternatives to Full-Time Employment: Paid internships
Company-Paid Training is available.
International Assignments are available to employees.
Contact: For college students and recent graduates: Mr. Geoff Gailey, Vice President of Human Resources; Phone: 317-841-9999. For individuals with previous experience: Human Resources Department; Phone: 317-841-9999; Fax: 317-841-2650.

CTB, INC.
STATE ROAD 15 NORTH, PO BOX 2000
MILFORD, INDIANA 46542-2000

Overview: Manufactures fabricated plate work (boiler shop). Established: 1995. Ownership: Private.
Key Statistics: Annual Sales: $138 million. Number of Employees: 900.
Hiring History: Number of professional employees hired in 1997: 20. 1996: 15.
Expertise/Education Sought: Finance, accounting, administration, engineering, agricultural engineering, manufacturing, plastics engineering
Alternatives to Full-Time Employment: Paid internships
Company-Paid Training is available.

CTB, Inc. (continued)

International Assignments are available to employees.

Contact: Mr. Rick Van Puffelen, Human Resources Manager; Phone: 219-658-4191; Fax: 219-658-4133.

Company URL: http://www.ctbinc.com

CUMMINS ENGINE COMPANY

PO BOX 3005
COLUMBUS, INDIANA 47202-3005

Overview: Manufactures diesel engines for buses, trucks, and pickups. Established: 1919. Ownership: Public.

Key Statistics: Annual Sales: $5.3 billion. Number of Employees: 24,300.

Expertise/Education Sought: Electrical engineering, accounting, mechanical engineering, research and development, finance

Alternatives to Full-Time Employment: Paid internships

Contact: Mr. Dwight McMillan, Director of Corporate Research; Phone: 812-377-7512; Fax: 812-377-3294.

DAMON CORPORATION

PO BOX 1107
ELKHART, INDIANA 46515-1107

Overview: Manufactures mobile homes, travel trailers, and park model mobile homes. Established: 1988. Ownership: Private.

Key Statistics: Annual Sales: $177.8 million. Number of Employees: 437.

Expertise/Education Sought: Mechanical engineering, electrical engineering, research and development, construction engineering, accounting, computer-aided design

Company-Paid Training is available.

Contact: For individuals with previous experience: Mr. Nigel Clark, Vice President of Human Resources; Phone: 219-266-0443; Fax: 219-266-0543.

DELPHI DELCO ELECTRONICS SYSTEMS

PO BOX 9005, CT17A
KOKOMO, INDIANA 46904-9005

Overview: Manufactures automotive electronics products. Established: 1985. Parent Company: GM. Ownership: Public.

Key Statistics: Annual Sales: $2.7 billion. Number of Employees: 29,000.

Hiring History: Number of professional employees hired in 1996: 120.

Expertise/Education Sought: Mechanical engineering, engineering, electronics/electronics engineering, automotive engineering, software engineering/development

Alternatives to Full-Time Employment: Part-time jobs, internships

Company-Paid Training is available.

International Assignments are available to employees.

Contact: For individuals with previous experience: Ms. Cynthia KCampbell, Manager of Staffing and Diversity; Phone: 765-451-0789; Fax: 765-451-0108.

DELTA FAUCET COMPANY

55 EAST 111TH STREET
INDIANAPOLIS, INDIANA 46280-1071

Overview: Manufactures hardware and home improvement products. Established: 1973. Parent Company: Masco Corporation. Ownership: Public.

Key Statistics: Annual Sales: $273.2 million. Number of Employees: 3,110.

Hiring History: Number of professional employees hired in 1997: 35. 1996: 35.

Expertise/Education Sought: Sales, mechanical engineering, finance, operations, accounting, marketing, management information systems, industrial design

Alternatives to Full-Time Employment: Paid internships

International Assignments are available to employees.

Contact: Employee Relations Manager; Phone: 317-848-1812, ext. 4687; Fax: 317-571-6513.

DEPUY, INC.

700 ORTHOPEDIC DRIVE
WARSAW, INDIANA 46580

Mailing Address: PO Box 988, Warsaw, IN 46580

Overview: Manufactures orthopedic products. Established: 1992. Ownership: Public.

Key Statistics: Annual Sales: $697.3 million. Number of Employees: 2,930.

Expertise/Education Sought: Manufacturing engineering, mechanical engineering, materials science, accounting, marketing, packaging engineering

Alternatives to Full-Time Employment: Paid internships

Company-Paid Training is available.

Contact: Mr. Mark Ardington, Human Resources Manager; Phone: 219-267-8143; Fax: 219-267-2284. Job Hotline: 219-372-7187.

Company URL: http://www.depuy.com

DOW AGROSCIENCES LLC

9330 ZIONSVILLE ROAD
INDIANAPOLIS, INDIANA 46268

Overview: Manufactures herbicides and pesticides. Established: 1989. Parent Company: The Dow Chemical Company. Ownership: Public.

Key Statistics: Annual Sales: $1.2 billion. Number of Employees: 3,100.

Expertise/Education Sought: Agricultural engineering, chemical engineering, accounting, environmental engineering, finance, chemistry, law, biological engineering, computer programming

Alternatives to Full-Time Employment: Paid internships

Company-Paid Training is available.

International Assignments are available to employees.

Contact: Human Resources Department; Phone: 317-337-3000; Fax: 317-337-4166.

ELI LILLY AND COMPANY

LILLY CORPORATE CENTER, DC 1811
INDIANAPOLIS, INDIANA 46285

Overview: Creates and delivers innovative pharmaceutical-based health-care products. Established: 1876. Ownership: Public.

Key Statistics: Annual Sales: $8.5 billion. Number of Employees: 31,100.

Hiring History: Number of professional employees hired in 1997: 500. 1996: 300.

Expertise/Education Sought: Engineering, information technology

Alternatives to Full-Time Employment: Paid internships

Company-Paid Training is available.

International Assignments are available to employees.

Contact: Manager of U.S. Recruiting and Staffing; Phone: 800-892-9121; Fax: 317-277-9721.

Company URL: http://www.lilly.com

EXCEL INDUSTRIES, INC.

1120 NORTH MAIN STREET
ELKHART, INDIANA 46514

Overview: Manufactures automotive doors and windows. Established: 1928. Ownership: Public.

Key Statistics: Annual Sales: $887.7 million. Number of Employees: 6,107.

Expertise/Education Sought: Finance, sales, marketing, engineering, design engineering, product engineering, manufacturing engineering, research and development

Contact: Mr. Michael Paquette, Corporate Vice President of Human Resources; Phone: 219-264-2131; Fax: 219-266-5418.

FOREST RIVER, INC.

3010 COLLEGE AVENUE
GOSHEN, INDIANA 46526-5050

Overview: Builds tent-type travel trailers. Established: 1996. Ownership: Private.

Key Statistics: Annual Sales: $200 million. Number of Employees: 900.

Hiring History: Number of professional employees hired in 1996: 200.

Expertise/Education Sought: Accounting, engineering, plant management, operations, management information systems, purchasing

Contact: For individuals with previous experience: Human Relations Department; Phone: 219-533-5934; Fax: 219-537-1211.

FRANKENTHAL GROUP LLC

5425 HIGHWAY 31 EAST
CLARKSVILLE, INDIANA 47129-9228

Overview: Auctions automobiles and related motor vehicles. Established: 1996. Ownership: Private.

Key Statistics: Annual Sales: $81.8 million. Number of Employees: 285.

Hiring History: Number of professional employees hired in 1997: 30. 1996: 20.

Expertise/Education Sought: Automotive engineering

Contact: For individuals with previous experience: Human Resources Department; Phone: 812-283-0734; Fax: 812-280-8458.

Company URL: http://www.frankenthalgroup.com

GECOM CORPORATION

1025 BARACHEL LANE
GREENSBURG, INDIANA 47240-1269

Overview: Manufactures motor vehicle parts and accessories. Established: 1987. Parent Company: Mitsui Mining. Ownership: Private.

Key Statistics: Annual Sales: $148 million. Number of Employees: 1,300.

Hiring History: Number of professional employees hired in 1997: 9.

Expertise/Education Sought: Manufacturing engineering, automotive engineering, mechanical engineering, design engineering

Company-Paid Training is available.

Contact: For individuals with previous experience: Mr. Larry Mohnkern, Human Resources Manager; Phone: 812-663-1724; Fax: 812-663-2230.

GREAT LAKES CHEMICAL CORPORATION

1 GREAT LAKES BOULEVARD
WEST LAFAYETTE, INDIANA 47906

Overview: Manufactures specialty chemicals. Established: 1932. Ownership: Public.

Key Statistics: Annual Sales: $2.2 billion. Number of Employees: 7,000.

Hiring History: Number of professional employees hired in 1997: 63. 1996: 63.

Expertise/Education Sought: Chemistry, chemical engineering, accounting, information services

Alternatives to Full-Time Employment: Paid internships

International Assignments are available to employees.

Contact: Ms. Diane M. Foster, Human Resource Generalist. E-mail: dfoster@glcc.com. Job Hotline: 765-497-6377.

Company URL: http://www.greatlakeschem.com

HAYNES INTERNATIONAL, INC.

1020 WEST PARK AVENUE
KOKOMO, INDIANA 46901-6330

Overview: Manufactures high performance alloys. Established: 1989. Ownership: Private.

Key Statistics: Annual Sales: $226.4 million. Number of Employees: 931.

Hiring History: Number of professional employees hired in 1997: 50.

Expertise/Education Sought: Research, design engineering, manufacturing engineering, metallurgical engineering, mechanical engineering, electrical engineering

Alternatives to Full-Time Employment: Paid internships

Company-Paid Training is available.

International Assignments are available to employees.

Contact: Ms. Jane Ortman, Employment Representative; Phone: 765-456-6402; Fax: 765-456-6155. E-mail: jortman@haynesintl.com.

Company URL: http://www.haynesintl.com

JAYCO, INC.

903 SOUTH MAIN STREET
MIDDLEBURY, INDIANA 46540-9706

Overview: Manufactures motor homes. Established: 1968. Ownership: Private.

Key Statistics: Annual Sales: $251.4 million. Number of Employees: 1,500.

Hiring History: Number of professional employees hired in 1997: 200. 1996: 200.

Expertise/Education Sought: Sales, management information systems, marketing, accounting, finance, electrical engineering, manufacturing, interior design, product engineering

Alternatives to Full-Time Employment: Paid internships

Company-Paid Training is available.

Contact: Mr. Chuck Owens, Manager of Employee Relations; Phone: 219-825-5861; Fax: 219-825-0674.

Company URL: http://www.jayco.com

GEORGE KOCH SONS, INC.

10 SOUTH 11TH AVENUE
EVANSVILLE, INDIANA 47744

Overview: Manufactures industrial fabrications. Established: 1873. Ownership: Private.

Key Statistics: Annual Sales: $412 million. Number of Employees: 2,700.

Expertise/Education Sought: Business management, estimating, mechanical engineering

Alternatives to Full-Time Employment: Paid internships

Company-Paid Training is available.

International Assignments are available to employees.

Contact: For individuals with previous experience: Mr. Thomas Utley, Manager of Human Resources; Phone: 812-465-9735; Fax: 812-465-9737.

MEAD JOHNSON NUTRITIONAL

2400 WEST LLOYD EXPRESSWAY
EVANSVILLE, INDIANA 47721

Overview: Manufactures nutritional and pharmaceutical products. Established: 1967. Parent Company: Bristol-Myers Squibb Company. Ownership: Public.

Key Statistics: Annual Sales: $569.4 million. Number of Employees: 6,000.

Hiring History: Number of professional employees hired in 1997: 50.

Expertise/Education Sought: Marketing, finance, accounting, chemical engineering, packaging engineering, process engineering

Alternatives to Full-Time Employment: Paid internships

Company-Paid Training is available.

Mead Johnson Nutritional (continued)

International Assignments are available to employees.

Contact: Mr. E. J. Meier, Staffing Manager; Phone: 812-429-5000; Fax: 812-429-7148.

Company URL: http://www.meadjohnson.com

NIBCO, INC.

1516 MIDDLEBURY STREET
ELKHART, INDIANA 46514

Overview: Manufactures plumbing fixtures, piping, and valve fittings. Established: 1976. Ownership: Private.

Key Statistics: Annual Sales: $461.3 million. Number of Employees: 3,518.

Hiring History: Number of professional employees hired in 1996: 40.

Expertise/Education Sought: Production management, information systems

Alternatives to Full-Time Employment: Paid internships

Company-Paid Training is available.

International Assignments are available to employees.

Contact: For individuals with previous experience: Ms. Cathy Nafe, Staffing Manager.

Company URL: http://www.nibco.com

NIPSCO INDUSTRIES, INC.

801 EAST 86TH AVENUE
MERRILLVILLE, INDIANA 46410-6272

Overview: Energy/Utility-based holding company. Established: 1853. Ownership: Public.

Key Statistics: Annual Sales: $2.5 billion. Number of Employees: 6,400.

Expertise/Education Sought: Finance, accounting, management, engineering, marketing, sales

Alternatives to Full-Time Employment: Part-time jobs, paid internships

Company-Paid Training is available.

International Assignments are available to employees.

Contact: Ms. Violet G. Sistovaris, Organizational Development Executive; Phone: 219-647-5603; Fax: 219-647-5589. E-mail: vsistovaris@nipsco.com.

Company URL: http://www.nipsco.com

CORPORATE STATEMENT

NIPSCO Industries, Inc. (NI), is an energy/utility-based holding company that provides electric energy, natural gas, and water to the public through its six wholly-owned regulated subsidiaries. Building on the strength of its core business expertise, NI is moving into new, diversified activities with an eye on an ever-changing, competitive environment.

ONE CALL COMMUNICATIONS, INC.

801 CONGRESSIONAL BOULEVARD
CARMEL, INDIANA 46032

Overview: Provides long distance telephone communications services. Established: 1982. Ownership: Private.

Key Statistics: Annual Sales: $83.8 million. Number of Employees: 410.

Expertise/Education Sought: Sales, data processing, computer programming

Alternatives to Full-Time Employment: Paid internships

Company-Paid Training is available.

Contact: For individuals with previous experience: Ms. Linda Brummett, Human Resource Manager; Phone: 317-843-1300; Fax: 317-580-7491. Job Hotline: 800-313-3160.

Company URL: http://www.onecall.net

PRECISION PLASTICS, INC.

900 WEST CONNEXION WAY
COLUMBIA CITY, INDIANA 46725-2751

Overview: Manufactures injection molded plastics. Established: 1981. Ownership: Private.

Key Statistics: Annual Sales: $30 million. Number of Employees: 415.

Expertise/Education Sought: Chemistry, accounting, finance, data processing, injection molding

Alternatives to Full-Time Employment: Paid internships

Company-Paid Training is available.

Contact: For individuals with previous experience: Ms. Claudia Keirn, Human Resources Manager; Phone: 219-244-6114; Fax: 219-244-5995. E-mail: ckeirn@pplastic.com.

REILLY INDUSTRIES, INC.

300 NORTH MERIDIAN STREET, SUITE 1500
INDIANAPOLIS, INDIANA 46229-3745

Overview: Manufactures asphalt, synthetic rubber, and chemicals. Established: 1905. Ownership: Private.

Key Statistics: Annual Sales: $207.1 million. Number of Employees: 1,019.

Hiring History: Number of professional employees hired in 1997: 15. 1996: 8.

Expertise/Education Sought: Chemistry, environmental engineering, chemical engineering, industrial engineering

Alternatives to Full-Time Employment: Paid internships

Company-Paid Training is available.

International Assignments are available to employees.

Contact: Mr. Brent Schoolcraft, Personnel Administrator; Phone: 317-247-8141, ext. 6551; Fax: 317-248-6545.

Company URL: http://www.reillyind.com

REINKE CONSTRUCTION CORPORATION

1216 WEST SAMPLE STREET
SOUTH BEND, INDIANA 46619

Overview: Provides general contractor services for schools and businesses. Established: 1946. Ownership: Private.

Key Statistics: Annual Sales: $20 million. Number of Employees: 100.

Hiring History: Number of professional employees hired in 1997: 3.

Expertise/Education Sought: Construction management

Alternatives to Full-Time Employment: Paid internships

Contact: Mr. James T. Hiss, Chief Operating Officer; Phone: 219-287-1561; Fax: 219-232-1239.

RIETH-RILEY CONSTRUCTION COMPANY, INC.

PO BOX 477
GOSHEN, INDIANA 46527

Overview: Provides highway, bridge, asphalt, and concrete construction services.

Expertise/Education Sought: Civil engineering, construction engineering, accounting, construction management, quality control

Alternatives to Full-Time Employment: Paid internships

Company-Paid Training is available.

Contact: Mr. James W. Eaton, Human Resources Manager; Phone: 219-875-5183; Fax: 219-875-8405.

ROCHE DIAGNOSTICS BOEHRINGER MANNHEIM CORPORATION
9115 HAGUE ROAD
INDIANAPOLIS, INDIANA 46250

OVERVIEW: Manufactures medical and surgical equipment and pharmaceuticals.

EXPERTISE/EDUCATION SOUGHT: Chemical engineering, mechanical engineering, software engineering/development, medical technology, manufacturing engineering

ALTERNATIVES TO FULL-TIME EMPLOYMENT: Part-time jobs

CONTACT: For individuals with previous experience: Human Resources; Phone: 317-845-2000.

SELMER COMPANY, INC.
600 INDUSTRIAL PARKWAY
ELKHART, INDIANA 46516-5414

OVERVIEW: Manufactures brass and woodwind musical instruments and parts and pianos. Established: 1993. Parent Company: Steinway, Inc. Ownership: Public.

KEY STATISTICS: Annual Sales: $257.9 million. Number of Employees: 1,959.

HIRING HISTORY: Number of professional employees hired in 1997: 40. 1996: 50.

EXPERTISE/EDUCATION SOUGHT: Engineering, manufacturing engineering

ALTERNATIVES TO FULL-TIME EMPLOYMENT: Paid internships

COMPANY-PAID TRAINING is available.

CONTACT: Ms. Judy Miller, Employee Relations Supervisor; Phone: 219-522-1675, ext. 315; Fax: 219-522-0334.

SHOE CARNIVAL, INC.
8233 BAUMGART ROAD
EVANSVILLE, INDIANA 47711-1599

OVERVIEW: Operates chain of shoe and accessory stores. Established: 1978. Ownership: Public.

KEY STATISTICS: Annual Sales: $233.9 million. Number of Employees: 1,880.

HIRING HISTORY: Number of professional employees hired in 1997: 196. 1996: 160.

EXPERTISE/EDUCATION SOUGHT: Retail management, information systems

ALTERNATIVES TO FULL-TIME EMPLOYMENT: Paid internships

COMPANY-PAID TRAINING is available.

CONTACT: Mr. Jim Merold, Director of Recruiting; Phone: 812-867-4578; Fax: 812-867-4247. E-mail: recruit@scbl.com. Job Hotline: 800-746-3462.

COMPANY URL: http://www.shoecarnival.com

STURGES COMPANY, INC.
3466 STELLHORN ROAD
FORT WAYNE, INDIANA 46815-4630

OVERVIEW: Operates supermarkets. Established: 1995. Ownership: Private.

KEY STATISTICS: Annual Sales: $50 million. Number of Employees: 220.

EXPERTISE/EDUCATION SOUGHT: Marketing, sales, manufacturing engineering

CONTACT: For individuals with previous experience: Human Resources Department; Phone: 219-485-5861; Fax: 219-485-3637.

TRI-MANUFACTURING, INC.
333 SOUTH 3RD STREET
TERRE HAUTE, INDIANA 47807-3410

OVERVIEW: Manufactures aerospace components. Established: 1988. Parent Company: UNC, Inc. Ownership: Public.

KEY STATISTICS: Annual Sales: $30.2 million. Number of Employees: 325.

HIRING HISTORY: Number of professional employees hired in 1997: 10.

EXPERTISE/EDUCATION SOUGHT: Mechanical engineering, electrical engineering

COMPANY-PAID TRAINING is available.

CONTACT: Mr. David Heavin, Director of Human Resources; Phone: 812-234-1591.

WABASH NATIONAL CORPORATION
1000 SAGAMORE PARKWAY SOUTH
LAFAYETTE, INDIANA 47905

MAILING ADDRESS: PO Box 6129, Lafayette, IN 47903

OVERVIEW: Manufactures truck trailers, vans, bimodal vehicles, and related equipment. Established: 1985. Ownership: Public.

KEY STATISTICS: Annual Sales: $631.5 million. Number of Employees: 4,000.

HIRING HISTORY: Number of professional employees hired in 1997: 7. 1996: 7.

EXPERTISE/EDUCATION SOUGHT: Computer-aided design, mechanical engineering, industrial engineering, computer programming

ALTERNATIVES TO FULL-TIME EMPLOYMENT: Part-time jobs, paid internships

COMPANY-PAID TRAINING is available.

INTERNATIONAL ASSIGNMENTS are available to employees.

CONTACT: Mr. Chuck Fish, Vice President Human Resources; Phone: 765-448-1591; Fax: 765-477-9405.

COMPANY URL: http://www.wncwabash.com/wabash

WPTA-TV, INC.
3401 BUTLER ROAD
FORT WAYNE, INDIANA 46808

OVERVIEW: Operates TV broadcasting stations. Established: 1989. Parent Company: Granite Broadcasting. Ownership: Public.

KEY STATISTICS: Annual Sales: $9.1 million. Number of Employees: 100.

HIRING HISTORY: Number of professional employees hired in 1997: 12. 1996: 15.

EXPERTISE/EDUCATION SOUGHT: Sales, engineering, production, journalism, production, production

ALTERNATIVES TO FULL-TIME EMPLOYMENT: Internships

CONTACT: For individuals with previous experience: Human Resources; Phone: 219-483-0584; Fax: 219-482-8750. E-mail: jand@wpta.com.

ZIMMER, INC.
PO BOX 708
WARSAW, INDIANA 46581-0708

OVERVIEW: Manufactures orthopedic products. Established: 1972. Parent Company: Bristol-Myers Squibb Company. Ownership: Public.

KEY STATISTICS: Annual Sales: $340 million. Number of Employees: 4,450.

EXPERTISE/EDUCATION SOUGHT: Design engineering, sales, computer programming, marketing, finance

ALTERNATIVES TO FULL-TIME EMPLOYMENT: Paid internships

COMPANY-PAID TRAINING is available.

INTERNATIONAL ASSIGNMENTS are available to employees.

CONTACT: For individuals with previous experience: Salaried Recruiting; Phone: 219-267-6131. Job Hotline: 800-267-JOBS.

IOWA

ACCESS AIR, INC.
601 LOCUST STREET, SUITE 330
DES MOINES, IOWA 50309-3741

OVERVIEW: Provides scheduled air transportation service. Established: 1996. Ownership: Private.

KEY STATISTICS: Annual Sales: $58 million.

Access Air, Inc. (continued)

Expertise/Education Sought: Aviation, aeronautical engineering
Contact: Human Resource Department; Phone: 515-245-3745; Fax: 515-247-2216. E-mail: accessair@netins.com.

COMMUNICATIONS DATA SERVICES, INC.
1901 BELL AVENUE
DES MOINES, IOWA 50315-1067

Overview: Provides information retrieval, data processing and magazine/newspaper subscription fulfillment services. Established: 1982. Parent Company: Hearst Corporation. Ownership: Private.
Key Statistics: Annual Sales: $84.7 million. Number of Employees: 2,700.
Hiring History: Number of professional employees hired in 1997: 50.
Expertise/Education Sought: Computer programming, marketing, management
Alternatives to Full-Time Employment: Paid internships
Company-Paid Training is available.
Contact: Ms. Brooke Tyler, Recruitment Specialist; Phone: 515-280-4130; Fax: 515-246-6687. E-mail: btyler@cdsfulfillment.com. Job Hotline: 800-290-1996.
Company URL: http://www.fulfillment.com

CUSTOM-PAK, INC.
86 16TH AVENUE NORTH
CLINTON, IOWA 52732-3083

Mailing Address: PO Box 3083, Clinton, IA 52732-3083
Overview: Blow molds, designs and manufactures industrial parts. Established: 1974. Ownership: Private.
Key Statistics: Annual Sales: $44.9 million. Number of Employees: 600.
Hiring History: Number of professional employees hired in 1997: 98. 1996: 15.
Expertise/Education Sought: Electrical engineering, mechanical engineering, hydraulic engineering, robotic engineering
Alternatives to Full-Time Employment: Paid internships
Company-Paid Training is available.
Contact: For individuals with previous experience: Ms. Barb Soukup, Human Resources Development; Phone: 319-242-1801; Fax: 319-244-5362. E-mail: soukup@custom-pak.com.
Company URL: http://www.custom-pak.com

DOERFER ENGINEERING CORPORATION
201 WASHINGTON STREET
CEDAR FALLS, IOWA 50613-2729

Overview: Provides engineering services for ship building. Established: 1995. Ownership: Private.
Key Statistics: Annual Sales: $7.8 million. Number of Employees: 115.
Hiring History: Number of professional employees hired in 1997: 5. 1996: 6.
Expertise/Education Sought: Engineering
Contact: For individuals with previous experience: Ms. Julie Setaro, Human Resources Representative; Phone: 319-277-3110.

HY-VEE, INC.
5820 WESTOWN PARKWAY
WEST DES MOINES, IOWA 50266

Overview: Operates chain of food stores. Established: 1930. Ownership: Private.
Key Statistics: Annual Sales: $2.8 billion. Number of Employees: 35,000.
Hiring History: Number of professional employees hired in 1997: 2,000. 1996: 280.
Expertise/Education Sought: Retail, sales, business administration
Alternatives to Full-Time Employment: Part-time jobs, paid internships
Company-Paid Training is available.
Contact: For individuals with previous experience: Mr. Jerry Willis, Manager of Human Resources; Phone: 515-226-2800; Fax: 515-267-2935.
Company URL: http://www.hy-vee.com

IES INDUSTRIES, INC.
200 1ST STREET, SE
CEDAR RAPIDS, IOWA 52401-1409

Overview: Provides electric and natural gas services. Established: 1986. Ownership: Public.
Key Statistics: Annual Sales: $755 million. Number of Employees: 2,016.
Expertise/Education Sought: Nuclear engineering, mechanical engineering, electrical engineering, civil engineering
Alternatives to Full-Time Employment: Paid internships
Contact: For individuals with previous experience: Human Resources Department; Phone: 800-275-2099.

INTERMEC TECHNOLOGIES CORPORATION
550 2ND STREET, SE
CEDAR RAPIDS, IOWA 52401-2023

Overview: Designs, manufactures, and markets computerized systems for inventory control. Established: 1997. Ownership: Private.
Key Statistics: Annual Sales: $250 million. Number of Employees: 954.
Expertise/Education Sought: Administration, engineering, software engineering/development
Alternatives to Full-Time Employment: Paid internships
International Assignments are available to employees.
Contact: For individuals with previous experience: Human Resources Department; Phone: 319-369-3659; Fax: 319-369-3791.
Company URL: http://www.intermec.com

KEOKUK STEEL CASTINGS COMPANY
600 MORGAN STREET
KEOKUK, IOWA 52632

Overview: Manufactures steel and related products. Established: 1991. Parent Company: NACO, Inc. Ownership: Private.
Key Statistics: Annual Sales: $50 million. Number of Employees: 450.
Hiring History: Number of professional employees hired in 1996: 15.
Expertise/Education Sought: Metallurgy, industrial technician
Alternatives to Full-Time Employment: Paid internships
Company-Paid Training is available.
International Assignments are available to employees.
Contact: Mr. William Bartlett, Personnel Director; Phone: 319-524-2661; Fax: 319-524-1382.

MCLEOD USA, INC.
6400 C STREET, SW, PO BOX 3177
CEDAR RAPIDS, IOWA 52406-3177

Overview: Provides local and long-distance telecommunications services. Established: 1996. Ownership: Private.
Key Statistics: Annual Sales: $69.1 million. Number of Employees: 543.
Hiring History: Number of professional employees hired in 1997: 700.
Expertise/Education Sought: Sales, management information systems, engineering
Alternatives to Full-Time Employment: Paid internships
Company-Paid Training is available.
Contact: Human Resources Department; Phone: 319-369-0000; Fax: 319-298-6942. E-mail: resumes@mcleodusa.com.
Company URL: http://www.mcleodusa.com

MIDAMERICAN ENERGY
666 GRAND AVENUE
DES MOINES, IOWA 50309

OVERVIEW: Public utility.
HIRING HISTORY: Number of professional employees hired in 1996: 300.
EXPERTISE/EDUCATION SOUGHT: Electrical engineering, mechanical engineering, accounting, information technology
ALTERNATIVES TO FULL-TIME EMPLOYMENT: Paid internships
COMPANY-PAID TRAINING is available.
CONTACT: For individuals with previous experience: Human Relations Department. Job Hotline: 888-257-1369.
COMPANY URL: http://www.midamerican.com

SHIVE-HATTERY, INC.
115 3RD STREET, SE, 9TH FLOOR
CEDAR RAPIDS, IOWA 52401

OVERVIEW: Provides engineering and architectural consulting services. Established: 1896. Ownership: Private.
KEY STATISTICS: Number of Employees: 240.
HIRING HISTORY: Number of professional employees hired in 1997: 17. 1996: 13.
EXPERTISE/EDUCATION SOUGHT: Architecture, electrical engineering, mechanical engineering, process engineering, structural engineering
ALTERNATIVES TO FULL-TIME EMPLOYMENT: Part-time jobs, paid internships
COMPANY-PAID TRAINING is available.
CONTACT: Ms. Gloria J. Frost, Vice President, Director of Human Resources, PO Box 1599, Cedar Rapids, Iowa, 52406-1599; Phone: 319-362-0313; Fax: 319-362-2883. E-mail: gfrost@shive-hattery.com.

CORPORATE STATEMENT

Shive-Hattery, Inc., is a 220-person multidiscipline consulting, engineering, and architectural firm with in-house disciplines required to fully serve the industrial, retail, development, environmental, traffic/transportation, education, medical, and municipal client sectors.

SMITHWAY MOTOR XPRESS CORPORATION
PO BOX 404
FORT DODGE, IOWA 50501

OVERVIEW: Transports building materials. Established: 1995. Ownership: Public.
KEY STATISTICS: Annual Sales: $93.7 million. Number of Employees: 550.
HIRING HISTORY: Number of professional employees hired in 1997: 500. 1996: 500.
EXPERTISE/EDUCATION SOUGHT: Transportation, computer science
ALTERNATIVES TO FULL-TIME EMPLOYMENT: Paid internships
COMPANY-PAID TRAINING is available.
CONTACT: Mr. Tom Nelson, Human Resource Director; Phone: 515-576-7418; Fax: 515-576-8794. E-mail: tomn@smxc.com.
COMPANY URL: http://www.smxc.com

TAYLOR BALL, LLC
6000 WESTOWN PARKWAY, SUITE 200W
WEST DES MOINES, IOWA 50266-7711

OVERVIEW: Provides government construction services. Established: 1980. Ownership: Private.
KEY STATISTICS: Annual Sales: $133.6 million. Number of Employees: 400.
HIRING HISTORY: Number of professional employees hired in 1996: 30.
EXPERTISE/EDUCATION SOUGHT: Construction engineering, construction, project management
ALTERNATIVES TO FULL-TIME EMPLOYMENT: Paid internships
COMPANY-PAID TRAINING is available.
CONTACT: For individuals with previous experience: Ms. Laura Hanke, Human Resources Manager; Phone: 515-267-6004; Fax: 515-267-6068. E-mail: lhanke@aol.com.
COMPANY URL: http://www.taylorball.com

VERMEER MANUFACTURING COMPANY
2411 HIGHWAY 102
PELLA, IOWA 50219-7814

MAILING ADDRESS: PO Box 200, Pella, IA 50219
OVERVIEW: Manufactures heavy industrial and farm equipment. Established: 1948. Ownership: Private.
KEY STATISTICS: Annual Sales: $450 million. Number of Employees: 2,200.
HIRING HISTORY: Number of professional employees hired in 1997: 25. 1996: 50.
EXPERTISE/EDUCATION SOUGHT: Mechanical engineering, design engineering, computer programming, marketing, sales, accounting, computer-aided design, information systems, technical writing, environmental engineering
ALTERNATIVES TO FULL-TIME EMPLOYMENT: Paid internships
COMPANY-PAID TRAINING is available.
CONTACT: For individuals with previous experience: Mr. Larry Nesset, Human Resource Manager; Phone: 515-621-7459; Fax: 515-621-7722. E-mail: lnesset@vermeermfg.com.
COMPANY URL: http://www.vermeermfg.com

V-T INDUSTRIES, INC.
1000 INDUSTRIAL PARK
HOLSTEIN, IOWA 51025

OVERVIEW: Manufactures household doors and related products. Established: 1956. Ownership: Private.
KEY STATISTICS: Annual Sales: $39.9 million. Number of Employees: 534.
HIRING HISTORY: Number of professional employees hired in 1997: 12. 1996: 6.
EXPERTISE/EDUCATION SOUGHT: Product management, manufacturing
ALTERNATIVES TO FULL-TIME EMPLOYMENT: Paid internships
COMPANY-PAID TRAINING is available.
CONTACT: Mr. James Cooney, Director of Human Resources; Phone: 712-368-4381; Fax: 712-368-4667. E-mail: jccooney@pionet.com.

WATERLOO INDUSTRIES, INC.
PO BOX 2095
WATERLOO, IOWA 50704

OVERVIEW: Manufactures metal tool boxes. Parent Company: FortuneBrand Home And Office.
EXPERTISE/EDUCATION SOUGHT: Manufacturing engineering, design engineering, mechanical engineering, industrial engineering, marketing, management
ALTERNATIVES TO FULL-TIME EMPLOYMENT: Paid internships
CONTACT: Human Resources Department; Fax: 319-232-3924.

KANSAS

Cessna

CESSNA AIRCRAFT COMPANY

PO BOX 7704
WICHITA, KANSAS 67277

OVERVIEW: Manufactures aircraft. Established: 1927. Parent Company: Textron, Inc. Ownership: Public.

KEY STATISTICS: Annual Sales: $1.3 billion. Number of Employees: 10,000.

EXPERTISE/EDUCATION SOUGHT: Aeronautical engineering, mechanical engineering, electrical engineering, manufacturing engineering, industrial engineering, quality engineering, computer science, computer programming

CONTACT: Mr. John Dandurand, Employment Representative; Phone: 316-941-7753; Fax: 316-941-7865.

COMPANY URL: http://www.cessna.textron.com

CORPORATE STATEMENT

These are extraordinary times for Cessna. Though the name Cessna has always been synonymous with outstanding aircraft, the company's products now dominate the market more than ever. Today, more than half of the business jets being sold worldwide are Cessna Citations. The introduction of eight new business jets in the last ten years, the restart of the single-engine piston line, ten years of continued growth in sales and employment, domination of market share, and a commitment to employees make Cessna "simply the best."

DILLON COMPANIES, INC.

700 EAST 30TH AVENUE
HUTCHINSON, KANSAS 67502-8435

OVERVIEW: Operates chain of retail stores. Established: 1983. Parent Company: The Kroger Company. Ownership: Public.

KEY STATISTICS: Annual Sales: $6.4 billion. Number of Employees: 46,000.

EXPERTISE/EDUCATION SOUGHT: Retail

CONTACT: For individuals with previous experience: Mr. Harold Ryan, Vice President of Human Resources; Phone: 316-663-6801; Fax: 316-663-3378.

KOCH GLITSCH

4111 EAST 37TH STREET NORTH
WICHITA, KANSAS 67220-3203

OVERVIEW: Provides engineering services to the oil and gas industries. Established: 1945. Parent Company: Koch Industries, Inc. Ownership: Private.

KEY STATISTICS: Annual Sales: $246.8 million. Number of Employees: 2,500.

HIRING HISTORY: Number of professional employees hired in 1997: 350. 1996: 100.

EXPERTISE/EDUCATION SOUGHT: Chemical engineering, mechanical engineering

ALTERNATIVES TO FULL-TIME EMPLOYMENT: Paid internships

INTERNATIONAL ASSIGNMENTS are available to employees.

CONTACT: For individuals with previous experience: Ms. Trish Calvert, Personnel Director; Phone: 316-828-5110; Fax: 316-828-8093. E-mail: calvertt@kochind.com.

LEARJET, INC.

1 LEARJET WAY, PO BOX 7707
WICHITA, KANSAS 67209

OVERVIEW: Manufactures jets. Established: 1990. Parent Company: Bombardier Corporation. Ownership: Private.

KEY STATISTICS: Annual Sales: $551.9 million. Number of Employees: 4,975.

EXPERTISE/EDUCATION SOUGHT: Aeronautical engineering, design engineering, avionics, mechanical engineering, industrial engineering

ALTERNATIVES TO FULL-TIME EMPLOYMENT: Internships

COMPANY-PAID TRAINING is available.

INTERNATIONAL ASSIGNMENTS are available to employees.

CONTACT: Ms. Sandy Brown, Employment Department, PO Box 7707, Wichita, Kansas, 67277-7707; Phone: 316-946-2200, ext. 3578; Fax: 316-946-2163. Job Hotline: 316-946-2562.

SEABOARD CORPORATION

9000 WEST 67TH STREET
SHAWNEE MISSION, KANSAS 66202-3631

OVERVIEW: Diversified international agribusiness and transportation company engaged in poultry and pork processing, commodity merchandising, baking, shipping, produce storage, and distribution. Established: 1946. Ownership: Private.

KEY STATISTICS: Annual Sales: $1.5 billion. Number of Employees: 11,699.

HIRING HISTORY: Number of professional employees hired in 1996: 150.

EXPERTISE/EDUCATION SOUGHT: Finance, business management, data processing, engineering

ALTERNATIVES TO FULL-TIME EMPLOYMENT: Internships

COMPANY-PAID TRAINING is available.

INTERNATIONAL ASSIGNMENTS are available to employees.

CONTACT: Mr. Peter Mirakian, Senior Human Resources Manager; Phone: 913-676-8800; Fax: 913-676-8872.

WOLF CREEK NUCLEAR OPERATING CORPORATION

1550 OXEN LANE, NE
BURLINGTON, KANSAS 66839

MAILING ADDRESS: PO Box 411, Burlington, KS 66839-0411

OVERVIEW: Nuclear power utility. Established: 1986. Ownership: Private.

KEY STATISTICS: Annual Sales: $155.5 million. Number of Employees: 1,006.

EXPERTISE/EDUCATION SOUGHT: Mechanical engineering, electrical engineering, nuclear engineering, civil engineering, chemical engineering

ALTERNATIVES TO FULL-TIME EMPLOYMENT: Paid internships

CONTACT: Mr. David Reynolds, Human Resources Specialist; Phone: 316-364-8831, ext. 8673. E-mail: dareyno@wcnoc.com.

YELLOW SERVICES, INC.

10990 ROE AVENUE
OVERLAND PARK, KANSAS 66214

OVERVIEW: Provides information systems services for Yellow Corporation. Parent Company: Yellow Corporation.

EXPERTISE/EDUCATION SOUGHT: Computer science, business, information systems

ALTERNATIVES TO FULL-TIME EMPLOYMENT: Paid internships

COMPANY-PAID TRAINING is available.

INTERNATIONAL ASSIGNMENTS are available to employees.

CONTACT: For individuals with previous experience: Mr. Kip Esry, Human Resources Manager, MS B105, Overland Park, Kansas, 66214; Phone: 913-344-3411; Fax: 913-344-4871. E-mail: kip.esry@yellowcorp.com. Job Hotline: 913-344-5900.

COMPANY URL: http://www.yellowcorp.com

KENTUCKY

ATLANTIS PLASTICS, INC.
390 COMMUNITY DRIVE
HENDERSON, KENTUCKY 42420

OVERVIEW: Manufactures injection molded plastics. Established: 1985. Ownership: Public.

KEY STATISTICS: Annual Sales: $70 million. Number of Employees: 800.

HIRING HISTORY: Number of professional employees hired in 1997: 200. 1996: 50.

EXPERTISE/EDUCATION SOUGHT: Mechanical engineering, engineering

COMPANY-PAID TRAINING is available.

CONTACT: For individuals with previous experience: Human Resources Department, 1870 The Exchange, Suite 200, Henderson, Kentucky, 42420; Phone: 502-827-9801; Fax: 502-827-9841.

ATR WIRE & CABLE, INC.
1857 SOUTH DANVILLE BYPASS
DANVILLE, KENTUCKY 40422-9499

OVERVIEW: Manufactures tire cord. Established: 1981. Parent Company: Tokyo Rope. Ownership: Private.

KEY STATISTICS: Annual Sales: $85 million. Number of Employees: 574.

HIRING HISTORY: Number of professional employees hired in 1997: 50. 1996: 125.

EXPERTISE/EDUCATION SOUGHT: Manufacturing

COMPANY-PAID TRAINING is available.

CONTACT: For individuals with previous experience: Mr. Michael Bewley, Manager of Human Resources; Phone: 606-236-9220, ext. 254; Fax: 606-236-9489.

BROWN & WILLIAMSON
401 SOUTH 4TH AVENUE, SUITE 200
LOUISVILLE, KENTUCKY 40202

MAILING ADDRESS: PO Box 35090, Louisville, KY 40232

OVERVIEW: Manufactures and markets cigarettes and other tobacco products. Established: 1927. Parent Company: BAT. Ownership: Private.

KEY STATISTICS: Annual Sales: $1.3 billion. Number of Employees: 4,800.

EXPERTISE/EDUCATION SOUGHT: Accounting, marketing, business management, data processing, human resources, auditing

ALTERNATIVES TO FULL-TIME EMPLOYMENT: Internships

INTERNATIONAL ASSIGNMENTS are available to employees.

CONTACT: Ms. Carolyn Craul, Recruiter; Phone: 502-568-7000; Fax: 502-568-7120.

CLARION MANUFACTURING CORPORATION OF AMERICA
237 BEAVER ROAD
WALTON, KENTUCKY 41094

OVERVIEW: Manufactures car stereos and cellular phones. Established: 1986. Parent Company: Clarion, Ltd. Ownership: Private.

KEY STATISTICS: Annual Sales: $72.8 million. Number of Employees: 228.

EXPERTISE/EDUCATION SOUGHT: Accounting, finance, data processing, electrical engineering, electronics/electronics engineering, purchasing

ALTERNATIVES TO FULL-TIME EMPLOYMENT: Paid internships

COMPANY-PAID TRAINING is available.

CONTACT: Ms. Mary Schumer, Personnel Generalist; Phone: 606-485-6600; Fax: 606-485-4099.

COMPUTER SERVICES, INC.
3901 TECHNOLOGY DRIVE
PADUCAH, KENTUCKY 42001-2503

OVERVIEW: Provides data processing services. Established: 1965. Ownership: Private.

KEY STATISTICS: Annual Sales: $25.1 million. Number of Employees: 400.

EXPERTISE/EDUCATION SOUGHT: Data processing

ALTERNATIVES TO FULL-TIME EMPLOYMENT: Part-time jobs

CONTACT: For individuals with previous experience: Mrs. Ann Carney, Human Resources Department Manager; Phone: 502-442-7361, ext. 10103; Fax: 502-442-7361. E-mail: acarney@csiweb.com.

COMPANY URL: http://www.csiweb.com

COURTAULDS COATINGS
400 SOUTH 13TH STREET
LOUISVILLE, KENTUCKY 40203

OVERVIEW: Manufactures fibers, coatings, and packaging materials. Established: 1980. Parent Company: Courtaulds, PLC. Ownership: Private.

KEY STATISTICS: Annual Sales: $850 million. Number of Employees: 4,200.

HIRING HISTORY: Number of professional employees hired in 1997: 500. 1996: 500.

EXPERTISE/EDUCATION SOUGHT: Sales, chemistry, accounting, information technology

ALTERNATIVES TO FULL-TIME EMPLOYMENT: Paid internships

COMPANY-PAID TRAINING is available.

INTERNATIONAL ASSIGNMENTS are available to employees.

CONTACT: For individuals with previous experience: Ms. Betty Garmon, Personnel Manager; Phone: 502-588-9292; Fax: 502-588-9788.

EQUITABLE BAG COMPANY, INC.
7600 EMPIRE DRIVE
FLORENCE, KENTUCKY 41042-1010

OVERVIEW: Manufactures paper, plastic, and foil bags. Established: 1994. Ownership: Private.

KEY STATISTICS: Annual Sales: $75.8 million. Number of Employees: 588.

HIRING HISTORY: Number of professional employees hired in 1997: 10. 1996: 15.

EXPERTISE/EDUCATION SOUGHT: Engineering, graphic arts, management information systems, industrial engineering

ALTERNATIVES TO FULL-TIME EMPLOYMENT: Paid internships

CONTACT: For individuals with previous experience: Human Resources; Phone: 606-371-2150; Fax: 606-647-5303.

HITACHI AUTOMOTIVE PRODUCTS USA
955 WARWICK ROAD, PO BOX 510
HARRODSBURG, KENTUCKY 40330-1067

OVERVIEW: Manufactures engine electrical equipment. Established: 1985. Parent Company: Hitachi America, Ltd. Ownership: Private.

KEY STATISTICS: Annual Sales: $70.7 million. Number of Employees: 1,140.

HIRING HISTORY: Number of professional employees hired in 1997: 100. 1996: 150.

EXPERTISE/EDUCATION SOUGHT: Electrical engineering, mechanical engineering

ALTERNATIVES TO FULL-TIME EMPLOYMENT: Paid internships

COMPANY-PAID TRAINING is available.

CONTACT: Ms. Stephanie Rogers, Recruiter; Phone: 606-734-9451; Fax: 606-734-5309. E-mail: stephanie.rogers@hap.com.

NATIONAL PROCESSING COMPANY
1231 DURRETT LANE
LOUISVILLE, KENTUCKY 40285

Overview: Provides data processing services. Established: 1981. Parent Company: National Processing, Inc. Ownership: Public.

Key Statistics: Annual Sales: $373.7 million. Number of Employees: 8,200.

Hiring History: Number of professional employees hired in 1997: 6. 1996: 4.

Expertise/Education Sought: Data processing, finance, accounting, computer programming

Alternatives to Full-Time Employment: Paid internships

Company-Paid Training is available.

International Assignments are available to employees.

Contact: Mr. Ron Tomes, Vice President of Human Resources; Phone: 502-326-7041; Fax: 502-364-2041. E-mail: rtomes@mail.npc.net.

Company URL: http://www.npc.net

SCOTTY'S CONTRACTING STONE COMPANY
2323 BARREN RIVER ROAD
BOWLING GREEN, KENTUCKY 42101-9483

Overview: Provides highway paving and drainage system construction services. Established: 1972. Ownership: Private.

Key Statistics: Annual Sales: $51 million. Number of Employees: 65.

Hiring History: Number of professional employees hired in 1997: 15.

Expertise/Education Sought: Construction engineering, surveying, accounting, finance, civil engineering

Contact: For individuals with previous experience: Mr. Brad Walker, Human Resources Manager; Phone: 502-781-3998; Fax: 502-781-3690.

STEEL TECHNOLOGIES, INC.
15415 SHELBYVILLE ROAD
LOUISVILLE, KENTUCKY 40245

Overview: Processes flat rolled steel for manufacturing purposes. Established: 1971. Ownership: Public.

Key Statistics: Annual Sales: $294.2 million. Number of Employees: 424.

Expertise/Education Sought: Mechanical engineering, electrical engineering, metallurgical engineering, accounting, data processing, engineering technology

Alternatives to Full-Time Employment: Paid internships

Contact: Mrs. Connie Glendenning, Associate Manager of Human Resources; Phone: 502-245-2110, ext. 249; Fax: 502-244-8730. E-mail: cglendenning@steeltechnologies.com.

TOKICO (USA), INC.
301 MAYDE ROAD
BEREA, KENTUCKY 40403-9777

Overview: Manufactures suspension and brake systems. Established: 1987. Parent Company: TOKICO, Ltd. Ownership: Private.

Key Statistics: Annual Sales: $210.7 million. Number of Employees: 590.

Hiring History: Number of professional employees hired in 1997: 25. 1996: 10.

Expertise/Education Sought: Mechanical engineering, industrial engineering, electrical technology, electrical engineering, materials management, management information systems

Alternatives to Full-Time Employment: Paid internships

Company-Paid Training is available.

Contact: Mr. Bill Wildman, Director of General Affairs; Phone: 606-986-2359; Fax: 606-986-7114.

TOPY CORPORATION
980 CHENAULT ROAD
FRANKFORT, KENTUCKY 40601

Mailing Address: PO Box 1010, Frankfort, KY 40602-1010

Overview: Manufactures automotive parts. Established: 1985. Parent Company: Topy Industries, Ltd. Ownership: Private.

Key Statistics: Annual Sales: $46.1 million. Number of Employees: 500.

Hiring History: Number of professional employees hired in 1996: 10.

Expertise/Education Sought: Mechanical engineering, accounting, finance, data processing

Company-Paid Training is available.

Contact: Mr. Tak Fujibayashi, Executive Vice President; Phone: 502-695-6163, ext. 230; Fax: 502-875-5424.

UNION UNDERWEAR COMPANY, INC.
1 FRUIT OF THE LOOM DRIVE
BOWLING GREEN, KENTUCKY 42103

Overview: Manufactures undergarments and socks. Established: 1985. Parent Company: Fruit Of The Loom. Ownership: Public.

Key Statistics: Annual Sales: $2.4 billion. Number of Employees: 30,000.

Expertise/Education Sought: Accounting, marketing, finance, sales, manufacturing engineering, electrical engineering, graphic arts, design, market research

Alternatives to Full-Time Employment: Part-time jobs, paid internships

International Assignments are available to employees.

Contact: For individuals with previous experience: Human Resources; Phone: 502-781-6400.

Company URL: http://www.fruit.com

VOGT-NEM, INC.
4000 DUPONT CIRCLE, SUITE 400
LOUISVILLE, KENTUCKY 40207

Overview: Provides pipe fitting, platework, and boiler shop services. Established: 1880. Parent Company: NEM, Bv, The Netherlands. Ownership: Private.

Key Statistics: Annual Sales: $30 million. Number of Employees: 450.

Hiring History: Number of professional employees hired in 1997: 27. 1996: 14.

Expertise/Education Sought: Manufacturing engineering, industrial engineering, structural engineering, electrical engineering, project engineering, mechanical engineering

Alternatives to Full-Time Employment: Paid internships

Company-Paid Training is available.

International Assignments are available to employees.

Contact: For individuals with previous experience: Ms. Virginia McCombs, Director of Human Resources; Phone: 502-899-4604; Fax: 502-899-4690. E-mail: hr@vogt-nem.com.

Company URL: http://www.vogtnem.com

VOGT VALVE COMPANY
901 WEST MAGNOLIA AVENUE
LOUISVILLE, KENTUCKY 40208-2225

Overview: Manufactures plumbing and heating valves and pipe fittings. Established: 1996. Parent Company: Edward Valves. Ownership: Private.

Key Statistics: Annual Sales: $48 million. Number of Employees: 280.

Hiring History: Number of professional employees hired in 1997: 5. 1996: 355.

Expertise/Education Sought: Manufacturing

Company-Paid Training is available.

International Assignments are available to employees.

Contact: Mr. Mark Cain, Director of Human Resources; Phone: 502-635-3305; Fax: 502-637-7344.

ZEON CHEMICALS, INC.
4100 BELLS LANE
LOUISVILLE, KENTUCKY 40211-2127

Overview: Manufactures synthetic rubber. Established: 1989. Parent Company: Nippon Zeon. Ownership: Private.

Key Statistics: Annual Sales: $150 million. Number of Employees: 400.

Expertise/Education Sought: Mechanical engineering, chemical engineering, electrical engineering

Company-Paid Training is available.

Contact: Mr. Mark Myers, Compensation and Staffing Specialist; Phone: 502-775-7700.

LOUISIANA

BABCOCK & WILCOX
1450 POYDRAS STREET, 22ND FLOOR
NEW ORLEANS, LOUISIANA 70112-6050

Overview: Manufactures steam generating equipment and associated products. Established: 1867. Parent Company: McDermott International. Ownership: Public.

Key Statistics: Annual Sales: $1.6 billion. Number of Employees: 18,000.

Hiring History: Number of professional employees hired in 1996: 25.

Expertise/Education Sought: Chemical engineering, mechanical engineering

Contact: Mr. Robert Jongbloed, Manager of College Recruiting and Equal Employment Opportunity; Phone: 504-587-6400; Fax: 504-587-6150. E-mail: bob.jongbloed@mcdermott.com.

Company URL: http://www.mcdermott.com

CLECO
2030 DONAHUE FERRY ROAD
PINEVILLE, LOUISIANA 71360

Overview: Electric utility. Established: 1934. Ownership: Public.

Key Statistics: Assets: $435.4 million. Number of Employees: 1,200.

Hiring History: Number of professional employees hired in 1996: 62.

Expertise/Education Sought: Civil engineering, electrical engineering, mechanical engineering, accounting, computer programming, program analysis, marketing

Alternatives to Full-Time Employment: Paid internships

Company-Paid Training is available.

International Assignments are available to employees.

Contact: Mr. Richard Sasser, Group Leader of Core Human Resources Service Group. E-mail: richard.sasser@cleco.com. Job Hotline: 318-484-7628.

Company URL: http://www.cleco.com

ENGINEERS & CONSTRUCTORS INTERNATIONAL
2638 SOUTH SHERWOOD FOREST BOULEVARD (SUITE 100)
BATON ROUGE, LOUISIANA 70816

Overview: Provides engineering services. Established: 1981. Parent Company: MES Engineering, Inc. Ownership: Private.

Key Statistics: Annual Sales: $10 million. Number of Employees: 30.

Hiring History: Number of professional employees hired in 1997: 125. 1996: 65.

Expertise/Education Sought: Agricultural engineering

Alternatives to Full-Time Employment: Paid internships

International Assignments are available to employees.

Contact: Mr. Gary Mooney, Marketing Coordinator; Phone: 504-293-7768, ext. 157; Fax: 504-292-8364. E-mail: ecilhome@aol.com.

ENTERGY CORPORATION
639 LOYOLA AVENUE
NEW ORLEANS, LOUISIANA 70113-3121

Overview: Electric utility. Established: 1993. Ownership: Public.

Key Statistics: Annual Sales: $1.8 billion. Number of Employees: 756.

Hiring History: Number of professional employees hired in 1997: 246.

Expertise/Education Sought: Finance, sales, information technology, accounting, data processing, computer science, management information systems, human resources

Alternatives to Full-Time Employment: Part-time jobs, paid internships

Company-Paid Training is available.

International Assignments are available to employees.

Contact: For college students and recent graduates: Ms. Linda Wilson, College Relations Manager; Phone: 504-576-4885; Fax: 504-576-4428. For individuals with previous experience: Mr. Gerald Bailey, Manager of Human Resources; Phone: 504-576-4885; Fax: 504-576-4428. Job Hotline: 504-576-4885.

Company URL: http://www.entergy.com

EPCON, INC.
120 MALLARD STREET, SUITE 200
SAINT ROSE, LOUISIANA 70087-1452

Overview: Provides engineering services. Established: 1996. Ownership: Private.

Key Statistics: Annual Sales: $34.5 million. Number of Employees: 270.

Hiring History: Number of professional employees hired in 1997: 200. 1996: 200.

Expertise/Education Sought: Engineering, electrical engineering, computer-aided design, drafting, chemical engineering, construction management, mechanical engineering

Contact: For individuals with previous experience: Ms. Mary Noble, Director of Personnel; Phone: 504-468-8900.

Company URL: http://www.epconinc.com

FALCON INLAND, INC.
24 CONCORD ROAD
HOUMA, LOUISIANA 70360-7561

Overview: Drills oil and gas wells. Established: 1995. Ownership: Public.

Key Statistics: Annual Sales: $57.6 million. Number of Employees: 800.

Expertise/Education Sought: Machinery, electrical technology

Alternatives to Full-Time Employment: Paid internships

Company-Paid Training is available.

International Assignments are available to employees.

Contact: Mr. Rick Melancon, Personnel Director; Phone: 504-876-6987.

HEAD & ENGQUIST EQUIPMENT, LLC
6600 AIRLINE HIGHWAY
BATON ROUGE, LOUISIANA 70805-4701

Overview: Wholesales construction machinery. Established: 1994. Ownership: Private.

Key Statistics: Annual Sales: $100 million. Number of Employees: 315.

Hiring History: Number of professional employees hired in 1997: 100. 1996: 47.

Alternatives to Full-Time Employment: Paid internships

Company-Paid Training is available.

Contact: For individuals with previous experience: Mr. Richard Knight, Human Resources Manager; Phone: 504-354-2254; Fax: 504-354-2398.

INTRALOX, INC.
201 LAITRAM LANE
HARAHAN, LOUISIANA 70123-5395

Overview: Manufactures plastic conveyor belting. Established: 1973. Parent Company: Laitram Corporation. Ownership: Private.
Key Statistics: Annual Sales: $70 million. Number of Employees: 686.
Hiring History: Number of professional employees hired in 1996: 100.
Expertise/Education Sought: Mechanical engineering, accounting, finance, data processing, sales, plastics engineering
Alternatives to Full-Time Employment: Paid internships
International Assignments are available to employees.
Contact: For individuals with previous experience: Mr. James Evans, Manager of Human Resources; Phone: 504-733-6000; Fax: 504-733-2143. Job Hotline: 504-733-6000.

LAFOURCHE TELEPHONE COMPANY
112 WEST 10TH STREET
LAROSE, LOUISIANA 70373

Overview: Provides telephone services. Established: 1983. Parent Company: SJI, Inc. Ownership: Private.
Key Statistics: Annual Sales: $69.3 million. Number of Employees: 300.
Hiring History: Number of professional employees hired in 1997: 4. 1996: 6.
Expertise/Education Sought: Accounting, finance, electrical engineering, data processing
Alternatives to Full-Time Employment: Part-time jobs
Company-Paid Training is available.
Contact: Ms. Donna Plaisance, Human Resources; Phone: 504-693-4567; Fax: 504-693-3897. E-mail: donnap@latelco.com. Job Hotline: 504-691-JOBS.
Company URL: http://www.mobiletel.com/latelco.htm

MAINTENANCE ENTERPRISES, INC.
52410 CLARKE ROAD
WHITE CASTLE, LOUISIANA 70788-4914

Overview: Builds and remodels industrial buildings and warehouses. Established: 1967. Parent Company: CROWN ENTERPRISES, Inc. Ownership: Private.
Key Statistics: Annual Sales: $57.5 million. Number of Employees: 450.
Hiring History: Number of professional employees hired in 1997: 750. 1996: 450.
Expertise/Education Sought: Construction, estimating
Company-Paid Training is available.
International Assignments are available to employees.
Contact: Mr. Lewis Jordan, Human Resource Manager; Phone: 504-545-3040, ext. 145; Fax: 504-545-4141. Job Hotline: 504-545-4142.
Company URL: http://www.crownenterprises.com

J. RAY MCDERMOTT, SA
1450 POYDRAS STREET
NEW ORLEANS, LOUISIANA 70112

Overview: Provides offshore construction services. Established: 1994. Parent Company: McDermott, Inc. Ownership: Public.
Key Statistics: Annual Sales: $1.4 billion. Number of Employees: 10,800.
Hiring History: Number of professional employees hired in 1997: 25. 1996: 30.
Expertise/Education Sought: Civil engineering
Alternatives to Full-Time Employment: Part-time jobs
Company-Paid Training is available.
International Assignments are available to employees.
Contact: Mr. Robert Jongbloed, College Recruiter.
Company URL: http://www.mcdermott.com

SUNBELT PLASTICS
4611 CENTRAL AVENUE
MONROE, LOUISIANA 71203-6032

Overview: Manufactures plastics. Established: 1979. Ownership: Private.
Key Statistics: Annual Sales: $55 million. Number of Employees: 350.
Hiring History: Number of professional employees hired in 1996: 150.
Expertise/Education Sought: Chemical engineering, electrical engineering, mechanical engineering, plastics, plastics engineering, accounting, finance, data processing
Alternatives to Full-Time Employment: Paid internships
International Assignments are available to employees.
Contact: For individuals with previous experience: Mrs. Barabara McLawchlin, Director of Human Resources; Phone: 318-388-2200; Fax: 318-388-6955.

TIDEWATER, INC.
1440 CANAL STREET, SUITE 2100
NEW ORLEANS, LOUISIANA 70112-2700

Overview: Distributes compressors and petroleum products. Established: 1956. Ownership: Public.
Key Statistics: Annual Sales: $803 million. Number of Employees: 8,400.
Hiring History: Number of professional employees hired in 1996: 16.
Expertise/Education Sought: Business administration, law, marketing, finance, accounting, engineering
Contact: For individuals with previous experience: Ms. Mary-Elizabeth Torrens, Director of Human Resources; Phone: 504-566-4550; Fax: 504-566-4514.

WALK HAYDEL AND ASSOCIATES, INC.
600 CARONDELET STREET
NEW ORLEANS, LOUISIANA 70130-3511

Overview: Provides project engineering and construction management services. Established: 1995. Parent Company: Dames And Moore. Ownership: Public.
Key Statistics: Annual Sales: $51 million. Number of Employees: 600.
Expertise/Education Sought: Architecture, petroleum/petrochemical engineering, chemical engineering, process engineering, drafting
International Assignments are available to employees.
Contact: For college students and recent graduates: Ms. Lisa Reitman, Human Resources Administrator; Phone: 504-586-8111; Fax: 504-599-5224. For individuals with previous experience: Mr. E. Garin Danner, Director of Human Resources; Phone: 504-586-8111; Fax: 504-599-5224.

MAINE

BATH IRON WORKS CORPORATION
700 WASHINGTON STREET
BATH, MAINE 04530-2574

Overview: Provides ship building services. Established: 1995. Parent Company: General Dynamics Corporation. Ownership: Public.
Key Statistics: Annual Sales: $850 million. Number of Employees: 8,000.
Hiring History: Number of professional employees hired in 1997: 200. 1996: 100.
Expertise/Education Sought: Mechanical engineering, electrical engineering, marine engineering, naval architecture
Alternatives to Full-Time Employment: Paid internships

Company-Paid Training is available.
International Assignments are available to employees.
Contact: Ms. Nancy Zugehoer, Manager of Personnel; Phone: 207-442-4278; Fax: 207-442-5288.

DEXTER SHOE COMPANY
71 RAILROAD AVENUE
DEXTER, MAINE 04930-1234

Overview: Manufactures men's and women's footwear. Established: 1993. Parent Company: Berkshire Hathaway. Ownership: Public.
Key Statistics: Annual Sales: $243.5 million. Number of Employees: 3,200.
Hiring History: Number of professional employees hired in 1996: 200.
Expertise/Education Sought: Accounting, manufacturing, computer programming, customer service and support
International Assignments are available to employees.
Contact: For individuals with previous experience: Human Resources; Phone: 207-924-7341; Fax: 207-924-5589.

FAIRCHILD SEMICONDUCTOR CORPORATION
333 WESTERN AVENUE
SOUTH PORTLAND, MAINE 04106-1705

Overview: Manufactures semiconductors. Established: 1997. Ownership: Public.
Key Statistics: Annual Sales: $691.3 million. Number of Employees: 6,180.
Expertise/Education Sought: Manufacturing, software engineering/development, engineering, electrical engineering
Alternatives to Full-Time Employment: Paid internships
Company-Paid Training is available.
International Assignments are available to employees.
Contact: For individuals with previous experience: Mr. Robert Carlino, Staffing Supervisor; Phone: 207-775-8563; Fax: 207-761-6431.

GEIGER BROTHERS
MOUNT HOPE AVENUE
LEWISTON, MAINE 04240

Overview: Wholesales advertising novelties. Established: 1878. Ownership: Private.
Key Statistics: Annual Sales: $82 million. Number of Employees: 550.
Hiring History: Number of professional employees hired in 1997: 98. 1996: 48.
Expertise/Education Sought: Manufacturing engineering, sales, marketing
Alternatives to Full-Time Employment: Paid internships
Contact: For individuals with previous experience: Human Relations Department. E-mail: jobs@geiger.com.
Company URL: http://www.geiger.com

HERBERT C. HAYNES, INC.
RR 2
WINN, MAINE 04495

Mailing Address: PO Box 96, Winn, ME 04495
Overview: Produces pulpwood and wood chips. Established: 1963. Ownership: Private.
Key Statistics: Annual Sales: $86.1 million. Number of Employees: 100.
Expertise/Education Sought: Equipment operation, forestry
Contact: Ms. Ginger Maxwell, Office Manager; Fax: 207-736-2900. E-mail: hchaynesinc@telplus.net.

HUSSEY SEATING COMPANY
38 DYER STREET EXTENSION
NORTH BERWICK, MAINE 03906

Overview: Manufactures stadium and bleacher seating. Established: 1835. Ownership: Private.
Key Statistics: Annual Sales: $50 million. Number of Employees: 400.
Hiring History: Number of professional employees hired in 1997: 100. 1996: 25.
Expertise/Education Sought: Drafting, accounting, sales, customer service and support, manufacturing, marketing
Company-Paid Training is available.
Contact: For individuals with previous experience: Mr. Richard Hussey, Employee Coordinator; Phone: 207-676-2271; Fax: 207-676-9506.

IRVING TANNING COMPANY
MAIN STREET
HARTLAND, MAINE 04943

Overview: Operates leather tanning and finishing facility. Established: 1962. Ownership: Public.
Key Statistics: Annual Sales: $107.8 million. Number of Employees: 550.
Hiring History: Number of professional employees hired in 1996: 150.
Expertise/Education Sought: Manufacturing
Contact: For individuals with previous experience: Human Resources; Phone: 207-938-4491, ext. 259; Fax: 207-938-2977.

JORDAN'S MEATS
38 INDIA STREET
PORTLAND, MAINE 04112-0588

Overview: Produces sausage and other prepared meats. Established: 1990. Ownership: Private.
Key Statistics: Annual Sales: $203.6 million. Number of Employees: 750.
Expertise/Education Sought: Manufacturing engineering, industrial engineering
Alternatives to Full-Time Employment: Paid internships
Contact: For individuals with previous experience: Employment Coordinator; Phone: 207-772-5411, ext. 331; Fax: 207-772-4814.

NORTHLAND TELEPHONE COMPANY
2401 CONGRESS STREET
PORTLAND, MAINE 04102-1932

Overview: Provides telecommunications services. Established: 1994. Parent Company: NJD Communications. Ownership: Private.
Key Statistics: Annual Sales: $15.5 million. Number of Employees: 40.
Hiring History: Number of professional employees hired in 1997: 10.
Expertise/Education Sought: Telecommunications engineering, customer service and support, management, communications
Contact: For individuals with previous experience: Human Resources; Fax: 207-874-6012. E-mail: susanj@northlandlink.com.

H.E. SARGENT, INC.
101 BENNOCH ROAD
STILIWATER, MAINE 04489

Overview: Provides site preparation and construction services. Established: 1996. Parent Company: Fru-Con Construction Corporation. Ownership: Private.
Key Statistics: Annual Sales: $80 million. Number of Employees: 175.
Hiring History: Number of professional employees hired in 1997: 16. 1996: 10.
Expertise/Education Sought: Construction, project management
Alternatives to Full-Time Employment: Paid internships
Company-Paid Training is available.
Contact: For individuals with previous experience: Human Relations Department; Phone: 207-827-4435; Fax: 207-827-2945.

SPENCER PRESS, INC.

90 SPENCER DRIVE
WELLS, MAINE 04090

Overview: Provides commercial lithographic and offset printing services. Established: 1940. Ownership: Private.

Key Statistics: Annual Sales: $75 million. Number of Employees: 675.

Hiring History: Number of professional employees hired in 1997: 5. 1996: 5.

Expertise/Education Sought: Machinery, sales

Company-Paid Training is available.

Contact: For individuals with previous experience: Ms. Denise Hamilton, Human Resources Manager; Phone: 207-646-9926, ext. 362; Fax: 207-646-5021. E-mail: dhamilton@spencerpress.com.

Company URL: http://www.spencerpress.com

WAUSAU PAPERS OTIS MILL

1 MILL STREET
JAY, MAINE 04239-1897

Overview: Manufactures specialty paper. Established: 1993. Parent Company: Wausau—Mosinee Paper Corporation. Ownership: Public.

Key Statistics: Annual Sales: $41.7 million. Number of Employees: 290.

Hiring History: Number of professional employees hired in 1996: 30.

Expertise/Education Sought: Management, customer service and support, manufacturing engineering

Alternatives to Full-Time Employment: Paid internships

Company-Paid Training is available.

International Assignments are available to employees.

Contact: For individuals with previous experience: Human Resources; Phone: 207-897-7200; Fax: 207-897-7207.

MARYLAND

ALLIEDSIGNAL TECHNICAL SERVICES

1 BENDIX ROAD
COLUMBIA, MARYLAND 21045

Overview: Provides technical services to the aerospace industry. Established: 1985. Parent Company: AlliedSignal, Inc. Ownership: Public.

Key Statistics: Annual Sales: $852.2 million. Number of Employees: 7,000.

Hiring History: Number of professional employees hired in 1997: 700.

Expertise/Education Sought: Electrical engineering, computer engineering, mechanical engineering, aerospace engineering, computer science, mathematics

Alternatives to Full-Time Employment: Part-time jobs, paid internships

International Assignments are available to employees.

Contact: For college students and recent graduates: Staffing Department; Phone: 410-964-7000; Fax: 410-964-7741. For individuals with previous experience: Staffing Department; Phone: 410-964-7000; Fax: 410-964-7741. Job Hotline: 800-638-7816.

Company URL: http://www.alliedsignal.com/aerospace

ALPHA SCIENTIFIC, INC.

416 HUNGERFORD DRIVE
ROCKVILLE, MARYLAND 20850-4127

Overview: Provides professional, scientific, and engineering equipment. Established: 1996. Ownership: Private.

Key Statistics: Number of Employees: 600.

Hiring History: Number of professional employees hired in 1997: 200.

Expertise/Education Sought: Research and development

Alternatives to Full-Time Employment: Paid internships

Company-Paid Training is available.

International Assignments are available to employees.

Contact: Human Resource Department; Phone: 301-738-7155; Fax: 301-738-3277.

The Johns Hopkins University
Applied Physics Laboratory

THE APPLIED PHYSICS LABORATORY OF THE JOHNS HOPKINS UNIVERSITY

11100 JOHNS HOPKINS ROAD
LAUREL, MARYLAND 20723

Overview: Provides research and development laboratory services. Established: 1942.

Key Statistics: Number of Employees: 2,800.

Expertise/Education Sought: Analysis, systems engineering, software engineering/development, communications engineering, electronic design and development, space physics, defense systems

Contact: Employment Office Department PG-99; Fax: 240-228-5274.

CORPORATE STATEMENT

As a technical resource, APL makes contributions in defense, space, and biomedicine. APL's primary role involves national security issues. The technology developed through defense research is also applied to civil sector problems.

AVATECH SOLUTIONS, INC.

11403 CRON HILL DRIVE, SUITE A
OWINGS MILLS, MARYLAND 21117-6217

Overview: Designs computer systems. Established: 1997. Ownership: Private.

Key Statistics: Annual Sales: $12.2 million. Number of Employees: 120.

Expertise/Education Sought: Engineering

Company-Paid Training is available.

Contact: For individuals with previous experience: Mrs. Jamie Elyar, Administrative Manager; Phone: 410-902-6900; Fax: 410-902-8324. E-mail: jelyar@avat.com.

Company URL: http://www.avat.com

BALTIMORE AIRCOIL COMPANY

7595 MONTEVIDEO ROAD
JESSUP, MARYLAND 20794-9323

Mailing Address: PO Box 7322, Baltimore, MD 21227-0322

Overview: Manufactures refrigeration and air conditioning equipment. Established: 1985. Parent Company: Amsted Industries, Inc. Ownership: Private.

Key Statistics: Annual Sales: $108.7 million. Number of Employees: 1,225.

Hiring History: Number of professional employees hired in 1997: 10. 1996: 10.

Expertise/Education Sought: Chemical engineering, mechanical engineering, manufacturing engineering, marketing, research, design engineering, industrial engineering

International Assignments are available to employees.

Contact: Ms. Marion Thompson, Manager of Human Resources; Phone: 410-799-6203; Fax: 410-799-6564.

CABLE TV MONTGOMERY

20 WEST GUDE DRIVE
ROCKVILLE, MARYLAND 20850

Overview: Provides cable television services. Established: 1994. Parent Company: SBC Media Ventures. Ownership: Public.

Key Statistics: Annual Sales: $64.6 million. Number of Employees: 484.
Expertise/Education Sought: Communications, customer service and support, accounting, finance, data processing, electrical engineering
Company-Paid Training is available.
Contact: For individuals with previous experience: Ms. Grace Killelea, Director of Human Resources; Phone: 301-294-7600. Job Hotline: 301-294-7679.
Company URL: http://www.ctvm.com

CENTURY TECHNOLOGIES, INC.

8405 COLESVILLE ROAD, SUITE 400
SILVER SPRING, MARYLAND 20910-3317

Overview: Provides systems integration telecommunications and network services. Established: 1978. Ownership: Private.
Key Statistics: Annual Sales: $15.9 million. Number of Employees: 350.
Hiring History: Number of professional employees hired in 1997: 169. 1996: 130.
Expertise/Education Sought: Networking, software engineering/development
Alternatives to Full-Time Employment: Part-time jobs, paid internships
Company-Paid Training is available.
Contact: Mr. Earl Richmond, Human Resources Manager; Phone: 301-585-4800; Fax: 301-588-1619. E-mail: erichmond@centech.com.
Company URL: http://www.centech.com

COMPUTER DATA SYSTEMS, INC.

1 CURIE COURT
ROCKVILLE, MARYLAND 20850-4310

Overview: Provides computer systems integration and consulting services. Established: 1968. Parent Company: Affiliated Computer Services, Inc. Ownership: Public.
Key Statistics: Annual Sales: $304.4 million. Number of Employees: 3,400.
Hiring History: Number of professional employees hired in 1997: 1,400. 1996: 1,200.
Expertise/Education Sought: Information systems, computer programming, computer science
Alternatives to Full-Time Employment: Part-time jobs, paid internships
Company-Paid Training is available.
International Assignments are available to employees.
Contact: Mr. James Hollister, Director of Employment; Phone: 301-921-7000; Fax: 301-921-0795. E-mail: james.hollister@cdsi.com. Job Hotline: 800-422-7721.
Company URL: http://www.cdsi.com

COMSAT CORPORATION

6560 ROCK SPRING DRIVE
BETHESDA, MARYLAND 20817

Overview: Provides satellite-communications services. Established: 1963. Ownership: Public.
Key Statistics: Annual Sales: $1 billion. Number of Employees: 3,766.
Expertise/Education Sought: Communications, engineering, computer science, telecommunications
Alternatives to Full-Time Employment: Paid internships
Company-Paid Training is available.
International Assignments are available to employees.
Contact: Ms. Marian Hyder, Professional Recruiter; Phone: 301-214-3000; Fax: 301-214-7134.
Company URL: http://www.comsat.com

CROWN CENTRAL PETROLEUM CORPORATION

1 NORTH CHARLES STREET
BALTIMORE, MARYLAND 21201-3740

Overview: Petroleum refining and petrochemical production company. Established: 1923. Ownership: Public.
Key Statistics: Annual Sales: $1.6 billion. Number of Employees: 2,904.
Hiring History: Number of professional employees hired in 1997: 30. 1996: 30.
Expertise/Education Sought: Marketing, finance, accounting, management information systems, computer programming
Alternatives to Full-Time Employment: Paid internships
Company-Paid Training is available.
Contact: For individuals with previous experience: Human Relations Department; Phone: 410-539-7400; Fax: 410-659-4875.

CTA, INC.

6116 EXECUTIVE BOULEVARD, SUITE 800
ROCKVILLE, MARYLAND 20852

Overview: Provides computer programming services and furthers commercial physical research. Established: 1979. Ownership: Private.
Key Statistics: Annual Sales: $179.7 million. Number of Employees: 1,250.
Expertise/Education Sought: Computer engineering, electrical engineering, systems engineering, management information systems, computer science, business management, aerospace engineering
Contact: For individuals with previous experience: Ms. Nancy Nelkin, Human Resources Manager; Phone: 301-816-1200; Fax: 301-816-1367.

DISCLOSURE, INC.

5161 RIVER ROAD
BETHESDA, MARYLAND 20816

Overview: Provides public information retrieval services. Established: 1995. Parent Company: Primark. Ownership: Public.
Key Statistics: Annual Sales: $67.5 million. Number of Employees: 950.
Hiring History: Number of professional employees hired in 1997: 100. 1996: 150.
Expertise/Education Sought: Computer programming, sales, technology, production, marketing, accounting, finance, law
Alternatives to Full-Time Employment: Part-time jobs, paid internships
Company-Paid Training is available.
International Assignments are available to employees.
Contact: Human Resources Department; Phone: 301-951-1300; Fax: 301-718-2340. E-mail: karen.palisin@disclosure.com. Job Hotline: 301-951-1300, ext. 5.
Company URL: http://www.disclosure.com

GENERAL PHYSICS CORPORATION

6700 ALEXANDER BELL DRIVE
COLUMBIA, MARYLAND 21046-2100

Overview: Provides management and training and development services. Established: 1994. Parent Company: GP Strategies. Ownership: Public.
Key Statistics: Annual Sales: $16.9 million. Number of Employees: 300.
Hiring History: Number of professional employees hired in 1997: 400.
Expertise/Education Sought: Business administration, engineering, training and development
Alternatives to Full-Time Employment: Paid internships
Company-Paid Training is available.
International Assignments are available to employees.
Contact: For individuals with previous experience: Human Resource Department; Phone: 800-727-6677; Fax: 410-290-2426. E-mail: hr@columbia.genphysics.com.
Company URL: http://www.genphysics.com

HECHINGER INVESTMENT COMPANY

1801 MCCORMICK DRIVE
LARGO, MARYLAND 20774

Overview: Operates chain of home improvement stores. Established: 1911. Ownership: Private.
Key Statistics: Annual Sales: $4.5 billion. Number of Employees: 36,000.
Hiring History: Number of professional employees hired in 1996: 80.

Hechinger Investment Company (continued)

Expertise/Education Sought: Business administration, marketing, retail, sales
Alternatives to Full-Time Employment: Paid internships
Company-Paid Training is available.
Contact: For individuals with previous experience: Staffing and Development; Phone: 301-341-1000; Fax: 301-925-3906.

HONEYWELL MEASUREX DMC CORPORATION

15810 GAITHERS DRIVE
GAITHERSBURG, MARYLAND 20877-1404

Overview: Manufactures measurement and control devices. Established: 1996. Ownership: Public.
Key Statistics: Annual Sales: $29.2 million. Number of Employees: 333.
Hiring History: Number of professional employees hired in 1997: 30. 1996: 30.
Expertise/Education Sought: Electrical engineering, software engineering/development, electrical technology
Alternatives to Full-Time Employment: Paid internships
Company-Paid Training is available.
International Assignments are available to employees.
Contact: Human Resources Department; Phone: 301-948-2450; Fax: 301-670-0506. E-mail: resumes@dc.hodes.com.
Company URL: http://www.honeywell.com

MCI TELECOMMUNICATIONS CORPORATION

PO BOX 10130
GAITHERSBURG, MARYLAND 20898

Overview: Provides telecommunications services. Established: 1968. Parent Company: MCI Telecommunications, Inc. Ownership: Public.
Key Statistics: Annual Sales: $18.5 billion. Number of Employees: 55,285.
Hiring History: Number of professional employees hired in 1997: 4,500. 1996: 4,500.
Expertise/Education Sought: Electrical engineering, computer programming, software engineering/development, management information systems, computer science, telecommunications, electronics/electronics engineering
Alternatives to Full-Time Employment: Paid internships
Contact: Recruiting Headquarters; Fax: 800-691-6982. E-mail: mci@alexus.com.
Company URL: http://www.mci.com

NATIONAL INSTITUTE OF STANDARDS AND TECHNOLOGY

OFFICE OF HUMAN RESOURCES MANAGEMENT
GAITHERSBURG, MARYLAND 20899

Overview: Promotes public health, public safety, and environmental issues. Division of the U.S. Department of Commerce.
Hiring History: Number of professional employees hired in 1996: 300.
Expertise/Education Sought: Physics, mechanical engineering, civil engineering, mathematics, computer science
Alternatives to Full-Time Employment: Part-time jobs, paid internships
Company-Paid Training is available.
International Assignments are available to employees.
Contact: Ms. Deborah Martin, Staffing Specialist; Phone: 301-975-3007; Fax: 301-948-6107. Job Hotline: 301-926-4851.
Company URL: http://www.nist.gov

NATIONAL INSTITUTES OF HEALTH

9000 ROCKVILLE PIKE
BETHESDA, MARYLAND 20892

Overview: Conducts studies for the purpose of preventing, detecting, diagnosing, and treating diseases and disabilities.
Expertise/Education Sought: Biology, behavioral research, biotechnology, physics, chemistry
Alternatives to Full-Time Employment: Part-time jobs, paid internships
International Assignments are available to employees.
Contact: For individuals with previous experience: Mr. Herbert Casey, Director Division Career Resources; Phone: 301-496-5979; Fax: 301-496-1209. Job Hotline: 301-496-2403.
Company URL: http://www.nih.gov

NATIONAL SECURITY AGENCY

S232R (AWE), 9800 SAVAGE ROAD, SUITE 6840
FORT GEORGE G. MEADE, MARYLAND 20755-6840

Overview: Provides national security in communications. Established: 1952.
Expertise/Education Sought: Asian, Middle Eastern, and Slavic (except Russian) languages, computer science, mathematics, advanced degreed mathematicians
Contact: Mr. Kenneth Acosta, Diversity and College Recruitment Manager; Phone: 800-255-8415; Fax: 410-859-4593.

For more information, see full organization profile, pg. 286.

NAVAL AVIATION SYSTEMS TEAM

HUMAN RESOURCES OFFICE, PROFESSIONAL RECRUITMENT, BUILDING 1489
PATUXENT RIVER, MARYLAND 20670-1139

Overview: Conducts aviation research and development. Established: 1945.
Hiring History: Number of professional employees hired in 1996: 525.
Expertise/Education Sought: Aerospace engineering, electrical engineering, electronics/electronics engineering, computer engineering, mechanical engineering, operations research, computer science, materials engineering
Alternatives to Full-Time Employment: Part-time jobs, internships
Company-Paid Training is available.
Contact: Mrs. Jean V. Lawrie-Parker, Personnel Specialist; Phone: 301-342-5501; Fax: 301-342-6682.
Company URL: http://www.nawcad.navy.mil

NIRO, INC.

9165 RUMSEY ROAD
COLUMBIA, MARYLAND 21045-1943

Overview: Provides engineering services. Established: 1995. Parent Company: GEA. Ownership: Private.
Key Statistics: Annual Sales: $106 million. Number of Employees: 550.
Hiring History: Number of professional employees hired in 1997: 125. 1996: 20.
Expertise/Education Sought: Chemical engineering, electrical engineering, mechanical engineering, process engineering, pharmaceutical science, control engineering
Alternatives to Full-Time Employment: Paid internships
Company-Paid Training is available.
International Assignments are available to employees.
Contact: For individuals with previous experience: Mr. Ronald Matzek, Human Resources Manager; Phone: 410-997-8700; Fax: 410-997-5021.

OAO TECHNOLOGY SOLUTIONS, INC.

7500 GREENWAY CENTER DRIVE
GREENBELT, MARYLAND 20770-3502

Overview: Provides business services. Established: 1996. Ownership: Private.

Key Statistics: Annual Sales: $57.9 million. Number of Employees: 1,684.

Expertise/Education Sought: Hardware engineering, software engineering/development

Alternatives to Full-Time Employment: Paid internships

Company-Paid Training is available.

Contact: For individuals with previous experience: Human Resources; Phone: 301-486-0400; Fax: 301-486-0415.

Company URL: http://www.oaot.com

ORANGE TECHNOLOGIES, INC.

13 FIRSTFIELD ROAD
GAITHERSBURG, MARYLAND 20878-1700

Overview: Provides computer integrated systems services. Established: 1996. Ownership: Private.

Key Statistics: Annual Sales: $9.8 million. Number of Employees: 108.

Expertise/Education Sought: Computer programming, networking, computer-aided design

Alternatives to Full-Time Employment: Part-time jobs, paid internships

Company-Paid Training is available.

International Assignments are available to employees.

Contact: For individuals with previous experience: Mr. Paul Hutchens, Manager; Phone: 305-840-3220; Fax: 305-577-5362. E-mail: jtcox@orange.com.

Company URL: http://www.orange.com

POLY-SEAL CORPORATION

1810 PORTAL STREET
BALTIMORE, MARYLAND 21224-6512

Overview: Manufactures polymer seals. Established: 1969. Ownership: Private.

Key Statistics: Annual Sales: $50.9 million. Number of Employees: 535.

Hiring History: Number of professional employees hired in 1997: 200.

Expertise/Education Sought: Plastics engineering, mechanical engineering, product design and development, computer science, marketing, sales

Alternatives to Full-Time Employment: Part-time jobs

Company-Paid Training is available.

International Assignments are available to employees.

Contact: Ms. Maria Hayward, Senior Human Resources Manager; Phone: 410-633-1990, ext. 1340; Fax: 410-633-6311. E-mail: maria-h@polyseal.usa.com.

Company URL: http://www.poly-seal.com

PRESTON TRUCKING COMPANY, INC.

151 EASTON BOULEVARD
PRESTON, MARYLAND 21655

Overview: Provides truck freight and shipment services. Established: 1993. Parent Company: Yellow Corporation. Ownership: Public.

Key Statistics: Annual Sales: $635.3 million. Number of Employees: 8,895.

Expertise/Education Sought: Management, transportation

Contact: Ms. Alison Ball, Director of Human Resources; Phone: 410-673-7151. E-mail: ptchr@dmv.com.

Raytheon

RAYTHEON STX CORPORATION

4400 FORBES BOULEVARD
LANHAM, MARYLAND 20706

Overview: Provides services in earth, space, and environmental applications; transportation, navigation, and control; and information and data management. Established: 1973. Parent Company: The Raytheon Company. Ownership: Public.

Key Statistics: Number of Employees: 1,400.

Hiring History: Number of professional employees hired in 1997: 300. 1996: 210.

Expertise/Education Sought: Computer science, software engineering/development, physics, computer programming, data processing

Alternatives to Full-Time Employment: Part-time jobs, paid internships

Contact: For college students and recent graduates: Ms. Kathy Maguire, Recruiter; Phone: 301-794-5000; Fax: 301-794-5242. E-mail: recruiting@stx.com. For individuals with previous experience: Ms. Jennifer Lynne Freeland, Principal Technical Recruiter; Phone: 301-794-5000; Fax: 301-794-5242. E-mail: recruiting@stx.com.

Company URL: http://www.stx.com

CORPORATE PROFILE

Raytheon STX (RSTX) is an innovative, results-oriented technology leader providing support services and systems development to government, commercial, and international clients. The company specializes in Earth, space, and environmental applications; transportation, navigation, and control; and information and data management.

Specific areas of engineering expertise include scientific system development for the collection and analysis of remotely sensed data; instrument engineering; scientific data center management, including systems engineering, software development, relational database design and management, and scientific analysis; network engineering and systems administration; and other advanced computing technologies.

Opportunities exist for all levels of personnel who have an appropriate degree combined with academic or professional experience in software development or systems engineering in two or more of the following areas: UNIX/C, VMS, FORTRAN, X-Windows, GUI, RDBMS, (Oracle, Sybase, Ingres), object-oriented DBMS, object-oriented programming (C++), Web development (Perl, Java), LAN/WAN, supercomputing, parallel processing, and systems administration. A combination of both scientific and computer studies or experience is highly desirable.

The company is primarily located in the Washington, D.C., area, with offices nationwide. Raytheon STX is an Equal Opportunity Employer.

RMS INFORMATIONS SYSTEMS

4221 FORBES BOULEVARD
LANHAM, MARYLAND 20706-4343

Overview: Offers data processing and preparation services. Established: 1996. Ownership: Private.

Key Statistics: Annual Sales: $98 million. Number of Employees: 1,000.

Hiring History: Number of professional employees hired in 1997: 100.

Expertise/Education Sought: Telecommunications, computer engineering, information technology

Alternatives to Full-Time Employment: Paid internships

Contact: For individuals with previous experience: Human Resources Department; Phone: 301-306-4460; Fax: 301-459-2018.

Company URL: http://www.rmstech.com

TRACOR SYSTEMS TECHNOLOGIES, INC.

1601 RESEARCH BOULEVARD
ROCKVILLE, MARYLAND 20850

Overview: Provides computer engineering services. Established: 1947. Parent Company: Tracor, Inc. Ownership: Public.

Key Statistics: Annual Sales: $270 million. Number of Employees: 3,200.

Hiring History: Number of professional employees hired in 1997: 400. 1996: 25.

Expertise/Education Sought: Engineering, computer science, mathematics

Alternatives to Full-Time Employment: Paid internships

International Assignments are available to employees.

Contact: For individuals with previous experience: Mr. Ben Gotkin, College Recruiter; Phone: 301-838-6320; Fax: 301-838-6303. E-mail: gotkin@tst.tracor.com.

Company URL: http://www.tst.tracor.com

WANG GLOBAL
6700 ROCKLEDGE DRIVE
BETHESDA, MARYLAND 20817-1804

Overview: Manufactures software. Established: 1984. Parent Company: Wang Laboratories, Inc. Ownership: Public.

Key Statistics: Annual Sales: $246.8 million. Number of Employees: 2,880.

Hiring History: Number of professional employees hired in 1996: 1,200.

Expertise/Education Sought: Software engineering/development, electrical engineering, networking, communication engineering

Alternatives to Full-Time Employment: Part-time jobs

Company-Paid Training is available.

International Assignments are available to employees.

Contact: Miss Stephenie Orso, Staffing Specialist, 7900 Westpark Drive, McLain, Virginia, 22102-1804; Phone: 703-827-3695; Fax: 703-827-3390.

Company URL: http://www.wang.com

WEINSCHEL CORPORATION
5305 SPECTRUM DRIVE
FREDERICK, MARYLAND 21703-7362

Overview: Manufactures microwave test equipment. Established: 1995. Parent Company: Microwave Components Enterprises, Inc. Ownership: Private.

Key Statistics: Annual Sales: $6.1 million. Number of Employees: 110.

Hiring History: Number of professional employees hired in 1997: 12.

Expertise/Education Sought: Accounting, finance, electrical engineering, manufacturing engineering, information systems, computer programming

Alternatives to Full-Time Employment: Paid internships

Company-Paid Training is available.

Contact: Human Resources; Phone: 301-831-4701; Fax: 301-831-4571. E-mail: weinschel@aol.com.

Company URL: http://www.weinschel.com

MASSACHUSETTS

ADDISON WESLEY LONGMAN
ONE JACOB WAY
READING, MASSACHUSETTS 01867

Overview: Publisher of textbooks, multimedia, and learning resources. Established: 1942. Parent Company: Pearson, PLC. Ownership: Private.

Key Statistics: Annual Sales: $950 million. Number of Employees: 4,000.

Hiring History: Number of professional employees hired in 1997: 300.

Expertise/Education Sought: Technology

Contact: Mr. Jeffrey H. Kline, Divisional Human Resources Manager; Phone: 617-944-3700; Fax: 617-944-5390.

Company URL: http://www.aw.com

ALDEN PRODUCTS COMPANY
117 NORTH MAIN STREET
BROCKTON, MASSACHUSETTS 02401-3908

Overview: Produces electronic connectors. Established: 1930. Ownership: Private.

Key Statistics: Annual Sales: $15 million. Number of Employees: 135.

Hiring History: Number of professional employees hired in 1996: 6.

Expertise/Education Sought: Engineering, manufacturing engineering, accounting, research

Alternatives to Full-Time Employment: Part-time jobs, paid internships

Company-Paid Training is available.

Contact: For individuals with previous experience: Ms. Rose Penney, Operations Clerk; Phone: 508-583-0160; Fax: 508-583-0164.

Company URL: http://www.aldenproducts.com

ALLEGRO MICROSYSTEMS, INC.
115 NORTHEAST CUTOFF
WORCESTER, MASSACHUSETTS 01606-1224

Overview: Manufactures semiconductors and related electronic devices. Established: 1990. Parent Company: Sanken Electric. Ownership: Private.

Key Statistics: Annual Sales: $204.5 million. Number of Employees: 1,900.

Expertise/Education Sought: Design engineering, product engineering, electrical engineering, engineering

Alternatives to Full-Time Employment: Part-time jobs, paid internships

Contact: For individuals with previous experience: Human Resource Department, PO Box 15036, Worcester, Massachusetts, 01615-0036; Phone: 508-853-5000; Fax: 508-853-8378.

Company URL: http://www.allegromicrosystems.com

ALTRON, INC.
1 JEWEL DRIVE
WILMINGTON, MASSACHUSETTS 01887-3390

Overview: Manufactures printed circuit boards. Established: 1970. Ownership: Public.

Key Statistics: Annual Sales: $165.2 million. Number of Employees: 1,085.

Expertise/Education Sought: Engineering, sales management, purchasing, production

Alternatives to Full-Time Employment: Part-time jobs, paid internships

Company-Paid Training is available.

Contact: For college students and recent graduates: Ms. Debbie Dubois, Human Resources Supervisor; Phone: 978-658-5800; Fax: 978-933-1588. For individuals with previous experience: Ms. Angela Culot, Human Resources Manager; Phone: 978-658-5800; Fax: 978-933-1588. E-mail: aculot@altron.com.

Company URL: http://www.altron.com

AMERICAN ELECTRIC CABLE COMPANY
181 APPLETON STREET
HOLYOKE, MASSACHUSETTS 01040-5793

Overview: Manufactures insulated wire cables, cords and wire specialties. Established: 1981. Ownership: Private.

Key Statistics: Annual Sales: $11.6 million. Number of Employees: 90.

Hiring History: Number of professional employees hired in 1997: 100. 1996: 60.

Expertise/Education Sought: Engineering, quality assurance

Contact: For individuals with previous experience: Human Resources; Phone: 413-539-9893; Fax: 413-533-0164.

AMERICAN SAW AND MANUFACTURING COMPANY
301 CHESTNUT STREET
EAST LONGMEADOW, MASSACHUSETTS 01028-5601

Overview: Manufactures blades for hand and power saws. Established: 1918. Ownership: Private.

Key Statistics: Annual Sales: $75 million. Number of Employees: 650.

Expertise/Education Sought: Manufacturing

Alternatives to Full-Time Employment: Part-time jobs, paid internships

Company-Paid Training is available.

International Assignments are available to employees.
Contact: For individuals with previous experience: Human Resources Department; Phone: 413-525-3961; Fax: 413-525-9655. Job Hotline: 413-525-1488.
Company URL: http://www.lenoxsaw.com

AMERICAN SCIENCE AND ENGINEERING
829 MIDDLESEX TURNPIKE
BILLERICA, MASSACHUSETTS 01821-3907

Overview: Provides commercial physical research services. Established: 1958. Ownership: Public.
Key Statistics: Annual Sales: $28.5 million. Number of Employees: 150.
Hiring History: Number of professional employees hired in 1997: 50. 1996: 100.
Expertise/Education Sought: Engineering, science, finance, sales
Alternatives to Full-Time Employment: Part-time jobs, paid internships
International Assignments are available to employees.
Contact: Mr. Marji Holdenen, Contract Recruiter; Phone: 978-262-8777; Fax: 978-262-8896. E-mail: mholden@as-e.com.
Company URL: http://www.as-e.com

ANALOG DEVICES, INC.
1 TECHNOLOGY WAY
NORWOOD, MASSACHUSETTS 02062-2634

Overview: Manufactures integrated circuits, semiconductors, and related devices. Established: 1965. Ownership: Public.
Key Statistics: Annual Sales: $1.2 billion. Number of Employees: 6,900.
Hiring History: Number of professional employees hired in 1996: 1,000.
Expertise/Education Sought: Management information systems, mechanical engineering, electrical engineering, engineering
Alternatives to Full-Time Employment: Paid internships
Contact: Ms. Marsha Terrelonge, Senior Human Resource Representative; Phone: 617-329-4700. E-mail: mterrelonge@analog.com.

APPLIED RECYCLING, INC.
430 BEDFORD STREET
LEXINGTON, MASSACHUSETTS 02173

Overview: Manufactures machine tool accessories. Established: 1996. Ownership: Private.
Key Statistics: Annual Sales: $15 million.
Hiring History: Number of professional employees hired in 1997: 56. 1996: 7.
Expertise/Education Sought: Environmental engineering, manufacturing engineering, manufacturing, chemistry, chemical engineering
Alternatives to Full-Time Employment: Paid internships
International Assignments are available to employees.
Contact: For college students and recent graduates: Ms. Moon Kim, Analyst; Phone: 781-863-2800; Fax: 781-674-9997. For individuals with previous experience: Mr. Richard Rosen, Chairman and Chief Executive Officer; Phone: 781-863-2800; Fax: 781-674-9997.

ARANEX
7 OAK PARK
BEDFORD, MASSACHUSETTS 01730

Overview: Manufactures communications products for midrange and mainframe computers. Established: 1982. Parent Company: IDE Corporation. Ownership: Private.
Key Statistics: Annual Sales: $20.3 million. Number of Employees: 150.
Hiring History: Number of professional employees hired in 1997: 20.
Expertise/Education Sought: Computer programming, computer science, technical engineering, communications
Alternatives to Full-Time Employment: Part-time jobs, paid internships
Company-Paid Training is available.
Contact: For individuals with previous experience: Ms. Janif Schelfhaudt, Human Resources Manager; Phone: 781-533-0520; Fax: 781-533-0500. E-mail: jschelfh@aranex.com.
Company URL: http://www.aranex.com

ASA INTERNATIONAL, LTD.
10 SPEEN STREET
FRAMINGHAM, MASSACHUSETTS 01701-4661

Overview: Develops software and provides hardware support, software support and network systems services. Established: 1969. Ownership: Public.
Key Statistics: Annual Sales: $25.5 million. Number of Employees: 230.
Hiring History: Number of professional employees hired in 1997: 50. 1996: 50.
Expertise/Education Sought: Computer programming, program analysis, software engineering/development, analysis
Alternatives to Full-Time Employment: Part-time jobs, internships
Company-Paid Training is available.
International Assignments are available to employees.
Contact: Human Resources Department; Phone: 508-626-2727; Fax: 508-626-0645.
Company URL: http://www.asaint.com

BBN
10 MOULTON STREET
CAMBRIDGE, MASSACHUSETTS 02138

Overview: Provides software, systems, and network engineering. Established: 1948. Parent Company: GTE. Ownership: Public.
Key Statistics: Annual Sales: $234.3 million. Number of Employees: 2,000.
Hiring History: Number of professional employees hired in 1997: 805. 1996: 350.
Expertise/Education Sought: Telecommunications, business management, management information systems, software engineering/development, computer engineering
Alternatives to Full-Time Employment: Part-time jobs, paid internships
International Assignments are available to employees.
Contact: Mr. Bob Hale, Director of Employment; Phone: 617-873-2506; Fax: 617-873-3099. E-mail: bhale@bbn.com.

BIRD, INC.
1077 PLEASANT STREET
NORWOOD, MASSACHUSETTS 02062-4609

Overview: Manufactures concrete products. Established: 1990. Parent Company: Bird Corporation. Ownership: Public.
Key Statistics: Annual Sales: $60 million. Number of Employees: 164.
Expertise/Education Sought: Manufacturing
Alternatives to Full-Time Employment: Paid internships
Company-Paid Training is available.
Contact: For individuals with previous experience: Human Resources Department; Phone: 781-551-0656, ext. 125; Fax: 781-551-9507.

BIRD MACHINE COMPANY, INC.
100 NEPONSET STREET, PO BOX 9103
SOUTH WALPOLE, MASSACHUSETTS 02071-9103

Overview: Designs and manufactures machinery for liquid/solid separation problems. Established: 1989. Parent Company: Baker Hughes, Inc. Ownership: Public.
Key Statistics: Annual Sales: $110 million. Number of Employees: 500.
Hiring History: Number of professional employees hired in 1997: 18.
Expertise/Education Sought: Accounting, engineering, manufacturing engineering, drafting
Alternatives to Full-Time Employment: Paid internships

Bird Machine Company, Inc. (continued)

Company-Paid Training is available.

International Assignments are available to employees.

Contact: Mr. Tom Daly, Human Resources Manager; Phone: 508-668-0400; Fax: 508-660-1385. E-mail: tom.daly@bakerhughes.com.

BIW CABLE SYSTEMS, INC.

9 FORGE PARK
FRANKLIN, MASSACHUSETTS 02038-3135

Overview: Manufactures nonferrous wiredrawn, insulated cable. Established: 1988. Parent Company: Draka USA Corporation. Ownership: Private.

Key Statistics: Annual Sales: $37.8 million. Number of Employees: 300.

Hiring History: Number of professional employees hired in 1997: 12.

Expertise/Education Sought: Accounting, engineering

Contact: For individuals with previous experience: Human Resources; Phone: 508-520-1200; Fax: 508-520-1929.

BOSE CORPORATION

THE MOUNTAIN
FRAMINGHAM, MASSACHUSETTS 01701

Overview: Manufactures and markets audio speakers. Established: 1964. Ownership: Private.

Key Statistics: Annual Sales: $395.9 million. Number of Employees: 3,500.

Expertise/Education Sought: Sales, law, accounting, marketing, finance

Alternatives to Full-Time Employment: Internships

Contact: For individuals with previous experience: Central Recruiting Department; Phone: 508-879-7330; Fax: 508-766-7499.

BOSTON EDISON COMPANY

800 BOYLSTON STREET
BOSTON, MASSACHUSETTS 02199-8003

Overview: Electric utility. Established: 1886. Ownership: Public.

Key Statistics: Annual Sales: $1.7 billion. Number of Employees: 3,362.

Expertise/Education Sought: Sales, marketing, mechanical engineering, marketing, industrial engineering, information technology, civil engineering, structural engineering, electrical engineering

Alternatives to Full-Time Employment: Paid internships

Company-Paid Training is available.

Contact: For individuals with previous experience: Ms. Jo Anne Van Derveer, Senior Human Resources Specialist, 800 Boylston Street, P-305, Boston, Massachusetts, 02199-8003; Phone: 617-424-2000; Fax: 617-424-2629. E-mail: joanne_vanderveer@bedison.com. Job Hotline: 617-424-2640.

H.L. BOUTON COMPANY

11 KENDRICK ROAD
WAREHAM, MASSACHUSETTS 02571-1077

Mailing Address: PO Box 840, Buzzard's Bay, MA 02532-1940

Overview: Manufactures goggles, spectacles, and eyeglass lenses and frames. Established: 1982. Ownership: Public.

Key Statistics: Annual Sales: $13.2 million. Number of Employees: 120.

Hiring History: Number of professional employees hired in 1997: 12. 1996: 12.

Expertise/Education Sought: Machinery

Company-Paid Training is available.

Contact: For individuals with previous experience: Personnel Department, 320 Main Street, Buzzards Bay, Massachusetts, 02532-1077; Phone: 508-759-8000; Fax: 508-759-9727.

BROOKFIELD ENGINEERING LABS

240 CUSHING STREET
STOUGHTON, MASSACHUSETTS 02072-2398

Overview: Manufactures analytical instruments. Established: 1934. Ownership: Private.

Key Statistics: Annual Sales: $26.5 million. Number of Employees: 190.

Hiring History: Number of professional employees hired in 1997: 30. 1996: 30.

Expertise/Education Sought: Electronics/electronics engineering, electrical engineering, mechanical engineering

Company-Paid Training is available.

Contact: Human Resources; Phone: 508-946-6200; Fax: 508-946-6262. E-mail: hr@brookfieldengineering.com.

Company URL: http://www.brookfieldengineering.com

BULL HN WORLDWIDE INFORMATION

300 CONCORD ROAD
BILLERICA, MASSACHUSETTS 01821

Overview: Manufactures large computer systems and communications equipment. Established: 1992. Ownership: Private.

Key Statistics: Annual Sales: $372.5 million. Number of Employees: 1,350.

Hiring History: Number of professional employees hired in 1997: 117.

Expertise/Education Sought: Data processing, computer programming, systems integration, project management, technical, sales, management

Alternatives to Full-Time Employment: Paid internships

Company-Paid Training is available.

International Assignments are available to employees.

Contact: Ms. Debra Mitrano, Staffing Specialist; Phone: 978-294-4390; Fax: 978-294-6601. E-mail: debra.mitrano@bull.com.

Company URL: http://www.us.bull.com

CABOT CORPORATION

75 STATE STREET
BOSTON, MASSACHUSETTS 02109-1817

Overview: Refines metals and manufactures carbon black, silica, electronic materials, and plastics. Established: 1882. Ownership: Public.

Key Statistics: Annual Sales: $1.6 billion. Number of Employees: 4,700.

Expertise/Education Sought: Finance, marketing, engineering, human resources

Alternatives to Full-Time Employment: Paid internships

International Assignments are available to employees.

Contact: Human Resources; Phone: 617-345-0100; Fax: 617-342-6103.

Company URL: http://www.cabot-corp.com

CAMP DRESSER & MCKEE, INC.

1 CAMBRIDGE CENTER
CAMBRIDGE, MASSACHUSETTS 02142-1601

Overview: A full service consulting engineering, construction and operations firm. Established: 1947. Ownership: Private.

Key Statistics: Annual Sales: $321 million. Number of Employees: 2,300.

Hiring History: Number of professional employees hired in 1996: 225.

Expertise/Education Sought: Environmental engineering, civil engineering, chemical engineering, construction engineering, waste management, electrical engineering

Alternatives to Full-Time Employment: Part-time jobs, paid internships

International Assignments are available to employees.
Contact: For college students and recent graduates: Ms. Roberta Cabral, College Recruiter; Phone: 617-252-8000; Fax: 617-621-2565. For individuals with previous experience: Mr. Edward Hagan, Corporate Human Resources Manager; Phone: 617-621-8181; Fax: 617-577-7501.
Company URL: http://www.cdm.com

C & K COMPONENTS, INC.
57 STANLEY AVENUE
WATERTOWN, MASSACHUSETTS 02172-4802
Overview: Manufactures electromechanical components. Established: 1957. Ownership: Private.
Key Statistics: Annual Sales: $49.3 million.
Expertise/Education Sought: Electronics/electronics engineering, manufacturing
Alternatives to Full-Time Employment: Part-time jobs, paid internships
Company-Paid Training is available.
Contact: For individuals with previous experience: Human Resources; Phone: 617-926-6400; Fax: 617-964-2686. E-mail: hrdept@ckcor.com.
Company URL: http://www.ckcorp.com

DRAPER LABORATORY

THE CHARLES STARK DRAPER LABORATORY, INC.
555 TECHNOLOGY SQUARE
CAMBRIDGE, MASSACHUSETTS 02139-3563
Overview: Nonprofit scientific research and development corporation. Established: 1973. Ownership: Private.
Key Statistics: Annual Sales: $247.8 million. Number of Employees: 1,300.
Hiring History: Number of professional employees hired in 1996: 70.
Expertise/Education Sought: Aeronautical/astronautical engineering, computer science, electrical engineering, mechanical engineering, ocean engineering, systems engineering
Alternatives to Full-Time Employment: Paid internships
Company-Paid Training is available.
Contact: Professional Employment, MS#44; Phone: 617-258-2678; Fax: 617-258-1113. E-mail: hr@draper.com.
Company URL: http://www.draper.com
For more information, see full corporate profile, pg. 268.

CHASE-WALTON ELASTOMERS, INC.
29 APSLEY STREET
HOUSTON, MASSACHUSETTS 01749-1543
Overview: Manufactures silicone rubber products. Established: 1974. Ownership: Private.
Key Statistics: Annual Sales: $8 million. Number of Employees: 100.
Expertise/Education Sought: Manufacturing engineering, engineering, polymers
Contact: For individuals with previous experience: Ms. Kathy O'Toole, Human Resources Manager; Phone: 508-485-5600; Fax: 978-562-5178.

CHIRON DIAGNOSTICS CORPORATION
63 NORTH STREET
MEDFIELD, MASSACHUSETTS 02052
Overview: Manufactures medical diagnostic apparatus. Established: 1995. Parent Company: Chiron Corporation. Ownership: Public.
Key Statistics: Annual Sales: $600 million. Number of Employees: 3,200.
Hiring History: Number of professional employees hired in 1996: 22.
Expertise/Education Sought: Software engineering/development, chemistry, biology, biochemistry, medical technology, information technology
Alternatives to Full-Time Employment: Part-time jobs, paid internships
Company-Paid Training is available.
International Assignments are available to employees.
Contact: Human Resources and Staffing Department; Fax: 508-359-3887. E-mail: hrjobs@kirondiag.com.
Company URL: http://www.chiron.com

CIVIGENICS, INC.
200 EAST MAIN STREET
MILFORD, MASSACHUSETTS 01757-2808
Overview: Provides management services. Established: 1995. Ownership: Private.
Key Statistics: Annual Sales: $27 million. Number of Employees: 500.
Hiring History: Number of professional employees hired in 1997: 350. 1996: 200.
Expertise/Education Sought: Social service, management, human resources, administration
Alternatives to Full-Time Employment: Paid internships
Company-Paid Training is available.
Contact: Human Resources; Phone: 508-634-1877; Fax: 508-634-1875.

COMMERCIAL UNION INSURANCE COMPANY
1 BEACON STREET
BOSTON, MASSACHUSETTS 02108-3100
Overview: Provides Insurance. Established: 1969. Ownership: Private.
Key Statistics: Annual Sales: $2.2 billion. Number of Employees: 6,600.
Hiring History: Number of professional employees hired in 1997: 200. 1996: 120.
Expertise/Education Sought: Risk management
Company-Paid Training is available.
Contact: Mr. William Queenan, Human Resources Consultant; Phone: 617-725-6000; Fax: 617-725-6262.

COMMONWEALTH ENERGY SYSTEM
1 MAIN STREET
CAMBRIDGE, MASSACHUSETTS 02142-9150
Overview: Provides accounting services. Established: 1927. Ownership: Public.
Key Statistics: Annual Sales: $35.6 million. Number of Employees: 198.
Expertise/Education Sought: Computer science, engineering
Company-Paid Training is available.
Contact: For individuals with previous experience: Human Resources; Phone: 617-225-4000; Fax: 617-354-5325. E-mail: resourcesh@comelectric.com.

CONTINENTAL RESOURCES, INC.
175 MIDDLESEX TURNPIKE
BEDFORD, MASSACHUSETTS 01730-1409
Overview: Provides computers, peripherals, and software. Established: 1962. Ownership: Private.
Key Statistics: Annual Sales: $125.7 million. Number of Employees: 231.
Hiring History: Number of professional employees hired in 1997: 100.
Expertise/Education Sought: Sales, systems engineering
Alternatives to Full-Time Employment: Paid internships
Company-Paid Training is available.
Contact: For individuals with previous experience: Human Resources Department; Phone: 781-533-0332; Fax: 781-533-0212.

CSC
29 SAWYER ROAD
WALTHAM, MASSACHUSETTS 02154-3423

OVERVIEW: Consulting firm providing data management, telecommunications, and outsourcing. Established: 1986. Parent Company: Computer Sciences Corporation. Ownership: Public.

KEY STATISTICS: Annual Sales: $125.8 million. Number of Employees: 2,300.

HIRING HISTORY: Number of professional employees hired in 1996: 858.

EXPERTISE/EDUCATION SOUGHT: Management consulting, systems integration, information systems

ALTERNATIVES TO FULL-TIME EMPLOYMENT: Part-time jobs, internships

COMPANY-PAID TRAINING is available.

INTERNATIONAL ASSIGNMENTS are available to employees.

CONTACT: Ms. Patricia Romeo, Director of Recruiting; Phone: 513-624-2040.

COMPANY URL: http://www.csc.com

CUMBERLAND FARMS, INC.
777 DEDHAM STREET
CANTON, MASSACHUSETTS 02021-1484

OVERVIEW: Operates chain of convenience stores. Established: 1938. Ownership: Private.

KEY STATISTICS: Annual Sales: $1.4 billion. Number of Employees: 6,900.

HIRING HISTORY: Number of professional employees hired in 1997: 2,000. 1996: 2,000.

EXPERTISE/EDUCATION SOUGHT: Manufacturing engineering, retail management

ALTERNATIVES TO FULL-TIME EMPLOYMENT: Paid internships

COMPANY-PAID TRAINING is available.

CONTACT: Ms. Michelle Gouveia, Employment Specialist; Phone: 781-828-4900, ext. 5018; Fax: 781-828-9012. E-mail: cfijobs@aol.com. Job Hotline: 800-444-JOBS.

COMPANY URL: http://www.cumberlandfarms.com

DALE MEDICAL PRODUCTS, INC.
7 CROSS STREET
PLAINVILLE, MASSACHUSETTS 02762-0056

MAILING ADDRESS: PO Box 1556, Plainville, MA 02762-0056

OVERVIEW: Manufactures ankle, knee, and abdominal supports. Established: 1961. Ownership: Private.

KEY STATISTICS: Annual Sales: $10 million. Number of Employees: 86.

HIRING HISTORY: Number of professional employees hired in 1997: 5. 1996: 10.

EXPERTISE/EDUCATION SOUGHT: Manufacturing engineering

ALTERNATIVES TO FULL-TIME EMPLOYMENT: Part-time jobs

CONTACT: For individuals with previous experience: Human Relations Department; Phone: 800-343-3980; Fax: 800-752-1230.

COMPANY URL: http://www.dalemed.com

DANIELS PRINTING, LP
40 COMMERCIAL STREET
EVERETT, MASSACHUSETTS 02149

OVERVIEW: Provides commercial lithographic and offset printing services. Established: 1958. Ownership: Private.

KEY STATISTICS: Annual Sales: $70 million. Number of Employees: 425.

EXPERTISE/EDUCATION SOUGHT: Accounting, sales, manufacturing engineering

COMPANY-PAID TRAINING is available.

CONTACT: Human Resource Department; Phone: 617-389-7900; Fax: 617-389-5520.

DATA GENERAL CORPORATION
4400 COMPUTER DRIVE
WESTBOROUGH, MASSACHUSETTS 01580

OVERVIEW: Manufactures computers. Established: 1968. Ownership: Public.

KEY STATISTICS: Annual Sales: $1.5 billion. Number of Employees: 5,000.

EXPERTISE/EDUCATION SOUGHT: Marketing, software engineering/development, hardware engineering

ALTERNATIVES TO FULL-TIME EMPLOYMENT: Paid internships

INTERNATIONAL ASSIGNMENTS are available to employees.

CONTACT: Ms. Sandy Bradshaw, Corporate Staffing Manager; Fax: 508-898-4686.

COMPANY URL: http://www.dg.com

DATA TECHNOLOGY, INC.
260A FORDHAM ROAD
WILMINGTON, MASSACHUSETTS 01888-1721

OVERVIEW: Manufactures computer peripheral equipment. Established: 1971. Ownership: Private.

KEY STATISTICS: Annual Sales: $13.4 million. Number of Employees: 85.

EXPERTISE/EDUCATION SOUGHT: Engineering, electrical technology

CONTACT: For individuals with previous experience: Human Resources Department; Phone: 781-935-8820; Fax: 781-935-9656.

DAVID CLARK COMPANY, INC.
360 FRANKLIN STREET
WORCESTER, MASSACHUSETTS 01604-4916

MAILING ADDRESS: PO Box 15054, Worchester, MA 01615-0054

OVERVIEW: Manufactures radio broadcasting equipment. Established: 1975. Ownership: Private.

KEY STATISTICS: Annual Sales: $40.9 million. Number of Employees: 340.

HIRING HISTORY: Number of professional employees hired in 1997: 70. 1996: 70.

EXPERTISE/EDUCATION SOUGHT: Engineering, computer programming, computer science, technical

ALTERNATIVES TO FULL-TIME EMPLOYMENT: Part-time jobs, paid internships

COMPANY-PAID TRAINING is available.

CONTACT: For individuals with previous experience: Mr. Peter Kasparson, Assistant Personnel Manager; Phone: 508-756-6216. E-mail: humanresources@davidclark.com. Job Hotline: 508-751-5861.

COMPANY URL: http://www.davidclark.com

DAYMARC, INC.
1 MONARCH DRIVE
LITTLETON, MASSACHUSETTS 01460-1425

OVERVIEW: Manufactures electronic measure instruments. Established: 1994. Parent Company: Cohu. Ownership: Public.

KEY STATISTICS: Annual Sales: $19.2 million. Number of Employees: 225.

EXPERTISE/EDUCATION SOUGHT: Engineering

ALTERNATIVES TO FULL-TIME EMPLOYMENT: Paid internships

COMPANY-PAID TRAINING is available.

INTERNATIONAL ASSIGNMENTS are available to employees.

CONTACT: For individuals with previous experience: Human Resources Department; Phone: 978-486-1060; Fax: 978-486-1070.

DIGITAL EQUIPMENT CORPORATION
200 FOREST STREET, MS-MRO1-3/L32
MARLBORO, MASSACHUSETTS 01752-3085

OVERVIEW: Provides open client/server computing solutions from personal computers to integrated information systems. Established: 1957. Ownership: Public.

KEY STATISTICS: Annual Sales: $13 billion. Number of Employees: 54,900.

HIRING HISTORY: Number of professional employees hired in 1997: 70.

EXPERTISE/EDUCATION SOUGHT: Software engineering/development, electrical engineering, computer engineering, mechanical engineering, computer science

Alternatives to Full-Time Employment: Paid internships

Company-Paid Training is available.

International Assignments are available to employees.

Contact: Mrs. Cindy Gallerani, College Relations Manager; Phone: 508-467-5608; Fax: 508-467-5661.

Company URL: http://www.digital.com

DODGE COMPANY, INC.

165 CAMBRIDGEPARK DRIVE
CAMBRIDGE, MASSACHUSETTS 02140-0193

Overview: Processes chemicals for embalming and cosmetics. Established: 1957. Ownership: Private.

Key Statistics: Annual Sales: $16 million. Number of Employees: 105.

Hiring History: Number of professional employees hired in 1996: 6.

Expertise/Education Sought: Manufacturing

Contact: For individuals with previous experience: Ms. Donna Hetherington, Comptroller; Fax: 617-661-1428.

DOLAN-JENNER INDUSTRIES, INC.

678 ANDOVER STREET
LAWRENCE, MASSACHUSETTS 01843

Overview: Manufactures glass, quartz, coherent, and high temperature optical fibers for industry and laboratory. Established: 1996. Parent Company: Danaher Corporation. Ownership: Public.

Key Statistics: Annual Sales: $15.4 million.

Hiring History: Number of professional employees hired in 1997: 12. 1996: 6.

Expertise/Education Sought: Manufacturing engineering, mechanical engineering, marketing, sales, customer service and support

Alternatives to Full-Time Employment: Paid internships

Company-Paid Training is available.

Contact: Human Resources Department; Phone: 978-681-8000; Fax: 978-687-6011.

DURO INDUSTRIES, INC.

110 CHACE STREET
FALL RIVER, MASSACHUSETTS 02724-1416

Overview: Manufactures finishes for military, industrial, and commercial fabrics. Established: 1985. Ownership: Private.

Key Statistics: Annual Sales: $82.8 million. Number of Employees: 1,400.

Hiring History: Number of professional employees hired in 1997: 50.

Expertise/Education Sought: Engineering, computer science, chemistry, accounting, finance

Alternatives to Full-Time Employment: Paid internships

Contact: For individuals with previous experience: Mrs. Eva Custadio, Director of Human Resources; Phone: 508-675-0101; Fax: 508-673-1945.

DYNAMICS RESEARCH CORPORATION

60 FRONTAGE ROAD
ANDOVER, MASSACHUSETTS 01810-5423

Overview: Provides data processing and preparation services. Established: 1955. Ownership: Public.

Key Statistics: Annual Sales: $130.2 million. Number of Employees: 1,249.

Expertise/Education Sought: Engineering

Alternatives to Full-Time Employment: Part-time jobs, paid internships

Contact: For individuals with previous experience: Mr. Matt McKethmie, Senior Human Resources Representative; Phone: 508-475-9090, ext. 2855; Fax: 508-474-8948. E-mail: mmckethmie@drc.com.

Company URL: http://www.drc.com

DYNATECH CORPORATION

3 NEW ENGLAND EXECUTIVE PARK
BURLINGTON, MASSACHUSETTS 01803-5006

Overview: Provides voice data and communications services. Established: 1957. Ownership: Public.

Key Statistics: Annual Sales: $362.4 million. Number of Employees: 1,500.

Expertise/Education Sought: Accounting, finance, electronics/electronics engineering, sales, communications engineering

Alternatives to Full-Time Employment: Paid internships

International Assignments are available to employees.

Contact: For individuals with previous experience: Mr. Mario Alosco, Director of Executive Staffing; Phone: 617-272-6100; Fax: 617-272-2304.

Company URL: http://www.dynatech.com

ECC CORPORATION

156 PRINCETON STREET
HOLDEN, MASSACHUSETTS 01520

Mailing Address: PO Box 176, Jefferson, MA 01522-0176

Overview: Manufactures printed circuit boards. Established: 1957. Ownership: Private.

Key Statistics: Number of Employees: 130.

Hiring History: Number of professional employees hired in 1997: 30. 1996: 10.

Expertise/Education Sought: Engineering, electrical engineering, test engineering, imaging technology, process engineering, computer-aided design, marketing, sales

Alternatives to Full-Time Employment: Paid internships

Company-Paid Training is available.

Contact: Mr. Russell Vickstrom, Human Resources Manager; Phone: 508-829-5301; Fax: 508-829-9663. E-mail: hr@ecccorp.com.

EG&G, INC.

45 WILLIAM STREET
WELLESLEY, MASSACHUSETTS 02181-4004

Overview: Provides technologically advanced products and services. Established: 1947. Ownership: Public.

Key Statistics: Annual Sales: $1.4 billion. Number of Employees: 13,000.

Hiring History: Number of professional employees hired in 1996: 250.

Expertise/Education Sought: Finance, human resources, law

Alternatives to Full-Time Employment: Part-time jobs, paid internships

Company-Paid Training is available.

International Assignments are available to employees.

Contact: Ms. Deborah Burk, Director of Human Resources; Phone: 781-431-4249. E-mail: dburke@egginc.com.

Company URL: http://www.egginc.com

EMC CORPORATION

171 SOUTH STREET
HOPKINTON, MASSACHUSETTS 01748-2208

Overview: Manufactures computer memory boards. Established: 1979. Ownership: Public.

Key Statistics: Annual Sales: $2.3 billion. Number of Employees: 4,800.

Expertise/Education Sought: Hardware engineering, software engineering/development

Alternatives to Full-Time Employment: Paid internships

Company-Paid Training is available.

International Assignments are available to employees.

Contact: For college students and recent graduates: Mrs. Patti McCrossan, College Recruiter; Phone: 508-435-1000; Fax: 508-435-8884. For individuals with previous experience: Ms. Lisa Benjamin, Employment Manager, 35 Parkwood Drive, Hopkinton, Massachusetts, 01748-2208; Phone: 508-435-1000; Fax: 508-435-8884. E-mail: benjamin_lisa@isusemc.com.

Company URL: http://www.emc.com

EPEC, INC.

174 DUCHAINE BOULEVARD
NEW BEDFORD, MASSACHUSETTS 02745

Overview: Manufactures printed circuit boards. Established: 1981. Ownership: Private.

Key Statistics: Annual Sales: $7 million. Number of Employees: 95.

Hiring History: Number of professional employees hired in 1996: 6.

Expertise/Education Sought: Manufacturing

Company-Paid Training is available.

Contact: For individuals with previous experience: Human Relations Department.

FITZPATRICK COMPANIES, INC.

RR 102
LEE, MASSACHUSETTS 01238

Overview: Operates mail order business. Established: 1996. Ownership: Private.

Key Statistics: Annual Sales: $90.9 million. Number of Employees: 530.

Expertise/Education Sought: Accounting, manufacturing engineering, business management, finance, design engineering, retail management, hospital administration

Alternatives to Full-Time Employment: Part-time jobs, internships

Company-Paid Training is available.

Contact: For individuals with previous experience: Ms. Rebecca Riordan, Human Resources Manager; Phone: 413-243-1474, ext. 223; Fax: 413-243-1067.

Company URL: http://www.countrycurtains.com

FMC-CROSBY VALVE, INC.

43 KENDRICK STREET
WRENTHAM, MASSACHUSETTS 02093-1544

Overview: Manufactures pressure relief devices, valves, test benches, and accessories. Established: 1995. Parent Company: FMC. Ownership: Public.

Key Statistics: Annual Sales: $83.7 million. Number of Employees: 540.

Hiring History: Number of professional employees hired in 1997: 15. 1996: 15.

Expertise/Education Sought: Sales, accounting, mechanical engineering

Alternatives to Full-Time Employment: Paid internships

Company-Paid Training is available.

International Assignments are available to employees.

Contact: Mr. William Larocque, Manager of Employee Relations; Phone: 508-384-3121; Fax: 508-384-8675. E-mail: bill_larocque@fmc.com.

FOXBORO

A SIEBE COMPANY

THE FOXBORO COMPANY

33 COMMERCIAL STREET
FOXBORO, MASSACHUSETTS 02035

Overview: Manufactures industrial process control equipment. Established: 1908. Parent Company: Siebe, PLC.

Key Statistics: Annual Sales: $1.2 billion. Number of Employees: 3,400.

Hiring History: Number of professional employees hired in 1997: 300. 1996: 250.

Expertise/Education Sought: Electrical engineering, chemical engineering, computer science

Contact: Mr. Paul Fanning, Manager of Human Resources; Phone: 508-549-6149; Fax: 508-549-4486. E-mail: pfanning@foxboro.com.

Company URL: http://www.foxboro.com

CORPORATE PROFILE

The Foxboro Company is a world leader in process automation, manufacturing, and marketing a range of control products for continuous, batch, and sequential industrial processes. The company is also the manufacturer of advanced measurement and intelligent monitoring instruments, both electronic and pneumatic. The company also offers extensive engineering and application services, education and training, and postsale technical support. Foxboro customers include the chemical and petrochemical, oil and gas, pulp and paper, food, pharmaceutical, mining and minerals, electric power, textile, water, and wastewater treatment industries. The company's instruments and systems are used primarily to measure and control the condition of processes and the composition of materials being processed. These products also generate information used for the total management of the process plant. Headquartered in Foxboro, Massachusetts, the Foxboro Company is committed to excellence and to the total quality process. This is reflected in the company's receiving ISO9001 certification and a number of quality awards.

Building on their education and experience, employees of the Foxboro Company can carve out a rewarding career either in a specific function or through movement in a number of business areas. Besides on-the-job training, employees can avail themselves of a broad range of formal training and development offered through internal and external means.

The company offers its employees a very competitive and comprehensive benefits package that includes tuition assistance.

The company, in hiring new college graduates, looks traditionally to those possessing a degree in engineering—specifically electrical or chemical—or computer science.

FTP SOFTWARE, INC.

2 HIGH STREET
NORTH ANDOVER, MASSACHUSETTS 01845

Overview: Provides custom computer programming services. Established: 1986. Ownership: Public.

Key Statistics: Annual Sales: $101.1 million. Number of Employees: 804.

Hiring History: Number of professional employees hired in 1997: 40. 1996: 70.

Expertise/Education Sought: Technical, marketing, finance, accounting, sales, marketing, software engineering/development

Alternatives to Full-Time Employment: Part-time jobs, paid internships

Company-Paid Training is available.

Contact: Ms. Elaine LaCrois, Senior Technical Recruiter; Phone: 978-685-4000; Fax: 978-684-6557. E-mail: elacroix@ftp.com.

Company URL: http://www.ftp.com

GENRAD, INC.

7 TECHNOLOGY PARK DRIVE
WESTFORD, MASSACHUSETTS 01886-3141

Overview: Manufactures electronic measure instruments. Established: 1915. Ownership: Public.

Key Statistics: Annual Sales: $183.5 million. Number of Employees: 1,239.

Hiring History: Number of professional employees hired in 1997: 100.

Expertise/Education Sought: Engineering, software engineering/development, accounting

Alternatives to Full-Time Employment: Paid internships

International Assignments are available to employees.

Contact: Human Resources Department; Phone: 978-589-7079; Fax: 978-589-2091.

Company URL: http://www.genrad.com

GEORGE MCQUESTEN COMPANY, INC.

600 IRON HORSE PARK
NORTH BILLERICA, MASSACHUSETTS 01862-1618

Overview: Wholesales lumber. Established: 1937. Ownership: Private.

Key Statistics: Annual Sales: $49.7 million. Number of Employees: 98.

Hiring History: Number of professional employees hired in 1996: 8.

Expertise/Education Sought: Building science

Contact: Mr. Richard Starrak, President; Phone: 978-663-3435; Fax: 978-667-0934.

LOUIS M. GERSON COMPANY, INC.

15 SPROAT STREET
MIDDLEBORO, MASSACHUSETTS 02346

Overview: Manufactures household furnishings. Established: 1945. Ownership: Private.

Key Statistics: Annual Sales: $35 million. Number of Employees: 30.

Hiring History: Number of professional employees hired in 1996: 30.

Expertise/Education Sought: Accounting, engineering, marketing, management, sales, production, quality control, warehousing, maintenance

Company-Paid Training is available.

Contact: Mrs. Jackie Clifford, Human Resources Supervisor; Phone: 508-947-4000, ext. 2103; Fax: 508-947-8668.

THE GILLETTE COMPANY

PRUDENTIAL TOWER
BOSTON, MASSACHUSETTS 02199-8004

Mailing Address: Human Resources, 40th Floor, Boston, MA 02199-8004

Overview: Manufactures razors and blades, toiletries, dental products, stationery products, and small appliances. Established: 1901. Ownership: Public.

Key Statistics: Annual Sales: $9.7 billion. Number of Employees: 44,100.

Expertise/Education Sought: Business administration, law, marketing, sales

Alternatives to Full-Time Employment: Paid internships

Company-Paid Training is available.

International Assignments are available to employees.

Contact: Senior Recruiter, Prudential Tower, 40th Floor, Boston, Massachusetts, 02199-8004; Phone: 617-421-7700; Fax: 617-421-7617. Job Hotline: 617-421-7567.

HAARTZ CORPORATION

87 HAYWARD ROAD
ACTON, MASSACHUSETTS 01720-3005

Overview: Manufactures coated fabrics. Established: 1922. Ownership: Private.

Key Statistics: Annual Sales: $80 million. Number of Employees: 300.

Hiring History: Number of professional employees hired in 1997: 66.

Expertise/Education Sought: Engineering

Contact: For individuals with previous experience: Human Resource Department; Phone: 978-264-2600; Fax: 978-264-2624. E-mail: hrussell@haartz.com.

Company URL: http://www.haartz.com

HADCO TECH EAST

46 ROGERS ROAD
WARD HILL, MASSACHUSETTS 01835-6969

Overview: Manufactures printed circuit boards. Established: 1996. Parent Company: Capitol Corporations. Ownership: Public.

Key Statistics: Annual Sales: $13.6 million. Number of Employees: 160.

Hiring History: Number of professional employees hired in 1996: 75.

Expertise/Education Sought: Manufacturing engineering, electrical engineering, electronics/electronics engineering, design engineering, design

Contact: For individuals with previous experience: Ms. Nancy Langevin, Manager of Personnel; Phone: 978-556-6233; Fax: 978-556-6111. E-mail: nlangevin@hadco.com.

HAEMONETICS CORPORATION

400 WOOD ROAD
BRAINTREE, MASSACHUSETTS 02184

Overview: Manufactures blood processing machines and related disposables. Established: 1971. Ownership: Public.

Key Statistics: Annual Sales: $309.8 million. Number of Employees: 1,526.

Expertise/Education Sought: Electrical engineering, software engineering/development, mechanical engineering, polymers, biomedical engineering, finance, accounting, life sciences, general nursing

Alternatives to Full-Time Employment: Part-time jobs, paid internships

Company-Paid Training is available.

International Assignments are available to employees.

Contact: Human Resources Department; Phone: 781-356-9259; Fax: 781-848-9959.

Company URL: http://www.haemonetics.com

HARCO, INC.

215 MASSACHUSETTS AVENUE
ARLINGTON, MASSACHUSETTS 02174-8607

Overview: Provides engineering services. Established: 1994. Ownership: Private.

Key Statistics: Annual Sales: $12 million.

Hiring History: Number of professional employees hired in 1997: 7. 1996: 4.

Expertise/Education Sought: Marine engineering, marine research, marine biology, marine science

Company-Paid Training is available.

International Assignments are available to employees.

Contact: For individuals with previous experience: Human Resources Department; Phone: 781-643-2690; Fax: 781-643-2252.

HARVARD APPARATUS, INC.

84 OCTOBER HILL ROAD
HOLLISTON, MASSACHUSETTS 01746

Overview: Manufactures laboratory apparatus and furniture. Established: 1996. Ownership: Private.

Key Statistics: Annual Sales: $10.3 million.

Hiring History: Number of professional employees hired in 1997: 5.

Expertise/Education Sought: Technology, science

Alternatives to Full-Time Employment: Paid internships

Company-Paid Training is available.

International Assignments are available to employees.

Contact: For individuals with previous experience: Mr. Chane Graziano, Chief Executive Officer; Phone: 508-655-7000; Fax: 508-655-6029. E-mail: chaneg@aol.com.

HARVEY INDUSTRIES, INC.

43 EMERSON ROAD
WALTHAM, MASSACHUSETTS 02154

Overview: Wholesales roofing and siding materials. Established: 1961. Ownership: Private.

Key Statistics: Annual Sales: $152.8 million. Number of Employees: 750.

Expertise/Education Sought: Marketing, manufacturing engineering, sales, manufacturing

Alternatives to Full-Time Employment: Paid internships

Company-Paid Training is available.

Contact: For individuals with previous experience: Human Resources; Fax: 617-899-4912.

Company URL: http://www.harveyind.com

HAZEN PAPER COMPANY

240 SOUTH WATER STREET
HOLYOKE, MASSACHUSETTS 01040-5979

Mailing Address: PO Box 189, Holyoke, MA 01041

Overview: Manufactures coated and laminated paper. Established: 1925. Ownership: Private.

Key Statistics: Annual Sales: $58.2 million. Number of Employees: 185.

Hiring History: Number of professional employees hired in 1997: 20.

Expertise/Education Sought: Accounting, administration, chemical engineering, industrial engineering, manufacturing engineering

Alternatives to Full-Time Employment: Paid internships

Company-Paid Training is available.

Contact: For individuals with previous experience: Mr. Phillip McDonald, Director of Human Resources; Phone: 413-538-8204; Fax: 413-552-0794.

HIGH VOLTAGE ENGINEERING CORPORATION

401 EDGEWATER PLACE, SUITE 680
WAKEFIELD, MASSACHUSETTS 01880-6219

Overview: Manufactures process control instruments. Established: 1988. Ownership: Private.

Key Statistics: Annual Sales: $135 million. Number of Employees: 1,300.

Expertise/Education Sought: Accounting, administration, computer science, engineering

Contact: Mr. Arthur Burdett, Vice President of Human Resources; Phone: 781-224-1001, ext. 109; Fax: 781-224-1011.

HNU SYSTEMS, INC.

160 CHARLEMONT STREET
NEWTON, MASSACHUSETTS 02161-1999

Overview: Manufactures scales and detectors. Established: 1973. Ownership: Private.

Key Statistics: Annual Sales: $5.3 million. Number of Employees: 55.

Hiring History: Number of professional employees hired in 1997: 6. 1996: 10.

Expertise/Education Sought: Sales, marketing, mechanical engineering, electrical engineering, software engineering/development, chemistry

Alternatives to Full-Time Employment: Internships

International Assignments are available to employees.

Contact: Mr. Edward McFarland, Controller; Phone: 617-964-6690, ext. 237; Fax: 617-558-0056.

Company URL: http://www.hnu.com

HONEYWELL CONSUMER PRODUCTS

250 TURNPIKE ROAD
SOUTHBOROUGH, MASSACHUSETTS 01772-1742

Overview: Manufactures household fans, heaters, and humidifiers. Established: 1996. Parent Company: Honeywell, Inc. Ownership: Public.

Key Statistics: Annual Sales: $180.4 million. Number of Employees: 2,230.

Hiring History: Number of professional employees hired in 1997: 30.

Expertise/Education Sought: Manufacturing engineering, mechanical engineering, electrical engineering, HVAC engineering, sales, marketing, design engineering

International Assignments are available to employees.

Contact: Mrs. Mary Hassan, Vice President of Human Resources; Phone: 508-490-7000; Fax: 508-460-8056.

H.P. HOOD, INC.

90 EVERETT AVENUE, SUITE 200
CHELSEA, MASSACHUSETTS 02150-2301

Overview: Processes milk and dairy products. Established: 1995. Parent Company: Catamount. Ownership: Private.

Key Statistics: Annual Sales: $271.4 million. Number of Employees: 1,196.

Expertise/Education Sought: Accounting, engineering, marketing, management, sales, management information systems, computer programming, computer science, administration, customer service and support

Contact: For individuals with previous experience: Human Resources; Phone: 617-887-3000; Fax: 617-887-8369.

HYDE MANUFACTURING COMPANY

54 EASTFORD ROAD
SOUTHBRIDGE, MASSACHUSETTS 01550-3604

Overview: Manufactures hand and edge tools. Established: 1959. Ownership: Private.

Key Statistics: Annual Sales: $39.5 million. Number of Employees: 650.

Hiring History: Number of professional employees hired in 1997: 41. 1996: 41.

Expertise/Education Sought: Manufacturing

Company-Paid Training is available.

Contact: For individuals with previous experience: Mr. Richard Ayres, Director of Human Resources.

IDEAL TAPE COMPANY

1400 MIDDLESEX STREET
LOWELL, MASSACHUSETTS 01851-2097

Overview: Produces rubber floorcoverings, pressure-sensitive tape, and soles and heels for footwear. Established: 1908. Parent Company: American Biltrite, Inc. Ownership: Public.

Key Statistics: Annual Sales: $418 million. Number of Employees: 2,770.

Hiring History: Number of professional employees hired in 1997: 12. 1996: 7.

Expertise/Education Sought: Manufacturing

Alternatives to Full-Time Employment: Paid internships

International Assignments are available to employees.

Contact: Mrs. Bonnie Posnak, Vice President of Human Resources; Phone: 508-458-6833; Fax: 508-458-0302.

INSTRON CORPORATION

100 ROYAL STREET
CANTON, MASSACHUSETTS 02021-1089

Overview: Manufactures physical property testing equipment. Established: 1946. Ownership: Public.

Key Statistics: Annual Sales: $153.1 million. Number of Employees: 1,095.

Hiring History: Number of professional employees hired in 1996: 71.

Expertise/Education Sought: Mechanical engineering, software engineering/development, electrical engineering

Alternatives to Full-Time Employment: Paid internships

Company-Paid Training is available.

International Assignments are available to employees.

Contact: For individuals with previous experience: Human Resources; Phone: 781-828-2500; Fax: 781-575-5776.

Company URL: http://www.instron.com

IRIS GRAPHICS, INC.

6 CROSBY DRIVE
BEDFORD, MASSACHUSETTS 01730-1402

Overview: Manufactures computer graphics printers. Established: 1990. Parent Company: Scitex Development Corporation. Ownership: Private.

Key Statistics: Annual Sales: $75 million. Number of Employees: 277.

Expertise/Education Sought: Manufacturing engineering, mechanical engineering, electrical engineering, design engineering, software engineering/development

Alternatives to Full-Time Employment: Paid internships

Company-Paid Training is available.

Contact: Miss Caren Jenkins, Recruiter; Phone: 617-275-8777; Fax: 617-275-8590.

Company URL: http://www.irisgraphics.com

ISI SYSTEMS, INC.
2 TECH DRIVE
ANDOVER, MASSACHUSETTS 01810-2434

Overview: Provides software and processing services to the property and casualty insurance market. Established: 1989. Parent Company: CGI. Ownership: Private.

Key Statistics: Annual Sales: $120 million. Number of Employees: 900.

Hiring History: Number of professional employees hired in 1997: 150.

Expertise/Education Sought: Customer service and support, accounting, training and development, computer programming

Alternatives to Full-Time Employment: Part-time jobs

Company-Paid Training is available.

Contact: Employment Coordinator; Phone: 978-682-5500; Fax: 978-687-4305.

Company URL: http://www.isisys.com

JAMESBURY, INC.
640 LINCOLN STREET
WORCESTER, MASSACHUSETTS 01605-2011

Overview: Manufactures industrial valves. Established: 1988. Parent Company: Neles Controls. Ownership: Private.

Key Statistics: Annual Sales: $180 million. Number of Employees: 900.

Expertise/Education Sought: Accounting, administration, business management, engineering, finance, manufacturing engineering

Company-Paid Training is available.

International Assignments are available to employees.

Contact: For individuals with previous experience: Ms. Marcia Sirat, Employment Training Manager; Phone: 508-852-0215; Fax: 508-852-8207.

JAY CASHMAN, INC.
285 DORCHESTER AVENUE
BOSTON, MASSACHUSETTS 02017

Overview: Heavy construction. Established: 1997. Ownership: Private.

Key Statistics: Annual Sales: $72.3 million.

Hiring History: Number of professional employees hired in 1997: 35. 1996: 25.

Expertise/Education Sought: Civil engineering, construction engineering

Contact: For individuals with previous experience: Human Resources; Phone: 617-268-1300; Fax: 617-561-7116.

Company URL: http://www.jaycashman.com

KAYE INSTRUMENTS, INC.
15 DEANGELO DRIVE
BEDFORD, MASSACHUSETTS 01730-2281

Overview: Manufactures industrial process measuring equipment. Established: 1993. Parent Company: Bowthorpe, PLC. Ownership: Private.

Key Statistics: Annual Sales: $12.7 million. Number of Employees: 80.

Hiring History: Number of professional employees hired in 1996: 20.

Expertise/Education Sought: Customer service and support, engineering

Company-Paid Training is available.

International Assignments are available to employees.

Contact: For individuals with previous experience: Human Resources; Phone: 781-275-0300; Fax: 781-275-9024.

Company URL: http://www.kayeinc.com

KAYEM FOODS, INC.
75 ARLINGTON STREET
CHELSEA, MASSACHUSETTS 02150-2365

Overview: Operates meat packing plant. Established: 1909. Ownership: Private.

Key Statistics: Annual Sales: $110 million. Number of Employees: 440.

Hiring History: Number of professional employees hired in 1997: 5.

Expertise/Education Sought: Food services, production, warehousing

Alternatives to Full-Time Employment: Paid internships

Company-Paid Training is available.

Contact: For individuals with previous experience: Human Resources; Phone: 617-889-1600; Fax: 617-889-5931.

KEANE, INC.
10 CITY SQUARE
BOSTON, MASSACHUSETTS 02129

Overview: Manages, analyzes, designs, programs, markets, and installs computer based information systems. Established: 1965. Ownership: Public.

Key Statistics: Annual Sales: $467.1 million. Number of Employees: 5,338.

Hiring History: Number of professional employees hired in 1997: 3,000. 1996: 300.

Expertise/Education Sought: Accounting, finance, data processing, software engineering/development, consulting, computer programming

Alternatives to Full-Time Employment: Part-time jobs

Company-Paid Training is available.

International Assignments are available to employees.

Contact: For college students and recent graduates: Mr. Chuck Dazis, College Recruiting Manager; Phone: 617-241-9200; Fax: 617-241-8027. For individuals with previous experience: Mr. David Pollard, Manager of Human Resources; Phone: 617-241-9200; Fax: 617-241-8027. Job Hotline: 800-73K-EANE, ext. 1335.

KENS FOODS, INC.
1 DANGELO DRIVE
MARLBOROUGH, MASSACHUSETTS 01752-3066

Mailing Address: PO Box 849, Marlboro, MA 01752

Overview: Manufactures food products. Established: 1958. Ownership: Private.

Key Statistics: Annual Sales: $131.6 million. Number of Employees: 300.

Hiring History: Number of professional employees hired in 1997: 65.

Expertise/Education Sought: Production, manufacturing

Alternatives to Full-Time Employment: Paid internships

Company-Paid Training is available.

Contact: Mr. Ray Felice, Director of Human Resources; Phone: 508-485-7540; Fax: 508-485-6882.

KEVLIN CORPORATION
5 CORNELL PLACE
WILMINGTON, MASSACHUSETTS 01887-2129

Overview: Designs, manufactures, and sells microwave equipment for radar systems in air traffic control and defense industries. Established: 1996. Ownership: Private.

Key Statistics: Annual Sales: $12 million. Number of Employees: 78.

Hiring History: Number of professional employees hired in 1997: 5. 1996: 4.

Expertise/Education Sought: Mechanical engineering, electrical engineering

Contact: Ms. Marge Fitzgerald, Director of Human Resources; Phone: 978-657-3900; Fax: 978-658-5170.

KEY POLYMER CORPORATION
17 SHEPARD STREET, LAWRENCE INDUSTRIAL PARK
LAWRENCE, MASSACHUSETTS 01842

Overview: Manufactures adhesives. Established: 1959. Ownership: Private.

Key Statistics: Annual Sales: $11.2 million. Number of Employees: 75.

Key Polymer Corporation (continued)

Expertise/Education Sought: Chemical engineering
Contact: Human Relations Department; Phone: 978-683-9411; Fax: 978-686-7729.

KOLLMORGEN CORPORATION
1601 TRAPELO ROAD
WALTHAM, MASSACHUSETTS 02154-7333

Overview: Manufactures motors and generators. Established: 1907. Ownership: Public.
Key Statistics: Annual Sales: $230.4 million. Number of Employees: 1,850.
Expertise/Education Sought: Manufacturing
Alternatives to Full-Time Employment: Part-time jobs, internships
International Assignments are available to employees.
Contact: For individuals with previous experience: Human Resources; Phone: 781-890-5655; Fax: 781-190-7150.
Company URL: http://www.kollmorgen.com

KVAERNER US, INC.
116 RODDY AVENUE
ATTLEBORO, MASSACHUSETTS 02703-7900

Overview: Provides chemical engineering services. Established: 1990. Parent Company: Kvaerner. Ownership: Private.
Key Statistics: Annual Sales: $320 million. Number of Employees: 5,000.
Hiring History: Number of professional employees hired in 1997: 20. 1996: 20.
Expertise/Education Sought: Manufacturing
Contact: Human Resources; Phone: 508-399-6400.

LEACH AND GARNER COMPANY
57 JOHN L DIETSCH SQUARE
NORTH ATTLEBORO, MASSACHUSETTS 02761-0200

Overview: Processes precious metals. Established: 1899. Ownership: Private.
Key Statistics: Annual Sales: $48 million. Number of Employees: 400.
Hiring History: Number of professional employees hired in 1997: 40.
Expertise/Education Sought: Manufacturing engineering
Alternatives to Full-Time Employment: Part-time jobs, paid internships
Company-Paid Training is available.
Contact: Ms. Christine LaChance, Director of Human Resources; Phone: 508-695-7800, ext. 243; Fax: 508-695-4093.
Company URL: http://www.genfind.com

LINCOLN LABORATORY
244 WOOD STREET
LEXINGTON, MASSACHUSETTS 02173-9108

Overview: Researches and develops advanced electronics and technologies. Established: 1951. Parent Company: Massachusetts Institute of Technology.
Key Statistics: Number of Employees: 2,200.
Expertise/Education Sought: Electrical engineering, physics, mathematics, computer science
Contact: Human Resources Office, JO99.
Company URL: http://www.ll.mit.edu

CORPORATE STATEMENT

MIT Lincoln Laboratory is a research and development center located in Lexington, Massachusetts. The Laboratory's fundamental mission continues to be to apply science, by means of advanced technology, to critical problems of national security with emphasis on air defense, space surveillance, radar system development, communication systems, airport weather surveillance, and air traffic control.

For more information, see full corporate profile, pg. 280.

LOJACK CORPORATION
333 ELM STREET
DEDHAM, MASSACHUSETTS 02026

Overview: Develops computer tracking devices to locate stolen vehicles. Established: 1978. Ownership: Public.
Key Statistics: Annual Sales: $61.7 million. Number of Employees: 290.
Expertise/Education Sought: Accounting, sales, marketing
Contact: Ms. Cindy Hood, Office Manager; Phone: 781-326-4700; Fax: 781-326-7255.
Company URL: http://www.lojack.com

THE LS STARRETT COMPANY
121 CRESCENT STREET
ATHOL, MASSACHUSETTS 01331

Overview: Manufactures precision measuring tools. Established: 1880. Ownership: Public.
Key Statistics: Annual Sales: $250.5 million. Number of Employees: 2,740.
Expertise/Education Sought: Sales, production
Alternatives to Full-Time Employment: Paid internships
Company-Paid Training is available.
International Assignments are available to employees.
Contact: For individuals with previous experience: Mr. Joel Shaughnessy, Personnel Manager; Phone: 978-249-3551; Fax: 978-249-8495.

M/A-COM, INC.
1011 PAWTUCKET BOULEVARD
LOWELL, MASSACHUSETTS 01854-1040

Overview: Produces diodes, integrated circuits, solid state transistors, and related devices. Established: 1995. Parent Company: AMP, Inc. Ownership: Public.
Key Statistics: Annual Sales: $418.8 million. Number of Employees: 4,000.
Hiring History: Number of professional employees hired in 1997: 125.
Expertise/Education Sought: Manufacturing, management, engineering technology
Alternatives to Full-Time Employment: Paid internships
Company-Paid Training is available.
International Assignments are available to employees.
Contact: Mr. Jon Curley, Manager of Employment and College Relations; Phone: 978-442-5000; Fax: 978-442-4443. E-mail: careers@macom.com. Job Hotline: 888-622-6656.
Company URL: http://www.macom.com

MASSACHUSETTS DEPARTMENT OF ENVIRONMENTAL PROTECTION
1 WINTER STREET
BOSTON, MASSACHUSETTS 02108-4747

Overview: State government office. Established: 1980. Ownership: Private.
Key Statistics: Number of Employees: 600.
Expertise/Education Sought: Environmental engineering, environmental planning
Alternatives to Full-Time Employment: Paid internships
Company-Paid Training is available.
Contact: Recruiting Office; Phone: 617-292-5000; Fax: 617-556-1049.
Company URL: http://www.state.ma.us/deb/bas/hire.htm

MASSACHUSETTS HEAVY INDUSTRIES

115 EAST HOWARD STREET
QUINCY, MASSACHUSETTS 02169-8711

Overview: Builds and repairs ships. Established: 1995. Ownership: Private.

Key Statistics: Annual Sales: $102 million.

Alternatives to Full-Time Employment: Paid internships

Contact: Mr. Kevin Fennessey, Manager of Human Resources.

MASSACHUSETTS PORT AUTHORITY

1 HARBORSIDE DRIVE, SUITE 200 SOUTH
EAST BOSTON, MASSACHUSETTS 02128

Overview: Operates airport and transportation terminals. Established: 1956. Ownership: Private.

Key Statistics: Annual Sales: $247.2 million. Number of Employees: 2,520.

Expertise/Education Sought: Accounting, finance, information systems, marketing, human resources, aviation, management, law

Alternatives to Full-Time Employment: Paid internships

Company-Paid Training is available.

International Assignments are available to employees.

Contact: For college students and recent graduates: Mrs. Julie Wasson, Human Resources; Phone: 617-973-5300. E-mail: employment@massport.com. For individuals with previous experience: Mrs. Tonda Lumley, Human Resources; Phone: 617-568-3900; Fax: 617-568-3940. E-mail: employment@massport.com. Job Hotline: 617-568-3901.

Company URL: http://www.massport.com

MASSACHUSETTS TURNPIKE AUTHORITY

10 PARK PLAZA, SUITE 4160
BOSTON, MASSACHUSETTS 02116-1933

Overview: Operates toll road. Established: 1952. Ownership: Private.

Key Statistics: Annual Sales: $169.7 million. Number of Employees: 1,315.

Expertise/Education Sought: Computer programming, civil engineering, accounting, finance

Contact: For individuals with previous experience: Personnel Director; Phone: 617-248-2800; Fax: 617-248-2903.

MATRIX-ONE

2 EXECUTIVE DRIVE
CHELMSFORD, MASSACHUSETTS 01824-2558

Overview: Provides computer programming services and software development. Established: 1997. Ownership: Private.

Key Statistics: Annual Sales: $20 million. Number of Employees: 60.

Expertise/Education Sought: Engineering, engineering technology

Company-Paid Training is available.

International Assignments are available to employees.

Contact: Human Resource Department; Phone: 978-454-0444; Fax: 978-441-0071.

Company URL: http://www.matrix-one.com

MEDICAL INFORMATION TECHNOLOGY

MEDITECH CIRCLE
WESTWOOD, MASSACHUSETTS 02090

Overview: Develops hospital and administrative software. Established: 1969. Ownership: Private.

Key Statistics: Annual Sales: $116.5 million. Number of Employees: 1,500.

Hiring History: Number of professional employees hired in 1997: 250.

Expertise/Education Sought: Marketing, software engineering/development, electrical engineering, accounting, finance, clinical nursing, pharmacy, medical technology, radiology

Alternatives to Full-Time Employment: Part-time jobs

Company-Paid Training is available.

Contact: For individuals with previous experience: Recruiting; Phone: 617-821-3000, ext. 4351. E-mail: jobs@meditech.com.

Company URL: http://www.meditech.com

MESTEK, INC.

260 NORTH ELM STREET
WESTFIELD, MASSACHUSETTS 01085-1614

Overview: Manufactures heating equipment. Established: 1986. Ownership: Public.

Key Statistics: Annual Sales: $299.5 million. Number of Employees: 2,411.

Hiring History: Number of professional employees hired in 1996: 85.

Expertise/Education Sought: Drafting, sales, electrical engineering, mechanical engineering, manufacturing engineering

Alternatives to Full-Time Employment: Paid internships

Contact: Mr. Dennis Punch, Recruiter; Phone: 413-568-9571; Fax: 413-562-7630.

MICRO NETWORKS CORPORATION

324 CLARK STREET
WORCESTER, MASSACHUSETTS 01606-1214

Overview: Manufactures semiconductors and integrated circuits. Established: 1997. Parent Company: Sawgrass Electronics Group, Inc. Ownership: Private.

Key Statistics: Annual Sales: $15 million. Number of Employees: 80.

Expertise/Education Sought: Electronic design and development, electronics/electronics engineering

Company-Paid Training is available.

International Assignments are available to employees.

Contact: Mr. Tom Vangel, Human Resources Director; Phone: 508-852-5400; Fax: 508-853-8296.

Company URL: http://www.mnc.com

MICROWAVE DEVELOPMENT LABS, INC.

10 MICHIGAN DRIVE
NATICK, MASSACHUSETTS 01760

Overview: Produces electronic components. Established: 1948. Ownership: Private.

Key Statistics: Annual Sales: $5.2 million. Number of Employees: 80.

Hiring History: Number of professional employees hired in 1996: 3.

Expertise/Education Sought: Engineering, system test

Alternatives to Full-Time Employment: Part-time jobs

Contact: For individuals with previous experience: Personnel; Phone: 508-655-0060; Fax: 508-655-6972.

Company URL: http://www.micro-dev-labs.com

MILLIPORE CORPORATION

80 ASHBY ROAD
BEDFORD, MASSACHUSETTS 01730

Overview: Manufactures specialty application filters. Established: 1954. Ownership: Public.

Key Statistics: Annual Sales: $618.7 million. Number of Employees: 3,200.

Hiring History: Number of professional employees hired in 1996: 190.

Expertise/Education Sought: Mechanical engineering, data processing, chemical engineering, manufacturing engineering, chemistry, microbiology, materials science, materials engineering

Alternatives to Full-Time Employment: Part-time jobs, paid internships

Company-Paid Training is available.

International Assignments are available to employees.

Contact: For college students and recent graduates: Ms. Patti Zambarano, Employment Specialist; Phone: 781-533-2219; Fax: 781-533-3130. E-mail: patti_zambarano@millipore.com. For individuals with previous experience: Human Resources Department; Phone: 781-275-9200; Fax: 781-275-5550.

Company URL: http://www.millipore.com

MITRE

THE MITRE CORPORATION

202 BURLINGTON ROAD
BEDFORD, MASSACHUSETTS 01730-1420

Mailing Address: PO Box 208, Bedford, MA 01730-0208

Overview: Technical advisor to the Air Force, other branches of the Department of Defense, the FAA, and NASA. Established: 1958. Ownership: Private.

Key Statistics: Number of Employees: 4,200.

Expertise/Education Sought: Electrical engineering, computer engineering, computer science, mathematics, physics, telecommunications

Alternatives to Full-Time Employment: Part-time jobs, paid internships

Company-Paid Training is available.

International Assignments are available to employees.

Contact: Corporate Recruitment, Department PO 98; Fax: 781-271-3402. E-mail: jobs@mitre.org.

Company URL: http://www.mitre.org

For more information, see full corporate profile, pg. 284.

MODUS MEDIA INTERNATIONAL

690 CANTON STREET
WESTWOOD, MASSACHUSETTS 02090-2321

Overview: Designs computer systems and offers system software development services. Established: 1997. Ownership: Private.

Key Statistics: Annual Sales: $17.2 million. Number of Employees: 148.

Expertise/Education Sought: Software engineering/development, administration

Alternatives to Full-Time Employment: Internships

Company-Paid Training is available.

International Assignments are available to employees.

Contact: Human Resources; Phone: 781-407-2000.

Company URL: http://www.modusmedia.com

NEW ENGLAND FROZEN FOODS, INC.

1 HARVEST LANE
SOUTHBOROUGH, MASSACHUSETTS 01772

Overview: Wholesales packaged frozen foods. Established: 1931. Ownership: Private.

Expertise/Education Sought: Sales, transportation

Contact: For individuals with previous experience: Personnel; Phone: 508-481-0300; Fax: 508-481-6456.

NORTHERN RESEARCH & ENGINEERING CORPORATION

39 OLYMPIA AVENUE
WOBURN, MASSACHUSETTS 01801-2073

Overview: Provides commercial and industrial design services. Established: 1978. Parent Company: Ingersoll-Rand. Ownership: Public.

Key Statistics: Annual Sales: $16 million. Number of Employees: 100.

Hiring History: Number of professional employees hired in 1997: 6. 1996: 20.

Expertise/Education Sought: Mechanical engineering

Alternatives to Full-Time Employment: Paid internships

Company-Paid Training is available.

Contact: For individuals with previous experience: Human Resources; Phone: 781-935-9050; Fax: 781-935-9052.

Company URL: http://www.nrec.com

NORTON COMPANY

1 NEW BOND STREET
WORCESTER, MASSACHUSETTS 01606-0008

Mailing Address: Box Number 15008, Worcester, MA 01615

Overview: Manufactures abrasive wheels and sandpaper. Established: 1990. Parent Company: St. Gobain. Ownership: Private.

Key Statistics: Annual Sales: $1.4 billion. Number of Employees: 11,000.

Expertise/Education Sought: Manufacturing

Alternatives to Full-Time Employment: Paid internships

Company-Paid Training is available.

International Assignments are available to employees.

Contact: Human Resources. E-mail: 5. Job Hotline: 800-562-8742.

THE OPEN GROUP

11 CAMBRIDGE CENTER
CAMBRIDGE, MASSACHUSETTS 02142-1401

Overview: Provides custom computer programming services. Established: 1988. Ownership: Private.

Key Statistics: Annual Sales: $38.9 million. Number of Employees: 307.

Hiring History: Number of professional employees hired in 1997: 85. 1996: 98.

Expertise/Education Sought: Software engineering/development, computer programming, computer science, marketing, research

Alternatives to Full-Time Employment: Part-time jobs, paid internships

International Assignments are available to employees.

Contact: Mr. Phil Sheerin, Human Resources Department; Phone: 617-621-8869; Fax: 617-621-8781. E-mail: P.SHEERIN@OPENGROUP.ORG.

Company URL: http://www.opengroup.org

OSRAM SYLVANIA, INC.

100 ENDICOTT STREET
DANVERS, MASSACHUSETTS 01923-3623

Overview: Manufactures light bulbs. Established: 1993. Parent Company: Siemens. Ownership: Private.

Key Statistics: Annual Sales: $866.7 million. Number of Employees: 13,000.

Expertise/Education Sought: Engineering, accounting, production

Alternatives to Full-Time Employment: Paid internships

Company-Paid Training is available.

International Assignments are available to employees.

Contact: Human Resources; Phone: 800-328-4775; Fax: 978-750-2106.

Company URL: http://www.sylvania.com

PARAMETRIC TECHNOLOGY CORPORATION

128 TECHNOLOGY DRIVE
WALTHAM, MASSACHUSETTS 02154

Overview: Develops and markets computer aided design and computer aided manufacturing software. Established: 1972. Ownership: Public.

Key Statistics: Annual Sales: $477.2 million. Number of Employees: 1,500.

Hiring History: Number of professional employees hired in 1997: 1,200. 1996: 800.

Expertise/Education Sought: Sales, product design and development, marketing, finance, accounting, technical writing, mechanical engineering, software engineering/development, administration supervision

Alternatives to Full-Time Employment: Paid internships

Company-Paid Training is available.

International Assignments are available to employees.

Contact: For individuals with previous experience: Human Resource Representative; Phone: 781-398-5000; Fax: 781-398-6000. E-mail: careers@ptcjobs.com.

Company URL: http://www.ptc.com

PERINI CORPORATION
73 MOUNT WAYTE AVENUE
FRAMINGHAM, MASSACHUSETTS 01701-5800

Overview: Provides construction and real estate development services. Established: 1894. Ownership: Public.

Key Statistics: Annual Sales: $1.3 billion. Number of Employees: 2,400.

Hiring History: Number of professional employees hired in 1997: 100. 1996: 100.

Expertise/Education Sought: Civil engineering, accounting, finance

Company-Paid Training is available.

Contact: For individuals with previous experience: Mr. Douglas Mure, Vice President of Human Resources; Phone: 508-628-2223; Fax: 508-628-2821. E-mail: dmure@perini.com.

PICTURETEL CORPORATION
100 MINUTEMAN ROAD
ANDOVER, MASSACHUSETTS 01810

Overview: Manufactures video conferencing equipment. Established: 1984. Ownership: Public.

Key Statistics: Annual Sales: $482.5 million. Number of Employees: 1,496.

Hiring History: Number of professional employees hired in 1997: 150. 1996: 150.

Expertise/Education Sought: Sales, marketing, accounting, research and development, hardware engineering, software engineering/development, computer science, electronics/electronics engineering

Alternatives to Full-Time Employment: Paid internships

International Assignments are available to employees.

Contact: For college students and recent graduates: Miss Jane Cotreau, Human Resources Representative; Phone: 978-623-4383; Fax: 978-623-4853. For individuals with previous experience: Miss Winnie Leung, Human Resources Consultant; Phone: 978-623-4383; Fax: 978-623-4853. E-mail: leungw@pictel.com.

Company URL: http://www.picturetel.com

PILOT SOFTWARE, INC.
1 CANAL PARK
CAMBRIDGE, MASSACHUSETTS 02141

Overview: Develops executive information software. Established: 1996. Parent Company: Platinum. Ownership: Public.

Key Statistics: Annual Sales: $22.7 million. Number of Employees: 290.

Hiring History: Number of professional employees hired in 1997: 100. 1996: 100.

Expertise/Education Sought: Accounting, finance, software engineering/development, data processing, computer programming

Alternatives to Full-Time Employment: Part-time jobs

Company-Paid Training is available.

International Assignments are available to employees.

Contact: For individuals with previous experience: Senior Human Resources Representative; Phone: 617-374-1123; Fax: 617-374-6687.

Company URL: http://www.pilotsw.com

PLYMOUTH RUBBER COMPANY, INC.
104 REVERE STREET
CANTON, MASSACHUSETTS 02021-2964

Overview: Manufactures rubber tape. Established: 1894. Ownership: Public.

Key Statistics: Annual Sales: $57.2 million. Number of Employees: 450.

Expertise/Education Sought: Manufacturing

Contact: Mr. Donald Amdrejczyk, Director of Human Resources; Phone: 781-828-0220, ext. 1207; Fax: 781-828-6041.

POLAROID CORPORATION
549 TECHNOLOGY SQUARE
CAMBRIDGE, MASSACHUSETTS 02139-3539

Mailing Address: 565 Technology Square, Cambridge, MA 02139-3539

Overview: Manufactures photographic and imaging equipment. Established: 1937. Ownership: Public.

Key Statistics: Annual Sales: $2.3 billion. Number of Employees: 10,046.

Hiring History: Number of professional employees hired in 1997: 10.

Expertise/Education Sought: Mechanical engineering, optics, electrical engineering, information systems, chemical engineering, chemistry, computer science

Alternatives to Full-Time Employment: Part-time jobs, paid internships

International Assignments are available to employees.

Contact: Ms. Peg Kercher, Relations Manager, 565 Technology Square, Cambridge, Massachusetts, 02139-3539; Phone: 617-386-3970; Fax: 617-386-3925. E-mail: kerchem@polaroid.com.

PRIMARK CORPORATION
1000 WINTER STREET, SUITE 4300 NORTH
WALTHAM, MASSACHUSETTS 02154-1248

Overview: Provides information technology services. Established: 1981. Ownership: Public.

Key Statistics: Annual Sales: $767.1 million. Number of Employees: 5,938.

Expertise/Education Sought: Human resources, law, accounting, finance, administration

Alternatives to Full-Time Employment: Part-time jobs

Company-Paid Training is available.

International Assignments are available to employees.

Contact: Ms. Diane Robson, Manager of Human Resources; Phone: 781-466-6611; Fax: 781-890-6187.

Company URL: http://www.primark.com

PRINTED CIRCUIT CORPORATION
10 MICRO DRIVE
WOBURN, MASSACHUSETTS 01801-5702

Overview: Manufactures printed circuit boards. Established: 1961. Ownership: Private.

Key Statistics: Annual Sales: $23 million. Number of Employees: 300.

Expertise/Education Sought: Sales, marketing, process engineering, quality assurance, marketing

Contact: Human Resource Department; Phone: 617-935-9570; Fax: 617-932-6403.

PROGRESS SOFTWARE CORPORATION
14 OAK PARK DRIVE
BEDFORD, MASSACHUSETTS 01730-1414

Overview: Develops application development software. Established: 1981. Ownership: Public.

Key Statistics: Annual Sales: $176.7 million. Number of Employees: 1,105.

Expertise/Education Sought: Software engineering/development, accounting, finance, marketing

Alternatives to Full-Time Employment: Part-time jobs, internships

International Assignments are available to employees.

Contact: For individuals with previous experience: Human Resource Department; Phone: 617-280-4000; Fax: 617-280-4350. E-mail: hr@bedford.progress.com.

Company URL: http://www.progress.com

RAID, INC.
184 PLEASANT VALLEY STREET
METHUEN, MASSACHUSETTS 01844-5855

Overview: Develops computer software systems. Established: 1997. Ownership: Private.

Key Statistics: Annual Sales: $14 million.

Hiring History: Number of professional employees hired in 1997: 10. 1996: 5.

Expertise/Education Sought: Engineering, sales

Alternatives to Full-Time Employment: Paid internships

RAID, Inc. (continued)

Company-Paid Training is available.

Contact: For individuals with previous experience: Human Resources; Phone: 978-683-6444.

Raytheon
EXPECT GREAT THINGS

RAYTHEON COMPANY

141 SPRING STREET
LEXINGTON, MASSACHUSETTS 02173

Overview: Manufactures electronic and aircraft products and appliances and provides energy and environmental services. Established: 1922. Ownership: Public.

Key Statistics: Annual Sales: $15 billion. Number of Employees: 110,000.

Expertise/Education Sought: Electrical engineering, electronics/electronics engineering, accounting, data processing

Company-Paid Training is available.

International Assignments are available to employees.

Contact: Mr. Robert Murphy, Corporate Manager of Staffing; Phone: 781-860-2824; Fax: 781-860-2512. E-mail: robert_r_murphy@raytheon.com.

Company URL: http://www.raytheon.com

For more information, see full corporate profile, pg. 294.

RAYTHEON OPTICAL SYSTEMS, INC.

4 HARTWELL PLACE
LEXINGTON, MASSACHUSETTS 02173

Overview: Researches and develops highly complex advanced optics, electro-optical and sensor systems. Parent Company: Raytheon.

Hiring History: Number of professional employees hired in 1997: 12. 1996: 12.

Expertise/Education Sought: Mechanical engineering, electrical engineering, software engineering/development, optical engineering, systems engineering

Company-Paid Training is available.

International Assignments are available to employees.

Contact: Ms. Judy Woodbury, Human Resources Specialist; Phone: 781-402-2183; Fax: 781-402-2296.

RAYTHEON SERVICE COMPANY

2 WAYSIDE ROAD
BURLINGTON, MASSACHUSETTS 01803-4607

Overview: Provides engineering services. Established: 1994. Parent Company: Raytheon Company. Ownership: Public.

Key Statistics: Annual Sales: $251.8 million. Number of Employees: 5,981.

Hiring History: Number of professional employees hired in 1996: 1,000.

Expertise/Education Sought: Technical writing, electrical engineering, software engineering/development

Company-Paid Training is available.

International Assignments are available to employees.

Contact: For individuals with previous experience: Mr. James Dempsey, Manager of Human Resources; Phone: 617-238-2263; Fax: 617-238-2744.

Company URL: http://www.raytheon.com

REMEDIATION TECHNOLOGIES, INC.

9 POND LANE, SUITE 3A
CONCORD, MASSACHUSETTS 01742-2842

Overview: Provides hazardous waste removal; water and soil remediation services. Established: 1995. Parent Company: Thermo Remediation, Inc. Ownership: Public.

Key Statistics: Annual Sales: $50 million. Number of Employees: 325.

Expertise/Education Sought: Civil engineering, biological engineering, chemical engineering, environmental engineering, geology, hydrogeology

Company-Paid Training is available.

Contact: For individuals with previous experience: Mr. Mark O'Brien, Human Resources Manager; Phone: 978-371-1422; Fax: 978-369-9279.

SCHNEIDER AUTOMATION, INC.

1 HIGH STREET
NORTH ANDOVER, MASSACHUSETTS 01845

Overview: Manufactures programmable logic control instruments. Established: 1996. Parent Company: Groupe Schneider—North America. Ownership: Private.

Key Statistics: Number of Employees: 775.

Hiring History: Number of professional employees hired in 1996: 95.

Expertise/Education Sought: Software engineering/development, hardware engineering, applied engineering, accounting, training and development, manufacturing engineering

Alternatives to Full-Time Employment: Part-time jobs

Company-Paid Training is available.

International Assignments are available to employees.

Contact: Staffing Manager; Phone: 800-225-0962.

Company URL: http://www.modicon.com

E.J. SCIABA CONTRACTING COMPANY, INC.

18 WOLCOTT STREET
READVILLE, MASSACHUSETTS 02136

Overview: Provides construction services. Established: 1980. Ownership: Private.

Key Statistics: Annual Sales: $45 million. Number of Employees: 50.

Expertise/Education Sought: Structural engineering, construction engineering, mechanical engineering, electrical engineering, civil engineering, environmental engineering

Contact: Mr. Edward Schaber, President; Phone: 617-364-6150; Fax: 617-364-1055.

SCULLY SIGNAL COMPANY

70 INDUSTRIAL WAY
WILMINGTON, MASSACHUSETTS 01887-3479

Overview: Manufactures process control instruments. Established: 1964. Ownership: Private.

Key Statistics: Annual Sales: $16.7 million. Number of Employees: 174.

Hiring History: Number of professional employees hired in 1997: 12. 1996: 10.

Expertise/Education Sought: Engineering, industrial engineering, sales

Company-Paid Training is available.

International Assignments are available to employees.

Contact: Personnel Department; Phone: 617-692-8600; Fax: 617-692-8620. E-mail: pbettano@scully.com.

SIMPLEX

1 SIMPLEX PLAZA
GARDNER, MASSACHUSETTS 01440

Overview: Manufactures data systems, monitors, and control systems for buildings. Established: 1967. Ownership: Private.

Key Statistics: Annual Sales: $591 million. Number of Employees: 5,300.

Hiring History: Number of professional employees hired in 1997: 280.

Expertise/Education Sought: Software engineering/development, information technology

Alternatives to Full-Time Employment: Paid internships

Company-Paid Training is available.

International Assignments are available to employees.
Contact: For individuals with previous experience: Mr. Michael Stacey, Manager of Human Resources; Phone: 508-632-2500; Fax: 508-632-9062.
Company URL: http://www.simplexnet.com

SPRINGFIELD WIRE, INC.

243 COTTAGE STREET
SPRINGFIELD, MASSACHUSETTS 01104-4004

Overview: Manufactures electric heating units. Established: 1921. Ownership: Private.
Key Statistics: Annual Sales: $35.6 million. Number of Employees: 822.
Hiring History: Number of professional employees hired in 1997: 90.
Expertise/Education Sought: Electrical engineering, mechanical engineering
Contact: For individuals with previous experience: Mr. Paul Dunn, Personnel Manager; Phone: 413-781-6950.

STANDARD-THOMSON CORPORATION

152 GROVE STREET
WALTHAM, MASSACHUSETTS 02154-9109

Overview: Manufactures measuring and controlling devices. Established: 1988. Parent Company: Stant Corporation. Ownership: Private.
Key Statistics: Annual Sales: $32.9 million. Number of Employees: 375.
Hiring History: Number of professional employees hired in 1997: 70. 1996: 20.
Expertise/Education Sought: Mechanical engineering, quality engineering, environmental engineering, design engineering, product engineering
Alternatives to Full-Time Employment: Paid internships
Company-Paid Training is available.
Contact: Mr. Luis Mollindo, Senior Human Resources Administrator; Phone: 781-894-7310, ext. 384; Fax: 781-894-2235. E-mail: lmolline@stdthom.com.

STANMAR, INC.

130 BOSTON POST ROAD
SUDBURY, MASSACHUSETTS 01776-2406

Overview: Operates ski lodge. Established: 1965. Ownership: Private.
Key Statistics: Annual Sales: $25 million. Number of Employees: 30.
Hiring History: Number of professional employees hired in 1997: 4.
Expertise/Education Sought: Accounting, administration, business management, civil engineering, data processing, design engineering
Contact: For individuals with previous experience: Ms. Irene Barney, Administrative Assistant; Phone: 978-443-9922, ext. 122; Fax: 978-443-0479.

STD MANUFACTURING, INC.

1063 TURNPIKE STREET
STOUGHTON, MASSACHUSETTS 02072-1160

Mailing Address: PO Box 420, Stoughton, MA 02072-1160
Overview: Manufactures metal stampings machine parts. Established: 1953. Ownership: Private.
Key Statistics: Annual Sales: $6 million. Number of Employees: 75.
Hiring History: Number of professional employees hired in 1997: 10.
Expertise/Education Sought: Manufacturing engineering
Company-Paid Training is available.
Contact: For individuals with previous experience: Human Relations Department; Phone: 781-828-4400; Fax: 781-344-5895. E-mail: stdmed@thecia.net.

THE STOP & SHOP SUPERMARKET COMPANY

1385 HANCOCK STREET
QUINCY, MASSACHUSETTS 02169

Overview: Operates chain of food stores. Established: 1996. Parent Company: Royal Ahold. Ownership: Private.
Key Statistics: Annual Sales: $4.1 billion. Number of Employees: 36,000.
Hiring History: Number of professional employees hired in 1997: 100. 1996: 100.
Expertise/Education Sought: Human resources
Alternatives to Full-Time Employment: Paid internships
Company-Paid Training is available.
International Assignments are available to employees.
Contact: Ms. Julia Clark, Employment and Central Staff Manager, PO Box 1942, Quincy, Massachusetts, 02105; Phone: 617-770-8880; Fax: 617-770-8190.
Company URL: http://www.stopandshop.com

STRIDE RITE CORPORATION

191 SPRING STREET
LEXINGTON, MASSACHUSETTS 02173-9191

Overview: Manufactures children's footwear. Established: 1919. Ownership: Public.
Key Statistics: Annual Sales: $448.3 million. Number of Employees: 3,500.
Hiring History: Number of professional employees hired in 1997: 250. 1996: 250.
Expertise/Education Sought: Financial analysis, marketing, product management, computer programming, information systems
Alternatives to Full-Time Employment: Paid internships
International Assignments are available to employees.
Contact: For college students and recent graduates: Mr. Robert Moons, College Recruiter; Phone: 617-824-6000. E-mail: rmoons@striderite.com. For individuals with previous experience: Human Resources; Phone: 617-824-6000, ext. 4400; Fax: 617-824-6549.
Company URL: http://www.striderite.com

TAD TECHNICAL SERVICES CORPORATION

3 SPEEN STREET
FRAMINGHAM, MASSACHUSETTS 01701

Overview: Provides engineering recruitment services. Established: 1956. Parent Company: TAD Resources International, Inc. Ownership: Private.
Key Statistics: Annual Sales: $1.1 billion. Number of Employees: 40,000.
Hiring History: Number of professional employees hired in 1996: 300.
Expertise/Education Sought: Electrical engineering, mechanical engineering, drafting, design engineering, technical, software engineering/development
Company-Paid Training is available.
Contact: For individuals with previous experience: Mr. Mark Mahoney, Sales Recruiter; Phone: 508-875-8030; Fax: 508-875-7975.
Company URL: http://www.tadresources.com

TASC

Shaping *things to come.*

TASC

55 WALKERS BROOK DRIVE
READING, MASSACHUSETTS 01867

Overview: Provides integrated systems and information services. Established: 1966. Parent Company: Litton Industries, Inc.. Ownership: Public.
Key Statistics: Annual Sales: $385 million. Number of Employees: 2,700.
Hiring History: Number of professional employees hired in 1997: 541. 1996: 150.

TASC (continued)

Expertise/Education Sought: Information systems, management information systems, computer programming, computer science, electrical engineering

Alternatives to Full-Time Employment: Part-time jobs

Company-Paid Training is available.

Contact: Ms. Mary Beth Nason, Technical Recruiter; Phone: 781-942-2000; Fax: 781-942-2461. E-mail: mbnason@tasc.com.

Company URL: http://www.tasc.com

For more information, see full corporate profile, pg. 300.

TELCO SYSTEMS, INC.

63 NAHATAN STREET
NORWOOD, MASSACHUSETTS 02062-5732

Overview: Manufactures fiber optic transmission products, network access equipment, and Internetworking products. Established: 1972. Ownership: Public.

Key Statistics: Annual Sales: $94 million. Number of Employees: 388.

Hiring History: Number of professional employees hired in 1996: 100.

Expertise/Education Sought: Software engineering/development, hardware engineering

Alternatives to Full-Time Employment: Part-time jobs, paid internships

Company-Paid Training is available.

International Assignments are available to employees.

Contact: Ms. Kim Turner, Human Resources Representative; Phone: 617-551-0300, ext. 2239; Fax: 617-255-2385.

Company URL: http://www.telco.com

TERADYNE, INC.

321 HARRISON AVENUE
BOSTON, MASSACHUSETTS 02118-2238

Overview: Manufactures electronic systems and software. Established: 1960. Ownership: Public.

Key Statistics: Annual Sales: $1.2 billion. Number of Employees: 5,000.

Expertise/Education Sought: Mechanical engineering, electrical engineering, design engineering, accounting, marketing, sales, computer science

Alternatives to Full-Time Employment: Part-time jobs, paid internships

International Assignments are available to employees.

Contact: For individuals with previous experience: Human Resources; Phone: 617-422-2700; Fax: 617-422-2080.

Company URL: http://www.teradyne.com

THERMO ELECTRON CORPORATION

81 WYMAN STREET
WALTHAM, MASSACHUSETTS 02154-1223

Overview: Manufactures and distributes plastics, engines, and electrical products. Established: 1956. Ownership: Public.

Key Statistics: Annual Sales: $2.9 billion. Number of Employees: 17,760.

Expertise/Education Sought: Accounting, finance, sales, marketing

Alternatives to Full-Time Employment: Part-time jobs

Company-Paid Training is available.

Contact: For individuals with previous experience: Ms. Darlene Foster, Human Resources Representative; Phone: 781-622-1000; Fax: 781-622-1261.

Company URL: http://www.thermo.com

THINKING MACHINES CORPORATION

16 NEW ENGLAND EXECUTIVE PARK
BURLINGTON, MASSACHUSETTS 01803

Overview: Manufactures supercomputers and parallel processors. Established: 1983. Parent Company: New Valley. Ownership: Private.

Key Statistics: Annual Sales: $7 million. Number of Employees: 55.

Expertise/Education Sought: Software engineering/development

Alternatives to Full-Time Employment: Part-time jobs

Contact: Ms. Colleen Humphreys, Human Resources Director; Phone: 781-238-3437; Fax: 781-238-3430. E-mail: cos@think.com.

Company URL: http://www.think.com

TITLEIST AND FOOT-JOY WORLDWIDE

333 BRIDGE STREET
FAIRHAVEN, MASSACHUSETTS 02719-4905

Overview: Manufactures sporting and athletic goods. Established: 1976. Parent Company: Fortune Brands. Ownership: Public.

Key Statistics: Annual Sales: $334.2 million. Number of Employees: 3,198.

Expertise/Education Sought: Sales, engineering, information systems

Alternatives to Full-Time Employment: Part-time jobs

Contact: For individuals with previous experience: Mr. Jim McKenna, Human Resources Manager; Phone: 508-979-2000; Fax: 508-979-3900.

Company URL: http://www.titleist.com

TOWN AND COUNTRY FINE JEWELRY GROUP

25 UNION STREET
CHELSEA, MASSACHUSETTS 02150-1214

Overview: Manufactures precious metal jewelry. Established: 1955. Parent Company: Town And Country Corporation. Ownership: Public.

Key Statistics: Annual Sales: $209.2 million. Number of Employees: 2,500.

Expertise/Education Sought: Manufacturing

Company-Paid Training is available.

International Assignments are available to employees.

Contact: Human Resources Representative; Phone: 617-884-8500; Fax: 617-884-8297.

TREVI ICOS CORPORATION

475 C STREET
BOSTON, MASSACHUSETTS 02210-1900

Overview: Bridge, tunnel, and elevated highway construction. Established: 1997. Ownership: Private.

Key Statistics: Annual Sales: $50 million. Number of Employees: 130.

Expertise/Education Sought: Construction, engineering

Alternatives to Full-Time Employment: Paid internships

International Assignments are available to employees.

Contact: Human Resources; Phone: 617-357-5333; Fax: 617-357-5999.

TUBED PRODUCTS

44 O'NEILL STREET
EASTHAMPTON, MASSACHUSETTS 01027

Overview: Manufactures plastic squeeze tubes. Established: 1968. Parent Company: McCormick & Company, Inc. Ownership: Public.

Key Statistics: Annual Sales: $50 million. Number of Employees: 1,100.

Hiring History: Number of professional employees hired in 1997: 6. 1996: 10.

Expertise/Education Sought: Plastics engineering, industrial engineering, accounting, computer science

Alternatives to Full-Time Employment: Paid internships

Contact: For individuals with previous experience: Ms. Virginia Smith, Manager of Human Relations; Phone: 413-527-1250, ext. 301; Fax: 413-529-1214.

UTILICOM NETWORKS, INC.

124 WASHINGTON STREET, SUITE 201
FOXBORO, MASSACHUSETTS 02035-1368

Overview: Provides telecommunications services. Established: 1995. Ownership: Private.

Hiring History: Number of professional employees hired in 1997: 5. 1996: 10.

Expertise/Education Sought: Networking, technology, marketing

Alternatives to Full-Time Employment: Paid internships
Contact: For individuals with previous experience: Human Resources; Phone: 508-698-7000; Fax: 508-698-7100. E-mail: ucn@utilicom.net.
Company URL: http://www.utilicom.net

VALPEY-FISHER CORPORATION
75 SOUTH STREET
HOPKINTON, MASSACHUSETTS 01748

Overview: Produces electronic crystals. Established: 1982. Parent Company: MATEC Corporation. Ownership: Public.
Key Statistics: Annual Sales: $5.8 million. Number of Employees: 90.
Expertise/Education Sought: Engineering
Company-Paid Training is available.
International Assignments are available to employees.
Contact: For individuals with previous experience: Human Resources Department; Phone: 508-435-6831; Fax: 508-435-5289.

VOLEX, INC.
358 FAUNCE CORNER ROAD
DARTMOUTH, MASSACHUSETTS 02747

Overview: Provides cable assembly services. Established: 1992. Ownership: Private.
Key Statistics: Annual Sales: $115.9 million. Number of Employees: 1,300.
Hiring History: Number of professional employees hired in 1996: 50.
Expertise/Education Sought: Electrical engineering, mechanical engineering, data processing
Company-Paid Training is available.
International Assignments are available to employees.
Contact: For individuals with previous experience: Ms. Debra Campos, Administrative Assistant; Phone: 508-995-9500, ext. 3065; Fax: 508-998-5446.
Company URL: http://www.volex.com

WALKER MAGNETICS GROUP, INC.
17 ROCKDALE STREET
WORCESTER, MASSACHUSETTS 01606-1921

Overview: Manufactures process control instruments. Established: 1920. Ownership: Private.
Key Statistics: Annual Sales: $60.4 million. Number of Employees: 550.
Hiring History: Number of professional employees hired in 1997: 15. 1996: 15.
Expertise/Education Sought: Mechanical engineering, electrical engineering, accounting, production management, product engineering
International Assignments are available to employees.
Contact: For individuals with previous experience: Human Resources; Phone: 508-853-3232; Fax: 508-852-8649.

WANG LABORATORIES, INC.
600 TECHNOLOGY PARK DRIVE
BILLERICA, MASSACHUSETTS 01821

Overview: Manufactures software and provides system integration services. Established: 1951. Ownership: Public.
Key Statistics: Annual Sales: $1.3 billion. Number of Employees: 9,300.
Expertise/Education Sought: Computer programming, finance, marketing, software engineering/development
Alternatives to Full-Time Employment: Part-time jobs, paid internships
Company-Paid Training is available.
International Assignments are available to employees.
Contact: For college students and recent graduates: Human Resources; Phone: 978-967-5000. For individuals with previous experience: Human Resources; Phone: 978-967-5000.
Company URL: http://www.wang.com

WEST LYNN CREAMER, INC.
626 LYNNWAY
LYNN, MASSACHUSETTS 01905-3030

Overview: Processes milk and dairy products. Established: 1976. Ownership: Private.
Key Statistics: Annual Sales: $191.6 million. Number of Employees: 850.
Hiring History: Number of professional employees hired in 1997: 300.
Expertise/Education Sought: Industrial distribution, electrical engineering
Alternatives to Full-Time Employment: Paid internships
Company-Paid Training is available.
Contact: For individuals with previous experience: Human Resource Department; Phone: 781-599-1300; Fax: 781-599-7810.

W.A. WILDE COMPANY
200 SUMMER STREET
HOLLISTON, MASSACHUSETTS 01746-2299

Overview: Provides direct mail advertising services. Established: 1868. Ownership: Private.
Key Statistics: Annual Sales: $21.2 million. Number of Employees: 400.
Hiring History: Number of professional employees hired in 1997: 100.
Expertise/Education Sought: Accounting, production, application developer, software engineering/development
Alternatives to Full-Time Employment: Part-time jobs
Company-Paid Training is available.
Contact: Human Resources; Phone: 508-429-5515; Fax: 508-893-0375. E-mail: resume@wilde.com.
Company URL: http://www.wilde.com

XYPLEX NETWORKS, INC.
295 FOSTER STREET
LITTLETON, MASSACHUSETTS 01460-1708

Overview: Manufactures data communications equipment. Established: 1996. Parent Company: M. R. V. Communications. Ownership: Public.
Key Statistics: Annual Sales: $100 million. Number of Employees: 400.
Expertise/Education Sought: Accounting, finance, data processing, software engineering/development, hardware engineering
Alternatives to Full-Time Employment: Part-time jobs, internships
International Assignments are available to employees.
Contact: For individuals with previous experience: Human Resources Department.
Company URL: http://www.xyplex.com

MICHIGAN

AETNA INDUSTRIES, INC.
24331 SHERWOOD AVENUE
CENTER LINE, MICHIGAN 48015-1060

Overview: Manufactures auto body parts. Established: 1996. Ownership: Private.
Key Statistics: Annual Sales: $165.1 million. Number of Employees: 1,500.
Expertise/Education Sought: Production
Contact: For individuals with previous experience: Human Resources; Phone: 810-759-2200; Fax: 810-756-8674.

ALBERT KAHN ASSOCIATES, INC.

7430 SECOND AVENUE, SUITE 700
DETROIT, MICHIGAN 48202-2798

Overview: Provides architectural and engineering services. Established: 1895. Ownership: Private.

Key Statistics: Number of Employees: 330.

Expertise/Education Sought: Civil engineering, mechanical engineering, electrical engineering, structural engineering, architecture

Alternatives to Full-Time Employment: Paid internships

International Assignments are available to employees.

Contact: Ms. Diane Gill, Human Resources Coordinator; Phone: 313-202-7609; Fax: 313-871-8553. E-mail: diane.gill@akahn.com.

Company URL: http://www.albertkahn.com

CORPORATE PROFILE

Albert Kahn Associates (AKA), ranked among the top 25 architectural engineering firms in the United States, designs precedent-setting industrial, commercial, and institutional buildings throughout this country and abroad. Its clients include some of the most familiar names in the business world—Avon Products, BMW AG, Budd Company, Caterpillar, Chrysler Corporation, Colgate-Palmolive, Ethyl Corporation, Ford Motor Company, General Electric, General Motors, Henry Ford Health System, Eli Lilly, Mack Trucks, Max Factor, Mercedes-Benz, Nippondenso, Nissan, Providence Hospital, Toyota Motor Manufacturing, United Technologies, University of Michigan, and Whirlpool.

AKA offers a stimulating work environment for graduating and experienced architects and engineers. Thanks to a diverse client base and building type, AKA architects and engineers design facilities that incorporate such contemporary techniques as cogeneration, heat/energy recovery, energy management, thermal storage, peak-load shedding, air-conditioning stratification in industrial HVAC systems, power distribution and lighting systems, monitoring and controlling electrical power systems, and integrated fire, security, and access control systems using card readers, CCTV, intercom with hard wire, broadband, and fiber-optic cabling techniques. An in-house computer network integrates the design process.

AKA has captured numerous local, state, and national honors, including the national and Michigan American Institute of Architects' prestigious Architectural Firm Awards for "consistently producing distinguished architecture through the collaboration of related architectural disciplines." The firm was also honored with AIA's 1994 Intern Development Program Outstanding Firm Award.

Benefits include a liberal vacation policy, hospitalization, employee life insurance, disability insurance, profit-sharing, and pension and 401k plans.

Professionally trained employees may advance to management positions in the employee-owned corporation.

Albert Kahn Associates is an Equal Opportunity Employer, M/F/D/V.

ALLIED STAFFING, INC.

17187 NORTH LAUREL PARK DRIVE
LIVONIA, MICHIGAN 48152-3940

Overview: Provides employment services. Established: 1994. Parent Company: Acro Service Corp.. Ownership: Private.

Key Statistics: Annual Sales: $6.1 million. Number of Employees: 500.

Expertise/Education Sought: Engineering

International Assignments are available to employees.

Contact: For individuals with previous experience: Technical and Clerical Department; Phone: 734-591-1100; Fax: 734-591-1217.

ALPS AUTOMOTIVE, INC.

1500 ATLANTIC BOULEVARD
AUBURN HILLS, MICHIGAN 48326-1500

Overview: Manufactures automobile supplies and parts. Established: 1981. Ownership: Private.

Key Statistics: Annual Sales: $155.6 million. Number of Employees: 132.

Expertise/Education Sought: Engineering

Contact: For individuals with previous experience: Human Resources Department.

AMERITECH ADVERTISING SERVICES

100 EAST BIG BEAVER ROAD, SUITE 920
TROY, MICHIGAN 48083

Overview: Publishes telephone directories. Established: 1983. Parent Company: Ameritech Corporation. Ownership: Public.

Key Statistics: Annual Sales: $223.7 million. Number of Employees: 2,955.

Expertise/Education Sought: Customer service and support

Alternatives to Full-Time Employment: Paid internships

Company-Paid Training is available.

Contact: For college students and recent graduates: Mrs. Denise King, Project Manager; Phone: 810-524-7800; Fax: 810-740-5329. For individuals with previous experience: Human Resources Department; Phone: 810-524-7800; Fax: 810-740-5329. Job Hotline: 248-524-7800.

Company URL: http://www.ameritech.com

AMIGO MOBILITY INTERNATIONAL

6693 DIXIE HIGHWAY
BRIDGEPORT, MICHIGAN 48722

Overview: Supplies surgical and hospital equipment. Established: 1968. Ownership: Private.

Key Statistics: Annual Sales: $12 million. Number of Employees: 130.

Expertise/Education Sought: Manufacturing engineering, mechanical engineering

Alternatives to Full-Time Employment: Paid internships

Company-Paid Training is available.

International Assignments are available to employees.

Contact: For individuals with previous experience: Human Resources; Phone: 517-777-0910; Fax: 517-777-8184. E-mail: amigoint@concentric.net.

Company URL: http://www.myamigo.com

AMWAY CORPORATION

7575 EAST FULTO STREET
ADA, MICHIGAN 49355

Overview: Markets consumer and household products through independent distributors. Established: 1949. Ownership: Private.

Key Statistics: Annual Sales: $2.5 billion. Number of Employees: 13,000.

Expertise/Education Sought: Marketing, accounting, chemical engineering

Alternatives to Full-Time Employment: Paid internships

Company-Paid Training is available.

International Assignments are available to employees.

Contact: For college students and recent graduates: Mrs. Pam Powers, Advisor For College Relations; Phone: 616-787-6777; Fax: 616-787-5675. For individuals with previous experience: Mr. Ron Schut, Manager Professional Recruiting; Phone: 616-787-6777; Fax: 616-787-5675.

Company URL: http://www.amway.com

ARAMARK FACILITY SERVICES, INC.

2911 RESEARCH DRIVE
ROCHESTER HILLS, MICHIGAN 48309-3579

Overview: Provides facility management services. Established: 1987. Parent Company: Aramark Corporation. Ownership: Private.

Key Statistics: Annual Sales: $17.5 million. Number of Employees: 500.

Expertise/Education Sought: Facilities management

Alternatives to Full-Time Employment: Paid internships

Company-Paid Training is available.

INTERNATIONAL ASSIGNMENTS are available to employees.

CONTACT: For college students and recent graduates: Mr. Tom Horn, Director of Staffing and College Relations, 1101 Market Street, Philadelphia, Pennsylvania, 19107; Phone: 215-238-3496; Fax: 215-238-8170. For individuals with previous experience: Human Resources; Phone: 248-852-1011, ext. 115; Fax: 248-852-4007.

ASC, INC.

ONE SUNROOF CENTER
SOUTHGATE, MICHIGAN 48195-3044

OVERVIEW: Manufactures automotive sunroofs, convertible systems, and specialty vehicles. Established: 1965.

KEY STATISTICS: Annual Sales: $400 million. Number of Employees: 2,000.

EXPERTISE/EDUCATION SOUGHT: Mechanical engineering, manufacturing engineering, quality engineering, design engineering, computer programming

CONTACT: Staffing Department; Phone: 313-246-0665; Fax: 313-246-2609.

COMPANY URL: http://www.ascglobal.com

CORPORATE STATEMENT

Since its humble beginning in 1965 as a small start-up entrepreneurial company, ASC Incorporated has grown at a remarkable rate. In the span of three decades, the company has become an internationally respected, world-class, Tier 1 leader in the specialized field of automotive enhancement technology and open-air systems. Not content to rest on past accomplishments, the company is forging ahead, building on its thirty-year tradition of craftsmanship, quality, and product integrity.

AVL NORTH AMERICA CORPORATION

47519 HALYARD DRIVE
PLYMOUTH, MICHIGAN 48170-2438

OVERVIEW: Manufactures process control instruments. Established: 1993. Parent Company: AVL. Ownership: Private.

KEY STATISTICS: Annual Sales: $11.4 million. Number of Employees: 120.

HIRING HISTORY: Number of professional employees hired in 1996: 10.

EXPERTISE/EDUCATION SOUGHT: Engineering, software engineering/ development, mechanical engineering, electrical engineering

ALTERNATIVES TO FULL-TIME EMPLOYMENT: Part-time jobs

INTERNATIONAL ASSIGNMENTS are available to employees.

CONTACT: For individuals with previous experience: Human Resources; Phone: 313-414-9600; Fax: 313-414-9643.

COMPANY URL: http://www.avlna.com

THE BUDD COMPANY

3155 WEST BIG BEAVER ROAD
TROY, MICHIGAN 48007-2601

OVERVIEW: Manufactures automotive parts and accessories. Established: 1978. Parent Company: Thyssen, AG. Ownership: Private.

KEY STATISTICS: Annual Sales: $1.8 billion. Number of Employees: 9,000.

EXPERTISE/EDUCATION SOUGHT: Administration, accounting, finance, law, electrical engineering, mechanical engineering

ALTERNATIVES TO FULL-TIME EMPLOYMENT: Internships

COMPANY-PAID TRAINING is available.

CONTACT: Ms. Linda Bayly, Corporate Personnel Manager; Phone: 248-643-3613; Fax: 248-643-3687.

COMPANY URL: http://www.buddcompany.com

CADE INDUSTRIES, INC.

2365 WOODLAKE DRIVE, SUITE 120
OKEMOS, MICHIGAN 48864

OVERVIEW: Designs, manufactures, and repairs high-technology composite components. Established: 1980. Ownership: Public.

KEY STATISTICS: Annual Sales: $34.9 million. Number of Employees: 515.

EXPERTISE/EDUCATION SOUGHT: Engineering, management

ALTERNATIVES TO FULL-TIME EMPLOYMENT: Internships

COMPANY-PAID TRAINING is available.

CONTACT: For individuals with previous experience: Human Resources; Phone: 517-347-6185; Fax: 517-347-1333. E-mail: sherylmull@acd.net.

CADILLAC PRODUCTS, INC.

5800 CROOKS ROAD
TROY, MICHIGAN 48098-2100

OVERVIEW: Manufactures plastics products. Established: 1942. Ownership: Private.

KEY STATISTICS: Annual Sales: $110 million. Number of Employees: 925.

HIRING HISTORY: Number of professional employees hired in 1997: 50.

EXPERTISE/EDUCATION SOUGHT: Mechanical engineering, plastics

ALTERNATIVES TO FULL-TIME EMPLOYMENT: Paid internships

COMPANY-PAID TRAINING is available.

CONTACT: For college students and recent graduates: Ms. Debbie Fulton, Recruiting Coordinator; Phone: 810-879-5000; Fax: 810-879-7420. For individuals with previous experience: Recruiting Coordinator; Phone: 248-879-5000; Fax: 248-879-7420.

CASCADE DIE CASTING GROUP, INC.

3040 CHARLEVOIX DRIVE, SE
GRAND RAPIDS, MICHIGAN 49546-7000

OVERVIEW: Provides die casting services. Established: 1985. Parent Company: TCH Industries, Inc. Ownership: Private.

KEY STATISTICS: Annual Sales: $67.5 million. Number of Employees: 625.

EXPERTISE/EDUCATION SOUGHT: Accounting, finance, payroll, data processing, electrical technology, electrical engineering, project engineering, process engineering, benefits administration

ALTERNATIVES TO FULL-TIME EMPLOYMENT: Paid internships

COMPANY-PAID TRAINING is available.

CONTACT: Mr. Paul Dieterle, Director of Human Resources; Phone: 616-956-0966; Fax: 616-956-9149.

CELLO-FOIL PRODUCTS, INC.

155 BROOK STREET
BATTLE CREEK, MICHIGAN 49017-3031

OVERVIEW: Manufactures flexible packaging converters. Established: 1949. Ownership: Private.

KEY STATISTICS: Annual Sales: $60 million. Number of Employees: 265.

EXPERTISE/EDUCATION SOUGHT: Computer science

COMPANY-PAID TRAINING is available.

CONTACT: Mr. Terry A. McFall, Director of Human Resources; Phone: 616-964-7137; Fax: 616-968-2194.

CHRYSLER CORPORATION

1000 CHRYSLER DRIVE, CIMS: 485-01-99
AUBURN HILLS, MICHIGAN 48326-2766

OVERVIEW: Manufactures cars, trucks, and automotive components. Established: 1925. Ownership: Public.

KEY STATISTICS: Annual Sales: $61.1 billion. Number of Employees: 130,000.

HIRING HISTORY: Number of professional employees hired in 1997: 1,400. 1996: 1,000.

EXPERTISE/EDUCATION SOUGHT: Mechanical engineering, electrical engineering, computer science, supply chain management, finance, human resources

INTERNATIONAL ASSIGNMENTS are available to employees.

CONTACT: College Relations; Phone: 248-512-2187.

COMPANY URL: http://www.chrysler.com

Chrysler Corporation (continued)

CORPORATE PROFILE

Chrysler Corporation designs, engineers, manufactures, and distributes cars, light trucks, sport-utility vehicles, minivans, automotive components and systems, and aftermarket parts.

Numerous milestones mark the company's 74-year history, and the most recent 15-year period has included many benchmarks such as the invention of the minivan, acquisition of American Motors Corporation and the Jeep® line of sport-utility vehicles, platform teams, the dedication of Chrysler's World Headquarters and Chrysler Technology Center, and the introduction of award-winning and internationally recognized products like the Dodge Viper and the Plymouth Prowler.

Chrysler is an international company and has opened three major facilities within the last two years: the Chrysler World Headquarters in Auburn Hills, Michigan; the Chrysler Europe office in Brussels; and Chrysler Japan in Tokyo. Most engineering, manufacturing, and administrative activities are concentrated in southeast Michigan.

Organized in product platform teams, Chrysler seeks people who can work in teams both as members and leaders. Ideal candidates are innovative problem solvers with professional commitment and a strong work ethic. Academic achievement and leadership ability are also important.

New graduates typically enter Chrysler either by direct assignment or in a rotational program. For additional information, Chrysler Corporation's home page on the Web is at http://www.chrysler.com.

CMS ENERGY CORPORATION
330 TOWN CENTER DRIVE
DEARBORN, MICHIGAN 48126-2709

Overview: Supplies nonregulated energy. Established: 1987. Ownership: Public.

Key Statistics: Assets: $4.3 billion. Number of Employees: 9,663.

Hiring History: Number of professional employees hired in 1997: 100. 1996: 200.

Expertise/Education Sought: Accounting, law, finance, human resources, computer programming, program analysis, mechanical engineering, electrical engineering

Alternatives to Full-Time Employment: Paid internships

Company-Paid Training is available.

International Assignments are available to employees.

Contact: Ms. Paulette Boggs, Staffing Director; Phone: 517-788-1497.

Company URL: http://www.cmsenergy.com

CMS MARKETING SERVICES & TRADING COMPANY
330 TOWN CENTER DRIVE, FAIRLANE PLAZA SOUTH
DEARBORN, MICHIGAN 48126-2711

Overview: Provides electric and natural gas service. Established: 1997. Parent Company: CMS Energies Corporation. Ownership: Public.

Key Statistics: Annual Sales: $300 million. Number of Employees: 150.

Expertise/Education Sought: Project engineering, construction engineering

Alternatives to Full-Time Employment: Paid internships

Company-Paid Training is available.

International Assignments are available to employees.

Contact: Human Resource Department; Phone: 313-436-9493; Fax: 313-441-0402. E-mail: lmmurphy@cmsenergy.com.

Company URL: http://www.cmsenergy.com

COMPLETE BUSINESS SOLUTIONS, INC.
32605 WEST 12 MILE ROAD
FARMINGTON HILLS, MICHIGAN 48334

Overview: Provides system integration services. Established: 1985. Ownership: Private.

Key Statistics: Annual Sales: $83.2 million. Number of Employees: 1,600.

Hiring History: Number of professional employees hired in 1997: 400. 1996: 300.

Expertise/Education Sought: Data processing, software engineering/development

Company-Paid Training is available.

International Assignments are available to employees.

Contact: Mr. Pankaj Ved, Director Corporate Recruiting; Phone: 248-848-2220; Fax: 248-488-1949. E-mail: pved@cbsinc.com.

Company URL: http://www.cbsinc.com

COMPUTER METHODS CORPORATION
13740 MERRIMAN ROAD
LIVONIA, MICHIGAN 48150

Overview: Develops computer software. Established: 1977. Ownership: Private.

Key Statistics: Annual Sales: $26.9 million. Number of Employees: 400.

Hiring History: Number of professional employees hired in 1997: 175. 1996: 175.

Expertise/Education Sought: Computer programming, computer science, electrical engineering, management information systems, software engineering/development

Company-Paid Training is available.

Contact: Ms. Kim Clute, Director of Human Resources; Phone: 313-522-5187, ext. 1113; Fax: 313-522-2705. E-mail: kclute@cmethods.com.

Company URL: http://www.cmethods.com

CONSUMERS ENERGY
212 WEST MICHIGAN AVENUE
JACKSON, MICHIGAN 49201

Overview: Electric and gas utility. Established: 1987. Parent Company: CMS Energy Corporation. Ownership: Public.

Key Statistics: Assets: $3.8 billion. Number of Employees: 8,873.

Expertise/Education Sought: Mechanical engineering, electrical engineering, accounting, management information systems, computer science

Alternatives to Full-Time Employment: Paid internships

International Assignments are available to employees.

Contact: Mrs. Paulette E. Boggs, Director, Staffing and Diversity; Phone: 517-788-1497; Fax: 517-788-0258.

Company URL: http://www.consumersenergy.com

CONTINENTAL PLASTICS COMPANY
33525 GROESBECK HIGHWAY
FRASER, MICHIGAN 48026-4205

Overview: Manufactures injection molded plastics. Established: 1957. Ownership: Private.

Key Statistics: Annual Sales: $102 million. Number of Employees: 520.

Hiring History: Number of professional employees hired in 1996: 75.

Expertise/Education Sought: Mechanical engineering, injection molding

Contact: For individuals with previous experience: Mr. Pat Luckino, Human Resources Director; Phone: 810-294-4600; Fax: 810-294-0187.

CP INTERNATIONAL, INC.
512 WASHINGTON AVENUE
GRAND HAVEN, MICHIGAN 49417-1455

Overview: Manufactures aluminum die castings. Established: 1996. Ownership: Private.

Key Statistics: Annual Sales: $45 million.

Hiring History: Number of professional employees hired in 1997: 5.

Expertise/Education Sought: Mechanical engineering, sales

Alternatives to Full-Time Employment: Paid internships

International Assignments are available to employees.

Contact: For individuals with previous experience: Human Resource Department; Phone: 800-486-8866; Fax: 888-233-9934.

DELPHI AUTOMOTIVE SYSTEMS
ONE PONTIAC PLAZA
PONTIAC, MICHIGAN 48340-2952

Overview: Designs, develops, and manufactures automotive systems and components. Established: 1903. Ownership: Public.

Key Statistics: Annual Sales: $26.4 billion. Number of Employees: 216,000.

Expertise/Education Sought: Mechanical engineering, industrial engineering, manufacturing engineering, electrical engineering, materials engineering

Contact: For college students and recent graduates: Divisional College Relations and Professional Staffing, 4800 South Saginaw Street, Flint, Michigan, 48501-1360; Fax: 810-257-6342. For individuals with previous experience: Divisional College Relations and Professional Staffing, 4800 South Saginaw Street, Flint, Michigan, 48501-1360; Fax: 810-257-6342.

Company URL: http://www.delphiauto.com

CORPORATE PROFILE

Delphi Automotive Systems is a diversified supplier of automotive components, systems, and modules. The company employs 216,000 people and operates 208 manufacturing facilities, forty-six joint ventures in thirty-six countries, and twenty-seven state-of-the-art technical centers.

The global headquarters of Delphi Automotive Systems is located in Troy, Michigan, with regional headquarters in Paris, Tokyo, and Sao Paulo. Delphi consists of seven operating divisions, each of which is a world leader in system design, engineering, manufacturing, delivery, and service. The divisions are as follows: **Chassis Systems,** *which focuses on intelligent and innovative ride and handling systems;* **Delco Electronics Systems,** *which is responsible for innovative safety, security, audio, communication, and control technologies that are designed to make vehicles safer, smarter, more comfortable, and more convenient;* **Energy & Engine Management Systems,** *which develops environmentally responsible fuel systems with enhanced performance and fuel economy;* **Harrison Thermal Systems,** *which focuses on powertrain performance and passenger comfort, optimizing engine and transmission cooling and interior climate control systems;* **Interior Systems,** *which develops products that dramatically improve driver and passenger safety, security, comfort, and convenience;* **Packard Electric Systems,** *which focuses on the electrical/electronic architecture for reliable complex systems; and* **Saginaw Steering Systems,** *which develops steering systems that enhance vehicle control, improve fuel economy, and simplify assembly.*

For more information about Delphi Automotive Systems, prospective applicants should visit the World Wide Web (http://www.delphiauto.com). Delphi Automotive Systems and its divisions are proud to be Equal Opportunity Employers.

DENSO INTERNATIONAL AMERICA, INC.
24777 DENSO DRIVE
SOUTHFIELD, MICHIGAN 48034-5133

Overview: Manufactures automotive parts and supplies. Established: 1985. Ownership: Private.

Key Statistics: Annual Sales: $2.3 billion. Number of Employees: 3,600.

Hiring History: Number of professional employees hired in 1997: 71.

Expertise/Education Sought: Sales, design engineering, accounting, applications engineering, information systems

Alternatives to Full-Time Employment: Paid internships

Contact: For individuals with previous experience: Mrs. Gail Combs, Section Manager Employment; Phone: 248-350-7792; Fax: 248-350-7772.

Company URL: http://www.denso-int.com

DENSO MANUFACTURING MICHIGAN, INC.
1 DENSO ROAD
BATTLE CREEK, MICHIGAN 49015-1056

Overview: Manufactures automotive parts and supplies. Established: 1985. Parent Company: Denso Corporation. Ownership: Private.

Key Statistics: Annual Sales: $695 million. Number of Employees: 1,550.

Hiring History: Number of professional employees hired in 1996: 28.

Expertise/Education Sought: Sales, marketing, accounting, finance, data processing, computer programming, design engineering, manufacturing engineering

Alternatives to Full-Time Employment: Part-time jobs, paid internships

Company-Paid Training is available.

Contact: Mr. Brad Vanvyl, Supervisor of Human Resources; Phone: 616-965-3322, ext. 2541; Fax: 616-966-2127.

DIVERSEY LEVER, INC.
14496 SHELDON ROAD, SUITE 210
PLYMOUTH, MICHIGAN 48170-2475

Overview: Manufactures polishes. Established: 1996. Parent Company: Unilever Corporation. Ownership: Private.

Key Statistics: Annual Sales: $166.9 million. Number of Employees: 1,400.

Hiring History: Number of professional employees hired in 1997: 100. 1996: 75.

Expertise/Education Sought: Accounting, engineering, science, computer programming

Alternatives to Full-Time Employment: Paid internships

Company-Paid Training is available.

International Assignments are available to employees.

Contact: Ms. Stacey Woodman, Human Resources Generalist; Phone: 734-414-3182; Fax: 734-414-3369.

DONNELLY CORPORATION
414 EAST 40TH STREET
HOLLAND, MICHIGAN 49423-5313

Overview: Manufactures automotive glass and plastic products. Established: 1905. Ownership: Public.

Key Statistics: Annual Sales: $671.3 million. Number of Employees: 2,966.

Expertise/Education Sought: Sales, accounting, marketing

Alternatives to Full-Time Employment: Paid internships

International Assignments are available to employees.

Contact: Mr. James Wujkowski, Manager of Professional Recruitment; Phone: 616-786-6200; Fax: 616-786-6236. E-mail: jameswujkowski@donnelly.com.

Company URL: http://www.donnelly.com

THE DOW CHEMICAL COMPANY
2030 WILLARD H DOW CENTER
MIDLAND, MICHIGAN 48674

Mailing Address: Applicant Center, PO Box 16, Midland, MI 48641-1655

Overview: Manufactures chemical, plastic, and consumer products. Established: 1897. Ownership: Public.

Key Statistics: Annual Sales: $20.1 billion. Number of Employees: 40,289.

Hiring History: Number of professional employees hired in 1997: 200. 1996: 200.

Expertise/Education Sought: Marketing, finance, sales, human resources, technical

Alternatives to Full-Time Employment: Paid internships

Company-Paid Training is available.

International Assignments are available to employees.

Contact: Ms. Diane Dittenhafer, Recruitment Manager; Phone: 517-636-1288; Fax: 517-636-3674.

Company URL: http://www.dow.com

DOW CORNING CORPORATION

2200 WEST SALZBURG ROAD
MIDLAND, MICHIGAN 48686-0994

Overview: Manufactures silicone products and lubricants. Established: 1943. Parent Company: Dow Chemical Company. Ownership: Private.

Key Statistics: Annual Sales: $2.5 billion. Number of Employees: 8,700.

Expertise/Education Sought: Chemistry, chemical engineering, mechanical engineering, electrical engineering

Alternatives to Full-Time Employment: Paid internships

Company-Paid Training is available.

International Assignments are available to employees.

Contact: For college students and recent graduates: Mr. Joseph Mathis-Lilley, Staffing Manager; Phone: 517-496-4000, ext. 6376; Fax: 517-499-6109. For individuals with previous experience: Mr. Doug Dawson, Director of Human Resources; Phone: 517-496-4000, ext. 8463; Fax: 517-499-6109.

DTE ENERGY COMPANY

2000 2ND AVENUE
DETROIT, MICHIGAN 48226-1203

Overview: Generates, transmits, and distributes electrical service. Established: 1995. Ownership: Public.

Key Statistics: Annual Sales: $3.6 billion. Number of Employees: 8,400.

Expertise/Education Sought: Engineering

Alternatives to Full-Time Employment: Part-time jobs, paid internships

Company-Paid Training is available.

Contact: Mr. Milton Hill, Human Resources Consultant; Phone: 313-235-9495; Fax: 313-235-9499.

Company URL: http://www.detroitedison.com

FLOWSERVE CORPORATION FLUID SEALING DIVISION

2100 FACTORY STREET
KALAMAZOO, MICHIGAN 49001-4161

Overview: Designs, manufactures, and distributes mechanical sealing systems. Established: 1995. Parent Company: Flowserve Corporation. Ownership: Public.

Key Statistics: Annual Sales: $84.7 million. Number of Employees: 1,100.

Expertise/Education Sought: Mechanical engineering, design engineering, industrial distribution

Alternatives to Full-Time Employment: Paid internships

Company-Paid Training is available.

International Assignments are available to employees.

Contact: Ms. Cyndi Haaf, Human Resources Representative; Phone: 616-381-2650; Fax: 616-381-3249.

Company URL: http://www.flowserve.com

FORD MOTOR COMPANY

AMERICAN ROAD
DEARBORN, MICHIGAN 48121

Overview: Manufactures automobiles and trucks. Established: 1903. Ownership: Public.

Key Statistics: Annual Sales: $154 billion. Number of Employees: 363,000.

Hiring History: Number of professional employees hired in 1997: 2,000.

Expertise/Education Sought: Engineering, finance, computer science, marketing, manufacturing management

Alternatives to Full-Time Employment: Part-time jobs, paid internships

Company-Paid Training is available.

International Assignments are available to employees.

Contact: Human Resources Customer Operations, Salaried Recruitment, PO Box 0520, Department Code: USAUPTJO, Allen Park, Michigan, 48101-0520. E-mail: recruit2@ford.com. Job Hotline: 888-621-1723.

Company URL: http://www.ford.com/careercenter

For more information, see full corporate profile, pg. 276.

FREUDENBERG-NOK

47690 EAST ANCHOR COURT
PLYMOUTH, MICHIGAN 48170-2455

Overview: Manufactures automotive seals and molded products. Established: 1989. Parent Company: Freudenberg And Company. Ownership: Private.

Key Statistics: Annual Sales: $628.6 million. Number of Employees: 5,000.

Hiring History: Number of professional employees hired in 1996: 50.

Expertise/Education Sought: Marketing, sales, accounting, engineering

Alternatives to Full-Time Employment: Paid internships

Company-Paid Training is available.

International Assignments are available to employees.

Contact: Ms. Ashley Ault, Human Resources Coordinator; Phone: 313-451-0020, ext. 217; Fax: 313-451-0125.

GENERAL DYNAMICS LAND SYSTEMS

38500 MOUND ROAD
STERLING HEIGHTS, MICHIGAN 48310-3200

Overview: Manufactures military vehicles. Established: 1982. Parent Company: General Dynamics Corporation. Ownership: Public.

Key Statistics: Annual Sales: $871 million. Number of Employees: 3,000.

Expertise/Education Sought: Systems engineering, hardware engineering, mechanical engineering

Alternatives to Full-Time Employment: Paid internships

Company-Paid Training is available.

International Assignments are available to employees.

Contact: Ms. Milli Miller, Manager of Staffing; Phone: 810-825-4000; Fax: 810-825-4546.

GENERAL HOST CORPORATION

6501 EAST NEVADA STREET
DETROIT, MICHIGAN 48234-9977

Overview: Owns and operates retail nursery and craft stores. Established: 1911. Ownership: Public.

Key Statistics: Annual Sales: $530.8 million. Number of Employees: 5,000.

Hiring History: Number of professional employees hired in 1997: 200. 1996: 200.

Expertise/Education Sought: Accounting, finance, data processing, marketing

Contact: Ms. Tracy Williams, Human Resources Manager; Phone: 313-564-2244; Fax: 313-366-4253.

GENERAL MOTORS CORPORATION

3044 WEST GRAND BOULEVARD, ROOM 482108162
DETROIT, MICHIGAN 48202-3091

Overview: Manufactures cars, trucks, and automotive parts. Established: 1908. Ownership: Public.

Key Statistics: Annual Sales: $164.1 billion. Number of Employees: 647,000.

Expertise/Education Sought: Accounting, finance, computer programming, computer science, sales, chemical engineering, electrical engineering, metallurgical engineering, manufacturing, mechanical engineering

Alternatives to Full-Time Employment: Paid internships

Company-Paid Training is available.

International Assignments are available to employees.

Contact: Ms. Betty Anderson, Director of Educational Relations; Phone: 313-556-3569.

GENTEX CORPORATION

600 NORTH CENTENNIAL STREET
ZEELAND, MICHIGAN 49464-1318

Overview: Manufactures automotive mirrors, smoke detectors, and signaling devices. Established: 1974. Ownership: Public.

Key Statistics: Annual Sales: $148.7 million. Number of Employees: 1,075.

Hiring History: Number of professional employees hired in 1997: 1,100. 1996: 200.

Expertise/Education Sought: Chemical engineering, design engineering, electrical engineering, mechanical engineering, electrical technology

Alternatives to Full-Time Employment: Paid internships

International Assignments are available to employees.

Contact: Mr. Kurt Wassink, Staffing Specialist; Phone: 616-772-1800; Fax: 616-772-7348.

Company URL: http://www.gentex.com

GERBER PRODUCTS COMPANY

445 STATE STREET
FREMONT, MICHIGAN 49413

Overview: Produces processed baby foods, children's apparel, and child care products. Established: 1994. Parent Company: Novartis. Ownership: Private.

Key Statistics: Annual Sales: $1 billion. Number of Employees: 5,100.

Expertise/Education Sought: Accounting, data processing, finance, sales, marketing, law, computer programming, business administration

Alternatives to Full-Time Employment: Paid internships

Contact: For individuals with previous experience: Ms. Melissa Redinger, Senior Manager Human Resources; Phone: 616-928-2658, ext. 2658; Fax: 616-928-2819.

GKN AUTOMOTIVE, INC.

3300 UNIVERSITY DRIVE
AUBURN HILLS, MICHIGAN 48326-2362

Overview: Manufactures universal joints and drive shafts for motor vehicles. Established: 1985. Parent Company: GKN. Ownership: Private.

Key Statistics: Annual Sales: $186.4 million. Number of Employees: 2,000.

Expertise/Education Sought: Engineering

Alternatives to Full-Time Employment: Paid internships

Company-Paid Training is available.

International Assignments are available to employees.

Contact: For individuals with previous experience: Human Resources; Phone: 810-377-1200.

HAWORTH, INC.

ONE HAWORTH CENTER
HOLLAND, MICHIGAN 49423-9576

Overview: Manufactures office furniture. Established: 1948. Ownership: Private.

Key Statistics: Annual Sales: $1.2 billion. Number of Employees: 8,800.

Expertise/Education Sought: Mechanical engineering, industrial engineering, information systems, marketing, drafting, environmental, production

Alternatives to Full-Time Employment: Part-time jobs, internships

International Assignments are available to employees.

Contact: Staffing and Diversity; Phone: 616-393-3000; Fax: 616-393-1551. Job Hotline: 616-393-3900.

Company URL: http://www.haworth-furn.com

HILLSDALE TOOL AND MANUFACTURING

135 EAST SOUTH STREET
HILLSDALE, MICHIGAN 49242-1807

Overview: Manufactures machine and engine parts. Established: 1971. Parent Company: Eagle-Picher Industries, Inc. Ownership: Public.

Key Statistics: Annual Sales: $165 million. Number of Employees: 1,000.

Hiring History: Number of professional employees hired in 1996: 70.

Expertise/Education Sought: Mechanical engineering, manufacturing engineering, tool and die engineering

Alternatives to Full-Time Employment: Paid internships

Company-Paid Training is available.

International Assignments are available to employees.

Contact: Mr. Sam Trego, Vice President of Human Resources; Phone: 517-439-9381, ext. 587; Fax: 517-439-5953.

HY-LIFT, LLC

1185 EAST KEATING AVENUE
MUSKEGON, MICHIGAN 49442-6018

Overview: Manufactures carburetors, pistons, rings, and valves. Established: 1996. Parent Company: W.A. Thomas. Ownership: Private.

Key Statistics: Annual Sales: $20.3 million. Number of Employees: 325.

Hiring History: Number of professional employees hired in 1997: 15. 1996: 4.

Expertise/Education Sought: Electrical engineering, manufacturing engineering, industrial engineering

Alternatives to Full-Time Employment: Paid internships

Company-Paid Training is available.

Contact: For individuals with previous experience: Human Resources.

INMET CORPORATION

300 DINO DRIVE
ANN ARBOR, MICHIGAN 48103-9457

Overview: Produces microwave components. Established: 1995. Parent Company: MCE Companies, Inc. Ownership: Private.

Key Statistics: Annual Sales: $36 million. Number of Employees: 550.

Hiring History: Number of professional employees hired in 1996: 20.

Expertise/Education Sought: Microwave engineering

Alternatives to Full-Time Employment: Part-time jobs, paid internships

Company-Paid Training is available.

Contact: For individuals with previous experience: Human Resources Department; Phone: 313-426-5553; Fax: 313-426-5557.

Company URL: http://www.inmet.com

INTERMET CORPORATION

5445 CORPORATE DRIVE, SUITE 200
TROY, MICHIGAN 48098

Overview: Foundry. Established: 1984. Ownership: Public.

Key Statistics: Annual Sales: $109.4 million. Number of Employees: 1,500.

Expertise/Education Sought: Engineering, accounting, management information systems, manufacturing

Alternatives to Full-Time Employment: Paid internships

International Assignments are available to employees.

Contact: For individuals with previous experience: Human Resources; Phone: 248-952-2500; Fax: 248-952-2501.

JOHNSON CONTROLS, INC.

49200 HALYARD DRIVE
PLYMOUTH, MICHIGAN 48170

Overview: Manufactures batteries, automotive seats, building and industrial controls, and plastics. Established: 1985. Ownership: Public.

Key Statistics: Annual Sales: $3.5 billion. Number of Employees: 22,864.

Johnson Controls, Inc. (continued)

Hiring History: Number of professional employees hired in 1997: 300. 1996: 240.

Expertise/Education Sought: Mechanical engineering, industrial engineering, aerospace engineering

Alternatives to Full-Time Employment: Paid internships

Company-Paid Training is available.

International Assignments are available to employees.

Contact: For college students and recent graduates: Mr. Bernie Spraggins, Senior Staffing Specialist; Phone: 734-254-5848; Fax: 734-254-5843. For individuals with previous experience: Ms. Ingrid Joris, Recruiting Manager; Phone: 734-254-5000; Fax: 734-254-5843. E-mail: ingridjoris@jci.com.

LA-Z-BOY, INC.

1284 NORTH TELEGRAPH ROAD
MONROE, MICHIGAN 48162-5138

Overview: Manufactures chairs, sofas, and other furniture. Established: 1929. Ownership: Public.

Key Statistics: Annual Sales: $1 billion. Number of Employees: 10,733.

Hiring History: Number of professional employees hired in 1997: 20. 1996: 20.

Expertise/Education Sought: Mechanical engineering, manufacturing engineering, marketing, finance, sales

Alternatives to Full-Time Employment: Part-time jobs, paid internships

Company-Paid Training is available.

Contact: Mr. Mark Lohman, Manager of Human Resources; Phone: 313-242-1444, ext. 4434; Fax: 313-241-3452.

Company URL: http://www.La-Z-Boy.com

LEAR CORPORATION

2998 WATERVIEW DRIVE
ROCHESTER HILLS, MICHIGAN 48309

Overview: Manufactures automobile seats. Established: 1995. Ownership: Public.

Key Statistics: Annual Sales: $55.2 million. Number of Employees: 764.

Hiring History: Number of professional employees hired in 1996: 25.

Expertise/Education Sought: Manufacturing engineering, research, design engineering, test engineering

Alternatives to Full-Time Employment: Paid internships

Contact: Human Resources Administrator; Phone: 810-371-8400.

Company URL: http://www.lear.com

LECO CORPORATION, INC.

3000 LAKEVIEW AVENUE
SAINT JOSEPH, MICHIGAN 49085-2319

Overview: Manufactures analytical instruments. Established: 1936. Ownership: Private.

Key Statistics: Annual Sales: $70.4 million. Number of Employees: 800.

Hiring History: Number of professional employees hired in 1997: 72. 1996: 50.

Expertise/Education Sought: Software engineering/development, electrical engineering, hardware engineering, marketing, finance, computer programming, sales, field service, mechanical engineering

Alternatives to Full-Time Employment: Part-time jobs, paid internships

International Assignments are available to employees.

Contact: Mr. John Mattix, Manager of Human Resources; Phone: 616-983-5533, ext. 218; Fax: 616-983-3850.

Company URL: http://www.leco.com

LESLIE METAL ARTS, INC.

3225 32ND STREET, SE
GRAND RAPIDS, MICHIGAN 49512-1870

Overview: Manufactures small automobile parts. Established: 1945. Ownership: Private.

Key Statistics: Annual Sales: $178.7 million. Number of Employees: 1,250.

Expertise/Education Sought: Automotive engineering

Alternatives to Full-Time Employment: Paid internships

Company-Paid Training is available.

Contact: For individuals with previous experience: Human Resources; Phone: 616-949-1851; Fax: 616-949-1761.

LIBRALTER PLASTICS, INC.

3175 MARTIN ROAD
WALLED LAKE, MICHIGAN 48390-1628

Overview: Manufactures injection molded plastics. Established: 1977. Ownership: Private.

Key Statistics: Annual Sales: $105 million. Number of Employees: 755.

Hiring History: Number of professional employees hired in 1997: 35. 1996: 25.

Expertise/Education Sought: Project engineering, electronics/electronics engineering, plastics engineering, finance, accounting, information systems

Alternatives to Full-Time Employment: Paid internships

Company-Paid Training is available.

Contact: For individuals with previous experience: Mr. Rick Felstow, Director of Human Resources; Phone: 248-669-4900, ext. 241; Fax: 248-669-5372.

MARK I MOLDED PLASTICS

1640 EAST CHICAGO ROAD
JONESVILLE, MICHIGAN 49250-9749

Mailing Address: PO Box 35, Jonesville, MI 49250

Overview: Manufactures injection molded plastics. Established: 1987. Parent Company: Excel Industries. Ownership: Public.

Key Statistics: Annual Sales: $37.4 million. Number of Employees: 500.

Expertise/Education Sought: Product engineering, process engineering, management

Contact: For individuals with previous experience: Ms. Carol Moody, Human Resources Manager; Phone: 517-849-9911; Fax: 517-849-9740.

MASCO CORPORATION

21001 VAN BORN ROAD
TAYLOR, MICHIGAN 48180-1340

Overview: Manufactures automotive parts, plumbing fixtures, hardware, and furnishings. Established: 1929. Ownership: Public.

Key Statistics: Annual Sales: $3.2 billion. Number of Employees: 22,800.

Expertise/Education Sought: Product engineering, manufacturing engineering, operations, sales, marketing, design engineering, automotive engineering, finance

Alternatives to Full-Time Employment: Paid internships

International Assignments are available to employees.

Contact: Mr. Dave Kinsella, Director of Executive Recruiting; Phone: 313-274-7400, ext. 411; Fax: 313-374-6657.

Company URL: http://www.masco.com

MCKECHNIE VEHICLE COMPONENTS

5750 NEW KING DRIVE, SUITE 100
TROY, MICHIGAN 48098-2696

Overview: Manufactures motor vehicle wheels and parts. Established: 1989. Ownership: Private.

Key Statistics: Annual Sales: $46.1 million. Number of Employees: 500.

Hiring History: Number of professional employees hired in 1997: 7. 1996: 5.

Expertise/Education Sought: Engineering, sales

Alternatives to Full-Time Employment: Paid internships

Company-Paid Training is available.

International Assignments are available to employees.

Contact: Ms. Bonnie Perry, Recruiter; Phone: 606-276-5823; Fax: 606-277-0665.

MICHIGAN AUTOMOTIVE COMPRESSOR

2400 NORTH DEARING ROAD
PARMA, MICHIGAN 49269-9719

Overview: Manufactures air conditioning equipment for motor vehicles. Established: 1989. Parent Company: Denso Corporation And Toyota Automatic Loom. Ownership: Private.

Key Statistics: Annual Sales: $53.3 million. Number of Employees: 600.

Hiring History: Number of professional employees hired in 1996: 80.

Expertise/Education Sought: Mechanical engineering, accounting, business, logistics, packaging engineering, information systems, quality engineering

Alternatives to Full-Time Employment: Paid internships

Company-Paid Training is available.

Contact: Ms. Robin Hicks, Human Resources Specialist; Phone: 517-531-5500; Fax: 517-531-1711. E-mail: hicksr@michauto.com.

Company URL: http://www.michauto.com

MICHIGAN DEPARTMENT OF TRANSPORTATION

425 WEST OTTAWA STREET
LANSING, MICHIGAN 48913

Overview: Provides transportation services and facilities.

Expertise/Education Sought: Urban and regional planning, land surveying, real state management, civil engineering, environmental engineering

Contact: For individuals with previous experience: Recruiter, PO Box 30050, Lansing, Michigan, 48909; Phone: 517-373-1620; Fax: 517-373-8214.

NICHOLAS PLASTICS, INC.

11700 48TH AVENUE
ALLENDALE, MICHIGAN 49401

Overview: Provides extrusion products for automobiles and furniture. Established: 1978. Ownership: Private.

Key Statistics: Annual Sales: $50 million. Number of Employees: 360.

Hiring History: Number of professional employees hired in 1997: 10.

Expertise/Education Sought: Plastics engineering, chemical engineering, statistics, mathematics, business, mechanical engineering, manufacturing engineering

Alternatives to Full-Time Employment: Paid internships

Company-Paid Training is available.

Contact: Ms. Cathy Wilson, Human Resources Director; Phone: 616-892-4100; Fax: 616-892-4225.

NISSAN RESEARCH AND DEVELOPMENT

39001 SUNRISE DRIVE
FARMINGTON HILLS, MICHIGAN 48331-3487

Overview: Provides automobile design engineering services. Established: 1977. Parent Company: Nissan Moter Corporation. Ownership: Private.

Key Statistics: Annual Sales: $27.9 million. Number of Employees: 440.

Hiring History: Number of professional employees hired in 1996: 40.

Expertise/Education Sought: Mechanical engineering, electrical engineering

Alternatives to Full-Time Employment: Part-time jobs, paid internships

Company-Paid Training is available.

International Assignments are available to employees.

Contact: For individuals with previous experience: Ms. Holly Bracco, Department of Human Resources, PO Box 9200, Farmington Hills, Michigan, 48333-3487; Phone: 248-488-4189; Fax: 248-488-3904.

Company URL: http://www.nissan-nrd.com

OGIHARA AMERICA CORPORATION

1480 WEST MCPHERSON PARK
HOWELL, MICHIGAN 48843

Overview: Manufactures automotive trim and accessories. Established: 1984. Ownership: Private.

Key Statistics: Annual Sales: $48.8 million. Number of Employees: 446.

Hiring History: Number of professional employees hired in 1997: 150.

Expertise/Education Sought: Mechanical engineering, manufacturing, automotive engineering

Alternatives to Full-Time Employment: Paid internships

Company-Paid Training is available.

International Assignments are available to employees.

Contact: For individuals with previous experience: Ms. Jean Smith, Human Resources Representative; Phone: 517-548-4900; Fax: 517-545-3806.

PENSKE MOTORSPORTS, INC.

13400 WEST OUTER DRIVE
DETROIT, MICHIGAN 48239-1309

Overview: Owns and operates auto racing facilities and related promotional business. Established: 1995. Parent Company: Penske Corporation. Ownership: Private.

Key Statistics: Annual Sales: $55.2 million. Number of Employees: 184.

Expertise/Education Sought: Engineering

Alternatives to Full-Time Employment: Paid internships

Company-Paid Training is available.

International Assignments are available to employees.

Contact: Human Resources Acquisitions; Phone: 313-592-5000; Fax: 313-762-1031.

PEREGRINE, INC.

25200 TELEGRAPH ROAD
SOUTHFIELD, MICHIGAN 48086

Overview: Manufactures automotive stampings and body parts. Established: 1996. Ownership: Private.

Key Statistics: Annual Sales: $330.7 million. Number of Employees: 3,000.

Expertise/Education Sought: Automotive engineering

Alternatives to Full-Time Employment: Paid internships

Contact: Human Resources; Phone: 248-354-2100.

Company URL: http://www.peregrineinc.com

PLASTIPAK PACKAGING, INC.

9135 GENERAL COURT
PLYMOUTH, MICHIGAN 48170-4621

Overview: Manufactures plastic packaging. Established: 1956. Parent Company: Absopure. Ownership: Private.

Key Statistics: Annual Sales: $455 million. Number of Employees: 2,400.

Hiring History: Number of professional employees hired in 1997: 35. 1996: 30.

Expertise/Education Sought: Sales, marketing, accounting, data processing, distribution, manufacturing, quality control

Company-Paid Training is available.

International Assignments are available to employees.

Contact: Mr. Matt Leslie, Human Resources Representative; Phone: 734-455-3600; Fax: 734-354-7391. E-mail: hr00mjl%ppak@mcimail.com.

PRINCE CORPORATION

1 PRINCE CENTER
HOLLAND, MICHIGAN 49423-5407

Overview: Manufactures and distributes automotive trim and accessories. Established: 1989. Parent Company: Johnson Controls, Inc. Ownership: Public.

Prince Corporation (continued)

KEY STATISTICS: Annual Sales: $373.3 million. Number of Employees: 4,000.

EXPERTISE/EDUCATION SOUGHT: Accounting, finance, data processing, sales, automotive engineering

ALTERNATIVES TO FULL-TIME EMPLOYMENT: Paid internships

INTERNATIONAL ASSIGNMENTS are available to employees.

CONTACT: Mr. Robert Farnum, Manager of Staffing; Phone: 616-392-5151; Fax: 616-394-1080.

COMPANY URL: http://www.johnsoncontrols.com/prince.htm

REXAIR, INC.

3221 WEST BIG BEAVER ROAD
TROY, MICHIGAN 48084-2803

OVERVIEW: Manufactures home vacuum cleaners. Established: 1987. Parent Company: US Industries. Ownership: Public.

KEY STATISTICS: Annual Sales: $142.3 million. Number of Employees: 600.

HIRING HISTORY: Number of professional employees hired in 1997: 4. 1996: 10.

EXPERTISE/EDUCATION SOUGHT: Accounting, sales, computer-aided design, product design and development

COMPANY-PAID TRAINING is available.

CONTACT: For individuals with previous experience: Ms. Ethel Madden, Human Resources Director, 3221 West Big Beaver Road, Suite 2, Troy, Michigan, 48084-2803; Phone: 810-643-7222; Fax: 810-643-7676.

RIETER AUTOMOTIVE NORTH AMERICA, INC.

38555 HILLS TECH DRIVE
FARMINGTON HILLS, MICHIGAN 48331

OVERVIEW: Manufactures sound deadening materials for automobiles. Established: 1986. Parent Company: Rieter Automotive Systems. Ownership: Private.

KEY STATISTICS: Annual Sales: $250 million. Number of Employees: 2,000.

HIRING HISTORY: Number of professional employees hired in 1997: 18.

EXPERTISE/EDUCATION SOUGHT: Design engineering, automotive engineering, acoustic engineering

ALTERNATIVES TO FULL-TIME EMPLOYMENT: Paid internships

INTERNATIONAL ASSIGNMENTS are available to employees.

CONTACT: Mrs. Paula Gleckler, Human Resources Manager; Phone: 248-848-0100, ext. 305; Fax: 248-848-5170.

RIVIERA TOOL COMPANY

5460 EXECUTIVE PARKWAY, SE
GRAND RAPIDS, MICHIGAN 49512-5507

OVERVIEW: Manufactures specialty dies, tools, and jigs. Established: 1988. Ownership: Private.

KEY STATISTICS: Annual Sales: $22 million. Number of Employees: 140.

EXPERTISE/EDUCATION SOUGHT: Engineering, management information systems

ALTERNATIVES TO FULL-TIME EMPLOYMENT: Paid internships

CONTACT: For individuals with previous experience: Manager; Phone: 616-698-2100; Fax: 616-698-2470.

ROUGE STEEL COMPANY

3001 MILLER ROAD
DEARBORN, MICHIGAN 48121-1699

MAILING ADDRESS: PO Box 1699, Dearborn, MI 48120-1699

OVERVIEW: Manufactures steel and flat rolled steel products. Established: 1989. Ownership: Private.

KEY STATISTICS: Annual Sales: $1.3 billion. Number of Employees: 3,119.

HIRING HISTORY: Number of professional employees hired in 1997: 50. 1996: 50.

EXPERTISE/EDUCATION SOUGHT: Accounting, finance, mechanical engineering, electrical engineering, metallurgical engineering

COMPANY-PAID TRAINING is available.

CONTACT: Mr. Mel Baggett, Manager of Personnel, PO Box 1639, Dearborn, Michigan, 48121-1699; Phone: 313-322-1572; Fax: 313-337-4185.

SLOVCAN STEEL, INC. (NV)

26211 CENTRAL PARK BOULEVARD
SOUTHFIELD, MICHIGAN 48076-4107

OVERVIEW: Provider of metal distribution services. Established: 1997. Ownership: Private.

KEY STATISTICS: Annual Sales: $100 million.

EXPERTISE/EDUCATION SOUGHT: Engineering, milling engineering

CONTACT: For individuals with previous experience: Human Resources; Phone: 248-357-4862; Fax: 248-357-4951.

SPARTON CORPORATION

2400 EAST GANSON STREET
JACKSON, MICHIGAN 49202-3772

OVERVIEW: Manufactures military electronic equipment. Established: 1900. Ownership: Public.

KEY STATISTICS: Annual Sales: $142.7 million. Number of Employees: 1,600.

EXPERTISE/EDUCATION SOUGHT: Materials management

COMPANY-PAID TRAINING is available.

CONTACT: Ms. Brenda Gettings, Human Resource Supervisor; Phone: 517-787-8600; Fax: 517-787-8046.

COMPANY URL: http://www.sparton.com

SPX CORPORATION

700 TERRACE POINT ROAD
MUSKEGON, MICHIGAN 49440-3301

OVERVIEW: Manufactures automotive parts. Established: 1911. Ownership: Public.

KEY STATISTICS: Annual Sales: $1.1 billion. Number of Employees: 5,100.

EXPERTISE/EDUCATION SOUGHT: Accounting, operations, finance, marketing, data processing, electrical engineering, mechanical engineering, materials engineering

ALTERNATIVES TO FULL-TIME EMPLOYMENT: Part-time jobs, paid internships

INTERNATIONAL ASSIGNMENTS are available to employees.

CONTACT: Mr. Steve Lison, Vice President of Human Resources; Phone: 616-724-5107; Fax: 616-724-5130.

COMPANY URL: http://www.spx.com

THE STANDARD PRODUCTS COMPANY

2401 SOUTH GULLEY ROAD
DEARBORN, MICHIGAN 48124-2440

OVERVIEW: Manufactures automotive products. Established: 1927. Ownership: Public.

KEY STATISTICS: Annual Sales: $1.1 billion. Number of Employees: 10,350.

HIRING HISTORY: Number of professional employees hired in 1997: 102. 1996: 120.

EXPERTISE/EDUCATION SOUGHT: Technical engineering, mechanical engineering, electrical engineering, manufacturing engineering, management

ALTERNATIVES TO FULL-TIME EMPLOYMENT: Paid internships

COMPANY-PAID TRAINING is available.

INTERNATIONAL ASSIGNMENTS are available to employees.

CONTACT: Ms. Theresa Boegler, Corporate Manager of Employment; Phone: 313-791-2145; Fax: 313-561-8806.

COMPANY URL: http://www.stdproducts.com

STATE OF MICHIGAN

400 SOUTH PINE
LANSING, MICHIGAN 48909

OVERVIEW: Provides public services to Michigan residents including human services, natural resource conservation, law enforcement, public health, and state highways.

KEY STATISTICS: Number of Employees: 58,000.

EXPERTISE/EDUCATION SOUGHT: Computer programming, information systems, nursing, civil engineering, environmental science

ALTERNATIVES TO FULL-TIME EMPLOYMENT: Part-time jobs, paid internships

COMPANY-PAID TRAINING is available.

CONTACT: Information Office, Michigan Department of Civil Service, PO Box 30002, Lansing, Michigan, 48909; Phone: 800-877-1766; Fax: 517-373-3175.

COMPANY URL: http://www.mdcs.state.mi.us/

CORPORATE PROFILE

The Employer *The state of Michigan provides its employees with the daily opportunity to make a difference in the lives of Michigan residents. Whether in computer science or social science, civil engineering or civil rights, state government provides opportunities for meaningful public service. While helping others, employees help themselves to develop professionally through training opportunities and professional affiliations. As they develop, employees can move up in an organization that has more than 58,000 positions, including entry-level through specialist- and managerial-level positions. Benefits include a competitive salary; group health, dental, vision, and life insurance; retirement; 401(k); and tuition assistance for continuing education.*

Career Opportunities *The state of Michigan is recruiting candidates with degrees in a variety of fields, including but not limited to agriculture; computer science; civil, electrical, and environmental engineering; environmental science; geology and surveying; health care; nursing; and civil, drafting, and electrical technologies. The state of Michigan seeks information technology professionals with expertise in COBOL, SQL, Oracle, PowerBuilder, Visual Basic, IBM, Unisys, Bull/Honeywell, Windows NT, client/servers, LAN management, and more.*

Interested applicants should call for a college graduate or information technology information packet. The state of Michigan is an Equal Opportunity Employer that values diversity.

STEELCASE, INC.

901 44TH STREET, SE
GRAND RAPIDS, MICHIGAN 49508-7575

OVERVIEW: Manufactures and markets office furniture. Established: 1912. Ownership: Private.

KEY STATISTICS: Annual Sales: $2.9 billion. Number of Employees: 19,200.

HIRING HISTORY: Number of professional employees hired in 1996: 300.

EXPERTISE/EDUCATION SOUGHT: Sales, finance, marketing, manufacturing engineering

ALTERNATIVES TO FULL-TIME EMPLOYMENT: Paid internships

COMPANY-PAID TRAINING is available.

CONTACT: Mr. Thomas Dryer, Manager of Employment; Phone: 616-247-2209; Fax: 616-247-2256.

STRYKER CORPORATION

2725 FAIRFIELD ROAD
KALAMAZOO, MICHIGAN 49002-4085

OVERVIEW: Manufactures medical equipment and instruments. Established: 1938. Ownership: Public.

KEY STATISTICS: Annual Sales: $910.1 million. Number of Employees: 4,629.

EXPERTISE/EDUCATION SOUGHT: Accounting, sales, mechanical engineering, design engineering, electrical engineering, computer-aided design, manufacturing, marketing

ALTERNATIVES TO FULL-TIME EMPLOYMENT: Internships

INTERNATIONAL ASSIGNMENTS are available to employees.

CONTACT: For individuals with previous experience: Mr. David Huisjin, Office of Human Resources.

COMPANY URL: http://www.strykercorp.com

TCI CABLEVISION OF WEST MICHIGAN, INC.

3500 PATTERSON AVENUE, SE, SUITE A
GRAND RAPIDS, MICHIGAN 49512

MAILING ADDRESS: PO Box 128901, Grand Rapids, MI 49512

OVERVIEW: Operates local cable service. Established: 1984. Parent Company: TCI Cablevision. Ownership: Public.

KEY STATISTICS: Annual Sales: $93.8 million. Number of Employees: 700.

EXPERTISE/EDUCATION SOUGHT: Customer service and support, marketing, technology

CONTACT: For individuals with previous experience: Employee Relations Department; Phone: 616-977-2200; Fax: 616-977-2212.

TECUMSEH PRODUCTS COMPANY

100 EAST PATTERSON STREET
TECUMSEH, MICHIGAN 49286-2041

OVERVIEW: Manufactures refrigeration compressors and engines for lawn mowers and snow blowers. Established: 1930. Ownership: Public.

KEY STATISTICS: Annual Sales: $1.8 billion. Number of Employees: 16,300.

EXPERTISE/EDUCATION SOUGHT: Finance, human resources, sales, process engineering, technical engineering, mechanical engineering, electrical engineering

ALTERNATIVES TO FULL-TIME EMPLOYMENT: Paid internships

COMPANY-PAID TRAINING is available.

INTERNATIONAL ASSIGNMENTS are available to employees.

CONTACT: For individuals with previous experience: Mr. Thomas Ciminillo, Corporate Staff Representative; Phone: 517-423-8411, ext. 551; Fax: 517-423-8760.

TRIPLE S PLASTICS, INC.

14320 SOUTH PORTAGE ROAD
VICKSBURG, MICHIGAN 49097

OVERVIEW: Manufactures plastics. Established: 1969. Ownership: Public.

KEY STATISTICS: Annual Sales: $64.6 million. Number of Employees: 850.

EXPERTISE/EDUCATION SOUGHT: Plastics engineering, mechanical engineering, sales, marketing, finance, accounting

ALTERNATIVES TO FULL-TIME EMPLOYMENT: Paid internships

CONTACT: For individuals with previous experience: Mr. Phil Weaver, Corporate Director of Human Resources; Phone: 616-649-0545, ext. 2232; Fax: 616-649-3427.

TROY DESIGN, INC.

2653 INDUSTRIAL ROW
TROY, MICHIGAN 48084

OVERVIEW: Provides engineering design services. Established: 1977. Ownership: Private.

KEY STATISTICS: Annual Sales: $50 million. Number of Employees: 800.

HIRING HISTORY: Number of professional employees hired in 1996: 150.

EXPERTISE/EDUCATION SOUGHT: Electrical engineering, mechanical engineering, automotive engineering

CONTACT: For individuals with previous experience: Human Resources; Phone: 248-689-4900; Fax: 248-689-1109.

COMPANY URL: http://www.troydesign.com

TSK AMERICA, INC.

39205 COUNTRY CLUB DRIVE, SUITE C-26
FARMINGTON HILLS, MICHIGAN 48331

Overview: Wholesales industrial machinery and equipment. Established: 1995. Parent Company: Tokyo Seimitsu, Inc. Ownership: Private.

Key Statistics: Annual Sales: $30.4 million. Number of Employees: 120.

Hiring History: Number of professional employees hired in 1997: 20.

Expertise/Education Sought: Engineering, business, electrical engineering, assurance

Alternatives to Full-Time Employment: Part-time jobs

Company-Paid Training is available.

Contact: Ms. Karen Mikowski, Corporate Administrative Assistant; Phone: 248-489-5500; Fax: 248-489-5503.

Company URL: http://www.tsk.co.jp

UNITED TECHNOLOGIES AUTOMOTIVE

5200 AUTO CLUB DRIVE
DEARBORN, MICHIGAN 48126

Overview: Manufactures aviation and automotive equipment. Established: 1978. Parent Company: United Technologies Corporation. Ownership: Public.

Key Statistics: Annual Sales: $2.7 billion. Number of Employees: 38,000.

Expertise/Education Sought: Automotive engineering, mechanical engineering, electrical engineering, design engineering, marketing, accounting, manufacturing engineering, electronics/electronics engineering

Alternatives to Full-Time Employment: Part-time jobs, paid internships

International Assignments are available to employees.

Contact: Recruiting and Staffing; Phone: 313-593-9000; Fax: 313-593-9587.

Company URL: http://www.uta.com

U.S. ARMY TANK AUTOMOTIVE AND ARMAMENTS COMMAND (TACOM RDE CENTER)

COMMANDER US ARMY—TANK AND ARMAMENTS, ATTENTION: AMSTA-RM-P
WARREN, MICHIGAN 48397-5000

Overview: Provides advanced automotive technology services.

Expertise/Education Sought: Software engineering/development, electronics/electronics engineering, electrical engineering, materials engineering, ceramic engineering, computer science, chemical engineering, chemistry, physics

Contact: For individuals with previous experience: Mr. James Long, Human Resources Manager; Phone: 810-574-6655; Fax: 810-574-7599.

VARITY KELSEY-HAYES

12025 TECH CENTER DRIVE
LIVONIA, MICHIGAN 48150

Overview: Manufactures wheels and brakes for the automotive aftermarket. Established: 1992. Parent Company: Lucas Varity PLC. Ownership: Public.

Key Statistics: Annual Sales: $778.2 million. Number of Employees: 4,200.

Hiring History: Number of professional employees hired in 1997: 100. 1996: 100.

Expertise/Education Sought: Manufacturing engineering, design engineering, electrical engineering, mechanical engineering, software engineering/development

Alternatives to Full-Time Employment: Part-time jobs, paid internships

Company-Paid Training is available.

International Assignments are available to employees.

Contact: Mr. John Kovach, Vice President of Human Resources; Phone: 313-266-2600; Fax: 313-266-4603. E-mail: john_kovach@kh.com.

VISTEON AUTOMOTIVE SYSTEMS

5500 AUTO CLUB DRIVE
DEARBORN, MICHIGAN 48126

Overview: Designs, produces, and markets automotive systems. Established: 1996. Parent Company: Ford Motor Company. Ownership: Public.

Key Statistics: Annual Sales: $17 billion. Number of Employees: 75,000.

Hiring History: Number of professional employees hired in 1997: 500.

Expertise/Education Sought: Software engineering/development, embedded systems, electrical engineering, electronics/electronics engineering, mechanical engineering, process engineering, management information systems, sales, marketing, manufacturing management

Alternatives to Full-Time Employment: Part-time jobs, paid internships

Company-Paid Training is available.

International Assignments are available to employees.

Contact: Salaried Recruitment, PO Box 0520; Fax: 313-337-2967. E-mail: recruit2@ford.com.

Company URL: http://www.visteonet.com

WHIRLPOOL CORPORATION

2000 M-63 NORTH
BENTON HARBOR, MICHIGAN 49022

Overview: Manufactures consumer appliances. Established: 1911. Ownership: Public.

Key Statistics: Annual Sales: $8.7 billion. Number of Employees: 48,163.

Expertise/Education Sought: Engineering, management information systems, systems analysis, marketing, finance

Alternatives to Full-Time Employment: Part-time jobs, paid internships

International Assignments are available to employees.

Contact: Mr. Kent Price, Director of Management Resources; Phone: 616-923-2472.

WOLLIN PRODUCTS, INC.

4295 NORTH ROOSEVELT ROAD
STEVENSVILLE, MICHIGAN 49127-0159

Overview: Manufactures injection molded plastics. Established: 1995. Ownership: Private.

Key Statistics: Annual Sales: $33.6 million. Number of Employees: 450.

Expertise/Education Sought: Computer-aided design, plastics engineering

Contact: Mr. Stephen Aykroyd, Personnel Director; Phone: 616-429-3201; Fax: 616-429-6060.

MINNESOTA

ACCESS CASH INTERNATIONAL

4105 LEXINGTON AVENUE NORTH, SUITE 305
ARDEN HILLS, MINNESOTA 55126

Overview: Wholesales office equipment. Established: 1994. Ownership: Private.

Key Statistics: Number of Employees: 25.

Hiring History: Number of professional employees hired in 1997: 14. 1996: 10.

Expertise/Education Sought: Computer programming, data processing, information systems, management, management information systems, software engineering/development

Contact: For individuals with previous experience: Ms. Jan Gillespie, Marketing Director; Phone: 612-490-0413, ext. 302; Fax: 612-490-0545.

ADC TELECOMMUNICATIONS, INC.
12501 WHITEWATER DRIVE
MINNETKA, MINNESOTA 66343

Mailing Address: PO Box 1101, Minneapolis, MN 55440

Overview: Designs and markets cable management, transmission, and networking products. Established: 1952. Ownership: Public.

Key Statistics: Annual Sales: $828 million. Number of Employees: 4,620.

Hiring History: Number of professional employees hired in 1997: 400.

Expertise/Education Sought: Sales, accounting, management, electrical engineering, information engineering, mechanical engineering

Alternatives to Full-Time Employment: Part-time jobs, paid internships

Company-Paid Training is available.

International Assignments are available to employees.

Contact: Human Resources Department; Phone: 612-946-8080; Fax: 612-946-3292.

Company URL: http://www.adc.com

ADVANCED FLEX, INC.
15115 MINNETONKA INDUSTRIAL ROAD
MINNETONKA, MINNESOTA 55345

Overview: Manufactures printed circuit boards. Established: 1982. Ownership: Private.

Key Statistics: Annual Sales: $20.7 million. Number of Employees: 320.

Hiring History: Number of professional employees hired in 1997: 97.

Expertise/Education Sought: Process engineering, information systems, chemical engineering

Alternatives to Full-Time Employment: Paid internships

Company-Paid Training is available.

Contact: For individuals with previous experience: Ms. Kellie McLaughlin, Human Resources Representative; Phone: 612-930-4800; Fax: 612-935-4236. Job Hotline: 612-930-4901.

Company URL: http://www.afiweb.com

ALLIANT TECHSYSTEMS
600 SECOND STREET, NE
HOPKINS, MINNESOTA 55343-8367

Overview: Manufactures military hardware and navigation equipment. Established: 1990. Ownership: Public.

Key Statistics: Annual Sales: $1.1 billion. Number of Employees: 6,800.

Expertise/Education Sought: Mechanical engineering, electrical engineering, computer science, accounting, finance, management information systems, design engineering, software engineering/development

Alternatives to Full-Time Employment: Paid internships

Company-Paid Training is available.

Contact: Ms. Jeanne Jakel, Manager of Human Resource Services; Phone: 612-931-5752. E-mail: jeanne_jakel@atk.com.

Company URL: http://www.atk.com

AMERICAN MEDICAL SYSTEMS, INC.
10700 BREN ROAD WEST
MINNETONKA, MINNESOTA 55343-9679

Overview: Designs, manufactures and markets implantable urology products. Established: 1985. Parent Company: Pfizer, Inc. Ownership: Public.

Key Statistics: Annual Sales: $90 million. Number of Employees: 560.

Expertise/Education Sought: Biomedical engineering, physicians, research and development, public administration, analysis

Alternatives to Full-Time Employment: Paid internships

Company-Paid Training is available.

International Assignments are available to employees.

Contact: Mr. Ned Vandellen, Human Resources Administrator; Phone: 612-930-6181; Fax: 612-930-6514.

ANALYSTS INTERNATIONAL CORPORATION
7615 METRO BOULEVARD
MINNEAPOLIS, MINNESOTA 55439

Overview: Provides computer consulting, systems analysis, and programming services. Established: 1966. Ownership: Public.

Key Statistics: Annual Sales: $439.5 million. Number of Employees: 4,650.

Hiring History: Number of professional employees hired in 1996: 350.

Expertise/Education Sought: Software engineering/development, program analysis, project management, computer programming

Alternatives to Full-Time Employment: Part-time jobs

Contact: For individuals with previous experience: Recruiting; Phone: 612-835-5900.

Company URL: http://www.analysts.com

ANDERSEN CORPORATION
100 4TH AVENUE NORTH
BAYPORT, MINNESOTA 55003-1046

Overview: Manufactures windows, doors, and related accessories. Established: 1903. Ownership: Private.

Key Statistics: Annual Sales: $1.1 billion. Number of Employees: 3,700.

Hiring History: Number of professional employees hired in 1996: 60.

Expertise/Education Sought: Marketing, chemical engineering, computer programming, finance, design engineering, management information systems, accounting, computer science, mechanical engineering, human resources

Alternatives to Full-Time Employment: Paid internships

Company-Paid Training is available.

Contact: For individuals with previous experience: Manager of Corporate Staffing Services; Phone: 612-439-5150.

ARCTIC CAT, INC.
600 BROOKS AVENUE SOUTH
THIEF RIVER FALLS, MINNESOTA 56701

Overview: Manufactures transportation equipment. Established: 1982. Ownership: Public.

Key Statistics: Annual Sales: $468.6 million. Number of Employees: 1,722.

Hiring History: Number of professional employees hired in 1997: 200. 1996: 50.

Expertise/Education Sought: Accounting, finance, data processing, electrical engineering, mechanical engineering

Company-Paid Training is available.

International Assignments are available to employees.

Contact: For individuals with previous experience: Human Resources Department; Phone: 218-681-8558; Fax: 218-681-9859.

BERNAFON-MAICO, INC.
9675 WEST 76TH STREET
EDEN PRAIRIE, MINNESOTA 55344

Overview: Supplies surgical appliances and hearing aids. Established: 1995. Parent Company: Oticon. Ownership: Private.

Key Statistics: Annual Sales: $10.5 million. Number of Employees: 74.

Hiring History: Number of professional employees hired in 1997: 10.

Expertise/Education Sought: Electrical technology

Alternatives to Full-Time Employment: Paid internships

Company-Paid Training is available.

Contact: Ms. Gina Harms, Human Resources Administrator; Phone: 612-941-4200; Fax: 612-903-4200.

CATERPILLAR PAVING PRODUCTS, INC.
9401 85TH AVENUE NORTH
MINNEAPOLIS, MINNESOTA 55445

Overview: Manufactures concrete products and construction machinery. Established: 1985. Parent Company: Caterpillar, Inc. Ownership: Public.

Key Statistics: Annual Sales: $78.1 million. Number of Employees: 679.

Expertise/Education Sought: Mechanical engineering

Alternatives to Full-Time Employment: Paid internships

Contact: For individuals with previous experience: Mr. William Jetson, Manager of Human Resources; Phone: 613-493-1384; Fax: 613-493-1345.

CENEX, INC.
5500 CENEX DRIVE
INVER GROVE HEIGHTS, MINNESOTA 55077-1733

Mailing Address: PO Box 64089, St. Paul, MN 55164-0089

Overview: Manufactures fertilizers, farm supplies, and feed products. Established: 1927. Ownership: Private.

Key Statistics: Annual Sales: $2.7 billion. Number of Employees: 3,300.

Hiring History: Number of professional employees hired in 1996: 100.

Expertise/Education Sought: Marketing, sales, sales

Company-Paid Training is available.

International Assignments are available to employees.

Contact: Mr. Richard Baldwin, Director of Human Resources; Phone: 612-451-4511; Fax: 612-451-4694. E-mail: dbald@cnxlol.com. Job Hotline: 612-481-2250.

CERIDIAN CORPORATION
8100 34TH AVENUE SOUTH
MINNEAPOLIS, MINNESOTA 55425-1640

Overview: Provides transaction processing, computer systems integration and payroll services. Established: 1957. Ownership: Public.

Key Statistics: Annual Sales: $1.5 billion. Number of Employees: 10,800.

Expertise/Education Sought: Engineering

Alternatives to Full-Time Employment: Part-time jobs, paid internships

Company-Paid Training is available.

Contact: For individuals with previous experience: Human Resource Department; Phone: 612-853-8100; Fax: 612-853-5495.

Company URL: http://www.ceridian.com

DELUXE CORPORATION
3680 VICTORIA STREET NORTH
SHOREVIEW, MINNESOTA 55126-2966

Overview: Prints checks and forms and provides electronic payment systems. Established: 1915. Ownership: Public.

Key Statistics: Annual Sales: $1.9 billion. Number of Employees: 19,600.

Expertise/Education Sought: Electrical engineering, mechanical engineering, customer service and support, telemarketing

Alternatives to Full-Time Employment: Part-time jobs, paid internships

Company-Paid Training is available.

International Assignments are available to employees.

Contact: Ms. Lorraine Fernandez, Employment Recruiter; Phone: 612-481-4100; Fax: 612-481-4357. Job Hotline: 612-481-4100.

Company URL: http://www.deluxe.com

DYNEON, LLC
6744 33RD STREET NORTH
OAKDALE, MINNESOTA 55128

Overview: Produces custom compound purchased resins. Established: 1996. Parent Company: 3M. Ownership: Private.

Key Statistics: Annual Sales: $350 million. Number of Employees: 650.

Expertise/Education Sought: Technical

Alternatives to Full-Time Employment: Part-time jobs, paid internships

International Assignments are available to employees.

Contact: For individuals with previous experience: Human Resource Department; Phone: 612-737-4108; Fax: 612-575-3737.

Company URL: http://www.dyneon.com

ECOLAB, INC.
370 NORTH WABASHA STREET
SAINT PAUL, MINNESOTA 55102-1349

Overview: Provides equipment and services for housekeeping, nursing, and sanitizing. Established: 1923. Ownership: Public.

Key Statistics: Annual Sales: $1.5 billion. Number of Employees: 9,000.

Expertise/Education Sought: Accounting, chemical engineering, financial analysis, sales, marketing

Alternatives to Full-Time Employment: Paid internships

Company-Paid Training is available.

International Assignments are available to employees.

Contact: For college students and recent graduates: Mr. Pat Caldie, Director of Managment Development; Phone: 612-293-2288. For individuals with previous experience: Ms. Sue Metcalf, Director of Human Resources; Phone: 612-293-2089. Job Hotline: 612-293-4473.

Company URL: http://www.ecolab.com

GE CAPITAL—IT SOLUTIONS
5121 WINNETKA AVENUE
MINNEAPOLIS, MINNESOTA 55428

Overview: Markets computer products and services and emergency alert systems. Established: 1996. Parent Company: General Electric. Ownership: Public.

Key Statistics: Annual Sales: $1.5 billion. Number of Employees: 3,500.

Hiring History: Number of professional employees hired in 1997: 1,000.

Expertise/Education Sought: Accounting, marketing, warehousing, purchasing, management information systems, human resources

Company-Paid Training is available.

International Assignments are available to employees.

Contact: For individuals with previous experience: Human Resources; Phone: 612-557-2500; Fax: 612-551-5566.

Company URL: http://www.gecits.ge.com

GME CONSULTANTS, INC.
14000 21ST AVENUE NORTH
MINNEAPOLIS, MINNESOTA 55447

Overview: Provides consulting engineering services. Established: 1981. Ownership: Private.

Key Statistics: Annual Sales: $6.3 million. Number of Employees: 100.

Hiring History: Number of professional employees hired in 1997: 20. 1996: 20.

Expertise/Education Sought: Environmental, geotechnology, engineering

Alternatives to Full-Time Employment: Paid internships

Company-Paid Training is available.

Contact: For individuals with previous experience: Mr. Tom Venema, Vice President; Phone: 612-559-1859; Fax: 612-559-0720. E-mail: gme@gmeconsultant.com.

HICKORY TECH CORPORATION
221 EAST HICKORY STREET
MANKATO, MINNESOTA 56001-3610

Overview: Provides telephone services. Established: 1985. Ownership: Public.

Key Statistics: Annual Sales: $66.6 million. Number of Employees: 417.

Expertise/Education Sought: Software engineering/development, accounting, finance, data processing, construction engineering

Alternatives to Full-Time Employment: Paid internships

Company-Paid Training is available.

Contact: Ms. Janet McCullough, Human Resources Manager; Phone: 507-387-1852; Fax: 507-386-0700. Job Hotline: 800-229-6448.

Company URL: http://www.hickorytech.com

Honeywell

HONEYWELL TECHNOLOGY CENTER

3660 TECHNOLOGY DRIVE
MINNEAPOLIS, MINNESOTA 55418

Overview: Researches and develops advanced technologies and products. Established: 1885. Parent Company: Honeywell, Inc. Ownership: Public.

Key Statistics: Number of Employees: 55,000.

Expertise/Education Sought: Electrical engineering, computer science, hardware engineering, systems engineering

Alternatives to Full-Time Employment: Paid internships

Contact: Ms. Esther Saarela, Human Resources Representative; Phone: 612-951-7436; Fax: 612-951-7181.

CORPORATE PROFILE

The Honeywell Technology Center (HTC) is the company's primary research and development organization, responsible for developing and transferring advanced technology, development and productivity tools, components, architectures, and product and system concepts to its operating divisions. These developments provide Honeywell and Honeywell customers a market advantage and drive future generations of Honeywell products, systems, and service.

HTC concentrates in the areas of sensors—microsensors, integrated optic sensors, and sensor systems; information processing—signal processing architectures and systems, RF, data acquisition, and photonics; displays—devices and systems; controls—environmental and industrial; navigation—aerospace navigation, and guidance and control systems; and software—architecture and modeling, embedded, expert systems, frameworks and tools, and automation.

In addition to internal funding, HTC has a contract base that includes activities with government laboratories and agencies, industry consortia, NASA, aerospace primes, and a variety of commercial companies that have technology objectives and requirements compatible with our internally driven efforts.

The Honeywell Technology Center seeks employees with master's degrees in aerospace engineering, computer science, electrical engineering, mathematics, optics, and physics. Doctoral specialties sought are aerospace engineering, computer science, electrical engineering, mathematics, optics, physics, software engineering, solid-state physics, material science, and chemical engineering.

HTC is headquartered in Minneapolis, Minnesota and has a facility in Phoenix, Arizona.

HUTCHINSON TECHNOLOGY, INC.

40 WEST HIGHLAND PARK DRIVE
HUTCHINSON, MINNESOTA 55350-9783

Overview: Manufactures suspension assemblies for disk drives. Established: 1965. Ownership: Public.

Key Statistics: Annual Sales: $453.2 million. Number of Employees: 5,479.

Hiring History: Number of professional employees hired in 1997: 2,000. 1996: 10,000.

Expertise/Education Sought: Mechanical engineering, electrical engineering, industrial engineering, chemical engineering, quality engineering

Alternatives to Full-Time Employment: Paid internships

Company-Paid Training is available.

Contact: Mr. Doug D. Anderson, Corporate Staffing; Phone: 320-587-1972; Fax: 320-587-1290.

IMATION CORPORATION

1 IMATION PLACE
OAKDALE, MINNESOTA 55128-3414

Overview: Manufactures magnetic disks, drums, and magnetic tape. Established: 1996. Ownership: Public.

Key Statistics: Annual Sales: $2.3 billion. Number of Employees: 9,400.

Hiring History: Number of professional employees hired in 1997: 200. 1996: 202.

Expertise/Education Sought: Engineering, chemical engineering, computer science, customer service and support, design engineering, electrical engineering, management information systems, manufacturing engineering, mechanical engineering, software engineering/development

Alternatives to Full-Time Employment: Part-time jobs, paid internships

International Assignments are available to employees.

Contact: For individuals with previous experience: Staffing Department; Phone: 612-704-4000; Fax: 612-704-4171. Job Hotline: 612-704-5627.

Company URL: http://www.imation.com

IMI CORNELIUS, INC.

1 CORNELIUS PLACE
ANOKA, MINNESOTA 55303-1583

Overview: Manufactures beverage dispensing equipment. Established: 1981. Parent Company: Imperial Metal, Inc. Ownership: Private.

Key Statistics: Annual Sales: $450 million. Number of Employees: 3,500.

Hiring History: Number of professional employees hired in 1996: 35.

Expertise/Education Sought: Accounting, marketing, finance, sales, electronics/electronics engineering, mechanical engineering, design engineering, information systems

Alternatives to Full-Time Employment: Paid internships

Company-Paid Training is available.

International Assignments are available to employees.

Contact: For individuals with previous experience: Staffing Administrator; Phone: 612-422-3076.

LECTEC CORPORATION

10701 RED CIRCLE DRIVE
MINNETONKA, MINNESOTA 55343-9136

Overview: Manufactures electrodes, polymers, and transdermal drug delivery systems. Established: 1977. Ownership: Public.

Key Statistics: Annual Sales: $13.1 million. Number of Employees: 88.

Expertise/Education Sought: Chemical engineering, engineering

Alternatives to Full-Time Employment: Part-time jobs

Contact: For individuals with previous experience: Ms. Alice Ong, Human Resources.

Company URL: http://www.lectec.com

MANFUFACTURERS SERVICES, LTD.

4300 ROUND LAKE ROAD WEST
SAINT PAUL, MINNESOTA 55112-3923

Overview: Manufactures computer peripherals. Established: 1994. Ownership: Public.

Key Statistics: Annual Sales: $43.1 million. Number of Employees: 400.

Hiring History: Number of professional employees hired in 1997: 50.

Expertise/Education Sought: Engineering, mechanical engineering, accounting

Alternatives to Full-Time Employment: Paid internships

Company-Paid Training is available.

International Assignments are available to employees.

Contact: For individuals with previous experience: Human Resource Department; Phone: 612-604-2476; Fax: 612-604-2420.

MILLER WASTE MILLS, INC.

580 EAST FRONT STREET
WINONA, MINNESOTA 55987-4256

OVERVIEW: Manufactures plastic pellets and reinforced thermoplastics. Established: 1927. Ownership: Private.

KEY STATISTICS: Annual Sales: $125 million. Number of Employees: 600.

EXPERTISE/EDUCATION SOUGHT: Plastics, marketing, sales

ALTERNATIVES TO FULL-TIME EMPLOYMENT: Internships

CONTACT: Mr. Richard Kulas, Personnel Manager; Phone: 507-454-6900; Fax: 507-454-2089.

MINNESOTA CORN PROCESSORS

901 NORTH HIGHWAY 59
MARSHALL, MINNESOTA 56258

OVERVIEW: Wholesales corn. Established: 1995. Ownership: Private.

HIRING HISTORY: Number of professional employees hired in 1997: 9. 1996: 125.

EXPERTISE/EDUCATION SOUGHT: Engineering, management, production management

ALTERNATIVES TO FULL-TIME EMPLOYMENT: Paid internships

COMPANY-PAID TRAINING is available.

CONTACT: For individuals with previous experience: Human Resources Department.

MINNESOTA DEPARTMENT OF TRANSPORTATION

395 JOHN IRELAND BOULEVARD, MS200
SAINT PAUL, MINNESOTA 55155-1899

OVERVIEW: Provides transportation management services. Established: 1871.

KEY STATISTICS: Number of Employees: 5,000.

HIRING HISTORY: Number of professional employees hired in 1997: 88.

EXPERTISE/EDUCATION SOUGHT: Civil engineering, environmental engineering, structural engineering, electrical engineering, research and development

COMPANY-PAID TRAINING is available.

CONTACT: Ms. Jolene Forman, Graduate Engineer Program Coordinator; Phone: 612-296-1719; Fax: 612-297-7944. E-mail: jef100@dot.state.mn.us.

COMPANY URL: http://www.dot.state.mn.us

CORPORATE STATEMENT

On the cutting edge of many technologies, Mn/DOT strives to provide high-quality transportation services for the citizens of Minnesota.

Mn/DOT is continually working to create a more diverse workplace by attracting more women and members of minority groups through engineering recruitment.

MINNESOTA MINING AND MANUFACTURING

3M CENTER, BUILDING 224-1W-02
SAINT PAUL, MINNESOTA 55144

OVERVIEW: Manufactures adhesives, polymers, coatings, surgical supplies, and digital information and imaging products. Established: 1902. Ownership: Public.

KEY STATISTICS: Annual Sales: $14.2 billion. Number of Employees: 74,289.

HIRING HISTORY: Number of professional employees hired in 1997: 1,000.

EXPERTISE/EDUCATION SOUGHT: Accounting, finance, computer science, information technology, marketing

ALTERNATIVES TO FULL-TIME EMPLOYMENT: Part-time jobs, paid internships

COMPANY-PAID TRAINING is available.

INTERNATIONAL ASSIGNMENTS are available to employees.

CONTACT: For individuals with previous experience: Staffing Department; Phone: 612-736-6006.

COMPANY URL: http://www.mmm.com

MTS SYSTEMS CORPORATION

14000 TECHNOLOGY DRIVE
EDEN PRAIRIE, MINNESOTA 55344-2290

OVERVIEW: Manufactures measuring and controlling devices. Established: 1963. Ownership: Public.

KEY STATISTICS: Annual Sales: $261 million. Number of Employees: 1,725.

EXPERTISE/EDUCATION SOUGHT: Software engineering/development, mechanical engineering, electrical engineering, computer science, systems engineering, applications engineering

ALTERNATIVES TO FULL-TIME EMPLOYMENT: Part-time jobs, paid internships

COMPANY-PAID TRAINING is available.

INTERNATIONAL ASSIGNMENTS are available to employees.

CONTACT: Mr. Charles Cox, Employment Coordinator; Phone: 612-937-4172; Fax: 612-937-4929. E-mail: infor@mts.com.

COMPANY URL: http://www.mts.com

NORTHWEST AIRLINES, INC.

SAINT PAUL INTERNATIONAL AIRPORT
SAINT PAUL, MINNESOTA 55111-3034

OVERVIEW: Provides passenger airline services. Established: 1989. Parent Company: NWA, Inc. Ownership: Public.

KEY STATISTICS: Annual Sales: $9.7 billion. Number of Employees: 45,000.

EXPERTISE/EDUCATION SOUGHT: Aeronautical engineering, aviation, electronics/electronics engineering, flight attendant

ALTERNATIVES TO FULL-TIME EMPLOYMENT: Paid internships

INTERNATIONAL ASSIGNMENTS are available to employees.

CONTACT: For individuals with previous experience: Mr. Perliss Johnson, Director of Corporate Human Resources, 5101 Northwest Drive, Saint Paul, Minnesota, 55111-3034; Phone: 612-726-7273; Fax: 612-726-6646.

PLASTICS, INC.

224 RYAN AVENUE
SAINT PAUL, MINNESOTA 55102-0830

OVERVIEW: Manufactures food storage containers, microwave cookware and other cookware products. Established: 1987. Parent Company: Newell Company. Ownership: Public.

KEY STATISTICS: Annual Sales: $33.2 million. Number of Employees: 445.

HIRING HISTORY: Number of professional employees hired in 1997: 31. 1996: 30.

EXPERTISE/EDUCATION SOUGHT: Accounting, finance, data processing, mechanical engineering, industrial engineering, marketing

ALTERNATIVES TO FULL-TIME EMPLOYMENT: Paid internships

COMPANY-PAID TRAINING is available.

CONTACT: For college students and recent graduates: Ms. Julie Gast, Human Resource Representative; Phone: 612-229-5349; Fax: 612-229-5477. For individuals with previous experience: Mr. Andy Killebrew, Vice President of Human Resources; Phone: 612-229-5374; Fax: 612-229-5477. E-mail: andy.killebrew@anchorhockingplastics.com.

POLARIS INDUSTRIES, INC.

1225 HIGHWAY 169 NORTH
MINNEAPOLIS, MINNESOTA 55441

OVERVIEW: Manufactures snowmobiles and all terrain vehicles for recreational and utility use. Established: 1994. Ownership: Public.

KEY STATISTICS: Annual Sales: $1.2 billion. Number of Employees: 3,500.

HIRING HISTORY: Number of professional employees hired in 1997: 200. 1996: 200.

EXPERTISE/EDUCATION SOUGHT: Mechanical engineering, industrial engineering, electrical engineering, program analysis, administration

Alternatives to Full-Time Employment: Part-time jobs, internships
Contact: Ms. Mary Zins, Manager of Human Resources; Phone: 612-542-0500; Fax: 612-542-0599.
Company URL: http://www.polarisindustries.com

RAMSEY TECHNOLOGY, INC.
501 90TH AVENUE, NW
MINNEAPOLIS, MINNESOTA 55433-8005

Overview: Provides inspection and testing services. Established: 1994. Ownership: Public.
Key Statistics: Annual Sales: $70 million. Number of Employees: 425.
Expertise/Education Sought: Engineering, customer service and support
Alternatives to Full-Time Employment: Paid internships
Company-Paid Training is available.
Contact: For individuals with previous experience: Human Resources Department; Phone: 612-783-2601; Fax: 612-783-2525.

SCIMED LIFE SYSTEMS, INC.
1 SCIMED PLACE
MAPLE GROVE, MINNESOTA 55311-1566

Overview: Manufactures medical devices. Established: 1995. Parent Company: Boston Scientific Corporation. Ownership: Public.
Key Statistics: Annual Sales: $300 million. Number of Employees: 1,848.
Hiring History: Number of professional employees hired in 1997: 100. 1996: 35.
Expertise/Education Sought: Engineering, mechanical engineering, chemical engineering, research and development, biology, manufacturing engineering, quality engineering
Alternatives to Full-Time Employment: Paid internships
Company-Paid Training is available.
Contact: Staffing Professions; Phone: 612-494-1700; Fax: 612-494-2290. Job Hotline: 612-494-2600, ext. 5627.
Company URL: http://www.bsci.com

SHORT-ELLIOTT-HENDRICKSON, INC.
3535 VADNAIS CENTER DRIVE
SAINT PAUL, MINNESOTA 55110

Overview: Provides architecture and construction engineering services. Established: 1979. Ownership: Private.
Key Statistics: Annual Sales: $25 million. Number of Employees: 290.
Hiring History: Number of professional employees hired in 1997: 25. 1996: 25.
Expertise/Education Sought: Civil engineering, environmental engineering, technical engineering, water resources/quality, waste management
Alternatives to Full-Time Employment: Paid internships
Company-Paid Training is available.
Contact: Ms. Joan Johnson, Manager of Human Resources; Phone: 612-490-2000; Fax: 612-490-2150.
Company URL: http://www.sehinc.com

SMARTE CARTE, INC.
4455 WHITE BEAR PARKWAY
SAINT PAUL, MINNESOTA 55110-7626

Overview: Provides cart, locker, and stroller services in airports, shopping centers, and entertainment facilities. Established: 1996. Parent Company: SMC Holdings Corporation. Ownership: Private.
Key Statistics: Annual Sales: $56.1 million. Number of Employees: 1,000.
Hiring History: Number of professional employees hired in 1996: 200.
Expertise/Education Sought: Administration, marketing
Alternatives to Full-Time Employment: Paid internships
Company-Paid Training is available.
International Assignments are available to employees.
Contact: For individuals with previous experience: Ms. Kay Reissner, Recruiting Specialist; Phone: 612-429-3614; Fax: 612-426-0927.

STARKEY LABORATORIES, INC.
6700 WASHINGTON AVENUE SOUTH
EDEN PRAIRIE, MINNESOTA 55344

Overview: Manufactures custom made hearing aids. Established: 1970. Ownership: Private.
Key Statistics: Annual Sales: $303.1 million. Number of Employees: 3,450.
Hiring History: Number of professional employees hired in 1996: 35.
Expertise/Education Sought: Electronics/electronics engineering, audio engineering, research and development, mechanical engineering, electrical engineering, process engineering, design engineering, product design and development, quality control
Alternatives to Full-Time Employment: Paid internships
Company-Paid Training is available.
Contact: For college students and recent graduates: Ms. Paulette Daniel, Human Resources; Phone: 612-941-6401; Fax: 612-828-6944. For individuals with previous experience: Mr. Larry Miller, Director of Human Resources; Phone: 612-941-6401; Fax: 612-828-6944. Job Hotline: 800-328-8604.

SUNGARD FINANCIAL SYSTEMS, INC.
601 2ND AVENUE SOUTH
HOPKINS, MINNESOTA 55343

Overview: Provides bookkeeping services. Established: 1990. Parent Company: Sungard Data, Inc. Ownership: Public.
Key Statistics: Annual Sales: $35.6 million. Number of Employees: 565.
Expertise/Education Sought: Accounting, finance, data processing, investment
Alternatives to Full-Time Employment: Part-time jobs, paid internships
Company-Paid Training is available.
Contact: For individuals with previous experience: Human Resources Department; Phone: 612-935-3300; Fax: 612-936-8888.
Company URL: http://www.sungard.com

THERMO KING CORPORATION
314 WEST 90TH STREET
MINNEAPOLIS, MINNESOTA 55420-3630

Overview: Designs, develops, and manufactures transport temperature control equipment. Established: 1961. Parent Company: Ingersoll-Rand Company. Ownership: Public.
Key Statistics: Annual Sales: $1.1 billion. Number of Employees: 4,300.
Hiring History: Number of professional employees hired in 1996: 100.
Expertise/Education Sought: Accounting, finance, marketing, sales, mechanical engineering, electrical engineering, industrial engineering
Alternatives to Full-Time Employment: Paid internships
Company-Paid Training is available.
International Assignments are available to employees.
Contact: Human Relations Department; Phone: 612-887-2200; Fax: 612-885-3404. E-mail: human_resources@thermoking.com.

THE TORO COMPANY
8111 LYNDALE AVENUE SOUTH
MINNEAPOLIS, MINNESOTA 55420-1136

Overview: Manufactures turf and lawn care machinery. Established: 1914. Ownership: Public.
Key Statistics: Annual Sales: $930.9 million. Number of Employees: 4,579.
Hiring History: Number of professional employees hired in 1997: 164. 1996: 100.
Expertise/Education Sought: Mechanical engineering, electrical engineering, marketing, accounting, finance, law, data processing
Alternatives to Full-Time Employment: Paid internships
Contact: For individuals with previous experience: Ms. Sherry Bromely, Senior Staffing Manager; Phone: 612-887-8943; Fax: 612-887-7961. E-mail: sherry.bromley@toro.com. Job Hotline: 612-887-7300.
Company URL: http://www.toro.com

TWIN CITY OPTICAL COMPANY
5205 HIGHWAY 169 NORTH
MINNEAPOLIS, MINNESOTA 55442-3900

Overview: Manufactures eyeglasses. Established: 1994. Parent Company: The Churchill Companies. Ownership: Private.
Key Statistics: Annual Sales: $22.1 million. Number of Employees: 350.
Expertise/Education Sought: Sales, marketing, laboratory technology
Alternatives to Full-Time Employment: Paid internships
Company-Paid Training is available.
Contact: For individuals with previous experience: Human Resources Department; Phone: 612-551-2000.

VALSPAR CORPORATION
1101 SOUTH 3RD STREET
MINNEAPOLIS, MINNESOTA 55415

Overview: Manufactures and distributes paints and coatings. Established: 1806. Ownership: Public.
Key Statistics: Annual Sales: $859.8 million. Number of Employees: 2,855.
Expertise/Education Sought: Sales, marketing, finance, chemistry, chemical engineering, mechanical engineering, accounting
Alternatives to Full-Time Employment: Paid internships
International Assignments are available to employees.
Contact: Mr. Paul Mason, Manager of Staffing; Phone: 612-332-7371.
Company URL: http://www.valspar.com

VISION-EASE LENS, INC.
7100 NORTHLAND CIRCLE NORTH
MINNEAPOLIS, MINNESOTA 55428-1500

Overview: Provides business services. Established: 1996. Parent Company: BMC Industries. Ownership: Public.
Key Statistics: Annual Sales: $16 million. Number of Employees: 500.
Expertise/Education Sought: Optical engineering, marketing, sales, research and development
Contact: Human Resources; Phone: 800-633-5829; Fax: 612-536-8789.

WEST GROUP
610 OPPERMAN DRIVE-PO BOX 64526
SAINT PAUL, MINNESOTA 55164-0526

Overview: Provides information services. Established: 1982. Parent Company: Thomson Corporation. Ownership: Private.
Key Statistics: Annual Sales: $32 million. Number of Employees: 450.
Expertise/Education Sought: Production, engineering
Alternatives to Full-Time Employment: Paid internships
Company-Paid Training is available.
International Assignments are available to employees.
Contact: Human Resources; Phone: 612-687-4357; Fax: 612-687-5827. Job Hotline: 612-687-8980.

MISSISSIPPI

AUTOMATED HEALTH SYSTEMS, INC.
3545 LAKELAND DRIVE
JACKSON, MISSISSIPPI 39208

Mailing Address: PO Box 1043, Jackson, MS 39215
Overview: Develops and markets computer software for health-care facilities. Parent Company: Blue Cross and Blue Shield of Mississippi.
Hiring History: Number of professional employees hired in 1997: 3.
Expertise/Education Sought: Software engineering/development
Company-Paid Training is available.
Contact: For individuals with previous experience: Ms. Kelly Helmick, Recruiting, PO Box 1043, Jackson, Mississippi, 39215; Phone: 601-532-3704; Fax: 601-536-5886.

BERNSTEIN AND SONS SHIRT CORPORATION
206 WEST RAILROAD AVENUE
CRYSTAL SPRINGS, MISSISSIPPI 39059

Overview: Manufactures women's blouses and shirts. Established: 1920. Ownership: Private.
Key Statistics: Annual Sales: $27.5 million. Number of Employees: 750.
Expertise/Education Sought: Manufacturing
Contact: For individuals with previous experience: Ms. Edith Barnes, Human Resources Assistant Manager; Phone: 601-892-4551, ext. 126; Fax: 601-892-1109.

HEATCRAFT, INC.
3984 HIGHWAY 51 SOUTH
GRENADA, MISSISSIPPI 38901

Mailing Address: PO Box 948, Grenada, MS 38902
Overview: Manufactures heating and air conditioning systems. Established: 1986. Parent Company: Lennox International, Inc. Ownership: Private.
Key Statistics: Annual Sales: $230 million. Number of Employees: 3,300.
Expertise/Education Sought: Accounting, computer programming, finance, sales, marketing, mechanical engineering, electrical engineering, design engineering, distribution
Alternatives to Full-Time Employment: Paid internships
Company-Paid Training is available.
Contact: For individuals with previous experience: Mr. Andy Carter, Human Resources Director, PO Box 799900, Grenada, Mississippi, 38901; Phone: 601-229-2000; Fax: 601-229-2288.

INGALLS SHIPBUILDING, INC.
1000 ACCESS ROAD
PASCAGOULA, MISSISSIPPI 39568

Overview: Constructs ships. Established: 1968. Parent Company: Litton Industries, Inc. Ownership: Public.
Key Statistics: Annual Sales: $1.1 billion. Number of Employees: 10,000.
Hiring History: Number of professional employees hired in 1997: 200.
Expertise/Education Sought: Mechanical engineering, electrical engineering, design engineering, industrial engineering, computer science
Alternatives to Full-Time Employment: Part-time jobs
Contact: For individuals with previous experience: Ms. Sandy Delia, Human Resources Recruiter; Phone: 601-935-1269; Fax: 601-935-2741. E-mail: DeliaSM@Ingalls.com.
Company URL: http://www.ingalls.com

PIPER IMPACT, INC.
HIGHWAY 15 NORTH
NEW ALBANY, MISSISSIPPI 38652

Overview: Manufactures metal extrusions. Established: 1986. Parent Company: Quanex. Ownership: Public.
Key Statistics: Annual Sales: $120 million. Number of Employees: 1,200.
Expertise/Education Sought: Accounting, data processing, mechanical engineering, electrical engineering, chemical engineering, design engineering
Contact: For individuals with previous experience: Ms. Nancy Carson, Director of Human Resources; Phone: 601-534-5046; Fax: 601-534-5016.

RAYTHEON AEROSPACE

555 INDUSTRIAL DRIVE SOUTH
MADISON, MISSISSIPPI 39110-9072

Overview: Provides aircraft cleaning and janitorial services. Established: 1977. Parent Company: Raytheon Aircraft Company. Ownership: Public.

Key Statistics: Annual Sales: $437.3 million. Number of Employees: 4,700.

Hiring History: Number of professional employees hired in 1997: 100. 1996: 150.

Expertise/Education Sought: Avionics

Alternatives to Full-Time Employment: Paid internships

International Assignments are available to employees.

Contact: For individuals with previous experience: Mr. Fred Hann, Personnel Specialist; Phone: 601-856-2274, ext. 422; Fax: 601-856-8006.

WORLDCOM, INC.

515 EAST AMITE STREET
JACKSON, MISSISSIPPI 39201

Overview: Provides transmission services. Established: 1994. Ownership: Public.

Key Statistics: Annual Sales: $221.8 million.

Expertise/Education Sought: Telecommunications, sales, information systems, engineering, finance

Alternatives to Full-Time Employment: Paid internships

Company-Paid Training is available.

International Assignments are available to employees.

Contact: For college students and recent graduates: Ms. Alana Rushing, College Recruiter; Phone: 601-360-8586; Fax: 601-974-8269. E-mail: alana.rushing@wcom.com. For individuals with previous experience: Ms. Derindea Phillips, Corporate Recruiter; Phone: 601-360-8069; Fax: 601-974-8269. E-mail: derindea.phillips@wcom.com. Job Hotline: 601-360-8987.

Company URL: http://www.wcom.com

MISSOURI

A.B. CHANCE COMPANY

210 NORTH ALLEN STREET
CENTRALIA, MISSOURI 65240-1302

Overview: Manufactures electrical switchgear apparatus. Established: 1994. Parent Company: Hubbell, Inc. Ownership: Public.

Key Statistics: Annual Sales: $160 million. Number of Employees: 1,100.

Expertise/Education Sought: Mechanical engineering, electrical engineering

Alternatives to Full-Time Employment: Paid internships

International Assignments are available to employees.

Contact: Ms. Areta Mayes, Manager of Employment and Benefits; Phone: 314-682-5521.

ACF INDUSTRIES, INC.

620 NORTH 2ND STREET
SAINT CHARLES, MISSOURI 63301-5418

Overview: Manufactures and leases railroad equipment and freight cars. Established: 1984. Ownership: Private.

Key Statistics: Annual Sales: $619.7 million. Number of Employees: 2,400.

Hiring History: Number of professional employees hired in 1996: 10.

Expertise/Education Sought: Accounting, engineering, computer programming, finance

Alternatives to Full-Time Employment: Paid internships

Contact: For individuals with previous experience: Ms. Dotty O'Brien, Employee Relations Specialist; Phone: 314-940-5156; Fax: 314-940-5169.

AMERENUE

1901 CHOUTEAU AVENUE
SAINT LOUIS, MISSOURI 63166-6149

Overview: Electric utility. Established: 1881. Parent Company: Ameren Corporation. Ownership: Public.

Key Statistics: Annual Sales: $2 billion. Number of Employees: 8,000.

Hiring History: Number of professional employees hired in 1997: 39.

Expertise/Education Sought: Electrical engineering, systems analysis, computer science, mechanical engineering, computer programming

Contact: Ms. Johnetta Carver, Staffing Supervisor; Phone: 314-554-2119; Fax: 314-206-0175. E-mail: johnetta_carver@ameren.com.

Company URL: http://www.ameren.com

CORPORATE PROFILE

For more than 100 years, Ameren has generated energy for millions of customers in Missouri and Illinois. The Ameren Corporation is a holding company created by the merger of the former Union Electric Company (now AmerenUE) and Central Illinois Public Service Company (now AmerenCIPS). Ameren owns and operates fourteen generating plants. Energy operations such as these enable Ameren to be one of the nation's largest investor-owned utilities. The Callaway County Nuclear Power Plant, currently in operation, emphasizes the company's commitment to supplying efficient energy for today and for future generations. Individuals interested in comprehensive engineering programs are offered a variety of assignments at Ameren. The **Engineering and Construction** *function comprises four major departments: Electrical Engineering, Construction and Services, Power Plant Design Engineering, and Transmission Planning. Engineers serving in these departments are personally involved in specific projects, often working on one particular project from inception to completion.*

Power Operations *is responsible for supplying Ameren customers with power. Duties include sales and economic operations and equipment maintenance. Power Operations departments are Power Plants, Energy Supply, and Fossil Fuels.*

Responsibilities in **Customer Services** *cover a wide spectrum of engineering work. Skilled professionals plan, design, operate, and maintain Ameren's transmission and distribution systems.*

Departments in the **Nuclear** *function are Nuclear Operations, Nuclear Engineering, Nuclear Services, and Licensing and Fuels. Skilled visionaries in this area provide the power of the future to effectively counter the increase in consumer demand.*

Corporate Planning *also covers a variety of different functions. These include Business Analysis, Corporate Analysis, Load Analysis and Forecasting, Energy Services, and Rate Engineering.*

Information Technology *provides for the operations programming, organizing, and planning of Ameren's computer network and telecommunications system. This includes the areas of Administration, Data Processing, Development, Planning and Support, and Telecommunications.*

As a private enterprise entrusted with an essential public service, Ameren has the responsibility to offer fair wages and job satisfaction, provide reliable electric service at a fair rate to customers, and make a fair return for investors. Professionals choose employment at Ameren because it provides comprehensive engineering and computer programs, excellent salaries and benefits, and a work environment that encourages individual input and application.

ANGELICA CORPORATION
424 SOUTH WOODSMILL ROAD
CHESTERFIELD, MISSOURI 63017

OVERVIEW: Manufactures uniforms, rents textiles for hospitals, and markets nursing uniforms and shoes.

EXPERTISE/EDUCATION SOUGHT: Manufacturing engineering, mechanical engineering, textiles, accounting, marketing, sales, manufacturing

ALTERNATIVES TO FULL-TIME EMPLOYMENT: Paid internships

COMPANY-PAID TRAINING is available.

INTERNATIONAL ASSIGNMENTS are available to employees.

CONTACT: For individuals with previous experience: Human Relations Department.

ASSOCIATED ELECTRIC COOPERATIVE
2814 SOUTH GOLDEN AVENUE
SPRINGFIELD, MISSOURI 65807-3213

OVERVIEW: Provides electric services. Established: 1962. Ownership: Private.

KEY STATISTICS: Annual Sales: $556.6 million. Number of Employees: 720.

HIRING HISTORY: Number of professional employees hired in 1997: 12. 1996: 12.

EXPERTISE/EDUCATION SOUGHT: Electrical engineering, accounting, finance, computer programming, mechanical engineering, auditing

ALTERNATIVES TO FULL-TIME EMPLOYMENT: Paid internships

COMPANY-PAID TRAINING is available.

CONTACT: Ms. Kristina Galbraith, Human Resources Generalist; Phone: 417-881-1204; Fax: 417-885-9252. E-mail: kgalbraith@aeci.org.

COMPANY URL: http://www.aeci.org

BLACK & VEATCH

BLACK & VEATCH
8400 WARD PARKWAY
KANSAS CITY, MISSOURI 64114

MAILING ADDRESS: PO Box 8405, Kansas City, MO 64114-0405

OVERVIEW: Provides full service engineering, procurement, and construction services. Established: 1915.

KEY STATISTICS: Annual Sales: $1.8 billion. Number of Employees: 8,000.

EXPERTISE/EDUCATION SOUGHT: Engineering, information technology

ALTERNATIVES TO FULL-TIME EMPLOYMENT: Internships

INTERNATIONAL ASSIGNMENTS are available to employees.

CONTACT: For college students and recent graduates: Ms. Katy May, College Recruiter. For individuals with previous experience: Mr. Joe Sewards, Employment Director.

COMPANY URL: http://www.bv.com

CORPORATE PROFILE

Founded in 1915, Black & Veatch is a partnership of professional engineers and architects that ranks among the largest design firms in the United States. Its professionals use their skills and talents to provide effective responses to environmental, economic, building, and energy needs. The firm is headquartered in Kansas City, Missouri, with offices around the world.

Because of the diverse nature of the firm, numerous career experiences exist worldwide in the power, infrastructure, and process businesses as well as with various administrative groups. Black & Veatch provides complete engineering, architectural, financial, and management services, including studies, facilities design, and construction management. Black & Veatch also provides hazardous waste management and building design and construction services and develops cogeneration and independent power projects for utilities and industry.

The Process Division (Black & Veatch Pritchard, Inc.) continues to dominate the market in sulfur recovery and maintains a significant market share in natural gas and natural gas liquids processing. Its low cost, highly flexible, and reliable gas processing facilities are in increasing demand as the process industries respond to intense competition.

Black & Veatch has completed many successful and award-winning projects, including the 120 MW natural gas-fired simple-cycle combustion turbine Gadong Power Station Extension in Brunei, which was recognized as "Project of the Year" by Power Engineering *and* Power Engineering International *magazines; an industrial process treatment facility in Chandler, Arizona, which received the American Academy of Environmental Engineers' "Best of Show"; and a groundwater treatment facility in Muskegon, Michigan, which received an "Award of Excellence" from the Associated Builders and Contractors, Inc., Western Michigan Chapter.*

Since its establishment, Black & Veatch has been entrusted with more than 30,000 projects by some 6,100 different clients. More than 350 of these assignments have been for clients in more than 65 countries and territories outside of the continental United States. Today the firm continues to emphasize the principles of integrity and quality engineering established by its founding partners and for the past 10 years has been ranked the U.S. and worldwide leader in project awards for electricity generation by the McCoy Power Reports.

Black & Veatch places a great deal of importance on the well-being of its employees. A comprehensive health and financial benefits program is available. Vacations, holidays, and firm-sponsored recreational activities are provided to help employees maintain a well-rounded lifestyle.

BURNS AND MCDONNELL, INC.
9400 WARD PARKWAY
KANSAS CITY, MISSOURI 64114-3319

OVERVIEW: Provides pollution control and sanitary engineering services. Established: 1991. Ownership: Private.

KEY STATISTICS: Annual Sales: $9.2 million. Number of Employees: 150.

HIRING HISTORY: Number of professional employees hired in 1997: 400. 1996: 300.

EXPERTISE/EDUCATION SOUGHT: Electrical engineering, environmental engineering, civil engineering, structural engineering, mechanical engineering, geotechnical engineering, chemical engineering, geologist, geography, environmental science

ALTERNATIVES TO FULL-TIME EMPLOYMENT: Paid internships

COMPANY-PAID TRAINING is available.

INTERNATIONAL ASSIGNMENTS are available to employees.

CONTACT: For college students and recent graduates: Ms. Melissa Hragyil, College Relations Coordinator; Phone: 816-822-3129; Fax: 816-822-3516. For individuals with previous experience: Mr. John R. Rice, Personnel Relations Manager; Phone: 816-822-3035; Fax: 816-822-3516.

COMPANY URL: http://www.burnsmcd.com

CERNER CORPORATION
2800 ROCK CREEK PARKWAY
KANSAS CITY, MISSOURI 64117-2551

OVERVIEW: Develops software and provides health-care and clinical information system services. Established: 1979. Ownership: Public.

KEY STATISTICS: Annual Sales: $189.1 million. Number of Employees: 1,400.

EXPERTISE/EDUCATION SOUGHT: Software engineering/development, computer programming

ALTERNATIVES TO FULL-TIME EMPLOYMENT: Paid internships

COMPANY-PAID TRAINING is available.

INTERNATIONAL ASSIGNMENTS are available to employees.

CONTACT: For college students and recent graduates: Mrs. Susan Woehrman, Manager of Recruiting; Phone: 816-221-1024, ext. 1660; Fax: 816-474-1742. For individuals with previous experience: Mrs. Colette Hunt, Recruiting Team Leader; Phone: 816-221-1024; Fax: 816-474-1742. E-mail: chunt@cerner.com.

COMPANY URL: http://www.cerner.com

COOPER AUTOMOTIVE
901 ROOSEVELT PARKWAY
CHESTERFIELD, MISSOURI 63017

Overview: Manufactures spark plugs. Established: 1989. Parent Company: Cooper Industries. Ownership: Public.

Key Statistics: Annual Sales: $350.5 million. Number of Employees: 5,645.

Hiring History: Number of professional employees hired in 1996: 25.

Expertise/Education Sought: Finance, management information systems, production

International Assignments are available to employees.

Contact: For individuals with previous experience: Employee Relations Department.

ENERGIZER
800 SOUTH CHAUTEAU
SAINT LOUIS, MISSOURI 63102

Overview: Manufactures batteries and battery chargers. Established: 1986. Parent Company: Ralston Purina Company. Ownership: Public.

Key Statistics: Annual Sales: $2.2 billion. Number of Employees: 17,000.

Expertise/Education Sought: Mechanical engineering, operations, quality engineering

Company-Paid Training is available.

International Assignments are available to employees.

Contact: For college students and recent graduates: Ms. Lynda Davis, Human Resource Recruiter; Phone: 314-982-2191; Fax: 314-982-3800. E-mail: lyndak.davis@energizer.com. For individuals with previous experience: Ms. Becky Gockel, Human Resources Recruiter; Phone: 314-982-5485; Fax: 314-982-3800. E-mail: beckys.gockel@energizer.com. Job Hotline: 314-982-2962.

Company URL: http://www.energizer.com

ENGINEERED PRODUCTS, INC.
12202 MISSOURI BOTTOM ROAD
HAZELWOOD, MISSOURI 63042-2318

Overview: Manufactures injection molded plastics. Established: 1984. Ownership: Private.

Key Statistics: Number of Employees: 300.

Expertise/Education Sought: Mechanical engineering

Contact: For individuals with previous experience: Employment Opportunities; Phone: 314-731-1220.

FABTECH, INC.
777 NORTHWEST BLUE PARKWAY, SUITE 350
LEES SUMMIT, MISSOURI 64086-5709

Overview: Produces semiconductor circuit networks. Established: 1996. Parent Company: Lite-On Power Semiconductor Corporation. Ownership: Private.

Key Statistics: Annual Sales: $18 million. Number of Employees: 100.

Hiring History: Number of professional employees hired in 1997: 100. 1996: 50.

Expertise/Education Sought: Manufacturing

Contact: Mr. Carol Haverpamp, Administrative Manager.

FARMLAND FOODS, INC.
10150 NORTH EXECUTIVE HILLS BOULEVARD
KANSAS CITY, MISSOURI 64153-0227

Overview: Produces and markets processed meats. Established: 1966. Parent Company: Farmland Industries, Inc. Ownership: Public.

Key Statistics: Annual Sales: $1.4 billion. Number of Employees: 4,000.

Expertise/Education Sought: Marketing, sales, accounting, finance, engineering, industrial engineering

Alternatives to Full-Time Employment: Paid internships

Company-Paid Training is available.

International Assignments are available to employees.

Contact: Mrs. Debbie Mesmer, Senior Employment Specialist; Phone: 816-891-1442; Fax: 816-891-1429. E-mail: damesmer@farmland.com.

FARMLAND INDUSTRIES, INC.
3315 NORTH OAK TRAFFICWAY
KANSAS CITY, MISSOURI 64116-2798

Overview: Manufactures and distributes feed, fertilizer, and chemicals; markets grain; processes beef and pork products. Established: 1929. Ownership: Public.

Key Statistics: Annual Sales: $9.1 billion. Number of Employees: 14,700.

Expertise/Education Sought: Data processing, accounting, finance, sales, marketing, advertising, economics, agricultural science

Alternatives to Full-Time Employment: Part-time jobs, paid internships

Company-Paid Training is available.

Contact: Ms. Pam Parker, Director of Human Resources, PO Box 7305, Kansas City, Missouri, 64116; Phone: 816-459-6363; Fax: 816-459-5954. E-mail: psparker@farmland.com. Job Hotline: 800-822-8263.

Company URL: http://www.farmland.com

GALAXY TELECOM, LP
1220 NORTH MAIN STREET
SIKESTON, MISSOURI 63801-4827

Overview: Provides cable TV services. Established: 1994. Parent Company: Galaxy Cablevision. Ownership: Private.

Key Statistics: Annual Sales: $62.3 million. Number of Employees: 500.

Hiring History: Number of professional employees hired in 1997: 250. 1996: 212.

Expertise/Education Sought: Accounting, electrical engineering

Alternatives to Full-Time Employment: Paid internships

Company-Paid Training is available.

Contact: Mr. Mark Anderson, Director of Technical Operations; Phone: 573-472-8255; Fax: 573-471-7281.

Company URL: http://www.enjoytv.com

GLOBAL SYSTEMS, INC.
8585 PAGE AVENUE
SAINT LOUIS, MISSOURI 63114-6008

Overview: Provides engineering services. Established: 1996. Ownership: Private.

Key Statistics: Annual Sales: $9.5 million. Number of Employees: 137.

Hiring History: Number of professional employees hired in 1997: 20.

Expertise/Education Sought: Electrical engineering, information systems

Company-Paid Training is available.

Contact: For individuals with previous experience: Human Resource Department; Phone: 314-567-5732; Fax: 314-428-6022. E-mail: gsi@inlink.com.

HNTB CORPORATION
PO BOX 412197
KANSAS CITY, MISSOURI 64141-2197

Overview: Provides architectural and engineering design and consulting services. Established: 1966. Ownership: Private.

Key Statistics: Annual Sales: $209.7 million. Number of Employees: 2,000.

Expertise/Education Sought: Architecture, urban and regional planning, interior design, registered nursing, allied health

Alternatives to Full-Time Employment: Part-time jobs, paid internships

Company-Paid Training is available.

Contact: Ms. B. A. Stein, Recruiting Specialist; Phone: 816-472-1201; Fax: 816-472-4064. E-mail: bstein@hntb.com.

Company URL: http://www.hntb.com

HOECHST MARION ROUSSEL
10236 MARION PARK DRIVE
KANSAS CITY, MISSOURI 64137-1405

Overview: Produces prescription drugs and consumer health-care products. Established: 1995. Parent Company: Hoechst, AG. Ownership: Private.

Key Statistics: Annual Sales: $307.7 million. Number of Employees: 2,400.

Expertise/Education Sought: Packaging engineering, process engineering

Alternatives to Full-Time Employment: Internships

Contact: For individuals with previous experience: Mrs. Judy McNish, Manager of Recruiting; Phone: 816-966-5332; Fax: 816-966-2769.

INTELLIMARK
701 EMERSON ROAD SUITE 300
SAINT LOUIS, MISSOURI 63141-5437

Overview: Provides data processing consulting services. Established: 1995. Parent Company: Staffmark. Ownership: Private.

Key Statistics: Annual Sales: $7.1 million. Number of Employees: 115.

Expertise/Education Sought: Technology

Alternatives to Full-Time Employment: Paid internships

Company-Paid Training is available.

International Assignments are available to employees.

Contact: Account Manager; Phone: 314-432-0018; Fax: 314-432-5198.

LACLEDE GAS COMPANY
720 OLIVE STREET
SAINT LOUIS, MISSOURI 63101-2338

Overview: Natural gas production, storage, and distribution public utility. Established: 1857. Ownership: Public.

Key Statistics: Annual Sales: $544.8 million. Number of Employees: 2,072.

Expertise/Education Sought: Accounting, mechanical engineering, electrical engineering

Contact: Mr. Bill Rearden, Manager of Employment; Phone: 314-342-0646; Fax: 314-342-3166. Job Hotline: 314-342-0762.

LEAR CORPORATION—MASLAND DIVISION
8485 PROSPECT AVENUE
KANSAS CITY, MISSOURI 64132-2376

Overview: Manufactures mechanical rubber automotive products. Established: 1996. Parent Company: Lear Corporation. Ownership: Public.

Key Statistics: Annual Sales: $45 million. Number of Employees: 350.

Hiring History: Number of professional employees hired in 1996: 90.

Expertise/Education Sought: Manufacturing

Alternatives to Full-Time Employment: Paid internships

International Assignments are available to employees.

Contact: For individuals with previous experience: Human Resources; Phone: 816-444-3611.

LEONARDS METAL, INC.
3600 MUELLER
SAINT CHARLES, MISSOURI 63302

Mailing Address: PO Box 900, St. Charles, MO 63302-1900

Overview: Manufactures parts used in the aerospace industry. Established: 1984. Ownership: Private.

Key Statistics: Annual Sales: $30 million. Number of Employees: 520.

Hiring History: Number of professional employees hired in 1997: 146. 1996: 200.

Expertise/Education Sought: Mechanical engineering

Alternatives to Full-Time Employment: Internships

Company-Paid Training is available.

Contact: Ms. Sarah Jamison, Recruiting Coordinator; Phone: 314-946-6525; Fax: 314-916-2127.

M AND A ELECTRIC POWER COOPERATIVE
HIGHWAY PP
POPLAR BLUFF, MISSOURI 63901

Mailing Address: PO Box 670, Poplar Bluff, MO 63902

Overview: Provides electrical services. Established: 1948. Ownership: Private.

Key Statistics: Annual Sales: $7.4 million. Number of Employees: 40.

Expertise/Education Sought: Sales

Contact: For individuals with previous experience: Human Relations Department.

MASSMAN CONSTRUCTION COMPANY
8901 STATE LINE ROAD
KANSAS CITY, MISSOURI 64114

Overview: Provides marine and heavy construction services. Established: 1917.

Key Statistics: Annual Sales: $40 million. Number of Employees: 400.

Expertise/Education Sought: Civil engineering, construction engineering, project management

Contact: Mr. Keith Jacobson, Vice President.

CORPORATE STATEMENT

Massman Construction Company is involved in the construction of difficult marine foundation projects on a nationwide basis. Preferred candidates are those who can function in a technical home office environment as well as a remote jobsite trailer. Technical aptitude as well as excellent interpersonal skills are required of successful employees.

MIDWEST RESEARCH INSTITUTE
425 VOLKER BOULEVARD
KANSAS CITY, MISSOURI 64110-2299

Overview: Provides nonprofit contract, biological, environmental, and engineering research services. Established: 1943. Ownership: Private.

Key Statistics: Annual Sales: $226.6 million. Number of Employees: 1,150.

Expertise/Education Sought: Chemical engineering, chemistry, chemical analysis, toxicology, molecular biology, microbiology, environmental engineering, environmental science

Alternatives to Full-Time Employment: Part-time jobs, paid internships

Contact: Ms. Rochelle Anderson, Employment Recruiter; Phone: 816-753-7600, ext. 1206; Fax: 816-753-2304. E-mail: randerson@mriresearch.org. Job Hotline: 816-753-7600, ext. 7506.

Company URL: http://www.mriresearch.org

NATIONAL IMAGERY AND MAPPING AGENCY
3200 SOUTH SECOND STREET, (HRRC/L-28)
SAINT LOUIS, MISSOURI 63118-3399

Overview: Provides timely, relevant and accurate imagery, imagery intelligence, and geospatial information in support of national security objectives. Established: 1996. Division of the Department of Defense.

Expertise/Education Sought: Geology, geoscience, computer science, mathematics, image processing, imaging technology

Alternatives to Full-Time Employment: Part-time jobs

Company-Paid Training is available.

International Assignments are available to employees.

Contact: Ms. Sheree Cannady, Personnel Management Specialist; Phone: 800-777-6104, ext. 120; Fax: 314-260-1235. Job Hotline: 800-777-6104.

Company URL: http://www.nima.mil

PAYLESS CASHWAYS, INC.

2300 MAIN STREET
KANSAS CITY, MISSOURI 64141

Overview: Operates chain of home improvement stores. Established: 1988. Ownership: Public.

Key Statistics: Annual Sales: $2.6 billion. Number of Employees: 16,664.

Expertise/Education Sought: Management, information systems, accounting, human resources

Alternatives to Full-Time Employment: Part-time jobs

Company-Paid Training is available.

Contact: Ms. Penny Trimble, Staffing Administrator; Phone: 816-234-6306; Fax: 816-234-6046. Job Hotline: 816-234-6474.

Company URL: http://www.payless.cashways.com

RALCORP HOLDINGS, INC.

PO BOX 618
SAINT LOUIS, MISSOURI 63188-0618

Overview: Manufactures cereals, baby foods, cookies, and crackers. Established: 1994. Ownership: Public.

Key Statistics: Annual Sales: $256 million. Number of Employees: 2,099.

Expertise/Education Sought: Accounting, marketing, customer service and support, sales

Alternatives to Full-Time Employment: Paid internships

Company-Paid Training is available.

Contact: Ms. Deana Spuhler, Human Resources Coordinator. Job Hotline: 314-877-7555.

Company URL: http://www.ralcorp.com

RALSTON PURINA COMPANY

801 CHOUTEAU AVENUE
SAINT LOUIS, MISSOURI 63102

Overview: Manufactures pet foods and baked goods. Established: 1894. Ownership: Public.

Key Statistics: Annual Sales: $6.1 billion. Number of Employees: 31,837.

Expertise/Education Sought: Computer programming, accounting, finance, computer science, management, business administration, auditing, program analysis, data processing

Alternatives to Full-Time Employment: Internships

International Assignments are available to employees.

Contact: For individuals with previous experience: Employment Department; Phone: 314-982-1000. Job Hotline: 314-982-2962.

Company URL: http://www.ralstonpurina.com

SBC COMMUNICATION, INC.

ONE BELL CENTER, SUITE 224
SAINT LOUIS, MISSOURI 63101

Overview: Provides telecommunications services. Established: 1984. Ownership: Public.

Key Statistics: Annual Sales: $9.7 billion. Number of Employees: 49,470.

Expertise/Education Sought: Telecommunications, electrical engineering

Alternatives to Full-Time Employment: Part-time jobs, paid internships

Company-Paid Training is available.

International Assignments are available to employees.

Contact: For college students and recent graduates: Mr. Mark Jacobi, Senior Recruiter; Phone: 314-235-2469; Fax: 314-235-4371. For individuals with previous experience: Ms. Janie Briggs, Manager of Recruiting Operations, 1010 Pine Street, Saint Louis, Missouri, 63101-3002; Phone: 314-235-9800; Fax: 314-235-4371. Job Hotline: 800-444-6303.

Company URL: http://www.sbc.com

SHERWOOD, DAVIS & GECK

1915 OLIVE STREET
SAINT LOUIS, MISSOURI 63103-1625

Overview: Manufactures and distributes disposable medical products. Established: 1982. Parent Company: American Home Products Corporation. Ownership: Public.

Key Statistics: Annual Sales: $1 billion. Number of Employees: 8,000.

Hiring History: Number of professional employees hired in 1997: 250. 1996: 100.

Expertise/Education Sought: Accounting, marketing, finance, chemical engineering, mechanical engineering, electrical engineering

Company-Paid Training is available.

International Assignments are available to employees.

Contact: For individuals with previous experience: Human Resource Manager.

SPEEDLINE TECHNOLOGIES

HIGHWAY 5 SOUTH
CAMDENTON, MISSOURI 65020

Overview: Manufactures machinery for making electronic components. Established: 1992. Parent Company: Cookson America. Ownership: Private.

Key Statistics: Annual Sales: $65.8 million. Number of Employees: 669.

Hiring History: Number of professional employees hired in 1997: 150. 1996: 150.

Expertise/Education Sought: Industrial technician, electrical technology, electrical engineering, mechanical engineering, software engineering/development

Alternatives to Full-Time Employment: Paid internships

International Assignments are available to employees.

Contact: Mr. David Campanini, Human Resources Manager; Phone: 573-346-3341; Fax: 573-346-0235. Job Hotline: 573-346-3341, ext. 403.

STORZ INSTRUMENT COMPANY

3365 TREE COURT INDUSTRIAL BOULEVARD
SAINT LOUIS, MISSOURI 63122-6611

Overview: Manufactures surgical instruments and equipment. Established: 1986. Parent Company: Bausch And Lomb. Ownership: Public.

Key Statistics: Annual Sales: $190 million. Number of Employees: 900.

Hiring History: Number of professional employees hired in 1996: 50.

Expertise/Education Sought: Sales, mechanical engineering, electrical engineering, marketing, manufacturing engineering, ophthalmological engineering, design engineering

International Assignments are available to employees.

Contact: Mr. Web Sant, Manager of Employment; Phone: 314-225-5051; Fax: 314-225-3032.

STUPP BROTHERS, INC.

3800 WEBER ROAD
SAINT LOUIS, MISSOURI 63125

Overview: Manufactures fabricated steel and pipe. Established: 1856. Ownership: Private.

Key Statistics: Annual Sales: $202 million. Number of Employees: 500.

Hiring History: Number of professional employees hired in 1997: 3. 1996: 6.

Expertise/Education Sought: Drafting, engineering

Company-Paid Training is available.

Contact: For individuals with previous experience: Mr. Ed Stolle, Jr., Director of Human Resources; Phone: 314-638-5000.

TRANSACTION TECHNOLOGIES, INC.

1 TRANS TECH DRIVE
UNION, MISSOURI 63084

Overview: Manufactures printed circuit boards. Ownership: Private.

Key Statistics: Annual Sales: $16.1 million. Number of Employees: 250.

Transaction Technologies, Inc. (continued)

Hiring History: Number of professional employees hired in 1997: 50. 1996: 50.

Expertise/Education Sought: Engineering

Company-Paid Training is available.

Contact: For individuals with previous experience: Human Resources; Phone: 314-583-1440; Fax: 314-583-1822.

Company URL: http://www.coinco.com

UTILICORP UNITED, INC.

21 NINTH STREET
KANSAS CITY, MISSOURI 64105

Overview: Electric and gas utility. Established: 1926. Ownership: Public.

Key Statistics: Annual Sales: $4.3 billion. Number of Employees: 4,746.

Expertise/Education Sought: Electrical engineering, finance, accounting, mechanical engineering, marketing, energy/energy management

Alternatives to Full-Time Employment: Paid internships

Company-Paid Training is available.

Contact: Mr. Bill Englehart, Recruiting; Phone: 816-467-3463; Fax: 816-467-9463. E-mail: bengleha@utilicorp.com.

Company URL: http://www.utilicorp.com

VENTURE STORES, INC.

2001 EAST TERRA LANE
O' FALLON, MISSOURI 63366

Overview: Operates chain of retail stores. Established: 1990. Ownership: Public.

Key Statistics: Annual Sales: $1.5 billion. Number of Employees: 12,600.

Expertise/Education Sought: Merchandising, computer programming, management information systems, accounting, retail, industrial engineering

Alternatives to Full-Time Employment: Paid internships

Company-Paid Training is available.

Contact: For individuals with previous experience: Executive Recruitment Manager; Phone: 314-281-5500; Fax: 314-281-6049.

MONTANA

BLACKFOOT TELEPHONE COOPERATIVE, INC.

1221 NORTH RUSSELL STREET
MISSOULA, MONTANA 59802-1898

Overview: Wholesales electronic parts. Established: 1993. Ownership: Private.

Key Statistics: Annual Sales: $21.2 million. Number of Employees: 110.

Hiring History: Number of professional employees hired in 1997: 15. 1996: 4.

Expertise/Education Sought: Telecommunications

Alternatives to Full-Time Employment: Part-time jobs

Contact: For individuals with previous experience: Ms. Joan Mandeville, Assistant Manager; Phone: 406-541-2121, ext. 5300; Fax: 406-541-5333. E-mail: jmandeville@blackfoot.net.

Company URL: http://www.blackfoot.net

NEBRASKA

AG PROCESSING, INC.

12700 WEST DODGE ROAD
OMAHA, NEBRASKA 68154-6103

Mailing Address: PO Box 2047, Omaha, NE 68103-2047

Overview: Produces soybean based products. Established: 1943. Ownership: Private.

Key Statistics: Annual Sales: $2.8 billion. Number of Employees: 3,100.

Expertise/Education Sought: Merchandising, actuarial, marketing, operations, agricultural science, project management, data processing

International Assignments are available to employees.

Contact: For individuals with previous experience: Ms. Linda Brown, Manager of Recruiting; Phone: 402-496-7809, ext. 2223.

BALDWIN FILTERS

4400 EAST HIGHWAY 30
KEARNEY, NEBRASKA 68847-6010

Overview: Manufactures motor vehicle parts. Established: 1981. Parent Company: Clarcor, Inc. Ownership: Public.

Key Statistics: Annual Sales: $74.2 million. Number of Employees: 800.

Expertise/Education Sought: Manufacturing engineering, industrial engineering, mechanical engineering

Alternatives to Full-Time Employment: Paid internships

Company-Paid Training is available.

Contact: Mr. Robert Bjerke, Vice President of Human Resources; Phone: 308-234-1951; Fax: 308-233-9294.

CONCRETE INDUSTRIES

6300 CORNHUSKER HIGHWAY, PO BOX 29529
LINCOLN, NEBRASKA 68529

Overview: Manufactures concrete products. Established: 1949. Parent Company: Nebco, Inc. Ownership: Private.

Key Statistics: Annual Sales: $60 million. Number of Employees: 400.

Expertise/Education Sought: Civil engineering, structural engineering

Alternatives to Full-Time Employment: Paid internships

Contact: For individuals with previous experience: Mr. James Schmersal, Engineering Manager; Phone: 402-434-1800; Fax: 402-434-1899.

LEO A DALY COMPANY

8600 INDIAN HILLS DRIVE
OMAHA, NEBRASKA 68114-4039

Overview: Provides architectural, engineering, and interior design services. Established: 1915. Ownership: Private.

Key Statistics: Annual Sales: $63.5 million. Number of Employees: 625.

Hiring History: Number of professional employees hired in 1997: 20. 1996: 12.

Expertise/Education Sought: Architecture, structural engineering, civil engineering, mechanical engineering, interior design, electrical engineering

Alternatives to Full-Time Employment: Paid internships

Contact: For individuals with previous experience: Mr. Frank Thompson, Vice President of Human Resources; Phone: 402-391-8111; Fax: 402-391-8564. E-mail: frthompson@leoadaly.com.

DATA TRANSMISSION NETWORK

9110 WEST DODGE ROAD, SUITE 200
OMAHA, NEBRASKA 68114-3324

Overview: Provides electronic agriculture and financial market information services. Established: 1986. Ownership: Public.

Key Statistics: Annual Sales: $98.4 million. Number of Employees: 900.

Expertise/Education Sought: Customer service and support, computer programming, project engineering, hardware engineering, software engineering/development

Alternatives to Full-Time Employment: Paid internships

Company-Paid Training is available.

International Assignments are available to employees.

Contact: Ms. Shealyn Becker, Human Resources Assistant; Phone: 402-390-2328; Fax: 402-255-8750. E-mail: sbecker@dtn.com.

Company URL: http://www.dtn.com

HDR, INC.

8404 INDIAN HILLS DRIVE
OMAHA, NEBRASKA 68114-4049

Overview: Provides engineering, architectural, and management consulting services. Established: 1996. Ownership: Private.

Key Statistics: Annual Sales: $199 million. Number of Employees: 1,751.

Hiring History: Number of professional employees hired in 1997: 485. 1996: 200.

Expertise/Education Sought: Civil engineering, structural engineering, mechanical engineering, electrical engineering, environmental engineering

Alternatives to Full-Time Employment: Part-time jobs

Contact: Mr. Rob Johnson, National Recruitment Director; Phone: 402-399-1205. E-mail: rojohnso@www.hdrinc.com.

Company URL: http://www.hdrinc.com

IBP, INC.

HIGHWAY 35, PO BOX 515
DAKOTA CITY, NEBRASKA 68731

Overview: Owns and operates beef and pork processing plants. Established: 1960. Ownership: Public.

Key Statistics: Annual Sales: $12.5 billion. Number of Employees: 34,000.

Expertise/Education Sought: Accounting, finance, data processing, law, economics, information systems

Alternatives to Full-Time Employment: Paid internships

Company-Paid Training is available.

International Assignments are available to employees.

Contact: Ms. Patricia Dinges, Manager of Recruitment; Phone: 402-241-2979; Fax: 402-241-2025.

Company URL: http://www.ibpinc.com

ITI MARKETING SERVICES, INC.

902 NORTH 91ST PLAZA
OMAHA, NEBRASKA 68114

Overview: Provides telemarketing services. Established: 1995. Ownership: Private.

Key Statistics: Annual Sales: $124.8 million. Number of Employees: 7,500.

Expertise/Education Sought: Telemarketing

Company-Paid Training is available.

Contact: For individuals with previous experience: Ms. Carmen Vanwetering, Human Resources Manager; Phone: 402-393-5730; Fax: 402-392-9090. Job Hotline: 402-393-5610.

Company URL: http://www.itimarketing.com

NEBRASKA PUBLIC POWER DISTRICT

1414 15TH STREET
COLUMBUS, NEBRASKA 68601-5226

Overview: Electric utility. Established: 1939. Ownership: Private.

Key Statistics: Annual Sales: $452.4 million. Number of Employees: 2,012.

Expertise/Education Sought: Electrical engineering, mechanical engineering, nuclear engineering

Alternatives to Full-Time Employment: Paid internships

Company-Paid Training is available.

Contact: Mr. Steve Albers, Employment Consultant, PO Box 499, Columbus, Nebraska, 68601; Phone: 402-563-5253; Fax: 402-563-5771. E-mail: scalber@nppd.com. Job Hotline: 402-563-5010.

Company URL: http://www.nppd.com

ONE/NET

9828 TAYLOR STREET
OMAHA, NEBRASKA 68134-3754

Overview: Provides computer integrated systems services. Established: 1993. Ownership: Private.

Key Statistics: Annual Sales: $10 million.

Hiring History: Number of professional employees hired in 1997: 3. 1996: 5.

Expertise/Education Sought: Computer science, networking

Alternatives to Full-Time Employment: Internships

Contact: For individuals with previous experience: Mr. Doug Petry, President; Phone: 402-572-6900; Fax: 402-572-7500. E-mail: onenet@msn.com.

Company URL: http://www.freeyellow.com/members2/onenet

PGLA 1, A GENERAL PARTNERSHIP

650 INDUSTRIAL ROAD
BLAIR, NEBRASKA 68008-2649

Overview: Manufactures chemical preparations. Established: 1997. Parent Company: Cargill & Purac. Ownership: Private.

Key Statistics: Annual Sales: $24.7 million. Number of Employees: 200.

Expertise/Education Sought: Chemical engineering

Contact: For individuals with previous experience: Ms. Lynette Jennings, Human Resource Manager; Phone: 402-533-4148; Fax: 402-533-4133.

Company URL: http://www.cargill.com

VALMONT INDUSTRIES, INC.

HIGHWAY 275, PO BOX 358
VALLEY, NEBRASKA 68064

Overview: Manufactures irrigation systems and light poles.

Expertise/Education Sought: Sales, manufacturing engineering, civil engineering, structural engineering, agricultural engineering, mechanical engineering, electrical engineering, marketing

Alternatives to Full-Time Employment: Paid internships

Company-Paid Training is available.

International Assignments are available to employees.

Contact: Mr. Terry McElain, Senior Vice President; Phone: 402-359-2201; Fax: 402-343-0661.

Company URL: http://www.valmont.com

NEVADA

CSR

7150 POLLOCK DRIVE
LAS VEGAS, NEVADA 89119

Overview: Manufactures concrete. Established: 1990. Parent Company: CSR America, Inc. Ownership: Private.

Key Statistics: Annual Sales: $70 million. Number of Employees: 400.

CSR (continued)

Hiring History: Number of professional employees hired in 1997: 10. 1996: 10.

Expertise/Education Sought: Manufacturing engineering, minerals engineering, mechanical engineering, sales, accounting, marketing, construction management

Contact: Human Resources.

EG&G SPECIAL PROJECTS

821 GRIER DRIVE
LAS VEGAS, NEVADA 89119-3717

Overview: Conducts commercial physical research. Established: 1983. Parent Company: EG&G, Inc. Ownership: Public.

Key Statistics: Annual Sales: $93.8 million. Number of Employees: 1,200.

Expertise/Education Sought: Software engineering/development, electrical engineering, electronics/electronics engineering

Alternatives to Full-Time Employment: Internships

Company-Paid Training is available.

Contact: For individuals with previous experience: Mr. Jack Pollock, Personnel Administrator; Phone: 702-361-1660, ext. 1010; Fax: 702-361-4879. E-mail: pollock@egginc.com.

IGT

9295 PROTOTYPE DRIVE
RENO, NEVADA 89511-8986

Overview: Manufactures and designs gaming machines and software. Established: 1971. Parent Company: International Game Technology. Ownership: Public.

Key Statistics: Annual Sales: $733.5 million. Number of Employees: 2,800.

Hiring History: Number of professional employees hired in 1997: 500. 1996: 800.

Expertise/Education Sought: Software engineering/development, hardware engineering, electrical engineering, mechanical engineering, management, manufacturing engineering, accounting, logistics, information systems, firmware engineering

Alternatives to Full-Time Employment: Paid internships

Company-Paid Training is available.

International Assignments are available to employees.

Contact: Ms. Kim Buschelman, Supervisor of Employment; Phone: 702-448-0355; Fax: 702-448-1600. E-mail: igthr@intgame.com. Job Hotline: 702-448-JOBS.

Company URL: http://www.igtgame.com

LIONHEART TECHNOLOGIES, INC.

2000 ARROWHEAD DRIVE
CARSON CITY, NEVADA 89706-0403

Overview: Manufactures electromedical equipment. Established: 1996. Ownership: Private.

Key Statistics: Annual Sales: $27.3 million. Number of Employees: 275.

Hiring History: Number of professional employees hired in 1996: 12.

Expertise/Education Sought: Electrical engineering

Contact: For individuals with previous experience: Human Resources; Phone: 702-883-3400.

NATIONAL SYSTEMS, INC.

3885 SOUTH DECATUR BOULEVARD
LAS VEGAS, NEVADA 89103-5855

Mailing Address: 121 Fisher Avenue, Staten Island, NY 10307

Overview: Provides professional engineering services. Established: 1996. Ownership: Private.

Key Statistics: Annual Sales: $10 million.

Hiring History: Number of professional employees hired in 1997: 8.

Expertise/Education Sought: Engineering, environmental engineering

Contact: For individuals with previous experience: Human Resource Department; Phone: 718-948-5945; Fax: 718-948-8078.

NEW HAMPSHIRE

BALZERS PROCESS SYSTEMS, INC.

25 SAGAMORE PARK ROAD
HOUSTON, NEW HAMPSHIRE 03051-4901

Overview: Manufactures air flow controls. Established: 1995. Ownership: Private.

Key Statistics: Annual Sales: $70 million. Number of Employees: 68.

Expertise/Education Sought: Technical

Alternatives to Full-Time Employment: Paid internships

International Assignments are available to employees.

Contact: For individuals with previous experience: Human Resources; Phone: 603-594-1500; Fax: 603-594-1538.

BEEDE ELECTRICAL INSTRUMENT COMPANY, INC.

175 SOUTH MAIN STREET
PENACOOK, NEW HAMPSHIRE 03303-1950

Mailing Address: PO Box 8910, Penacook, NH 03303

Overview: Manufactures electronic measure instruments. Established: 1946. Parent Company: Northfield New Hampshire. Ownership: Private.

Key Statistics: Annual Sales: $20 million. Number of Employees: 240.

Expertise/Education Sought: Electrical engineering, management information systems

Alternatives to Full-Time Employment: Paid internships

Company-Paid Training is available.

Contact: For individuals with previous experience: Ms. Lurie Methven, Director of Human Resources; Phone: 603-753-6362, ext. 202; Fax: 603-753-8314.

CONSOLIDATED PROCESS MACHINERY

9 EXECUTIVE PARK DRIVE
MERRIMACK, NEW HAMPSHIRE 03054-4058

Overview: Manufactures farm machinery and equipment. Established: 1996. Parent Company: Gencor Industries. Ownership: Public.

Key Statistics: Annual Sales: $60 million. Number of Employees: 490.

Hiring History: Number of professional employees hired in 1997: 12.

Expertise/Education Sought: Manufacturing engineering, warehousing

International Assignments are available to employees.

Contact: Human Resource Department; Phone: 603-423-6310; Fax: 603-423-6388. E-mail: swoodward@cpm.md.com.

Company URL: http://www.cpmroskamp.com

CUSHCRAFT CORPORATION

48 PERIMETER ROAD
MANCHESTER, NEW HAMPSHIRE 03103-3327

Mailing Address: PO Box 4680, Manchester, NH 03108

Overview: Manufactures radio and television communications equipment. Established: 1970. Ownership: Private.

Key Statistics: Annual Sales: $7.1 million. Number of Employees: 75.

Hiring History: Number of professional employees hired in 1997: 15. 1996: 10.

Expertise/Education Sought: Manufacturing engineering

ALTERNATIVES TO FULL-TIME EMPLOYMENT: Part-time jobs
INTERNATIONAL ASSIGNMENTS are available to employees.
CONTACT: For individuals with previous experience: Human Resources; Phone: 603-627-7877; Fax: 603-627-1764.
COMPANY URL: http://www.cushcraft.com

HADCO CORPORATION
12 MANOR PARKWAY
SALEM, NEW HAMPSHIRE 03079

OVERVIEW: Manufactures printed circuits. Established: 1966. Ownership: Public.
KEY STATISTICS: Annual Sales: $350.7 million. Number of Employees: 3,005.
EXPERTISE/EDUCATION SOUGHT: Electrical engineering, accounting, finance, data processing, purchasing, sales, process engineering, human resources, manufacturing engineering
ALTERNATIVES TO FULL-TIME EMPLOYMENT: Paid internships
COMPANY-PAID TRAINING is available.
INTERNATIONAL ASSIGNMENTS are available to employees.
CONTACT: Ms. Lorraine Peterson, Recruiter; Phone: 603-898-8000, ext. 2362; Fax: 603-890-1298. E-mail: lpeterson@hadco.com.
COMPANY URL: http://www.hadco.com

HARVARD INDUSTRIES/KINGSTON-WARREN CORPORATION
171 ROUTE 85
NEWFIELDS, NEW HAMPSHIRE 03856

MAILING ADDRESS: PO Box 169, Newfield, NH 03856
OVERVIEW: Produces rubber automotive products. Established: 1986. Ownership: Public.
KEY STATISTICS: Annual Sales: $144 million. Number of Employees: 1,500.
EXPERTISE/EDUCATION SOUGHT: Manufacturing engineering, automotive engineering
CONTACT: Mr. William Pickering, Manager of Human Resources; Phone: 603-772-3771; Fax: 603-772-0984.

HITCHINER MANUFACTURING COMPANY
ELM STREET, PO BOX 2001
MILFORD, NEW HAMPSHIRE 03055

OVERVIEW: Manufactures metal castings. Established: 1949. Ownership: Private.
KEY STATISTICS: Annual Sales: $138.3 million. Number of Employees: 3,000.
HIRING HISTORY: Number of professional employees hired in 1997: 250. 1996: 200.
EXPERTISE/EDUCATION SOUGHT: Process engineering, metallurgical engineering
ALTERNATIVES TO FULL-TIME EMPLOYMENT: Part-time jobs, paid internships
COMPANY-PAID TRAINING is available.
INTERNATIONAL ASSIGNMENTS are available to employees.
CONTACT: Mr. Richard Bickford, Corporate Manager of Human Resources; Phone: 603-673-1100; Fax: 603-672-2408.
COMPANY URL: http://www.hitchiner.com

KINGSBURY CORPORATION
80 LAUREL STREET
KEENE, NEW HAMPSHIRE 03431

OVERVIEW: Manufactures metal cutting machine tools. Established: 1984. Ownership: Private.
KEY STATISTICS: Annual Sales: $50 million. Number of Employees: 525.
HIRING HISTORY: Number of professional employees hired in 1997: 12.
EXPERTISE/EDUCATION SOUGHT: Electrical engineering, control engineering
ALTERNATIVES TO FULL-TIME EMPLOYMENT: Paid internships
INTERNATIONAL ASSIGNMENTS are available to employees.
CONTACT: For individuals with previous experience: Ms. Diane Tisdale, Manager of Human Resources; Phone: 603-352-5212, ext. 1211; Fax: 603-357-1955.

KOLLSMAN, INC.
220 DANIEL WEBSTER HIGHWAY
MERRIMACK, NEW HAMPSHIRE 03054

OVERVIEW: Manufactures search and navigation equipment. Established: 1995. Ownership: Private.
KEY STATISTICS: Annual Sales: $68.7 million. Number of Employees: 845.
HIRING HISTORY: Number of professional employees hired in 1997: 95. 1996: 45.
EXPERTISE/EDUCATION SOUGHT: Sales, marketing, management, engineering
CONTACT: Human Resources; Phone: 603-886-2124; Fax: 603-595-6166.
COMPANY URL: http://www.kollsman.com

MARKEM CORPORATION
PO BOX 2100, 150 CONGRESS STREET
KEENE, NEW HAMPSHIRE 03431-4307

OVERVIEW: Manufactures printing machinery. Established: 1911. Ownership: Private.
KEY STATISTICS: Annual Sales: $231.5 million. Number of Employees: 1,800.
EXPERTISE/EDUCATION SOUGHT: Engineering, marketing, chemistry, manufacturing engineering, software engineering/development
ALTERNATIVES TO FULL-TIME EMPLOYMENT: Part-time jobs, paid internships
COMPANY-PAID TRAINING is available.
INTERNATIONAL ASSIGNMENTS are available to employees.
CONTACT: For college students and recent graduates: Mr. Barry Mallis, Training and Development Manager; Phone: 603-352-1130, ext. 2232; Fax: 603-352-0525. E-mail: bmallis@markem.com. For individuals with previous experience: Mr. Peter Plante, Manager of Human Resources; Phone: 603-352-1130, ext. 2232; Fax: 603-352-0525. E-mail: pplante@markem.com.
COMPANY URL: http://www.markem.com

MONADNOCK PAPER MILLS, INC.
117 ANTRIM ROAD
BENNINGTON, NEW HAMPSHIRE 03442

OVERVIEW: Produces specialty papers. Established: 1948. Ownership: Private.
KEY STATISTICS: Annual Sales: $58.2 million. Number of Employees: 240.
HIRING HISTORY: Number of professional employees hired in 1997: 5. 1996: 25.
EXPERTISE/EDUCATION SOUGHT: Technical sales
COMPANY-PAID TRAINING is available.
CONTACT: Mr. Joseph Fletcher, Vice President Human Resources; Phone: 603-588-3311; Fax: 603-588-6289. E-mail: jfletcher@mpm.com.

MPB CORPORATION
PO BOX 547
KEENE, NEW HAMPSHIRE 03431

OVERVIEW: Manufactures ball bearings and disks for commercial use. Established: 1990. Parent Company: The Timken Company. Ownership: Public.
KEY STATISTICS: Annual Sales: $122.3 million. Number of Employees: 1,630.
EXPERTISE/EDUCATION SOUGHT: Manufacturing engineering, product engineering
ALTERNATIVES TO FULL-TIME EMPLOYMENT: Paid internships
COMPANY-PAID TRAINING is available.

MPB Corporation (continued)

International Assignments are available to employees.
Contact: For individuals with previous experience: Mr. Robert M. Rooney, Director of Human Resources; Phone: 603-352-0310, ext. 242; Fax: 603-355-4556. E-mail: rooneyr@timken.com.

NASHUA CORPORATION
44 FRANKLIN STREET
NASHUA, NEW HAMPSHIRE 03061

Mailing Address: PO Box 2002, Nashua, NH 03061
Overview: Manufactures paper adhesive labels. Established: 1904. Ownership: Public.
Key Statistics: Annual Sales: $389.7 million. Number of Employees: 2,340.
Hiring History: Number of professional employees hired in 1997: 25.
Expertise/Education Sought: Manufacturing, marketing
Company-Paid Training is available.
Contact: For individuals with previous experience: Human Resources Department; Phone: 603-880-2323.
Company URL: http://www.nashua.com

NATIONAL SEA PRODUCTS, INC.
PO BOX 839
PORTSMOUTH, NEW HAMPSHIRE 03802-0839

Overview: Processes fresh and frozen fish and seafood. Established: 1968. Parent Company: National Sea Products of Halifax (Canada). Ownership: Private.
Key Statistics: Annual Sales: $63 million. Number of Employees: 175.
Expertise/Education Sought: Manufacturing, finance, sales, human resources, production
Contact: For individuals with previous experience: Human Resources Department.

PRIME TANNING COMPANY, INC.
216 AIRPORT DRIVE
ROCHESTER, NEW HAMPSHIRE 03866

Overview: Operates leather tanning facility. Established: 1926. Ownership: Private.
Key Statistics: Annual Sales: $100 million. Number of Employees: 1,000.
Hiring History: Number of professional employees hired in 1996: 80.
Expertise/Education Sought: Manufacturing, industrial engineering
Company-Paid Training is available.
Contact: Mr. Richard Euler, Director of Human Resources; Phone: 603-330-3100; Fax: 603-330-2080.

PUBLIC SERVICE OF NEW HAMPSHIRE
1000 ELM STREET
MANCHESTER, NEW HAMPSHIRE 03101-1730

Mailing Address: PO Box 330, Manchester, NH 03105
Overview: Provides electric services. Established: 1992. Parent Company: Notheast Utility. Ownership: Public.
Key Statistics: Annual Sales: $1.1 billion. Number of Employees: 1,279.
Hiring History: Number of professional employees hired in 1996: 8.
Expertise/Education Sought: Electrical engineering
Contact: Mrs. Claire Guyette, Human Resource Coordinator; Phone: 603-634-2732; Fax: 603-634-2597.

RDF CORPORATION
23 ELM AVENUE
HUDSON, NEW HAMPSHiRE 03051-3224

Overview: Manufactures environmental control instruments. Established: 1955. Ownership: Private.
Key Statistics: Annual Sales: $5.6 million. Number of Employees: 82.
Hiring History: Number of professional employees hired in 1997: 27. 1996: 20.
Expertise/Education Sought: Administration, accounting, business management, customer service and support, design engineering, electrical engineering, finance
Company-Paid Training is available.
Contact: Ms. Tracy DeMonte, Human Resources Manager; Phone: 603-882-5195, ext. 213; Fax: 603-882-6925.

SHAER SHOE CORPORATION
10 COMMERCE PARK NORTH
BEDFORD, NEW HAMPSHIRE 03110-6905

Overview: Manufactures women's footwear. Established: 1990. Ownership: Private.
Key Statistics: Annual Sales: $32 million. Number of Employees: 600.
Hiring History: Number of professional employees hired in 1997: 200. 1996: 200.
Expertise/Education Sought: Manufacturing
Contact: Ms. Shirley Johnson, Human Resource Manager; Phone: 603-625-8566; Fax: 207-778-9240.

STANDEX INTERNATIONAL CORPORATION
6 MANOR PARKWAY
SALEM, NEW HAMPSHIRE 03079-2841

Overview: Manages manufacturing (hardware, food service equipment, electronic products, machinery), printing, and mail order fruit operations. Established: 1955. Ownership: Public.
Key Statistics: Annual Sales: $564.6 million. Number of Employees: 5,000.
Hiring History: Number of professional employees hired in 1997: 5.
Expertise/Education Sought: Administration, accounting, business management, auditing
Alternatives to Full-Time Employment: Part-time jobs, paid internships
Company-Paid Training is available.
Contact: Mr. James Mettling, Director of Human Resources; Phone: 603-893-9701, ext. 314; Fax: 603-893-7324. E-mail: Mettling@standex.com.
Company URL: http://www.standex.com

SUMMIT PACKAGING SYSTEMS, INC.
16 AMMON DRIVE
MANCHESTER, NEW HAMPSHIRE 03103

Mailing Address: PO Box 5304, Manchester, NH 03103
Overview: Operates plastics processing facility. Established: 1976. Ownership: Private.
Key Statistics: Annual Sales: $90 million. Number of Employees: 350.
Expertise/Education Sought: Accounting, administration, finance, industrial engineering
Company-Paid Training is available.
International Assignments are available to employees.
Contact: Ms. Lillian Filia, Human Resource Director; Phone: 603-669-5410, ext. 156; Fax: 603-644-2594.

TEXTRON AUTOMOTIVE COMPANY
ROUTE 11, PO BOX 547
FARMINGTON, NEW HAMPSHIRE 03835

Overview: Manufactures automotive parts. Established: 1964. Parent Company: Textron, Inc. Ownership: Public.
Key Statistics: Annual Sales: $500 million. Number of Employees: 5,000.
Expertise/Education Sought: Automotive engineering, mechanical engineering, accounting, electrical engineering, laboratory technology

Company-Paid Training is available.

Contact: Ms. Sandra Conley, Manager of Human Resources; Phone: 603-755-4200; Fax: 603-755-4825.

TIMBERLAND COMPANY, INC.

200 DOMAIN DRIVE
STRATHAM, NEW HAMPSHIRE 03885-2575

Overview: Markets shoes and related apparel. Established: 1933. Ownership: Public.

Key Statistics: Annual Sales: $690 million. Number of Employees: 5,500.

Hiring History: Number of professional employees hired in 1996: 250.

Expertise/Education Sought: Administration, transportation, design, accounting, marketing, sales, customer service and support

International Assignments are available to employees.

Contact: Staffing; Phone: 603-772-9500; Fax: 603-773-1662.

Company URL: http://www.timberland.com

UNITRODE CORPORATION

7 CONTINENTAL BOULEVARD
MERRIMACK, NEW HAMPSHIRE 03054-4303

Overview: Produces integrated microcircuits. Established: 1960. Ownership: Public.

Key Statistics: Annual Sales: $133.5 million. Number of Employees: 562.

Hiring History: Number of professional employees hired in 1997: 175. 1996: 145.

Expertise/Education Sought: Information systems, computer science, computer analysis, computer engineering, computer-aided design

Alternatives to Full-Time Employment: Internships

Company-Paid Training is available.

Contact: Ms. Laura Gasparini, Senior Human Resources Representative; Phone: 603-424-2410. E-mail: gasparini@unitrode.com.

Company URL: http://www.unitrode.com

WORTHEN INDUSTRIES, INC.

3 EAST SPIT BROOK ROAD
NASHUA, NEW HAMPSHIRE 03060-5710

Overview: Manufactures adhesives. Established: 1957. Ownership: Private.

Key Statistics: Annual Sales: $55 million. Number of Employees: 225.

Expertise/Education Sought: Manufacturing

Company-Paid Training is available.

Contact: For individuals with previous experience: Ms. Janet Kelliher, Benefits Manager; Fax: 603-888-3059. E-mail: jkelliher@worthenind.com.

Company URL: http://www.worthenind.com

NEW JERSEY

ALLIED BEVERAGE GROUP, LLC

625 RAHWAY AVENUE
UNION, NEW JERSEY 07083-6631

Overview: Wholesales wine. Established: 1996. Ownership: Private.

Key Statistics: Annual Sales: $81.7 million.

Expertise/Education Sought: Sales, warehousing, transportation

Contact: For individuals with previous experience: Human Resources; Phone: 908-810-4000; Fax: 908-687-6732.

ALLIED BUILDING PRODUCTS CORPORATION

15 EAST UNION AVENUE
EAST RUTHERFORD, NEW JERSEY 07073-2127

Overview: Wholesales lumber and building materials. Established: 1996. Parent Company: Oldcastle. Ownership: Private.

Key Statistics: Number of Employees: 1,300.

Hiring History: Number of professional employees hired in 1997: 500. 1996: 300.

Expertise/Education Sought: Marketing

Company-Paid Training is available.

Contact: For individuals with previous experience: Mr. Bill Stevenson, Director of Human Resources; Phone: 201-507-3815.

AMERICAN HOME PRODUCTS CORPORATION

5 GIRALDA FARMS
MADISON, NEW JERSEY 07940

Overview: Manufactures and distributes pharmaceuticals, consumer products, medical supplies, and food products. Established: 1926. Ownership: Public.

Key Statistics: Annual Sales: $14.1 billion. Number of Employees: 59,747.

Expertise/Education Sought: Biochemistry, marketing, public relations, sales, compliance

Alternatives to Full-Time Employment: Paid internships

International Assignments are available to employees.

Contact: Mr. John Nathan, Director of Human Resources; Phone: 973-660-5301; Fax: 973-660-7026. E-mail: nathanj@ahp.com.

Company URL: http://www.ahp.com

AMERICAN STANDARD, INC.

1 CENTENNIAL AVENUE
PISCATAWAY, NEW JERSEY 08854

Overview: Manufactures plumbing fixtures. Established: 1988. Ownership: Public.

Key Statistics: Annual Sales: $5.8 billion. Number of Employees: 44,000.

Expertise/Education Sought: Technical engineering, product design and development, quality control, mechanical engineering, accounting, administration

Alternatives to Full-Time Employment: Part-time jobs

Contact: Human Resources Department, PO Box 6820, Piscataway, New Jersey, 08855-6820; Phone: 908-980-3000.

Company URL: http://www.us.amstd.com

ANADIGICS, INC.

35 TECHNOLOGY DRIVE
WARREN, NEW JERSEY 07059-5148

Overview: Manufactures integrated circuits. Established: 1985. Ownership: Public.

Key Statistics: Annual Sales: $68.9 million. Number of Employees: 500.

Hiring History: Number of professional employees hired in 1996: 110.

Expertise/Education Sought: Manufacturing engineering, mechanical engineering, electrical engineering, product engineering, process engineering, product design and development, design engineering

International Assignments are available to employees.

Contact: For individuals with previous experience: Human Resources Department.

ARROW FASTENER COMPANY, INC.

271 MAYHILL STREET
SADDLE BROOK, NEW JERSEY 07663-5303

Overview: Manufactures fastening systems, supplies, and accessories. Established: 1929. Ownership: Private.

Key Statistics: Annual Sales: $10 million. Number of Employees: 700.

Arrow Fastener Company, Inc. (continued)

Expertise/Education Sought: Manufacturing engineering, design engineering, manufacturing, sales

Contact: For individuals with previous experience: Personnel Department.

AUTOMATIC DATA PROCESSING

1 ADP BOULEVARD
ROSELAND, NEW JERSEY 07068

Overview: Provides computerized-business services. Established: 1949. Ownership: Public.

Key Statistics: Annual Sales: $4.1 billion. Number of Employees: 30,000.

Hiring History: Number of professional employees hired in 1996: 800.

Expertise/Education Sought: Accounting, sales, marketing, finance, computer science

Alternatives to Full-Time Employment: Paid internships

International Assignments are available to employees.

Contact: For college students and recent graduates: Mrs. Patty Gallagher-Ryan, College Recruiter; Phone: 201-994-5000; Fax: 201-535-3834. For individuals with previous experience: Mrs. Mary Jane Patterson, Senior Technical Recruiter; Phone: 973-994-5000; Fax: 973-535-3834.

Company URL: http://www.adp.com

C.R. BARD, INC.

730 CENTRAL AVENUE
MURRAY HILL, NEW JERSEY 07974

Overview: Develops, manufactures, and markets health-care products and instruments. Established: 1907. Ownership: Public.

Key Statistics: Annual Sales: $1.2 billion. Number of Employees: 9,800.

Hiring History: Number of professional employees hired in 1997: 150. 1996: 150.

Expertise/Education Sought: Sales, biomedical engineering, quality control

Alternatives to Full-Time Employment: Paid internships

Company-Paid Training is available.

International Assignments are available to employees.

Contact: Mr. Michael Conforti, Corporate Human Resources Manager; Phone: 908-277-8000; Fax: 908-277-8412.

BASF CORPORATION

3000 CONTINENTAL DRIVE NORTH
MOUNT OLIVE, NEW JERSEY 07828-1234

Overview: Manufactures industrial and agricultural chemicals, plastics, fibers and pharmaceutical preparations. Established: 1985. Parent Company: BASF, AG. Ownership: Private.

Key Statistics: Annual Sales: $6.1 billion. Number of Employees: 14,400.

Expertise/Education Sought: Accounting, marketing, logistics, chemical engineering, computer science

Alternatives to Full-Time Employment: Part-time jobs, paid internships

International Assignments are available to employees.

Contact: For college students and recent graduates: Ms. Jakarta Robinson, Proffesional Staffing and Recruitment; Phone: 973-426-4714; Fax: 973-426-2438. For individuals with previous experience: Mr. Ronald A. Hadley, Manager of Corporate Staffing; Phone: 973-426-3115; Fax: 973-426-2412.

Company URL: http://www.basf.com

BEL FUSE, INC.

198-200 VAN VORST STREET
JERSEY CITY, NEW JERSEY 07302

Overview: Manufactures local area network components. Established: 1949. Ownership: Public.

Key Statistics: Annual Sales: $65.5 million. Number of Employees: 707.

Hiring History: Number of professional employees hired in 1997: 3.

Expertise/Education Sought: Electrical engineering, mechanical engineering, accounting, administration, finance

Company-Paid Training is available.

International Assignments are available to employees.

Contact: Ms. Miriam Martinez, Personnel Manager; Phone: 201-432-0463, ext. 211; Fax: 201-432-9542.

BENJAMIN MOORE AND COMPANY

51 CHESTNUT RIDGE ROAD
MONTVALE, NEW JERSEY 07645-1801

Overview: Manufacturers paints, stains, and finishes. Established: 1883. Ownership: Public.

Key Statistics: Annual Sales: $625.2 million. Number of Employees: 2,082.

Hiring History: Number of professional employees hired in 1996: 200.

Expertise/Education Sought: Chemistry, chemical engineering, sales, marketing, finance, data processing

Alternatives to Full-Time Employment: Internships

Company-Paid Training is available.

Contact: For individuals with previous experience: Corporate Human Resources; Phone: 201-573-9600; Fax: 201-573-6631.

BESTFOODS

700 SYLVAN AVENUE
ENGLEWOOD CLIFFS, NEW JERSEY 07632

Overview: Produces processed foods. Established: 1906. Ownership: Public.

Key Statistics: Annual Sales: $9.8 billion. Number of Employees: 51,000.

Expertise/Education Sought: Accounting, auditing, marketing, finance, sales, management information systems

Alternatives to Full-Time Employment: Paid internships

International Assignments are available to employees.

Contact: Ms. Linda Paulhamus, Manager of Staffing; Phone: 201-894-2323; Fax: 201-894-2742.

Company URL: http://www.bestfoods.com

BLUESTONE SOFTWARE, INC.

1000 BRIGGS ROAD
MOUNT LAUREL, NEW JERSEY 08054-4101

Overview: Provides custom computer programming services. Established: 1997. Ownership: Private.

Key Statistics: Annual Sales: $11 million. Number of Employees: 100.

Expertise/Education Sought: Software engineering/development

Alternatives to Full-Time Employment: Internships

Company-Paid Training is available.

Contact: For individuals with previous experience: Ms. Shelly Dockerty, Technical Recruiter; Phone: 609-727-4600; Fax: 609-727-0124. E-mail: hr@bluestone.com.

Company URL: http://www.bluestone.com

BOC GASES

575 MOUNTAIN AVENUE
MURRAY HILL, NEW JERSEY 07974

Overview: Manufacturers compressed gases, anesthetics, turbines, pumps, and flow measuring devices. Established: 1983. Parent Company: BOC Group, Inc. Ownership: Private.

Key Statistics: Annual Sales: $1.8 billion. Number of Employees: 13,000.

Hiring History: Number of professional employees hired in 1997: 250.

Expertise/Education Sought: Accounting, finance, engineering

ALTERNATIVES TO FULL-TIME EMPLOYMENT: Part-time jobs
CONTACT: Ms. Jackie Cohen, Manager of Employee Relations; Phone: 908-771-1853; Fax: 908-771-1702.
COMPANY URL: http://www.boc.com

BOGEN COMMUNICATIONS, INC.

50 SPRING STREET
RAMSEY, NEW JERSEY 07446

OVERVIEW: Manufactures paging, intercom, and public address equipment. Established: 1993. Ownership: Public.
KEY STATISTICS: Annual Sales: $46.3 million. Number of Employees: 137.
EXPERTISE/EDUCATION SOUGHT: Electrical engineering, management information systems, sales, electronics/electronics engineering
ALTERNATIVES TO FULL-TIME EMPLOYMENT: Paid internships
COMPANY-PAID TRAINING is available.
INTERNATIONAL ASSIGNMENTS are available to employees.
CONTACT: For individuals with previous experience: Human Resource Department; Phone: 201-934-8500; Fax: 201-934-9832.

BWD AUTOMOTIVE

US HIGHWAY 1 AND ROUTE 130
NORTH BRUNSWICK, NEW JERSEY 08902

OVERVIEW: Manufactures electronic control equipment. Established: 1933. Parent Company: Echlin Corporation. Ownership: Private.
KEY STATISTICS: Annual Sales: $50 million. Number of Employees: 600.
EXPERTISE/EDUCATION SOUGHT: Manufacturing engineering, mechanical engineering, industrial engineering, sales, accounting, materials management, engineering, marketing
ALTERNATIVES TO FULL-TIME EMPLOYMENT: Paid internships
COMPANY-PAID TRAINING is available.
CONTACT: For individuals with previous experience: Mr. Joseph Hammerman, Human Resources Director; Phone: 732-545-7000; Fax: 732-545-7201.

CABLEVISION OF HUDSON COUNTY

360 FIRST STREET
HOBOKEN, NEW JERSEY 07030-2454

OVERVIEW: Provides cable TV services. Established: 1995. Parent Company: Cablevision Systems Corporation. Ownership: Public.
KEY STATISTICS: Annual Sales: $5.4 million. Number of Employees: 40.
EXPERTISE/EDUCATION SOUGHT: Customer service and support, field service, field engineering, sales, administration, public affairs
ALTERNATIVES TO FULL-TIME EMPLOYMENT: Paid internships
COMPANY-PAID TRAINING is available.
CONTACT: Human Resources; Phone: 201-798-1614, ext. 8362; Fax: 201-798-4163.

CAMPBELL SOUP COMPANY

1 CAMPBELL PLACE
CAMDEN, NEW JERSEY 08103

OVERVIEW: Manufactures soups and processed foods. Established: 1869. Ownership: Public.
KEY STATISTICS: Annual Sales: $8 billion. Number of Employees: 37,000.
HIRING HISTORY: Number of professional employees hired in 1997: 200. 1996: 200.
EXPERTISE/EDUCATION SOUGHT: Accounting, finance, engineering, product design and development, law, marketing
ALTERNATIVES TO FULL-TIME EMPLOYMENT: Paid internships
COMPANY-PAID TRAINING is available.
INTERNATIONAL ASSIGNMENTS are available to employees.
CONTACT: Ms. Martha Landry, Corporate Recruiter; Phone: 609-342-6242; Fax: 609-342-3721.
COMPANY URL: http://www.campbellsoup.com

CESSI TECHNOLOGIES, INC.

43 RAILROAD AVENUE
BEACHWOOD, NEW JERSEY 08722-3806

OVERVIEW: Provides business consulting and systems analysis services. Established: 1996. Ownership: Private.
KEY STATISTICS: Annual Sales: $80.1 million.
EXPERTISE/EDUCATION SOUGHT: Control engineering, software engineering/development, nursing
COMPANY-PAID TRAINING is available.
CONTACT: President and Owner; Phone: 732-818-7715; Fax: 732-818-7716. E-mail: info@cessi.com.
COMPANY URL: http://www.cessi.com

CHECKPOINT SYSTEMS, INC.

101 WOLF DRIVE
THOROFARE, NEW JERSEY 08086

MAILING ADDRESS: PO Box 188, Thorofare, NJ 08086-0188
OVERVIEW: Manufactures electronic surveillance devices. Established: 1977. Ownership: Public.
KEY STATISTICS: Annual Sales: $291.8 million. Number of Employees: 2,628.
EXPERTISE/EDUCATION SOUGHT: Sales, research and development, software engineering/development, information services, radio frequency
ALTERNATIVES TO FULL-TIME EMPLOYMENT: Part-time jobs, paid internships
INTERNATIONAL ASSIGNMENTS are available to employees.
CONTACT: Ms. Theresa McHale, Manager of Human Resources; Phone: 609-848-1800; Fax: 609-848-0937.
COMPANY URL: http://www.checkpointsystems.com

CHURCH AND DWIGHT COMPANY, INC.

469 NORTH HARRISON STREET
PRINCETON, NEW JERSEY 08543

OVERVIEW: Manufactures baking soda and cleaning agents. Established: 1846. Ownership: Public.
KEY STATISTICS: Annual Sales: $527.8 million. Number of Employees: 966.
EXPERTISE/EDUCATION SOUGHT: Accounting, marketing, sales
INTERNATIONAL ASSIGNMENTS are available to employees.
CONTACT: Mr. Joseph Holland, Senior Manager Administration; Phone: 609-683-5900; Fax: 609-497-7269.

COOPERHEAT, INC.

1021 CENTENNIAL AVENUE
PISCATAWAY, NEW JERSEY 08854-4125

OVERVIEW: Provides metal heat treatment services to industrial companies. Established: 1992. Parent Company: I.I.S.I.. Ownership: Public.
KEY STATISTICS: Annual Sales: $51 million. Number of Employees: 450.
EXPERTISE/EDUCATION SOUGHT: Accounting, finance, mechanical engineering, metallurgical engineering, industrial engineering, marketing, sales
ALTERNATIVES TO FULL-TIME EMPLOYMENT: Part-time jobs, paid internships
COMPANY-PAID TRAINING is available.
INTERNATIONAL ASSIGNMENTS are available to employees.
CONTACT: Ms. Maria Marcelli, Corporate Market Coordinator; Phone: 732-981-0800, ext. 322; Fax: 732-981-0850.
COMPANY URL: http://www.cooperheat.com

DIALOGIC

1515 ROUTE 10
PARSIPPANY, NEW JERSEY 07054

Overview: Manufactures computer telephony and components. Established: 1983.

Key Statistics: Annual Sales: $261 million. Number of Employees: 1,000.

Hiring History: Number of professional employees hired in 1997: 300.

Expertise/Education Sought: Computer science, software engineering/development, computer engineering, computer programming, sales

Alternatives to Full-Time Employment: Paid internships

Company-Paid Training is available.

International Assignments are available to employees.

Contact: For college students and recent graduates: Mr. Frank T. Johnson, Manager of University Relations; Phone: 973-993-3000, ext. 6341; Fax: 973-993-5762. E-mail: johnsonf@dialogic.com. For individuals with previous experience: Mr. Terry Williams, Staffing Manager; Phone: 973-993-3000, ext. 6411; Fax: 973-993-5762. E-mail: twilliams@dialogic.com.

Company URL: http://www.dialogic.com

For more information, see full corporate profile, pg. 274.

DURO-TEST CORPORATION

9 LAW DRIVE
FAIRFIELD, NEW JERSEY 07004-3206

Overview: Manufactures lighting fixtures. Established: 1995. Ownership: Private.

Key Statistics: Annual Sales: $126.7 million. Number of Employees: 1,900.

Expertise/Education Sought: Sales, accounting, finance, marketing

International Assignments are available to employees.

Contact: For individuals with previous experience: Ms. Shirley Benson, Employment Manager; Phone: 973-808-1800, ext. 5311; Fax: 973-472-5220.

ELECTRO-CATHETER CORPORATION

2100 FELVER COURT
RAHWAY, NEW JERSEY 07065

Overview: Manufactures surgical and medical equipment. Established: 1961. Ownership: Public.

Key Statistics: Annual Sales: $7.4 million. Number of Employees: 98.

Hiring History: Number of professional employees hired in 1996: 9.

Expertise/Education Sought: Engineering, research and development

Company-Paid Training is available.

Contact: Ms. Emily Gordon, Personnel Administrator; Phone: 732-382-5600; Fax: 732-382-7107. E-mail: elecath@earthlink.com.

FOSTER WHEELER CORPORATION

PERRYVILLE CORPORATE PARK
CLINTON, NEW JERSEY 08809

Overview: Provides industrial engineering and construction services. Established: 1900. Ownership: Public.

Key Statistics: Annual Sales: $4 billion. Number of Employees: 12,085.

Expertise/Education Sought: Accounting, finance, computer programming, marketing, mechanical engineering, electrical engineering, structural engineering, environmental engineering, chemical engineering

Alternatives to Full-Time Employment: Paid internships

Company-Paid Training is available.

International Assignments are available to employees.

Contact: For college students and recent graduates: Mr. Robert Austin, Manager of Human Resources; Phone: 908-713-3450; Fax: 908-730-4980. For individuals with previous experience: Mr. Richard Lively, Director of Human Resources; Phone: 908-713-3450; Fax: 908-713-3490.

FRANCO MANUFACTURING COMPANY, INC.

555 PROSPECT STREET
METUCHEN, NEW JERSEY 08840-2271

Overview: Processes textiles. Established: 1958. Ownership: Private.

Key Statistics: Annual Sales: $42.9 million. Number of Employees: 600.

Expertise/Education Sought: Manufacturing

International Assignments are available to employees.

Contact: For individuals with previous experience: Human Resources; Phone: 732-494-0500; Fax: 732-906-0591.

GAS MATERIAL CORPORATION

1361 ALPS ROAD
WAYNE, NEW JERSEY 07470-3700

Overview: Manufactures industrial organic chemicals. Established: 1996. Parent Company: ISP Corporation. Ownership: Private.

Key Statistics: Annual Sales: $612.4 million. Number of Employees: 2,700.

Expertise/Education Sought: Administration, engineering, computer science, customer service and support

Alternatives to Full-Time Employment: Paid internships

Company-Paid Training is available.

International Assignments are available to employees.

Contact: For individuals with previous experience: Human Resource Department; Phone: 973-628-3000; Fax: 973-628-3119.

Company URL: http://www.gas.com

GEMINI INDUSTRIES, INC.

215 ENTIN ROAD
CLIFTON, NEW JERSEY 07014-1424

Overview: Manufactures electronic accessories. Established: 1988. Ownership: Private.

Key Statistics: Annual Sales: $172 million. Number of Employees: 450.

Hiring History: Number of professional employees hired in 1996: 302.

Expertise/Education Sought: Business administration, marketing, office management, computer science, electronics/electronics engineering

Company-Paid Training is available.

Contact: For individuals with previous experience: Mrs. Dee Kean, Employment Coordinator, Building 179, 215 Entin Road, Clifton, New Jersey, 07014-1424; Phone: 201-471-9050, ext. 536; Fax: 201-547-7215. E-mail: deekean@gemini-usa.com.

GENERAL FOAM CORPORATION

W100 CENTURY ROAD
PARAMUS, NEW JERSEY 07652

Overview: Produces foam rubber. Established: 1994. Parent Company: PMC, Inc. Ownership: Private.

Key Statistics: Annual Sales: $56.1 million. Number of Employees: 727.

Expertise/Education Sought: Sales, manufacturing, laboratory assistance

International Assignments are available to employees.

Contact: For individuals with previous experience: Personnel Department; Phone: 201-262-7500.

GPU ENERGY

300 MADISON AVENUE
MORRISTOWN, NEW JERSEY 07960

Overview: Electric utility. Established: 1941. Parent Company: GPU, Inc. Ownership: Public.

Key Statistics: Assets: $2.1 billion. Number of Employees: 2,538.

Expertise/Education Sought: Electrical engineering, mechanical engineering, marketing

Company-Paid Training is available.

Contact: Ms. Lois Theis, Senior Administrator, 915 Valley Road, Morristown, New Jersey, 07962-6108; Phone: 973-455-8502; Fax: 973-455-8582. Job Hotline: 800-893-5781.

GPU, INC.

300 MADISON AVENUE
MORRISTOWN, NEW JERSEY 07960

Overview: Public utility. Established: 1906. Ownership: Public.

Key Statistics: Assets: $3.9 billion. Number of Employees: 9,300.

Expertise/Education Sought: Nuclear engineering, mechanical engineering, electrical engineering, accounting, management information systems, human resources

Alternatives to Full-Time Employment: Part-time jobs, paid internships

Company-Paid Training is available.

International Assignments are available to employees.

Contact: For individuals with previous experience: Human Resources; Phone: 973-455-8200; Fax: 973-644-4107.

Company URL: http://www.gpu.com

GRAVER WATER SYSTEMS

750 WALNUT AVENUE
CRANFORD, NEW JERSEY 07016

Overview: Designs wastewater treatment systems. Established: 1983. Parent Company: Marmon Industrial Corporation. Ownership: Private.

Key Statistics: Annual Sales: $35 million. Number of Employees: 120.

Hiring History: Number of professional employees hired in 1997: 3. 1996: 6.

Expertise/Education Sought: Design engineering, mechanical engineering, chemical engineering, environmental engineering, data processing

Contact: For individuals with previous experience: Mr. Michael Coyne, Controller Vice President; Phone: 908-653-4200; Fax: 908-653-4320.

THE GREAT ATLANTIC AND PACIFIC TEA COMPANY, INC.

2 PARAGON DRIVE
MONTVALE, NEW JERSEY 07645

Overview: Operates chain of food stores. Established: 1859. Ownership: Public.

Key Statistics: Annual Sales: $10.1 billion. Number of Employees: 84,000.

Expertise/Education Sought: Computer science, accounting, finance, retail

Alternatives to Full-Time Employment: Paid internships

International Assignments are available to employees.

Contact: Ms. Corine Blake, Director of Personnel, PO Box 418, Montvale, New Jersey, 07645-1718; Phone: 201-930-4416; Fax: 201-930-8144.

HAYWARD INDUSTRIES, INC.

620 DIVISION STREET
ELIZABETH, NEW JERSEY 07201-2012

Overview: Manufactures pool filtration equipment. Established: 1964. Ownership: Public.

Key Statistics: Annual Sales: $234.5 million. Number of Employees: 1,500.

Expertise/Education Sought: Management information systems, sales, management, finance, product engineering, marketing

Company-Paid Training is available.

International Assignments are available to employees.

Contact: For individuals with previous experience: Human Resources; Phone: 908-351-5400; Fax: 908-351-0604.

HOFFMANN-LA ROCHE, INC.

340 KINGSLAND STREET
NUTLEY, NEW JERSEY 07110

Overview: Manufactures pharmaceutical products. Established: 1905. Ownership: Private.

Key Statistics: Annual Sales: $2.5 billion. Number of Employees: 5,000.

Expertise/Education Sought: Accounting, finance, marketing, data processing, computer science, business administration, pharmacy, pharmaceutical, biochemistry

Alternatives to Full-Time Employment: Paid internships

Company-Paid Training is available.

International Assignments are available to employees.

Contact: Mr. Alvin L. Vinson, Director of Human Resources; Phone: 973-235-5000. Job Hotline: 973-235-5000.

K. HOVNANIAN ENTERPRISES, INC.

10 STATE HIGHWAY 35
RED BANK, NEW JERSEY 07701

Overview: Develops real estate. Established: 1959. Ownership: Public.

Key Statistics: Annual Sales: $807.5 million. Number of Employees: 1,100.

Hiring History: Number of professional employees hired in 1996: 200.

Expertise/Education Sought: Architecture, civil engineering, planning, environmental engineering

Alternatives to Full-Time Employment: Part-time jobs

Contact: Mr. Bill Carpitella, Senior Vice President; Phone: 732-747-7800; Fax: 732-747-0913. Job Hotline: 888-287-0517.

INGERSOLL-RAND COMPANY

200 CHESTNUT RIDGE ROAD
WOODCLIFF LAKE, NEW JERSEY 07675-7703

Overview: Manufactures industrial machinery and equipment. Established: 1871. Ownership: Public.

Key Statistics: Annual Sales: $6.7 billion. Number of Employees: 41,874.

Expertise/Education Sought: Sales, accounting, human resources, finance, mechanical engineering, industrial engineering

Alternatives to Full-Time Employment: Part-time jobs, paid internships

Company-Paid Training is available.

International Assignments are available to employees.

Contact: Human Resource; Phone: 201-573-0123; Fax: 201-573-3168.

Company URL: http://www.ingersoll-rand.com

ISP/GAF CORPORATION

1361 ALPS ROAD
WAYNE, NEW JERSEY 07470-3700

Overview: Manufactures specialty chemicals and building materials. Established: 1989. Ownership: Private.

Key Statistics: Annual Sales: $1.3 billion. Number of Employees: 5,000.

Expertise/Education Sought: Chemical engineering, computer programming, accounting, law, marketing, sales, research

Alternatives to Full-Time Employment: Paid internships

International Assignments are available to employees.

Contact: For individuals with previous experience: Mr. Gary Schneid, Director of Employment; Phone: 973-628-3083; Fax: 973-628-3577. E-mail: gschneid@ispcorp.com.

Company URL: http://www.ispcorp.com

ITT AVIONICS

100 KINGSLAND ROAD
CLIFTON, NEW JERSEY 07014

Overview: Designs, develops and supports electronic warfare systems. Established: 1967. Parent Company: ITT Industries.

Hiring History: Number of professional employees hired in 1997: 80. 1996: 71.

ITT Avionics (continued)

Expertise/Education Sought: Software engineering/development, ASCI/MMIC design, threat analysis, systems engineering
Alternatives to Full-Time Employment: Paid internships
Company-Paid Training is available.
Contact: Mr. Thomas J. Gildea, Human Resource Manager; Phone: 973-284-2665; Fax: 973-284-5220. E-mail: tgilda@avionics.itt.com.

KEPTEL, INC.
56 PARK ROAD
PINTON FALLS, NEW JERSEY 07724

Overview: Manufactures telephone network interface systems. Established: 1994. Parent Company: Antec Corporation. Ownership: Public.
Key Statistics: Annual Sales: $50 million. Number of Employees: 776.
Hiring History: Number of professional employees hired in 1996: 20.
Expertise/Education Sought: Manufacturing engineering, mechanical engineering, sales, warehousing
Alternatives to Full-Time Employment: Part-time jobs, paid internships
Company-Paid Training is available.
Contact: Human Resources; Phone: 908-389-8800.
Company URL: http://www.antec.com

Lucent Technologies
Bell Labs Innovations

We make the things that make communications work.

LUCENT TECHNOLOGIES
600 MOUNTAIN AVENUE
MURRAY HILL, NEW JERSEY 07974

Overview: Designs and builds telecommunication systems. Established: 1996. Ownership: Public.
Key Statistics: Annual Sales: $280 billion. Number of Employees: 130,000.
Expertise/Education Sought: Radio frequency engineering, software engineering/development, systems engineering, circuit design, marketing
Contact: For college students and recent graduates: Mr. Donald Jensen, University Relations, 101 Crawfords Corner Road, Room 1B212, Holmdel, New Jersey, 07733; Phone: 908-949-4633; Fax: 908-949-5900. E-mail: dmjensen@lucent.com. For individuals with previous experience: Mr. Reginald Barefield, District Manager of Human Resources, 283 King George Road, Room B1C-40, Warren, New Jersey, 07059; Phone: 908-559-4735; Fax: 908-559-2084. E-mail: rbarefield@lucent.com.
Company URL: http://www.lucent.com

For more information, see full corporate profile, pg. 282.

LUCENT TECHNOLOGIES CORPORATION
600-700 MOUNTAIN AVENUE
MURRAY HILL, NEW JERSEY 07974

Overview: Manufactures industrial inorganic chemicals. Established: 1989. Ownership: Public.
Key Statistics: Annual Sales: $50 million. Number of Employees: 100.
Hiring History: Number of professional employees hired in 1996: 5.
Expertise/Education Sought: Engineering, environmental
Alternatives to Full-Time Employment: Internships
International Assignments are available to employees.
Contact: Personnel Department; Phone: 908-582-8500; Fax: 908-508-2576.
Company URL: http://www.lucent.com

MALLINCKRODT BAKER, INC.
222 RED SCHOOL LANE
PHILLIPSBURG, NEW JERSEY 08865-2219

Overview: Manufactures industrial inorganic chemicals. Established: 1995. Parent Company: Mallinckrodt, Inc. Ownership: Public.
Key Statistics: Annual Sales: $144.8 million. Number of Employees: 900.
Expertise/Education Sought: Manufacturing
Alternatives to Full-Time Employment: Paid internships
Company-Paid Training is available.
International Assignments are available to employees.
Contact: Human Resources; Phone: 908-859-2151; Fax: 908-859-9368. E-mail: hpkilpa@mkg.com.

MARCAL PAPER MILLS, INC.
1 MARKET STREET
ELMWOOD PARK, NEW JERSEY 07407-1401

Overview: Manufactures paper napkins, tissue, and toilet paper. Established: 1932. Ownership: Private.
Key Statistics: Number of Employees: 1,150.
Hiring History: Number of professional employees hired in 1997: 75. 1996: 75.
Contact: For individuals with previous experience: Mr. Jim Nelson, Director of Human Resources; Phone: 201-703-6242; Fax: 201-703-6451.
Company URL: http://www.marcalpaper.com

MERRILL LYNCH—PRIVATE CLIENT GROUP
800 SCUDDERS MILL ROAD
PLAINSBORO, NEW JERSEY 08536

Overview: Investment firm. Established: 1914. Parent Company: Merrill Lynch & Company. Ownership: Public.
Key Statistics: Assets: $865 billion. Number of Employees: 50,000.
Expertise/Education Sought: Technology, information systems
Contact: Ms. Amanda Nealley, College Recruitment Manager; Phone: 609-282-1450.
Company URL: http://www.ml.com

METEX CORPORATION
970 NEW DURHAM ROAD
EDISON, NEW JERSEY 08818-4020

Overview: Manufactures wire mesh. Established: 1991. Parent Company: United Capital Corporation. Ownership: Public.
Key Statistics: Annual Sales: $55.6 million. Number of Employees: 500.
Hiring History: Number of professional employees hired in 1996: 4.
Expertise/Education Sought: Manufacturing, customer service and support
Company-Paid Training is available.
Contact: For individuals with previous experience: Human Resources Department; Phone: 908-287-0800; Fax: 908-248-8739.

MINOLTA CORPORATION
101 WILLIAMS DRIVE
RAMSEY, NEW JERSEY 07446-1217

Overview: Manufactures photocopying machines and photographic equipment and supplies. Established: 1959. Ownership: Private.
Key Statistics: Annual Sales: $1.1 billion. Number of Employees: 3,880.
Hiring History: Number of professional employees hired in 1997: 75.
Expertise/Education Sought: Sales, accounting, marketing, software engineering/development, credit management
Contact: Ms. Linda Loveland, Employment and Compensation Manager; Phone: 201-934-5251; Fax: 201-825-7567. E-mail: LLoveland@Minolta.com.

MINOLTA INFORMATION SYSTEMS

1 INTERNATIONAL BOULEVARD, FLOOR 9
MAHWAH, NEW JERSEY 07495-0025

Overview: Designs computer systems and offers computer integration and software development services. Established: 1997. Parent Company: Minolta Corporation. Ownership: Private.

Key Statistics: Annual Sales: $13 million. Number of Employees: 52.

Expertise/Education Sought: Software engineering/development, product management

Company-Paid Training is available.

Contact: Human Resources; Phone: 201-825-4000.

Company URL: http://www.misi.minolta.com

MISTRAS HOLDINGS CORPORATION

195 CLARKSVILLE ROAD
LAWRENCEVILLE, NEW JERSEY 08648-2301

Mailing Address: PO Box 3135, Princeton, NJ 08543-3135

Overview: Manufactures measuring and controlling devices. Established: 1994. Ownership: Private.

Key Statistics: Annual Sales: $18 million. Number of Employees: 193.

Hiring History: Number of professional employees hired in 1997: 26. 1996: 25.

Expertise/Education Sought: Electrical engineering, engineering, manufacturing engineering

Company-Paid Training is available.

International Assignments are available to employees.

Contact: Ms. Linda Azzinaro, Human Resources Director; Phone: 609-716-4016; Fax: 609-716-0706.

NATIONAL STARCH AND CHEMICAL COMPANY

10 FINDERNE AVENUE
BRIDGEWATER, NEW JERSEY 08807-3355

Overview: Manufactures starch, adhesives, and resins. Established: 1989. Parent Company: ICI. Ownership: Private.

Key Statistics: Annual Sales: $2.4 billion. Number of Employees: 7,700.

Hiring History: Number of professional employees hired in 1996: 25.

Expertise/Education Sought: Chemistry, food science, sales, finance, chemical engineering, materials science, polymers

Alternatives to Full-Time Employment: Part-time jobs, paid internships

Company-Paid Training is available.

International Assignments are available to employees.

Contact: Miss Colleen Twill, Employment Supervisor; Phone: 800-366-4031; Fax: 908-685-6956. E-mail: nsc.recruit@nstarch.com.

Company URL: http://www.nationalstarch.com

NER DATA PRODUCTS, INC.

307 SOUTH DELSEA DRIVE
GLASSBORO, NEW JERSEY 08028

Mailing Address: PO Box 524, Glassboro, NJ 08028

Overview: Manufactures typewriter ribbons and plastic cartridges. Established: 1985. Parent Company: Hargro Associates. Ownership: Private.

Key Statistics: Annual Sales: $64.6 million. Number of Employees: 600.

Expertise/Education Sought: Accounting, finance, marketing, sales, data processing, engineering

Alternatives to Full-Time Employment: Part-time jobs, paid internships

Company-Paid Training is available.

Contact: For individuals with previous experience: Ms. Kathie Kane, Manager of Human Resources; Phone: 609-881-5524; Fax: 609-881-2393.

Company URL: http://www.nerdata.com

NETWORK PROGRAMS LLC

255 OLD NEW BRUNSWICK ROAD
PISCATAWAY, NEW JERSEY 08854-3734

Overview: Develops computer software and applications. Established: 1996. Ownership: Private.

Key Statistics: Annual Sales: $25 million. Number of Employees: 160.

Expertise/Education Sought: Software engineering/development, systems analysis, systems administration

Alternatives to Full-Time Employment: Paid internships

Company-Paid Training is available.

International Assignments are available to employees.

Contact: Human Resources; Phone: 732-562-1111; Fax: 732-562-0395. E-mail: resumes@mpi.www.com.

NEW JERSEY NATURAL GAS COMPANY

1415 WYCKOFF ROAD
WALL, NEW JERSEY 07719

Overview: Provides natural gas distribution services. Established: 1995. Parent Company: NJ Resources Corporation. Ownership: Public.

Expertise/Education Sought: Accounting, electrical engineering, engineering, management, manufacturing engineering, chemical engineering

Alternatives to Full-Time Employment: Internships

Contact: For individuals with previous experience: Ms. Hollis Cooper, Human Resources Consultant; Phone: 732-938-1262; Fax: 732-938-2134. Job Hotline: 732-938-1000.

NEW JERSEY TURNPIKE AUTHORITY

ROUTE 18 AND EXIT 9
EAST BRUNSWICK, NEW JERSEY 08816

Mailing Address: PO Box 1121, New Brunswick, NJ 08903-1121

Overview: Operates toll road. Established: 1948. Ownership: Private.

Key Statistics: Annual Sales: $444.1 million. Number of Employees: 2,395.

Expertise/Education Sought: Transportation

Alternatives to Full-Time Employment: Paid internships

Company-Paid Training is available.

Contact: Human Resources Department, PO Box 1121, East Brunswick, New Jersey, 08903; Phone: 732-247-0900; Fax: 732-828-4508.

NORTHERN BUILDING PRODUCTS

1 REMSEN PLACE
RIDGEFIELD, NEW JERSEY 07657-2321

Overview: Manufactures metal window and door frames. Established: 1978. Ownership: Private.

Key Statistics: Annual Sales: $6.8 million. Number of Employees: 80.

Hiring History: Number of professional employees hired in 1997: 4.

Expertise/Education Sought: Accounting, administration, design engineering, industrial engineering

Contact: For individuals with previous experience: Human Resources Department; Phone: 201-943-6400; Fax: 201-943-9282.

OHM CORPORATION

16406 US ROUTE 224 EAST
TRENTON, NEW JERSEY 08691-1904

Overview: Provides site environmental remediation and construction services. Established: 1969. Ownership: Public.

Key Statistics: Annual Sales: $550 million. Number of Employees: 3,100.

Expertise/Education Sought: Construction management, civil engineering, environmental engineering, geologist, hydrogeologist

Contact: For college students and recent graduates: Mr. Tom Young, Staffing Specialist; Phone: 609-588-6492; Fax: 609-588-6399. For individuals with previous experience: Mr. Jay Earl, Corporate Recruiting Manager; Phone: 609-588-6430; Fax: 609-588-6399.

Company URL: http://www.ohm.com

ORTHO-CLINICAL DIAGNOSTICS
1001 US HIGHWAY 202
RARITAN, NEW JERSEY 08869-0606

Overview: Provides blood diagnostic services. Established: 1973. Parent Company: Johnson & Johnson Corporation. Ownership: Public.

Key Statistics: Annual Sales: $132.1 million. Number of Employees: 1,600.

Hiring History: Number of professional employees hired in 1997: 70. 1996: 50.

Expertise/Education Sought: Laboratory technology, biology, microbiology, chemistry, quality engineering

Company-Paid Training is available.

International Assignments are available to employees.

Contact: For individuals with previous experience: Ms. Jean L. Messina, Human Resources Director; Phone: 908-218-8525; Fax: 908-704-3649.

PATHMARK STORES, INC.
200 MILIK STREET
CARTERET, NEW JERSEY 07008

Overview: Operates chain of food stores. Established: 1987. Ownership: Private.

Key Statistics: Annual Sales: $3.7 billion. Number of Employees: 29,700.

Expertise/Education Sought: Accounting, finance, retail management

Alternatives to Full-Time Employment: Paid internships

Company-Paid Training is available.

Contact: Ms. Renee Perlman, Human Resources Recruiter; Phone: 908-499-3000, ext. 4192; Fax: 908-499-4961.

PFISTER CHEMICAL, INC.
1098 US HIGHWAY 46
RIDGEFIELD, NEW JERSEY 07657-1016

Mailing Address: PO Box 15, Ridgefield, NJ 07657

Overview: Produces dyes and cyclic organic intermediates. Established: 1936. Ownership: Private.

Key Statistics: Annual Sales: $20 million. Number of Employees: 130.

Hiring History: Number of professional employees hired in 1997: 4. 1996: 4.

Expertise/Education Sought: Manufacturing, chemistry

Company-Paid Training is available.

Contact: For individuals with previous experience: Ms. Sally Bowen, Human Resources; Phone: 201-945-5400; Fax: 201-945-0159.

Company URL: http://www.pfisterchemical.com

PRESIDENT CONTAINER, INC.
200 WEST COMMERCIAL AVENUE
MOONACHIE, NEW JERSEY 07074-1610

Overview: Manufactures corrugated boxes. Established: 1955. Ownership: Private.

Key Statistics: Annual Sales: $50 million. Number of Employees: 300.

Hiring History: Number of professional employees hired in 1997: 80.

Expertise/Education Sought: Manufacturing, distribution, customer service and support

Contact: Mr. Jon Van Skiver, Director of Human Resources; Phone: 201-933-7500, ext. 580; Fax: 201-842-1452.

PUBLIC SERVICE ELECTRIC AND GAS
80 PARK PLAZA
NEWARK, NEW JERSEY 07101

Mailing Address: PO Box 570, Newark, NJ 07101-0570

Overview: Public utility providing gas and electric service in the state of New Jersey. Established: 1903. Parent Company: Public Service Enterprise Group, Inc. Ownership: Public.

Key Statistics: Annual Sales: $6 billion. Number of Employees: 11,600.

Expertise/Education Sought: Computer science, electrical engineering, mechanical engineering, nuclear engineering, information systems

Contact: Mr. Al Branca, Employment Manager, 234 Pierson Avenue, Edison, New Jersey, 08837; Phone: 908-417-2771; Fax: 908-417-1159. E-mail: abranca@pseg.com.

CORPORATE STATEMENT

PSE&G is a regulated utility providing electric and energy services to more than 2 million customers and more than 5.5 million residents of New Jersey. It is the state's largest utility and one of America's largest combined electric and gas energy companies.

RAPIDFORMS, INC.
301 GROVE ROAD
THOROFARE, NEW JERSEY 08086-9400

Overview: Wholesales business forms. Established: 1971. Parent Company: NEBS. Ownership: Public.

Key Statistics: Annual Sales: $89 million. Number of Employees: 600.

Expertise/Education Sought: Accounting, finance, sales, graphic arts, graphic arts, graphics engineering, print production, printing

Company-Paid Training is available.

Contact: For individuals with previous experience: Human Resources Department; Phone: 609-384-1144; Fax: 609-384-1797.

RCN CORPORATION
105 CARNEGIE CENTER
PRINCETON, NEW JERSEY 08540-6215

Overview: Provides long distance phone services. Established: 1990. Ownership: Private.

Key Statistics: Annual Sales: $20.9 million. Number of Employees: 130.

Hiring History: Number of professional employees hired in 1997: 600.

Expertise/Education Sought: Customer service and support, communication engineering, communications, telecommunications, construction, networks, network management systems

Alternatives to Full-Time Employment: Paid internships

Company-Paid Training is available.

International Assignments are available to employees.

Contact: Mr. Ed Jackenthal, Employment Manager; Phone: 609-734-7504; Fax: 609-734-3777.

Company URL: http://www.rcn.com

RICOH CORPORATION
5 DEDRICK PLACE
CALDWELL, NEW JERSEY 07006-6304

Overview: Manufactures copiers, facsimile machines, cameras, and other office equipment. Established: 1962. Parent Company: Ricoh Company, Ltd. Ownership: Private.

Key Statistics: Annual Sales: $1.5 billion. Number of Employees: 2,800.

Expertise/Education Sought: Software engineering/development, computer programming, data processing, finance, digital technology, accounting, marketing

Alternatives to Full-Time Employment: Part-time jobs

Contact: Ms. Joanne Ehman, Manager of Human Resources; Phone: 973-882-2000, ext. 2233; Fax: 973-882-2199.

Company URL: http://www.ricoh.com

SANDVIK COROMANT COMPANY
1702 NEVINS ROAD
FAIR LAWN, NEW JERSEY 07410-2886

Overview: Produces cold-rolled steel strip and wire. Established: 1919. Parent Company: AB Sandvik Coromant. Ownership: Private.

Key Statistics: Annual Sales: $638.6 million. Number of Employees: 2,858.

Hiring History: Number of professional employees hired in 1997: 151.

Expertise/Education Sought: Manufacturing engineering, sales, marketing

Alternatives to Full-Time Employment: Paid internships
Company-Paid Training is available.
International Assignments are available to employees.
Contact: For individuals with previous experience: Human Relations Department; Phone: 201-794-5152; Fax: 201-794-5165.
Company URL: http://www.sandvik.com

SEALED AIR CORPORATION
PARK 80 EAST
SADDLE BROOK, NEW JERSEY 07663

Overview: Manufactures protective packaging for consumer and commercial use. Established: 1960. Ownership: Public.
Key Statistics: Annual Sales: $789.6 million. Number of Employees: 4,200.
Expertise/Education Sought: Chemical engineering, design engineering, manufacturing engineering
Alternatives to Full-Time Employment: Part-time jobs, paid internships
Company-Paid Training is available.
International Assignments are available to employees.
Contact: For individuals with previous experience: Benefits Department; Phone: 201-791-7600; Fax: 201-703-4205.

SHARP ELECTRONICS CORPORATION
PO BOX 650
MAHWAH, NEW JERSEY 07430

Overview: Manufactures and markets consumer electronics products, consumer appliances, and office equipment. Established: 1962. Parent Company: Sharp Corporation Japan. Ownership: Private.
Key Statistics: Annual Sales: $3.1 billion. Number of Employees: 2,300.
Expertise/Education Sought: Marketing, sales
Alternatives to Full-Time Employment: Paid internships
Company-Paid Training is available.
International Assignments are available to employees.
Contact: Mr. Steve Ness, Manager of Employment; Phone: 201-529-8442; Fax: 201-529-8869.

SIEMENS HEARING INSTRUMENTS, INC.
10 CONSTITUTION AVENUE
PISCATAWAY, NEW JERSEY 08854-6145

Overview: Manufactures hearing aids. Established: 1991. Parent Company: Siemans AG. Ownership: Private.
Key Statistics: Annual Sales: $63 million. Number of Employees: 370.
Expertise/Education Sought: Production, sales, marketing, electrical engineering
Company-Paid Training is available.
Contact: For individuals with previous experience: Ms. Crystal Ruiz, Manager of Human Resources; Phone: 732-562-6600; Fax: 732-562-6617.

SONY ELECTRONICS, INC.
1 SONY DRIVE
PARK RIDGE, NEW JERSEY 07656-3211

Overview: Manufactures and markets electronic equipment, office equipment, and home appliances. Established: 1988. Parent Company: Sony Corporation. Ownership: Private.
Key Statistics: Annual Sales: $19.2 billion. Number of Employees: 42,000.
Expertise/Education Sought: Computer science, manufacturing engineering, accounting, electronics/electronics engineering, finance, electrical engineering, marketing, sales
Contact: For individuals with previous experience: Recruitment; Phone: 201-930-1000; Fax: 201-930-1001.

SOUTH JERSEY INDUSTRIES, INC.
1 SOUTH JERSEY PLAZA
FOLSOM, NEW JERSEY 08037

Overview: Distributes and transmits natural gas; mines industrial sand and gravel. Established: 1969. Ownership: Public.
Key Statistics: Annual Sales: $355.5 million. Number of Employees: 1,100.
Hiring History: Number of professional employees hired in 1996: 20.
Expertise/Education Sought: Accounting, finance, auditing, marketing, computer programming
Alternatives to Full-Time Employment: Paid internships
Company-Paid Training is available.
Contact: For individuals with previous experience: Mrs. Patricia Wilus, Human Resources Manager; Phone: 609-561-9000; Fax: 609-561-8225. Job Hotline: 609-561-9000, ext. 4453.

SPRINGFIELD PRECISION INSTRUMENTS
76 PASSAIC STREET
WOOD RIDGE, NEW JERSEY 07075-1004

Overview: Manufactures industrial control and process control instruments. Established: 1997. Ownership: Public.
Key Statistics: Annual Sales: $14.8 million. Number of Employees: 150.
Hiring History: Number of professional employees hired in 1997: 20.
Expertise/Education Sought: Engineering, manufacturing
Contact: For individuals with previous experience: Human Resources Department; Phone: 973-777-2900.

TERUMO MEDICAL CORPORATION
2100 COTTONTAIL LANE
SOMERSET, NEW JERSEY 08873-1271

Overview: Manufactures medical supplies. Established: 1972. Parent Company: Terumo Corporation. Ownership: Private.
Key Statistics: Annual Sales: $162 million. Number of Employees: 500.
Expertise/Education Sought: Sales, accounting, marketing
Alternatives to Full-Time Employment: Paid internships
Company-Paid Training is available.
Contact: For individuals with previous experience: Ms. Helene Katz, Corporate Human Resources Manager; Phone: 732-302-4900; Fax: 732-302-3083.

TOYS R US, INC.
461 FROM ROAD
PARAMUS, NEW JERSEY 07652-3524

Overview: Operates chain of toy stores. Established: 1928. Ownership: Public.
Key Statistics: Annual Sales: $9.9 billion. Number of Employees: 60,000.
Hiring History: Number of professional employees hired in 1997: 1,400. 1996: 1,300.
Expertise/Education Sought: Management
Alternatives to Full-Time Employment: Paid internships
Company-Paid Training is available.
International Assignments are available to employees.
Contact: For college students and recent graduates: Ms. Lisa Slater, Manager of Field Recruitment; Phone: 201-599-8808. For individuals with previous experience: Mr. James Gorenc, Director of Employment; Phone: 201-599-8808.
Company URL: http://www.toysrus.com

TURTLE & HUGHES, INC.
1900 LOWER ROAD
LINDEN, NEW JERSEY 07036-6519

Overview: Wholesales electrical apparatus and equipment. Established: 1964. Ownership: Private.
Key Statistics: Annual Sales: $133.6 million. Number of Employees: 310.

Turtle & Hughes, Inc. (continued)

Hiring History: Number of professional employees hired in 1996: 5.
Expertise/Education Sought: Manufacturing engineering, electrical engineering, accounting
Alternatives to Full-Time Employment: Part-time jobs, paid internships
Company-Paid Training is available.
Contact: Human Resources Department; Phone: 732-574-3600; Fax: 732-388-4471.
Company URL: http://www.turtle.com

TYCO SUBMARINE SYSTEMS, LTD.
340 MOUNT KEMBLE AVENUE
MORRISTOWN, NEW JERSEY 07960-6656

Overview: Manufactures current-carrying wiring devices. Established: 1997. Parent Company: Tyco International, Ltd. Ownership: Public.
Key Statistics: Annual Sales: $56.3 million. Number of Employees: 800.
Expertise/Education Sought: Engineering
Company-Paid Training is available.
International Assignments are available to employees.
Contact: For individuals with previous experience: Mr. Christopher Cornwell, Senior Manager of Staffing and Employment; Phone: 973-326-5727; Fax: 973-326-2748. E-mail: ccornwel@submarinesystems.com.
Company URL: http://www.submarinesystems.com

UNION CAMP CORPORATION
1600 VALLEY ROAD
WAYNE, NEW JERSEY 07470-2043

Overview: Manufactures paper products. Established: 1861. Ownership: Public.
Key Statistics: Annual Sales: $4 billion. Number of Employees: 19,268.
Hiring History: Number of professional employees hired in 1996: 200.
Expertise/Education Sought: Chemical engineering, mechanical engineering, information systems, sales, marketing, accounting, safety management
Alternatives to Full-Time Employment: Paid internships
Company-Paid Training is available.
International Assignments are available to employees.
Contact: Mrs. Bathsheba Sams, Director of Staffing; Phone: 973-628-2000, ext. 2870; Fax: 973-628-2848.

VENICE MAID FOODS, INC.
270 NORTH MILL ROAD
VINELAND, NEW JERSEY 08360

Mailing Address: PO Box 1505, Vineland, NJ 08360
Overview: Canned specialties, including soup and broth. Established: 1997. Ownership: Private.
Key Statistics: Annual Sales: $40 million. Number of Employees: 190.
Hiring History: Number of professional employees hired in 1997: 41.
Expertise/Education Sought: Food services, administration, quality control, research and development, engineering, management information systems
Contact: Human Resource Department; Phone: 609-691-2100; Fax: 609-696-1295.

WARNER-LAMBERT COMPANY
201 TABOR ROAD
MORRIS PLAINS, NEW JERSEY 07950-2693

Overview: Manufactures pharmaceutical preparations. Established: 1866. Ownership: Public.
Key Statistics: Annual Sales: $7 billion. Number of Employees: 40,000.
Expertise/Education Sought: Pharmaceutical science, chemistry, biochemistry, life sciences, computer science, engineering
Alternatives to Full-Time Employment: Paid internships
Company-Paid Training is available.
International Assignments are available to employees.
Contact: Resume Processing Center, Department 98107, 6 Century Drive, 3rd floor, Parsippany, New Jersey, 07054. E-mail: wlstaffing@mops.wl.com.
Company URL: http://www.warner-lambert.com

WESTFALIA SEPARATOR, INC.
100 FAIRWAY COURT
NORTHVALE, NEW JERSEY 07647-2401

Overview: Wholesales industrial machinery and equipment. Established: 1950. Parent Company: GEA, AG. Ownership: Private.
Key Statistics: Annual Sales: $75.6 million. Number of Employees: 233.
Hiring History: Number of professional employees hired in 1997: 25. 1996: 20.
Expertise/Education Sought: Electrical engineering, mechanical engineering, chemical engineering
Alternatives to Full-Time Employment: Paid internships
Company-Paid Training is available.
Contact: For individuals with previous experience: Personnel Department; Phone: 201-767-3900; Fax: 201-767-3416.

WISE FOODS, INC.
8 CAMPUS DRIVE
PARSIPPANY, NEW JERSEY 07054-4493

Overview: Manufactures and markets a variety of food, nonfood consumer, and industrial products. Established: 1995. Parent Company: KKR, Inc. Ownership: Private.
Key Statistics: Annual Sales: $3.7 billion. Number of Employees: 11,200.
Hiring History: Number of professional employees hired in 1996: 25.
Expertise/Education Sought: Marketing, finance, sales, accounting, chemical engineering
Alternatives to Full-Time Employment: Paid internships
Company-Paid Training is available.
Contact: For individuals with previous experience: Ms. Leslie Drake, Director of Human Resources; Phone: 973-993-8911; Fax: 973-898-1860.

A. ZEREGAS SONS, INC.
20-01 BROADWAY
FAIR LAWN, NEW JERSEY 07410-2058

Overview: Manufactures macaroni, noodles and spaghetti . Established: 1848. Ownership: Private.
Key Statistics: Annual Sales: $60 million. Number of Employees: 200.
Hiring History: Number of professional employees hired in 1997: 185.
Expertise/Education Sought: Manufacturing
Alternatives to Full-Time Employment: Paid internships
Company-Paid Training is available.
Contact: For individuals with previous experience: Human Resources Department; Phone: 201-797-1400; Fax: 201-797-0148.

NEW MEXICO

COMMODORE ADVANCED SCIENCES, INC.
2340 MENAUL BOULEVARD, NE, SUITE 400
ALBUQUERQUE, NEW MEXICO 87107

Overview: Researches and develops environmental and nuclear studies. Established: 1977. Parent Company: Commodore. Ownership: Private.
Key Statistics: Annual Sales: $40 million. Number of Employees: 300.

Hiring History: Number of professional employees hired in 1997: 25. 1996: 25.

Expertise/Education Sought: Environmental engineering, nuclear engineering, civil engineering, design engineering, chemical engineering, electrical engineering, health and safety, health physics

Contact: For individuals with previous experience: Human Resources Office; Phone: 505-872-3508; Fax: 505-872-6829. E-mail: vtinaje@adv-sci.com.

CONTACT NEW MEXICO, INC.

10820 CENTRAL AVENUE SE
ALBUQUERQUE, NEW MEXICO 87123

Mailing Address: PO Box 11818, Albuquerque, NM 87192

Overview: Provides paging services. Established: 1996. Ownership: Private.

Key Statistics: Annual Sales: $10.7 million. Number of Employees: 130.

Hiring History: Number of professional employees hired in 1996: 10.

Expertise/Education Sought: Communications

Company-Paid Training is available.

Contact: For individuals with previous experience: Ms. Betty Swanger, Human Resource Manager; Phone: 505-275-4600; Fax: 505-275-4646.

PHELPS DODGE TYRONE, INC.

NEW MEXICO STATE HIGHWAY 90 SOUTH
TYRONE, NEW MEXICO 88065

Overview: Mines copper ore. Established: 1995. Parent Company: Phelps Dodge, Inc. Ownership: Public.

Key Statistics: Annual Sales: $50.4 million. Number of Employees: 500.

Hiring History: Number of professional employees hired in 1996: 4.

Expertise/Education Sought: Mining engineering, metallurgical engineering

Alternatives to Full-Time Employment: Internships

International Assignments are available to employees.

Contact: For individuals with previous experience: Ms. Liz Gary, Employee Relations Manager; Phone: 505-538-5331; Fax: 505-388-2021.

REFRIGERATION TECHNOLOGY, INC.

301 ANTONE STREET
SUNLAND PARK, NEW MEXICO 88063

Overview: Manufactures environmental controls for air conditioning and refrigeration systems. Established: 1997. Parent Company: RTI, Inc. Ownership: Public.

Key Statistics: Annual Sales: $8 million. Number of Employees: 100.

Hiring History: Number of professional employees hired in 1997: 20.

Expertise/Education Sought: Mechanical engineering

Alternatives to Full-Time Employment: Paid internships

International Assignments are available to employees.

Contact: For individuals with previous experience: Human Resource Department; Phone: 505-589-5431; Fax: 505-589-0710. E-mail: rti@whc.net.

Sandia National Laboratories

SANDIA NATIONAL LABORATORIES

PO BOX 5800, (MS1023)
ALBUQUERQUE, NEW MEXICO 87185-1023

Overview: Engaged in research and development of nuclear technology, the environment, energy, manufacturing, space, the nation's economy, and global security. Established: 1948. Parent Company: Lockheed Martin Corporation. Ownership: Private.

Key Statistics: Number of Employees: 7,500.

Hiring History: Number of professional employees hired in 1997: 125. 1996: 80.

Expertise/Education Sought: Management information systems, computer science, electrical engineering, mechanical engineering, software engineering/development

Alternatives to Full-Time Employment: Paid internships

Company-Paid Training is available.

Contact: For college students and recent graduates: Mr. Ed Gullick, Recruiting Coordinator; Phone: 505-844-2559; Fax: 505-844-6636. E-mail: emgulli@sandia.gov. For individuals with previous experience: Ms. B. J. Jones, Staffing Manager; Phone: 505-845-0096; Fax: 505-844-6636. E-mail: bjjones@sandia.gov.

Company URL: http://www.sandia.gov

For more information, see full corporate profile, pg. 296.

TECHNOLOGY MANAGEMENT LIMITED COMPANY

2500 LOUISIANA BOULEVARD, NE SUITE 513
ALBUQUERQUE, NEW MEXICO 87110-4343

Overview: Provides consulting engineering services. Established: 1993. Ownership: Private.

Key Statistics: Number of Employees: 219.

Hiring History: Number of professional employees hired in 1997: 4.

Expertise/Education Sought: Accounting, engineering

International Assignments are available to employees.

Contact: Ms. Elizabeth Walker, Human Resources Director; Phone: 505-872-3146; Fax: 505-872-3246.

Company URL: http://www.tmcservices.com

NEW YORK

AARQUE MANAGEMENT CORPORATION

111 WEST 2ND STREET
JAMESTOWN, NEW YORK 14702-0310

Overview: Produces cold finished steel shapes. Established: 1994. Ownership: Private.

Key Statistics: Annual Sales: $236 million. Number of Employees: 800.

Expertise/Education Sought: Accounting, metallurgy, manufacturing engineering

Contact: For individuals with previous experience: Mr. Donald Proctor, Treasurer; Phone: 716-664-6014; Fax: 716-664-5057.

ABC, INC.

77 WEST 66TH STREET
NEW YORK, NEW YORK 10023

Overview: Operates radio and television stations and networks. Established: 1996. Parent Company: Walt Disney Company. Ownership: Public.

Key Statistics: Annual Sales: $919 million. Number of Employees: 8,000.

Expertise/Education Sought: Finance, accounting, law, marketing, computer programming, electrical engineering

Alternatives to Full-Time Employment: Internships

Company-Paid Training is available.

International Assignments are available to employees.

Contact: Mr. Brendan Burke, Director of Employee Relations; Phone: 212-456-7777.

ADL DATA SYSTEMS, INC.
20 LIVINGSTONE AVENUE, 2ND FLOOR
DOBBS FERRY, NEW YORK 10522

Overview: Provides computer programming services.

Hiring History: Number of professional employees hired in 1997: 8.

Expertise/Education Sought: Marketing, software engineering/development

Company-Paid Training is available.

Contact: For individuals with previous experience: Ms. Shelly Pollack, Controller; Phone: 914-591-1800; Fax: 914-591-1818. E-mail: shelley@mail.adldata.com.

AGRILINK FOODS, INC.
90 LINDEN OAKS
ROCHESTER, NEW YORK 14625

Overview: Distributes canned fruits and vegetables, sauces, and snack foods. Established: 1994. Parent Company: Pro-Fac Cooperative. Ownership: Public.

Key Statistics: Annual Sales: $730.8 million. Number of Employees: 3,363.

Hiring History: Number of professional employees hired in 1996: 3.

Expertise/Education Sought: Finance, business management, accounting

Alternatives to Full-Time Employment: Internships

Contact: Ms. Lois Warlick-Jarvie, Vice President of Human Resources, PO Box 20670, Rochester, New York, 14602-0670; Phone: 716-383-1850, ext. 136; Fax: 716-383-1568.

AGWAY, INC.
333 BUTTERNUT DRIVE
DE WITT, NEW YORK 13214-1879

Overview: Manufactures and distributes agricultural supplies and products. Established: 1964. Ownership: Public.

Key Statistics: Annual Sales: $1.7 billion. Number of Employees: 7,100.

Expertise/Education Sought: Agricultural economics

Alternatives to Full-Time Employment: Paid internships

Contact: For individuals with previous experience: Human Relations Department; Phone: 315-449-7203; Fax: 315-449-6289.

ALBANY INTERNATIONAL CORPORATION
1373 BROADWAY
ALBANY, NEW YORK 12204

Overview: Manufactures felt. Established: 1895. Ownership: Public.

Key Statistics: Annual Sales: $692.8 million. Number of Employees: 5,854.

Hiring History: Number of professional employees hired in 1996: 35.

Expertise/Education Sought: Marketing, finance, sales, process engineering, chemical engineering, mechanical engineering, electrical engineering, information systems

Alternatives to Full-Time Employment: Paid internships

Company-Paid Training is available.

International Assignments are available to employees.

Contact: For individuals with previous experience: Mr. Barry Jessee, Director of Human Resources; Phone: 518-445-2200.

AMERADA HESS CORPORATION
1185 AVENUE OF THE AMERICAS
NEW YORK, NEW YORK 10036-2601

Overview: Produces oil, gas, and petroleum based products. Established: 1920. Ownership: Public.

Key Statistics: Annual Sales: $8.3 billion. Number of Employees: 9,085.

Expertise/Education Sought: Marketing, finance, sales, accounting, management information systems, chemical engineering, mechanical engineering, geology, geophysics

Alternatives to Full-Time Employment: Paid internships

Company-Paid Training is available.

International Assignments are available to employees.

Contact: Mr. Larry Fox, Manager of Employee Relations; Phone: 212-997-8500; Fax: 212-536-8318. E-mail: lfox@hess.com.

AMERICAN EXPRESS COMPANY
3 WORLD FINANCIAL CENTER
NEW YORK, NEW YORK 10285

Overview: Provides credit cards, travelers checks, and other financial services. Established: 1850. Ownership: Public.

Key Statistics: Annual Sales: $16.2 billion. Number of Employees: 72,229.

Hiring History: Number of professional employees hired in 1996: 100.

Expertise/Education Sought: Marketing, finance, accounting, auditing, data processing, human resources, operations, sales, technology

Alternatives to Full-Time Employment: Paid internships

International Assignments are available to employees.

Contact: For college students and recent graduates: Ms. Deborah Hickey, College Recruiter; Fax: 212-619-9770. For individuals with previous experience: Mr. Neil Begley, Senior Director of Human Resources, American Express Tower, New York, New York, 10285; Phone: 212-640-5050. Job Hotline: 212-640-2171.

AMERICAN INTERNATIONAL TELEPHONE, INC.
287 BOWMAN AVENUE
PURCHASE, NEW YORK 10577-2517

Overview: Provides long distance phone services. Established: 1995. Ownership: Private.

Key Statistics: Annual Sales: $30 million. Number of Employees: 45.

Hiring History: Number of professional employees hired in 1996: 10.

Expertise/Education Sought: Engineering

Alternatives to Full-Time Employment: Internships

Company-Paid Training is available.

Contact: Mr. Charles Eisenberg, President; Phone: 914-251-1460, ext. 246; Fax: 914-251-1460. E-mail: charles@aitoasis.com.

AMETEK ROTRON
9 HASBROUCK LANE
WOODSTOCK, NEW YORK 12498-1894

Overview: Manufactures blowers and fans. Established: 1976. Parent Company: Ametek. Ownership: Public.

Key Statistics: Annual Sales: $63 million. Number of Employees: 450.

Hiring History: Number of professional employees hired in 1996: 12.

Expertise/Education Sought: Electrical engineering, engineering, mechanical engineering, industrial engineering

Contact: For individuals with previous experience: Human Resource Department; Phone: 914-679-2401; Fax: 914-679-1852.

AMMANN & WHITNEY
96 MORTON STREET
NEW YORK, NEW YORK 10014-3388

Overview: Engineering consulting firm. Established: 1949. Parent Company: Lewis Berger International, Inc. Ownership: Private.

Key Statistics: Annual Sales: $10.3 million. Number of Employees: 150.

Hiring History: Number of professional employees hired in 1997: 50.

Expertise/Education Sought: Architecture, structural engineering, mechanical engineering, design engineering, electrical engineering, engineering, civil engineering, construction engineering

Alternatives to Full-Time Employment: Internships

Contact: For individuals with previous experience: Human Resources Department; Phone: 212-524-7200.

AMPACET CORPORATION
660 WHITE PLAINS ROAD
TARRYTOWN, NEW YORK 10591-5107

Overview: Produces custom compound purchased resins. Established: 1954. Ownership: Private.

Key Statistics: Annual Sales: $335 million. Number of Employees: 700.

Hiring History: Number of professional employees hired in 1997: 5.

Expertise/Education Sought: Manufacturing engineering

Contact: For individuals with previous experience: Human Resources; Fax: 914-333-1616.

ANDREA ELECTRONICS CORPORATION
1140 45TH ROAD
LONG ISLAND CITY, NEW YORK 11101-5213

Overview: Manufactures radio communications equipment. Established: 1934. Ownership: Public.

Key Statistics: Annual Sales: $9.2 million. Number of Employees: 86.

Hiring History: Number of professional employees hired in 1997: 8. 1996: 4.

Expertise/Education Sought: Accounting, engineering, marketing, electrical engineering, design engineering, sales

Alternatives to Full-Time Employment: Part-time jobs

Contact: Executive Management; Phone: 718-729-8500; Fax: 718-729-8500.

Company URL: http://www.andreaelectronics.com

ANVIL NETWARE
228 EAST 45TH STREET
NEW YORK, NEW YORK 10017-3303

Overview: Wholesales women's and children's clothing. Established: 1994. Ownership: Private.

Key Statistics: Annual Sales: $204.2 million. Number of Employees: 2,301.

Expertise/Education Sought: Accounting, sales, marketing, manufacturing

Contact: For individuals with previous experience: Ms. Cari Wiemokly, Office Manager; Phone: 212-476-0300; Fax: 212-476-0323.

THE APPAREL GROUP, LTD.
1370 AVENUE OF THE AMERICAS
NEW YORK, NEW YORK 10019-4602

Mailing Address: PO Box 32100, Louisville, KY 40232

Overview: Manufactures men's and boys' shirts. Established: 1990. Parent Company: TAL. Ownership: Private.

Key Statistics: Annual Sales: $37 million. Number of Employees: 1,200.

Hiring History: Number of professional employees hired in 1997: 20. 1996: 25.

Expertise/Education Sought: Engineering, information systems

Contact: For individuals with previous experience: Human Resources; Phone: 502-473-6300; Fax: 502-473-6118.

I. APPEL CORPORATION
149 MADISON AVENUE
NEW YORK, NEW YORK 10016-6711

Overview: Manufactures women's and children's underwear and nightwear. Established: 1976. Ownership: Private.

Key Statistics: Annual Sales: $58.5 million. Number of Employees: 1,750.

Expertise/Education Sought: Manufacturing engineering, sales

Contact: For individuals with previous experience: Mr. Milton Schneider, President; Phone: 212-592-8600; Fax: 212-592-8799.

APPLETON PAPERS, INC. NEWTON FALLS MILL
875 COUNTY ROUTE 60
NEWTON FALLS, NEW YORK 13666

Overview: Manufactures specialty papers for business and industry. Established: 1995. Parent Company: Appleton. Ownership: Private.

Key Statistics: Annual Sales: $70 million. Number of Employees: 400.

Hiring History: Number of professional employees hired in 1996: 50.

Expertise/Education Sought: Manufacturing engineering

Alternatives to Full-Time Employment: Paid internships

Company-Paid Training is available.

International Assignments are available to employees.

Contact: Human Resources Department; Phone: 315-848-3321, ext. 322; Fax: 315-848-5908.

ARROW ELECTRONICS, INC.
25 HUB DRIVE
MELVILLE, NEW YORK 11747-3509

Overview: Distributes computer and electronics products. Established: 1935. Ownership: Public.

Key Statistics: Annual Sales: $6.5 billion. Number of Employees: 7,900.

Expertise/Education Sought: Management information systems, finance, human resources, accounting, marketing, electrical engineering, sales

Alternatives to Full-Time Employment: Internships

Company-Paid Training is available.

International Assignments are available to employees.

Contact: Mrs. M. J. Wheble, Staffing Manager; Phone: 516-391-1300; Fax: 516-391-1501. E-mail: mwheble@arrow.com.

Company URL: http://www.arrow.com

ART LEATHER MANUFACTURING COMPANY
4510 94TH STREET
ELMHURST, NEW YORK 11373-2800

Overview: Manufactures notebooks, albums, and binders. Established: 1990. Ownership: Private.

Key Statistics: Annual Sales: $31.8 million. Number of Employees: 575.

Expertise/Education Sought: Manufacturing

Contact: Director of Human Resources; Phone: 718-699-6300; Fax: 718-699-9339.

AVNET, INC.
80 CUTTER MILL ROAD
GREAT NECK, NEW YORK 11021-3108

Overview: Distributes computer products and provides computer maintenance services. Established: 1955. Ownership: Public.

Key Statistics: Annual Sales: $5.4 billion. Number of Employees: 9,400.

Expertise/Education Sought: Accounting, law, marketing, sales, software engineering/development

Alternatives to Full-Time Employment: Part-time jobs, paid internships

Company-Paid Training is available.

International Assignments are available to employees.

Contact: Mr. Robert Zierk, Vice President of Human Resources; Phone: 310-665-2600, ext. 2674; Fax: 310-642-1839.

Company URL: http://www.avnet.com

BAROUH EATON ALLEN CORPORATION
67 KENT AVENUE
BROOKLYN, NEW YORK 11211-1926

Overview: Manufactures carbon paper and typewriter ribbons. Established: 1955. Ownership: Private.

Key Statistics: Annual Sales: $40 million. Number of Employees: 146.

Barouh Eaton Allen Corporation (continued)

Expertise/Education Sought: Manufacturing, production

Contact: For college students and recent graduates: Human Resources Department; Phone: 718-782-2601; Fax: 718-486-6340. For individuals with previous experience: Mr. Theo Ulgar, Personnel Manager; Phone: 718-782-2601, ext. 219; Fax: 718-486-6340.

MATTHEW BENDER AND COMPANY, INC.

2 PARK AVENUE
NEW YORK, NEW YORK 10016

Overview: Book publisher. Established: 1963. Parent Company: Times Mirror Company. Ownership: Public.

Key Statistics: Annual Sales: $162.3 million. Number of Employees: 1,400.

Hiring History: Number of professional employees hired in 1997: 350. 1996: 300.

Expertise/Education Sought: Marketing, manufacturing engineering, information systems, sales

Alternatives to Full-Time Employment: Part-time jobs, paid internships

Contact: Human Resources; Phone: 212-448-2071; Fax: 212-448-2339.

BICC CABLES CORPORATION

1 CROSFIELD AVENUE
WEST NYACK, NEW YORK 10994

Overview: Manufactures electrical cables. Established: 1988. Parent Company: BICC Group. Ownership: Private.

Key Statistics: Annual Sales: $513.7 million. Number of Employees: 2,000.

Hiring History: Number of professional employees hired in 1997: 70. 1996: 70.

Expertise/Education Sought: Accounting, finance, sales, manufacturing engineering, electrical engineering

Company-Paid Training is available.

Contact: Ms. Maureen Bailey, Manager of Employee Relations; Phone: 914-353-4000, ext. 241; Fax: 914-353-5070.

BIOPHARMACEUTICS, INC.

990 STATION ROAD
BELLPORT, NEW YORK 11713

Overview: Manufactures generic pharmaceuticals, consumer products, and cancer treatments.

Expertise/Education Sought: Manufacturing

Contact: For individuals with previous experience: Ms. Carol Massimo, Personnel Manager.

BLEYER INDUSTRIES, INC.

260 WEST SUNRISE HIGHWAY
VALLEY STREAM, NEW YORK 11581

Overview: Manufactures paper cups, paper plates, and paperboard. Established: 1981. Ownership: Private.

Key Statistics: Annual Sales: $37.4 million. Number of Employees: 310.

Expertise/Education Sought: Accounting, finance, marketing

Company-Paid Training is available.

Contact: For individuals with previous experience: Employment Opportunities; Phone: 516-374-3800; Fax: 516-374-3980.

BROOKLYN UNION GAS COMPANY, INC.

1 METROTECH CENTER
BROOKLYN, NEW YORK 11201

Overview: Gas utility. Established: 1849. Parent Company: KeySpan Energy Corporation. Ownership: Private.

Key Statistics: Annual Sales: $811.3 million. Number of Employees: 3,000.

Expertise/Education Sought: Human resources, accounting, marketing, computer science, finance, chemical engineering

Alternatives to Full-Time Employment: Part-time jobs, paid internships

Company-Paid Training is available.

Contact: Mr. Lou Donaldson, Director of Recruitment; Phone: 718-403-2810; Fax: 718-488-1762. Job Hotline: 718-403-6300.

Company URL: http://www.bug.com

BUFFALO CHINA, INC.

658 BAILEY AVENUE
BUFFALO, NEW YORK 14206-3003

Overview: Manufactures china products. Established: 1983. Parent Company: Oneida, Ltd. Ownership: Public.

Key Statistics: Annual Sales: $39.1 million. Number of Employees: 800.

Expertise/Education Sought: Sales, marketing

Alternatives to Full-Time Employment: Paid internships

Company-Paid Training is available.

Contact: For individuals with previous experience: Ms. Mary Lou Domino, Manager of Human Resources, 51 Hayes Place, Buffalo, New York, 14210-3003; Phone: 716-824-8515; Fax: 716-824-2736.

CANON USA, INC.

1 CANON PLAZA
LAKE SUCCESS, NEW YORK 11042

Overview: Manufactures office equipment and cameras. Established: 1966. Parent Company: Canon, Inc, Tokyo. Ownership: Private.

Key Statistics: Annual Sales: $7.5 billion. Number of Employees: 7,000.

Hiring History: Number of professional employees hired in 1997: 300. 1996: 9,999.

Expertise/Education Sought: Marketing, sales, finance, accounting, auditing, merchandising, business management

Alternatives to Full-Time Employment: Paid internships

Company-Paid Training is available.

Contact: Human Resources; Phone: 516-328-5050, ext. 5050; Fax: 516-328-4669.

Company URL: http://www.usa.canon.com

CARL ZEISS, INC.

1 ZEISS DRIVE
THORNWOOD, NEW YORK 10594-1939

Overview: Manufactures medical and surgical equipment, microscopes, measuring machines, and other precision instruments. Established: 1925. Ownership: Private.

Key Statistics: Annual Sales: $300 million. Number of Employees: 890.

Expertise/Education Sought: Electronics/electronics engineering, software engineering/development, design engineering, physics, biology, chemistry

Alternatives to Full-Time Employment: Paid internships

Company-Paid Training is available.

Contact: For individuals with previous experience: Mr. Lawrence Hart, Vice President of Human Resources; Phone: 914-747-1800; Fax: 914-681-7482.

CARTER-WALLACE, INC.

1345 AVENUE OF THE AMERICAS
NEW YORK, NEW YORK 10105-0302

Overview: Manufactures pharmaceuticals, diagnostics, toiletries, and pet products. Established: 1880. Ownership: Public.

Key Statistics: Annual Sales: $648.8 million. Number of Employees: 3,460.

Expertise/Education Sought: Chemical engineering, mechanical engineering, accounting, finance, marketing, sales, management information systems, pharmacy

Alternatives to Full-Time Employment: Paid internships

Company-Paid Training is available.
Contact: Mrs. Christine Glasco, Director of Staffing and Development; Phone: 609-655-6000.

CITY OF ROCHESTER
30 CHURCH STREET, CITY HALL 103 A
ROCHESTER, NEW YORK 14614

Overview: Local government. Ownership: Private.
Key Statistics: Number of Employees: 175.
Expertise/Education Sought: Municipal planning, municipal engineering
Alternatives to Full-Time Employment: Paid internships
Company-Paid Training is available.
Contact: Human Resources; Phone: 716-428-6573; Fax: 716-428-6902. Job Hotline: 706-428-7115.

COLGATE-PALMOLIVE COMPANY
300 PARK AVENUE
NEW YORK, NEW YORK 10022-7402

Overview: Manufactures personal care and household products. Established: 1806. Ownership: Public.
Key Statistics: Annual Sales: $8.7 billion. Number of Employees: 37,300.
Expertise/Education Sought: Sales, treasury management, marketing, finance, accounting, public relations
Alternatives to Full-Time Employment: Paid internships
International Assignments are available to employees.
Contact: Mr. Bob Joy, Vice President Global Human Resources; Phone: 212-310-2000; Fax: 212-000-0000.
Company URL: http://www.colgate.com

COMPUTER® ASSOCIATES
Software superior by design.

COMPUTER ASSOCIATES INTERNATIONAL, INC.
ONE COMPUTER ASSOCIATES PLAZA
ISLANDIA, NEW YORK 11788-7000

Overview: Multiplatform software company. Established: 1976. Ownership: Public.
Key Statistics: Annual Sales: $4.5 billion. Number of Employees: 11,000.
Hiring History: Number of professional employees hired in 1997: 500.
Expertise/Education Sought: Computer programming, software engineering/development, project management
Alternatives to Full-Time Employment: Paid internships
Company-Paid Training is available.
International Assignments are available to employees.
Contact: Mr. Andy Press, Human Resources; Phone: 516-342-4402; Fax: 800-962-9224. E-mail: resumes@cai.com.
Company URL: http://www.cai.com

For more information, see full corporate profile, pg. 272.

COMPUTER TASK GROUP, INC.
800 DELAWARE AVENUE
BUFFALO, NEW YORK 14209-2006

Overview: Provides software development and consulting services for information and communication systems. Established: 1966. Ownership: Public.
Key Statistics: Annual Sales: $365.1 million. Number of Employees: 5,000.
Expertise/Education Sought: Computer science, computer programming, software engineering/development, data processing, information systems
Alternatives to Full-Time Employment: Part-time jobs, paid internships
Company-Paid Training is available.
International Assignments are available to employees.
Contact: Director of Recruiting; Phone: 800-992-5350; Fax: 716-882-8000.
Company URL: http://www.ctg.com

CONSOLIDATED EDISON COMPANY OF NEW YORK
4 IRVING PLACE
NEW YORK, NEW YORK 10003-3598

Overview: Electric and gas utility. Established: 1884. Ownership: Public.
Key Statistics: Assets: $7 billion. Number of Employees: 15,801.
Expertise/Education Sought: Accounting, computer science, finance
Alternatives to Full-Time Employment: Paid internships
Company-Paid Training is available.
Contact: For college students and recent graduates: Ms. Sharon Gay, Specialist; Phone: 212-460-2033; Fax: 212-228-9439. For individuals with previous experience: Mrs. Lorretta Vanacore, Manager of Recruiting and Staffing; Phone: 212-460-4600; Fax: 212-228-9439.

CURTIS INSTRUMENTS, INC.
200 KISCO AVENUE
MOUNT KISCO, NEW YORK 10549-1407

Overview: Manufactures instruments to test electricity and fluids. Established: 1960. Ownership: Private.
Key Statistics: Annual Sales: $69.6 million. Number of Employees: 660.
Hiring History: Number of professional employees hired in 1997: 5.
Expertise/Education Sought: Electrical engineering, electronics/electronics engineering, mechanical engineering, design engineering, analog engineering
Alternatives to Full-Time Employment: Paid internships
International Assignments are available to employees.
Contact: Ms. Lisa Cristantiello, Human Resources Generalist; Phone: 914-666-2971, ext. 161; Fax: 914-666-2188. E-mail: curtismtk1@aol.com.
Company URL: http://www.curtisinst.com

DATAMEDIC CORPORATION
20 OSER AVENUE
HAUPPAUGE, NEW YORK 11788-3825

Overview: Provides computer billing services for dentists and physicians. Established: 1972. Ownership: Private.
Key Statistics: Annual Sales: $21.2 million. Number of Employees: 300.
Hiring History: Number of professional employees hired in 1997: 25. 1996: 7.
Expertise/Education Sought: Software engineering/development, computer programming, accounting
Alternatives to Full-Time Employment: Paid internships
Contact: Human Resources Department; Phone: 516-435-8880; Fax: 516-435-1062.

DAVIS CONSTRUCTION COMPANY, INC.
400 PARK AVENUE
NEW YORK, NEW YORK 10022-4406

Overview: Builds nonresidential construction. Established: 1994. Ownership: Private.
Key Statistics: Annual Sales: $5.2 million. Number of Employees: 30.
Hiring History: Number of professional employees hired in 1997: 15. 1996: 5.
Expertise/Education Sought: Accounting, finance, project management, architecture, purchasing, contracts administration/management
Contact: For individuals with previous experience: Human Resources; Phone: 212-308-7365; Fax: 212-308-8602.

DEL GLOBAL TECHNOLOGIES CORPORATION
1 COMMERCE PARK
VALHALLA, NEW YORK 10595-1455

Overview: Produces electronic components. Established: 1954. Ownership: Public.

Del Global Technologies Corporation (continued)

Key Statistics: Annual Sales: $54.7 million. Number of Employees: 455.
Hiring History: Number of professional employees hired in 1997: 10.
Expertise/Education Sought: Manufacturing, medical technology, marketing, design engineering, imaging technology
International Assignments are available to employees.
Contact: For individuals with previous experience: Human Resources; Phone: 914-686-3600; Fax: 914-686-5424.

DIGITEC 2000, INC.
8 WEST 38TH STREET, FLOOR 5
NEW YORK, NEW YORK 10018

Overview: Provides telecommunications services. Established: 1987. Ownership: Private.
Key Statistics: Annual Sales: $26 million. Number of Employees: 39.
Expertise/Education Sought: Telecommunications, customer service and support, data processing
Contact: Human Resources; Phone: 212-944-8888, ext. 867; Fax: 212-391-4984.

DORSEY GAGE COMPANY, INC.
253 MANSION STREET
POUGHKEEPSIE, NEW YORK 12601-2623

Mailing Address: PO Box D, Poughkeepsie, NY 12602
Overview: Manufactures precision measuring instruments. Established: 1955. Ownership: Private.
Key Statistics: Annual Sales: $10.5 million. Number of Employees: 78.
Hiring History: Number of professional employees hired in 1997: 15. 1996: 38.
Expertise/Education Sought: Computer programming, mechanical engineering, machinery
Alternatives to Full-Time Employment: Paid internships
Company-Paid Training is available.
Contact: For individuals with previous experience: Human Relations Department; Phone: 914-454-3111; Fax: 914-454-3140. E-mail: gwlut@aol.com.

DOW JONES AND COMPANY
200 LIBERTY STREET
NEW YORK, NEW YORK 10281-1003

Overview: Provides on line and print news services. Established: 1882. Ownership: Public.
Key Statistics: Annual Sales: $2.5 billion. Number of Employees: 11,844.
Expertise/Education Sought: Journalism, finance, computer programming, marketing, accounting, economics, printing, sales
Alternatives to Full-Time Employment: Paid internships
Company-Paid Training is available.
International Assignments are available to employees.
Contact: Staffing; Phone: 212-416-2000.
Company URL: http://www.dowjones.com

DUGGAL COLOR PROJECTS, INC.
9 WEST 20TH STREET
NEW YORK, NEW YORK 10011-3701

Overview: Provides photofinishing laboratory services. Established: 1961. Ownership: Private.
Key Statistics: Annual Sales: $17.6 million. Number of Employees: 170.
Expertise/Education Sought: Photo technicians
Alternatives to Full-Time Employment: Paid internships
Contact: For individuals with previous experience: Personnel Department; Phone: 212-242-7000, ext. 232; Fax: 212-633-1266.

DUNLOP TIRE CORPORATION
200 JOHN JAMES AUDUBON PARK
AMHERST, NEW YORK 14228

Overview: Manufactures tires. Established: 1986. Ownership: Private.
Key Statistics: Annual Sales: $772.1 million. Number of Employees: 3,276.
Expertise/Education Sought: Mechanical engineering, sales, marketing
Alternatives to Full-Time Employment: Paid internships
Company-Paid Training is available.
Contact: For individuals with previous experience: Mr. Charles Cookson, Corporate Manager Human Resources; Phone: 716-639-5236; Fax: 716-639-5017. E-mail: Ccookson@dunloptire.com.

EAGLE ELECTRIC MANUFACTURING COMPANY, INC.
45-31 COURT SQUARE
LONG ISLAND CITY, NEW YORK 11101

Overview: Manufactures electric plugs, sockets, and related devices. Established: 1920. Ownership: Private.
Key Statistics: Annual Sales: $200 million. Number of Employees: 2,400.
Expertise/Education Sought: Manufacturing engineering, customer service and support, electrical engineering, design engineering, administration
Alternatives to Full-Time Employment: Paid internships
Contact: Human Resource Department; Phone: 718-361-4853; Fax: 718-784-2792.

EASTCO INDUSTRIAL SAFETY CORPORATION
130 WEST 10TH STREET
HUNTINGTON STATION, NEW YORK 11746

Overview: Manufactures fire resistant protective clothing. Established: 1940. Ownership: Public.
Key Statistics: Annual Sales: $28 million. Number of Employees: 200.
Hiring History: Number of professional employees hired in 1996: 10.
Expertise/Education Sought: Manufacturing, production, sales
Company-Paid Training is available.
Contact: For individuals with previous experience: Human Relations Department; Phone: 516-427-1802; Fax: 516-427-1840.

EASTMAN KODAK COMPANY
343 STATE STREET
ROCHESTER, NEW YORK 14650-1139

Overview: Produces imaging systems and photographic equipment and accessories. Established: 1880. Ownership: Public.
Key Statistics: Annual Sales: $16 billion. Number of Employees: 94,800.
Expertise/Education Sought: Electrical engineering, chemical engineering, optical engineering, mechanical engineering, software engineering/development, computer engineering, industrial engineering, computer science, biochemistry, management information systems
Alternatives to Full-Time Employment: Part-time jobs, paid internships
Company-Paid Training is available.
International Assignments are available to employees.
Contact: Staffing Department; Phone: 716-724-4000; Fax: 716-724-9416. E-mail: staffing@kodak.com.
Company URL: http://www.kodak.com

EBY ELECTRO COMPANY, INC.
140 ADAMS BOULEVARD
FARMINGDALE, NEW YORK 11735-6614

Overview: Produces electronic connectors. Established: 1980. Ownership: Private.
Key Statistics: Annual Sales: $8.2 million. Number of Employees: 75.
Expertise/Education Sought: Electronics/electronics engineering
Alternatives to Full-Time Employment: Part-time jobs

International Assignments are available to employees.

Contact: For individuals with previous experience: Mr. Sid Salmon, President; Phone: 516-694-8989; Fax: 516-694-1155.

Company URL: http://www.ebynet.com

EDER ASSOCIATES, DIVISION OF GANNETT FLEMING

480 FOREST AVENUE
LOCUST VALLEY, NEW YORK 11560-2118

Overview: Provides environmental engineering services. Established: 1970. Parent Company: Gannett Fleming, Inc. Ownership: Private.

Key Statistics: Annual Sales: $13.5 million. Number of Employees: 165.

Hiring History: Number of professional employees hired in 1997: 25. 1996: 50.

Expertise/Education Sought: Environmental engineering, engineering

Alternatives to Full-Time Employment: Paid internships

Contact: For individuals with previous experience: Human Resources; Phone: 516-671-8440; Fax: 516-671-3349.

ELLANEF MANUFACTURING CORPORATION

9711 50TH AVENUE
CORONA, NEW YORK 11368-2740

Overview: Manufactures aircraft parts and assemblies. Established: 1941. Ownership: Private.

Key Statistics: Annual Sales: $55.7 million. Number of Employees: 600.

Hiring History: Number of professional employees hired in 1997: 50. 1996: 40.

Expertise/Education Sought: Computer programming, engineering, production control

Company-Paid Training is available.

Contact: For individuals with previous experience: Mr. Ernest Constantine, Human Resources Manager; Phone: 718-699-4000, ext. 227; Fax: 718-592-0722.

EMERSON POWER TRANSMISSION CORPORATION

620 SOUTH AURORA STREET
ITHACA, NEW YORK 14850-5726

Overview: Manufactures power transmissions. Established: 1983. Parent Company: Emerson Electric Company. Ownership: Public.

Key Statistics: Number of Employees: 4,600.

Hiring History: Number of professional employees hired in 1997: 30. 1996: 30.

Expertise/Education Sought: Sales, business administration, management, engineering

Alternatives to Full-Time Employment: Paid internships

Company-Paid Training is available.

International Assignments are available to employees.

Contact: Mr. Lou Sposigo, Director of Training; Phone: 607-274-6105; Fax: 607-272-7239. Job Hotline: 800-354-9825.

ESPEY MANUFACTURING AND ELECTRONICS CORPORATION

233 BALLSTON AVENUE
SARATOGA SPRINGS, NEW YORK 12866-4755

Overview: Produces electronic components. Established: 1928. Ownership: Public.

Key Statistics: Annual Sales: $16.8 million. Number of Employees: 190.

Hiring History: Number of professional employees hired in 1996: 20.

Expertise/Education Sought: Manufacturing, design engineering, data processing, administration, business management, manufacturing engineering

Contact: Personnel Manager; Phone: 518-584-4100, ext. 211; Fax: 518-584-5701.

Company URL: http://www.espey.com

FINK BAKING COMPANY, LLC

535 54TH AVENUE
LONG ISLAND CITY, NEW YORK 11101-5924

Overview: Manufactures bakery products. Established: 1966. Ownership: Private.

Key Statistics: Annual Sales: $65 million. Number of Employees: 700.

Hiring History: Number of professional employees hired in 1996: 450.

Expertise/Education Sought: Sales, transportation

Company-Paid Training is available.

Contact: For individuals with previous experience: Human Resources; Phone: 718-392-8300; Fax: 718-729-1303.

FISHER-PRICE, INC.

636 GIRARD AVENUE
EAST AURORA, NEW YORK 14052-1824

Overview: Manufactures and distributes toys. Established: 1993. Parent Company: Mattel. Ownership: Public.

Key Statistics: Annual Sales: $1.2 billion. Number of Employees: 4,000.

Expertise/Education Sought: Marketing, mechanical engineering, sales, accounting, manufacturing engineering

Alternatives to Full-Time Employment: Paid internships

Company-Paid Training is available.

Contact: For individuals with previous experience: Ms. Jacqueline Evans, Employment Coordinator; Phone: 716-687-3000, ext. 3622; Fax: 716-687-3238. E-mail: careers@fisher-price.com.

FOREST LABORATORIES, INC.

909 THIRD AVENUE
NEW YORK, NEW YORK 10022

Overview: Manufactures pharmaceuticals. Established: 1956. Ownership: Public.

Key Statistics: Annual Sales: $280.7 million. Number of Employees: 1,663.

Hiring History: Number of professional employees hired in 1996: 25.

Expertise/Education Sought: Product design and development

Alternatives to Full-Time Employment: Paid internships

Contact: For college students and recent graduates: Ms. Michelle Schachter, College Recruiter; Phone: 212-421-7850; Fax: 212-750-9152. E-mail: staffing@frx.com. For individuals with previous experience: Mr. Bernard McGovern, Director of Human Resources; Phone: 212-421-7850; Fax: 212-750-9152. E-mail: staffing@frx.com.

FREQUENCY ELECTRONICS, INC.

55 CHARLES LINDBERGH BOULEVARD
UNIONDALE, NEW YORK 11553-3699

Overview: Manufactures frequency standards, generators, filters, and clocks. Established: 1961. Ownership: Public.

Key Statistics: Annual Sales: $27.9 million. Number of Employees: 212.

Expertise/Education Sought: Engineering

Company-Paid Training is available.

Contact: Mr. Robert Klomp, Director of Human Resources; Phone: 516-794-4500, ext. 2066; Fax: 516-794-3889.

GAYMAR INDUSTRIES, INC.

10 CENTRE DRIVE
ORCHARD PARK, NEW YORK 14127-2295

Overview: Manufactures medical equipment. Established: 1956. Ownership: Private.

Key Statistics: Annual Sales: $40 million. Number of Employees: 240.

Expertise/Education Sought: Manufacturing engineering, design engineering, electrical engineering, mechanical engineering, marketing, computer programming, quality engineering

Alternatives to Full-Time Employment: Internships

Contact: For individuals with previous experience: Mr. Paul Kunkemoeller, Director of Human Resources; Phone: 716-662-2551, ext. 741; Fax: 716-662-8785. E-mail: paulkunkemoeller@gamar.com.

GENERAL MECHATRONICS CORPORATION
60 MILBAR BOULEVARD
FARMINGDALE, NEW YORK 11735

Overview: Manufactures aircraft parts and components. Established: 1961. Ownership: Private.

Key Statistics: Annual Sales: $18.3 million. Number of Employees: 100.

Expertise/Education Sought: Manufacturing engineering, administration

Contact: Human Resources Department; Phone: 516-249-7900.

GOODTIMES HOME VIDEO CORPORATION
16 EAST 40TH STREET
NEW YORK, NEW YORK 10016

Overview: Manufactures magnetic and optical recording media equipment. Established: 1984. Ownership: Private.

Key Statistics: Annual Sales: $62.4 million. Number of Employees: 650.

Expertise/Education Sought: Accounting, finance, data processing, sales, marketing

Alternatives to Full-Time Employment: Internships

Contact: Ms. Lily Rettis, Director of Human Resources; Phone: 212-951-3040; Fax: 212-213-9319.

GRANITE BROADCASTING CORPORATION
767 3RD AVENUE, 34TH FLOOR
NEW YORK, NEW YORK 10017-2023

Overview: Operates television broadcasting stations. Established: 1988. Ownership: Public.

Key Statistics: Annual Sales: $129.2 million. Number of Employees: 1,170.

Expertise/Education Sought: Accounting, computer programming, finance, communications

Alternatives to Full-Time Employment: Paid internships

Company-Paid Training is available.

Contact: For individuals with previous experience: Ms. Dolores Perez, Director of Human Resources; Phone: 212-826-2530; Fax: 212-826-2858.

GRETAGMACBETH LLC
617 LITTLE BRITAIN ROAD
NEW WINDSOR, NEW YORK 12553-7209

Overview: Manufactures electric equipment and supplies. Established: 1997. Ownership: Private.

Key Statistics: Annual Sales: $37 million. Number of Employees: 200.

Hiring History: Number of professional employees hired in 1997: 20.

Expertise/Education Sought: Engineering, marketing

Company-Paid Training is available.

International Assignments are available to employees.

Contact: For individuals with previous experience: Human Resources Department; Phone: 914-565-7660; Fax: 914-565-1155.

GUARDIAN LIFE INSURANCE COMPANY
201 PARK AVENUE SOUTH
NEW YORK, NEW YORK 10003-1601

Overview: Provides life, health, and disability insurance. Established: 1860. Ownership: Private.

Key Statistics: Annual Sales: $6.9 billion. Number of Employees: 5,155.

Expertise/Education Sought: Marketing, sales, finance, actuarial, accounting, underwriting, computer programming, law, industrial engineering

Alternatives to Full-Time Employment: Part-time jobs, paid internships

Company-Paid Training is available.

Contact: Ms. Lisa Engelson, Employment Supervisor; Phone: 212-598-8407; Fax: 212-533-5156.

Company URL: http://www.theguardian.com

HALSEY DRUG COMPANY, INC.
1825-1839 PACIFIC STREET
BROOKLYN, NEW YORK 11233

Overview: Manufactures pharmaceutical preparations, including antihistamines and antibiotics. Established: 1935. Ownership: Public.

Key Statistics: Annual Sales: $12.4 million. Number of Employees: 360.

Expertise/Education Sought: Manufacturing, chemical engineering, electrical engineering

Contact: For individuals with previous experience: Human Resources; Phone: 718-467-7500; Fax: 718-467-4261.

HAZEN AND SAWYER, PC
730 BROADWAY, 6TH FLOOR
NEW YORK, NEW YORK 10003-9594

Overview: Provides environmental and scientific engineering services. Established: 1951. Ownership: Private.

Key Statistics: Annual Sales: $55 million. Number of Employees: 425.

Hiring History: Number of professional employees hired in 1997: 15. 1996: 8.

Expertise/Education Sought: Environmental engineering, civil engineering

Alternatives to Full-Time Employment: Paid internships

Company-Paid Training is available.

International Assignments are available to employees.

Contact: Ms. Lianne Rubbo, Human Resource Manager; Phone: 212-539-7105; Fax: 212-614-9049.

Company URL: http://www.hazenandsawyer.com

HENDRICKSON/SCALAMANDRE/POSILLICO
1610 NEW HIGHWAY
FARMINGDALE, NEW YORK 11735-1510

Overview: General contractor. Established: 1995. Ownership: Private.

Key Statistics: Annual Sales: $100 million. Number of Employees: 500.

Expertise/Education Sought: Engineering

Contact: For individuals with previous experience: Mr. Steve Bongiorno, Controller; Phone: 516-249-1872; Fax: 516-249-8124.

HERAEUS, INC.
540 MADISON AVENUE
NEW YORK, NEW YORK 10022

Overview: Produces metal pastes. Established: 1977. Parent Company: Heraeus Holding, GMBH. Ownership: Private.

Key Statistics: Annual Sales: $550 million. Number of Employees: 500.

Expertise/Education Sought: Trading, precious metals, manufacturing

Contact: For individuals with previous experience: Employment; Phone: 212-752-2705, ext. 31; Fax: 212-752-4686.

HOFFMAN MILLS, INC.
470 PARK AVENUE SOUTH
NEW YORK, NEW YORK 10016-6819

Overview: Produces broadwoven synthetic fabric. Established: 1948. Ownership: Private.

Key Statistics: Annual Sales: $46 million. Number of Employees: 600.

Expertise/Education Sought: Manufacturing

Alternatives to Full-Time Employment: Internships

International Assignments are available to employees.

Contact: For individuals with previous experience: Mr. Joseph Leone, Chief Financial Officer; Phone: 212-684-3700; Fax: 212-779-1299.

HOST APPAREL, INC.
1430 BROADWAY, FLOOR 10
NEW YORK, NEW YORK 10018-3397

Overview: Manufactures men's and boys' underwear and nightwear. Established: 1959. Ownership: Private.

Key Statistics: Number of Employees: 60.

EXPERTISE/EDUCATION SOUGHT: Manufacturing, sales, merchandising
CONTACT: For individuals with previous experience: Human Resources; Phone: 212-302-0800.

HOUSE OF PERFECTION, INC.
131 WEST 33RD STREET, 7TH FLOOR
NEW YORK, NEW YORK 10001-2988

OVERVIEW: Manufactures girls' and children's clothing. Established: 1945. Ownership: Private.
KEY STATISTICS: Annual Sales: $39 million. Number of Employees: 500.
HIRING HISTORY: Number of professional employees hired in 1997: 6.
EXPERTISE/EDUCATION SOUGHT: Manufacturing
CONTACT: For individuals with previous experience: Human Resource Department; Phone: 212-239-9780; Fax: 212-268-1431.

HRH CONSTRUCTION INTERIORS
1 PARK AVENUE
NEW YORK, NEW YORK 10016

OVERVIEW: Provides construction management services. Established: 1977. Parent Company: Starrett Corporation. Ownership: Public.
KEY STATISTICS: Annual Sales: $17.5 million. Number of Employees: 300.
EXPERTISE/EDUCATION SOUGHT: Construction management
ALTERNATIVES TO FULL-TIME EMPLOYMENT: Paid internships
CONTACT: For individuals with previous experience: Human Resource; Phone: 212-751-3100; Fax: 212-696-5458.

IDF INTERNATIONAL, INC.
330 WEST 42ND STREET
NEW YORK, NEW YORK 10036-6902

OVERVIEW: Provides engineering services. Established: 1991. Ownership: Private.
KEY STATISTICS: Annual Sales: $15.6 million. Number of Employees: 190.
HIRING HISTORY: Number of professional employees hired in 1997: 4.
EXPERTISE/EDUCATION SOUGHT: Engineering, electrical engineering, structural engineering
INTERNATIONAL ASSIGNMENTS are available to employees.
CONTACT: For individuals with previous experience: Mr. John Nesi, Personnel Manager; Phone: 212-563-6900; Fax: 212-563-0076.

IEH CORPORATION
140 58TH STREET, SUITE 8E-F
BROOKLYN, NEW YORK 11220-2521

OVERVIEW: Produces electronic connectors. Established: 1943. Ownership: Public.
KEY STATISTICS: Number of Employees: 90.
HIRING HISTORY: Number of professional employees hired in 1997: 4. 1996: 4.
EXPERTISE/EDUCATION SOUGHT: Electrical engineering, quality control
CONTACT: For individuals with previous experience: Human Resources Department; Phone: 718-492-9673; Fax: 718-492-9898.

IMPERIAL SCHRADE CORPORATION
7 SCHRADE COURT
ELLENVILLE, NEW YORK 12428-1041

OVERVIEW: Manufactures cutlery. Established: 1947. Ownership: Private.
KEY STATISTICS: Annual Sales: $44.6 million. Number of Employees: 850.
HIRING HISTORY: Number of professional employees hired in 1997: 150. 1996: 150.
EXPERTISE/EDUCATION SOUGHT: Mechanical engineering, project engineering
ALTERNATIVES TO FULL-TIME EMPLOYMENT: Paid internships
COMPANY-PAID TRAINING is available.
CONTACT: For individuals with previous experience: Corporate Recruiter; Phone: 914-647-7600, ext. 349; Fax: 914-647-8702.

INCO UNITED STATES, INC.
1 NEW YORK PLAZA
NEW YORK, NEW YORK 10004-1901

OVERVIEW: Mines nickel and extracts precious metals. Established: 1975. Parent Company: Inco, Ltd. Ownership: Private.
KEY STATISTICS: Annual Sales: $1.2 billion. Number of Employees: 2,429.
EXPERTISE/EDUCATION SOUGHT: Computer programming, systems analysis, accounting, finance, data processing, computer programming, human resources
CONTACT: For individuals with previous experience: Mr. Ted Von Hoff, Manager of Human Resources; Phone: 212-612-5500; Fax: 212-612-5879.

INDUSTRIAL ACOUSTICS COMPANY, INC.
1160 COMMERCE AVENUE
BRONX, NEW YORK 10462-5506

OVERVIEW: Manufactures prefabricated metal components. Established: 1949. Ownership: Public.
KEY STATISTICS: Annual Sales: $73.6 million. Number of Employees: 730.
HIRING HISTORY: Number of professional employees hired in 1997: 6. 1996: 7.
EXPERTISE/EDUCATION SOUGHT: Engineering
ALTERNATIVES TO FULL-TIME EMPLOYMENT: Paid internships
COMPANY-PAID TRAINING is available.
INTERNATIONAL ASSIGNMENTS are available to employees.
CONTACT: Mr. Raymond Svana, Professional Recruiter; Phone: 718-931-8000, ext. 725; Fax: 718-430-4750.
COMPANY URL: http://www.industrialacoustics.com

INFORMATION BUILDERS, INC.
2 PENN PLAZA
NEW YORK, NEW YORK 10121

OVERVIEW: Manufactures and markets computer software. Established: 1975. Ownership: Private.
KEY STATISTICS: Annual Sales: $261 million. Number of Employees: 1,800.
HIRING HISTORY: Number of professional employees hired in 1997: 200.
EXPERTISE/EDUCATION SOUGHT: Computer programming, marketing, sales
ALTERNATIVES TO FULL-TIME EMPLOYMENT: Paid internships
COMPANY-PAID TRAINING is available.
INTERNATIONAL ASSIGNMENTS are available to employees.
CONTACT: Mr. Kier McAulay, Technical Recruiter; Phone: 212-736-4433, ext. 3596; Fax: 212-239-6674. E-mail: kier_mcaulay@ibi.com.
COMPANY URL: http://www.ibi.com

INTERNATIONAL FLAVORS AND FRAGRANCES
521 WEST 57TH STREET
NEW YORK, NEW YORK 10019-2905

OVERVIEW: Manufactures synthetic flavors and fragrances for commercial products. Established: 1909. Ownership: Public.
KEY STATISTICS: Annual Sales: $1.4 billion. Number of Employees: 4,629.
HIRING HISTORY: Number of professional employees hired in 1997: 20. 1996: 10.
EXPERTISE/EDUCATION SOUGHT: Finance, marketing, law, sales, chemical engineering, chemistry, chemical analysis, manufacturing engineering, product design and development
ALTERNATIVES TO FULL-TIME EMPLOYMENT: Paid internships
COMPANY-PAID TRAINING is available.
INTERNATIONAL ASSIGNMENTS are available to employees.
CONTACT: Ms. Kashmira Palkhivala, Employee Relations Manager; Phone: 212-765-5500, ext. 7217; Fax: 212-708-7119.

KAHN-LUCAS-LANCASTER, INC.
100 WEST 33RD STREET, ROOM 921
NEW YORK, NEW YORK 10001-2912

OVERVIEW: Manufactures girls' clothing. Established: 1964. Ownership: Private.

KEY STATISTICS: Annual Sales: $20 million. Number of Employees: 500.

EXPERTISE/EDUCATION SOUGHT: Manufacturing engineering

ALTERNATIVES TO FULL-TIME EMPLOYMENT: Paid internships

CONTACT: For individuals with previous experience: Human Resources Department; Phone: 717-684-6911; Fax: 717-684-0229.

KEY SERVICES CORPORATION
17 CORPORATE WOODS BOULEVARD
ALBANY, NEW YORK 12211

OVERVIEW: Provides data processing services for financial institutions. Established: 1985. Parent Company: Key Corporation. Ownership: Public.

KEY STATISTICS: Annual Sales: $269.6 million. Number of Employees: 4,230.

EXPERTISE/EDUCATION SOUGHT: Computer programming, computer analysis, software engineering/development, program analysis, management information systems

ALTERNATIVES TO FULL-TIME EMPLOYMENT: Internships

COMPANY-PAID TRAINING is available.

CONTACT: Ms. Darcy Batzold, Technical Recruiter; Phone: 518-391-1558; Fax: 518-391-1579. Job Hotline: 888-539-2562.

KEYSPAN ENERGY CORPORATION
1 METROTECH CENTER
BROOKLYN, NEW YORK 11201-3831

OVERVIEW: Natural gas distributor. Established: 1996. Ownership: Private.

KEY STATISTICS: Annual Sales: $902.3 million. Number of Employees: 2,816.

EXPERTISE/EDUCATION SOUGHT: Systems administration, engineering

ALTERNATIVES TO FULL-TIME EMPLOYMENT: Paid internships

COMPANY-PAID TRAINING is available.

CONTACT: Mr. Louis Donaldson, Employment Recruiter; Phone: 718-403-2000; Fax: 718-488-1762. Job Hotline: 718-403-2810.

COMPANY URL: http://www.bug.com

KEYSTONE ELECTRONICS CORPORATION
3107 20TH ROAD
ASTORIA, NEW YORK 11105-2017

OVERVIEW: Produces electronic connectors. Established: 1957. Ownership: Private.

KEY STATISTICS: Annual Sales: $10 million. Number of Employees: 130.

EXPERTISE/EDUCATION SOUGHT: Electronics/electronics engineering

CONTACT: For individuals with previous experience: Personnel Department; Phone: 718-956-8900; Fax: 718-956-9040.

LAND N SEA, INC.
1375 BROADWAY
NEW YORK, NEW YORK 10018-7001

OVERVIEW: Manufactures women's blouses and shirts. Established: 1958. Ownership: Private.

KEY STATISTICS: Annual Sales: $100 million. Number of Employees: 900.

EXPERTISE/EDUCATION SOUGHT: Manufacturing, sales

COMPANY-PAID TRAINING is available.

CONTACT: For individuals with previous experience: Human Resources; Phone: 212-444-6000; Fax: 212-444-6018.

LECROY CORPORATION
700 CHESTNUT RIDGE ROAD
SPRING VALLEY, NEW YORK 10977-6437

OVERVIEW: Manufactures oscillographs and oscilloscopes. Established: 1963. Ownership: Public.

KEY STATISTICS: Annual Sales: $117.1 million. Number of Employees: 555.

HIRING HISTORY: Number of professional employees hired in 1997: 65. 1996: 65.

EXPERTISE/EDUCATION SOUGHT: Engineering, marketing, management, sales, manufacturing

ALTERNATIVES TO FULL-TIME EMPLOYMENT: Part-time jobs

INTERNATIONAL ASSIGNMENTS are available to employees.

CONTACT: For individuals with previous experience: Human Relations Department; Phone: 914-425-2000; Fax: 914-425-8967.

COMPANY URL: http://www.lecroy.com

LEVITON MANUFACTURING COMPANY, INC.
5925 LITTLE NECK PARKWAY
LITTLE NECK, NEW YORK 11362-2591

OVERVIEW: Manufactures wiredrawn nonferrous cable. Established: 1941. Ownership: Private.

KEY STATISTICS: Annual Sales: $200 million. Number of Employees: 1,150.

EXPERTISE/EDUCATION SOUGHT: Engineering, data processing, customer service and support, accounting, finance, management, administration

ALTERNATIVES TO FULL-TIME EMPLOYMENT: Internships

CONTACT: Ms. Janet Howitt, Human Resources Manager; Phone: 718-281-6386; Fax: 718-281-6129.

LIFETIME TELEVISION
309 WEST 49TH STREET
NEW YORK, NEW YORK 10019-7316

OVERVIEW: Television broadcasting company. Established: 1994. Parent Company: Hearst Corporation and Disney/ABC. Ownership: Private.

KEY STATISTICS: Annual Sales: $46.5 million. Number of Employees: 350.

EXPERTISE/EDUCATION SOUGHT: Data processing, finance, electrical engineering, television broadcasting, communications, accounting, marketing, computer programming

ALTERNATIVES TO FULL-TIME EMPLOYMENT: Part-time jobs, internships

COMPANY-PAID TRAINING is available.

CONTACT: Ms. Debra Henderson, Director of Recruitment; Phone: 212-424-7000; Fax: 212-424-7155.

COMPANY URL: http://www.lifetimetv.com

LOEWS CORPORATION
667 MADISON AVENUE
NEW YORK, NEW YORK 10021-8029

OVERVIEW: Manages operations in the financial and hotel industries. Established: 1969. Parent Company: Lorillard Tobacco Company. Ownership: Public.

KEY STATISTICS: Assets: $20.4 billion. Number of Employees: 34,700.

EXPERTISE/EDUCATION SOUGHT: Accounting, marketing, finance, management information systems, hospitality/hotel and restaurant management

ALTERNATIVES TO FULL-TIME EMPLOYMENT: Paid internships

COMPANY-PAID TRAINING is available.

INTERNATIONAL ASSIGNMENTS are available to employees.

CONTACT: For individuals with previous experience: Mr. Jerry Champness, Manager of Employment, 1 Park Avenue, New York, New York, 10016-8029; Phone: 212-545-2512; Fax: 212-545-2525.

LONG ISLAND LIGHTING COMPANY
175 EAST OLD COUNTRY ROAD
HICKSVILLE, NEW YORK 11801-4257

OVERVIEW: Electric utility. Established: 1910. Ownership: Public.

KEY STATISTICS: Annual Sales: $3.2 billion. Number of Employees: 5,413.

HIRING HISTORY: Number of professional employees hired in 1997: 3.

EXPERTISE/EDUCATION SOUGHT: Information systems

ALTERNATIVES TO FULL-TIME EMPLOYMENT: Internships

COMPANY-PAID TRAINING is available.

CONTACT: Ms. Caroline Rubin, Employment Administrator; Phone: 516-545-3151; Fax: 516-545-3180.

THE LONG ISLAND RAILROAD COMPANY

JAMAICA STATION
JAMAICA STATION, NEW YORK 11435

OVERVIEW: Operates railroad. Established: 1966. Parent Company: Metropolitan Transportation Authority. Ownership: Private.

KEY STATISTICS: Annual Sales: $345.2 million. Number of Employees: 6,100.

EXPERTISE/EDUCATION SOUGHT: Finance, accounting, auditing, transportation

CONTACT: For individuals with previous experience: Employment Office; Phone: 718-558-7400. Job Hotline: 718-558-7560.

LOUIS HORNICK & COMPANY, INC.

261 5TH AVENUE
NEW YORK, NEW YORK 10016-7701

OVERVIEW: Manufactures curtains. Established: 1918. Ownership: Private.

KEY STATISTICS: Annual Sales: $50 million. Number of Employees: 525.

EXPERTISE/EDUCATION SOUGHT: Engineering

ALTERNATIVES TO FULL-TIME EMPLOYMENT: Paid internships

COMPANY-PAID TRAINING is available.

CONTACT: For individuals with previous experience: Human Resources Department; Phone: 212-679-2448; Fax: 212-779-7098.

COMPANY URL: http://www.louishornick.com

LUMEX, DIVISION OF GRAHAM-FIELD

100 SPENCE STREET
BAYSHORE, NEW YORK 11706-2206

OVERVIEW: Manufactures canes, orthopedic crutches, walkers, and wheelchairs. Established: 1996. Parent Company: Graham-Field Inc. Ownership: Public.

KEY STATISTICS: Annual Sales: $34.3 million. Number of Employees: 450.

EXPERTISE/EDUCATION SOUGHT: Accounting, administration, engineering, marketing, mechanical engineering, design engineering, sales, home health care

ALTERNATIVES TO FULL-TIME EMPLOYMENT: Paid internships

COMPANY-PAID TRAINING is available.

CONTACT: Ms. Aida Malavey, Director of Human Resources; Phone: 516-273-2200, ext. 404.

MATERIALS RESEARCH CORPORATION

560 ROUTE 303
ORANGEBURG, NEW YORK 10962-1314

OVERVIEW: Manufactures industrial equipment. Established: 1989. Parent Company: Sony Corporation of America. Ownership: Private.

KEY STATISTICS: Annual Sales: $68.8 million. Number of Employees: 700.

HIRING HISTORY: Number of professional employees hired in 1996: 50.

EXPERTISE/EDUCATION SOUGHT: Physics, materials science, field engineering, process engineering, software engineering/development

ALTERNATIVES TO FULL-TIME EMPLOYMENT: Paid internships

CONTACT: For individuals with previous experience: Human Resources, 2120 West Guadeloupe, Orangeburg, New York, 10962-1314; Phone: 914-268-7979.

MEDICAL LABORATORY AUTOMATION

270 MARBLE AVENUE
PLEASANTVILLE, NEW YORK 10570-3411

OVERVIEW: Manufactures blood coagulation devices. Established: 1963. Ownership: Private.

KEY STATISTICS: Annual Sales: $59.2 million. Number of Employees: 400.

HIRING HISTORY: Number of professional employees hired in 1996: 50.

EXPERTISE/EDUCATION SOUGHT: Design engineering, mechanical engineering, electrical engineering, industrial engineering

COMPANY-PAID TRAINING is available.

CONTACT: For individuals with previous experience: Ms. Pat McDonough, Human Resources Representative; Phone: 914-747-3020, ext. 329.

METROPOLITAN TRANSPORTATION AUTHORITY

347 MADISON AVENUE
NEW YORK, NEW YORK 10017-3706

OVERVIEW: Provides local transit services. Established: 1965. Ownership: Private.

KEY STATISTICS: Annual Sales: $3.2 billion. Number of Employees: 64,465.

HIRING HISTORY: Number of professional employees hired in 1997: 100.

EXPERTISE/EDUCATION SOUGHT: Public relations

ALTERNATIVES TO FULL-TIME EMPLOYMENT: Paid internships

COMPANY-PAID TRAINING is available.

CONTACT: Ms. Jessie Crawford, Assistant Manager; Phone: 212-878-7000; Fax: 212-878-0166. Job Hotline: 212-878-7490.

COMPANY URL: http://www.mta.nyc.ny.us

MITSUBISHI CHEMICAL AMERICA, INC.

1 NORTH LEXINGTON
WHITE PLAINS, NEW YORK 10601-1711

OVERVIEW: Manufactures chemicals and pharmaceutical products. Established: 1981. Parent Company: Mitsubishi Chemical Corporation. Ownership: Private.

KEY STATISTICS: Annual Sales: $850 million. Number of Employees: 2,491.

HIRING HISTORY: Number of professional employees hired in 1996: 7.

EXPERTISE/EDUCATION SOUGHT: Accounting, finance, marketing, sales, pharmacology

COMPANY-PAID TRAINING is available.

INTERNATIONAL ASSIGNMENTS are available to employees.

CONTACT: For individuals with previous experience: Ms. Takuya Wada, Director of Resources; Phone: 914-286-3611; Fax: 914-761-0108. E-mail: takuya_wada@m-chem.com.

COMPANY URL: http://www.m-kagaku.co.jp

MONOFRAX, INC.

1870 NEW YORK AVENUE
FALCONER, NEW YORK 14733-1797

OVERVIEW: Produces nonclay refractors. Established: 1996. Parent Company: Vesuvius America/Cookson Group. Ownership: Private.

KEY STATISTICS: Annual Sales: $50 million. Number of Employees: 370.

EXPERTISE/EDUCATION SOUGHT: Ceramic engineering

COMPANY-PAID TRAINING is available.

CONTACT: For individuals with previous experience: Mr. Sal Cantali, Manager of Human Resources; Phone: 716-483-7200, ext. 206; Fax: 716-665-2478.

MOOG, INC.

SENECA STREET AT JAMISON
EAST AURORA, NEW YORK 14052

OVERVIEW: Manufactures high performance controls for commercial, industrial, aerospace, and defense applications. Established: 1951. Ownership: Public.

KEY STATISTICS: Annual Sales: $407.2 million. Number of Employees: 3,229.

EXPERTISE/EDUCATION SOUGHT: Mechanical engineering

ALTERNATIVES TO FULL-TIME EMPLOYMENT: Paid internships

COMPANY-PAID TRAINING is available.

INTERNATIONAL ASSIGNMENTS are available to employees.

CONTACT: For college students and recent graduates: Ms. Kathy Boni, Human Resources Plant 23; Phone: 716-652-2000; Fax: 716-687-4820. For individuals with previous experience: Human Resources Plant 23; Phone: 716-652-2000; Fax: 716-687-4820.

NALGE NUNC INTERNATIONAL
75 PANORAMA CREEK DRIVE
ROCHESTER, NEW YORK 14625

OVERVIEW: Manufactures and distributes plastic products. Established: 1994. Parent Company: Sybron International. Ownership: Public.

KEY STATISTICS: Annual Sales: $220.4 million. Number of Employees: 1,584.

EXPERTISE/EDUCATION SOUGHT: Mechanical engineering, computer science, marketing, plastics engineering

ALTERNATIVES TO FULL-TIME EMPLOYMENT: Part-time jobs

COMPANY-PAID TRAINING is available.

INTERNATIONAL ASSIGNMENTS are available to employees.

CONTACT: For individuals with previous experience: Mr. Martin J. Wheeler, Manager of Employment and Compensation; Phone: 716-264-3961; Fax: 716-264-3969. E-mail: mwheeler@nalgenunc.com.

COMPANY URL: http://www.nalgenunc.com

NAPCO SECURITY SYSTEMS, INC.
333 BAYVIEW AVENUE
AMITYVILLE, NEW YORK 11701

OVERVIEW: Manufactures electronic security devices and alarm products. Established: 1969. Ownership: Public.

KEY STATISTICS: Annual Sales: $53.3 million. Number of Employees: 1,000.

HIRING HISTORY: Number of professional employees hired in 1997: 45.

EXPERTISE/EDUCATION SOUGHT: Design engineering, computer analysis, electronics/electronics engineering

COMPANY-PAID TRAINING is available.

CONTACT: Human Resources Department; Phone: 516-842-9400; Fax: 516-789-8176. E-mail: sbeebe@napcosecurity.com.

COMPANY URL: http://www.napcosecurity.com

NATIONAL BROADCASTING COMPANY, INC.
30 ROCKEFELLER PLAZA
NEW YORK, NEW YORK 10112

OVERVIEW: Operates radio and television stations and networks. Established: 1986. Parent Company: General Electric. Ownership: Public.

KEY STATISTICS: Annual Sales: $5.2 billion. Number of Employees: 5,000.

EXPERTISE/EDUCATION SOUGHT: Television broadcasting, telecommunications, public relations, sales, computer science, multimedia

ALTERNATIVES TO FULL-TIME EMPLOYMENT: Internships

COMPANY-PAID TRAINING is available.

INTERNATIONAL ASSIGNMENTS are available to employees.

CONTACT: NBC Employment Office; Phone: 212-664-2277; Fax: 212-664-5761.

NATIONAL FUEL GAS COMPANY
10 LAFAYETTE SQUARE
BUFFALO, NEW YORK 14203-1824

OVERVIEW: Distributes petroleum and natural gas products. Established: 1902. Ownership: Public.

KEY STATISTICS: Annual Sales: $1.2 billion. Number of Employees: 2,843.

EXPERTISE/EDUCATION SOUGHT: Accounting, finance

ALTERNATIVES TO FULL-TIME EMPLOYMENT: Paid internships

COMPANY-PAID TRAINING is available.

CONTACT: Mr. Paul Pavlovic, Program Manager; Phone: 716-857-7421; Fax: 716-857-7195. Job Hotline: 716-857-7821.

COMPANY URL: http://www.natfuel.com

NEW YORK POWER AUTHORITY
1633 BROADWAY
NEW YORK, NEW YORK 10019-6708

OVERVIEW: Utility company. Established: 1931. Ownership: Private.

KEY STATISTICS: Annual Sales: $1.4 billion. Number of Employees: 3,500.

HIRING HISTORY: Number of professional employees hired in 1997: 70.

EXPERTISE/EDUCATION SOUGHT: Electrical engineering, law, civil engineering, computer programming, mechanical engineering, nuclear engineering, accounting, marketing, human resources, public relations

ALTERNATIVES TO FULL-TIME EMPLOYMENT: Paid internships

COMPANY-PAID TRAINING is available.

CONTACT: Ms. Ann Tivenan, Employee Services Specialist; Phone: 212-468-6691; Fax: 212-468-6695. E-mail: tivenan.a@nypa.gov.

NEW YORK STATE ELECTRIC AND GAS
4500 VESTAL PARKWAY EAST
BINGHAMTON, NEW YORK 13902

OVERVIEW: Electric and gas utility. Established: 1852. Ownership: Public.

KEY STATISTICS: Annual Sales: $2.1 billion. Number of Employees: 4,000.

EXPERTISE/EDUCATION SOUGHT: Marketing, information systems

ALTERNATIVES TO FULL-TIME EMPLOYMENT: Paid internships

COMPANY-PAID TRAINING is available.

CONTACT: Mr. Ronald Thorpe, Professional Employment Specialist; Phone: 607-762-4106; Fax: 607-762-4004. E-mail: rthorpe@nyseg.com.

COMPANY URL: http://www.nyseg.com

NIAGARA MOHAWK POWER CORPORATION
1304 BUCKLEY ROAD
SYRACUSE, NEW YORK 13212-4201

OVERVIEW: Electric and gas utility. Established: 1937. Ownership: Public.

KEY STATISTICS: Annual Sales: $4 billion. Number of Employees: 8,600.

EXPERTISE/EDUCATION SOUGHT: Computer science

ALTERNATIVES TO FULL-TIME EMPLOYMENT: Paid internships

CONTACT: Ms. Dianne Thompson, Manager Organizational Staffing; Phone: 315-460-4551; Fax: 315-460-4574. Job Hotline: 800-867-5222.

COMPANY URL: http://www.nimo.com

NORTH AMERICAN SPECIALTIES
120-12 28TH AVENUE
FLUSHING, NEW YORK 11354-1049

OVERVIEW: Manufactures dies and die holders. Established: 1981. Ownership: Private.

KEY STATISTICS: Annual Sales: $10 million. Number of Employees: 300.

EXPERTISE/EDUCATION SOUGHT: Tool and die engineering

CONTACT: For individuals with previous experience: Controller; Phone: 718-961-6210; Fax: 718-886-0573.

O'BRIEN AND GERE ENGINEERS, INC.
5000 BRITTONFIELD PARKWAY
EAST SYRACUSE, NEW YORK 13057

OVERVIEW: Provides engineering consulting services. Established: 1985. Parent Company: O'Brien and Gere, Ltd. Ownership: Private.

KEY STATISTICS: Annual Sales: $71.5 million. Number of Employees: 600.

HIRING HISTORY: Number of professional employees hired in 1997: 28. 1996: 38.

EXPERTISE/EDUCATION SOUGHT: Chemical engineering, civil engineering, environmental engineering, mechanical engineering, structural engineering, electrical engineering

ALTERNATIVES TO FULL-TIME EMPLOYMENT: Part-time jobs, paid internships

COMPANY-PAID TRAINING is available.

CONTACT: For individuals with previous experience: Ms. Lisa Thompson, Personnel Administrator; Phone: 315-437-6100; Fax: 315-463-7554. E-mail: thompsoa@obg.com.

COMPANY URL: http://www.obg.com

ODD JOB TRADING
66 WEST 48TH STREET
NEW YORK, NEW YORK 10036-1702

OVERVIEW: Operates department stores. Established: 1996. Parent Company: Mazel Stores. Ownership: Public.

Key Statistics: Annual Sales: $32.3 million. Number of Employees: 500.
Expertise/Education Sought: Retail
Company-Paid Training is available.
Contact: For individuals with previous experience: Mr. Jerry Lacher, Vice President of Operations; Phone: 212-575-0477; Fax: 212-302-6673.

OGDEN SERVICES CORPORATION
2 PENNSYLVANIA PLAZA
NEW YORK, NEW YORK 10121

Overview: Provides office cleaning services. Established: 1939. Ownership: Public.
Key Statistics: Annual Sales: $2 billion. Number of Employees: 30,000.
Expertise/Education Sought: Aviation, energy/energy management, entertainment
Alternatives to Full-Time Employment: Paid internships
Company-Paid Training is available.
International Assignments are available to employees.
Contact: Human Resources; Phone: 212-868-6000; Fax: 212-868-8116.
Company URL: http://www.ogdencorp.com

PACE INDUSTRIES, INC.
405 LEXINGTON AVENUE
NEW YORK, NEW YORK 10174

Overview: Manufactures air conditioning systems. Established: 1988. Ownership: Private.
Key Statistics: Annual Sales: $470 million. Number of Employees: 5,300.
Hiring History: Number of professional employees hired in 1996: 4.
Expertise/Education Sought: Management, accounting
Company-Paid Training is available.
Contact: For individuals with previous experience: Mr. William Ostan, Vice President of Human Resources; Phone: 212-916-8100; Fax: 212-916-8109.

PAGING NETWORK OF NEW YORK
990 STEWART AVENUE
GARDEN CITY, NEW YORK 11530

Overview: Provides paging services. Established: 1991. Ownership: Public.
Key Statistics: Annual Sales: $6.9 million. Number of Employees: 80.
Hiring History: Number of professional employees hired in 1997: 30. 1996: 30.
Expertise/Education Sought: Customer service and support, sales, technical
Alternatives to Full-Time Employment: Part-time jobs
International Assignments are available to employees.
Contact: For individuals with previous experience: Mr. Marc Steiner, Human Resources Manager; Phone: 516-745-0300, ext. 312; Fax: 516-745-0922. E-mail: marc_steiner@pagenet.com.
Company URL: http://www.pagenet.com

PALL CORPORATION
2200 NORTHERN BOULEVARD
EAST HILLS, NEW YORK 11548

Overview: Manufactures filters. Established: 1946. Ownership: Public.
Key Statistics: Annual Sales: $960.4 million. Number of Employees: 7,700.
Expertise/Education Sought: Accounting, industrial engineering, mechanical engineering, finance, marketing, human resources, computer programming, laboratory technology
Alternatives to Full-Time Employment: Part-time jobs, paid internships
Company-Paid Training is available.
International Assignments are available to employees.
Contact: For individuals with previous experience: Ms. Karen Elkins, Corporate Human Resources Manager; Phone: 516-484-5400, ext. 4281; Fax: 516-621-4534. E-mail: karen_elkins@pall.com. Job Hotline: 516-484-5400, ext. 5555.
Company URL: http://www.pall.com

PARSONS AND WHITTEMORE, INC.
4 INTERNATIONAL DRIVE
RYE BROOK, NEW YORK 10573

Overview: Manufactures pulp and paper. Established: 1980. Ownership: Private.
Key Statistics: Annual Sales: $180 million. Number of Employees: 1,800.
Expertise/Education Sought: Computer science, data processing, pulp and paper science, forestry, accounting, finance, sales, marketing
International Assignments are available to employees.
Contact: For individuals with previous experience: Mr. Arthur Schwartz, President; Phone: 914-937-9009; Fax: 914-937-2259.

PAWLING CORPORATION
157 CHARLES COLMAN BOULEVARD
PAWLING, NEW YORK 12564-1188

Overview: Produces rubber floorcoverings and rubber automotive products. Established: 1945. Ownership: Private.
Key Statistics: Annual Sales: $43 million. Number of Employees: 315.
Hiring History: Number of professional employees hired in 1997: 45. 1996: 25.
Expertise/Education Sought: Manufacturing
Alternatives to Full-Time Employment: Paid internships
Company-Paid Training is available.
Contact: For individuals with previous experience: Ms. Dawn Giordano, Human Resources Manager; Phone: 914-855-1000; Fax: 914-855-1937.

PAYCHEX, INC.
911 PANORAMA TRAIL SOUTH
ROCHESTER, NEW YORK 14625-2320

Overview: Provides payroll processing services and software. Established: 1970. Ownership: Public.
Key Statistics: Annual Sales: $734.7 million. Number of Employees: 4,440.
Hiring History: Number of professional employees hired in 1997: 1,500.
Expertise/Education Sought: Quality control, data processing, computer programming, finance, accounting, human resources, marketing, software engineering/development, sales
Alternatives to Full-Time Employment: Part-time jobs, internships
Company-Paid Training is available.
Contact: For college students and recent graduates: Ms. Sarah Brown, Recruiter; Phone: 716-383-3430; Fax: 716-383-3449. For individuals with previous experience: Ms. Michelle Haugh, Recruiter; Phone: 716-383-3430; Fax: 716-383-3449. Job Hotline: 800-896-5010.
Company URL: http://www.paychex.com

PEARSON, INC.
30 ROCKEFELLER PLAZA
NEW YORK, NEW YORK 10112-5095

Overview: Manages entertainment businesses. Established: 1981. Parent Company: Pearson, PLC. Ownership: Private.
Key Statistics: Annual Sales: $990.9 million. Number of Employees: 8,500.
Hiring History: Number of professional employees hired in 1997: 9. 1996: 16.
Expertise/Education Sought: Accounting, finance, management information systems, law, human resources
Alternatives to Full-Time Employment: Paid internships
Contact: For individuals with previous experience: Mr. Randall C. Keller, Vice President of Human Resources; Phone: 212-641-2400.

PEPSICO, INC.

700 ANDERSON HILL ROAD
PURCHASE, NEW YORK 10577-1403

Overview: Manufactures soft drinks and snack foods; operates fast food restaurants. Established: 1919. Ownership: Public.

Key Statistics: Annual Sales: $31.6 billion. Number of Employees: 140,000.

Expertise/Education Sought: Finance, strategic planning, auditing, management information systems

Alternatives to Full-Time Employment: Part-time jobs, paid internships

Company-Paid Training is available.

International Assignments are available to employees.

Contact: Mr. Charles Rogers, Vice President of Executive Staffing; Phone: 914-253-2000; Fax: 914-253-2070.

Company URL: http://www.pepsico.com

PERIPHONICS CORPORATION

4000 VETERAN'S MEMORIAL HIGHWAY
BOHEMIA, NEW YORK 11716-1024

Overview: Manufactures electronic voice response systems. Established: 1995. Ownership: Public.

Key Statistics: Annual Sales: $111.2 million. Number of Employees: 733.

Expertise/Education Sought: Software engineering/development, computer science, application developer

Alternatives to Full-Time Employment: Paid internships

Company-Paid Training is available.

International Assignments are available to employees.

Contact: Ms. Janet Anderson, Senior Director of Human Resources; Phone: 516-468-9000; Fax: 516-467-1755. E-mail: hr@peri.com.

Company URL: http://www.peri.com

PETRY MEDIA CORPORATION

1290 AVENUE OF THE AMERICAS
NEW YORK, NEW YORK 10104-0101

Overview: Provides advertising services. Established: 1995. Ownership: Private.

Key Statistics: Annual Sales: $126.5 million. Number of Employees: 1,265.

Hiring History: Number of professional employees hired in 1997: 60.

Expertise/Education Sought: Information services, accounting, finance, human resources, sales, research, sales, television broadcasting

Alternatives to Full-Time Employment: Paid internships

Company-Paid Training is available.

Contact: For individuals with previous experience: Human Resources; Phone: 212-603-5000; Fax: 212-603-5712.

Company URL: http://www.petrymedia.com

PFIZER, INC.

235 EAST 42ND STREET
NEW YORK, NEW YORK 10017

Overview: Manufactures medical, consumer, and food science products, veterinary drugs, and pharmaceuticals. Established: 1849. Ownership: Public.

Key Statistics: Annual Sales: $11.3 billion. Number of Employees: 43,800.

Expertise/Education Sought: Sales, chemistry, computer science, marketing, business administration, human resources, finance, manufacturing, market research

Alternatives to Full-Time Employment: Part-time jobs, paid internships

Company-Paid Training is available.

International Assignments are available to employees.

Contact: Manager of Corporate Recruiting; Phone: 212-573-7252; Fax: 212-808-6399. Job Hotline: 212-573-4150.

Company URL: http://www.pfizer.com

PHILIP MORRIS USA

120 PARK AVENUE
NEW YORK, NEW YORK 10017-5592

Overview: Manufactures tobacco products, beer, and packaged foods. Established: 1985. Parent Company: Philip Morris Companies, Inc. Ownership: Public.

Key Statistics: Annual Sales: $12.5 billion. Number of Employees: 20,000.

Expertise/Education Sought: Information technology

Alternatives to Full-Time Employment: Paid internships

Company-Paid Training is available.

International Assignments are available to employees.

Contact: Mr. John J. Tucker, Recruiter; Phone: 212-880-5000; Fax: 212-878-2167.

PHILIPS ELECTRONICS NORTH AMERICA CORPORATION

100 EAST 42ND STREET, FLOOR 23
NEW YORK, NEW YORK 10017-5613

Overview: Manufactures household audio and video equipment. Established: 1959. Ownership: Private.

Key Statistics: Annual Sales: $4 billion. Number of Employees: 34,900.

Expertise/Education Sought: Electrical engineering

Alternatives to Full-Time Employment: Part-time jobs, paid internships

International Assignments are available to employees.

Contact: Ms. Denise Townsend, Supervisor of Human Resources; Phone: 212-850-5354; Fax: 212-850-7342.

Company URL: http://www.philips.com

PHILIPS RESEARCH

345 SCARBOROUGH ROAD
BRIARCLIFF MANOR, NEW YORK 10510

Overview: Provides research and development for consumer electronics. Established: 1945. Parent Company: Philips International Corporate Research. Ownership: Public.

Key Statistics: Annual Sales: $8 million. Number of Employees: 240.

Hiring History: Number of professional employees hired in 1997: 20. 1996: 45.

Expertise/Education Sought: Software engineering/development, IC design, digital communications, power system optics

Alternatives to Full-Time Employment: Paid internships

Company-Paid Training is available.

International Assignments are available to employees.

Contact: Ms. Sandie Spadaro, Manager, Human Resources; Phone: 914-945-6106; Fax: 914-945-6400. E-mail: sacs@philabs.research.philips.com.

Company URL: http://www.philips.usa

CORPORATE PROFILE

Philips Research offers its employees challenging projects and an exhilarating intellectual atmosphere. Philips is the U.S. research and development group of Philips Electronics, one of the world's largest and most innovative electronics companies. The parent organization, Philips Electronics North America, is a Fortune 100 leader with more than $7 million in annual sales and a full spectrum of high-quality products, including consumer electronics, electronic components, scientific instrumentation, and lighting systems.

Currently, Philips is looking for talented team players to work at its state-of-the-art research and development center, which is located on a 100-acre campus, 1 hour north of New York City, in Westchester County. The diversified opportunities include positions in such fields as Lighting and Projection Display Systems, Microelectronic Devices, Video Communications, Advanced

Television Systems, Integrated Power Circuits, Software Engineering, and Image Processing. Advanced degrees (M.S. or Ph.D.) are required.

As a member of the Philips team, you'll enjoy a forward-thinking environment, professional enhancement opportunities, an attractive salary, and a comprehensive benefits plan.

Philips is an Equal Opportunity Employer.

PHONE PROGRAMS, INC.

40 ELMONT ROAD
ELMONT, NEW YORK 11003

OVERVIEW: Provides business telephone services. Established: 1968. Ownership: Private.

KEY STATISTICS: Number of Employees: 40.

HIRING HISTORY: Number of professional employees hired in 1996: 5.

EXPERTISE/EDUCATION SOUGHT: Telecommunications

ALTERNATIVES TO FULL-TIME EMPLOYMENT: Internships

CONTACT: For individuals with previous experience: Mr. Pete Walker, Director of Operations; Phone: 516-775-5410; Fax: 516-775-5827.

PC Photocircuits CORPORATION

PHOTOCIRCUITS CORPORATION

31 SEA CLIFF AVENUE
GLEN COVE, NEW YORK 11542

OVERVIEW: Manufactures printed circuit boards. Established: 1951. Ownership: Private.

KEY STATISTICS: Annual Sales: $320 million. Number of Employees: 3,200.

EXPERTISE/EDUCATION SOUGHT: Chemical engineering, mechanical engineering, materials science, production, waste treatment, facilities engineering, quality assurance, customer service and support

ALTERNATIVES TO FULL-TIME EMPLOYMENT: Summer internship positions are available to qualified candidates.

CONTACT: Robert F. Potorski, Director of Human Resources; Fax: 516-674-1080. Brenda Niehaus, Manager–Human Resources, Peachtree City, GA 30269; Fax: 770-487-7746.

CORPORATE PROFILE

The Company *Photocircuits Corporation, with facilities in Glen Cove (Long Island), New York, and Peachtree City, Georgia, is one of the largest independent producers of printed circuit boards in North America and one of the leading PCB manufacturers in the world.*

THE PIKE COMPANY, INC.

1 CIRCLE STREET
ROCHESTER, NEW YORK 14607-1007

OVERVIEW: Provides general contracting services. Established: 1985. Ownership: Private.

KEY STATISTICS: Annual Sales: $82.3 million. Number of Employees: 200.

HIRING HISTORY: Number of professional employees hired in 1997: 200. 1996: 100.

EXPERTISE/EDUCATION SOUGHT: Mechanical engineering, construction engineering, structural engineering, estimating, project management

ALTERNATIVES TO FULL-TIME EMPLOYMENT: Paid internships

COMPANY-PAID TRAINING is available.

CONTACT: Mr. Carl H. Tannert, Jr., Director of Corporate Development; Phone: 716-271-5256; Fax: 716-271-3101. E-mail: tannc@pikeco.com.

COMPANY URL: http://www.pikeco.com

PORTA SYSTEMS CORPORATION

575 UNDERHILL BOULEVARD
SYOSSET, NEW YORK 11791

OVERVIEW: Manufactures telecommunications test equipment. Established: 1969. Ownership: Public.

KEY STATISTICS: Annual Sales: $58 million. Number of Employees: 354.

HIRING HISTORY: Number of professional employees hired in 1997: 25. 1996: 20.

EXPERTISE/EDUCATION SOUGHT: Electronics/electronics engineering, electrical engineering, computer science, mechanical engineering

ALTERNATIVES TO FULL-TIME EMPLOYMENT: Paid internships

INTERNATIONAL ASSIGNMENTS are available to employees.

CONTACT: Ms. Donna Landeck, Manager of Human Resources; Phone: 516-364-9300; Fax: 516-682-4640. E-mail: dlandeck@compuserve.com.

THE PORT AUTHORITY OF NEW YORK AND NEW JERSEY

1 WORLD TRADE CENTER, 61ST EAST
NEW YORK, NEW YORK 10048

OVERVIEW: Operates airports and transportation terminals. Established: 1921. Ownership: Private.

KEY STATISTICS: Annual Sales: $2.2 billion. Number of Employees: 8,400.

EXPERTISE/EDUCATION SOUGHT: Engineering, transportation, public policy and planning, aviation

ALTERNATIVES TO FULL-TIME EMPLOYMENT: Part-time jobs, paid internships

INTERNATIONAL ASSIGNMENTS are available to employees.

CONTACT: Human Resources Department; Phone: 212-435-4700; Fax: 212-435-4660.

COMPANY URL: http://www.panynj.gov

PROTECTIVE CLOSURES COMPANY, INC.

2150 ELMWOOD AVENUE
BUFFALO, NEW YORK 14207-1910

OVERVIEW: Manufactures plastic products. Established: 1983. Parent Company: Mark IV Industries. Ownership: Public.

KEY STATISTICS: Annual Sales: $15.4 million. Number of Employees: 210.

EXPERTISE/EDUCATION SOUGHT: Engineering

ALTERNATIVES TO FULL-TIME EMPLOYMENT: Internships

CONTACT: For individuals with previous experience: Human Resource Department; Phone: 716-876-9855; Fax: 716-874-1680.

RICH, INC.

76 BEAVER STREET
NEW YORK, NEW YORK 10005-3402

OVERVIEW: Provides data processing and preparation services. Established: 1994. Ownership: Private.

KEY STATISTICS: Annual Sales: $5.1 million. Number of Employees: 75.

HIRING HISTORY: Number of professional employees hired in 1996: 10.

EXPERTISE/EDUCATION SOUGHT: Computer science

COMPANY-PAID TRAINING is available.

CONTACT: For individuals with previous experience: Ms. Nibya Molina, Payroll Manager; Phone: 212-747-0200; Fax: 212-747-0451.

RICH PRODUCTS CORPORATION

1150 NIAGARA STREET
BUFFALO, NEW YORK 14213-1714

OVERVIEW: Manufactures frozen foods. Established: 1944. Ownership: Private.

KEY STATISTICS: Annual Sales: $1 billion. Number of Employees: 5,988.

EXPERTISE/EDUCATION SOUGHT: Sales, human resources, marketing, food services, manufacturing engineering, accounting, public relations, international relations

ALTERNATIVES TO FULL-TIME EMPLOYMENT: Paid internships

COMPANY-PAID TRAINING is available.

Rich Products Corporation (continued)

International Assignments are available to employees.

Contact: Ms. Karen Macris, Employment Manager, PO Box 245, Buffalo, New York, 14240; Phone: 716-878-8000; Fax: 716-878-8534. E-mail: kmacris@rich.com.

ROCHESTER GAS AND ELECTRIC CORPORATION
89 EAST AVENUE
ROCHESTER, NEW YORK 14649

Overview: Distributes electric and gas power. Established: 1848. Ownership: Public.

Key Statistics: Assets: $1.1 billion. Number of Employees: 1,960.

Hiring History: Number of professional employees hired in 1997: 45. 1996: 20.

Expertise/Education Sought: Accounting, finance, engineering, information systems

Company-Paid Training is available.

Contact: Mr. LeRoy Valentine, Manager of Diversity and Staffing; Phone: 716-724-8982; Fax: 716-771-2895.

Company URL: http://www.rge.com

RUAL MANUFACTURING COMPANY, INC.
4411 ROUTE 9
HUDSON, NEW YORK 12534-3104

Overview: Manufactures electric heating pads and vaporizers. Established: 1955. Parent Company: Kaz, Inc. Ownership: Private.

Key Statistics: Annual Sales: $28.4 million. Number of Employees: 304.

Hiring History: Number of professional employees hired in 1996: 300.

Expertise/Education Sought: Machinery

Contact: For individuals with previous experience: Human Relations Department; Phone: 518-828-0450; Fax: 518-828-0450.

SABIN METAL CORPORATION
300 PANTIGO PLACE, SUITE 102
EAST HAMPTON, NEW YORK 11937-2630

Overview: Produces secondary precious metals. Established: 1945. Ownership: Private.

Key Statistics: Annual Sales: $59.3 million. Number of Employees: 220.

Expertise/Education Sought: Refining, precious metals

Alternatives to Full-Time Employment: Paid internships

International Assignments are available to employees.

Contact: For individuals with previous experience: Mr. Richard DeSantis, Assistant to the President; Phone: 516-329-1717, ext. 12; Fax: 516-329-1985. E-mail: sabincorp@aol.com.

Company URL: http://www.sabinmetal.com

S&S X-RAY PRODUCTS, INC.
1101 LINWOOD STREET
BROOKLYN, NEW YORK 11208

Overview: Manufactures X-ray products. Established: 1959. Ownership: Private.

Key Statistics: Annual Sales: $29 million. Number of Employees: 300.

Expertise/Education Sought: Accounting, finance, data processing, electrical engineering

Alternatives to Full-Time Employment: Part-time jobs

International Assignments are available to employees.

Contact: For individuals with previous experience: Mr. Norman Shoenfeld, President; Phone: 718-649-8500; Fax: 718-257-0219. E-mail: info@ssxray.com.

SCHLEGEL SYSTEMS, INC.
1555 JEFFERSON ROAD
ROCHESTER, NEW YORK 14623

Overview: Manufactures rubber strips and belting and plastic hardware and building products. Established: 1986. Parent Company: Unipoly Company. Ownership: Private.

Key Statistics: Annual Sales: $516.6 million. Number of Employees: 5,000.

Hiring History: Number of professional employees hired in 1997: 60.

Expertise/Education Sought: Design engineering, mechanical engineering

Company-Paid Training is available.

Contact: Mr. Jim Falkner, Human Resources Manager; Phone: 716-427-7200, ext. 5289; Fax: 716-427-7216.

SCHRODER AND COMPANY, INC.
787 7TH AVENUE
NEW YORK, NEW YORK 10019-6016

Overview: Provides securities brokerage services. Established: 1986. Parent Company: Schroder Plc. Ownership: Private.

Key Statistics: Annual Sales: $123.5 million. Number of Employees: 1,282.

Expertise/Education Sought: Institution and equity trading, research, finance, operations

Alternatives to Full-Time Employment: Paid internships

Company-Paid Training is available.

International Assignments are available to employees.

Contact: Human Resources Department; Phone: 212-492-6000; Fax: 212-492-7188.

F. SCHUMACHER AND COMPANY
79 MADISON AVENUE
NEW YORK, NEW YORK 10016-7803

Overview: Wholesales woven and upholstered fabric. Established: 1889. Ownership: Private.

Key Statistics: Annual Sales: $278 million. Number of Employees: 1,550.

Hiring History: Number of professional employees hired in 1997: 300. 1996: 284.

Expertise/Education Sought: Sales, manufacturing, advertising

Alternatives to Full-Time Employment: Paid internships

Contact: For individuals with previous experience: Human Resources Department; Phone: 212-213-7888; Fax: 212-213-7711. E-mail: gmaddox@ssco.com.

SEQUA CORPORATION
200 PARK AVENUE
NEW YORK, NEW YORK 10166

Overview: Manufactures aerospace machinery. Established: 1929. Ownership: Public.

Key Statistics: Annual Sales: $1.5 billion. Number of Employees: 9,350.

Hiring History: Number of professional employees hired in 1996: 8.

Expertise/Education Sought: Environmental science, manufacturing engineering, accounting, finance, marketing, human resources, aerospace engineering

Company-Paid Training is available.

Contact: Ms. Carolyn Velechko, Human Resource Specialist. E-mail: carolyn_velechko@sequa.com.

HENRY I. SIEGEL COMPANY, INC.
1372 BROADWAY, FLOOR 12
NEW YORK, NEW YORK 10018-6106

Overview: Manufactures men's and boys' trousers and slacks. Established: 1988. Ownership: Public.

Key Statistics: Annual Sales: $212 million. Number of Employees: 4,000.

Expertise/Education Sought: Manufacturing

Contact: For individuals with previous experience: Personnel Department; Phone: 212-302-6452; Fax: 212-819-9172.

SIEMENS CORPORATION
1301 AVENUE OF THE AMERICAS
NEW YORK, NEW YORK 10019-6022

Overview: Manufactures electronics and related equipment. Established: 1954. Parent Company: Siemens AG. Ownership: Private.
Key Statistics: Annual Sales: $8.1 billion. Number of Employees: 45,757.
Hiring History: Number of professional employees hired in 1997: 50.
Expertise/Education Sought: Communications, law, management information systems, finance
Alternatives to Full-Time Employment: Paid internships
Company-Paid Training is available.
International Assignments are available to employees.
Contact: Ms. Karen Mounday, Manager of Recruiting; Phone: 212-258-4198; Fax: 212-258-4476.

SMITH BARNEY, INC.
388 GREENWICH STREET
NEW YORK, NEW YORK 10013-2468

Overview: Provides brokerage and investment services. Established: 1993. Parent Company: Travelers. Ownership: Public.
Key Statistics: Annual Sales: $2.8 billion. Number of Employees: 28,000.
Expertise/Education Sought: Finance, accounting, underwriting, systems analysis, computer programming
Alternatives to Full-Time Employment: Paid internships
Company-Paid Training is available.
International Assignments are available to employees.
Contact: For individuals with previous experience: Corporate Recruiting; Phone: 212-816-6000. Job Hotline: 212-816-7735.

SUPERIOR PRINTING INK COMPANY, INC.
70 BETHUNE STREET
NEW YORK, NEW YORK 10014

Overview: Manufactures printing ink. Established: 1961. Ownership: Private.
Key Statistics: Annual Sales: $75 million. Number of Employees: 525.
Expertise/Education Sought: Technical
Alternatives to Full-Time Employment: Paid internships
Company-Paid Training is available.
Contact: Mr. Bob Valente, Director of Human Resources; Phone: 212-741-3600; Fax: 212-633-8283.

SYMBOL TECHNOLOGIES, INC.
ONE SYMBOL PLAZA
HOLTSVILLE, NEW YORK 11742

Overview: Manufactures bar code equipment. Established: 1973. Ownership: Public.
Key Statistics: Annual Sales: $656.7 million. Number of Employees: 2,900.
Expertise/Education Sought: Accounting, marketing, sales, engineering, information systems
Alternatives to Full-Time Employment: Part-time jobs, paid internships
Contact: Ms. Carole DeMayo, Director of Human Resources; Phone: 516-738-2400. E-mail: demayo@symbol.com.
Company URL: http://www.symbol.com

TATA, INC.
101 PARK AVENUE, FLOOR 26
NEW YORK, NEW YORK 10178-2699

Overview: Provides wholesale metals services. Established: 1990. Ownership: Private.
Key Statistics: Annual Sales: $153.1 million.
Expertise/Education Sought: Industrial distribution, manufacturing
International Assignments are available to employees.
Contact: For individuals with previous experience: Human Resources; Phone: 212-557-7979; Fax: 212-557-7987. E-mail: tatainc@aol.com.

TELEPORT COMMUNICATIONS GROUP
1 TELEPORT DRIVE
STATEN ISLAND, NEW YORK 10311-1001

Overview: Provides local telecommunications services. Established: 1983. Ownership: Private.
Key Statistics: Annual Sales: $192.1 million. Number of Employees: 1,500.
Hiring History: Number of professional employees hired in 1997: 800. 1996: 800.
Expertise/Education Sought: Electrical engineering, telecommunications, accounting, finance, business administration, computer science
Alternatives to Full-Time Employment: Paid internships
Contact: For individuals with previous experience: Mr. Wayne Balnicki, Staffing Manager; Phone: 718-355-2000; Fax: 718-355-2783. E-mail: hr@tcg.com.
Company URL: http://www.tcg.com

TENNECO PACKAGING
18 PECK AVENUE
GLENS FALLS, NEW YORK 12801-3833

Overview: Manufactures plastic and plastic foam products. Established: 1987. Ownership: Private.
Key Statistics: Annual Sales: $175 million. Number of Employees: 890.
Expertise/Education Sought: Engineering
Contact: For individuals with previous experience: Ms. Barbara Dido, Human Resources Manager.

TIAA-CREF FUND
730 3RD AVENUE
NEW YORK, NEW YORK 10017-3206

Overview: Provides insurance and manages pensions for educational and research organizations. Established: 1918. Ownership: Private.
Key Statistics: Annual Sales: $12 billion. Number of Employees: 5,000.
Hiring History: Number of professional employees hired in 1996: 650.
Expertise/Education Sought: Management information systems, computer programming
Alternatives to Full-Time Employment: Part-time jobs, paid internships
Company-Paid Training is available.
International Assignments are available to employees.
Contact: Human Resources—Staffing; Phone: 212-490-9000; Fax: 212-916-5883.
Company URL: http://www.tiaa-cref.org

TIME, INC.
1271 AVENUE OF THE AMERICAS
NEW YORK, NEW YORK 10020

Overview: Publishes magazines and books. Established: 1988. Parent Company: Time-Warner. Ownership: Private.
Key Statistics: Annual Sales: $4 billion. Number of Employees: 10,000.
Expertise/Education Sought: Analog engineering, marketing, finance, customer service and support, editing
Alternatives to Full-Time Employment: Paid internships
Company-Paid Training is available.
International Assignments are available to employees.
Contact: Human Resources; Phone: 212-522-3908; Fax: 212-522-4510.

TIMES MIRROR MAGAZINES, INC.
2 PARK AVENUE, FLOOR 9
NEW YORK, NEW YORK 10016-5679

Overview: Magazine publisher. Established: 1967. Parent Company: Times Mirror Company. Ownership: Public.
Key Statistics: Annual Sales: $225 million. Number of Employees: 550.

Times Mirror Magazines, Inc. (continued)

Expertise/Education Sought: Manufacturing, sales, advertising
Alternatives to Full-Time Employment: Paid internships
Company-Paid Training is available.
Contact: For individuals with previous experience: Ms. Kim Putnan, Employment Manager; Phone: 212-779-5000.
Company URL: http://www.tm.com

TOSHIBA AMERICA, INC.
1251 AVENUE OF THE AMERICAS
NEW YORK, NEW YORK 10020

Overview: Manufactures electronic, audio, and video equipment. Established: 1965. Parent Company: Toshiba Corporation. Ownership: Private.
Key Statistics: Annual Sales: $6.3 billion. Number of Employees: 7,500.
Hiring History: Number of professional employees hired in 1997: 10. 1996: 15.
Expertise/Education Sought: Finance, accounting, public relations, law, computer programming, audio engineering, audiovisual technology, electronics/electronics engineering, mechanical engineering
Alternatives to Full-Time Employment: Part-time jobs
Contact: For individuals with previous experience: Ms. Sarah Heinz, Human Resources Generalist; Phone: 212-596-0682; Fax: 212-593-3875.
Company URL: http://www.toshiba.com

TOSHIBA DISPLAY DEVICES, INC.
100 WESTINGHOUSE CIRCLE
HORSEHEADS, NEW YORK 14845

Overview: Manufactures television picture tubes. Established: 1990. Parent Company: Toshiba Corporation. Ownership: Private.
Key Statistics: Annual Sales: $360 million. Number of Employees: 1,500.
Hiring History: Number of professional employees hired in 1997: 25.
Expertise/Education Sought: Electrical engineering, process engineering, software engineering/development, program analysis, information systems
Alternatives to Full-Time Employment: Paid internships
Company-Paid Training is available.
Contact: For college students and recent graduates: Ms. Amy Hathaway, Employee Relations Specialist; Phone: 607-796-3500; Fax: 607-796-3814. E-mail: hathaway@servtech.com. For individuals with previous experience: Mr. Brian Spencer, Employee Relations Manager; Phone: 607-796-3500; Fax: 607-796-3814. E-mail: hathaway@servtech.com.

TOWN OF HAMBURG
6100 SOUTH PARK AVENUE
HAMBURG, NEW YORK 14075-3766

Overview: Local government. Established: 1812. Ownership: Private.
Key Statistics: Number of Employees: 532.
Expertise/Education Sought: Civil site engineering
Contact: For individuals with previous experience: Supervisor Office; Phone: 716-649-6111; Fax: 716-649-4087.

TURNER CONSTRUCTION COMPANY
375 HUDSON STREET
NEW YORK, NEW YORK 10014

Overview: Provides general contracting and construction management services. Established: 1902. Ownership: Public.
Key Statistics: Annual Sales: $534 million. Number of Employees: 2,400.
Expertise/Education Sought: Mechanical engineering, civil engineering, architectural engineering, construction management, accounting, building science
Alternatives to Full-Time Employment: Paid internships
Company-Paid Training is available.
International Assignments are available to employees.
Contact: Mr. Bruce Ruthven, Director of Human Resources; Phone: 212-229-6013; Fax: 212-229-6024.
Company URL: http://www.turnerconstruction.com

ULTRA ELECTRONICS DEFENSE, INC.
7525 COUNTY ROAD 42
FISHERS, NEW YORK 14453

Overview: Manufactures electronic measure instruments. Established: 1997. Parent Company: Ultra Electronics. Ownership: Private.
Key Statistics: Annual Sales: $19.6 million. Number of Employees: 190.
Expertise/Education Sought: Engineering, software engineering/development
Company-Paid Training is available.
International Assignments are available to employees.
Contact: For individuals with previous experience: Human Resources Department; Phone: 716-924-4000; Fax: 716-924-5732.

UNIFRAX CORPORATION
2351 WHIRLPOOL STREET
NIAGARA FALLS, NEW YORK 14305-2413

Overview: Producer of high performance, heat-management materials. Established: 1996. Ownership: Private.
Key Statistics: Annual Sales: $91.6 million. Number of Employees: 500.
Expertise/Education Sought: Engineering, marketing
Alternatives to Full-Time Employment: Internships
Company-Paid Training is available.
Contact: For individuals with previous experience: Human Resources; Phone: 716-278-3800.
Company URL: http://www.inifrax.com

UNIVERSAL INTERNATIONAL
160 BROADWAY, ROOM 901
NEW YORK, NEW YORK 10038-4201

Overview: Provides engineering services. Established: 1991. Ownership: Public.
Key Statistics: Annual Sales: $10 million.
Hiring History: Number of professional employees hired in 1997: 100. 1996: 125.
Expertise/Education Sought: Administration, finance
Company-Paid Training is available.
Contact: For individuals with previous experience: Mr. Andrew Thomas, Vice President; Phone: 212-587-9513; Fax: 212-587-9524.

US HEALTH MANAGEMENT CORPORATION
110 MARCUS DRIVE
MELVILLE, NEW YORK 11747-4228

Overview: Medical laboratory. Established: 1997. Parent Company: Fonar Corporation. Ownership: Public.
Key Statistics: Annual Sales: $30 million. Number of Employees: 90.
Hiring History: Number of professional employees hired in 1997: 50. 1996: 50.
Expertise/Education Sought: Engineering, machinery
Company-Paid Training is available.
International Assignments are available to employees.
Contact: For individuals with previous experience: Mr. Fred Peipman, Personnel Director; Phone: 516-694-2929, ext. 220.

VOLT SERVICES GROUP
1221 AVENUE OF THE AMERICAS
NEW YORK, NEW YORK 10020-1001

Overview: Provides employment services. Established: 1950. Ownership: Public.
Key Statistics: Annual Sales: $1 billion. Number of Employees: 31,400.

EXPERTISE/EDUCATION SOUGHT: Technical, computer science, manufacturing, computer operations
ALTERNATIVES TO FULL-TIME EMPLOYMENT: Paid internships
INTERNATIONAL ASSIGNMENTS are available to employees.
CONTACT: Mr. Joseph Musacchio, Senior Vice President; Phone: 212-719-7800, ext. 7841; Fax: 212-719-7850. E-mail: jobs@volteast.com. Job Hotline: 800-975-8658.
COMPANY URL: http://www.volteast.com

WESTVACO CORPORATION
299 PARK AVENUE
NEW YORK, NEW YORK 10171

OVERVIEW: Manufactures paper. Established: 1889. Ownership: Public.
KEY STATISTICS: Annual Sales: $3 billion. Number of Employees: 13,430.
HIRING HISTORY: Number of professional employees hired in 1996: 30.
EXPERTISE/EDUCATION SOUGHT: Sales, chemical engineering, process engineering, marketing, finance, physics
COMPANY-PAID TRAINING is available.
INTERNATIONAL ASSIGNMENTS are available to employees.
CONTACT: Ms. Cheryl Miller, Professional Recruiter; Phone: 804-327-6236; Fax: 804-327-5277. E-mail: wvcorecruit@aol.com.
COMPANY URL: http://www.westvaco.com

WHITING DOOR MANUFACTURING COPORATION
113 CEDAR STREET
AKRON, NEW YORK 14001-1038

MAILING ADDRESS: PO Box 388, Akron, NY 14001
OVERVIEW: Manufactures roll up doors. Established: 1976. Ownership: Private.
KEY STATISTICS: Annual Sales: $80 million. Number of Employees: 400.
EXPERTISE/EDUCATION SOUGHT: Mechanical engineering, manufacturing engineering, computer programming, accounting, finance, sales, marketing
CONTACT: For individuals with previous experience: Mr. Peter Gullo, Personnel Manager; Phone: 716-542-5427; Fax: 716-542-5947.
COMPANY URL: http://www.whitingdoor.com

WORLD WIDE TRADE RESOURCE
70 WEST 36TH STREET, ROOM 702
NEW YORK, NEW YORK 10018-8007

OVERVIEW: Provides computer related consulting services. Established: 1991. Ownership: Private.
KEY STATISTICS: Annual Sales: $10 million. Number of Employees: 100.
HIRING HISTORY: Number of professional employees hired in 1996: 100.
EXPERTISE/EDUCATION SOUGHT: Information systems, computer science, software engineering/development, computer programming, management
ALTERNATIVES TO FULL-TIME EMPLOYMENT: Paid internships
INTERNATIONAL ASSIGNMENTS are available to employees.
CONTACT: For individuals with previous experience: Ms. Kristyn Im, Vice President of Operations; Phone: 212-868-3341, ext. 111; Fax: 212-947-0572. E-mail: recruiter@it-inc.net.

YONKERS CONTRACTING COMPANY, INC.
969 MIDLAND AVENUE
YONKERS, NEW YORK 10704-1027

OVERVIEW: Provides heavy commercial construction services. Established: 1945. Ownership: Private.
KEY STATISTICS: Annual Sales: $270.6 million. Number of Employees: 2,500.
HIRING HISTORY: Number of professional employees hired in 1997: 20. 1996: 20.
EXPERTISE/EDUCATION SOUGHT: Civil engineering, construction, business administration, construction management, purchasing
CONTACT: Mr. John Chrysogelos, Jr., Human Resources; Phone: 914-565-1500; Fax: 914-378-8841.

ZIFF-DAVIS
500 CANAL VIEW BOULEVARD
ROCHESTER, NEW YORK 14623

OVERVIEW: Publishes technical manuals. Established: 1991. Ownership: Private.
KEY STATISTICS: Annual Sales: $42 million. Number of Employees: 240.
EXPERTISE/EDUCATION SOUGHT: Accounting, computer programming, technical writing, editing, finance, operations
COMPANY-PAID TRAINING is available.
INTERNATIONAL ASSIGNMENTS are available to employees.
CONTACT: For individuals with previous experience: Staffing Manager; Phone: 716-240-7500. E-mail: careerzde@zd.com. Job Hotline: 716-240-7632.
COMPANY URL: http://www.zdeducation.com

NORTH CAROLINA

AMITAL SPINNING CORPORATION
197 BOSCH BOULEVARD
NEW BERN, NORTH CAROLINA 28562-6924

OVERVIEW: Manufactures acrylic yarn. Established: 1986. Ownership: Private.
KEY STATISTICS: Annual Sales: $50.7 million. Number of Employees: 340.
HIRING HISTORY: Number of professional employees hired in 1997: 60. 1996: 75.
EXPERTISE/EDUCATION SOUGHT: Accounting, management
COMPANY-PAID TRAINING is available.
CONTACT: For individuals with previous experience: Mr. David Jones, Director of Personnel; Phone: 919-636-3435; Fax: 919-637-8043.

BELLSOUTH TELECOMMUNICATION
5715 GLENWOOD AVENUE, ROOM 319
RALEIGH, NORTH CAROLINA 27612-3252

OVERVIEW: Provides local telephone services. Ownership: Private.
KEY STATISTICS: Annual Sales: $11.8 million. Number of Employees: 100.
EXPERTISE/EDUCATION SOUGHT: Communications, engineering
ALTERNATIVES TO FULL-TIME EMPLOYMENT: Paid internships
COMPANY-PAID TRAINING is available.
CONTACT: For individuals with previous experience: Human Resources.

BURLINGTON INDUSTRIES, INC.
3330 WEST FRIENDLY AVENUE
GREENSBORO, NORTH CAROLINA 27410

OVERVIEW: Manufactures textiles. Established: 1937. Ownership: Public.
KEY STATISTICS: Annual Sales: $2.2 billion. Number of Employees: 21,000.
EXPERTISE/EDUCATION SOUGHT: Management information systems, electrical engineering, industrial engineering
ALTERNATIVES TO FULL-TIME EMPLOYMENT: Paid internships
COMPANY-PAID TRAINING is available.
INTERNATIONAL ASSIGNMENTS are available to employees.
CONTACT: Mr. Tony Michaels, Recruiting; Phone: 910-379-4588; Fax: 910-379-3329.
COMPANY URL: http://www.burlington.com

CAROTEK, INC.
700 SAM NEWELL ROAD
MATTHEWS, NORTH CAROLINA 28105-4515

OVERVIEW: Supplies industrial machinery and equipment. Established: 1990. Parent Company: Ad Bel, Ltd. Ownership: Private.

KEY STATISTICS: Annual Sales: $60 million. Number of Employees: 150.

EXPERTISE/EDUCATION SOUGHT: Chemical engineering, mechanical engineering, environmental engineering, industrial engineering

ALTERNATIVES TO FULL-TIME EMPLOYMENT: Part-time jobs, paid internships

COMPANY-PAID TRAINING is available.

CONTACT: For individuals with previous experience: Mr. Rod Holland, Operations Manager; Phone: 704-847-4406; Fax: 704-845-5077.

COMPANY URL: http://www.carotek.com

COGENTRIX ENERGY, INC.
9405 ARROW POINT BOULEVARD
CHARLOTTE, NORTH CAROLINA 28273-8167

OVERVIEW: Generates electric service. Established: 1995. Ownership: Private.

KEY STATISTICS: Annual Sales: $109 million. Number of Employees: 393.

EXPERTISE/EDUCATION SOUGHT: Engineering

CONTACT: For individuals with previous experience: Human Resource Department.

DUFF-NORTON, INC.
9415 PIONEER AVENUE
CHARLOTTE, NORTH CAROLINA 28273-6318

MAILING ADDRESS: PO Box 7010, Charlotte, NC 28241

OVERVIEW: Manufactures fluid control equipment. Established: 1996. Parent Company: Columbus McKinnon. Ownership: Public.

KEY STATISTICS: Annual Sales: $106.9 million. Number of Employees: 1,050.

EXPERTISE/EDUCATION SOUGHT: Accounting

ALTERNATIVES TO FULL-TIME EMPLOYMENT: Paid internships

COMPANY-PAID TRAINING is available.

INTERNATIONAL ASSIGNMENTS are available to employees.

CONTACT: Ms. Betty Martin, Manager of Human Resources; Phone: 704-588-0510; Fax: 704-588-8180.

FIELDCREST CANNON, INC.
1 LAKE CIRCLE DRIVE
KANNAPOLIS, NORTH CAROLINA 28081-3523

OVERVIEW: Manufactures sheets, bedding, and towels. Established: 1953. Parent Company: Pillowtex Corporation. Ownership: Public.

KEY STATISTICS: Annual Sales: $1.1 billion. Number of Employees: 11,312.

HIRING HISTORY: Number of professional employees hired in 1996: 30.

EXPERTISE/EDUCATION SOUGHT: Business management, chemical analysis, product management, marketing, finance, textiles

ALTERNATIVES TO FULL-TIME EMPLOYMENT: Paid internships

COMPANY-PAID TRAINING is available.

INTERNATIONAL ASSIGNMENTS are available to employees.

CONTACT: For individuals with previous experience: Human Resources; Phone: 704-939-2088; Fax: 704-939-4031.

FOOD LION, INC.
2110 EXECUTIVE DRIVE, PO BOX 1330
SALISBURY, NORTH CAROLINA 28145-1330

OVERVIEW: Operates chain of food stores. Established: 1976. Ownership: Public.

KEY STATISTICS: Annual Sales: $9 billion. Number of Employees: 73,170.

HIRING HISTORY: Number of professional employees hired in 1997: 200.

EXPERTISE/EDUCATION SOUGHT: Management information systems, drafting

ALTERNATIVES TO FULL-TIME EMPLOYMENT: Paid internships

COMPANY-PAID TRAINING is available.

CONTACT: For individuals with previous experience: Ms. Kay Wagoner, Recruitment Manager, PO Box 1330, Salisbury, North Carolina, 28145-1330; Phone: 704-633-8250, ext. 2231; Fax: 704-637-2581. Job Hotline: 704-633-8250, ext. 2144.

COMPANY URL: http://www.foodlion.com

GALEY AND LORD INDUSTRIES, INC.
7736 MCCLOUD ROAD, SUITE 300
GREENSBORO, NORTH CAROLINA 27409-9324

OVERVIEW: Produces broadwoven cotton fabric. Established: 1988. Ownership: Public.

KEY STATISTICS: Annual Sales: $405.6 million. Number of Employees: 5,451.

HIRING HISTORY: Number of professional employees hired in 1996: 75.

EXPERTISE/EDUCATION SOUGHT: Textiles, engineering, marketing

COMPANY-PAID TRAINING is available.

CONTACT: For individuals with previous experience: Ms. Dixie Pugh, Personnel Director; Phone: 910-665-3090; Fax: 910-665-3110.

HALL CONTRACTING CORPORATION
6415 LAKEVIEW ROAD
CHARLOTTE, NORTH CAROLINA 28269-2602

MAILING ADDRESS: PO Box 560218, Charlotte, NC 28256

OVERVIEW: Provides general contractor services. Established: 1954. Ownership: Private.

KEY STATISTICS: Annual Sales: $39.9 million. Number of Employees: 300.

HIRING HISTORY: Number of professional employees hired in 1996: 7.

EXPERTISE/EDUCATION SOUGHT: Structural engineering, mechanical engineering, electrical engineering, civil engineering

ALTERNATIVES TO FULL-TIME EMPLOYMENT: Internships

CONTACT: For individuals with previous experience: Mr. Mike Hall, President; Phone: 704-598-0818; Fax: 704-598-3855.

HOME INNOVATIONS, INC.
346 EAST PLAZA DRIVE
MOORESVILLE, NORTH CAROLINA 28115

MAILING ADDRESS: PO Box 297, Moorseville, NC 28115

OVERVIEW: Manufactures household furnishings. Established: 1995. Parent Company: Decorative Home Accents, Inc. Ownership: Private.

KEY STATISTICS: Annual Sales: $107 million. Number of Employees: 1,900.

EXPERTISE/EDUCATION SOUGHT: Accounting, administration, materials management, computer science, computer programming, computer science, management information systems, marketing, graphic arts

ALTERNATIVES TO FULL-TIME EMPLOYMENT: Paid internships

COMPANY-PAID TRAINING is available.

CONTACT: For individuals with previous experience: Ms. Patricia McCue, Coordinator of Corporate Human Resources; Phone: 704-664-2711, ext. 204; Fax: 704-664-4200.

INTERNATIONAL RESISTIVE COMPANY, INC.
736 GREENWAY ROAD
BOONE, NORTH CAROLINA 28607-4830

OVERVIEW: Manufactures electronic resistors. Established: 1990. Parent Company: TT Group. Ownership: Private.

KEY STATISTICS: Annual Sales: $65 million. Number of Employees: 726.

HIRING HISTORY: Number of professional employees hired in 1997: 10. 1996: 15.

EXPERTISE/EDUCATION SOUGHT: Maintenance, quality control, management, production, product engineering, production engineering, process engineering, management information systems, production management

Alternatives to Full-Time Employment: Paid internships
International Assignments are available to employees.
Contact: For individuals with previous experience: Ms. Sandra Reese, Human Resources Manager, PO Box 1860, Boone, North Carolina, 28607-1860; Phone: 704-264-8861; Fax: 704-264-8865.

KONICA MANUFACTURING USA, INC.

6900 KONICA DRIVE
WHITSETT, NORTH CAROLINA 27377

Overview: Manufactures photographic equipment and supplies. Established: 1987. Parent Company: Konica (Japan). Ownership: Private.
Key Statistics: Annual Sales: $192 million. Number of Employees: 375.
Hiring History: Number of professional employees hired in 1997: 50. 1996: 75.
Expertise/Education Sought: Mechanical engineering, chemical engineering, electrical engineering, computer engineering
Company-Paid Training is available.
Contact: Ms. Debbie Brown, Human Resource Specialist; Phone: 336-449-8000, ext. 253; Fax: 336-449-6330. E-mail: brownd@konica.com.

LIGGETT GROUP, INC.

700 WEST MAIN STREET
DURHAM, NORTH CAROLINA 27701-2801

Overview: Manufactures cigarettes. Established: 1986. Parent Company: Brooke Group, Ltd. Ownership: Public.
Key Statistics: Annual Sales: $401.1 million. Number of Employees: 570.
Hiring History: Number of professional employees hired in 1997: 50.
Expertise/Education Sought: Sales, production, administration
Contact: Human Resources; Phone: 919-683-8940; Fax: 919-683-9927.

NORTEL

PO BOX 13010, DEPARTMENT 1175 YRK, U.S. RESOURCING CENTER
RESEARCH TRIANGLE PARK, NORTH CAROLINA 27709

Overview: Global provider of communications products, systems, and networks. Established: 1895.
Expertise/Education Sought: Design engineering, software engineering/development, research and development, computer engineering, telecommunications
Alternatives to Full-Time Employment: Part-time jobs, paid internships
Company-Paid Training is available.
International Assignments are available to employees.
Contact: Mr. George Moore, University Relations Recruiter; Phone: 888-682-4473; Fax: 800-546-8092. Job Hotline: 888-682-4473.
Company URL: http://www.nortel.com

NUCOR CORPORATION

2100 REXFORD ROAD
CHARLOTTE, NORTH CAROLINA 28211

Overview: Manufactures steel. Established: 1904. Ownership: Public.
Key Statistics: Annual Sales: $3.6 billion. Number of Employees: 6,600.
Expertise/Education Sought: Metallurgical engineering
Contact: For individuals with previous experience: Human Relations Department; Phone: 704-366-7000.

PCS PHOSPHATE COMPANY, INC.

PO BOX 30321
RALEIGH, NORTH CAROLINA 27622

Overview: Potash mining company; manufactures fertilizers. Established: 1995. Parent Company: PCS Saskatoon, SK Canada. Ownership: Private.
Key Statistics: Annual Sales: $373.3 million. Number of Employees: 1,600.
Expertise/Education Sought: Chemical engineering, electrical engineering
Company-Paid Training is available.
Contact: For individuals with previous experience: Ms. Rhonda Downum, Human Resource Associate.

PERFECT FIT INDUSTRIES, INC.

201 CUTHBERTSON STREET
MONROE, NORTH CAROLINA 28110-3803

Overview: Manufactures household furnishings. Established: 1996. Ownership: Private.
Key Statistics: Annual Sales: $110 million. Number of Employees: 1,200.
Expertise/Education Sought: Manufacturing, sales
Contact: For individuals with previous experience: Mr. David Weather, Vice President of Facilities; Phone: 704-289-1531, ext. 214; Fax: 704-283-8856.

PIEDMONT NATURAL GAS COMPANY

1915 REXFORD ROAD
CHARLOTTE, NORTH CAROLINA 28211-3480

Overview: Distributes natural gas products. Established: 1950. Ownership: Public.
Key Statistics: Annual Sales: $685.1 million. Number of Employees: 2,000.
Expertise/Education Sought: Computer programming, data processing, accounting, marketing, finance
Alternatives to Full-Time Employment: Part-time jobs
Contact: For individuals with previous experience: Mr. Darryl Rabon, Supervisor of Employment.
Company URL: http://www.piedmontng.com

RESEARCH TRIANGLE INSTITUTE

PO BOX 12194
RESEARCH TRIANGLE PARK, NORTH CAROLINA 27709

Overview: Research institution. Established: 1958. Ownership: Private.
Key Statistics: Annual Sales: $143.4 million. Number of Employees: 1,480.
Hiring History: Number of professional employees hired in 1997: 150. 1996: 110.
Expertise/Education Sought: Chemistry, computer science, economics, statistics
Alternatives to Full-Time Employment: Part-time jobs, paid internships
Company-Paid Training is available.
Contact: Mr. Gerard Collins, Supervisor of Employment Services; Phone: 919-541-6200; Fax: 919-541-6506. E-mail: gwc@rti.org.
Company URL: http://www.rti.org

R.J. REYNOLDS TOBACCO COMPANY

401 NORTH MAIN STREET
WINSTON-SALEM, NORTH CAROLINA 27101-3818

Overview: Manufactures and markets cigarettes. Established: 1970. Parent Company: RJR Nabisco, Inc. Ownership: Public.
Key Statistics: Annual Sales: $5 billion. Number of Employees: 9,700.
Hiring History: Number of professional employees hired in 1997: 100. 1996: 200.
Expertise/Education Sought: Data processing, computer programming, accounting, sales, business management
Alternatives to Full-Time Employment: Paid internships
Contact: For individuals with previous experience: Mr. John C. Lovett, Director of Human Resources and Development; Phone: 910-741-6895; Fax: 910-741-1608.

SAS INSTITUTE, INC.
SAS CAMPUS DRIVE
CARY, NORTH CAROLINA 27513

OVERVIEW: Develops software. Established: 1976. Ownership: Private.

KEY STATISTICS: Annual Sales: $653.2 million. Number of Employees: 4,412.

HIRING HISTORY: Number of professional employees hired in 1996: 500.

EXPERTISE/EDUCATION SOUGHT: Software engineering/development, training and development, computer engineering, accounting

ALTERNATIVES TO FULL-TIME EMPLOYMENT: Part-time jobs

INTERNATIONAL ASSIGNMENTS are available to employees.

CONTACT: For individuals with previous experience: Human Resources Department; Phone: 919-677-8000; Fax: 919-677-4444. Job Hotline: 919-677-8000, ext. 5441.

COMPANY URL: http://www.sas.com

SEA-LAND SERVICE, INC.
6000 CARNEGIE BOULEVARD
CHARLOTTE, NORTH CAROLINA 28209-4637

OVERVIEW: Provides containerized shipping services. Established: 1986. Parent Company: CSX Corporation. Ownership: Public.

KEY STATISTICS: Annual Sales: $4.1 billion. Number of Employees: 8,892.

EXPERTISE/EDUCATION SOUGHT: Accounting, business, finance, marketing

ALTERNATIVES TO FULL-TIME EMPLOYMENT: Part-time jobs, paid internships

COMPANY-PAID TRAINING is available.

INTERNATIONAL ASSIGNMENTS are available to employees.

CONTACT: Mr. Steven Frey, Director of Human Resources; Phone: 704-571-2000; Fax: 704-571-4942.

COMPANY URL: http://www.csx.com

SHELBY YARN COMPANY
1100 BUFFALO STREET
SHELBY, NORTH CAROLINA 28150-4050

OVERVIEW: Mills synthetic fabrics. Established: 1997. Ownership: Private.

KEY STATISTICS: Annual Sales: $61.7 million. Number of Employees: 802.

EXPERTISE/EDUCATION SOUGHT: Business administration, sales, marketing, engineering

COMPANY-PAID TRAINING is available.

CONTACT: Mr. Wayne Walton, Human Resources; Phone: 704-487-2000.

SIECOR CORPORATION
800 17TH STREET, NW
HICKORY, NORTH CAROLINA 28601-3336

MAILING ADDRESS: PO Box 489, Hickory, NC 28603

OVERVIEW: Manufactures fiber optic cable and hardware. Established: 1980. Parent Company: Corning and Siemens. Ownership: Private.

KEY STATISTICS: Annual Sales: $458.1 million. Number of Employees: 3,500.

HIRING HISTORY: Number of professional employees hired in 1996: 90.

EXPERTISE/EDUCATION SOUGHT: Hardware engineering, electrical engineering, information services, mechanical engineering, materials engineering

ALTERNATIVES TO FULL-TIME EMPLOYMENT: Part-time jobs, paid internships

COMPANY-PAID TRAINING is available.

INTERNATIONAL ASSIGNMENTS are available to employees.

CONTACT: Ms. Diana Williams, Human Resources Consultant; Phone: 704-327-5000; Fax: 704-327-6408. Job Hotline: 704-327-5878.

COMPANY URL: http://www.siecor.com

SOUTHERN OPTICAL COMPANY
1909 NORTH CHURCH STREET
GREENSBORO, NORTH CAROLINA 27405-5631

OVERVIEW: Manufactures optical products. Established: 1996. Parent Company: Essilor of America. Ownership: Private.

KEY STATISTICS: Annual Sales: $69.8 million. Number of Employees: 710.

HIRING HISTORY: Number of professional employees hired in 1997: 8. 1996: 35.

EXPERTISE/EDUCATION SOUGHT: Optics

COMPANY-PAID TRAINING is available.

CONTACT: For individuals with previous experience: Mr. Jim Price, Vice President of Human Resources; Phone: 910-272-8146, ext. 255.

SUNSTATES MAINTENANCE CORPORATION
3400 WEST WENDOVER AVENUE, SUITE C
GREENSBORO, NORTH CAROLINA 27407-1523

OVERVIEW: Provides office cleaning services. Established: 1974. Ownership: Private.

KEY STATISTICS: Annual Sales: $25.2 million. Number of Employees: 2,000.

HIRING HISTORY: Number of professional employees hired in 1997: 3,000.

EXPERTISE/EDUCATION SOUGHT: Waste management, waste processing

COMPANY-PAID TRAINING is available.

CONTACT: For individuals with previous experience: Human Resources Department; Phone: 910-294-9411; Fax: 910-294-2971.

TAKATA RESTRAINT SYSTEMS, INC.
629 GREEN VALLEY ROAD
GREENSBORO, NORTH CAROLINA 27408-7726

OVERVIEW: Manufactures broadwoven synthetic fabric. Established: 1995. Parent Company: Takata, Inc. Ownership: Private.

KEY STATISTICS: Annual Sales: $65.4 million. Number of Employees: 850.

EXPERTISE/EDUCATION SOUGHT: Engineering, textiles, manufacturing

ALTERNATIVES TO FULL-TIME EMPLOYMENT: Paid internships

COMPANY-PAID TRAINING is available.

INTERNATIONAL ASSIGNMENTS are available to employees.

CONTACT: Human Resources; Phone: 336-547-1600; Fax: 336-547-1636.

TEK (AMERICA), INC.
2100 REXFORD ROAD, SUITE 216
CHARLOTTE, NORTH CAROLINA 28211-3484

OVERVIEW: Provides consulting engineering services. Established: 1995. Parent Company: Toray Engineering Co. (of Japan). Ownership: Private.

KEY STATISTICS: Annual Sales: $60 million.

HIRING HISTORY: Number of professional employees hired in 1997: 4. 1996: 6.

EXPERTISE/EDUCATION SOUGHT: Engineering, chemical engineering

INTERNATIONAL ASSIGNMENTS are available to employees.

CONTACT: For individuals with previous experience: Ms. Judy Rollings, Accounting Manager; Phone: 704-364-1009; Fax: 704-364-3321.

UNITED DOMINION INDUSTRIES
2300 FIRST UNION CENTER, 301 SOUTH COLLEGE STREET
CHARLOTTE, NORTH CAROLINA 28202-6039

OVERVIEW: Manufactures specialty products for commercial and industrial use. Established: 1909. Ownership: Public.

KEY STATISTICS: Annual Sales: $230.2 million. Number of Employees: 1,565.

EXPERTISE/EDUCATION SOUGHT: Accounting, finance, administration

INTERNATIONAL ASSIGNMENTS are available to employees.

CONTACT: Human Resources; Phone: 704-347-6800; Fax: 704-347-6900.

VERBATIM TAPE CORPORATION
1200 WEST HARRIS BOULEVARD
CHARLOTTE, NORTH CAROLINA 28262

OVERVIEW: Manufactures computer storage devices. Established: 1992. Parent Company: Mitsubishi Chemical. Ownership: Private.

KEY STATISTICS: Annual Sales: $48.6 million. Number of Employees: 450.
EXPERTISE/EDUCATION SOUGHT: Engineering, manufacturing
ALTERNATIVES TO FULL-TIME EMPLOYMENT: Paid internships
COMPANY-PAID TRAINING is available.
INTERNATIONAL ASSIGNMENTS are available to employees.
CONTACT: Human Resources; Phone: 704-547-6794; Fax: 704-547-6692.
COMPANY URL: http://www.verbatimcorp.com

VF JEANSWEAR, INC.
400 NORTH ELM STREET
GREENSBORO, NORTH CAROLINA 27401

OVERVIEW: Manufactures clothing and apparel. Established: 1982. Parent Company: VF Corporation. Ownership: Public.
KEY STATISTICS: Annual Sales: $332.3 million. Number of Employees: 10,300.
EXPERTISE/EDUCATION SOUGHT: Industrial engineering, manufacturing, computer engineering, distribution, finance, engineering
INTERNATIONAL ASSIGNMENTS are available to employees.
CONTACT: Human Resources; Phone: 336-332-3400; Fax: 336-332-4280.

WELLMAN, INC.
PO BOX 31331
CHARLOTTE, NORTH CAROLINA 28231

OVERVIEW: Manufactures polyester and nylon fibers. Established: 1927. Ownership: Public.
KEY STATISTICS: Annual Sales: $1.1 billion. Number of Employees: 3,600.
HIRING HISTORY: Number of professional employees hired in 1996: 75.
EXPERTISE/EDUCATION SOUGHT: Chemical engineering, finance, management, materials science
ALTERNATIVES TO FULL-TIME EMPLOYMENT: Paid internships
CONTACT: For individuals with previous experience: Human Resources Manager; Phone: 704-357-2000; Fax: 704-357-2253.

NORTH DAKOTA

AGGROW OILS LLC
780 NORTH 11TH STREET
CARRINGTON, NORTH DAKOTA 58421

OVERVIEW: Manufactures medicinal, botanical, and vegetable oils. Established: 1996. Ownership: Private.
KEY STATISTICS: Annual Sales: $12 million.
HIRING HISTORY: Number of professional employees hired in 1997: 12.
EXPERTISE/EDUCATION SOUGHT: Process engineering, engineering
COMPANY-PAID TRAINING is available.
CONTACT: For individuals with previous experience: Human Resources; Phone: 701-652-1990; Fax: 701-652-1991. E-mail: aggrow@daktel.com.

BASIN ELECTRIC POWER COOPERATIVE
1717 EAST INTERSTATE AVENUE
BISMARCK, NORTH DAKOTA 58501-0542

OVERVIEW: Generates and transmits electricity, produces natural gas, and mines coal. Established: 1961. Ownership: Private.
KEY STATISTICS: Assets: $400.3 million. Number of Employees: 1,996.
HIRING HISTORY: Number of professional employees hired in 1996: 10.
EXPERTISE/EDUCATION SOUGHT: Computer analysis, electrical engineering, mechanical engineering, data processing, finance, human resources, management, accounting
ALTERNATIVES TO FULL-TIME EMPLOYMENT: Part-time jobs, paid internships
CONTACT: Mrs. Sharon Klein, Employment and Equal Opportunity Office; Phone: 701-223-0441, ext. 2430; Fax: 701-255-5110. Job Hotline: 701-255-5175.

OHIO

AIRBORNE EXPRESS
145 HUNTER DRIVE
WILMINGTON, OHIO 45177-9390

OVERVIEW: Provides package delivery and courier services. Established: 1980. Parent Company: Airborne Freight Corporation. Ownership: Public.
KEY STATISTICS: Annual Sales: $782 million. Number of Employees: 6,833.
EXPERTISE/EDUCATION SOUGHT: Computer science, electrical engineering, mechanical engineering, systems analysis, program analysis
ALTERNATIVES TO FULL-TIME EMPLOYMENT: Paid internships
COMPANY-PAID TRAINING is available.
CONTACT: For individuals with previous experience: Mr. Howard Moore, Recruitment Manager; Phone: 937-382-5591, ext. 2430; Fax: 937-383-3838. E-mail: howard.moore@airborne.com.

AK STEEL CORPORATION
703 CURTIS STREET
MIDDLETOWN, OHIO 45043

OVERVIEW: Manufactures stainless steel and related products. Established: 1993. Ownership: Public.
KEY STATISTICS: Annual Sales: $1.1 billion. Number of Employees: 5,700.
HIRING HISTORY: Number of professional employees hired in 1997: 80. 1996: 60.
EXPERTISE/EDUCATION SOUGHT: Metallurgical engineering, electrical engineering, mechanical engineering, finance, marketing, purchasing, operations, research and development
ALTERNATIVES TO FULL-TIME EMPLOYMENT: Part-time jobs, paid internships
COMPANY-PAID TRAINING is available.
CONTACT: Mr. Michael Lehman, Senior Human Resources Representative, 1801 Crawford Street, Middletown, Ohio, 45043. E-mail: mike_lehman@aksteel.globalibmmail.com.
COMPANY URL: http://www.aksteel.com

ALPHA TECHNOLOGIES US, LP
2689 WINGATE AVENUE
AKRON, OHIO 44314

OVERVIEW: Manufactures process control instruments. Established: 1996. Ownership: Private.
KEY STATISTICS: Annual Sales: $35 million. Number of Employees: 230.
HIRING HISTORY: Number of professional employees hired in 1997: 10. 1996: 100.
EXPERTISE/EDUCATION SOUGHT: Sales, manufacturing, field technology, accounting, administration, management information systems, marketing, human resources, engineering, information technology
CONTACT: Mr. Chester Dawson, Director of Human Resources; Phone: 330-848-7211; Fax: 330-848-7326.

AMERICAN TRIM, LLC
1005 WEST GRAND AVENUE
LIMA, OHIO 45801-3429

Overview: Manufactures porcelain enameled products. Established: 1996. Ownership: Private.

Key Statistics: Annual Sales: $259.5 million. Number of Employees: 2,986.

Expertise/Education Sought: Manufacturing

Alternatives to Full-Time Employment: Paid internships

Company-Paid Training is available.

Contact: For individuals with previous experience: Human Resources; Phone: 419-228-1145; Fax: 419-221-4850.

ANCHOR HOCKING GLASS COMPANY
1115 WEST FIFTH AVENUE
LANCASTER, OHIO 43130

Overview: Manufactures consumer glassware, plastics, and household hardware. Established: 1990. Parent Company: Newell Company. Ownership: Public.

Key Statistics: Annual Sales: $62.4 million. Number of Employees: 8,900.

Expertise/Education Sought: Accounting, marketing, finance, human resources, production

Alternatives to Full-Time Employment: Paid internships

Contact: Mr. Karl Salmon, Vice President of Human Resources; Phone: 614-681-6144; Fax: 614-681-6076.

APPLE AMERICAN, LP INDIANA
8905 LAKE AVENUE
CLEVELAND, OHIO 44102-1221

Overview: Operates family restaurants. Established: 1996. Parent Company: Strang Corporation. Ownership: Private.

Key Statistics: Annual Sales: $40.7 million. Number of Employees: 1,435.

Expertise/Education Sought: Food chemistry, food engineering, food science, food service management, food services, franchising, restaurant management, hospitality/hotel and restaurant management

Alternatives to Full-Time Employment: Paid internships

Company-Paid Training is available.

Contact: Ms. Michelle Hoffman, Management Recruiter; Phone: 216-961-1227, ext. 287; Fax: 216-961-1966. E-mail: michellem@strangcorp.com.

APPLIED INDUSTRIAL TECHNOLOGIES
1 APPLIED PLAZA
CLEVELAND, OHIO 44115-2515

Overview: Manufactures bearings, tools, sealants, lubricants, and industrial equipment. Established: 1923. Ownership: Public.

Key Statistics: Annual Sales: $1.2 billion. Number of Employees: 5,000.

Hiring History: Number of professional employees hired in 1997: 500. 1996: 700.

Expertise/Education Sought: Distribution, sales, marketing

Alternatives to Full-Time Employment: Part-time jobs, paid internships

Company-Paid Training is available.

Contact: For individuals with previous experience: Ms. Carol Pavlik, Employment Supervisor; Phone: 216-426-4329; Fax: 216-426-4811. E-mail: cpavlik@appz-applied.com.

Company URL: http://www.appliedindustrial.com

THE AUSTIN COMPANY
3650 MAYFIELD ROAD
CLEVELAND, OHIO 44121-1734

Overview: Provides architectural and engineering services. Established: 1984. Ownership: Private.

Key Statistics: Annual Sales: $409.6 million. Number of Employees: 900.

Hiring History: Number of professional employees hired in 1997: 120. 1996: 120.

Expertise/Education Sought: Architecture, drafting, accounting, engineering

Contact: Ms. Jennifer Troy, Senior Persnel Services Administrator; Phone: 216-382-6600; Fax: 216-291-6684. E-mail: humanres@theaustin.com.

Company URL: http://www.theaustin.com

BEKAERT CORPORATION
3200 WEST MARKET STREET SOUTH
AKRON, OHIO 44333-3367

Overview: Manufactures steel and wire. Established: 1956. Parent Company: N.B. Bekaert. Ownership: Private.

Key Statistics: Annual Sales: $575 million. Number of Employees: 1,495.

Expertise/Education Sought: Human resources, finance, sales, data processing, accounting, auditing, engineering

Contact: For individuals with previous experience: Mr. Miguel Merzau, Corporate Manager of Human Resources; Phone: 330-867-3325.

BELCAN CORPORATION
10200 ANDERSON WAY
CINCINNATI, OHIO 45242-4718

Overview: Provides engineering services. Established: 1980. Ownership: Private.

Key Statistics: Annual Sales: $75 million. Number of Employees: 750.

Expertise/Education Sought: Machinery, engineering

Alternatives to Full-Time Employment: Paid internships

International Assignments are available to employees.

Contact: Human Resources; Phone: 800-423-5226; Fax: 513-985-7276. E-mail: belcan@belcan.tech.com.

Company URL: http://www.belcan.com

THE B. F. GOODRICH COMPANY
4020 KINROSS LAKES PARKWAY
RICHFIELD, OHIO 44286

Overview: Manufactures and markets chemical and aerospace products.

Expertise/Education Sought: Aeronautical engineering, mechanical engineering, electrical engineering, chemical engineering

Alternatives to Full-Time Employment: Paid internships

Company-Paid Training is available.

International Assignments are available to employees.

Contact: For individuals with previous experience: Vice President Human Resources, 9911 Brecksville Road, Richfield, Ohio, 44141; Phone: 330-659-7600; Fax: 330-659-7850.

BGP SERVICES
11501 NORTHLAKE DRIVE
CINCINNATI, OHIO 45249

Overview: Provides engineering support services. Established: 1986. Parent Company: Belcan Corporation. Ownership: Private.

Key Statistics: Annual Sales: $141.6 million. Number of Employees: 330.

Expertise/Education Sought: Electrical engineering, mechanical engineering, design engineering, manufacturing engineering

Alternatives to Full-Time Employment: Paid internships

Company-Paid Training is available.

Contact: Ms. Jeanne Bandewberg, Manager of Personnel; Phone: 513-247-0336; Fax: 513-605-5695.

BIRD ELECTRONIC CORPORATION

30303 AURORA ROAD
SOLON, OHIO 44139-2794

OVERVIEW: Manufactures electronic measuring test equipment. Established: 1995. Parent Company: Bird Technologies Group. Ownership: Private.

KEY STATISTICS: Annual Sales: $31.9 million. Number of Employees: 354.

HIRING HISTORY: Number of professional employees hired in 1997: 20.

EXPERTISE/EDUCATION SOUGHT: Engineering, electrical technology, radio frequency engineering

COMPANY-PAID TRAINING is available.

INTERNATIONAL ASSIGNMENTS are available to employees.

CONTACT: For individuals with previous experience: Mr. Michael Snyder, Manager of Human Resources; Phone: 440-248-1200, ext. 218; Fax: 440-248-0485.

BORDEN CHEMICAL, INC.

180 EAST BROAD STREET
COLUMBUS, OHIO 43215-3707

OVERVIEW: Manufactures industrial chemicals and packaging products. Established: 1996. Parent Company: Borden, Inc. Ownership: Private.

KEY STATISTICS: Annual Sales: $1.2 billion. Number of Employees: 3,000.

EXPERTISE/EDUCATION SOUGHT: Chemical engineering, information systems, industrial engineering, engineering

INTERNATIONAL ASSIGNMENTS are available to employees.

CONTACT: For individuals with previous experience: Human Relations Department; Phone: 614-225-4000.

BORDEN DECORATIVE PRODUCTS, INC.

1280 NORTH GRANT AVENUE
COLUMBUS, OHIO 43201-2871

OVERVIEW: Wholesales wallcoverings. Established: 1995. Parent Company: Borden, Inc. Ownership: Private.

KEY STATISTICS: Annual Sales: $132.3 million. Number of Employees: 830.

HIRING HISTORY: Number of professional employees hired in 1996: 25.

EXPERTISE/EDUCATION SOUGHT: Engineering, chemistry, inventory control

ALTERNATIVES TO FULL-TIME EMPLOYMENT: Paid internships

INTERNATIONAL ASSIGNMENTS are available to employees.

CONTACT: For individuals with previous experience: Mr. Tom Ness, Employee Relations Manager; Phone: 614-297-6198; Fax: 614-297-2996. Job Hotline: 614-225-4000.

BP CHEMICALS, INC.

4440 WARRENSVILLE CENTER
CLEVELAND, OHIO 44128

OVERVIEW: Manufactures industrial and agricultural chemicals. Established: 1963. Parent Company: British Petroleum. Ownership: Private.

KEY STATISTICS: Annual Sales: $637 million. Number of Employees: 2,600.

EXPERTISE/EDUCATION SOUGHT: Chemical engineering, mechanical engineering, marketing

ALTERNATIVES TO FULL-TIME EMPLOYMENT: Paid internships

INTERNATIONAL ASSIGNMENTS are available to employees.

CONTACT: Mr. Tom Hanich, Senior Recruiter; Phone: 216-586-3490; Fax: 216-586-6300.

BUCKEYE STEEL CASTINGS COMPANY

2211 PARSONS AVENUE
COLUMBUS, OHIO 43207-2386

OVERVIEW: Manufactures steel castings. Established: 1980. Parent Company: Worthington Industries, Inc. Ownership: Public.

KEY STATISTICS: Annual Sales: $130.6 million. Number of Employees: 1,200.

EXPERTISE/EDUCATION SOUGHT: Electrical engineering, metallurgical engineering, design engineering, mechanical engineering, manufacturing engineering, industrial engineering, customer service and support

CONTACT: Mr. Lawrence Jones, Manager of Personnel; Phone: 614-445-2154.

COMPANY URL: http://www.buckeyesteel.com

CALIBER TECHNOLOGY, INC.

PO BOX 3558
AKRON, OHIO 44309-3558

OVERVIEW: Provides trucking and shipping services. Established: 1930. Parent Company: FDX Corporation. Ownership: Public.

KEY STATISTICS: Annual Sales: $2.7 billion. Number of Employees: 38,167.

EXPERTISE/EDUCATION SOUGHT: Technology, information systems

ALTERNATIVES TO FULL-TIME EMPLOYMENT: Paid internships

COMPANY-PAID TRAINING is available.

CONTACT: Mrs. Sally Polen, Professional Recruiter; Phone: 330-643-6351; Fax: 330-384-2308. E-mail: spolen@calibersys.com.

COMPANY URL: http://www.fdxcorp.com

CINCINNATI BELL TELEPHONE COMPANY

201 EAST 4TH STREET
CINCINNATI, OHIO 45202

OVERVIEW: Provides telecommunications services. Established: 1873. Parent Company: Cin Bell, Inc. Ownership: Public.

KEY STATISTICS: Annual Sales: $617.1 million. Number of Employees: 3,500.

HIRING HISTORY: Number of professional employees hired in 1996: 650.

EXPERTISE/EDUCATION SOUGHT: Finance, accounting, marketing, electrical engineering, software engineering/development

ALTERNATIVES TO FULL-TIME EMPLOYMENT: Part-time jobs, paid internships

COMPANY-PAID TRAINING is available.

CONTACT: Ms. Lynda Breen, Professional Recruiter; Phone: 513-397-1265; Fax: 513-421-5694. E-mail: lbreen@cinbell.com.

COMPANY URL: http://www.cinbelltel.com

CINCINNATI MILACRON, INC.

4701 MARBURG AVENUE
CINCINNATI, OHIO 45209

OVERVIEW: Produces manufacturing systems, machinery, and related industrial products for the metalworking and plastics processing industries. Established: 1884. Ownership: Public.

KEY STATISTICS: Annual Sales: $1.7 billion. Number of Employees: 12,581.

EXPERTISE/EDUCATION SOUGHT: Mechanical engineering, business administration, electrical engineering, manufacturing engineering, industrial engineering, chemical engineering

ALTERNATIVES TO FULL-TIME EMPLOYMENT: Paid internships

COMPANY-PAID TRAINING is available.

INTERNATIONAL ASSIGNMENTS are available to employees.

CONTACT: For college students and recent graduates: Mr. Joe Roberts, Human Resources Administration; Phone: 513-841-7000; Fax: 513-841-8991. For individuals with previous experience: Mr. Steve Kuebbing, Human Resources Administration; Phone: 513-841-8100; Fax: 513-841-8991.

CINCINNATI SPECIALTIES, INC.

501 MURRAY ROAD
CINCINNATI, OHIO 45217-1014

OVERVIEW: Manufactures industrial inorganic chemicals. Established: 1994. Parent Company: PMC, Inc. Ownership: Private.

Cincinnati Specialties, Inc. (continued)

Key Statistics: Annual Sales: $30 million. Number of Employees: 200.
Hiring History: Number of professional employees hired in 1997: 10. 1996: 10.
Expertise/Education Sought: Chemistry, chemical engineering
Alternatives to Full-Time Employment: Paid internships
Company-Paid Training is available.
Contact: For individuals with previous experience: Human Relations Department.

CINTAS CORPORATION
6800 CINTAS BOULEVARD
CINCINNATI, OHIO 45262

Overview: Manufactures, rents, and markets industrial uniforms. Established: 1929. Ownership: Public.
Key Statistics: Annual Sales: $839.9 million. Number of Employees: 11,996.
Hiring History: Number of professional employees hired in 1997: 1,000. 1996: 500.
Expertise/Education Sought: Management, operations
Alternatives to Full-Time Employment: Paid internships
Company-Paid Training is available.
International Assignments are available to employees.
Contact: Mr. Kenneth May, National Recruitment; Phone: 513-573-4872; Fax: 513-573-4159.
Company URL: http://www.cintas-corp.com

COOPER TIRE AND RUBBER COMPANY
701 LIMA AVENUE
FINDLAY, OHIO 45840

Overview: Manufactures tires and other rubber products. Established: 1914. Ownership: Public.
Key Statistics: Annual Sales: $1.6 billion. Number of Employees: 8,932.
Hiring History: Number of professional employees hired in 1997: 50. 1996: 50.
Expertise/Education Sought: Design engineering, mechanical engineering, plant engineering, chemical engineering
Alternatives to Full-Time Employment: Paid internships
Company-Paid Training is available.
Contact: Ms. Linda Gallant, Employment Administrator; Phone: 419-424-4343; Fax: 419-424-7320.
Company URL: http://www.coopertire.com

CORE MATERIALS CORPORATION
800 MANOR PARK DRIVE
COLUMBUS, OHIO 43228-9640

Overview: Manufactures motor vehicle parts and accessories. Established: 1979. Ownership: Private.
Key Statistics: Annual Sales: $60 million. Number of Employees: 450.
Hiring History: Number of professional employees hired in 1997: 4. 1996: 5.
Expertise/Education Sought: Engineering, chemical engineering, industrial engineering
Company-Paid Training is available.
Contact: For individuals with previous experience: Human Resources; Phone: 614-870-5000; Fax: 614-870-5051.

CRANE PUMPS AND SYSTEMS, INC.
420 3RD STREET
PIQUA, OHIO 45356-3918

Overview: Manufactures pumps. Established: 1993. Parent Company: Crane Company. Ownership: Public.
Key Statistics: Annual Sales: $52.8 million. Number of Employees: 480.
Expertise/Education Sought: Industrial engineering, mechanical engineering
Alternatives to Full-Time Employment: Paid internships
Company-Paid Training is available.
International Assignments are available to employees.
Contact: Mr. Chuck Brooks, Personnel Manager, 420 3rd Street, PO Box 603, Piqua, Ohio, 45356-3918; Phone: 937-773-2442; Fax: 937-773-2238. Job Hotline: 937-773-2772, ext. 888.

CROWN EQUIPMENT CORPORATION
40 SOUTH WASHINGTON STREET
NEW BREMEN, OHIO 45869

Overview: Manufactures lift trucks and antenna rotators. Established: 1945. Ownership: Private.
Key Statistics: Annual Sales: $781.1 million. Number of Employees: 5,000.
Hiring History: Number of professional employees hired in 1996: 200.
Expertise/Education Sought: Mechanical engineering, design engineering, manufacturing engineering, accounting, data processing, finance
International Assignments are available to employees.
Contact: For individuals with previous experience: Mr. Max Fledderjohann, Personnel Director; Phone: 419-629-2311.

DAYCO PRODUCTS, INC.
1 PRESTIGE PLAZA
MIAMISBURG, OHIO 45401-1004

Overview: Manufactures belts, hoses, and rubber products for industrial and automotive industries. Established: 1988. Parent Company: Mark IV Industries, Inc. Ownership: Public.
Key Statistics: Annual Sales: $1.6 billion. Number of Employees: 16,000.
Hiring History: Number of professional employees hired in 1997: 75. 1996: 45.
Expertise/Education Sought: Sales, accounting, computer programming, program analysis, marketing, finance
Alternatives to Full-Time Employment: Paid internships
Company-Paid Training is available.
International Assignments are available to employees.
Contact: Mr. Charles Wolffis, Director of Human Resources, PO Box 1004, Miamisburg, Ohio, 45401-1004; Phone: 937-226-7000, ext. 5954; Fax: 937-226-8698.

THE DAYTON POWER AND LIGHT COMPANY
1065 WOODMAN DRIVE
DAYTON, OHIO 45401

Mailing Address: PO Box 8825, Dayton, OH 45401
Overview: Electric utility. Established: 1986. Parent Company: DPL, Inc. Ownership: Public.
Key Statistics: Assets: $1.3 billion. Number of Employees: 2,722.
Hiring History: Number of professional employees hired in 1997: 45. 1996: 45.
Expertise/Education Sought: Electrical engineering, engineering, mechanical engineering
Alternatives to Full-Time Employment: Part-time jobs, paid internships
Company-Paid Training is available.
Contact: For college students and recent graduates: Mr. Cory Pearson, Corporate Recruiter; Phone: 937-259-7258; Fax: 937-259-7290. E-mail: dtlhire@eri.com. For individuals with previous experience: Mr. Troy Washington, Supervisor of Employment; Phone: 937-259-7148; Fax: 937-259-7290. E-mail: dtlhire@eri.com.
Company URL: http://www.waytogo.com

ELSAG BAILEY PROCESS AUTOMATION

29801 EUCLID AVENUE
WICKLIFFE, OHIO 44092

Overview: Manufactures and distributes automotive parts and components. Established: 1993. Parent Company: Fin Mechanica. Ownership: Private.

Key Statistics: Annual Sales: $196.1 million. Number of Employees: 2,275.

Expertise/Education Sought: Marketing, finance, accounting, data processing, sales, electrical engineering, computer science, chemical engineering, mechanical engineering

Alternatives to Full-Time Employment: Paid internships

Company-Paid Training is available.

International Assignments are available to employees.

Contact: Ms. Marion Armburst, Senior Recruiter; Phone: 216-585-8500, ext. 8712; Fax: 216-585-5101. E-mail: marion.armburst@bailey.com.

Company URL: http://www.bailey.com

FIGGIE INTERNATIONAL, INC.

5875 LANDERBROOK DRIVE, SUITE 250
MAYFIELD HEIGHTS, OHIO 44124-7938

Overview: Manufactures fire fighting and protection equipment. Established: 1910. Ownership: Public.

Key Statistics: Annual Sales: $385.7 million. Number of Employees: 2,468.

Hiring History: Number of professional employees hired in 1997: 10. 1996: 5.

Expertise/Education Sought: Accounting

Alternatives to Full-Time Employment: Paid internships

International Assignments are available to employees.

Contact: For individuals with previous experience: Mr. William Sickman, Director of Human Resources and Administration; Phone: 440-446-1333; Fax: 440-442-9297.

FIRSTENERGY CORPORATION

76 SOUTH MAIN STREET
AKRON, OHIO 44308-1812

Overview: Provides electric service. Established: 1996. Ownership: Private.

Key Statistics: Annual Sales: $2.6 billion. Number of Employees: 9,100.

Expertise/Education Sought: Electrical engineering

Contact: For individuals with previous experience: Human Resources; Phone: 330-384-5100; Fax: 330-384-2455.

THE GENERAL CASTING COMPANY

550 LIBERTY ROAD, PO BOX 220
DELAWARE, OHIO 43015

Overview: Manufactures gray and ductile iron castings. Established: 1990. Ownership: Private.

Key Statistics: Annual Sales: $65 million. Number of Employees: 800.

Hiring History: Number of professional employees hired in 1997: 20. 1996: 20.

Expertise/Education Sought: Data processing, accounting, foundry engineering

Contact: Mr. Steve Richards, Vice President of Human Resources, PO Box 218, Delaware, Ohio, 43015; Phone: 614-363-1941; Fax: 614-363-1874.

M.A. HANNA COMPANY

200 PUBLIC SQUARE, SUITE 36-5000
CLEVELAND, OHIO 44114-2304

Overview: Manufactures specialty chemicals, plastics, and polymers. Established: 1885. Ownership: Public.

Key Statistics: Annual Sales: $2.1 billion. Number of Employees: 6,068.

Expertise/Education Sought: Chemical engineering, chemistry, data processing, accounting, marketing

Company-Paid Training is available.

International Assignments are available to employees.

Contact: For individuals with previous experience: Mr. L. L. Beach, Vice President of Human Resources; Phone: 216-589-4000; Fax: 216-589-4371.

Company URL: http://www.mahanna.com

HEIDELBERG FINISHING SYSTEMS

4900 WEBSTER STREET
DAYTON, OHIO 45414-4831

Overview: Manufactures office machines for binding. Established: 1996. Ownership: Private.

Key Statistics: Annual Sales: $46.1 million. Number of Employees: 500.

Hiring History: Number of professional employees hired in 1997: 130.

Expertise/Education Sought: Engineering, electrical engineering, manufacturing engineering

Alternatives to Full-Time Employment: Paid internships

Contact: Ms. Shuronda Morgan, Human Resources Representative; Phone: 937-278-2655; Fax: 937-278-5721.

HORIZON TELCOM, INC.

68 EAST MAIN STREET
CHILLICOTHE, OHIO 45601-0480

Overview: Provides local telephone services. Established: 1996. Ownership: Private.

Key Statistics: Annual Sales: $38.1 million. Number of Employees: 517.

Hiring History: Number of professional employees hired in 1997: 50. 1996: 50.

Expertise/Education Sought: Telecommunications, communications

Alternatives to Full-Time Employment: Paid internships

Company-Paid Training is available.

Contact: Human Resources Department; Phone: 740-772-8200, ext. 218; Fax: 740-775-7606.

HUFFY CORPORATION

225 BYERS ROAD
MIAMISBURG, OHIO 45342-3614

Overview: Manufactures bicycles and sporting goods. Established: 1924. Ownership: Public.

Key Statistics: Annual Sales: $701.9 million. Number of Employees: 8,020.

Expertise/Education Sought: Marketing, finance, human resources, accounting, computer programming

Alternatives to Full-Time Employment: Paid internships

Company-Paid Training is available.

Contact: For individuals with previous experience: Mr. Stan Davis, Vice President of Human Resorces and Organizational Development, 7701 Byers Road, Miamisburg, Ohio, 45342-3657; Phone: 937-866-6251, ext. 412; Fax: 937-865-5470.

INVACARE CORPORATION

1 INVACARE WAY
ELYRIA, OHIO 44035

Overview: Manufactures wheelchairs. Established: 1971. Ownership: Public.

Key Statistics: Annual Sales: $619.5 million. Number of Employees: 4,470.

Hiring History: Number of professional employees hired in 1997: 50.

Expertise/Education Sought: Sales, purchasing, finance, marketing, accounting, mechanical engineering, electrical engineering

Alternatives to Full-Time Employment: Part-time jobs, paid internships

Company-Paid Training is available.

Invacare Corporation (continued)

International Assignments are available to employees.
Contact: Ms. Ruth Conway, Manager of Human Resources; Phone: 216-329-6786; Fax: 216-365-7480.
Company URL: http://www.invacare.com

JAY PLASTICS, INC.
150 LONGVIEW AVENUE EAST
MANSFIELD, OHIO 44903-4206

Overview: Manufactures automobile parts. Established: 1946. Ownership: Private.
Key Statistics: Annual Sales: $74.4 million. Number of Employees: 650.
Expertise/Education Sought: Manufacturing engineering, quality control, computer science
Contact: For individuals with previous experience: Human Resources Administrator; Phone: 419-524-3778, ext. 3120; Fax: 419-526-3239.

JOHN MORRELL AND COMPANY, INC.
805 EAST KEMPER ROAD
CINCINNATI, OHIO 45246-4119

Overview: Packages meat. Established: 1995. Ownership: Public.
Key Statistics: Annual Sales: $1.5 billion. Number of Employees: 5,000.
Expertise/Education Sought: Industrial engineering, environmental engineering, computer science
Alternatives to Full-Time Employment: Paid internships
Company-Paid Training is available.
Contact: Ms. Aggie Carlson, Human Relations Department.

THE KROGER COMPANY
1014 VINE STREET
CINCINNATI, OHIO 45202

Overview: Operates chain of food stores. Established: 1883. Ownership: Public.
Key Statistics: Annual Sales: $25.2 billion. Number of Employees: 212,000.
Expertise/Education Sought: Law, sales, marketing, advertising, accounting, business management, retail, finance, economics
Contact: For individuals with previous experience: Department of Human Resources; Phone: 513-782-3451.

LENSCRAFTERS, INC.
8650 GOVERNORS HILL DRIVE
CINCINNATI, OHIO 45249

Overview: Operates chain of eyeglass stores. Established: 1995. Parent Company: Luxottica. Ownership: Private.
Key Statistics: Annual Sales: $849.7 million. Number of Employees: 13,500.
Hiring History: Number of professional employees hired in 1997: 200. 1996: 200.
Expertise/Education Sought: Marketing, information systems, merchandising, accounting, finance, human resources, operations
Alternatives to Full-Time Employment: Part-time jobs
Company-Paid Training is available.
Contact: Mr. Micheal Vacchiano, Manager of Recruitment; Phone: 513-583-6199; Fax: 513-583-6635. E-mail: mvacchia@lenscrafters.com. Job Hotline: 513-583-6884.
Company URL: http://www.lenscrafters.com

LIBBEY GLASS, INC.
300 MADISON AVENUE
TOLEDO, OHIO 43604-1204

Overview: Manufactures glassware. Established: 1993. Parent Company: Libbey, Inc. Ownership: Public.
Key Statistics: Annual Sales: $400.4 million. Number of Employees: 3,500.
Expertise/Education Sought: Electrical engineering, mechanical engineering, design engineering, industrial engineering, accounting, marketing, finance, computer programming
Contact: Human Resources Manager; Phone: 419-325-2100; Fax: 419-325-2137.

LIBBEY-OWENS-FORD COMPANY
811 MADISON AVENUE
TOLEDO, OHIO 43697-0799

Overview: Manufactures glass and glass products. Established: 1986. Parent Company: Pilkington Holdings, Inc. Ownership: Private.
Key Statistics: Annual Sales: $1.1 billion. Number of Employees: 7,565.
Expertise/Education Sought: Accounting, finance, marketing, auditing
Alternatives to Full-Time Employment: Paid internships
Company-Paid Training is available.
International Assignments are available to employees.
Contact: Ms. Jan Sterling, Supervisor of Personnel Services; Phone: 419-247-4887; Fax: 419-247-4815.

LIEBERT CORPORATION
1050 DEARBORN DRIVE
COLUMBUS, OHIO 43085

Overview: Manufactures air conditioning equipment. Established: 1987. Parent Company: Emerson Electric Company. Ownership: Public.
Key Statistics: Annual Sales: $337 million. Number of Employees: 3,800.
Hiring History: Number of professional employees hired in 1996: 100.
Expertise/Education Sought: Sales, management information systems, product management, engineering, sales, finance, information systems
Alternatives to Full-Time Employment: Part-time jobs, paid internships
Company-Paid Training is available.
International Assignments are available to employees.
Contact: For individuals with previous experience: Mr. Terry Martin, Human Resources Manager; Phone: 614-841-5753; Fax: 614-841-5890.
Company URL: http://www.Liebert.com

THE LIMITED STORES, INC.
3 LIMITED PARKWAY
COLUMBUS, OHIO 43230

Overview: Operates chain of women's apparel stores. Established: 1982. Parent Company: Limited, Inc. Ownership: Public.
Key Statistics: Annual Sales: $850 million. Number of Employees: 18,000.
Expertise/Education Sought: Marketing, product design and development, warehousing, information systems, sales, planning, finance, merchandising
Alternatives to Full-Time Employment: Paid internships
Company-Paid Training is available.
Contact: Human Resources Department; Phone: 614-479-2000; Fax: 614-479-2009.

THE LITTLE TIKES COMPANY
2180 BARLOW ROAD
HUDSON, OHIO 44236-4199

Overview: Manufactures children's vehicles and toys. Established: 1996. Parent Company: Rubbermaid. Ownership: Public.
Key Statistics: Annual Sales: $102.9 million. Number of Employees: 1,367.
Hiring History: Number of professional employees hired in 1997: 20.
Expertise/Education Sought: Manufacturing engineering

Alternatives to Full-Time Employment: Part-time jobs

Contact: For individuals with previous experience: Human Relations Department; Phone: 330-650-3000; Fax: 330-650-3819. Job Hotline: 330-650-3090.

Company URL: http://www.rubbermaid.com

LTV STEEL CORPORATION

200 PUBLIC SQUARE
CLEVELAND, OHIO 44114

Overview: Manufactures steel and oil field-drilling equipment. Established: 1953. Ownership: Public.

Key Statistics: Annual Sales: $4.1 billion. Number of Employees: 13,800.

Expertise/Education Sought: Accounting, finance, marketing

Alternatives to Full-Time Employment: Paid internships

Company-Paid Training is available.

International Assignments are available to employees.

Contact: Mr. Charles E. Butters, Director of Employee Development; Phone: 216-622-5234; Fax: 216-622-1066.

Company URL: http://www.ltvsteel.com

LUBRIZOL CORPORATION

29400 LAKELAND BOULEVARD
WICKLIFFE, OHIO 44092-2201

Overview: Manufactures plant genetics products and provides services for the agricultural industry. Established: 1928. Ownership: Public.

Key Statistics: Annual Sales: $1.6 billion. Number of Employees: 4,601.

Hiring History: Number of professional employees hired in 1997: 60. 1996: 93.

Expertise/Education Sought: Mechanical engineering, chemical engineering, chemistry

Company-Paid Training is available.

Contact: For college students and recent graduates: Ms. Laura Phelps, Human Resources Recruiter; Phone: 440-943-1200; Fax: 440-943-4228. E-mail: llp@lubrizol.com. For individuals with previous experience: Ms. Liz Martin, Manager of Staffing and Recruiting; Phone: 216-943-4200; Fax: 216-943-4228. E-mail: elc@lubrizol.com.

Company URL: http://www.lubrizol.com

MANSFIELD PLUMBING PRODUCTS

150 EAST 1ST STREET
PERRYSVILLE, OHIO 44864-9788

Overview: Manufactures sinks, toilets, bathtubs, and faucets. Established: 1986. Parent Company: Falcon Industries. Ownership: Private.

Key Statistics: Annual Sales: $150 million. Number of Employees: 1,500.

Expertise/Education Sought: Ceramic engineering, mechanical engineering, technical engineering

Contact: Mrs. Cathy Lance, Plant Employment and Benefits Supervisor; Phone: 419-938-5211; Fax: 419-938-1427.

MARION POWER SHOVEL COMPANY

617 WEST CENTER STREET
MARION, OHIO 43302-3509

Overview: Manufactures mining machinery. Established: 1995. Parent Company: Global Industrial Technology. Ownership: Public.

Key Statistics: Annual Sales: $45.8 million. Number of Employees: 456.

Expertise/Education Sought: Engineering

Contact: For individuals with previous experience: Ms. Terry Lynch, Manager Personnel Services; Phone: 614-383-5211, ext. 2176; Fax: 614-382-2054.

MASTERS BUILDERS, INC.

23700 CHAGRIN BOULEVARD
CLEVELAND, OHIO 44122-5506

Overview: Manufactures adhesives, sealants, and concrete products. Established: 1985. Parent Company: SKW. Ownership: Private.

Key Statistics: Annual Sales: $131.7 million. Number of Employees: 925.

Expertise/Education Sought: Research and development, chemical engineering, civil engineering

Company-Paid Training is available.

International Assignments are available to employees.

Contact: Ms. Radine Brown, Manager of Human Resources; Phone: 216-831-5500; Fax: 216-831-0043.

MEAD CORPORATION

COURTHOUSE PLAZA, NE
DAYTON, OHIO 45463

Overview: Manufactures paper, office supplies and packaging; provides on line database services. Established: 1846. Ownership: Public.

Key Statistics: Annual Sales: $4.7 billion. Number of Employees: 20,000.

Hiring History: Number of professional employees hired in 1997: 20. 1996: 50.

Expertise/Education Sought: Finance, human resources, information services, engineering

Alternatives to Full-Time Employment: Paid internships

International Assignments are available to employees.

Contact: Mr. Chuck Aardema, Director of Executive Staffing; Phone: 937-495-4314; Fax: 937-495-4317.

Company URL: http://www.mead.com

MEDEX, A FURON COMPANY

3637 LACON ROAD
HILLIARD, OHIO 43026

Overview: Distributes medical equipment and manufactures disposable medical supplies. Established: 1959. Parent Company: A Furon Company. Ownership: Public.

Key Statistics: Annual Sales: $92.5 million. Number of Employees: 1,100.

Expertise/Education Sought: Product design engineering

Alternatives to Full-Time Employment: Paid internships

Company-Paid Training is available.

International Assignments are available to employees.

Contact: For individuals with previous experience: Mr. Allan Fermier, Vice President of Human Resources; Phone: 614-876-2413; Fax: 614-876-8356. Job Hotline: 614-529-3838.

MEYER TOOL, INC.

3064 COLERAIN AVENUE
CINCINNATI, OHIO 45225-0098

Overview: Manufactures aircraft engine parts. Established: 1973. Ownership: Private.

Key Statistics: Annual Sales: $30 million. Number of Employees: 299.

Hiring History: Number of professional employees hired in 1997: 5. 1996: 5.

Expertise/Education Sought: Manufacturing engineering

Alternatives to Full-Time Employment: Paid internships

Company-Paid Training is available.

Contact: Ms. Diana Key, Human Resources Director; Phone: 513-853-4400, ext. 4317; Fax: 513-853-4495. E-mail: meyerhr@one.net.

MTD PRODUCTS, INC.
700 LIVERPOOL DRIVE
VALLEY CITY, OHIO 44280-9717

Overview: Manufactures automobile supplies and parts. Established: 1968. Ownership: Private.

Key Statistics: Annual Sales: $57 million. Number of Employees: 450.

Expertise/Education Sought: Engineering, accounting, manufacturing

Alternatives to Full-Time Employment: Internships

Company-Paid Training is available.

Contact: For individuals with previous experience: Human Resources; Phone: 330-225-2600.

NEATON AUTO PRODUCTS MANUFACTURING
975 SOUTH FRANKLIN STREET
EATON, OHIO 45320

Overview: Manufactures motor vehicle parts and accessories. Established: 1985. Parent Company: Nihomplast. Ownership: Private.

Key Statistics: Annual Sales: $96 million. Number of Employees: 520.

Hiring History: Number of professional employees hired in 1997: 105. 1996: 60.

Expertise/Education Sought: Engineering, industrial engineering, electrical engineering, mechanical engineering, plastics engineering

Alternatives to Full-Time Employment: Paid internships

Company-Paid Training is available.

International Assignments are available to employees.

Contact: For individuals with previous experience: Human Resources Department; Phone: 513-456-7103; Fax: 513-456-1437.

NESTLE USA—FOOD DIVISION
30003 BAINBRIDGE ROAD
SOLON, OHIO 44139-2205

Overview: Manufactures frozen foods. Established: 1985. Parent Company: Nestle USA, Inc. Ownership: Private.

Key Statistics: Number of Employees: 13,000.

Hiring History: Number of professional employees hired in 1997: 150.

Expertise/Education Sought: Administration, accounting, marketing, operations, management information systems, mechanical engineering, industrial engineering, finance, logistics

Alternatives to Full-Time Employment: Part-time jobs, paid internships

International Assignments are available to employees.

Contact: Mr. Craig Boerman, Human Resources Manager; Phone: 440-349-5757; Fax: 440-498-7730. Job Hotline: 888-NES-TLE1.

Company URL: http://www.nestle.com

NORCOLD, INC.
600 SOUTH KUTHER ROAD
SIDNEY, OHIO 45365-8840

Overview: Manufactures refrigerators and freezers. Established: 1997. Parent Company: Thetford Corporation. Ownership: Private.

Key Statistics: Annual Sales: $25.1 million. Number of Employees: 350.

Hiring History: Number of professional employees hired in 1997: 60. 1996: 10.

Expertise/Education Sought: Engineering, production, technical

Alternatives to Full-Time Employment: Paid internships

Company-Paid Training is available.

Contact: For individuals with previous experience: Human Resources; Phone: 937-497-3080; Fax: 937-497-3085.

NORDSON CORPORATION
28601 CLEMENS ROAD
WESTLAKE, OHIO 44145-1119

Overview: Manufactures industrial equipment for applying paints, adhesives, and coatings. Established: 1909. Ownership: Public.

Key Statistics: Annual Sales: $609.4 million. Number of Employees: 3,823.

Expertise/Education Sought: Finance, accounting, administration, business management

Alternatives to Full-Time Employment: Paid internships

Company-Paid Training is available.

International Assignments are available to employees.

Contact: For individuals with previous experience: Human Resource Associate; Phone: 440-414-5454; Fax: 440-892-9507. E-mail: mbush@nordson.com.

Company URL: http://www.nordson.com

NORTH AMERICAN REFRACTORIES COMPANY
1228 EUCLID AVENUE, SUITE 500
CLEVELAND, OHIO 44115

Overview: Manufactures steel. Established: 1978. Parent Company: Veitsch-Radex. Ownership: Private.

Key Statistics: Annual Sales: $250 million. Number of Employees: 1,500.

Hiring History: Number of professional employees hired in 1997: 20.

Expertise/Education Sought: Ceramic engineering, geology, manufacturing engineering, life sciences

International Assignments are available to employees.

Contact: For individuals with previous experience: Department of Human Resources; Phone: 216-621-5200; Fax: 216-621-0141.

NUCLEAR COMPUTERS
6333 NORTH DIXIE DRIVE
DAYTON, OHIO 45414-3305

Overview: Markets and repairs used computer electronic components. Established: 1990. Parent Company: Schindler Group, Inc. Ownership: Private.

Key Statistics: Annual Sales: $144.2 million.

Expertise/Education Sought: Engineering

Contact: For individuals with previous experience: Personnel; Phone: 937-264-8260; Fax: 937-264-8261.

OWENS CORNING
ONE OWENS CORNING PARKWAY
TOLEDO, OHIO 43659

Overview: Manufactures glass fiber materials and polyester resins. Established: 1938. Ownership: Public.

Key Statistics: Annual Sales: $3.8 billion. Number of Employees: 19,000.

Expertise/Education Sought: Mechanical engineering, chemical engineering, electrical engineering, materials management, analytical chemistry

Alternatives to Full-Time Employment: Part-time jobs, paid internships

International Assignments are available to employees.

Contact: For college students and recent graduates: Mrs. Maureen Comes, Human Resources Consultant; Phone: 419-248-8000. E-mail: maureen.comes@owenscorning.com. For individuals with previous experience: Human Resources Consultant; Phone: 419-248-8000.

Company URL: http://www.owenscorning.com

OWENS-ILLINOIS, INC.
1 SEAGATE
TOLEDO, OHIO 43666

Overview: Manufactures glass and plastic containers. Established: 1987. Ownership: Public.

Key Statistics: Annual Sales: $3.8 billion. Number of Employees: 30,800.

Hiring History: Number of professional employees hired in 1996: 50.

Expertise/Education Sought: Ceramic engineering, packaging engineering

Company-Paid Training is available.

International Assignments are available to employees.

Contact: For individuals with previous experience: Human Resource Department; Phone: 419-247-1427; Fax: 419-247-2216.

PARKER HANNIFIN CORPORATION
6035 PARKLAND BOULEVARD
CLEVELAND, OHIO 44124

Overview: Manufactures hydraulic and pneumatic components, aircraft and aerospace equipment, and automotive products. Established: 1924. Parent Company: Parker Hannifin Company. Ownership: Public.

Key Statistics: Annual Sales: $4.1 billion. Number of Employees: 34,927.

Hiring History: Number of professional employees hired in 1997: 120.

Expertise/Education Sought: Mechanical engineering, electrical engineering, accounting, aerospace engineering, agricultural engineering, medical technology, electrical technology

Alternatives to Full-Time Employment: Paid internships

Company-Paid Training is available.

International Assignments are available to employees.

Contact: Mr. Duane Crockrom, Manager of Human Resources; Phone: 216-896-3000; Fax: 216-896-4031.

Company URL: http://www.parker.com

PATHLORE SOFTWARE CORPORATION
7965 NORTH HIGH STREET, SUITE 300
COLUMBUS, OHIO 43235-4631

Overview: Provides computer software development services. Established: 1995. Ownership: Private.

Key Statistics: Annual Sales: $15 million. Number of Employees: 100.

Hiring History: Number of professional employees hired in 1996: 40.

Expertise/Education Sought: Software engineering/development, training and development

Alternatives to Full-Time Employment: Part-time jobs

Company-Paid Training is available.

International Assignments are available to employees.

Contact: For individuals with previous experience: Ms. Mariann Stopyra, Human Resources Administrator; Phone: 614-781-0036; Fax: 614-781-7200. E-mail: mstopyra@pathlore.com.

Company URL: http://www.pathlore.com

PERFECTION CORPORATION
222 NORTH LAKE STREET
MADISON, OHIO 44057-3118

Overview: Manufactures fluid systems products, and pipe fittings. Established: 1991. Parent Company: American Meter Corporation. Ownership: Private.

Key Statistics: Annual Sales: $47.1 million. Number of Employees: 300.

Hiring History: Number of professional employees hired in 1997: 4. 1996: 50.

Expertise/Education Sought: Mechanical engineering

Alternatives to Full-Time Employment: Paid internships

Company-Paid Training is available.

Contact: For individuals with previous experience: Mr. Christopher Jochum, Director of Human Resources; Phone: 216-428-1171, ext. 4323; Fax: 216-428-6765.

PIONEER INDUSTRIAL COMPONENTS, INC.
100 SOUTH PIONEER BOULEVARD
SPRINGBORO, OHIO 45066

Overview: Manufactures household audio and video equipment. Established: 1981. Parent Company: Pioneer Electronics Corporation. Ownership: Private.

Key Statistics: Annual Sales: $34.4 million. Number of Employees: 315.

Hiring History: Number of professional employees hired in 1997: 5. 1996: 5.

Expertise/Education Sought: Electrical engineering, mechanical engineering, information systems, computer programming, design engineering, software engineering/development, production engineering

Alternatives to Full-Time Employment: Paid internships

Contact: For individuals with previous experience: Human Resources.

PREFORMED LINE PRODUCTS COMPANY
660 BETA DRIVE
MAYFIELD VILLAGE, OHIO 44143

Overview: Manufactures connectors and splice cases for cable. Established: 1947. Ownership: Public.

Key Statistics: Annual Sales: $184.8 million. Number of Employees: 1,450.

Hiring History: Number of professional employees hired in 1997: 10. 1996: 10.

Expertise/Education Sought: Sales, accounting, marketing

Alternatives to Full-Time Employment: Part-time jobs

Contact: For individuals with previous experience: Ms. Marian Lawson, Manager of Personnel; Phone: 440-461-5200, ext. 9261; Fax: 440-442-6048.

Company URL: http://www.preformed.com

PRICE BROTHERS COMPANY
367 WEST 2ND STREET
DAYTON, OHIO 45402-1432

Overview: Provides concrete construction services. Established: 1899. Ownership: Private.

Key Statistics: Annual Sales: $102.8 million. Number of Employees: 610.

Expertise/Education Sought: Sales

Contact: For individuals with previous experience: Mr. Donn Lorenz, Vice President of Finance; Phone: 937-226-8700; Fax: 937-226-8903.

Company URL: http://www.pipesite.com

THE PROCTER & GAMBLE COMPANY
1 PROCTER & GAMBLE PLAZA
CINCINNATI, OHIO 45202-3315

Overview: Manufactures consumer, health and household products. Established: 1837. Ownership: Public.

Key Statistics: Annual Sales: $35.8 billion. Number of Employees: 106,000.

Hiring History: Number of professional employees hired in 1997: 800. 1996: 700.

Expertise/Education Sought: Sales, advertising, marketing, finance, accounting, auditing, systems integration, manufacturing engineering, chemical engineering

Alternatives to Full-Time Employment: Paid internships

Contact: Mr. Robert Pike, College Relations Director, 2 Procter & Gamble Plaza, Cincinnati, Ohio, 45202-3315; Phone: 513-983-3788; Fax: 513-983-4091.

Company URL: http://www.pg.com/careers

PROGRESSIVE PLASTICS, INC.
14801 EMERY AVENUE
CLEVELAND, OHIO 44135

Overview: Manufactures plastic containers. Established: 1966. Ownership: Private.

Key Statistics: Annual Sales: $30.2 million. Number of Employees: 325.

Hiring History: Number of professional employees hired in 1997: 280. 1996: 380.

Expertise/Education Sought: Marketing, computer programming, customer service and support, plastics engineering, product engineering, manufacturing

Progressive Plastics, Inc. (continued)

Alternatives to Full-Time Employment: Paid internships

Company-Paid Training is available.

Contact: For individuals with previous experience: Mr. Dennis George, Human Resource Recruiter; Phone: 800-252-0053; Fax: 216-252-6578. E-mail: ppi@en.com.

ROCKWELL AUTOMATION/RELIANCE ELECTRIC

6065 PARKLAND BOULEVARD
CLEVELAND, OHIO 44124

Overview: Manufactures industrial motors and power supplies for telecommunications equipment. Established: 1994. Parent Company: Rockwell International. Ownership: Public.

Key Statistics: Annual Sales: $716.7 million. Number of Employees: 10,282.

Hiring History: Number of professional employees hired in 1996: 30.

Expertise/Education Sought: Systems engineering, technical engineering, electrical engineering

Alternatives to Full-Time Employment: Paid internships

Company-Paid Training is available.

International Assignments are available to employees.

Contact: For individuals with previous experience: Ms. Constance S. Boykin, Manager of Human Resources; Phone: 216-266-7695; Fax: 216-266-1090.

ROYAL APPLIANCE MANUFACTURING COMPANY

650 ALPHA DRIVE
CLEVELAND, OHIO 44143-2123

Overview: Manufactures vacuum cleaners. Established: 1905. Ownership: Public.

Key Statistics: Annual Sales: $286.1 million. Number of Employees: 615.

Hiring History: Number of professional employees hired in 1997: 30. 1996: 25.

Expertise/Education Sought: Finance, accounting, sales, data processing, marketing, purchasing

Alternatives to Full-Time Employment: Paid internships

Company-Paid Training is available.

Contact: Ms. Gerri Jaffrey, Human Resources and Staffing; Phone: 440-449-6150; Fax: 440-449-7806.

Company URL: http://www.dirtdevil.com

RYDER PUBLIC TRANSPORTATION SERVICES

705 CENTRAL AVENUE, SUITE 500
CINCINNATI, OHIO 45202-6623

Overview: Provides highway transportation services. Established: 1934. Parent Company: Ryder System, Inc. Ownership: Public.

Key Statistics: Annual Sales: $5.5 billion. Number of Employees: 44,765.

Expertise/Education Sought: Project management

Alternatives to Full-Time Employment: Part-time jobs

Company-Paid Training is available.

Contact: Mr. Mike Ericsson, Division Recruiter, 77 West Port Plaza Drive, Suite 30, Cincinnati, Ohio, 45202; Phone: 314-579-1718; Fax: 314-275-7866.

Company URL: http://www.ryder.com

A. SCHULMAN, INC.

3550 WEST MARKET STREET
AKRON, OHIO 44333-2658

Overview: Manufactures and markets plastic resins and products. Established: 1928. Ownership: Public.

Key Statistics: Annual Sales: $996.4 million. Number of Employees: 1,939.

Expertise/Education Sought: Sales, marketing, accounting, management

Alternatives to Full-Time Employment: Paid internships

Company-Paid Training is available.

International Assignments are available to employees.

Contact: Ms. Toy Friedberg, Director of Human Resources; Phone: 330-666-3751; Fax: 330-668-7204.

SCITEX DIGITAL PRINTING, INC.

3000 RESEARCH BOULEVARD
DAYTON, OHIO 45420-4005

Overview: Manufactures electronic prepress systems for the printing and publishing industries. Established: 1993. Parent Company: Scitex Corporation, Ltd. Ownership: Private.

Key Statistics: Annual Sales: $155 million. Number of Employees: 637.

Hiring History: Number of professional employees hired in 1997: 80. 1996: 70.

Expertise/Education Sought: Mechanical engineering, electrical engineering, software engineering/development, computer science, applications engineering, electrical engineering

Alternatives to Full-Time Employment: Part-time jobs, paid internships

Company-Paid Training is available.

International Assignments are available to employees.

Contact: For individuals with previous experience: Mr. Joe Weil, Recruitment Director; Phone: 937-259-3302; Fax: 937-259-3190.

Company URL: http://www.scitex.com

THE SCOTT FETZER COMPANY

28800 CLEMENS ROAD
WESTLAKE, OHIO 44145-1134

Overview: Manufactures household and electrical products. Established: 1986. Parent Company: Berkshire Hathaway, Inc. Ownership: Public.

Key Statistics: Annual Sales: $954 million. Number of Employees: 9,540.

Hiring History: Number of professional employees hired in 1997: 9. 1996: 9.

Expertise/Education Sought: Law, administration, tax, accounting

Alternatives to Full-Time Employment: Paid internships

Company-Paid Training is available.

Contact: For individuals with previous experience: Ms. Robin Sabbarese, Benefits Administrator; Phone: 440-892-3000; Fax: 440-892-3060.

SEQUENTIA, INC.

15900 FOLTZ INDUSTRIAL PARK
STRONGSVILLE, OHIO 44136-5531

Overview: Manufactures plastic panels. Established: 1995. Ownership: Private.

Key Statistics: Annual Sales: $22.2 million. Number of Employees: 300.

Expertise/Education Sought: Manufacturing, plastics

Contact: For individuals with previous experience: Human Resources; Phone: 440-238-2400; Fax: 440-846-2137.

THE SHERWIN-WILLIAMS COMPANY

101 PROSPECT AVENUE NW
CLEVELAND, OHIO 44115-1075

Overview: Manufactures paints, lacquers and spray paints; operates retail stores. Established: 1866. Ownership: Public.

Key Statistics: Annual Sales: $4.1 billion. Number of Employees: 20,768.

Expertise/Education Sought: Management, chemical engineering, marketing, finance, business management, purchasing

Alternatives to Full-Time Employment: Paid internships

International Assignments are available to employees.
Contact: Human Resources; Phone: 216-566-2000.

SIA ADHESIVES, INC.
123 WEST BARTGES STREET
AKRON, OHIO 44311-1081

Overview: Manufactures adhesives and sealants. Established: 1996. Parent Company: Sovereign Specialty Chemicals, LP. Ownership: Private.
Key Statistics: Annual Sales: $22 million. Number of Employees: 105.
Hiring History: Number of professional employees hired in 1997: 10. 1996: 10.
Expertise/Education Sought: Sales, engineering, accounting, administration, adhesives
Company-Paid Training is available.
Contact: Ms. Jean Hoff, Human Resource Specialist; Phone: 330-374-3067; Fax: 330-374-2860.

STERIS CORPORATION
5960 HEISLEY ROAD
MENTOR, OHIO 44060-1834

Overview: Manufactures operating room tables, sterilization equipment, and lights. Established: 1996. Ownership: Public.
Hiring History: Number of professional employees hired in 1997: 400. 1996: 500.
Expertise/Education Sought: Management information systems, mechanical engineering, electrical engineering, medical technology, design engineering, product design engineering, manufacturing engineering, accounting, computer programming
Alternatives to Full-Time Employment: Paid internships
Company-Paid Training is available.
Contact: For individuals with previous experience: Director of Human Resources; Phone: 216-354-2600; Fax: 216-354-7043.

STRUCTURAL DYNAMICS RESEARCH CORPORATION
2000 EASTMAN DRIVE
MILFORD, OHIO 45150-2740

Overview: Develops engineering software and provides engineering services. Established: 1967. Ownership: Public.
Key Statistics: Annual Sales: $285.3 million. Number of Employees: 2,036.
Hiring History: Number of professional employees hired in 1997: 400. 1996: 400.
Expertise/Education Sought: Mechanical engineering, database management, marketing, human resources
Alternatives to Full-Time Employment: Part-time jobs
Company-Paid Training is available.
International Assignments are available to employees.
Contact: For college students and recent graduates: Ms. Carol Sorenson, Student Employment Specialist; Phone: 513-576-2400. E-mail: resume@sbrc.com. For individuals with previous experience: Mr. Derek Jackson, Director, Employment Services; Phone: 513-576-2400; Fax: 513-576-2922. E-mail: resume@sdrc.com.
Company URL: http://www.sdrc.com

TECHNEGLAS, INC.
727 EAST JENKINS AVENUE
COLUMBUS, OHIO 43207

Overview: Manufactures television glass funnels. Established: 1988. Parent Company: Nippon Electric Glass. Ownership: Private.
Key Statistics: Annual Sales: $650 million. Number of Employees: 3,500.
Expertise/Education Sought: Mechanical engineering, electrical engineering, ceramic engineering
Alternatives to Full-Time Employment: Part-time jobs, paid internships
International Assignments are available to employees.
Contact: For individuals with previous experience: Employment Manager; Phone: 614-443-6551; Fax: 614-445-1570.
Company URL: http://www.techneglas.com

TELXON CORPORATION
3330 WEST MARKET STREET
AKRON, OHIO 44333

Overview: Manufactures hand held, wireless, and portable computers. Established: 1969. Ownership: Public.
Key Statistics: Annual Sales: $466 million. Number of Employees: 1,600.
Expertise/Education Sought: Software engineering/development, electrical engineering, design engineering
Alternatives to Full-Time Employment: Paid internships
International Assignments are available to employees.
Contact: Miss Lisa Ross, Manager of Human Resources; Phone: 330-664-2967; Fax: 330-664-2813. E-mail: lross@telxon.com.

THE TIMKEN COMPANY
1835 DUEBER AVENUE, SW
CANTON, OHIO 44706-0927

Overview: Manufactures roller bearings and steel. Established: 1898. Ownership: Public.
Key Statistics: Annual Sales: $2.4 billion. Number of Employees: 19,130.
Hiring History: Number of professional employees hired in 1997: 100. 1996: 90.
Expertise/Education Sought: Mechanical engineering, metallurgical engineering, physics, chemistry, marketing, computer science, materials science, mathematics, general management, organizational behavior
Alternatives to Full-Time Employment: Paid internships
Company-Paid Training is available.
International Assignments are available to employees.
Contact: For college students and recent graduates: Ms. Arlene Zovath, Corporate Recruitment Analyst; Phone: 330-471-4022; Fax: 330-471-4462. E-mail: zovath@timken.com. For individuals with previous experience: Ms. Debra Rankin, Section Manager; Phone: 330-471-3933; Fax: 330-471-4462. E-mail: rankin@timken.com.
Company URL: http://www.timken.com

TRW, INC.
1900 RICHMOND ROAD
CLEVELAND, OHIO 44124

Overview: Manufactures automotive, aerospace, and defense products and information systems. Established: 1901. Ownership: Public.
Key Statistics: Annual Sales: $9.9 billion. Number of Employees: 65,200.
Expertise/Education Sought: Finance, law, human resources, business management, planning
Alternatives to Full-Time Employment: Paid internships
Company-Paid Training is available.
International Assignments are available to employees.
Contact: For college students and recent graduates: Ms. Faye Phillips, Human Resources Representative; Phone: 216-291-7000. For individuals with previous experience: Ms. Carol Dezzutti, Manager of Human Resources; Phone: 216-291-7000; Fax: 216-291-7321.
Company URL: http://www.trw.com

UNIVERSAL TUBULAR SYSTEMS
525 BEER ROAD
MANSFIELD, OHIO 44906-1214

Overview: Manufactures parts for gas stoves. Established: 1952. Parent Company: Burners Systems International. Ownership: Private.

Universal Tubular Systems (continued)

KEY STATISTICS: Annual Sales: $50 million. Number of Employees: 470.

HIRING HISTORY: Number of professional employees hired in 1997: 50. 1996: 50.

EXPERTISE/EDUCATION SOUGHT: Mechanical engineering, metallurgical engineering, design engineering, production engineering, accounting, finance, sales, marketing

ALTERNATIVES TO FULL-TIME EMPLOYMENT: Paid internships

INTERNATIONAL ASSIGNMENTS are available to employees.

CONTACT: For individuals with previous experience: Personnel Department; Phone: 419-529-5513; Fax: 419-529-4462.

UNIVERSITY OF DAYTON RESEARCH INSTITUTE

300 COLLEGE PARK
DAYTON, OHIO 45469-0105

OVERVIEW: Institute conducting sponsored research over a broad range of scientific and technical disciplines. Parent Company: University of Dayton.

HIRING HISTORY: Number of professional employees hired in 1997: 60. 1996: 40.

EXPERTISE/EDUCATION SOUGHT: Engineering, chemistry, environmental science, materials science, polymers, optics, physics

COMPANY-PAID TRAINING is available.

CONTACT: For individuals with previous experience: Ms. Paula L. Miller, Associate Director Human Resources; Phone: 513-229-2039; Fax: 513-229-3222. Job Hotline: 937-229-3377.

COMPANY URL: http://www.udri.udayton.edu

WAYNE-DALTON CORPORATION

ONE DOOR DRIVE
MOUNT HOPE, OHIO 44660

OVERVIEW: Manufactures doors, trim, and millwork products. Established: 1954. Ownership: Private.

KEY STATISTICS: Annual Sales: $300 million. Number of Employees: 2,483.

HIRING HISTORY: Number of professional employees hired in 1996: 300.

EXPERTISE/EDUCATION SOUGHT: Sales, accounting, purchasing, engineering

ALTERNATIVES TO FULL-TIME EMPLOYMENT: Paid internships

COMPANY-PAID TRAINING is available.

CONTACT: Ms. Kristi Loar, Employment Coordinator; Phone: 330-674-7015, ext. 4131; Fax: 330-674-5940. Job Hotline: 330-674-7015, ext. 4153.

WEASTEC, INC.

1600 NORTH HIGH STREET
HILLSBORO, OHIO 45133

OVERVIEW: Manufactures automotive electrical switches. Established: 1986. Parent Company: Toyo Denso. Ownership: Private.

KEY STATISTICS: Annual Sales: $80 million. Number of Employees: 500.

HIRING HISTORY: Number of professional employees hired in 1997: 100. 1996: 70.

EXPERTISE/EDUCATION SOUGHT: Electrical engineering, design engineering, mechanical engineering

ALTERNATIVES TO FULL-TIME EMPLOYMENT: Paid internships

COMPANY-PAID TRAINING is available.

CONTACT: Ms. Debbie Everhart, Supervisor of Human Resources; Phone: 513-393-6800; Fax: 513-393-3020.

COMPANY URL: http://www.weastec.com

WHITE CONSOLIDATED INDUSTRIES

11770 BEREA ROAD
CLEVELAND, OHIO 44111

OVERVIEW: Manufactures refrigerators and household appliances. Established: 1986. Parent Company: Electrolux, AB. Ownership: Private.

KEY STATISTICS: Annual Sales: $4.9 billion. Number of Employees: 29,000.

EXPERTISE/EDUCATION SOUGHT: Marketing, computer engineering, human resources, finance, mechanical engineering, electrical engineering

COMPANY-PAID TRAINING is available.

INTERNATIONAL ASSIGNMENTS are available to employees.

CONTACT: Mr. Joseph Burke, Vice President of Human Resources; Phone: 216-252-8393; Fax: 216-252-8160.

OKLAHOMA

ANSON GAS CORPORATION

4005 NORTHWEST EXPRESS WAY, SUITE 400
OKLAHOMA CITY, OKLAHOMA 73116

OVERVIEW: Explores for petroleum and gas and produces related products. Established: 1984. Ownership: Private.

KEY STATISTICS: Annual Sales: $17.9 million. Number of Employees: 85.

HIRING HISTORY: Number of professional employees hired in 1996: 20.

EXPERTISE/EDUCATION SOUGHT: Accounting, data processing

COMPANY-PAID TRAINING is available.

CONTACT: Ms. Karen Drennan, Personnel Director; Phone: 405-848-0525, ext. 3853; Fax: 405-879-3855.

BSW INTERNATIONAL

1 WEST 3RD STREET, SUITE 100
TULSA, OKLAHOMA 74103-3532

OVERVIEW: Provides architecture and engineering services. Established: 1989. Ownership: Private.

KEY STATISTICS: Annual Sales: $35.9 million. Number of Employees: 480.

HIRING HISTORY: Number of professional employees hired in 1996: 80.

EXPERTISE/EDUCATION SOUGHT: Marketing, finance, architecture, construction engineering, mechanical engineering, civil engineering, electrical engineering, structural engineering, construction

ALTERNATIVES TO FULL-TIME EMPLOYMENT: Paid internships

COMPANY-PAID TRAINING is available.

INTERNATIONAL ASSIGNMENTS are available to employees.

CONTACT: Ms. Chrisie Bedsworth, Human Resources Specialist; Phone: 918-582-8771, ext. 3290; Fax: 918-584-2656. E-mail: chrisie_bedsworth@bnwintl.com.

COMPANY URL: http://www.bswintl.com

CHEROKEE NATION

3.5 MILES SOUTH HIGHWAY 62
TAHLEQUAH, OKLAHOMA 74464-0948

MAILING ADDRESS: PO Box 948, Tahlequah, OK 74465-0948

OVERVIEW: Manufactures electronic components. Established: 1907. Ownership: Private.

KEY STATISTICS: Annual Sales: $42.7 million. Number of Employees: 1,500.

EXPERTISE/EDUCATION SOUGHT: Accounting, communications, finance, marketing, sales

Alternatives to Full-Time Employment: Internships
Company-Paid Training is available.
Contact: Ms. Loretta McNac, Recruitment and Training Specialist; Phone: 918-456-0671, ext. 2600; Fax: 918-458-6253.

CITGO PETROLEUM CORPORATION
6100 SOUTH YALE AVENUE
TULSA, OKLAHOMA 74136-1905

Mailing Address: PO Box 3758, Tulsa, OK 74102-3758
Overview: Operates petroleum refinery. Established: 1996. Parent Company: PDVSA. Ownership: Private.
Key Statistics: Annual Sales: $838.4 million. Number of Employees: 1,400.
Expertise/Education Sought: Manufacturing engineering, business administration
Alternatives to Full-Time Employment: Paid internships
Company-Paid Training is available.
Contact: Human Resources; Phone: 918-495-4000; Fax: 918-495-5458. Job Hotline: 918-495-4562.
Company URL: http://www.citgo.com

DOLESE BROTHERS COMPANY
20 NORTHWEST 13TH STREET
OKLAHOMA CITY, OKLAHOMA 73103-4806

Overview: Mines sand and crushed stone. Established: 1991. Ownership: Private.
Key Statistics: Annual Sales: $95 million. Number of Employees: 1,100.
Expertise/Education Sought: Drafting, transportation
Contact: For individuals with previous experience: Mr. Michael Rials, Director of Personnel; Phone: 405-297-8227; Fax: 405-297-8345.

THE FLINTCO COMPANIES, INC.
1624 WEST 21ST STREET
TULSA, OKLAHOMA 74107

Overview: Provides industrial and commercial construction services. Established: 1992. Parent Company: Flint Industries, Inc. Ownership: Private.
Key Statistics: Annual Sales: $403 million. Number of Employees: 3,000.
Hiring History: Number of professional employees hired in 1997: 30.
Expertise/Education Sought: Accounting, finance, sales
Alternatives to Full-Time Employment: Paid internships
Company-Paid Training is available.
International Assignments are available to employees.
Contact: Ms. Karen Hall, Director of Human Resources, PO Box 490, Tulsa, Oklahoma, 74101-0490; Phone: 918-587-8451, ext. 447; Fax: 918-586-0642. E-mail: khall@flintco.com.
Company URL: http://www.flintco.com

HILTI, INC.
5400 SOUTH 122ND EAST AVENUE
TULSA, OKLAHOMA 74146-6007

Overview: Produces metal fasteners. Established: 1981. Parent Company: Hilti AG. Ownership: Private.
Key Statistics: Annual Sales: $317.1 million. Number of Employees: 1,600.
Expertise/Education Sought: Manufacturing, maintenance
Alternatives to Full-Time Employment: Paid internships
Company-Paid Training is available.
Contact: For college students and recent graduates: Ms. Darylene Cunningham, Personnel Manager; Phone: 918-252-6000. For individuals with previous experience: Ms. Rosalia Cunningham, Personnel Manager; Phone: 918-252-6000. Job Hotline: 918-252-6406.
Company URL: http://www.hilti.com

PRO-QUIP CORPORATION
8522 EAST 61ST STREET
TULSA, OKLAHOMA 74133-1916

Overview: Provides engineering and design services. Established: 1986. Parent Company: Linde, AG. Ownership: Private.
Key Statistics: Annual Sales: $63.6 million. Number of Employees: 360.
Hiring History: Number of professional employees hired in 1997: 18. 1996: 12.
Expertise/Education Sought: Mechanical engineering, design engineering, drafting, engineering, construction engineering, chemical engineering
Company-Paid Training is available.
Contact: Mr. Ken Williams, Manager of Human Resources; Phone: 918-250-8522; Fax: 918-250-6915.

WILLBROS GROUP, INC.
2431 EAST 61ST STREET, SUITE 700
TULSA, OKLAHOMA 74136-1235

Overview: Water, sewer, utility line, and oil and gas pipeline construction. Established: 1976. Ownership: Private.
Key Statistics: Annual Sales: $150.4 million. Number of Employees: 2,000.
Expertise/Education Sought: Engineering, mechanical engineering
Company-Paid Training is available.
International Assignments are available to employees.
Contact: Human Resource Department; Phone: 918-748-7000; Fax: 918-748-7409. Job Hotline: 918-748-7702.

OREGON

AQUA GLASS WEST
5855 WASHBURN WAY
KLAMATH FALLS, OREGON 97603

Overview: Produces custom reinforced plastic and fiberglass products. Established: 1992. Parent Company: Masco. Ownership: Public.
Key Statistics: Annual Sales: $28 million. Number of Employees: 180.
Hiring History: Number of professional employees hired in 1997: 43.
Expertise/Education Sought: Manufacturing engineering
Alternatives to Full-Time Employment: Part-time jobs, paid internships
Company-Paid Training is available.
Contact: For individuals with previous experience: Mr. George Knight, Plant Manager; Phone: 541-885-3670; Fax: 541-885-5511.
Company URL: http://www.aquaglass.com

BONNEVILLE POWER ADMINISTRATION
905 NORTHEAST 11TH AVENUE
PORTLAND, OREGON 97232-4100

Mailing Address: PO Box 3621, Portland, OR 97208-3621
Overview: Electric utility. Established: 1937. Parent Company: U.S. Department of Energy. Ownership: Private.
Key Statistics: Annual Sales: $35.4 million. Number of Employees: 3,271.
Hiring History: Number of professional employees hired in 1997: 55. 1996: 50.
Expertise/Education Sought: Economics, finance, accounting, sales, auditing, engineering
Alternatives to Full-Time Employment: Part-time jobs, paid internships

Bonneville Power Administration (continued)

Company-Paid Training is available.

Contact: Ms. Linda Harris, Personnel Management Specialist; Phone: 503-230-3018; Fax: 503-230-4550. E-mail: llharris@bpa.gov. Job Hotline: 503-230-3055.

Company URL: http://www.bpa.gov

FREIGHTLINER TRUCKS
4747 NORTH CHANNEL AVENUE
PORTLAND, OREGON 97217

Overview: Provides engineering services for ship and boat building. Established: 1996. Parent Company: Daimler-Benz. Ownership: Private.

Key Statistics: Annual Sales: $8.9 million. Number of Employees: 125.

Hiring History: Number of professional employees hired in 1997: 1,700.

Expertise/Education Sought: Engineering, finance, computer programming, sales

Alternatives to Full-Time Employment: Paid internships

Company-Paid Training is available.

International Assignments are available to employees.

Contact: For individuals with previous experience: Human Resources Department; Phone: 503-735-8657; Fax: 503-735-6657. Job Hotline: 503-735-8657.

Company URL: http://www.freightliner.com

KOMATSU SLICON AMERICA, INC.
25300 NORTHWEST EVERGREEN
HILLSBORO, OREGON 97124

Overview: Manufactures semiconductors. Established: 1996. Parent Company: Komatsu Electronic Metals, Inc (Japan). Ownership: Private.

Key Statistics: Annual Sales: $120 million. Number of Employees: 233.

Hiring History: Number of professional employees hired in 1997: 150. 1996: 50.

Expertise/Education Sought: Process engineering, chemical engineering, equipment operation, maintenance, information systems

Company-Paid Training is available.

Contact: Ms. Susan Caswell, Staffing; Phone: 503-640-7000; Fax: 503-844-3297. E-mail: jobs@komsil.com.

MITSUBISHI SILICON AMERICA
1351 TANDEM AVENUE, NE
SALEM, OREGON 97303-4105

Overview: Manufactures polished silicon and epitaxial wafers. Established: 1986. Parent Company: Mitsubishi Materials Corporation. Ownership: Private.

Key Statistics: Annual Sales: $170 million. Number of Employees: 1,100.

Hiring History: Number of professional employees hired in 1997: 25. 1996: 25.

Expertise/Education Sought: Computer engineering, electrical engineering, chemical engineering, computer programming, information systems, mechanical engineering

Alternatives to Full-Time Employment: Part-time jobs, internships

International Assignments are available to employees.

Contact: Ms. Jan Wolfe, Senior Recruiter; Phone: 503-361-3536. E-mail: hr.recruiter@mmc-sil.com. Job Hotline: 503-315-6400.

Company URL: http://www.mmc-sil.com

MONACO COACH CORPORATION
91320 COBURG INDUSTRIAL WAY
COBURG, OREGON 97408-9492

Overview: Manufactures luxury motor homes. Established: 1996. Ownership: Public.

Expertise/Education Sought: Accounting, finance, marketing, engineering, industrial engineering, production management, production engineering

Alternatives to Full-Time Employment: Part-time jobs, paid internships

Contact: Ms. Deanna Ota, Personnel Manager; Phone: 541-686-8011; Fax: 541-302-3823.

Company URL: http://www.monaco-online.com

NIKE, INC.
1 BOWERMAN DRIVE
BEAVERTON, OREGON 97005-0979

Overview: Manufactures and markets athletic apparel and footwear. Established: 1964. Ownership: Public.

Key Statistics: Annual Sales: $9.2 billion. Number of Employees: 21,800.

Expertise/Education Sought: Research and development, administration, production, communications, finance, marketing, sales, management information systems, retail, design

Alternatives to Full-Time Employment: Paid internships

International Assignments are available to employees.

Contact: For college students and recent graduates: College Relations Department; Phone: 503-671-6453. For individuals with previous experience: Recruiting Resources; Phone: 503-671-6453; Fax: 888-767-9855.

Company URL: http://www.nike.com

PRAEGITZER INDUSTRIES, INC.
1270 SOUTHEAST MONMOUTH CUTOFF
DALLAS, OREGON 97338-9528

Overview: Manufactures printed circuit boards. Established: 1981. Ownership: Public.

Key Statistics: Annual Sales: $147.9 million. Number of Employees: 1,592.

Hiring History: Number of professional employees hired in 1996: 500.

Expertise/Education Sought: Computer programming, software engineering/development, data processing, electrical engineering, accounting, electronics/electronics engineering, hardware engineering, process engineering

Alternatives to Full-Time Employment: Part-time jobs, paid internships

Company-Paid Training is available.

International Assignments are available to employees.

Contact: Mr. Bob Schmelzer, Vice President of Human Resources; Phone: 503-623-1000, ext. 1327; Fax: 503-623-4434. E-mail: bob.schmelzer@pii.com.

Company URL: http://www.pii.com

SEIKO COMMUNICATIONS OF AMERICA
1625 NORTHWEST AMBERGLEN COURT
HILLSBORO, OREGON 97123

Overview: Provides radiotelephone communications services. Established: 1994. Parent Company: Seiko (Japan). Ownership: Private.

Key Statistics: Annual Sales: $8.5 million. Number of Employees: 100.

Expertise/Education Sought: Sales, marketing, engineering

International Assignments are available to employees.

Contact: For individuals with previous experience: Human Resources; Phone: 503-531-1703; Fax: 503-531-1575.

Company URL: http://www.messagewatch.com

U.S. FOREST INDUSTRIES, INC.
2611 WHITTLE AVENUE
MEDFORD, OREGON 97504-4719

Overview: Produces lumber products. Established: 1996. Ownership: Private.

KEY STATISTICS: Annual Sales: $117.8 million. Number of Employees: 527.

EXPERTISE/EDUCATION SOUGHT: Construction

CONTACT: For individuals with previous experience: Human Resources Department; Phone: 541-776-5750.

WACKER SILTRONIC CORPORATION

7200 NORTHWEST FRONT AVENUE
PORTLAND, OREGON 97210

MAILING ADDRESS: Po Box 83180, Portland, OR 97283

OVERVIEW: Manufactures silicon wafers. Established: 1978. Parent Company: Wacker Chemie. Ownership: Private.

KEY STATISTICS: Annual Sales: $107.8 million. Number of Employees: 1,400.

HIRING HISTORY: Number of professional employees hired in 1997: 200. 1996: 300.

EXPERTISE/EDUCATION SOUGHT: Chemical engineering, materials engineering, mechanical engineering, electrical engineering, chemistry, information technology

ALTERNATIVES TO FULL-TIME EMPLOYMENT: Paid internships

COMPANY-PAID TRAINING is available.

CONTACT: For individuals with previous experience: Mr. Robert Clay, Senior Corporate Recruiter; Phone: 503-219-7264; Fax: 503-241-7598. E-mail: bob_clay@siltronic.com. Job Hotline: 503-241-7547.

COMPANY URL: http://www.siltronic.com

WARN INDUSTRIES, INC.

13270 SOUTHEAST PHEASANT COURT
MILWAUKIE, OREGON 97222-1277

OVERVIEW: Manufactures four wheel drive accessories. Established: 1948. Ownership: Private.

KEY STATISTICS: Annual Sales: $116.6 million. Number of Employees: 634.

HIRING HISTORY: Number of professional employees hired in 1997: 40. 1996: 50.

EXPERTISE/EDUCATION SOUGHT: Manufacturing engineering, design engineering, management, technical

ALTERNATIVES TO FULL-TIME EMPLOYMENT: Part-time jobs, paid internships

COMPANY-PAID TRAINING is available.

INTERNATIONAL ASSIGNMENTS are available to employees.

CONTACT: For individuals with previous experience: Ms. Lisa Larson, Human Resources Administrator; Phone: 503-659-8750. E-mail: warn@teleport.com. Job Hotline: 503-786-4449.

COMPANY URL: http://www.warn.com

WILLAMETTE INDUSTRIES, INC.

1300 SOUTHWEST 5TH AVENUE
PORTLAND, OREGON 97201

OVERVIEW: Manufactures paper, packaging, lumber, and wood products. Established: 1906. Ownership: Public.

KEY STATISTICS: Annual Sales: $3.4 billion. Number of Employees: 13,700.

EXPERTISE/EDUCATION SOUGHT: Marketing, finance, accounting, electrical engineering, mechanical engineering, computer programming, chemical engineering

ALTERNATIVES TO FULL-TIME EMPLOYMENT: Paid internships

CONTACT: For individuals with previous experience: Mr. Dave Morthland, Vice President Human Resources; Phone: 503-227-5581, ext. 5645.

COMPANY URL: http://www.wii.com

PENNSYLVANIA

AALBORG KEYSTONE, INC.

1422 EAST AVENUE
ERIE, PENNSYLVANIA 16503-1561

OVERVIEW: Manufactures motors and generators. Established: 1997. Parent Company: AALBORG INDUSTRIES. Ownership: Private.

KEY STATISTICS: Annual Sales: $10 million. Number of Employees: 125.

EXPERTISE/EDUCATION SOUGHT: Engineering

CONTACT: For individuals with previous experience: Human Resources Department; Phone: 814-452-6421; Fax: 814-870-7460. E-mail: nmcmahon@aalborgkeystone.com.

ACCU-SORT SYSTEMS, INC.

511 SCHOOL HOUSE ROAD
TELFORD, PENNSYLVANIA 18969-1148

OVERVIEW: Manufactures bar code scanners and systems. Established: 1971. Ownership: Private.

KEY STATISTICS: Annual Sales: $48.5 million. Number of Employees: 450.

HIRING HISTORY: Number of professional employees hired in 1997: 149. 1996: 122.

EXPERTISE/EDUCATION SOUGHT: Manufacturing engineering, mechanical engineering, electrical engineering, software engineering/development, electronics/electronics engineering, computer science, electrical technology

ALTERNATIVES TO FULL-TIME EMPLOYMENT: Part-time jobs, paid internships

COMPANY-PAID TRAINING is available.

CONTACT: Ms. Donna Cuppett, Human Resources Manager; Phone: 215-721-5290; Fax: 215-721-5551. E-mail: hr@accusort.com.

COMPANY URL: http://www.accusort.com

ADELPHIA

MAIN AT WATER STREET
COUDERSPORT, PENNSYLVANIA 16915-1141

OVERVIEW: Provides cable television services. Established: 1952. Ownership: Public.

KEY STATISTICS: Annual Sales: $472.8 million. Number of Employees: 3,154.

EXPERTISE/EDUCATION SOUGHT: Communications, television broadcasting, human resources, marketing, finance, sales, electrical engineering, design engineering

CONTACT: For individuals with previous experience: Mr. Vincent Macdonald, Supervisor of Recruiting Department; Phone: 814-274-9830; Fax: 814-274-9623.

COMPANY URL: http://www.adelphia.net/career

AIR PRODUCTS AND CHEMICALS, INC.

7201 HAMILTON BOULEVARD
ALLENTOWN, PENNSYLVANIA 18195-1501

OVERVIEW: Manufactures industrial gases and chemicals. Established: 1940. Ownership: Public.

KEY STATISTICS: Annual Sales: $4 billion. Number of Employees: 15,200.

EXPERTISE/EDUCATION SOUGHT: Chemical engineering, information technology, finance, mechanical engineering

ALTERNATIVES TO FULL-TIME EMPLOYMENT: Internships

INTERNATIONAL ASSIGNMENTS are available to employees.

CONTACT: For college students and recent graduates: University Relations Recruiter; Phone: 610-481-7050; Fax: 610-481-8951. For individuals with previous experience: Ms. Barbara Paluszek, Human Resources; Phone: 610-481-4099; Fax: 610-706-6651.

COMPANY URL: http://www.airproducts.com

AMERICAN WESTERN REFINING, LP

PO BOX 8321
RADNOR, PENNSYLVANIA 19087-4515

Overview: Refines oil and gas. Established: 1981. Parent Company: Gadgil Western. Ownership: Public.

Key Statistics: Annual Sales: $68.7 million.

Hiring History: Number of professional employees hired in 1997: 3.

Expertise/Education Sought: Engineering, mechanical engineering, petroleum/petrochemical engineering, design engineering, computer programming

Contact: Mr. Chris Woods, Chief Financial Officer, PO Box 519, Radnor, Pennsylvania, 19087-4515; Phone: 610-964-9640; Fax: 610-995-9531.

AMETEK, INC.

STATION SQUARE
PAOLI, PENNSYLVANIA 19301

Overview: Manufactures motors and generators. Established: 1930. Ownership: Public.

Key Statistics: Annual Sales: $868.7 million. Number of Employees: 6,200.

Expertise/Education Sought: Engineering, business

Alternatives to Full-Time Employment: Paid internships

Company-Paid Training is available.

International Assignments are available to employees.

Contact: For individuals with previous experience: Human Resources; Phone: 610-647-2121; Fax: 610-296-3412.

Company URL: http://www.ametek.com

AMKOR ELECTRONICS, INC.

1345 ENTERPRISE DRIVE
WEST CHESTER, PENNSYLVANIA 19380-5964

Overview: Packages semiconductors. Established: 1970. Ownership: Private.

Key Statistics: Annual Sales: $960.8 million. Number of Employees: 480.

Expertise/Education Sought: Sales, customer service and support, data processing, electrical engineering, electronics/electronics engineering, computer analysis, computer programming

Alternatives to Full-Time Employment: Part-time jobs

Company-Paid Training is available.

International Assignments are available to employees.

Contact: For individuals with previous experience: Human Resources Department; Phone: 610-431-9600.

Company URL: http://www.amkor.com

AMP, INC.

441 FRIENDSHIP ROAD
HARRISBURG, PENNSYLVANIA 17105-3608

Overview: Manufactures electrical, electronic, and optic connection devices. Established: 1941. Ownership: Public.

Key Statistics: Annual Sales: $5.5 billion. Number of Employees: 44,985.

Hiring History: Number of professional employees hired in 1997: 500. 1996: 300.

Expertise/Education Sought: Computer science, electrical engineering, mechanical engineering, design engineering, sales, quality engineering, industrial engineering, automotive engineering

Alternatives to Full-Time Employment: Paid internships

Company-Paid Training is available.

International Assignments are available to employees.

Contact: For college students and recent graduates: Ms. Linda Spotts, Manager of College Relations; Phone: 717-592-6680; Fax: 717-592-2090. For individuals with previous experience: Professional Staffing; Phone: 717-564-0100; Fax: 717-592-2089.

Company URL: http://www.amp.com

ARAMARK CORPORATION

1101 MARKET STREET
PHILADELPHIA, PENNSYLVANIA 19107

Overview: Provides concessionaire services. Established: 1984. Ownership: Private.

Key Statistics: Annual Sales: $151.8 million. Number of Employees: 7,000.

Expertise/Education Sought: Accounting, administration, communications, computer programming, human resources, law, marketing, retail

Alternatives to Full-Time Employment: Part-time jobs, internships

Company-Paid Training is available.

International Assignments are available to employees.

Contact: Mr. Thomas W. Horn, Director of Staffing; Phone: 215-238-3225; Fax: 215-238-8170. E-mail: horn-tom@aramark.com.

Company URL: http://www.aramark.com

ARMCO, INC.

1 OXFORD CENTER, 301 GRANT STREET
PITTSBURGH, PENNSYLVANIA 15219-1415

Overview: Manufactures industrial equipment, steel and carbon products, and construction products; provides insurance. Established: 1899. Ownership: Public.

Key Statistics: Annual Sales: $1.7 billion. Number of Employees: 6,000.

Hiring History: Number of professional employees hired in 1996: 100.

Expertise/Education Sought: Chemical engineering, manufacturing engineering, mechanical engineering, quality control, chemistry, metallurgical engineering

Alternatives to Full-Time Employment: Paid internships

Company-Paid Training is available.

Contact: Human Resources; Phone: 412-255-9828; Fax: 412-255-9997. E-mail: Ninehauserla@armco.com.

Company URL: http://www.armco.com

ARMSTRONG WORLD INDUSTRIES, INC.

313 WEST LIBERTY STREET, PO BOX 3231
LANCASTER, PENNSYLVANIA 17604-3231

Mailing Address: PO Box 3001, Lancaster, PA 17601-3001

Overview: Manufacturer of quality, innovative flooring, ceiling, building and industrial products. Established: 1860. Ownership: Public.

Key Statistics: Number of Employees: 10,600.

Expertise/Education Sought: Chemical engineering, industrial engineering, electrical engineering, mechanical engineering, manufacturing science, process engineering, information systems

Alternatives to Full-Time Employment: Paid internships

International Assignments are available to employees.

Contact: Mr. Bing G. Spitler, Manager of Corporate Staffing. E-mail: resumes@armstrong.com.

Company URL: http://www.armstrong.com

CORPORATE PROFILE

From its humble beginnings in Pittsburgh, Pennsylvania, more than 137 years ago, Armstrong has continuously built on core competencies such as its famous floor coverings and today is an icon in building materials design and manufacturing. Armstrong's Building Products Operations were awarded the 1995 Malcolm Baldrige National Quality Award, regarded as the standard of business excellence.

Armstrong has also been recognized for its commitment to equal opportunity for minorities, women, veterans, and individuals with disabilities. People are hired from a variety of disciplines and depending on the position and requirements, college campus searches are frequently conducted.

By understanding and valuing the diverse skills and abilities of its employees, Armstrong is able to design and produce customer-focused products and services.

Competitive compensation and extensive benefits designed to attract goal-oriented professionals are offered.

ARROW INTERNATIONAL, INC.

2400 BERNVILLE ROAD
READING, PENNSYLVANIA 19605-9607

MAILING ADDRESS: PO Box 12888, Reading, PA 19612

OVERVIEW: Manufactures critical care and cardiovascular catheter kits. Established: 1975. Ownership: Public.

KEY STATISTICS: Annual Sales: $229.9 million. Number of Employees: 2,000.

EXPERTISE/EDUCATION SOUGHT: Mechanical engineering, biomedical engineering, sales

ALTERNATIVES TO FULL-TIME EMPLOYMENT: Paid internships

COMPANY-PAID TRAINING is available.

CONTACT: Ms. Linda Silva, Staffing Manager; Phone: 610-378-0131, ext. 3182; Fax: 610-478-3194.

COMPANY URL: http://www.arrowintl.com

ASPLUNDH TREE EXPERT COMPANY

708 BLAIR MILL ROAD
WILLOW GROVE, PENNSYLVANIA 19090-1701

OVERVIEW: Provides tree-trimming and removal services. Established: 1928. Ownership: Private.

KEY STATISTICS: Annual Sales: $935.5 million. Number of Employees: 20,000.

EXPERTISE/EDUCATION SOUGHT: Computer programming, accounting, finance

COMPANY-PAID TRAINING is available.

CONTACT: Mr. Michael Lynch, Director of Field Personnel; Phone: 215-784-4200; Fax: 215-784-4493.

AVO INTERNATIONAL

510 TOWNSHIP LINE ROAD
BLUE BELL, PENNSYLVANIA 19422-2701

OVERVIEW: Manufactures electronic measuring test equipment. Established: 1991. Parent Company: TBG. Ownership: Private.

KEY STATISTICS: Annual Sales: $28 million. Number of Employees: 170.

HIRING HISTORY: Number of professional employees hired in 1996: 15.

EXPERTISE/EDUCATION SOUGHT: Electrical technology, production

ALTERNATIVES TO FULL-TIME EMPLOYMENT: Part-time jobs

INTERNATIONAL ASSIGNMENTS are available to employees.

CONTACT: Ms. Nancy Cressman, Human Resource Manager; Phone: 215-646-9200; Fax: 215-628-9144.

COMPANY URL: http://www.avointl.com

BAYER CLOTHING GROUP, INC.

RD 4 BOX 91B
CLEARFIELD, PENNSYLVANIA 16830

OVERVIEW: Manufactures men's and boys' clothing. Established: 1986. Ownership: Private.

KEY STATISTICS: Annual Sales: $33.1 million. Number of Employees: 900.

EXPERTISE/EDUCATION SOUGHT: Manufacturing engineering

INTERNATIONAL ASSIGNMENTS are available to employees.

CONTACT: Miss Kris Young, Personnel Administration Manager; Phone: 814-765-7521; Fax: 814-765-0774.

BAYER CORPORATION

100 BAYER ROAD, BUILDING 16
PITTSBURGH, PENNSYLVANIA 15205-9741

OVERVIEW: Develops, manufactures, and distributes pharmaceuticals, chemicals, and imaging products. Established: 1954. Parent Company: Bayer, AG. Ownership: Private.

KEY STATISTICS: Annual Sales: $9 billion. Number of Employees: 24,000.

EXPERTISE/EDUCATION SOUGHT: Chemistry, engineering, international business (MBA), marketing, management information systems

CONTACT: Manager of Professional Staffing Programs; Phone: 412-777-5756; Fax: 412-777-5692.

COMPANY URL: http://www.bayer.com

CORPORATE PROFILE

Bayer Corporation's eight operating divisions develop, manufacture, and market more than 8,000 products that result in a comprehensive portfolio both creative and diverse. A $9-billion research-based company with 24,000 employees, its various businesses are concentrated in three core areas: health care and life sciences, chemicals, and imaging technologies.

A large part of Bayer's research and development efforts are dedicated to health care and life sciences, including treatments for cancer, dementia, metabolic disorders, and rheumatic diseases. Bayer is continuously strengthening the R&D facet of its operations to capitalize on opportunities in the North American market. However, Bayer's influence isn't limited by the boundaries of North America. The company is an integral part of the Bayer Group, a $34.4-billion company based in Leverkusen, Germany, that employs more than 144,000 people working for 350 companies in 150 countries.

Bayer Corporation strives to foster an environment that recognizes dedication and commitment toward its business goals, all the while gaining and retaining customer loyalty by anticipating and exceeding their requirements and expectations.

Its vast array of far-ranging activities is proof that Bayer Corporation is dedicated to its continuing commitment to the markets it serves.

BETHLEHEM STEEL CORPORATION

1170 8TH AVENUE
BETHLEHEM, PENNSYLVANIA 18016

OVERVIEW: Produces steel and manufactures steel products. Established: 1904. Ownership: Public.

KEY STATISTICS: Annual Sales: $4.7 billion. Number of Employees: 17,800.

HIRING HISTORY: Number of professional employees hired in 1997: 14. 1996: 48.

EXPERTISE/EDUCATION SOUGHT: Operations

ALTERNATIVES TO FULL-TIME EMPLOYMENT: Internships

COMPANY-PAID TRAINING is available.

CONTACT: For individuals with previous experience: Human Resources; Phone: 610-694-2424.

BINNEY AND SMITH, INC.

1100 CHURCH LANE
EASTON, PENNSYLVANIA 18044-0431

OVERVIEW: Manufactures, crayons, chalk, and art supplies. Established: 1984. Parent Company: Hallmark Cards. Ownership: Private.

KEY STATISTICS: Annual Sales: $562 million. Number of Employees: 2,300.

EXPERTISE/EDUCATION SOUGHT: Manufacturing

ALTERNATIVES TO FULL-TIME EMPLOYMENT: Internships

COMPANY-PAID TRAINING is available.

Binney and Smith, Inc. (continued)

International Assignments are available to employees.

Contact: For individuals with previous experience: Human Resources; Phone: 610-253-6271; Fax: 610-250-5768.

B. BRAUN MEDICAL, INC.

824 12TH AVENUE
BETHLEHEM, PENNSYLVANIA 18018-3524

Overview: Manufactures surgical and medical instruments. Established: 1979. Parent Company: B. Braun of America, Inc. Ownership: Private.

Key Statistics: Annual Sales: $200.4 million. Number of Employees: 1,600.

Expertise/Education Sought: Customer service and support, industrial engineering, materials engineering, packaging engineering, quality control, management, engineering, sales

Alternatives to Full-Time Employment: Paid internships

Company-Paid Training is available.

International Assignments are available to employees.

Contact: For individuals with previous experience: Mr. Peter Werley, Director of Human Resources; Phone: 610-691-5400, ext. 4255; Fax: 610-861-5991.

BUCKEYE PIPE LINE COMPANY

3900 HAMILTON BOULEVARD
ALLENTOWN, PENNSYLVANIA 18103-6122

Overview: Provides investment services. Established: 1996. Ownership: Private.

Key Statistics: Annual Sales: $49.6 million. Number of Employees: 600.

Expertise/Education Sought: Chemical engineering, electrical engineering, mechanical engineering, information systems

Alternatives to Full-Time Employment: Paid internships

Contact: For individuals with previous experience: Director of Human Resources; Phone: 610-770-4000; Fax: 610-770-4541.

Company URL: http://www.buckeye.com

BURNHAM CORPORATION

1241 HARRISBURG PIKE
LANCASTER, PENNSYLVANIA 17603-2515

Overview: Manufactures boilers. Established: 1856. Ownership: Public.

Key Statistics: Annual Sales: $159.9 million. Number of Employees: 1,600.

Expertise/Education Sought: Industrial engineering, marketing, engineering

Alternatives to Full-Time Employment: Internships

Company-Paid Training is available.

International Assignments are available to employees.

Contact: For individuals with previous experience: Mr. Kenneth Strutz, Vice President of Human Resources; Phone: 717-397-4701.

CDI CORPORATION

1717 ARCH STREET, 35TH FLOOR
PHILADELPHIA, PENNSYLVANIA 19103-2768

Overview: Provides employment and temporary personnel services. Established: 1950. Ownership: Public.

Key Statistics: Annual Sales: $1.4 billion. Number of Employees: 30,900.

Expertise/Education Sought: Civil engineering, computer engineering, data processing, construction engineering, design engineering, electrical engineering, electronics/electronics engineering, mechanical engineering, engineering technology

International Assignments are available to employees.

Contact: Human Resources Recruiting Department; Phone: 215-569-2200; Fax: 215-569-1750. E-mail: resumes@cdicorp.com.

CEGELEC AEG AUTOMATION SYSTEMS CORPORATION

701 TECHNOLOGY DRIVE
CANONSBURG, PENNSYLVANIA 15317-9587

Overview: Manufactures relays and industrial controls. Established: 1996. Parent Company: Cegelec Paris/France. Ownership: Private.

Key Statistics: Annual Sales: $80 million. Number of Employees: 220.

Hiring History: Number of professional employees hired in 1996: 10.

Expertise/Education Sought: Mechanical engineering, engineering technology

Contact: Ms. Vivian Semonian, Human Resources Supervisor; Phone: 412-873-9441; Fax: 412-873-9392.

CEGELEC AUTOMATION, INC.

301 ALPHA DRIVE
PITTSBURGH, PENNSYLVANIA 15238

Overview: Manufactures industrial motors and drive control equipment. Established: 1987. Parent Company: Alcatel Alsthom. Ownership: Private.

Key Statistics: Annual Sales: $70 million. Number of Employees: 361.

Hiring History: Number of professional employees hired in 1997: 20. 1996: 5.

Expertise/Education Sought: Electrical engineering, accounting, drafting, manufacturing

Alternatives to Full-Time Employment: Paid internships

Company-Paid Training is available.

International Assignments are available to employees.

Contact: Human Resources; Phone: 412-967-7696; Fax: 412-967-7660. E-mail: russo@cegelec.com.

Company URL: http://www.cegelecfr.com

COMCAST CORPORATION

1500 MARKET STREET
PHILADELPHIA, PENNSYLVANIA 19102-2148

Overview: Provides cable television services. Established: 1961. Ownership: Public.

Key Statistics: Annual Sales: $4 billion. Number of Employees: 16,400.

Expertise/Education Sought: Electrical engineering, design engineering, mechanical engineering, telecommunications, accounting, finance, construction engineering, sales, marketing

Alternatives to Full-Time Employment: Part-time jobs, paid internships

Contact: Mr. Al Peddrick, Director of Human Resources.

Company URL: http://www.comcast.com

COMPONENT INTERTECHNOLOGIES

2426 PERRY HIGHWAY
HADLEY, PENNSYLVANIA 16130-2924

Overview: Produces electronic circuits. Established: 1995. Ownership: Private.

Key Statistics: Annual Sales: $21.2 million. Number of Employees: 200.

Expertise/Education Sought: Engineering, electrical technology

Contact: For individuals with previous experience: Mrs. Robin Vaughn, Human Resources Manager; Phone: 724-253-3161, ext. 101; Fax: 724-253-3853. E-mail: rvaughn@cit-hadley.com.

COMPUTER AID, INC.

1209 HAUSMAN ROAD
ALLENTOWN, PENNSYLVANIA 18104-9300

Overview: Provides custom computer programming services. Established: 1981. Ownership: Private.

Key Statistics: Annual Sales: $39.2 million. Number of Employees: 840.

Hiring History: Number of professional employees hired in 1997: 750. 1996: 320.

Expertise/Education Sought: Computer programming, systems analysis

Company-Paid Training is available.

International Assignments are available to employees.

Contact: Mr. Michael Aloupis, Technical Recruiter Manager; Phone: 610-530-5000. E-mail: mike_aloupis@compaid.com.

Company URL: http://www.compaid.com

COMPUTERPEOPLE, INC.

4400 CAMPBELLS RUN ROAD
PITTSBURGH, PENNSYLVANIA 15205

Overview: Computer integrated systems design. Established: 1968. Ownership: Private.

Key Statistics: Annual Sales: $70 million. Number of Employees: 600.

Hiring History: Number of professional employees hired in 1997: 200.

Expertise/Education Sought: Computer programming, computer analysis, systems integration

Alternatives to Full-Time Employment: Internships

International Assignments are available to employees.

Contact: For individuals with previous experience: Mr. Chuck Bezilla, Marketing Manager; Phone: 412-494-9800, ext. 105; Fax: 412-494-9890.

Company URL: http://www.compeople.com

CONTRAVES BRASHEAR SYSTEMS, LP

615 EPSILON DRIVE
PITTSBURGH, PENNSYLVANIA 15238-2807

Overview: Manufactures optical instruments. Established: 1997. Ownership: Private.

Key Statistics: Annual Sales: $18 million. Number of Employees: 125.

Hiring History: Number of professional employees hired in 1997: 18.

Expertise/Education Sought: Optical engineering, software engineering/development, mechanical engineering, electrical engineering

Contact: Ms. Kathy Moretti, Human Resources Manager; Phone: 412-967-7700; Fax: 412-967-7050. E-mail: kmoretti@contraves-brashear.com.

Company URL: http://www.contraves-brashear.com

CROWN CORK AND SEAL

ONE CROWN WAY
PHILADELPHIA, PENNSYLVANIA 19154

Overview: Supplier of packaging products. Established: 1892.

Expertise/Education Sought: Chemical engineering, mechanical engineering, electrical engineering, project engineering, process engineering

Alternatives to Full-Time Employment: Part-time jobs, paid internships

Company-Paid Training is available.

International Assignments are available to employees.

Contact: Ms. Kim Douthip, Corporate Recruiter; Phone: 215-698-5379; Fax: 215-676-7245. E-mail: kdouthip@crowncork.com.

Company URL: http://www.crowncork.com

DAY & ZIMMERMANN, INC.

1818 MARKET STREET
PHILADELPHIA, PENNSYLVANIA 19103

Overview: Provides engineering, construction, and architectural services. Established: 1976. Ownership: Private.

Key Statistics: Annual Sales: $811.6 million. Number of Employees: 14,000.

Expertise/Education Sought: Electrical engineering, mechanical engineering, industrial engineering, civil engineering, accounting, finance, marketing, data processing, administration, information technology

Alternatives to Full-Time Employment: Part-time jobs

Company-Paid Training is available.

International Assignments are available to employees.

Contact: Mr. Anthony Natale, Corporate Vice President of Human Resources; Phone: 215-299-1574; Fax: 215-299-8326.

Company URL: http://www.dayzim.com

DENTSPLY INTERNATIONAL, INC.

570 WEST COLLEGE AVENUE
YORK, PENNSYLVANIA 17404-3880

Overview: Manufactures dental and health-care equipment and supplies. Established: 1983. Ownership: Public.

Key Statistics: Annual Sales: $656.6 million. Number of Employees: 5,100.

Expertise/Education Sought: Accounting, sales, data processing, manufacturing engineering

Alternatives to Full-Time Employment: Paid internships

Company-Paid Training is available.

International Assignments are available to employees.

Contact: Mr. Glenn Weingarth, Vice President Human Resources; Phone: 717-845-7511; Fax: 717-843-5989.

Company URL: http://www.dentsply.com

EICHLEAY CORPORATION

6585 PENN AVENUE
PITTSBURGH, PENNSYLVANIA 15206-4483

Overview: Provides construction and engineering consulting services. Established: 1875. Ownership: Private.

Key Statistics: Annual Sales: $120 million. Number of Employees: 2,000.

Hiring History: Number of professional employees hired in 1997: 70. 1996: 45.

Expertise/Education Sought: Engineering, mechanical engineering, civil engineering, environmental engineering, electrical engineering

Alternatives to Full-Time Employment: Paid internships

Contact: For individuals with previous experience: Human Resources Director; Phone: 412-363-9000; Fax: 412-365-3404.

Company URL: http://www.eichleay.com

EMI COMPANY

603 WEST 12TH STREET
ERIE, PENNSYLVANIA 16501-1508

Overview: Produces iron products. Established: 1988. Parent Company: Hitachi America, Ltd. Ownership: Private.

Key Statistics: Annual Sales: $50 million. Number of Employees: 285.

Expertise/Education Sought: Chemical engineering, mechanical engineering, finance

Alternatives to Full-Time Employment: Paid internships

Company-Paid Training is available.

Contact: Ms. Janice Butler, Human Resources Coordinator; Phone: 814-452-6431, ext. 6815; Fax: 814-453-3090.

FRAMATOME CONNECTORS USA, INC.

50 GRUMBACHER ROAD
YORK, PENNSYLVANIA 17402

Overview: Manufactures electric connectors and terminals. Established: 1989. Parent Company: Framatome Connectors International. Ownership: Private.

Key Statistics: Annual Sales: $348.6 million. Number of Employees: 2,200.

Hiring History: Number of professional employees hired in 1997: 70. 1996: 60.

Expertise/Education Sought: Design engineering, product engineering, management, sales, marketing, manufacturing engineering

International Assignments are available to employees.

Contact: For individuals with previous experience: Human Resources; Phone: 717-767-8000; Fax: 717-767-8057.

Company URL: http://www.fciconnect.com

FULLER COMPANY

2040 AVENUE C
BETHLEHEM, PENNSYLVANIA 18017-2188

Overview: Markets industrial machinery and equipment. Established: 1989. Parent Company: F. L. Smidth Company. Ownership: Private.

Fuller Company (continued)

Key Statistics: Number of Employees: 1,015.

Hiring History: Number of professional employees hired in 1997: 90. 1996: 97.

Expertise/Education Sought: Sales, accounting, computer science, management information systems, mechanical engineering, chemical engineering, electrical engineering

Alternatives to Full-Time Employment: Paid internships

Company-Paid Training is available.

International Assignments are available to employees.

Contact: Mr. Joseph Mangan, Manager of Employee Relations; Phone: 610-264-6579; Fax: 610-264-6170. E-mail: Resumes@Fullerco.com.

Company URL: http://www.fullerco.com

GPU ENERGY

1001 BROAD STREET
JOHNSTOWN, PENNSYLVANIA 15906-2437

Overview: Provides electrical services. Established: 1996. Ownership: Public.

Key Statistics: Annual Sales: $201 million. Number of Employees: 1,756.

Expertise/Education Sought: Accounting, finance, engineering, marketing

Contact: Human Resources.

GREENE TWEED AND COMPANY, INC.

2075 DETWILER ROAD
KULPSVILLE, PENNSYLVANIA 19443

Overview: Manufactures rubber O-rings. Established: 1964. Ownership: Private.

Key Statistics: Annual Sales: $100 million. Number of Employees: 800.

Expertise/Education Sought: Computer programming, accounting, sales, engineering

Alternatives to Full-Time Employment: Part-time jobs, paid internships

Company-Paid Training is available.

International Assignments are available to employees.

Contact: For individuals with previous experience: Ms. Wendy Willis, Manager of Human Resources, PO Box 305, Kulpsville, Pennsylvania, 19443-0305; Phone: 215-256-9521, ext. 429; Fax: 215-256-0189.

Company URL: http://www.gtweed.com

HALE PRODUCTS, INC.

700 SPRING MILL AVENUE
CONSHOHOCKEN, PENNSYLVANIA 19428-1954

Overview: Manufactures fire extinguishers and rescue tools. Established: 1994. Parent Company: Idex Corporation. Ownership: Public.

Key Statistics: Annual Sales: $50 million. Number of Employees: 525.

Expertise/Education Sought: Mechanical engineering

Alternatives to Full-Time Employment: Paid internships

Company-Paid Training is available.

International Assignments are available to employees.

Contact: For individuals with previous experience: Mr. Bill Kysor, Vice President Human Resource; Phone: 610-825-6300.

HARBISON-WALKER REFRACTORIES

600 GRANT STREET
PITTSBURGH, PENNSYLVANIA 15219

Overview: Processes clay refractors. Established: 1995. Parent Company: Global Industrial Technologies, Inc. Ownership: Public.

Key Statistics: Annual Sales: $263 million. Number of Employees: 1,615.

Hiring History: Number of professional employees hired in 1997: 12. 1996: 10.

Expertise/Education Sought: Chemical engineering, ceramic engineering, materials science

Company-Paid Training is available.

International Assignments are available to employees.

Contact: Mr. Marshall Stegall, Administrator of Pensions and Insurance; Phone: 412-562-6329; Fax: 412-562-6331. E-mail: stegallm@hwr.com.

Company URL: http://www.hwr.com

HUNT CORPORATION

1 COMMERCE SQUARE, 2005 MARKET STREET
PHILADELPHIA, PENNSYLVANIA 19103

Overview: Manufactures art supplies, office products, and office furniture. Established: 1899. Ownership: Public.

Key Statistics: Annual Sales: $327.5 million. Number of Employees: 1,980.

Expertise/Education Sought: Sales, finance, marketing, accounting, computer programming

Company-Paid Training is available.

International Assignments are available to employees.

Contact: For individuals with previous experience: Ms. Fran Lillo, Human Resources Generalist, 1 Commerce Square, Philadelphia, Pennsylvania, 19103; Phone: 215-841-2414; Fax: 215-656-3705.

INDSPEC TECHNOLOGIES, LTD.

411 7TH AVENUE, SUITE 300
PITTSBURGH, PENNSYLVANIA 15219-1923

Overview: Produces resorcinol. Established: 1993. Ownership: Private.

Key Statistics: Annual Sales: $78 million. Number of Employees: 388.

Hiring History: Number of professional employees hired in 1996: 3.

Expertise/Education Sought: Chemical engineering

Contact: For individuals with previous experience: Industrial Relations Manager.

INFINITY, A SUNGARD COMPANY

510 WALNUT STREET, SUITE 900
PHILADELPHIA, PENNSYLVANIA 19106-3619

Overview: Develops computer software. Established: 1987. Parent Company: Sungard Data Systems, Inc. Ownership: Public.

Key Statistics: Annual Sales: $40.3 million. Number of Employees: 400.

Expertise/Education Sought: Accounting, finance, data processing, software engineering/development

Alternatives to Full-Time Employment: Paid internships

International Assignments are available to employees.

Contact: For individuals with previous experience: Ms. Barbara Foley, Human Resources Administrator; Phone: 215-627-3800; Fax: 215-627-7009. E-mail: bfoley@sungard.com.

INTERMETRO INDUSTRIES CORPORATION

651 NORTH WASHINGTON STREET
WILKES BARRE, PENNSYLVANIA 18705-1707

Overview: Manufactures modular storage and material handling systems. Established: 1995. Parent Company: Emerson Electric Company. Ownership: Private.

Hiring History: Number of professional employees hired in 1997: 20. 1996: 40.

Expertise/Education Sought: Business management, accounting, engineering, mechanical engineering, industrial engineering, information systems

Alternatives to Full-Time Employment: Paid internships

Company-Paid Training is available.

International Assignments are available to employees.

Contact: Ms. Mary Ellen Petcavage, Employment Supervisor; Phone: 717-825-2741; Fax: 717-824-7520.

JOHNSTOWN WIRE TECHNOLOGY, INC.
124 LAUREL AVENUE
JOHNSTOWN, PENNSYLVANIA 15906-2246

Overview: Manufactures steel products. Established: 1995. Ownership: Private.

Key Statistics: Annual Sales: $43.4 million. Number of Employees: 250.

Hiring History: Number of professional employees hired in 1997: 15. 1996: 25.

Expertise/Education Sought: Electrical engineering, mechanical engineering

Company-Paid Training is available.

Contact: For individuals with previous experience: Manager of Human Resources; Phone: 814-532-5614; Fax: 814-532-5616.

KENNAMETAL, INC.
PO BOX 231, ROUTE 981 SOUTH
LATROBE, PENNSYLVANIA 15650

Overview: Manufactures metal-cutting machines and tools. Established: 1997. Ownership: Public.

Key Statistics: Annual Sales: $316.2 million. Number of Employees: 835.

Expertise/Education Sought: Engineering

Company-Paid Training is available.

International Assignments are available to employees.

Contact: For individuals with previous experience: Human Resources; Phone: 724-539-5000, ext. 6827; Fax: 724-539-6657.

KEYSTONE MERCY HEALTH PLAN
200 STEVENS DRIVE, SUITE 350
PHILADELPHIA, PENNSYLVANIA 19113-1522

Overview: Provides hospital and medical services plans. Established: 1996. Parent Company: Independence Blue Cross & Mercy Health System. Ownership: Private.

Key Statistics: Annual Sales: $400 million. Number of Employees: 760.

Expertise/Education Sought: Health care, management information systems, finance, nursing

Alternatives to Full-Time Employment: Paid internships

Company-Paid Training is available.

Contact: Ms. Kyna Kirkland, Manager of Human Resources Services; Phone: 215-937-8500; Fax: 215-937-5363.

KEYSTONE THERMOMETRICS CORPORATION
967 WINDFALL ROAD
SAINT MARY'S, PENNSYLVANIA 15857-3333

Overview: Manufactures thermistors. Established: 1995. Parent Company: Bowthorpe. Ownership: Private.

Key Statistics: Annual Sales: $20 million. Number of Employees: 360.

Hiring History: Number of professional employees hired in 1997: 3.

Expertise/Education Sought: Industrial engineering, sales, engineering

Company-Paid Training is available.

Contact: For individuals with previous experience: Human Resources; Phone: 814-834-9140; Fax: 814-834-1335.

KLEARFOLD HOLDINGS, INC.
110 GIBRALTAR ROAD
HORSHAM, PENNSYLVANIA 19044-2302

Overview: Investment holding company. Established: 1996. Ownership: Private.

Key Statistics: Annual Sales: $53.9 million. Number of Employees: 960.

Expertise/Education Sought: Packaging engineering, business administration

Company-Paid Training is available.

Contact: Human Resources; Phone: 215-343-9300; Fax: 215-343-0491.

KLING LINDQUIST
2301 CHESTNUT STREET
PHILADELPHIA, PENNSYLVANIA 19103-3035

Overview: Provides architectural services. Established: 1969. Ownership: Private.

Key Statistics: Annual Sales: $36.4 million. Number of Employees: 320.

Hiring History: Number of professional employees hired in 1997: 50. 1996: 50.

Expertise/Education Sought: Engineering, architecture, interior design

Alternatives to Full-Time Employment: Paid internships

Company-Paid Training is available.

International Assignments are available to employees.

Contact: Mr. Steve Huston, Technical Recruiter; Phone: 215-569-5279; Fax: 215-569-5963.

Company URL: http://www.tklp.com

LEHIGH PORTLAND CEMENT COMPANY
7660 IMPERIAL WAY
ALLENTOWN, PENNSYLVANIA 18195-1040

Overview: Produces Portland cement. Established: 1977. Parent Company: Heidelberg Zement, Inc. Ownership: Private.

Key Statistics: Annual Sales: $500 million. Number of Employees: 1,800.

Expertise/Education Sought: Engineering, technology, information technology, accounting, law, human resources

Alternatives to Full-Time Employment: Paid internships

Company-Paid Training is available.

International Assignments are available to employees.

Contact: Mr. Jeffrey Altmose, Human Resources Coordinator; Phone: 610-366-4600; Fax: 610-366-4885. E-mail: jaltamose@lehighcement.com.

MAC PLASTICS, INC.
250 WEST WYLIE AVENUE
WASHINGTON, PENNSYLVANIA 15301-2275

Overview: Manufactures finished plastic products. Established: 1961. Parent Company: Mac Plastics, Inc. Ownership: Private.

Key Statistics: Annual Sales: $32.3 million. Number of Employees: 350.

Hiring History: Number of professional employees hired in 1997: 55. 1996: 50.

Expertise/Education Sought: Industrial engineering, plastics engineering, mechanical engineering

Alternatives to Full-Time Employment: Paid internships

Company-Paid Training is available.

Contact: For individuals with previous experience: Ms. Karen Hess, Director of Human Resources; Phone: 412-222-2600, ext. 5008; Fax: 412-222-4585.

MANNESMANN DEMAG CORPORATION
345 ROUSER ROAD
CORAOPOLIS, PENNSYLVANIA 15108-4726

Overview: Provides engineering services. Established: 1994. Parent Company: Mannesmann Corporation. Ownership: Private.

Key Statistics: Annual Sales: $220 million. Number of Employees: 301.

Hiring History: Number of professional employees hired in 1997: 25.

Expertise/Education Sought: Administration, engineering

Alternatives to Full-Time Employment: Paid internships

Company-Paid Training is available.

Contact: Human Resources Department; Phone: 412-269-4600; Fax: 412-269-5571.

MARSULEX ENVIRONMENTAL TECHOLOGIES
200 NORTH 7TH STREET
LEBANON, PENNSYLVANIA 17046-5006

Overview: Manufactures industrial blowers and fans. Established: 1997. Parent Company: Marsulex LLC. Ownership: Private.

Marsulex Environmental Techologies (continued)

KEY STATISTICS: Annual Sales: $50 million. Number of Employees: 100.

EXPERTISE/EDUCATION SOUGHT: Engineering

COMPANY-PAID TRAINING is available.

CONTACT: For individuals with previous experience: Human Resource Department; Phone: 717-274-7000; Fax: 717-274-7015.

MASCOTECH SINTERED COMPONENTS

BOOT JACK HILL, STEIS DRIVE
RIDGWAY, PENNSYLVANIA 15853

MAILING ADDRESS: PO Box N, Ridgway, PA 15853

OVERVIEW: Manufactures powdered metal products. Established: 1972. Parent Company: MascoTech, Inc. Ownership: Public.

KEY STATISTICS: Annual Sales: $75 million. Number of Employees: 650.

HIRING HISTORY: Number of professional employees hired in 1997: 6. 1996: 8.

EXPERTISE/EDUCATION SOUGHT: Metallurgical engineering, product engineering, electrical engineering, automation, quality assurance, manufacturing engineering, industrial engineering, materials engineering

ALTERNATIVES TO FULL-TIME EMPLOYMENT: Paid internships

INTERNATIONAL ASSIGNMENTS are available to employees.

CONTACT: Mr. Michael Aiello, Human Resources Manager; Phone: 814-776-1141, ext. 292; Fax: 814-776-2440.

COMPANY URL: http://www.mascotech.com

MEDRAD, INC.

1 MEDRAD DRIVE
INDIANOLA, PENNSYLVANIA 15051-0780

OVERVIEW: Manufactures medical equipment. Established: 1995. Ownership: Private.

KEY STATISTICS: Annual Sales: $120 million. Number of Employees: 760.

HIRING HISTORY: Number of professional employees hired in 1997: 110. 1996: 80.

EXPERTISE/EDUCATION SOUGHT: Electrical engineering, mechanical engineering, industrial engineering, software engineering/development, information systems

ALTERNATIVES TO FULL-TIME EMPLOYMENT: Paid internships

COMPANY-PAID TRAINING is available.

INTERNATIONAL ASSIGNMENTS are available to employees.

CONTACT: For individuals with previous experience: Mr. Martin Resick, Director of Human Resources; Phone: 412-767-2400, ext. 2215; Fax: 412-767-4295.

MILLENNIA TECHNOLOGY, INC.

1105 PITTSBURGH STREET
CHESWICK, PENNSYLVANIA 15024-1330

OVERVIEW: Manufactures electronic components and circuits. Established: 1997. Ownership: Private.

KEY STATISTICS: Annual Sales: $6.7 million. Number of Employees: 175.

EXPERTISE/EDUCATION SOUGHT: Electrical engineering, computer programming

CONTACT: For individuals with previous experience: Human Resources; Phone: 412-274-7741; Fax: 412-828-5803.

MINE SAFETY APPLIANCES COMPANY

121 GAMMA DRIVE
PITTSBURGH, PENNSYLVANIA 15238

OVERVIEW: Manufactures and distributes mine safety appliances. Established: 1914. Ownership: Public.

KEY STATISTICS: Annual Sales: $501 million. Number of Employees: 4,300.

HIRING HISTORY: Number of professional employees hired in 1997: 20. 1996: 20.

EXPERTISE/EDUCATION SOUGHT: Finance, accounting, manufacturing engineering, sales, marketing

INTERNATIONAL ASSIGNMENTS are available to employees.

CONTACT: For individuals with previous experience: Mr. Kenneth Krall, Human Resources Manager, PO Box 426, Pittsburgh, Pennsylvania, 15230-0426; Phone: 412-967-3000.

COMPANY URL: http://www.msanet.com

MINE TECHNIK AMERICA, INC.

2045 WEST PIKE STREET
HOUSTON, PENNSYLVANIA 15342-1010

OVERVIEW: Wholesales mining machinery. Established: 1995. Parent Company: DBT. Ownership: Private.

KEY STATISTICS: Number of Employees: 200.

HIRING HISTORY: Number of professional employees hired in 1996: 3.

EXPERTISE/EDUCATION SOUGHT: Engineering

CONTACT: For individuals with previous experience: Human Resources.

MOORE PRODUCTS COMPANY

SUMNEYTOWN PIKE
SPRING HOUSE, PENNSYLVANIA 19477

OVERVIEW: Manufactures industrial temperature controllers. Established: 1940. Ownership: Public.

KEY STATISTICS: Annual Sales: $142.9 million. Number of Employees: 1,222.

HIRING HISTORY: Number of professional employees hired in 1996: 124.

EXPERTISE/EDUCATION SOUGHT: Industrial engineering, mechanical engineering, manufacturing engineering, electrical engineering, software engineering/development

ALTERNATIVES TO FULL-TIME EMPLOYMENT: Part-time jobs, paid internships

COMPANY-PAID TRAINING is available.

INTERNATIONAL ASSIGNMENTS are available to employees.

CONTACT: Mr. Michael Moran, Manager of Personnel; Phone: 215-646-7400, ext. 2470; Fax: 215-283-6341.

COMPANY URL: http://www.mooreproducts.com

MOTOR UP CORPORATION

1530 CHESTNUT STREET, SUITE 4
PHILADELPHIA, PENNSYLVANIA 19102-2728

OVERVIEW: Wholesales new auto parts. Established: 1995. Ownership: Private.

KEY STATISTICS: Annual Sales: $42 million. Number of Employees: 26.

HIRING HISTORY: Number of professional employees hired in 1997: 5. 1996: 5.

EXPERTISE/EDUCATION SOUGHT: Finance, sales, advertising, manufacturing

ALTERNATIVES TO FULL-TIME EMPLOYMENT: Part-time jobs, paid internships

COMPANY-PAID TRAINING is available.

INTERNATIONAL ASSIGNMENTS are available to employees.

CONTACT: For individuals with previous experience: Mr. Peter Landau, Financial Officer; Phone: 215-568-5500; Fax: 215-568-5503.

COMPANY URL: http://www.motorup.com

NAVAL CENTER FOR ACQUISITION TRAINING CAREER MANAGEMENT SITE

BOX 2033
MECHANICSBURG, PENNSYLVANIA 17055-0784

OVERVIEW: Responsible for the recruitment, hiring, training, and education of full-time civilians for the Navy. Established: 1974. Division of the Department of the Navy.

KEY STATISTICS: Number of Employees: 200,000.

HIRING HISTORY: Number of professional employees hired in 1997: 200. 1996: 180.

EXPERTISE/EDUCATION SOUGHT: Business administration, computer science, electrical engineering, mechanical engineering, aeronautical engineering

COMPANY-PAID TRAINING is available.

CONTACT: Mr. Sterling Phipps, Education and Training Specialist; Phone: 717-790-3980; Fax: 717-790-1980. E-mail: acqintern_don@fmso.navy.mil.

COMPANY URL: http://www.webcom/graphix/intern

ORGANIZATION PROFILE

The **Department of the Navy (DON) Acquisition Intern Program** *is a three-year training program that offers candidates rapid advancement, mobility, travel opportunities, and both formal and on-the-job training. Entry is at the GS-7 level with a target grade of GS-11; full-performance without further competition: GS-12. Academic background, professional experience, and the skills acquired through this program equip the candidate to advance into responsible, high-level positions in the Navy's acquisition professional community. As a civilian employee, benefits such as generous vacation time, a retirement plan with tax-deferred savings options, health and life insurance, tuition assistance, and others are offered.*

Engineers *plan, organize, monitor, oversee, and/or perform engineering activities that relate to the design, development, fabrication, installation, modification, or analysis of systems, systems components, ships, and aircraft. Chemical, mechanical, electrical, electronic, aerospace, computer, and industrial engineers must be college graduates with a degree in professional engineering (not engineering technology) from a school of engineering with at least one curriculum accredited by the Accreditation Board for Engineering and Technology (ABET) as an engineering curriculum and have a minimum 2.95 GPA.*

All positions require U.S. citizenship, Selective Service registration (men born after 12/31/59), security clearance (some positions may require passage of drug and polygraph tests) and a signed mobility agreement.

To apply, candidates must submit an Optional Form 612, a resume or other form that lists the candidate's qualifications, and a copy of college transcripts for all undergraduate course work. Applications or requests for additional information should be directed to: Naval Center for Acquisition Training, 5450 Carlisle Pike, PO Box 2033, Mechanicsburg, Pennsylvania 17055-0784, 717-790-3980, Fax: 717-790-1980, E-mail: acqintern_don@fmso.navy.mil

An Equal Opportunity Employer

NEW HOLLAND NORTH AMERICA, INC.

500 DILLER AVENUE
NEW HOLLAND, PENNSYLVANIA 17557-0903

OVERVIEW: Manufactures farm equipment. Established: 1991. Parent Company: Fiat. Ownership: Private.

KEY STATISTICS: Annual Sales: $341.1 million. Number of Employees: 3,793.

HIRING HISTORY: Number of professional employees hired in 1996: 70.

EXPERTISE/EDUCATION SOUGHT: Mechanical engineering, industrial engineering, agricultural engineering

ALTERNATIVES TO FULL-TIME EMPLOYMENT: Paid internships

COMPANY-PAID TRAINING is available.

INTERNATIONAL ASSIGNMENTS are available to employees.

CONTACT: Mr. Ted Lyon, Corporate Recruiting Manager; Phone: 717-355-3729; Fax: 717-355-3650. E-mail: tlyon1@newholland.com.

COMPANY URL: http://www.newholland.com

NOVACARE, INC.

1016 WEST NINTH AVENUE
KING OF PRUSSIA, PENNSYLVANIA 19406

OVERVIEW: Provides health-care and rehabilitation services and is a leading professional employer organization providing a broad array of employee-related services. Established: 1985. Ownership: Public.

KEY STATISTICS: Annual Sales: $22.8 million. Number of Employees: 750.

HIRING HISTORY: Number of professional employees hired in 1997: 500. 1996: 500.

EXPERTISE/EDUCATION SOUGHT: Information systems

ALTERNATIVES TO FULL-TIME EMPLOYMENT: Part-time jobs, paid internships

COMPANY-PAID TRAINING is available.

CONTACT: Clinical and Non-clinical Operations; Phone: 800-331-8840, ext. 7146; Fax: 610-992-3339. Job Hotline: 800-MOV-EJOB.

COMPANY URL: http://www.novacare.com

NOVA CHEMICAL, INC.

PARK WEST II, 2000 CLIFFMINE ROAD
PITTSBURG, PENNSYLVANIA 15275

OVERVIEW: Manufactures plastics. Established: 1937. Parent Company: NOVA Corporation. Ownership: Private.

KEY STATISTICS: Annual Sales: $755.7 million. Number of Employees: 500.

HIRING HISTORY: Number of professional employees hired in 1996: 25.

EXPERTISE/EDUCATION SOUGHT: Finance, chemical engineering, marketing, accounting, sales

COMPANY-PAID TRAINING is available.

INTERNATIONAL ASSIGNMENTS are available to employees.

CONTACT: For individuals with previous experience: Human Resources; Phone: 412-490-4000; Fax: 412-494-4861.

NTN-BCA CORPORATION

401 WEST LINCOLN AVENUE
LITITZ, PENNSYLVANIA 17543-8701

OVERVIEW: Manufactures ball and roller bearings. Established: 1996. Parent Company: NTN. Ownership: Private.

KEY STATISTICS: Annual Sales: $67.5 million. Number of Employees: 900.

EXPERTISE/EDUCATION SOUGHT: Manufacturing, engineering, human resources, customer service and support

INTERNATIONAL ASSIGNMENTS are available to employees.

CONTACT: For individuals with previous experience: Human Resources Department; Phone: 717-627-8303; Fax: 717-627-8306.

PARSONS ENERGY & CHEMICALS GROUP, INC.

2675 MORGANTOWN ROAD
READING, PENNSYLVANIA 19607

OVERVIEW: Provides engineering consulting services. Established: 1984. Parent Company: Parsons Corporation. Ownership: Public.

KEY STATISTICS: Annual Sales: $190.1 million. Number of Employees: 1,520.

HIRING HISTORY: Number of professional employees hired in 1997: 275.

EXPERTISE/EDUCATION SOUGHT: Finance, law

ALTERNATIVES TO FULL-TIME EMPLOYMENT: Paid internships

INTERNATIONAL ASSIGNMENTS are available to employees.

CONTACT: Ms. Denise Scibek, Senior Human Resources Representative; Phone: 610-855-2560; Fax: 610-855-2186. E-mail: denise_i_scibek@parsons.com.

COMPANY URL: http://www.parsons.com

PENNSYLVANIA DEPARTMENT OF TRANSPORTATION

555 WALNUT STREET, 9TH FLOOR FORUM PLACE
HARRISBURG, PENNSYLVANIA 17101-1900

OVERVIEW: Provides planning, design, construction, and maintenance of Pennsylvania's multimodal transportation system. Established: 1903.

KEY STATISTICS: Number of Employees: 12,000.

EXPERTISE/EDUCATION SOUGHT: Civil engineering

ALTERNATIVES TO FULL-TIME EMPLOYMENT: Part-time jobs, internships

COMPANY-PAID TRAINING is available.

CONTACT: Ms. Diana K. Hershey, Personnel Analyst; Phone: 717-783-2680; Fax: 717-772-1781.

Pennsylvania Department of Transportation (continued)

CORPORATE PROFILE

Background and Operations *Pennsylvania, the crossroads of the Northeast, has traditionally played an important part in our nation's growth. To effectively move people and goods within a highly industrialized region, the Pennsylvania Department of Transportation (PennDOT) has developed an extensive transportation system. Within this system the department is responsible for over 40,000 miles of state roadways and 24,000 bridges. Civil engineers, at the heart of this development, have constantly sought ways to improve and strengthen the system.*

Today, with the largest federal assistance program ever and full matching funding from the Commonwealth, Pennsylvania can look forward to years of strong new growth in the transportation industry. PennDOT is moving ahead with record-breaking construction and maintenance programs that will serve its citizens well into the 21st century. The programs require new, innovative planning and engineering to help keep people and commerce moving across the state and throughout the nation. On the road, highway emphasis is shifting from construction to maintenance. Traffic congestion and fuel conservation have made necessary the revitalization of mass transit in urban areas. Civil engineers are working at the forefront of these changes, creating the need for entry-level graduates who are able to assume progressively more responsible positions. The department foresees a need for an increasing number of civil engineers through the 1990s.

PennDOT's highway administration function is organized into Central Office Engineering Bureaus and eleven statewide Engineering Districts. The Central Office performs many planning and administrative functions, determining the state's overall policies and programs. Each district is responsible for carrying out these policies. Because the districts are nearer to the work site or problem location, they, in turn, recommend the necessary actions to the Central Office. This reciprocal relationship allows major areas of civil engineering, as well as many of the Central Office functions, to be represented in each district.

Summer, Co-op, and Internship Programs *The department has cooperative education and internship programs, ranging from three to six months, available for interested students. Participants have the opportunity to work with professional engineers in major areas of civil engineering.*

Summer internships are available in all eleven district offices and in Harrisburg. These positions are primarily inspectors assigned to actual construction projects. Summer experiences give students practical engineering knowledge and help them obtain a higher score on the Civil Service entrance examination.

Application and Information *The department recruits at civil engineering schools throughout the Northeast. Interested seniors should consult their placement or career services office for more information or write to the following address: Diana K. Hershey, College Relations Coordinator, Pennsylvania Department of Transportation, 555 Walnut Street, 9th Floor, Forum Place, Harrisburg, Pennsylvania 17101-1900, Phone: 717-783-2680.*

PHB, INC.
7900 WEST RIDGE ROAD
FAIRVIEW, PENNSYLVANIA 16415-1899

OVERVIEW: Aluminum and zinc die casting. Established: 1985. Ownership: Private.

KEY STATISTICS: Annual Sales: $79.9 million. Number of Employees: 914.

HIRING HISTORY: Number of professional employees hired in 1997: 60. 1996: 40.

EXPERTISE/EDUCATION SOUGHT: Tool and die engineering, die cast engineering, plastics engineering

ALTERNATIVES TO FULL-TIME EMPLOYMENT: Internships

COMPANY-PAID TRAINING is available.

CONTACT: Mr. Benjamin Hancock, Human Resources Manager; Phone: 814-474-5511, ext. 228; Fax: 814-474-2063.

PNC BANK
1 PNC PLAZA
PITTSBURGH, PENNSYLVANIA 15222-2707

OVERVIEW: Commercial banking and financial services company. Established: 1983. Ownership: Public.

KEY STATISTICS: Assets: $6.3 billion. Number of Employees: 25,020.

HIRING HISTORY: Number of professional employees hired in 1996: 600.

EXPERTISE/EDUCATION SOUGHT: Accounting, auditing, finance

ALTERNATIVES TO FULL-TIME EMPLOYMENT: Part-time jobs, paid internships

COMPANY-PAID TRAINING is available.

CONTACT: For college students and recent graduates: Ms. Karen Hardt, Manager of College Recruiting; Phone: 412-762-3528; Fax: 412-762-2256. For individuals with previous experience: Ms. Antonia Milke, Manager of Employment; Phone: 412-762-3255; Fax: 412-762-2256. Job Hotline: 412-762-3022.

COMPANY URL: http://www.pncbank.com

PPG INDUSTRIES, INC.
1 PPG PLACE
PITTSBURGH, PENNSYLVANIA 15272

OVERVIEW: Produces chemicals, glass, coatings, and resins. Established: 1883. Ownership: Public.

KEY STATISTICS: Annual Sales: $7.2 billion. Number of Employees: 31,300.

HIRING HISTORY: Number of professional employees hired in 1997: 114.

EXPERTISE/EDUCATION SOUGHT: Management information systems, accounting, chemistry, computer science, human resources, marketing, chemical engineering, electrical engineering, mechanical engineering

ALTERNATIVES TO FULL-TIME EMPLOYMENT: Part-time jobs, paid internships

COMPANY-PAID TRAINING is available.

INTERNATIONAL ASSIGNMENTS are available to employees.

CONTACT: For college students and recent graduates: Ms. Karen Bergman, Manager of Staffing; Phone: 412-434-3408; Fax: 412-434-2011. For individuals with previous experience: Ms. Karen Bergman, Manager of Staffing; Phone: 412-434-3408; Fax: 412-434-2011.

COMPANY URL: http://www.ppg.com

PQ CORPORATION
PO BOX 840
VALLEY FORGE, PENNSYLVANIA 19482

OVERVIEW: Supplies industrial inorganic chemicals. Established: 1994. Ownership: Private.

KEY STATISTICS: Annual Sales: $47 million. Number of Employees: 45.

EXPERTISE/EDUCATION SOUGHT: Administration, engineering, chemical engineering, electrical engineering

ALTERNATIVES TO FULL-TIME EMPLOYMENT: Paid internships

CONTACT: Staffing Group, 1200 West Swedesford Road, Valley Forge, Pennsylvania, 19312; Phone: 610-651-4200; Fax: 610-651-4435.

RESPIRONICS, INC.
1001 MURRY RIDGE DRIVE
MURRYSVILLE, PENNSYLVANIA 15668-8517

OVERVIEW: Manufactures respiratory equipment. Established: 1976. Ownership: Public.

KEY STATISTICS: Annual Sales: $178.6 million. Number of Employees: 1,565.

HIRING HISTORY: Number of professional employees hired in 1996: 70.

EXPERTISE/EDUCATION SOUGHT: Mechanical engineering, electrical engineering, software engineering/development, information systems, administration

ALTERNATIVES TO FULL-TIME EMPLOYMENT: Part-time jobs, paid internships

COMPANY-PAID TRAINING is available.

INTERNATIONAL ASSIGNMENTS are available to employees.

CONTACT: For individuals with previous experience: Human Resources; Phone: 724-733-0200; Fax: 724-387-5299. Job Hotline: 724-733-0200.

COMPANY URL: http://www.respironics.com

ROHM AND HAAS COMPANY
100 INDEPENDENCE MALL WEST
PHILADELPHIA, PENNSYLVANIA 19106-2399

OVERVIEW: Manufactures industrial chemicals, polymers, resins, and Plexiglas. Established: 1909. Ownership: Public.

KEY STATISTICS: Annual Sales: $4 billion. Number of Employees: 11,633.
HIRING HISTORY: Number of professional employees hired in 1997: 100.
EXPERTISE/EDUCATION SOUGHT: Chemical engineering, electrical engineering, mechanical engineering, data processing, accounting, finance, marketing
ALTERNATIVES TO FULL-TIME EMPLOYMENT: Part-time jobs, paid internships
COMPANY-PAID TRAINING is available.
INTERNATIONAL ASSIGNMENTS are available to employees.
CONTACT: Human Relations Department; Phone: 215-592-3000; Fax: 215-592-3377.
COMPANY URL: http://www.rohmhaas.com

SAP AMERICA, INC.
3999 WEST CHESTER PIKE
NEWTOWN SQUARE, PENNSYLVANIA 19073

OVERVIEW: Provides client/server business applications. Established: 1972. Parent Company: SAP, AG. Ownership: Public.
KEY STATISTICS: Annual Sales: $3.4 billion. Number of Employees: 13,000.
HIRING HISTORY: Number of professional employees hired in 1997: 900.
EXPERTISE/EDUCATION SOUGHT: Information systems, management information systems, finance, logistics, computer science
ALTERNATIVES TO FULL-TIME EMPLOYMENT: Part-time jobs, paid internships
COMPANY-PAID TRAINING is available.
INTERNATIONAL ASSIGNMENTS are available to employees.
CONTACT: For college students and recent graduates: Mr. Kirk Wilcox, College Recruiter, 950 Winter Street, Suite 3800, Waltham, Massachusetts, 02154; Phone: 888-727-5547; Fax: 888-672-6726. E-mail: campus.us@sap-ag.de. For individuals with previous experience: Staffing Manager, 950 Winter Street, Suite 3800, Waltham, Massachusetts, 02154; Fax: 888-672-6726. E-mail: staffing.america@sap-ag.de.
COMPANY URL: http://www.sap.com/usa

CORPORATE PROFILE

Background and Operations *As the world's leading provider of enterprise-wide software solutions, SAP America offers world-class customer service and support that confidently adds value to the customer's technology investment. With a commanding 31% share of the worldwide client/server enterprise application software market, SAP also leads the software industry in committing nearly 20% of revenues to R&D spending. SAP has helped more than 8,000 companies in more than 90 countries receive a better return on their information.*

Join SAP America in any one of the following positions:

Business Consultants *Business consultants work with clients to help them implement SAP products and improve their business operations using the vast capabilities of SAP software. Candidates must have an M.B.A. and a minimum of 2 years of work experience in finance, logistics, production, or HR.*

Technical Consultants *These professionals work with clients to address the technical issues involved with implementing SAP software. This position requires an M.S./B.S. in computer science or a related field; a master's is preferred. A background that includes working with networks, databases, programming, or operating systems is required.*

Customer Support Consultants *Consultants in this position provide direct support to customers and consultants who use SAP software. These positions require a B.S. or B.A. and strong communication skills; they are based in SAP's Customer Support Centers in Austin, Texas, or Philadelphia, Pennsylvania.*

Instructors/Trainers *These opportunities are for individuals who have experience presenting technical subject matter and facilitating student learning. Candidates should have a background teaching advanced software application or computer operating systems.*

Company Locations *Headquartered in Philadelphia, Pennsylvania, SAP America has offices in most major cities across the country, including Atlanta, Austin, Boston, Chicago, Cincinnati, Cleveland, Dallas, Detroit, Houston, Los Angeles, Minneapolis, New York, Pittsburgh, St. Louis, San Francisco, Seattle, and Washington, D.C.*

SCT CORPORATION
4 COUNTRY VIEW ROAD
MALVERN, PENNSYLVANIA 19355-1408

OVERVIEW: Develops applications software. Established: 1968. Ownership: Public.
KEY STATISTICS: Annual Sales: $215.3 million. Number of Employees: 2,200.
HIRING HISTORY: Number of professional employees hired in 1997: 1,200. 1996: 300.
EXPERTISE/EDUCATION SOUGHT: Program analysis, computer programming, network analysis, systems engineering
ALTERNATIVES TO FULL-TIME EMPLOYMENT: Part-time jobs, internships
COMPANY-PAID TRAINING is available.
INTERNATIONAL ASSIGNMENTS are available to employees.
CONTACT: For individuals with previous experience: Mr. Rick Rabinowitz, Recruiting Systems Administrator; Phone: 610-578-5194; Fax: 610-578-5162.
COMPANY URL: http://www.sctcorp.com

SKF USA, INC.
1100 1ST AVENUE
KING OF PRUSSIA, PENNSYLVANIA 19406-1312

OVERVIEW: Manufactures ball bearings. Established: 1933. Parent Company: AB SKF. Ownership: Private.
KEY STATISTICS: Annual Sales: $1.1 billion. Number of Employees: 5,800.
EXPERTISE/EDUCATION SOUGHT: Mechanical engineering, accounting, marketing
ALTERNATIVES TO FULL-TIME EMPLOYMENT: Paid internships
INTERNATIONAL ASSIGNMENTS are available to employees.
CONTACT: Ms. Mary Ellen Guarino, Manager of Staffing and Placement; Phone: 610-962-4547; Fax: 610-337-2924.

SMITHKLINE BEECHAM
1 FRANKLIN PLAZA
PHILADELPHIA, PENNSYLVANIA 19102-7929

OVERVIEW: Manufactures pharmaceuticals and consumer health products. Established: 1989. Ownership: Private.
KEY STATISTICS: Annual Sales: $6.7 billion. Number of Employees: 21,600.
HIRING HISTORY: Number of professional employees hired in 1997: 1,300. 1996: 1,000.
EXPERTISE/EDUCATION SOUGHT: Marketing, information systems, science
ALTERNATIVES TO FULL-TIME EMPLOYMENT: Part-time jobs, paid internships
INTERNATIONAL ASSIGNMENTS are available to employees.
CONTACT: Ms. Caroline DeMarco, Head of University Relations; Phone: 215-751-4000. E-mail: caroline.demarco@sb.com.
COMPANY URL: http://www.sb.com

SMITHS INDUSTRIES AEROSPACE
435 DEVON PARK DRIVE, SUITE 102
WAYNE, PENNSYLVANIA 19087-1937

OVERVIEW: Manufactures search and navigation equipment. Established: 1958. Parent Company: Smiths Industries. Ownership: Private.
KEY STATISTICS: Annual Sales: $691 million. Number of Employees: 6,000.
EXPERTISE/EDUCATION SOUGHT: Mechanical engineering, electronics/electronics engineering, computer science, electrical engineering, aerospace engineering
ALTERNATIVES TO FULL-TIME EMPLOYMENT: Part-time jobs, paid internships
COMPANY-PAID TRAINING is available.
INTERNATIONAL ASSIGNMENTS are available to employees.
CONTACT: Mr. Arthur Dekhayser, Human Resources Manager; Phone: 610-995-2055; Fax: 610-975-9443.
COMPANY URL: http://www.smithsind-aerospace.co.uk

SMS
51 VALLEY STREAM PARKWAY
MALVERN, PENNSYLVANIA 19355-1406

Overview: Develops and distributes software to medical fields. Established: 1969. Ownership: Public.

Key Statistics: Annual Sales: $767.4 million. Number of Employees: 4,420.

Hiring History: Number of professional employees hired in 1996: 600.

Expertise/Education Sought: Accounting, finance, marketing, sales, software engineering/development, computer programming

Alternatives to Full-Time Employment: Part-time jobs

Company-Paid Training is available.

International Assignments are available to employees.

Contact: For individuals with previous experience: Human Resources; Phone: 610-219-6300, ext. 3359; Fax: 610-219-8266.

Company URL: http://www.smed.com

STANLEY STORAGE SYSTEMS
11 GRAMMES ROAD
ALLENTOWN, PENNSYLVANIA 18103-4760

Overview: Manufactures metal partitions, cabinets, and display fixtures. Established: 1966. Parent Company: The Stanley Works. Ownership: Public.

Key Statistics: Annual Sales: $99.3 million. Number of Employees: 486.

Hiring History: Number of professional employees hired in 1997: 8.

Expertise/Education Sought: Engineering, marketing

Alternatives to Full-Time Employment: Part-time jobs, paid internships

International Assignments are available to employees.

Contact: Mr. George Carras, Human Resources Manager; Phone: 800-523-9462; Fax: 610-776-3836.

Company URL: http://www.industry.net/stanley.storage.systems

SUPERIOR GROUP, INC.
3 RADNOR CORPORATE CENTER
RADNOR, PENNSYLVANIA 19087

Mailing Address: PO Box 6760, Wayne, PA 19087-8760

Overview: Manufactures steel tubing. Established: 1982. Ownership: Private.

Key Statistics: Annual Sales: $500 million. Number of Employees: 3,900.

Hiring History: Number of professional employees hired in 1997: 50.

Expertise/Education Sought: Mechanical engineering, accounting, marketing

International Assignments are available to employees.

Contact: Mr. Thomas Pezick, Manager of Human Resources; Phone: 610-964-2032; Fax: 610-964-2037.

SUSQUEHANNA PFALTZGRAFF COMPANY
140 EAST MARKET STREET
YORK, PENNSYLVANIA 17401-1219

Overview: Manufactures stoneware and bone china. Established: 1941. Parent Company: Blaznet. Ownership: Private.

Key Statistics: Annual Sales: $78.1 million. Number of Employees: 3,200.

Expertise/Education Sought: Ceramic engineering, accounting, marketing

Alternatives to Full-Time Employment: Paid internships

Company-Paid Training is available.

International Assignments are available to employees.

Contact: For individuals with previous experience: Ms. Karen Brosey, Manager of Employee Recruitment; Phone: 717-848-5500.

Company URL: http://www.Pfaltzgraff.com

TRACOR AEROSPACE ELECTRONIC SYSTEMS
305 RICHARDSON ROAD
LANSDALE, PENNSYLVANIA 19446

Overview: Supplies electronic countermeasures and electronic warfare systems. Established: 1935. Parent Company: Tracor, Inc. Ownership: Public.

Key Statistics: Annual Sales: $886.9 million. Number of Employees: 9,200.

Expertise/Education Sought: Electrical engineering, computer science, mechanical engineering, manufacturing engineering, quality engineering

Alternatives to Full-Time Employment: Part-time jobs, paid internships

Contact: For college students and recent graduates: Ms. Marci Cavanaugh, Human Resources Representative; Phone: 215-996-2826; Fax: 215-996-2088. E-mail: jobs@aes.tracor.com. For individuals with previous experience: Ms. Karen Jones, Human Resources Representative; Phone: 215-996-2499; Fax: 215-996-2088. E-mail: jobs@aes.tracor.com.

Company URL: http://www.ael.com

UGI CORPORATION
460 NORTH GULPH ROAD
KING OF PRUSSIA, PENNSYLVANIA 19406-2815

Overview: Distributes natural gas, electricity, and propane. Established: 1991. Ownership: Public.

Key Statistics: Annual Sales: $1.6 billion. Number of Employees: 6,371.

Hiring History: Number of professional employees hired in 1997: 5. 1996: 10.

Expertise/Education Sought: Finance, law, information systems, computer programming, accounting, human resources

Alternatives to Full-Time Employment: Part-time jobs, paid internships

Company-Paid Training is available.

International Assignments are available to employees.

Contact: Ms. Lillian Early, Personnel Manager; Phone: 610-337-1000, ext. 3350; Fax: 610-992-3215.

Company URL: http://www.shareholder.com/ugi

UNIDYNE CORPORATION
118 PICKERING WAY, SUITE 104
EXTON, PENNSYLVANIA 19341-1325

Overview: Investment holding company. Established: 1996. Ownership: Private.

Key Statistics: Annual Sales: $16.6 million. Number of Employees: 136.

Expertise/Education Sought: Engineering, computer programming, automotive engineering, electronics/electronics engineering

Company-Paid Training is available.

International Assignments are available to employees.

Contact: For individuals with previous experience: Supervisor; Phone: 610-363-0314; Fax: 610-524-8715.

UNION PACIFIC CORPORATION
8TH AND EATON AVENUES
BETHLEHEM, PENNSYLVANIA 18018

Overview: Provides railroad and truck freight services. Established: 1969. Ownership: Public.

Key Statistics: Annual Sales: $8.8 billion. Number of Employees: 65,000.

Expertise/Education Sought: Finance, accounting, management information systems, computer programming, data processing

Company-Paid Training is available.

Contact: For individuals with previous experience: Human Resources; Phone: 610-861-3276; Fax: 610-861-3156.

UNISOURCE WORLDWIDE, INC.
PO BOX 3000
BERWYN, PENNSYLVANIA 19312

Mailing Address: 1100 Cassatt Road, Berwyn, PA 19312

Overview: Distributes paper and paper products. Established: 1968. Ownership: Public.

Key Statistics: Annual Sales: $7 billion. Number of Employees: 12,000.

Expertise/Education Sought: Accounting, finance, business management, computer programming, sales, marketing

Alternatives to Full-Time Employment: Paid internships

Company-Paid Training is available.

International Assignments are available to employees.

Contact: Corporate Manager Human Resources; Phone: 610-296-4470; Fax: 610-722-3400.

Company URL: http://www.unisource.com

UNITED REFINING COMPANY

15 BRADLEY STREET
WARREN, PENNSYLVANIA 16365-3224

Mailing Address: PO Box 780, Warren, PA 16365-0780

Overview: Manufactures and markets gasoline products. Established: 1986. Ownership: Private.

Key Statistics: Annual Sales: $800 million. Number of Employees: 2,800.

Expertise/Education Sought: Chemical engineering, project engineering

Company-Paid Training is available.

Contact: For individuals with previous experience: Mr. Lawrence Loughlin, Vice President of Human Resources; Phone: 814-723-1500, ext. 4670; Fax: 814-726-4602.

U.S. STEEL

600 GRANT STREET, SUITE 1959
PITTSBURGH, PENNSYLVANIA 15219-2702

Overview: Provides raw materials and manufactures steel and steel products. Established: 1901. Parent Company: USX Corporation. Ownership: Public.

Key Statistics: Annual Sales: $23.8 billion. Number of Employees: 42,195.

Hiring History: Number of professional employees hired in 1997: 100. 1996: 100.

Expertise/Education Sought: Metallurgical engineering, electrical engineering, mechanical engineering, marketing, accounting, materials science

Alternatives to Full-Time Employment: Paid internships

Company-Paid Training is available.

International Assignments are available to employees.

Contact: Mr. Richard Schinagl, Manager of Records and Corporate Affairs; Phone: 412-433-6688; Fax: 412-433-3561.

VISHAY INTERTECHNOLOGY, INC.

63 LINCOLN HIGHWAY
MALVERN, PENNSYLVANIA 19355

Overview: Manufactures passive electronic components. Established: 1962. Ownership: Public.

Key Statistics: Annual Sales: $1.1 billion. Number of Employees: 14,900.

Expertise/Education Sought: Electronics/electronics engineering, marketing, production management

Contact: For individuals with previous experience: Mr. William Spires, Vice President of Human Resources; Phone: 610-644-1300; Fax: 610-296-0657.

THE WEST COMPANY, INC.

101 GORDON DRIVE
LIONVILLE, PENNSYLVANIA 19341

Overview: Manufactures fabricated rubber, plastic, and glass products. Established: 1923. Ownership: Public.

Key Statistics: Annual Sales: $458.8 million. Number of Employees: 5,040.

Expertise/Education Sought: Plastics engineering, manufacturing engineering, industrial engineering, materials science

Company-Paid Training is available.

International Assignments are available to employees.

Contact: Ms. Deanne Price, Supervisor of Employment, North America, PO Box 645, Lionville, Pennsylvania, 19431-0645; Phone: 610-594-2900; Fax: 610-594-3011.

Company URL: http://www.thewestcompany.com

WILMAC CORPORATION

209 NORTH BEAVER STREET
YORK, PENNSYLVANIA 17403-5321

Mailing Address: PO Box 5047, York, PA 17405

Overview: Operate apartment buildings. Established: 1995. Ownership: Private.

Key Statistics: Number of Employees: 160.

Hiring History: Number of professional employees hired in 1997: 200. 1996: 300.

Expertise/Education Sought: Health care, real state management, pharmacy, therapy, physical therapy, occupational therapy, psychiatry, management

Alternatives to Full-Time Employment: Paid internships

Company-Paid Training is available.

Contact: For individuals with previous experience: Ms. Pat Reed, Corporate Recruiter; Phone: 717-854-7857; Fax: 717-846-2594.

XLCONNECT SOLUTIONS, INC.

411 EAGLEVIEW BOULEVARD
EXTON, PENNSYLVANIA 19341-1117

Overview: Provides computer related consulting services. Established: 1996. Parent Company: Intelligent Electronics, Inc. Ownership: Public.

Key Statistics: Annual Sales: $114.9 million. Number of Employees: 1,150.

Expertise/Education Sought: Accounting, information systems, engineering, management

Alternatives to Full-Time Employment: Paid internships

Company-Paid Training is available.

Contact: Human Resources; Phone: 610-458-5500.

YORK INTERNATIONAL CORPORATION

631 SOUTH RICHLAND AVENUE
YORK, PENNSYLVANIA 17403

Overview: Manufactures heating, refrigeration, and air conditioning equipment. Established: 1874. Ownership: Public.

Key Statistics: Annual Sales: $3.2 billion. Number of Employees: 20,100.

Hiring History: Number of professional employees hired in 1997: 87. 1996: 65.

Expertise/Education Sought: Mechanical engineering, electronics/electronics engineering, industrial engineering, manufacturing engineering, computer programming

Alternatives to Full-Time Employment: Part-time jobs, paid internships

Company-Paid Training is available.

International Assignments are available to employees.

Contact: Mr. John Tynan, Manager of Corporate Human Resources; Phone: 717-771-6580; Fax: 717-771-6567.

Company URL: http://www.york.com

YUASA-EXIDE, INC.

2366 BERNVILLE ROAD
READING, PENNSYLVANIA 19605

Mailing Address: PO Box 14145, Reading, PA 19612

Overview: Manufactures batteries. Established: 1965. Parent Company: Yuasa Corporation. Ownership: Private.

Key Statistics: Annual Sales: $240 million. Number of Employees: 2,400.

Expertise/Education Sought: Accounting, finance, data processing, chemical engineering, electrical engineering

Yuasa-Exide, Inc. (continued)

Alternatives to Full-Time Employment: Internships
Company-Paid Training is available.
International Assignments are available to employees.
Contact: For individuals with previous experience: Mr. Bill Litz, Director of Human Resources, PO Box 14415, Reading, Pennsylvania, 19612-4145; Phone: 610-208-1860; Fax: 610-372-8457.

RHODE ISLAND

ACS INDUSTRIES, INC.
160 HAMLET AVENUE
WOONSOCKET, RHODE ISLAND 02895-4835

Overview: Manufactures stainless steel wire, telephone cords, exhaust seal gaskets, polypropylene yarn, specialized filters. Established: 1989. Ownership: Private.
Key Statistics: Annual Sales: $157.1 million. Number of Employees: 2,200.
Expertise/Education Sought: Mechanical engineering, design engineering, metallurgical engineering, accounting, management information systems
International Assignments are available to employees.
Contact: For individuals with previous experience: Ms. Michelle Timmons, Manager of Personnel; Phone: 401-769-4700.

AMERICAN POWER CONVERSION
132 FAIRGROUNDS ROAD
WEST KINGSTON, RHODE ISLAND 02892-1511

Overview: Manufactures electronic components for personal computers, engineering work stations, and communications equipment. Established: 1981. Ownership: Public.
Key Statistics: Annual Sales: $706.9 million. Number of Employees: 2,650.
Expertise/Education Sought: Manufacturing, operations, sales, marketing, administration, management information systems, research and development, production, technical
Alternatives to Full-Time Employment: Part-time jobs, paid internships
International Assignments are available to employees.
Contact: Ms. Lisa DeFiuscio, Human Resources Manager; Phone: 401-789-5735, ext. 2444; Fax: 401-788-2710. Job Hotline: 401-789-5735, ext. 2940.
Company URL: http://www.apcc.com

AMTROL, INC.
1400 DIVISION ROAD
WEST WARWICK, RHODE ISLAND 02893-2300

Overview: Manufactures flow and expansion control technology. Established: 1996. Parent Company: The Cyprus Group. Ownership: Private.
Key Statistics: Annual Sales: $170.8 million. Number of Employees: 915.
Hiring History: Number of professional employees hired in 1997: 170. 1996: 15.
Expertise/Education Sought: Human resources, manufacturing engineering, accounting, manufacturing, computer programming, analysis
Alternatives to Full-Time Employment: Part-time jobs, paid internships
Company-Paid Training is available.
International Assignments are available to employees.
Contact: Mr. Michael Montigny, Director of Human Resources; Phone: 401-884-6300, ext. 233; Fax: 401-884-7816.
Company URL: http://www.amtrol.com

AQUIDNECK MANAGEMENT ASSOCIATES, LTD.
28 JACOME WAY
MIDDLETOWN, RHODE ISLAND 02842-5679

Overview: Provides computer software development services. Established: 1981. Ownership: Private.
Key Statistics: Annual Sales: $18 million. Number of Employees: 300.
Expertise/Education Sought: Computer programming, management, architecture, software engineering/development
Company-Paid Training is available.
Contact: For individuals with previous experience: Mr. Jeffrey Belden, Director of Administration; Phone: 401-849-8900; Fax: 401-848-0638.
Company URL: http://www.amaltd.com

DIMEO CONSTRUCTION COMPANY
75 CHAPMAN STREET
PROVIDENCE, RHODE ISLAND 02905-5405

Overview: Builds office buildings, hospitals, and schools. Established: 1932. Ownership: Private.
Key Statistics: Annual Sales: $64.5 million. Number of Employees: 170.
Hiring History: Number of professional employees hired in 1997: 20. 1996: 20.
Expertise/Education Sought: Construction
Alternatives to Full-Time Employment: Paid internships
Company-Paid Training is available.
Contact: For individuals with previous experience: Ms. Helena Langlais, Human Relations Department; Phone: 401-781-9800; Fax: 401-461-4580.

EG&G SEALOL, INC.
15 PIONEER AVENUE
WARWICK, RHODE ISLAND 02888

Overview: Manufactures gaskets, packing, and sealing devices. Established: 1968. Parent Company: EG&G, Inc. Ownership: Public.
Key Statistics: Annual Sales: $48.5 million. Number of Employees: 630.
Hiring History: Number of professional employees hired in 1997: 84. 1996: 84.
Expertise/Education Sought: Engineering, machinery, manufacturing, management information systems, accounting
Alternatives to Full-Time Employment: Paid internships
Company-Paid Training is available.
International Assignments are available to employees.
Contact: Ms. Christine Long, Senior Human Resource Generalist; Phone: 401-781-4700; Fax: 401-781-0930. E-mail: cdl@egginc.com.
Company URL: http://www.egginc.com

ELIZABETH WEBBING MILLS COMPANY
521 ROOSEVELT AVENUE
CENTRAL FALLS, RHODE ISLAND 02863-3138

Mailing Address: PO Box 1168, Pawtucket, RI 02862
Overview: Develops and manufactures narrow webbing fabric tapes. Established: 1959. Ownership: Private.
Key Statistics: Annual Sales: $104 million. Number of Employees: 600.
Hiring History: Number of professional employees hired in 1997: 50.
Expertise/Education Sought: Accounting, management, engineering, sales, purchasing, quality control
Contact: For individuals with previous experience: Human Relations Department; Phone: 401-723-0500; Fax: 401-724-5430.

ELMWOOD SENSORS, INC.
500 NARRAGANSETT PARK DRIVE
PAWTUCKET, RHODE ISLAND 02861-4325

OVERVIEW: Manufactures commercial and precision thermostats, thermal fuses, and temperature sensors. Established: 1981. Parent Company: Fasco Control. Ownership: Private.

KEY STATISTICS: Annual Sales: $47.9 million. Number of Employees: 650.

EXPERTISE/EDUCATION SOUGHT: Design engineering

CONTACT: For individuals with previous experience: Ms. Martha Norris, Human Resource Manager; Phone: 401-727-1300, ext. 304; Fax: 401-728-8818.

G-TECH CORPORATION
55 TECHNOLOGY WAY
WEST GREENWICH, RHODE ISLAND 02817-1711

OVERVIEW: Supplies and operates on line lottery systems and government benefits delivery systems. Established: 1990. Ownership: Public.

KEY STATISTICS: Annual Sales: $904.3 million. Number of Employees: 5,100.

EXPERTISE/EDUCATION SOUGHT: Software engineering/development, electrical engineering, hardware engineering, marketing, finance, administration, information systems

ALTERNATIVES TO FULL-TIME EMPLOYMENT: Part-time jobs, paid internships

INTERNATIONAL ASSIGNMENTS are available to employees.

CONTACT: Human Resources Department; Phone: 401-392-1000; Fax: 401-392-4950.

COMPANY URL: http://www.gtech.com

HASBRO, INC.
1027 NEWPORT AVENUE
PAWTUCKET, RHODE ISLAND 02862

OVERVIEW: Manufactures toys. Established: 1912. Ownership: Public.

KEY STATISTICS: Annual Sales: $3 billion. Number of Employees: 13,000.

HIRING HISTORY: Number of professional employees hired in 1997: 200. 1996: 200.

EXPERTISE/EDUCATION SOUGHT: Marketing, finance, accounting

ALTERNATIVES TO FULL-TIME EMPLOYMENT: Paid internships

COMPANY-PAID TRAINING is available.

INTERNATIONAL ASSIGNMENTS are available to employees.

CONTACT: Human Relations Department. Job Hotline: 401-727-5773.

KENNEY MANUFACTURING COMPANY
1000 JEFFERSON BOULEVARD
WARWICK, RHODE ISLAND 02886-2201

OVERVIEW: Manufactures drapery hardware and window blinds. Established: 1914. Ownership: Private.

KEY STATISTICS: Annual Sales: $100 million. Number of Employees: 550.

EXPERTISE/EDUCATION SOUGHT: Manufacturing

ALTERNATIVES TO FULL-TIME EMPLOYMENT: Internships

INTERNATIONAL ASSIGNMENTS are available to employees.

CONTACT: For individuals with previous experience: Personnel; Phone: 401-739-2200.

NEPTCO, INC.
30 HAMLET STREET
PAWTUCKET, RHODE ISLAND 02861-2827

OVERVIEW: Manufactures laminated plastics sheet. Established: 1988. Parent Company: Cookson America, Inc. Ownership: Private.

KEY STATISTICS: Annual Sales: $29.9 million. Number of Employees: 500.

HIRING HISTORY: Number of professional employees hired in 1996: 15.

EXPERTISE/EDUCATION SOUGHT: Manufacturing engineering, administration, finance

CONTACT: For individuals with previous experience: Human Resources Department; Phone: 401-722-5500, ext. 121; Fax: 401-722-6378.

AJ OSTER COMPANY
300 CENTERVILLE ROAD, 3RD FLOOR
WARWICK, RHODE ISLAND 02886-0211

OVERVIEW: Processors and distributors of metal foils and films. Established: 1991. Parent Company: Olin Corporation. Ownership: Public.

KEY STATISTICS: Annual Sales: $250 million. Number of Employees: 360.

EXPERTISE/EDUCATION SOUGHT: Manufacturing, marketing

INTERNATIONAL ASSIGNMENTS are available to employees.

CONTACT: For individuals with previous experience: Personnel; Phone: 401-736-2600.

T. SARDELLI & SONS, INC.
195 DUPONT DRIVE
PROVIDENCE, RHODE ISLAND 02907-3105

OVERVIEW: Manufactures precious metal jewelry. Established: 1949. Ownership: Private.

KEY STATISTICS: Annual Sales: $62.9 million. Number of Employees: 350.

HIRING HISTORY: Number of professional employees hired in 1997: 300.

EXPERTISE/EDUCATION SOUGHT: Manufacturing engineering, manufacturing management, manufacturing supervisors, distribution, sales

COMPANY-PAID TRAINING is available.

CONTACT: For individuals with previous experience: Ms. Sandy Morse, Employment Supervisor; Phone: 401-944-8510; Fax: 401-429-2170.

SIEBE NORTH, INC.
2000 PLAINFIELD PIKE
CRANSTON, RHODE ISLAND 02921

OVERVIEW: Manufactures utility and industrial gloves. Established: 1983. Parent Company: Siebe Gorman, Limited. Ownership: Private.

KEY STATISTICS: Annual Sales: $91.6 million. Number of Employees: 1,200.

HIRING HISTORY: Number of professional employees hired in 1997: 25.

EXPERTISE/EDUCATION SOUGHT: Research and development

INTERNATIONAL ASSIGNMENTS are available to employees.

CONTACT: Mrs. Pauline Nourse, Human Resources Director; Phone: 401-943-4400; Fax: 401-943-7974.

COMPANY URL: http://www.northsafety.com

STANLEY FASTENING SYSTEMS, LP
ROUTE 2
EAST GREENWHICH, RHODE ISLAND 02818

OVERVIEW: Manufactures office machines. Established: 1997. Parent Company: The Stanley Works In New Britain Connecticut. Ownership: Private.

KEY STATISTICS: Annual Sales: $102.1 million. Number of Employees: 1,100.

EXPERTISE/EDUCATION SOUGHT: Engineering

ALTERNATIVES TO FULL-TIME EMPLOYMENT: Paid internships

COMPANY-PAID TRAINING is available.

INTERNATIONAL ASSIGNMENTS are available to employees.

CONTACT: For individuals with previous experience: Human Resources; Phone: 401-884-2500; Fax: 401-884-3350.

TECHNIC, INC.
1 SPECTACLE STREET
CRANSTON, RHODE ISLAND 02910-1032

OVERVIEW: Produces metal treating compounds. Established: 1965. Ownership: Private.

Technic, Inc. (continued)

Key Statistics: Annual Sales: $120 million. Number of Employees: 295.
Hiring History: Number of professional employees hired in 1996: 50.
Expertise/Education Sought: Manufacturing engineering
Contact: For individuals with previous experience: Mr. Wayne Ganim, Director of Operations; Phone: 401-781-6100.

TEKNOR APEX COMPANY
505 CENTRAL AVENUE
PAWTUCKET, RHODE ISLAND 02861

Overview: Manufactures garden hoses and rubber. Established: 1968. Ownership: Private.
Key Statistics: Annual Sales: $300 million. Number of Employees: 2,000.
Hiring History: Number of professional employees hired in 1997: 30. 1996: 30.
Expertise/Education Sought: Chemical engineering, mechanical engineering, plastics engineering
Contact: For individuals with previous experience: Mr. Jack Yando, Director of Human Resources; Phone: 401-725-8000; Fax: 401-725-8095.
Company URL: http://www.teknorapex.com

TEXTRON, INC.
40 WESTMINSTER STREET
PROVIDENCE, RHODE ISLAND 02903

Overview: Manufactures aircraft systems and components, automotive parts, industrial products; provides financial services. Established: 1928. Ownership: Public.
Key Statistics: Annual Sales: $9.3 billion. Number of Employees: 57,000.
Hiring History: Number of professional employees hired in 1997: 65. 1996: 62.
Expertise/Education Sought: Business management, law, communications, finance, risk management, accounting
Alternatives to Full-Time Employment: Part-time jobs, paid internships
Company-Paid Training is available.
International Assignments are available to employees.
Contact: For individuals with previous experience: Mr. John Hambrick, Director of Human Resources; Phone: 401-421-2800.
Company URL: http://www.textron.com

TORAY PLASTICS AMERICA, INC.
50 BELVER AVENUE
NORTH KINGSTOWN, RHODE ISLAND 02852-7520

Overview: Manufactures plastics. Established: 1985. Parent Company: Toray Industries. Ownership: Private.
Key Statistics: Annual Sales: $249.5 million. Number of Employees: 600.
Hiring History: Number of professional employees hired in 1997: 43.
Expertise/Education Sought: Mechanical engineering, electrical engineering, chemical engineering
Company-Paid Training is available.
Contact: For individuals with previous experience: Mr. Harvey Greenhalgh, Human Resource Director; Phone: 401-294-4511; Fax: 401-294-2154.

SOUTH CAROLINA

AVX CORPORATION
801 17TH AVENUE SOUTH
MYRTLE BEACH, SOUTH CAROLINA 29577-4245

Overview: Manufactures electronic capacitors. Established: 1990. Ownership: Private.
Key Statistics: Annual Sales: $1.1 billion. Number of Employees: 13,000.
Expertise/Education Sought: Finance, marketing, data processing, engineering
Alternatives to Full-Time Employment: Paid internships
Contact: Mr. Ray Spitz, Director of Human Resources; Phone: 803-946-0231.

BLUE CROSS AND BLUE SHIELD OF SOUTH CAROLINA
I-20 EAST AT ALPINE ROAD
COLUMBIA, SOUTH CAROLINA 29219

Overview: Provides health insurance. Established: 1946. Ownership: Private.
Key Statistics: Annual Sales: $381.7 million. Number of Employees: 6,500.
Hiring History: Number of professional employees hired in 1997: 150. 1996: 300.
Expertise/Education Sought: Technical, health care, finance, program analysis, technology, systems analysis, telecommunications
Company-Paid Training is available.
Contact: Ms. Carolyn Clemens, Information Science Recruiter; Phone: 800-288-2227, ext. 44369; Fax: 803-419-8096. E-mail: bcbsis@mindspring.com.

BMW MANUFACTURING CORPORATION
1400 HIGHWAY 101 SOUTH
GREER, SOUTH CAROLINA 29651-8606

Mailing Address: PO Box 11000, Spartanburg, SC 29304-4100
Overview: Manufactures automobiles. Established: 1992. Parent Company: BMW, AG. Ownership: Private.
Key Statistics: Annual Sales: $236.7 million. Number of Employees: 1,700.
Expertise/Education Sought: Manufacturing engineering
Alternatives to Full-Time Employment: Internships
Contact: For individuals with previous experience: Human Resources; Phone: 864-968-6000.

BOWATER, INC.
55 EAST CAMPERDOWN WAY
GREENVILLE, SOUTH CAROLINA 29601-3511

Overview: Manufactures newsprint, coated paper, and computer forms. Established: 1964. Ownership: Public.
Key Statistics: Annual Sales: $1.7 billion. Number of Employees: 5,025.
Hiring History: Number of professional employees hired in 1996: 10.
Expertise/Education Sought: Finance, auditing, law, business management, human resources, sales, accounting, marketing, mechanical engineering
Alternatives to Full-Time Employment: Paid internships
Company-Paid Training is available.
International Assignments are available to employees.
Contact: Mr. Aaron Whitlock, Director, Compensation and Management; Phone: 864-271-7733; Fax: 864-282-9591. E-mail: whitlockad@bowater.com.
Company URL: http://www.bowater.com

CHICAGO PNEUMATIC TOOL COMPANY
1800 OVERVIEW DRIVE
ROCK HILL, SOUTH CAROLINA 29730-7463

Overview: Manufactures power-driven handtools. Established: 1987. Parent Company: Atlas Cop Company. Ownership: Private.

Key Statistics: Annual Sales: $95.3 million. Number of Employees: 2,200.

Expertise/Education Sought: Industrial engineering, sales, accounting

International Assignments are available to employees.

Contact: Human Resources; Phone: 803-817-7000.

CLARK-SCHWEBEL, INC.
2200 SOUTH MURRAY AVENUE
ANDERSON, SOUTH CAROLINA 29624-3139

Mailing Address: PO Box 2627, Anderson, SC 29622

Overview: Manufactures woven and non-woven industrial fabrics from fiberglass and other fibers. Established: 1996. Ownership: Private.

Key Statistics: Annual Sales: $108 million. Number of Employees: 1,400.

Expertise/Education Sought: Manufacturing

Alternatives to Full-Time Employment: Internships

International Assignments are available to employees.

Contact: Mr. David Barr, Marketing Manager; Phone: 864-224-3506, ext. 5174; Fax: 864-260-3377. Job Hotline: 864-225-7028.

DIXIE-NARCO, INC.
DIXIE-NARCO BOULEVARD
WILLISTON, SOUTH CAROLINA 29853

Overview: Manufactures vending machines. Established: 1988. Parent Company: Maytag Corporation. Ownership: Public.

Key Statistics: Annual Sales: $92 million. Number of Employees: 920.

Expertise/Education Sought: Mechanical engineering, information services, accounting, design engineering

International Assignments are available to employees.

Contact: Mr. Chris James, Manager of Human Resources Administration; Phone: 803-266-5000.

DUNLOP MAXFLI SPORTS CORPORATION
100 MAXFLI DRIVE
WESTMINSTER, SOUTH CAROLINA 29693

Overview: Manufactures golf equipment. Established: 1996. Parent Company: Dunlop Slazenger Group, Ltd. Ownership: Private.

Key Statistics: Annual Sales: $100 million. Number of Employees: 600.

Hiring History: Number of professional employees hired in 1996: 20.

Expertise/Education Sought: Marketing, finance, mechanical engineering, chemical engineering, electrical engineering, process engineering

Contact: Mr. Jim Bowers, Director of Human Resources; Phone: 864-647-4000; Fax: 864-647-4142.

FUJI PHOTO FILM, INC.
211 PUCKETT FERRY ROAD
GREENWOOD, SOUTH CAROLINA 29649

Mailing Address: PO Box 1306, Greenwood, SC 29648

Overview: Manufactures film and printing equipment. Established: 1989. Parent Company: Fuji Photo Film Company, Ltd. Ownership: Private.

Key Statistics: Annual Sales: $300 million. Number of Employees: 1,100.

Hiring History: Number of professional employees hired in 1997: 300. 1996: 300.

Expertise/Education Sought: Mechanical engineering, chemistry, industrial engineering, electrical engineering, accounting, management information systems, chemical engineering

Alternatives to Full-Time Employment: Paid internships

Company-Paid Training is available.

Contact: Ms. Candy Stack, Manager of Staffing and Organizational Development; Phone: 864-223-2888; Fax: 864-942-5203. E-mail: cstack@fujigreenwood.com.

GREENWOOD MILLS, INC.
104 MAXWELL AVENUE
GREENWOOD, SOUTH CAROLINA 29646-2641

Overview: Manufactures textiles. Established: 1889. Ownership: Private.

Key Statistics: Annual Sales: $421.9 million. Number of Employees: 6,500.

Expertise/Education Sought: Accounting, customer service and support, auditing, construction, quality control, finance, purchasing, human resources, data processing

Alternatives to Full-Time Employment: Paid internships

Company-Paid Training is available.

International Assignments are available to employees.

Contact: Mr. Warren Moore, Greig Division Personnel Manager; Phone: 864-229-2571; Fax: 864-941-4062.

KINGS ELECTRONICS COMPANY, INC.
WATERFORD INDUSTRIAL PARK
ROCK HILL, SOUTH CAROLINA 29730

Overview: Produces electronic connectors. Established: 1963. Ownership: Private.

Key Statistics: Annual Sales: $26.9 million. Number of Employees: 300.

Hiring History: Number of professional employees hired in 1997: 25. 1996: 20.

Expertise/Education Sought: Mechanical engineering, manufacturing engineering, industrial engineering, materials management, sales, product management, production management

Alternatives to Full-Time Employment: Part-time jobs, paid internships

Contact: Mr. Charles Robinson, Director of Human Resources.

Company URL: http://www.kingselectronics.com

MAYFAIR MILLS, INC.
1885 HAYNE STREET
ARCADIA, SOUTH CAROLINA 29320

Mailing Address: PO Box 850, Arcadia, SC 29320

Overview: Produces broadwoven cotton fabric. Established: 1934. Ownership: Private.

Key Statistics: Annual Sales: $120 million. Number of Employees: 1,850.

Expertise/Education Sought: Maintenance

Company-Paid Training is available.

Contact: For individuals with previous experience: Mrs. Thelma Canty, Assistant to Personnel Manager; Phone: 864-576-2610, ext. 1251; Fax: 864-576-6741.

MICHELIN NORTH AMERICA
1 PARKWAY SOUTH
GREENVILLE, SOUTH CAROLINA 29615-5022

Mailing Address: PO Box 19001, Greenville, SC 29602

Overview: Manufactures tires. Established: 1950. Ownership: Private.

Key Statistics: Annual Sales: $809.3 million. Number of Employees: 12,000.

Expertise/Education Sought: Mechanical engineering, industrial engineering, electrical engineering, chemical engineering, computer programming, management information systems

Alternatives to Full-Time Employment: Paid internships

Michelin North America (continued)

COMPANY-PAID TRAINING is available.

CONTACT: For college students and recent graduates: Ms. Audra Goodnough, Recruiting Coordinator; Phone: 864-458-6493; Fax: 864-458-5430. For individuals with previous experience: Mr. Greg Kerce, Director of Corporate Recruiting, PO Box 19001, Greenville, South Carolina, 29602-9001; Phone: 864-458-5286; Fax: 864-458-5430.

COMPANY URL: http://www.michelin.com

M-TEC CORPORATION

4854 O'HEAR AVENUE
NORTH CHARLESTON, SOUTH CAROLINA 29405-4972

OVERVIEW: Manufactures gaskets and sealing devices. Established: 1995. Ownership: Private.

KEY STATISTICS: Annual Sales: $18.8 million. Number of Employees: 244.

EXPERTISE/EDUCATION SOUGHT: Accounting, engineering, mechanical engineering

ALTERNATIVES TO FULL-TIME EMPLOYMENT: Paid internships

COMPANY-PAID TRAINING is available.

CONTACT: For individuals with previous experience: Human Resources Department; Phone: 803-744-6261; Fax: 803-554-5128.

COMPANY URL: http://www.rmengineeredproducts.com

P & B AUTOMOTIVE LLC

5100 OLD PEARMAN DAIRY ROAD
ANDERSON, SOUTH CAROLINA 29625-1314

MAILING ADDRESS: PO Box 227, Anderson, SC 29623-5227

OVERVIEW: Manufactures motor vehicle parts and accessories. Established: 1996. Parent Company: P & B Automotive. Ownership: Private.

KEY STATISTICS: Annual Sales: $41.4 million. Number of Employees: 450.

HIRING HISTORY: Number of professional employees hired in 1997: 70.

EXPERTISE/EDUCATION SOUGHT: Engineering, injection molding, management, business administration, accounting, sales, purchasing

ALTERNATIVES TO FULL-TIME EMPLOYMENT: Internships

COMPANY-PAID TRAINING is available.

INTERNATIONAL ASSIGNMENTS are available to employees.

CONTACT: Human Resource Department; Phone: 864-260-0000; Fax: 864-231-7537.

PIRELLI CABLE CORPORATION

705 INDUSTRIAL DRIVE
LEXINGTON, SOUTH CAROLINA 29072-3741

OVERVIEW: Manufactures cable. Established: 1978. Parent Company: Pirelli Tire International. Ownership: Private.

KEY STATISTICS: Annual Sales: $419.7 million. Number of Employees: 850.

EXPERTISE/EDUCATION SOUGHT: Mechanical engineering, accounting, finance, electrical engineering

ALTERNATIVES TO FULL-TIME EMPLOYMENT: Paid internships

INTERNATIONAL ASSIGNMENTS are available to employees.

CONTACT: Mr. Van Kent, Manager of Human Resources; Phone: 803-951-4800.

POLICY MANAGEMENT SYSTEMS CORPORATION

1 PMS CENTER, WILSON BOULEVARD
BLYTHEWOOD, SOUTH CAROLINA 29016

MAILING ADDRESS: PO Box 10, Columbia, SC 29202-0010

OVERVIEW: Provides insurance software systems support services. Established: 1974. Ownership: Public.

KEY STATISTICS: Annual Sales: $581.9 million. Number of Employees: 5,133.

HIRING HISTORY: Number of professional employees hired in 1996: 866.

EXPERTISE/EDUCATION SOUGHT: Accounting, marketing, sales, finance, data processing

ALTERNATIVES TO FULL-TIME EMPLOYMENT: Part-time jobs, internships

INTERNATIONAL ASSIGNMENTS are available to employees.

CONTACT: Manager of Recruiting; Phone: 803-735-4000; Fax: 803-714-3227.

COMPANY URL: http://www.pmsc.com

REEVES BROTHERS, INC.

US HIGHWAY 29 SOUTH
SPARTANBURG, SOUTH CAROLINA 29301

OVERVIEW: Produces broadwoven cotton fabric. Established: 1986. Ownership: Public.

KEY STATISTICS: Annual Sales: $116.7 million. Number of Employees: 1,800.

HIRING HISTORY: Number of professional employees hired in 1996: 5.

EXPERTISE/EDUCATION SOUGHT: Management, data processing, administration, sales

CONTACT: For individuals with previous experience: Human Resources Administrator; Phone: 864-576-1210; Fax: 864-595-2181.

RE PHELON COMPANY, INC.

895 UNIVERSITY PARKWAY
AIKEN, SOUTH CAROLINA 29801-2802

OVERVIEW: Manufactures engine electrical equipment. Established: 1944. Ownership: Private.

KEY STATISTICS: Annual Sales: $70 million. Number of Employees: 550.

HIRING HISTORY: Number of professional employees hired in 1996: 278.

EXPERTISE/EDUCATION SOUGHT: Die casting, die cast engineering

ALTERNATIVES TO FULL-TIME EMPLOYMENT: Internships

CONTACT: For individuals with previous experience: Mr. Robert Bowen, Director of Human Resources; Phone: 803-649-1381.

COMPANY URL: http://www.phelon.com

RUST ENVIRONMENT AND INFRASTRUCTURE

15 BRENDAN WAY
GREENVILLE, SOUTH CAROLINA 29615-3514

OVERVIEW: Manufactures environmental products. Established: 1993. Parent Company: Waste Management, Inc. Ownership: Public.

KEY STATISTICS: Annual Sales: $191.9 million. Number of Employees: 3,000.

HIRING HISTORY: Number of professional employees hired in 1997: 400. 1996: 250.

EXPERTISE/EDUCATION SOUGHT: Information technology, human resources, marketing, finance, computer-aided design, environmental engineering, civil engineering, chemical engineering, geology

ALTERNATIVES TO FULL-TIME EMPLOYMENT: Paid internships

COMPANY-PAID TRAINING is available.

INTERNATIONAL ASSIGNMENTS are available to employees.

CONTACT: For individuals with previous experience: Mr. William Kemp, Vice President of Human Resources; Phone: 864-234-3000. E-mail: Terry_Stevens@ccmail.rustei.com.

SONOCO PRODUCTS COMPANY

NORTH 2ND STREET
HARTSVILLE, SOUTH CAROLINA 29550-0160

MAILING ADDRESS: PO Box 160, Hartsville, SC 29551-0160

OVERVIEW: Manufactures packaging tubes, cans, and grocery bags. Established: 1899. Ownership: Public.

KEY STATISTICS: Annual Sales: $2.8 billion. Number of Employees: 15,000.

EXPERTISE/EDUCATION SOUGHT: Chemical engineering, mechanical engineering, electrical engineering, accounting, liberal arts

ALTERNATIVES TO FULL-TIME EMPLOYMENT: Part-time jobs, paid internships
COMPANY-PAID TRAINING is available.
INTERNATIONAL ASSIGNMENTS are available to employees.
CONTACT: For individuals with previous experience: Ms. Cindy Hartley, Vice President of Human Resources; Phone: 803-383-7247; Fax: 803-383-7478.
COMPANY URL: http://www.sonoco.com

SPAN-AMERICA MEDICAL SYSTEMS, INC.
70 COMMERCE CENTER
GREENVILLE, SOUTH CAROLINA 29615-5814

OVERVIEW: Manufactures polyurethane foam patient positioners, performs packaging and labeling services. Established: 1975. Ownership: Public.
KEY STATISTICS: Annual Sales: $31.5 million. Number of Employees: 270.
EXPERTISE/EDUCATION SOUGHT: Marketing, accounting, customer service and support, sales, research and development
ALTERNATIVES TO FULL-TIME EMPLOYMENT: Paid internships
COMPANY-PAID TRAINING is available.
CONTACT: Ms. Melinda Gage, Vice President of Human Resources; Phone: 864-288-8877; Fax: 864-288-8692.

WESTINGHOUSE SAVANNAH RIVER COMPANY
BUILDING 719-4A
AIKEN, SOUTH CAROLINA 29808

OVERVIEW: Nuclear production, waste management, and environmental restoration for the U.S. Department of Energy. Established: 1989. Parent Company: Westinghouse Electric Corporation. Ownership: Public.
KEY STATISTICS: Annual Sales: $1.4 billion. Number of Employees: 13,000.
HIRING HISTORY: Number of professional employees hired in 1997: 52. 1996: 38.
EXPERTISE/EDUCATION SOUGHT: Environmental engineering, environmental science, electrical engineering, geology, hydrogeology, chemical engineering, mechanical engineering, waste management
CONTACT: Mr. Willie R. Bell, Human Resources Department; Phone: 803-725-8455; Fax: 803-725-8781.

WHITTEN CENTER SCDDSN
3440 HARDEN STREET
COLUMBIA, SOUTH CAROLINA 29203-6835

MAILING ADDRESS: PO Box 4706, Columbia, SC 29240
OVERVIEW: Offers nursing and personal care. Established: 1920. Ownership: Private.
KEY STATISTICS: Annual Sales: $26.7 million. Number of Employees: 1,200.
HIRING HISTORY: Number of professional employees hired in 1997: 20. 1996: 25.
EXPERTISE/EDUCATION SOUGHT: Computer programming, accounting, software engineering/development, product management, product management
COMPANY-PAID TRAINING is available.
CONTACT: Human Resources; Phone: 803-737-6484.

SOUTH DAKOTA

MID COM
121 AIRPORT DRIVE
WATERTOWN, SOUTH DAKOTA 57201-5602

MAILING ADDRESS: PO Box 1330, Watertown, SD 57201
OVERVIEW: Manufactures electronic coils and transformers. Established: 1987. Parent Company: Amva, Inc. Ownership: Private.
KEY STATISTICS: Annual Sales: $75 million. Number of Employees: 1,200.
EXPERTISE/EDUCATION SOUGHT: Computer programming, electronics/electronics engineering, maintenance, electrical engineering, mechanical engineering, industrial engineering
ALTERNATIVES TO FULL-TIME EMPLOYMENT: Part-time jobs, paid internships
INTERNATIONAL ASSIGNMENTS are available to employees.
CONTACT: For college students and recent graduates: Ms. Kaye Nelson, Human Resource Coordinator; Phone: 605-886-4385; Fax: 605-882-0332. E-mail: knelson@midcom-inc.com. For individuals with previous experience: Ms. Carol J. Hinderaker, Vice President of Human Resources, 510 Tenth Avenue, SE, Watertown, South Dakota, 57201-5602; Phone: 605-886-4385; Fax: 605-886-3889.
COMPANY URL: http://www.midcom-inc.com

RAVEN INDUSTRIES, INC.
205 EAST 6TH STREET
SIOUX FALLS, SOUTH DAKOTA 57104

MAILING ADDRESS: PO Box 5107, Sioux Falls, SD 57117-5107
OVERVIEW: Manufactures plastics, electronic components, and sewn products. Established: 1956. Ownership: Public.
KEY STATISTICS: Annual Sales: $139.4 million. Number of Employees: 1,387.
HIRING HISTORY: Number of professional employees hired in 1997: 200. 1996: 200.
EXPERTISE/EDUCATION SOUGHT: Accounting, finance, management information systems, mechanical engineering, electrical engineering
ALTERNATIVES TO FULL-TIME EMPLOYMENT: Paid internships
COMPANY-PAID TRAINING is available.
CONTACT: Ms. Barbara Ohme, Manager of Human Resources; Phone: 605-336-2750, ext. 1200; Fax: 605-336-0165.

TENNESSEE

THE AREOSTRUCTURES CORPORATION
1431 VULTEE BOULEVARD
NASHVILLE, TENNESSEE 37217-2007

OVERVIEW: Manufactures aircraft engines and parts. Established: 1996. Ownership: Private.
KEY STATISTICS: Annual Sales: $126.4 million. Number of Employees: 1,500.
EXPERTISE/EDUCATION SOUGHT: Tool and die engineering
ALTERNATIVES TO FULL-TIME EMPLOYMENT: Paid internships
COMPANY-PAID TRAINING is available.
INTERNATIONAL ASSIGNMENTS are available to employees.
CONTACT: Mr. Johna Morganms, Employment Supervisor; Phone: 615-361-2000; Fax: 615-361-2531.

AUTOZONE, INC.

123 SOUTH FRONT STREET
MEMPHIS, TENNESSEE 38103

Mailing Address: PO Box 2198, Memphis, TN 38101-2198

Overview: Operates chain of retail automotive parts and supplies stores. Established: 1979. Ownership: Public.

Key Statistics: Annual Sales: $2.7 billion. Number of Employees: 28,700.

Hiring History: Number of professional employees hired in 1996: 300.

Expertise/Education Sought: Accounting, computer programming, finance, marketing, law, data processing, business management, construction, design, retail management

Alternatives to Full-Time Employment: Part-time jobs, paid internships

Company-Paid Training is available.

Contact: Staffing Services; Phone: 901-495-6500. E-mail: recruiters@autozone.com.

Company URL: http://www.autozone.com

BRIDGESTONE/FIRESTONE, INC.

50 CENTURY BOULEVARD
NASHVILLE, TENNESSEE 37214-3672

Overview: Manufactures tires. Established: 1988. Parent Company: Bridgestone. Ownership: Private.

Key Statistics: Annual Sales: $6.5 billion. Number of Employees: 45,000.

Expertise/Education Sought: Marketing, sales, mechanical engineering, chemical engineering

Alternatives to Full-Time Employment: Paid internships

Company-Paid Training is available.

International Assignments are available to employees.

Contact: For individuals with previous experience: Ms. Patti Brown, Director of Human Resources; Phone: 615-872-5000, ext. 1504; Fax: 615-872-1807.

DEVILBISS AIR POWER COMPANY

213 INDUSTRIAL DRIVE
JACKSON, TENNESSEE 38301-9615

Overview: Manufactures air and gas compressors Manufactures air compressors, generators, and pressure washers. Established: 1994. Parent Company: Investcorp. Ownership: Private.

Key Statistics: Annual Sales: $115.6 million. Number of Employees: 875.

Hiring History: Number of professional employees hired in 1996: 100.

Expertise/Education Sought: Accounting, administration, customer service and support, engineering, management, marketing, sales, management information systems, human resources

Alternatives to Full-Time Employment: Paid internships

Company-Paid Training is available.

International Assignments are available to employees.

Contact: For individuals with previous experience: Human Resources Department; Phone: 901-423-7000; Fax: 901-423-7900.

EASTMAN

EASTMAN CHEMICAL COMPANY

PO BOX 1975
KINGSPORT, TENNESSEE 37662-5215

Overview: Manufactures industrial chemicals, fibers, and plastics. Established: 1920. Ownership: Public.

Key Statistics: Annual Sales: $4.7 billion. Number of Employees: 16,100.

Hiring History: Number of professional employees hired in 1997: 71. 1996: 137.

Expertise/Education Sought: Chemical engineering, mechanical engineering, electrical engineering, chemistry, marketing, production engineering

Company-Paid Training is available.

Contact: Mr. Hoyt Denton, Manager of Employment; Phone: 423-229-2387; Fax: 423-224-0995.

Company URL: http://www.eastman.com

CORPORATE PROFILE

Company Overview. *Eastman Chemical Company is a leading manufacturer of industrial chemicals, fibers, and plastics. Eastman is respected in the chemical industry for its high quality and customer service*

Career Opportunties and Development. *Numerous entry-level positions are available due to Eastman's commitment to research and technical service, the demand for technical backgrounds in marketing, and the need for engineers and scientists in production. Career paths for professionally trained individuals are available in research and development, design, manufacturing, technical service, and technical sales. Depending upon interests and skills, employment with Eastman Chemical Company may lead to administrative and management responsibilities or a satisfying and rewarding career as a specialist in a particular field.*

Eastman encourages employees to continue personal and professional development. On-the-job training, tuition reimbursements, self-teach courses, and specialized training classes in management and personal skills offer employees growth and learning opportunities.

Company Locations. *Headquartered in Kingsport, Eastman maintains additional manufacturing facilities North America in Alvin, Texas; Batesville, Arkansas; Columbia, South Carolina; Longview, Texas; Rochester, New York; South Charleston, West Virginia; and Toronto, Canada. Manufacturing locations in Latin America are found in Cosoleacaque, Mexico; and *Zarate, Argentina. In Europe, Eastman has manufacturing facilities in Hartlepool and Workington, England; Rotterdam, Netherlands; San Roque, Spain; and Llangefni, Wales. Asian manufacturing plants are found in Hong Kong; Kuantan, Malaysia; and *Singapore. Eastman's products are marketed through sales offices in the United States, Europe, Asia, and South America. Interested persons should direct inquiries to Professional Employment, P.O. Box 1975, Kingsport, TN 37662-5215. All employment inquiries are referred to the appropriate unit for consideration. (*under construction)*

FEDEX CORPORATION

2605 NONCONNAH STREET, SUITE 150
MEMPHIS, TENNESSEE 38132-1702

Overview: Provides package delivery and courier services. Established: 1971. Parent Company: FDX Corporation. Ownership: Public.

Key Statistics: Annual Sales: $11.5 billion. Number of Employees: 126,000.

Expertise/Education Sought: Mechanical engineering, electrical engineering

Alternatives to Full-Time Employment: Paid internships

Company-Paid Training is available.

Contact: For individuals with previous experience: Mr. David Huffer, Senior Manager Recruitment, PO Box 727, Memphis, Tennessee, 38101; Phone: 901-922-1247; Fax: 901-922-4067.

Company URL: http://www.fedex.com

FLEETGUARD, INC.

402 BNA DRIVE, SUITE 500
NASHVILLE, TENNESSEE 37217-2526

Overview: Manufactures heavy duty filters. Established: 1974. Parent Company: Cummins Engine Company, Inc. Ownership: Public.

Key Statistics: Annual Sales: $233.1 million. Number of Employees: 2,500.

Expertise/Education Sought: Computer programming, data processing, finance, sales, marketing

Alternatives to Full-Time Employment: Part-time jobs, paid internships

Company-Paid Training is available.

Contact: Ms. Kathy Hibbs, Manager of Human Resources; Phone: 615-367-0040.

Company URL: http://www.fleetguard.com

INTERMEDIA MANAGEMENT, INC.

424 CHURCH STREET, SUITE 1600
NASHVILLE, TENNESSEE 37219-2308

Overview: Provides cable TV installation services. Established: 1988. Ownership: Private.

Key Statistics: Annual Sales: $161.3 million. Number of Employees: 1,200.

Hiring History: Number of professional employees hired in 1997: 135. 1996: 120.

Expertise/Education Sought: Operations, administration, law, engineering

Alternatives to Full-Time Employment: Paid internships

Company-Paid Training is available.

Contact: For individuals with previous experience: Ms. Donna Dearman, Human Resources Manager; Phone: 615-244-2300; Fax: 615-259-9957.

INTERNATIONAL COMFORT PRODUCTS CORPORATION

650 HEIL QUAKER AVENUE
LEWISBURG, TENNESSEE 37091

Overview: Manufactures refrigeration and heating equipment. Established: 1986. Ownership: Private.

Key Statistics: Annual Sales: $587.2 million. Number of Employees: 2,555.

Hiring History: Number of professional employees hired in 1996: 25.

Expertise/Education Sought: Accounting, computer programming, marketing, sales, industrial engineering, design engineering

Alternatives to Full-Time Employment: Paid internships

Company-Paid Training is available.

International Assignments are available to employees.

Contact: For college students and recent graduates: Mr. Ray Kliewer, College Recruiter; Phone: 931-359-3511. For individuals with previous experience: Human Resources Department; Phone: 931-359-3511; Fax: 931-270-3312.

INTERNATIONAL PAPER COMPANY

6400 POPLAR AVENUE
MEMPHIS, TENNESSEE 38197

Overview: Manufactures paper and wood products and specialty products. Established: 1898. Ownership: Public.

Key Statistics: Annual Sales: $20.1 billion. Number of Employees: 88,000.

Hiring History: Number of professional employees hired in 1997: 300. 1996: 300.

Expertise/Education Sought: Chemical engineering, electrical engineering, industrial engineering, mechanical engineering

Alternatives to Full-Time Employment: Part-time jobs, paid internships

Company-Paid Training is available.

International Assignments are available to employees.

Contact: Recruiting Coordinator; Phone: 901-763-5895; Fax: 901-763-6055.

Company URL: http://www.ipaper.com

KANTUS CORPORATION

201 GARRET PARKWAY
LEWISBURG, TENNESSEE 37091-3552

Overview: Manufactures motor vehicle parts and accessories. Established: 1984. Parent Company: Kansei Corporation. Ownership: Private.

Key Statistics: Annual Sales: $115 million. Number of Employees: 750.

Expertise/Education Sought: Mechanical engineering, electrical engineering, industrial engineering, data processing, accounting, marketing, finance

Company-Paid Training is available.

Contact: For individuals with previous experience: Ms. Lana Vanhooser, Senior Human Resource Representative; Phone: 931-359-0642; Fax: 931-359-2151.

KRAFT FOOD INGREDIENTS

8000 HORIZON CENTER
MEMPHIS, TENNESSEE 38133-4844

Overview: Produces and distributes food ingredients and specialty foods. Established: 1965. Parent Company: Kraft Foods. Ownership: Public.

Key Statistics: Annual Sales: $250 million. Number of Employees: 200.

Hiring History: Number of professional employees hired in 1996: 10.

Expertise/Education Sought: Marketing, food science, sales

Contact: For individuals with previous experience: Mrs. Vicky Odle, Associate Human Resources Manager; Phone: 901-381-6500.

LOCKHEED MARTIN ENERGY SYSTEMS, INC.

PO BOX 2009
OAK RIDGE, TENNESSEE 37831

Overview: Provides research in the areas of nuclear power and resource/energy recovery. Established: 1995. Parent Company: Lockheed Martin Corporation. Ownership: Public.

Key Statistics: Annual Sales: $705.6 million. Number of Employees: 14,700.

Hiring History: Number of professional employees hired in 1996: 9,999.

Expertise/Education Sought: Engineering science, environmental engineering, nuclear engineering, civil engineering, chemical engineering, science

Alternatives to Full-Time Employment: Part-time jobs, paid internships

Contact: For individuals with previous experience: Mr. Horace Moorman, Staffing Manager; Phone: 423-574-1553; Fax: 423-576-5549.

Company URL: http://www.ornl.gov/mmes.html

MAGNETEK

26 CENTURY BOULEVARD
NASHVILLE, TENNESSEE 37214-0159

Overview: Manufactures electrical equipment and components. Established: 1970. Ownership: Public.

Key Statistics: Annual Sales: $1.2 billion. Number of Employees: 13,800.

Expertise/Education Sought: Mechanical engineering, electrical engineering, design engineering, manufacturing engineering

Alternatives to Full-Time Employment: Paid internships

Company-Paid Training is available.

International Assignments are available to employees.

Contact: For individuals with previous experience: Human Resources Department; Phone: 615-316-5100; Fax: 615-316-5123.

MICRO CRAFT, INC.

207 BIG SPRINGS AVENUE
TULLAHOMA, TENNESSEE 37388-0370

Mailing Address: PO Box 370, Tullahoma, TN 37388

Overview: Manufactures aerospace testing equipment and provides engineering services. Established: 1958. Ownership: Private.

Key Statistics: Annual Sales: $50.4 million. Number of Employees: 620.

Expertise/Education Sought: Hardware engineering, aerospace engineering, field engineering

Alternatives to Full-Time Employment: Paid internships

Company-Paid Training is available.

International Assignments are available to employees.

Contact: For individuals with previous experience: Ms. Brenda Phillips, Director of Human Resources, PO Box 370, Tullahoma, Tennessee, 37388-0370; Phone: 931-455-2664; Fax: 931-455-9754.

Company URL: http://www.microcraft.com

MOHON INTERNATIONAL, INC.

1865 NORTH MARKET STREET
PARIS, TENNESSEE 38242

Mailing Address: PO Box 550, Paris, TN 38242

Mohon International, Inc. (continued)

Overview: Manufactures laboratory furniture for schools. Established: 1976. Ownership: Private.

Key Statistics: Annual Sales: $27.1 million. Number of Employees: 310.

Hiring History: Number of professional employees hired in 1997: 10. 1996: 10.

Expertise/Education Sought: Engineering, drafting

Alternatives to Full-Time Employment: Paid internships

Company-Paid Training is available.

Contact: Mr. Steve Mohon, President and Chief Executive Officer; Phone: 901-642-4251; Fax: 901-642-4262. E-mail: smohon@mohon.com.

Company URL: http://www.mohon.com

SEA RAY BOATS, INC.

2600 SEA RAY BOULEVARD
KNOXVILLE, TENNESSEE 37914

Overview: Builds and repairs inboard and outboard motorboats. Established: 1995. Parent Company: Brunswick Corporation. Ownership: Public.

Key Statistics: Annual Sales: $60.3 million. Number of Employees: 391.

Hiring History: Number of professional employees hired in 1997: 6.

Expertise/Education Sought: Mechanical engineering, sales, marketing, manufacturing engineering, accounting, management information systems, engineering, design engineering, finance

Alternatives to Full-Time Employment: Part-time jobs

Company-Paid Training is available.

Contact: Mrs. LuAnn Jarnagin, Vice President of Human Resources; Phone: 423-522-4181; Fax: 423-971-6423.

Company URL: http://www.searayboats.com

SERVICE MERCHANDISE COMPANY

7100 SERVICE MERCHANDISE DRIVE
BRENTWOOD, TENNESSEE 37027

Overview: Operates catalog showroom stores. Established: 1959. Ownership: Public.

Key Statistics: Annual Sales: $4 billion. Number of Employees: 26,261.

Expertise/Education Sought: Management information systems, accounting, purchasing, marketing, sales, management

Alternatives to Full-Time Employment: Part-time jobs

Contact: Ms. Judy Moore, Director of Recruiting; Phone: 616-660-4627. E-mail: jmoore@servicemerchandise.com. Job Hotline: 615-660-3199.

Company URL: http://www.servicemerchandise.com

SHONEY'S, INC.

1727 ELM HILL PIKE
NASHVILLE, TENNESSEE 37210-5701

Overview: Operates chain of family restaurants. Established: 1959. Ownership: Public.

Key Statistics: Annual Sales: $1.1 billion. Number of Employees: 38,000.

Expertise/Education Sought: Accounting, marketing, information systems, law, management

Alternatives to Full-Time Employment: Paid internships

Company-Paid Training is available.

Contact: Ms. Allison Griffith, Manager of Human Resources. E-mail: griffiam@shoneys.com. Job Hotline: 800-626-5630, ext. 223.

SMITH & NEPHEW, INC.

1450 EAST BROOKS ROAD
MEMPHIS, TENNESSEE 38116

Overview: Designs, markets and manufactures orthopedic implants, trauma, spine, and medical specialty products. Established: 1975. Parent Company: Smith & Newphews, PLC. Ownership: Private.

Key Statistics: Annual Sales: $326.1 million.

Expertise/Education Sought: Biology, mechanical engineering, materials engineering, biomedical engineering

International Assignments are available to employees.

Contact: For individuals with previous experience: Mr. Steve Ammons, Director of Human Resources; Phone: 901-396-2121; Fax: 901-348-6189.

Thomas&Betts

THOMAS & BETTS CORPORATION

8155 T & B BOULEVARD
MEMPHIS, TENNESSEE 38125

Overview: Manufactures and distributes electrical and electronic connectors, components, and systems. Established: 1898. Ownership: Public.

Key Statistics: Annual Sales: $2.1 billion. Number of Employees: 18,000.

Expertise/Education Sought: Accounting, customer service and support, auditing, logistics, manufacturing engineering, business management, information technology

Alternatives to Full-Time Employment: Paid internships

Contact: Ms. Angela Wilkinson, Recruiter; Phone: 901-252-5000; Fax: 901-252-1310.

CORPORATE PROFILE

Thomas & Betts recognizes the importance of giving its employees the opportunity for career development. The company's success depends greatly upon its ability to attract and retain talented people with various experiences and ambitions. In return, Thomas & Betts offers a challenging environment and the opportunity to participate in one of the most exciting companies in the industry.

It is very important that all employees have a common set of values and objectives that are understood, agreed upon, and practiced in their day-to-day work lives and a structure within which all employees can focus their energies on achieving core strategic objectives. The managing philosophy at Thomas & Betts governs how the organization and its resources are led, facilitated, and guided. This managing philosophy helps Thomas & Betts to successfully provide the product of choice in every market it chooses to serve.

With entry-level opportunities, Thomas & Betts offers competitive salaries and benefits, including medical and dental coverage, investment and pension plans, and a relocation program.

UNAKA COMPANY, INC.

102 COILE STREET
GREENEVILLE, TENNESSEE 37743

Mailing Address: PO Box 877, Greeneville, TN 37744

Overview: Manufactures cookware and household furniture. Established: 1989. Ownership: Private.

Key Statistics: Annual Sales: $96.8 million. Number of Employees: 1,000.

Expertise/Education Sought: Accounting, food science

Company-Paid Training is available.

Contact: For individuals with previous experience: Mr. Marion Bailey, General Manager, PO Box 877, Greeneville, Tennessee, 37744-0877; Phone: 423-639-1163; Fax: 423-639-7270. E-mail: mbailey587@aol.com.

Company URL: http://www.unaka.com

VP BUILDINGS, INC.

3200 PLAYERS CLUB CIRCLE
MEMPHIS, TENNESSEE 38187-0967

Overview: Manufactures prefabricated metal buildings. Established: 1997. Parent Company: LTV Steel. Ownership: Public.

Key Statistics: Annual Sales: $300 million. Number of Employees: 1,700.

Hiring History: Number of professional employees hired in 1997: 60. 1996: 60.

Expertise/Education Sought: Civil engineering

Company-Paid Training is available.

International Assignments are available to employees.

Contact: For individuals with previous experience: Human Resources Department; Phone: 901-748-8000; Fax: 901-748-9821.

Company URL: http://www.vp.com

TEXAS

ABB VETCO GRAY, INC.

10777 NORTHWEST FREEWAY
HOUSTON, TEXAS 77092-7325

Overview: Manufactures oil and gas drilling equipment. Established: 1991. Parent Company: ABB. Ownership: Private.

Key Statistics: Annual Sales: $450 million. Number of Employees: 2,200.

Hiring History: Number of professional employees hired in 1997: 200. 1996: 120.

Expertise/Education Sought: Manufacturing engineering, mechanical engineering

Alternatives to Full-Time Employment: Paid internships

Company-Paid Training is available.

International Assignments are available to employees.

Contact: For individuals with previous experience: Ms. Irene Duke, Director of Human Resources, 12221 North Houston Rosslyn Road, Houston, Texas, 77086-7325; Phone: 281-448-4410; Fax: 281-878-5155.

AIR FORCE RECRUITING SERVICE

550 D STREET WEST, SUITE 1
RANDOLPH AFB, TEXAS 78150-4527

Overview: Provides air and space force. Established: 1947. Division of the U.S. Air Force.

Key Statistics: Number of Employees: 370,000.

Expertise/Education Sought: Technical

International Assignments are available to employees.

Contact: Major Dori Johnson, Chief Line Officer Accessions; Phone: 210-652-2245.

Company URL: http://www.airforce.com

For more information, see full organization profile, pg. 304.

ALCON LABORATORIES

6201 SOUTH FREEWAY
FORT WORTH, TEXAS 76134-2099

Overview: Manufactures eye care products and ophthalmic instruments. Parent Company: Nestle, SA.

Expertise/Education Sought: Manufacturing engineering, mechanical engineering, ophthalmological engineering

Alternatives to Full-Time Employment: Paid internships

Company-Paid Training is available.

International Assignments are available to employees.

Contact: For individuals with previous experience: Placement Division, Mail Code T1-3, 6201 South Freeway, Fort Worth, Texas, 76134-2099; Phone: 817-551-0450; Fax: 817-551-4629. Job Hotline: 817-551-4575.

ALTURA ENERGY, LTD.

580 WESTLAKE PARK BOULEVARD
HOUSTON, TEXAS 77079-2662

Overview: Crude petroleum and natural gas products. Established: 1997. Parent Company: Amoco Production Company. Ownership: Private.

Key Statistics: Annual Sales: $110.4 million. Number of Employees: 450.

Expertise/Education Sought: Operations, engineering, geology

Contact: Human Resource Department; Phone: 281-552-1305; Fax: 281-552-1583.

AMERITRUCK DISTRIBUTION CORPORATION

301 COMMERCE, SUITE 1101
FORT WORTH, TEXAS 76102

Overview: Operates local trucking business. Established: 1995. Ownership: Public.

Key Statistics: Annual Sales: $224.3 million. Number of Employees: 3,200.

Expertise/Education Sought: Accounting, finance, marketing, management information systems, customer service and support, management, human resources, safety, maintenance

Contact: For individuals with previous experience: Mr. Ken Evans, Chief Financial Officer.

AMTECH CORPORATION

19111 DALLAS PARKWAY, SUITE 300
DALLAS, TEXAS 75287-3106

Overview: Manufactures electronic identification equipment. Established: 1983. Ownership: Public.

Key Statistics: Annual Sales: $116.5 million. Number of Employees: 860.

Hiring History: Number of professional employees hired in 1997: 70. 1996: 75.

Expertise/Education Sought: Computer science, software engineering/development, systems engineering

Company-Paid Training is available.

International Assignments are available to employees.

Contact: For individuals with previous experience: Ms. Cheryl Kucera, Human Resources Administrator; Phone: 972-733-6600; Fax: 972-733-6005.

Company URL: http://www.amtech.com

ASSEMBLY SERVICES, INC.

12058 ROJAS DRIVE
EL PASO, TEXAS 79936-7711

Overview: Wholesales industrial supplies. Established: 1966. Ownership: Private.

Key Statistics: Annual Sales: $48.3 million. Number of Employees: 310.

Hiring History: Number of professional employees hired in 1997: 350. 1996: 300.

Expertise/Education Sought: Manufacturing

Company-Paid Training is available.

Contact: For individuals with previous experience: Ms. Grace Hernandez, Personnel Administrator; Phone: 915-859-6680; Fax: 915-859-1927. E-mail: graciela.hernandez@quickiebrand.com.

AVO INTERNATIONAL

4271 BRONZE WAY
DALLAS, TEXAS 75237-1019

Overview: Manufactures electronic measure instruments. Established: 1987. Parent Company: TBG, Inc. Ownership: Private.

Key Statistics: Annual Sales: $30 million. Number of Employees: 275.

Expertise/Education Sought: Engineering, manufacturing

AVO International (continued)

International Assignments are available to employees.

Contact: For individuals with previous experience: Ms. Vickie Davis, Manager of Human Resources; Phone: 214-330-2205; Fax: 214-467-7341. Job Hotline: 214-333-3201.

Company URL: http://www.avointl.com

BAKER HUGHES, INC.

3900 ESSEX LANE, SUITE 1200
HOUSTON, TEXAS 77027-5112

Mailing Address: PO Box 4740, Houston, TX 77210-4740

Overview: Manufactures mining machinery. Established: 1982. Ownership: Public.

Key Statistics: Annual Sales: $553.2 million. Number of Employees: 5,500.

Hiring History: Number of professional employees hired in 1997: 3,000. 1996: 1,002.

Expertise/Education Sought: Manufacturing, information systems, chemical engineering, mechanical engineering, petroleum/petrochemical engineering, geologist

Alternatives to Full-Time Employment: Paid internships

Company-Paid Training is available.

International Assignments are available to employees.

Contact: Human Resources; Phone: 713-439-8600.

Company URL: http://www.bakerhughes.com

BANCTEC, INC.

4851 LBJ FREEWAY
DALLAS, TEXAS 75244

Overview: Manufactures image systems and process systems. Established: 1970. Ownership: Public.

Key Statistics: Annual Sales: $554 million. Number of Employees: 3,650.

Expertise/Education Sought: Software engineering/development, manufacturing engineering, networks

Alternatives to Full-Time Employment: Paid internships

Company-Paid Training is available.

International Assignments are available to employees.

Contact: Mr. Doug Johnson, Corporate Recruiter; Phone: 972-341-4000; Fax: 972-341-4867.

Company URL: http://www.banctec.com

BELMONT CONSTRUCTORS COMPANY, INC.

2925 BRIAR PARK DRIVE, SUITE 260
HOUSTON, TEXAS 77042

Overview: Provides construction services. Established: 1995. Ownership: Private.

Key Statistics: Annual Sales: $160.6 million. Number of Employees: 140.

Expertise/Education Sought: Civil engineering, construction engineering, accounting, finance

Company-Paid Training is available.

Contact: For individuals with previous experience: Mr. Wilson Yancey, Director of Human Resources; Phone: 713-435-6300; Fax: 713-435-6302.

BROWNING-FERRIS INDUSTRIES, INC.

757 NORTH ELDRIDGE PARKWAY
HOUSTON, TEXAS 77079-4435

Overview: Provides waste collection and disposal services. Established: 1970. Ownership: Public.

Key Statistics: Annual Sales: $5.8 billion. Number of Employees: 43,000.

Expertise/Education Sought: Accounting, law, sales, marketing, information systems, technical

Alternatives to Full-Time Employment: Paid internships

Company-Paid Training is available.

Contact: Recruitment; Phone: 713-870-7266; Fax: 713-584-8600.

CALTEX PETROLEUM CORPORATION

125 EAST JOHN CARPENTER FREEWAY
IRVING, TEXAS 75062-2709

Overview: Refines and markets petroleum products. Established: 1936. Parent Company: Texaco and Chevron. Ownership: Private.

Key Statistics: Annual Sales: $15 billion. Number of Employees: 8,000.

Expertise/Education Sought: Chemical engineering, mechanical engineering, accounting, finance

Contact: For individuals with previous experience: Ms. Nancy Wirth, Employment Specialist; Phone: 972-830-1000, ext. 3491; Fax: 972-830-1133.

CAMPBELL READY MIX, LP

105 EAST BOOTHE STREET
CLEVELAND, TEXAS 77327-4003

Overview: Manufactures ready-mix concrete. Established: 1996. Ownership: Private.

Key Statistics: Annual Sales: $41.5 million. Number of Employees: 315.

Expertise/Education Sought: Engineering

Company-Paid Training is available.

Contact: For individuals with previous experience: Human Resource Department; Phone: 281-443-8005; Fax: 281-592-1785.

CANRIG DRILLING TECHNOLOGY

14703 FM 1488 ROAD
MAGNOLIA, TEXAS 77354-1205

Overview: Manufactures oil and gas field machinery. Established: 1995. Parent Company: Nabors Industries. Ownership: Public.

Key Statistics: Annual Sales: $45 million. Number of Employees: 50.

Hiring History: Number of professional employees hired in 1997: 75.

Expertise/Education Sought: Engineering

Alternatives to Full-Time Employment: Paid internships

International Assignments are available to employees.

Contact: For individuals with previous experience: Human Resources Department; Phone: 281-259-8887; Fax: 281-259-8158.

CENTEX HOMES

2728 NORTH HARWOOD
DALLAS, TEXAS 75201-1516

Overview: Provides construction and real estate development services. Established: 1950. Parent Company: Centex Corporation. Ownership: Public.

Key Statistics: Annual Sales: $3.8 billion. Number of Employees: 8,926.

Hiring History: Number of professional employees hired in 1996: 200.

Expertise/Education Sought: Management, training and development, information systems, architecture

Alternatives to Full-Time Employment: Paid internships

Contact: For individuals with previous experience: Recruiter; Phone: 214-981-5000.

CENTRAL AND SOUTH WEST CORPORATION

1616 WOODALL RODGERS FREEWAY
DALLAS, TEXAS 75202-1234

Overview: Operates and manages utilities. Established: 1925. Ownership: Public.

Key Statistics: Assets: $5.2 billion. Number of Employees: 11,500.

Expertise/Education Sought: Marketing, computer programming, finance, mechanical engineering, electrical engineering, accounting

Alternatives to Full-Time Employment: Part-time jobs, paid internships

Company-Paid Training is available.

International Assignments are available to employees.

Contact: For individuals with previous experience: Ms. Denise Davis, Human Resource Manager. Job Hotline: 214-777-1877.

Company URL: http://www.csw.com

CHEMICAL LIME COMPANY

3700 HULEN STREET
FORT WORTH, TEXAS 76107-6816

MAILING ADDRESS: PO Box 985004, Fort Worth, TX 76185-5004
OVERVIEW: Manufactures lime and limestone products. Established: 1985. Parent Company: Lhoish. Ownership: Private.
KEY STATISTICS: Annual Sales: $98.1 million. Number of Employees: 858.
HIRING HISTORY: Number of professional employees hired in 1996: 30.
EXPERTISE/EDUCATION SOUGHT: Chemical engineering, mechanical engineering, manufacturing engineering
ALTERNATIVES TO FULL-TIME EMPLOYMENT: Internships
INTERNATIONAL ASSIGNMENTS are available to employees.
CONTACT: For individuals with previous experience: Director of Human Resources, PO Box 121874, Fort Worth, Texas, 76121-1874; Phone: 817-732-8164, ext. 132.
COMPANY URL: http://www.chemicallime.com

CONCORDE ADJUSTERS, INC.

669 AIRPORT FREEWAY, SUITE 400
HURST, TEXAS 76053-3964

OVERVIEW: Provides waste collection and disposal services. Established: 1973. Parent Company: Laidlaw, Inc. Ownership: Public.
KEY STATISTICS: Annual Sales: $607 million. Number of Employees: 6,070.
HIRING HISTORY: Number of professional employees hired in 1997: 7. 1996: 6.
EXPERTISE/EDUCATION SOUGHT: Claims processing/management, and administration
ALTERNATIVES TO FULL-TIME EMPLOYMENT: Paid internships
COMPANY-PAID TRAINING is available.
INTERNATIONAL ASSIGNMENTS are available to employees.
CONTACT: For individuals with previous experience: Mr. Russ Brown, Office Manager; Phone: 817-282-7580, ext. 277; Fax: 817-285-0401.

CONDEA VISTA COMPANY

900 THREADNEEDLE STREET
HOUSTON, TEXAS 77079

OVERVIEW: Manufactures chemicals. Established: 1991. Parent Company: RWE-DEA, AG. Ownership: Private.
KEY STATISTICS: Annual Sales: $817.7 million. Number of Employees: 1,400.
EXPERTISE/EDUCATION SOUGHT: Manufacturing
ALTERNATIVES TO FULL-TIME EMPLOYMENT: Part-time jobs
CONTACT: Human Resources Department.
COMPANY URL: http://www.condeavista.com

CONTINENTAL EMSCO

13111 NORTHWEST FREEWAY, SUITE 100
HOUSTON, TEXAS 77040-6318

OVERVIEW: Manufactures industrial machinery and equipment. Established: 1997. Parent Company: Conemsco. Ownership: Private.
KEY STATISTICS: Annual Sales: $230.9 million. Number of Employees: 1,500.
EXPERTISE/EDUCATION SOUGHT: Business administration, marketing, sales, engineering
ALTERNATIVES TO FULL-TIME EMPLOYMENT: Paid internships
COMPANY-PAID TRAINING is available.
INTERNATIONAL ASSIGNMENTS are available to employees.
CONTACT: Human Resources; Phone: 713-510-2320; Fax: 713-510-2371.
COMPANY URL: http://www.conemsco.com

COOPER INDUSTRIES, INC.

600 TRAVIS STREET, SUITE 5800
HOUSTON, TEXAS 77002

OVERVIEW: Manufactures power systems, tools, hardware, and petroleum, industrial, electrical, and automotive products;. Established: 1833. Ownership: Public.
KEY STATISTICS: Annual Sales: $5.3 billion. Number of Employees: 42,400.
EXPERTISE/EDUCATION SOUGHT: Industrial engineering, mechanical engineering, electrical engineering, human resources
COMPANY-PAID TRAINING is available.
CONTACT: For college students and recent graduates: Ms. Diane Easley, Manager of Recruitment and Development; Phone: 713-209-8400; Fax: 713-209-8987. E-mail: recruit@cooperindustries.com. For individuals with previous experience: Human Resources; Phone: 713-209-8427; Fax: 713-209-8987. E-mail: recruit@cooperindustries.com.

DALLAS SEMICONDUCTOR CORPORATION

4401 SOUTH BELTWOOD PARKWAY
DALLAS, TEXAS 75244-3292

OVERVIEW: Manufactures semiconductors. Established: 1984. Ownership: Public.
KEY STATISTICS: Annual Sales: $288.4 million. Number of Employees: 1,348.
HIRING HISTORY: Number of professional employees hired in 1997: 200.
EXPERTISE/EDUCATION SOUGHT: Design engineering, computer engineering, computer programming, design, product engineering, test engineering, process engineering
ALTERNATIVES TO FULL-TIME EMPLOYMENT: Part-time jobs, paid internships
COMPANY-PAID TRAINING is available.
INTERNATIONAL ASSIGNMENTS are available to employees.
CONTACT: For college students and recent graduates: Ms. Cindy Staley, College Recruiter; Phone: 972-371-3868. E-mail: recruiter@dalsemi.com. For individuals with previous experience: Mr. Pat Cockburn, Corporate Recruiter; Phone: 972-371-3868; Fax: 972-371-6337. E-mail: recruiter@dalsemi.com. Job Hotline: 972-371-4399.
COMPANY URL: http://www.dalsemi.com

DAVIS BROTHERS CONSTRUCTION COMPANY

2200 LOUETTA
SPRING, TEXAS 77388

OVERVIEW: Builds apartment buildings. Established: 1986. Ownership: Private.
KEY STATISTICS: Annual Sales: $40 million. Number of Employees: 68.
HIRING HISTORY: Number of professional employees hired in 1997: 6.
EXPERTISE/EDUCATION SOUGHT: Administration, construction
ALTERNATIVES TO FULL-TIME EMPLOYMENT: Paid internships
COMPANY-PAID TRAINING is available.
CONTACT: For individuals with previous experience: Ms. Sandy Satterwhite, Human Resources.

DAYBREAK SOFTWARE, INC.

4030 WEST BRAKER LANE, SUITE 555
AUSTIN, TEXAS 78759-4871

OVERVIEW: Provides computer software services. Established: 1996. Ownership: Private.
KEY STATISTICS: Annual Sales: $10 million.
EXPERTISE/EDUCATION SOUGHT: Computer programming, computer graphics
CONTACT: For individuals with previous experience: Human Resources Department; Phone: 512-794-6355; Fax: 512-794-6345.
COMPANY URL: http://www.daybreaksoft.com

DELL COMPUTER CORPORATION
1 DELL WAY
ROUND ROCK, TEXAS 78682

Overview: Manufactures personal computers. Established: 1984. Ownership: Public.

Key Statistics: Annual Sales: $7.8 billion. Number of Employees: 13,500.

Expertise/Education Sought: Electrical engineering, computer engineering, manufacturing engineering, transportation, program analysis, sales, customer service and support, production, network analysis

Alternatives to Full-Time Employment: Part-time jobs, internships

International Assignments are available to employees.

Contact: Human Resources Department; Phone: 512-338-4400.

Company URL: http://www.dell.com

DETEX CORPORATION
302 DETEX DRIVE
NEW BRAUNFELS, TEXAS 78130-3045

Overview: Manufactures security system devices. Established: 1923. Ownership: Private.

Key Statistics: Annual Sales: $14.5 million. Number of Employees: 126.

Hiring History: Number of professional employees hired in 1997: 25. 1996: 5.

Expertise/Education Sought: Electrical engineering, mechanical engineering, manufacturing engineering, information systems, administration, industrial engineering, engineering, management, design engineering, marketing

Alternatives to Full-Time Employment: Paid internships

Company-Paid Training is available.

International Assignments are available to employees.

Contact: Ms. Sandy Stoneking, Director of Human Resources; Phone: 830-629-2900, ext. 321; Fax: 830-620-6711. E-mail: detex@detex.com.

Company URL: http://www.detex.com

DOCUCORP, INC.
5910 NORTH CENTRAL EXPRESSWAY
DALLAS, TEXAS 75206-5125

Overview: Develops, markets, and supports document automation software. Established: 1997. Ownership: Private.

Key Statistics: Annual Sales: $18.8 million. Number of Employees: 240.

Hiring History: Number of professional employees hired in 1997: 10.

Expertise/Education Sought: Software engineering/development

International Assignments are available to employees.

Contact: For college students and recent graduates: Human Resources; Phone: 214-891-6500; Fax: 214-987-8187.

DRESSER INDUSTRIES, INC.
2001 ROSS AVENUE
DALLAS, TEXAS 75201-8001

Overview: Provides equipment and services for the petroleum and mining industries. Established: 1880. Ownership: Public.

Key Statistics: Annual Sales: $6.6 billion. Number of Employees: 32,100.

Expertise/Education Sought: Petroleum/petrochemical engineering, mechanical engineering, design engineering

Alternatives to Full-Time Employment: Paid internships

International Assignments are available to employees.

Contact: For individuals with previous experience: Ms. Tamme Gaddis, Human Resources Generalist; Phone: 214-740-6000; Fax: 214-760-6715.

Company URL: http://www.dresser.com

DSC COMMUNICATIONS CORPORATION
1000 COIT ROAD
PLANO, TEXAS 75075-5802

Overview: Manufactures telecommunications, electronic, and radio systems. Established: 1976. Ownership: Public.

Key Statistics: Annual Sales: $1.4 billion. Number of Employees: 6,500.

Hiring History: Number of professional employees hired in 1996: 750.

Expertise/Education Sought: Software engineering/development, design engineering, electrical engineering, marketing, sales, hardware engineering, test engineering, management information systems

Alternatives to Full-Time Employment: Paid internships

Company-Paid Training is available.

International Assignments are available to employees.

Contact: Staffing Department; Phone: 972-519-3000; Fax: 972-477-8149. Job Hotline: 972-477-JOBS.

Company URL: http://www.dsccc.com

DUKE ENERGY CORPORATION
5400 WESTHEIMER COURT
HOUSTON, TEXAS 77056-5310

Mailing Address: PO Box 1642, Houston, TX 77251

Overview: Operates natural gas pipelines. Established: 1989. Ownership: Public.

Key Statistics: Annual Sales: $1.5 billion.

Expertise/Education Sought: Accounting, finance, information systems, engineering, marketing, mechanical engineering, chemical engineering

Contact: Ms. Carrie Fowler, Human Resources Services Manager; Phone: 713-627-5885; Fax: 713-989-1503.

DYNEGY, INC.
1000 LOUISANNA
HOUSTON, TEXAS 77002-6000

Overview: Distributes natural gas products. Established: 1995. Ownership: Public.

Key Statistics: Annual Sales: $160.5 million. Number of Employees: 300.

Hiring History: Number of professional employees hired in 1996: 100.

Expertise/Education Sought: Accounting, finance, marketing

Alternatives to Full-Time Employment: Paid internships

International Assignments are available to employees.

Contact: Human Resources; Phone: 713-507-6400.

Company URL: http://www.ngccorp.com

ELECTRONIC DATA SYSTEMS CORPORATION
5400 LEGACY DRIVE
PLANO, TEXAS 75024-3105

Overview: Provides information technology services. Established: 1996. Ownership: Public.

Key Statistics: Annual Sales: $14.4 billion. Number of Employees: 100,000.

Expertise/Education Sought: Information systems, accounting, computer science, finance, business administration, electrical engineering, management information systems, mechanical engineering

Alternatives to Full-Time Employment: Part-time jobs, paid internships

Company-Paid Training is available.

International Assignments are available to employees.

Contact: Staffing Department; Phone: 972-605-2700; Fax: 972-605-0123.

Company URL: http://www.eds.com

ENSCO INTERNATIONAL, INC.
1445 ROSS AVENUE, SUITE 2700
DALLAS, TEXAS 75202-2737

Overview: Drills oil and gas wells. Established: 1994. Ownership: Public.

Key Statistics: Annual Sales: $62 million.

Hiring History: Number of professional employees hired in 1996: 8.
Expertise/Education Sought: Mechanical engineering, marine engineering, petroleum/petrochemical engineering
Alternatives to Full-Time Employment: Part-time jobs
International Assignments are available to employees.
Contact: For individuals with previous experience: Mr. Larry Eason, Manager of International Employment; Phone: 214-922-1500; Fax: 214-922-1554. E-mail: leason@enscous.com.
Company URL: http://www.enscous.com

ENSTAR NETWORKING
8304 ESTERS BOULEVARD, SUITE 840
IRVING, TEXAS 75063-2234

Mailing Address: 7450 Flying Cloud Drive, Eden Prairie, MN 55344
Overview: Designs computer systems and offers computer integration services. Established: 1983. Parent Company: ENStar, Inc. Ownership: Private.
Key Statistics: Annual Sales: $21.6 million. Number of Employees: 250.
Expertise/Education Sought: Engineering
Alternatives to Full-Time Employment: Paid internships
Company-Paid Training is available.
Contact: Mr. Chris Joerg, Corporate Recruiter; Phone: 612-942-3800; Fax: 612-944-8021. E-mail: cjoerg@enstar.com.
Company URL: http://www.enstar.com

ERICSSON, INC.
777 EAST CAMPBELL ROAD
DALLAS, TEXAS 75081

Overview: Researches and developments wireless communications equipment. Established: 1994. Ownership: Private.
Key Statistics: Annual Sales: $2.7 billion. Number of Employees: 8,000.
Expertise/Education Sought: Software engineering/development, radio frequency engineering, mechanical engineering, electronics/electronics engineering, research and development, digital technology
Alternatives to Full-Time Employment: Paid internships
Company-Paid Training is available.
International Assignments are available to employees.
Contact: Staffing Department; Phone: 972-583-0000; Fax: 972-583-1843.
Company URL: http://www.ericsson.com

J.C. EVANS CONSTRUCTION COMPANY
13505 BURNET ROAD
AUSTIN, TEXAS 78727-9647

Overview: Provides commercial building excavation services. Established: 1990. Ownership: Private.
Key Statistics: Annual Sales: $113.2 million. Number of Employees: 800.
Expertise/Education Sought: Accounting, finance, marketing
Alternatives to Full-Time Employment: Paid internships
Contact: For college students and recent graduates: Ms. Jan Fulford, Human Resource Director; Phone: 512-244-1400; Fax: 512-244-1900. For individuals with previous experience: Mr. Blake Coleman, President; Phone: 512-244-1400; Fax: 512-244-1900.

THE FAULKNER GROUP, INC.
3901 SOUTH LAMAR BOULEVARD, SUITE 100
AUSTIN, TEXAS 78704-7989

Overview: Provides commercial construction services. Established: 1984. Ownership: Private.
Key Statistics: Annual Sales: $300 million. Number of Employees: 2,100.
Hiring History: Number of professional employees hired in 1997: 75. 1996: 75.
Expertise/Education Sought: Structural engineering, civil engineering, mechanical engineering
Alternatives to Full-Time Employment: Paid internships
Company-Paid Training is available.
International Assignments are available to employees.
Contact: Mr. Denis Snyder, Human Resources Director; Phone: 512-448-9898; Fax: 512-448-9890. E-mail: dsnyder@faulkcomp.com.

FLS MILJO, INC.
100 GLENBOROUGH DRIVE
HOUSTON, TEXAS 77067-3600

Overview: Provides artists' agent and brokerage services. Established: 1996. Parent Company: FLS Miljo as Denmark. Ownership: Private.
Key Statistics: Annual Sales: $50 million. Number of Employees: 119.
Hiring History: Number of professional employees hired in 1996: 25.
Expertise/Education Sought: Mechanical engineering, structural engineering, design engineering, drafting, management, sales
Alternatives to Full-Time Employment: Part-time jobs
International Assignments are available to employees.
Contact: Mrs. Paula Chamoff, Human Resources Administrator; Phone: 281-539-3409; Fax: 281-539-3411. E-mail: mis@cottrell.com.
Company URL: http://www.cottrell.com

GEO SPACE CORPORATION
7334 NORTH GESSNER DRIVE
HOUSTON, TEXAS 77040-3144

Overview: Manufactures measuring, control, and geophysical electronic devices. Established: 1993. Parent Company: Oyo Geospace Corporation. Ownership: Private.
Key Statistics: Annual Sales: $17.1 million. Number of Employees: 175.
Hiring History: Number of professional employees hired in 1996: 10.
Expertise/Education Sought: Accounting, engineering, manufacturing engineering
Alternatives to Full-Time Employment: Part-time jobs
Company-Paid Training is available.
Contact: For individuals with previous experience: Mr. Pete Carlisle, Vice President of Human Resources; Phone: 713-939-7093; Fax: 713-937-8012.
Company URL: http://www.geospacecorp.com

GIFFORD-HILL
3500 MAPLE AVENUE
DALLAS, TEXAS 75219

Overview: Wholesales glass construction materials. Established: 1986. Parent Company: Hanson North America. Ownership: Private.
Key Statistics: Number of Employees: 2,300.
Expertise/Education Sought: Construction, construction, software engineering/development, computer science, application developer
Alternatives to Full-Time Employment: Paid internships
Company-Paid Training is available.
Contact: For individuals with previous experience: Human Resources; Phone: 214-525-5500; Fax: 214-525-5672.

GPM GAS COOPORATION
1300 POST OAK BOULEVARD, SUITE 800
HOUSTON, TEXAS 77056-3028

Overview: Produces petroleum. Established: 1992. Parent Company: Phillips Petroleum Company. Ownership: Public.
Key Statistics: Annual Sales: $362.9 million. Number of Employees: 1,157.
Hiring History: Number of professional employees hired in 1997: 10. 1996: 50.
Expertise/Education Sought: Marketing, safety, environmental
Alternatives to Full-Time Employment: Paid internships
Company-Paid Training is available.
Contact: Ms. Phyllis Beamguard, Human Resources Staffing; Phone: 713-297-6040; Fax: 713-297-6080. E-mail: pebeamg@bvemx.ppco.com.
Company URL: http://www.phillips66.com

GROUP MAINTENANCE AMERICA

8 EAST GREENWAY PLAZA, SUITE 1500
HOUSTON, TEXAS 77046-0801

Overview: Provides heating, ventilation, and air conditioning as well as plumbing and electrical services to residential and commercial customers. Established: 1996. Ownership: Private.

Key Statistics: Annual Sales: $66.9 million. Number of Employees: 711.

Expertise/Education Sought: Engineering

Contact: For individuals with previous experience: Human Resources Department; Phone: 713-860-0100; Fax: 713-626-4766.

Company URL: http://www.groupmac.com

GSE LINING TECHNOLOGY, INC.

19103 GUNDLE ROAD
HOUSTON, TEXAS 77073-3598

Overview: Manufactures polyethylene film. Established: 1995. Ownership: Public.

Key Statistics: Annual Sales: $92.2 million. Number of Employees: 650.

Expertise/Education Sought: Manufacturing

Alternatives to Full-Time Employment: Paid internships

Company-Paid Training is available.

International Assignments are available to employees.

Contact: For individuals with previous experience: Human Resources; Phone: 281-443-8564; Fax: 281-230-8607.

GTE DIRECTORIES CORPORATION

GTE PLACE, WEST AIRFIELD DRIVE
DFW AIRPORT TEXAS, TEXAS 75261

Overview: Publishes telephone books and Yellow Page directories. Established: 1936. Parent Company: GTE Corporation. Ownership: Public.

Key Statistics: Annual Sales: $659.9 million. Number of Employees: 4,692.

Expertise/Education Sought: Computer programming, systems analysis, telecommunications, accounting, sales, customer service and support, quality control, management information systems, planning

Alternatives to Full-Time Employment: Paid internships

Company-Paid Training is available.

International Assignments are available to employees.

Contact: Human Resources Department; Phone: 972-453-6000; Fax: 972-453-6888. E-mail: hr@gtebc.com. Job Hotline: 972-621-2200, ext. 9675.

Company URL: http://www.gte.com

GTE SERVICE CORPORATION

700 HIDDEN RIDGE
IRVING, TEXAS 75308

Overview: Provides telecommunications services. Established: 1926. Ownership: Public.

Key Statistics: Annual Sales: $1 billion. Number of Employees: 5,022.

Expertise/Education Sought: Computer programming, management information systems, telecommunications, sales, customer service and support, electrical engineering, computer science, finance, accounting

Alternatives to Full-Time Employment: Paid internships

Company-Paid Training is available.

International Assignments are available to employees.

Contact: For college students and recent graduates: Ms. Mary Jane Thompson, Staff Manager of Recruitment; Phone: 972-718-1515. For individuals with previous experience: Human Resources Department; Phone: 972-718-1220; Fax: 972-718-1633.

GULF COAST RICE AND TRADING

8018 SHANGRILA LANE
HOUSTON, TEXAS 77095-4133

Overview: Provides commodities brokerage services. Established: 1996. Ownership: Private.

Key Statistics: Annual Sales: $40 million.

Hiring History: Number of professional employees hired in 1997: 9.

Expertise/Education Sought: Manufacturing engineering

Contact: For individuals with previous experience: Human Relations Department; Fax: 281-855-2912.

HALLIBURTON COMPANY

PO BOX 3
HOUSTON, TEXAS 77001

Overview: Provides products and services in the exploration, development, and production of natural gas. Established: 1919. Ownership: Private.

Key Statistics: Annual Sales: $7.4 billion. Number of Employees: 60,000.

Expertise/Education Sought: Engineering

Contact: Human Resources Department; Phone: 713-676-5099; Fax: 713-676-8433.

Company URL: http://www.halliburton.com

HALLIBURTON ENERGY SERVICES

4100 CLINTON DRIVE
HOUSTON, TEXAS 75056

Overview: Provides oil field and construction services. Established: 1919. Ownership: Private.

Expertise/Education Sought: Civil engineering, industrial engineering

Alternatives to Full-Time Employment: Paid internships

International Assignments are available to employees.

Contact: For individuals with previous experience: Human Resources; Phone: 713-676-4415; Fax: 713-676-8433.

HITACHI SEMICONDUCTOR (AMERICA), INC.

6431 LONGHORN DRIVE
IRVING, TEXAS 75063-2738

Overview: Manufactures semiconductors. Established: 1987. Parent Company: Hitachi America, Ltd. Ownership: Private.

Key Statistics: Annual Sales: $205 million. Number of Employees: 872.

Hiring History: Number of professional employees hired in 1997: 300. 1996: 300.

Expertise/Education Sought: Mechanical engineering, electrical engineering, industrial engineering, accounting, marketing, management information systems, chemical engineering

Alternatives to Full-Time Employment: Paid internships

Company-Paid Training is available.

Contact: Human Resources; Phone: 972-756-3744; Fax: 972-756-3733. E-mail: staffing@hitachi-hisus.com.

Company URL: http://www.hitachi-hisus.com

HUNT OIL COMPANY

1445 ROSS AVENUE
DALLAS, TEXAS 75202

Overview: Produces and distributes petroleum and gas. Established: 1976. Ownership: Private.

Key Statistics: Annual Sales: $299 million. Number of Employees: 2,500.

Hiring History: Number of professional employees hired in 1997: 105.

Expertise/Education Sought: Geology, accounting, computer programming, petroleum/petrochemical engineering

Company-Paid Training is available.

International Assignments are available to employees.

Contact: Ms. Diane Brown, Recruiter; Phone: 214-978-8000; Fax: 214-978-8911. Job Hotline: 214-978-8022.

HUNT PETROLEUM CORPORATION

1601 ELM STREET, SUITE 3400
DALLAS, TEXAS 75201-7208

Overview: Produces crude petroleum and natural gas. Established: 1996. Ownership: Private.

KEY STATISTICS: Number of Employees: 85.
HIRING HISTORY: Number of professional employees hired in 1996: 20.
EXPERTISE/EDUCATION SOUGHT: Petroleum/petrochemical engineering
CONTACT: Ms. Kathy Smith, Personnel Administrator, Personnel Department, Dallas, Texas, 75201-7208; Phone: 214-880-8892; Fax: 214-880-8887.

INTEGRATED MEDICAL NETWORKS
511 EAST JOHN CARPENTER FREEWAY, SUITE 500
IRVING, TEXAS 75062

OVERVIEW: Develops software for hospital databases.
HIRING HISTORY: Number of professional employees hired in 1997: 20.
EXPERTISE/EDUCATION SOUGHT: Software engineering/development, technical engineering
COMPANY-PAID TRAINING is available.
CONTACT: For individuals with previous experience: Ms. Michelle Dennis, Manager of Human Resources. E-mail: michelle@imncorp.com.

INTELLICALL, INC.
2155 CHENNAULT DRIVE, SUITE 410
CARROLLTON, TEXAS 75006-5087

OVERVIEW: Provides telephone communication services. Established: 1985. Ownership: Public.
KEY STATISTICS: Annual Sales: $92.8 million. Number of Employees: 221.
EXPERTISE/EDUCATION SOUGHT: Electrical engineering, accounting, finance, data processing, software engineering/development
ALTERNATIVES TO FULL-TIME EMPLOYMENT: Part-time jobs
COMPANY-PAID TRAINING is available.
CONTACT: Ms. Rebecca Marks, Human Resources; Phone: 972-416-0022; Fax: 972-416-7213. E-mail: rmarks@intellicall.com.
COMPANY URL: http://www.intellicall.com

IXC COMMUNICATION, INC.
1122 CAPITAL OF TEXAS HIGHWAY SOUTH
AUSTIN, TEXAS 78746-1039

OVERVIEW: Provides data communications services. Established: 1995. Ownership: Private.
KEY STATISTICS: Annual Sales: $82.7 million. Number of Employees: 678.
HIRING HISTORY: Number of professional employees hired in 1997: 300.
EXPERTISE/EDUCATION SOUGHT: Telecommunications
ALTERNATIVES TO FULL-TIME EMPLOYMENT: Paid internships
COMPANY-PAID TRAINING is available.
INTERNATIONAL ASSIGNMENTS are available to employees.
CONTACT: Mr. Ron Boscamp, Manager of Staffing; Phone: 512-427-3917; Fax: 512-328-0624. E-mail: rboscamp@ixc-comm.com.

JC PENNEY COMPANY, INC.
6501 LEGACY DRIVE
PLANO, TEXAS 75024-3698

MAILING ADDRESS: PO Box 10001, Dallas, TX 75301
OVERVIEW: Operates national chain of department stores. Established: 1902. Ownership: Public.
KEY STATISTICS: Annual Sales: $23.6 billion. Number of Employees: 252,000.
EXPERTISE/EDUCATION SOUGHT: Information systems
CONTACT: Personnel; Phone: 972-431-2313; Fax: 972-431-2320.

KALDAIR, INC.
15810 PARK TEN PLAVE, SUITE 195
HOUSTON, TEXAS 77084-5134

OVERVIEW: Provides engineering services. Established: 1997. Ownership: Private.
KEY STATISTICS: Annual Sales: $10 million. Number of Employees: 32.
EXPERTISE/EDUCATION SOUGHT: Engineering, drafting, autocade, marketing
ALTERNATIVES TO FULL-TIME EMPLOYMENT: Paid internships
COMPANY-PAID TRAINING is available.
CONTACT: For individuals with previous experience: Office Manager, Human Resources; Phone: 281-492-2262; Fax: 281-492-2399.
COMPANY URL: http://www.kaldair.com

LIFE-TECH, INC.
4235 GREENBRIAR DRIVE
STAFFORD, TEXAS 77477

OVERVIEW: Manufactures medical and surgical equipment. Established: 1970. Ownership: Private.
KEY STATISTICS: Annual Sales: $11.2 million. Number of Employees: 113.
HIRING HISTORY: Number of professional employees hired in 1997: 5. 1996: 5.
EXPERTISE/EDUCATION SOUGHT: Engineering, accounting, sales
CONTACT: For individuals with previous experience: Human Resources Department; Phone: 281-491-6600.
COMPANY URL: http://www.lifetech.com

LYONDELL PETROCHEMICAL COMPANY
1221 MCKINNEY STREET, SUITE 1600
HOUSTON, TEXAS 77010

OVERVIEW: Manufactures and processes petroleum products and chemicals.
HIRING HISTORY: Number of professional employees hired in 1997: 300. 1996: 50.
EXPERTISE/EDUCATION SOUGHT: Chemical engineering, petroleum/petrochemical engineering, technical engineering
ALTERNATIVES TO FULL-TIME EMPLOYMENT: Paid internships
COMPANY-PAID TRAINING is available.
CONTACT: Ms. Karen Whipple, Human Resources Manager of Staffing; Phone: 713-652-7200; Fax: 713-652-4563.

MATRIX ENGINEERING, INC.
1725 WEST CARDINAL DRIVE
BEAUMONT, TEXAS 77705-6415

MAILING ADDRESS: PO Box 20958, Beaumont, TX 77720
OVERVIEW: Provides engineering services. Established: 1993. Parent Company: Waste Management. Ownership: Public.
KEY STATISTICS: Annual Sales: $39.6 million. Number of Employees: 622.
EXPERTISE/EDUCATION SOUGHT: Design engineering, construction engineering, architecture
CONTACT: For individuals with previous experience: Ms. Jan Ellender, Department of Human Resources; Phone: 409-842-5828; Fax: 409-842-2806.

MAXUS ENERGY CORPORATION
717 NORTH HARWOOD STREET
DALLAS, TEXAS 75201-6538

OVERVIEW: Produces crude petroleum and natural gas. Established: 1995. Parent Company: YPS, SA (Argentina). Ownership: Private.
KEY STATISTICS: Annual Sales: $738.1 million. Number of Employees: 2,027.
EXPERTISE/EDUCATION SOUGHT: Geology, petroleum/petrochemical engineering, production
ALTERNATIVES TO FULL-TIME EMPLOYMENT: Internships
INTERNATIONAL ASSIGNMENTS are available to employees.
CONTACT: Mr. Steve Ginsburg, Manager of Organizational Services; Fax: 214-979-1260.

M W KELLOGG
601 JEFFERSON STREET
HOUSTON, TEXAS 77002-7900

OVERVIEW: Heavy construction, including dams and waterways. Established: 1995. Parent Company: Dresser. Ownership: Public.
KEY STATISTICS: Annual Sales: $1.2 billion. Number of Employees: 4,000.

M W Kellogg (continued)

Expertise/Education Sought: Engineering

Contact: Human Resources; Phone: 713-753-5555. Job Hotline: 713-753-2225.

NORTH AMERICAN CONSTRUCTION CORPORATION

4754 SHAVANO OAK ROAD
SAN ANTONIO, TEXAS 78249-4009

Mailing Address: PO Box 690350, San Antonio, TX 78260

Overview: Provides construction services. Established: 1982. Ownership: Private.

Key Statistics: Annual Sales: $34.6 million. Number of Employees: 150.

Hiring History: Number of professional employees hired in 1996: 15.

Expertise/Education Sought: Mechanical engineering, construction

Contact: For individuals with previous experience: Ms. Elizabeth Hudson, Human Resources; Phone: 210-492-8944; Fax: 210-479-0553.

NSG

2010 MERRITT DRIVE
GARLAND, TEXAS 75041-6131

Overview: Operates sporting and recreational camps. Established: 1948. Ownership: Private.

Expertise/Education Sought: Manufacturing

Contact: Human Resources; Phone: 972-840-4043.

OCCIDENTAL CHEMICAL CORPORATION

5005 LBJ FREEWAY
DALLAS, TEXAS 75244-6119

Overview: Discovers and produces oil and gas, manufactures chemicals, and mines coal. Established: 1973. Parent Company: Occidental Petroleum Corporation. Ownership: Public.

Key Statistics: Annual Sales: $2.7 billion. Number of Employees: 8,300.

Hiring History: Number of professional employees hired in 1996: 28.

Expertise/Education Sought: Chemical engineering, accounting, sales, engineering, research and development, business management, marketing, safety management

Alternatives to Full-Time Employment: Part-time jobs, paid internships

Contact: For individuals with previous experience: Mr. James Reder, Manager of Staff Planning; Phone: 972-404-3921; Fax: 972-404-3414. E-mail: james_f._reder@oxy.com.

Company URL: http://www.oxychem.com

OMEGA OPTICAL COMPANY, INC.

13515 NORTH STEMMONS FREEWAY
DALLAS, TEXAS 75234-5765

Overview: Manufactures lenses for glasses. Established: 1994. Parent Company: ESSILOR. Ownership: Private.

Key Statistics: Annual Sales: $66.7 million. Number of Employees: 1,052.

Hiring History: Number of professional employees hired in 1997: 550.

Expertise/Education Sought: Accounting, marketing, data processing

International Assignments are available to employees.

Contact: Human Resources; Phone: 972-241-4141, ext. 310; Fax: 972-241-2947.

OPTEK TECHNOLOGY, INC.

1215 WEST CROSBY ROAD
CARROLLTON, TEXAS 75006-6905

Overview: Manufactures electronic components and semiconductors. Established: 1979. Ownership: Public.

Key Statistics: Annual Sales: $67.4 million. Number of Employees: 1,700.

Hiring History: Number of professional employees hired in 1997: 60. 1996: 20.

Expertise/Education Sought: Optics, sales, manufacturing engineering, mechanical engineering, automotive engineering, design, process engineering

Alternatives to Full-Time Employment: Part-time jobs, paid internships

Company-Paid Training is available.

International Assignments are available to employees.

Contact: Mr. Angie Crain, Human Resources Representative; Phone: 972-323-2363; Fax: 972-242-0818.

Company URL: http://www.optekinc.com

OVERHEAD DOOR CORPORATION

6750 LBJ FREEWAY
DALLAS, TEXAS 75240-6512

Overview: Manufactures metal garage doors and rolling doors. Established: 1996. Parent Company: Sanwa Shutter Corporation. Ownership: Private.

Key Statistics: Annual Sales: $304.5 million. Number of Employees: 3,969.

Hiring History: Number of professional employees hired in 1996: 50.

Expertise/Education Sought: Manufacturing, sales, accounting, marketing

Company-Paid Training is available.

International Assignments are available to employees.

Contact: Human Resource Department; Phone: 800-275-3290; Fax: 972-371-0322. E-mail: jobs@overheaddoor.com.

OYO CORPORATION, USA

9777 WEST GULF BANK ROAD, SUITE 5
HOUSTON, TEXAS 77040-3132

Overview: Produces electronic connectors. Established: 1983. Ownership: Private.

Key Statistics: Annual Sales: $71.3 million. Number of Employees: 650.

Expertise/Education Sought: Engineering

Company-Paid Training is available.

Contact: For individuals with previous experience: Human Resources, 9777 West Gulf Bank Road, Suite 10, Houston, Texas, 77040-3132; Phone: 713-849-0804; Fax: 713-849-4945.

PEROT SYSTEMS CORPORATION

12377 MERIT DRIVE, SUITE 1100
DALLAS, TEXAS 75251-3233

Overview: Provides outsourcing consulting services. Established: 1988. Ownership: Private.

Key Statistics: Annual Sales: $342.3 million. Number of Employees: 5,000.

Hiring History: Number of professional employees hired in 1997: 1,280.

Expertise/Education Sought: Technical, software engineering/development, systems integration, business development, management consulting

Alternatives to Full-Time Employment: Paid internships

Company-Paid Training is available.

International Assignments are available to employees.

Contact: Ms. Tony Uster, Senior Recruiting Manager; Phone: 214-383-5808; Fax: 800-607-3768.

Company URL: http://www.perotsystems.com

PILGRIM'S PRIDE CORPORATION

110 SOUTH TEXAS STREET
PITTSBURGH, TEXAS 75686-1532

Overview: Provides poultry processing services. Established: 1945. Ownership: Public.

Key Statistics: Annual Sales: $1.3 billion. Number of Employees: 12,354.

Expertise/Education Sought: Accounting, finance, computer programming, management information systems, data processing

Alternatives to Full-Time Employment: Paid internships

Company-Paid Training is available.

INTERNATIONAL ASSIGNMENTS are available to employees.
CONTACT: For individuals with previous experience: Ms. Goldie Harwell, Manager of Employee Relations; Phone: 903-855-4482; Fax: 903-855-4466.

PIZZA HUT WORLDWIDE, INC.
14841 DALLAS PARKWAY
DALLAS, TEXAS 75240-2100

OVERVIEW: Operates chain of pizza restaurants. Established: 1989. Ownership: Public.
KEY STATISTICS: Annual Sales: $54.2 million. Number of Employees: 2,500.
EXPERTISE/EDUCATION SOUGHT: Marketing, human resources, finance, operations
ALTERNATIVES TO FULL-TIME EMPLOYMENT: Paid internships
COMPANY-PAID TRAINING is available.
INTERNATIONAL ASSIGNMENTS are available to employees.
CONTACT: Staffing Department.
COMPANY URL: http://www.pizzahut.com

P.P.D. PHARMACO, INC.
4009 BANNISTER LANE
AUSTIN, TEXAS 78704-7792

OVERVIEW: Provides contract research services. Established: 1992. Parent Company: Applied Bioscience International. Ownership: Public.
KEY STATISTICS: Annual Sales: $98.9 million. Number of Employees: 1,105.
EXPERTISE/EDUCATION SOUGHT: Data processing, management, logistics, sales
COMPANY-PAID TRAINING is available.
INTERNATIONAL ASSIGNMENTS are available to employees.
CONTACT: Mr. Jerry Williams, Director of Human Resources; Phone: 512-440-5820; Fax: 512-440-2952. Job Hotline: 512-440-5900.
COMPANY URL: http://www.ppdpharmaco.com

QUEST MEDICAL, INC.
1 ALLENTOWN PARKWAY
ALLEN, TEXAS 75002

OVERVIEW: Manufactures medical and surgical products and equipment. Established: 1979. Ownership: Public.
KEY STATISTICS: Annual Sales: $26.1 million. Number of Employees: 370.
EXPERTISE/EDUCATION SOUGHT: Marketing
ALTERNATIVES TO FULL-TIME EMPLOYMENT: Paid internships
CONTACT: Human Resources Department; Phone: 972-390-9800; Fax: 972-390-2881.

RADIAN INTERNATIONAL
8501 NORTH MOPAC BOULEVARD
AUSTIN, TEXAS 78759-8341

OVERVIEW: Provides environmental research services. Established: 1975. Parent Company: Dow Chemical Company. Ownership: Public.
KEY STATISTICS: Annual Sales: $233.1 million. Number of Employees: 2,000.
HIRING HISTORY: Number of professional employees hired in 1997: 500.
EXPERTISE/EDUCATION SOUGHT: Meteorology, chemical engineering, environmental engineering, civil engineering, chemistry, geology, toxicology, environmental science, materials science, chemical safety
ALTERNATIVES TO FULL-TIME EMPLOYMENT: Part-time jobs, paid internships
COMPANY-PAID TRAINING is available.
INTERNATIONAL ASSIGNMENTS are available to employees.
CONTACT: For individuals with previous experience: Manager of Human Resources; Phone: 512-454-4797. E-mail: jp_kirksey@radian.com.
COMPANY URL: http://www.radian.com

RAYTHEON TRAINING
PO BOX 6171
ARLINGTON, TEXAS 76005-6171

OVERVIEW: Manufactures training systems. Established: 1988. Parent Company: Raytheon Company. Ownership: Private.
KEY STATISTICS: Annual Sales: $338 million. Number of Employees: 4,500.
EXPERTISE/EDUCATION SOUGHT: Electrical engineering, computer engineering, computer science, information systems, software engineering/development, systems engineering
ALTERNATIVES TO FULL-TIME EMPLOYMENT: Paid internships
INTERNATIONAL ASSIGNMENTS are available to employees.
CONTACT: Human Resources Department; Phone: 817-619-2000; Fax: 817-619-3576.
COMPANY URL: http://www.raytheon.com

REF-CHEM CONSTRUCTION CORPORATION
120 NORTH MUNGER STREET
PASADENA, TEXAS 77506-1326

OVERVIEW: Builds industrial buildings and warehouses. Established: 1995. Ownership: Private.
KEY STATISTICS: Annual Sales: $70.3 million. Number of Employees: 615.
EXPERTISE/EDUCATION SOUGHT: Design
COMPANY-PAID TRAINING is available.
CONTACT: Human Resources; Phone: 713-477-5065; Fax: 713-477-9615.
COMPANY URL: http://www.ref-chem.com

SANDVIK ROCK TOOLS, INC.
8760 CLAY ROAD
HOUSTON, TEXAS 77080-1859

OVERVIEW: Manufactures tools for the mining and construction industries. Established: 1987. Parent Company: Sandvik, AB. Ownership: Private.
KEY STATISTICS: Annual Sales: $75 million. Number of Employees: 380.
EXPERTISE/EDUCATION SOUGHT: Manufacturing
CONTACT: For individuals with previous experience: Human Resources Department.

SBC COMMUNICATIONS, INC.
175 EAST HOUSTON STREET
SAN ANTONIO, TEXAS 78205-2233

OVERVIEW: Provides telecommunications products and services. Established: 1984. Ownership: Public.
KEY STATISTICS: Annual Sales: $23.5 billion. Number of Employees: 110,000.
EXPERTISE/EDUCATION SOUGHT: Accounting, finance, sales, telecommunications, data processing
ALTERNATIVES TO FULL-TIME EMPLOYMENT: Part-time jobs, paid internships
INTERNATIONAL ASSIGNMENTS are available to employees.
CONTACT: For individuals with previous experience: Corporate Recruiter; Phone: 210-351-2480; Fax: 210-351-5122. Job Hotline: 800-444-6303.
COMPANY URL: http://www.sbc.com

Schlumberger

SCHLUMBERGER, LIMITED
300 SCHLUMBERGER DRIVE, DEPARTMENT PG 97/98
SUGAR LAND, TEXAS 77478

OVERVIEW: Provides oilfield services and measurement equipment. Established: 1927. Ownership: Public.
KEY STATISTICS: Annual Sales: $9 billion. Number of Employees: 55,000.
HIRING HISTORY: Number of professional employees hired in 1996: 500.

Schlumberger, Limited (continued)

Expertise/Education Sought: Technical engineering, chemical engineering, civil engineering, computer engineering, electrical engineering, mechanical engineering, petroleum/petrochemical engineering, computer science, geology, geophysics

Alternatives to Full-Time Employment: Paid internships

International Assignments are available to employees.

Contact: Miss Paula McCann, Manager of Recruiting, Coordination, and University Relations; Phone: 281-285-8447; Fax: 281-285-7173. E-mail: recruiting@slb.com.

Company URL: http://www.slb.com

For more information, see full corporate profile, pg. 298.

SEA LION TECHNOLOGY, INC.

5700 JOHNNY PALMER ROAD
TEXAS CITY, TEXAS 77590

Mailing Address: PO Box 18007, Texas City, TX 77592

Overview: Produces industrial organic chemicals. Established: 1993. Ownership: Private.

Key Statistics: Annual Sales: $10.5 million. Number of Employees: 55.

Hiring History: Number of professional employees hired in 1997: 5. 1996: 5.

Expertise/Education Sought: Manufacturing

Company-Paid Training is available.

Contact: For individuals with previous experience: Human Resources; Phone: 409-948-4351; Fax: 281-337-3694.

Company URL: http://www.sealiontech.com

SIEMENS ELECTROCOM, LP

2910 AVENUE F
ARLINGTON, TEXAS 76011-5214

Overview: Manufactures postal sorting machines. Established: 1994. Ownership: Private.

Key Statistics: Annual Sales: $402.8 million. Number of Employees: 882.

Hiring History: Number of professional employees hired in 1996: 200.

Expertise/Education Sought: Accounting, finance, business management, data processing, electrical engineering, mechanical engineering, software engineering/development, electrical technology

Alternatives to Full-Time Employment: Paid internships

Company-Paid Training is available.

International Assignments are available to employees.

Contact: Mr. Leonard Cameron, Human Resource Specialist; Phone: 817-695-7598; Fax: 817-695-3171.

SIGNATURE DATA SOLUTIONS

203 WEST WALL STREET, SUITE 1100
MIDLAND, TEXAS 79701-4516

Overview: Designs computer systems and offers computer integration services. Established: 1996. Ownership: Private.

Key Statistics: Annual Sales: $11 million.

Hiring History: Number of professional employees hired in 1997: 4.

Expertise/Education Sought: Computer programming, engineering

International Assignments are available to employees.

Contact: For individuals with previous experience: Mr. Erik C. Seay, Assistant Vice President; Phone: 915-570-6200; Fax: 915-570-6277. E-mail: eseay@sdsongone.com.

Company URL: http://www.sdsone.com

SKYTEAM AIRLINES, INC.

4201 NORTH MAIN STREET, SUITE 210
FORT WORTH, TEXAS 76106-2749

Overview: Operates airline. Established: 1996. Ownership: Private.

Key Statistics: Annual Sales: $36 million.

Expertise/Education Sought: Transportation

Contact: Mrs. Theresa Palmer-Foltz, Vice President; Phone: 817-625-2220; Fax: 817-625-2229.

SMT CENTRE OF TEXAS, INC.

15508 BRATTON LANE
AUSTIN, TEXAS 78728-3805

Overview: Manufactures printed circuit boards. Established: 1996. Parent Company: SMT Centre Of Ontario. Ownership: Private.

Key Statistics: Annual Sales: $25 million. Number of Employees: 160.

Hiring History: Number of professional employees hired in 1996: 20.

Expertise/Education Sought: Electrical engineering, manufacturing engineering

Company-Paid Training is available.

International Assignments are available to employees.

Contact: For individuals with previous experience: Human Resources; Phone: 512-310-4300; Fax: 512-310-4301.

SOLECTRON TEXAS, LP

12455 RESEARCH BOULEVARD
AUSTIN, TEXAS 78759-2314

Overview: Manufactures computers. Established: 1996. Parent Company: Solectron. Ownership: Private.

Key Statistics: Annual Sales: $600 million. Number of Employees: 1,400.

Expertise/Education Sought: Manufacturing

Alternatives to Full-Time Employment: Paid internships

International Assignments are available to employees.

Contact: For individuals with previous experience: Ms. Linda Caudel, Recruiter; Phone: 512-425-7145; Fax: 512-425-7347.

Company URL: http://www.solectron.com

SOUTHDOWN, INC.

1200 SMITH STREET, SUITE 2400
HOUSTON, TEXAS 77002-4402

Overview: Manufactures cement and concrete. Established: 1930. Ownership: Public.

Key Statistics: Annual Sales: $664.4 million. Number of Employees: 2,400.

Expertise/Education Sought: Chemical engineering, mining, process engineering

Alternatives to Full-Time Employment: Internships

Company-Paid Training is available.

Contact: Mr. Bob Osborn, Manager of Human Resources; Phone: 713-650-6200; Fax: 713-653-6815.

SOUTHWESTERN BELL MOBILE SYSTEMS

17330 PRESTON ROAD, 100A
DALLAS, TEXAS 75252

Overview: Provides cellular mobile telephone services. Established: 1983. Parent Company: SBC Corporation. Ownership: Public.

Key Statistics: Annual Sales: $334 million. Number of Employees: 4,810.

Expertise/Education Sought: Sales, electrical engineering, electronics/electronics engineering

Alternatives to Full-Time Employment: Paid internships

Company-Paid Training is available.

International Assignments are available to employees.

Contact: Human Resources; Phone: 972-733-2000; Fax: 972-733-5909.

SOUTHWEST FOUNDATION FOR BIOMEDICAL RESEARCH

7620 NORTHWEST LOOP 410 AT MILITARY DRIVE WEST
SAN ANTONIO, TEXAS 78227

Mailing Address: PO Box 760549, San Antonio, TX 78245-0549

Overview: Independent, nonprofit biomedical research institution. Established: 1941.
Hiring History: Number of professional employees hired in 1996: 3.
Expertise/Education Sought: Computer programming, systems analysis
Alternatives to Full-Time Employment: Part-time jobs
Contact: Ms. E. Gwen Bridgeford, Human Resources Specialist; Phone: 210-670-3256; Fax: 210-670-3328.
Company URL: http://www.sfbr.org

SOUTHWEST RESEARCH INSTITUTE
PO DRAWER 28510
SAN ANTONIO, TEXAS 78228-0510

Overview: Provides industrial and scientific research. Established: 1966. Ownership: Private.
Key Statistics: Annual Sales: $258.9 million. Number of Employees: 2,624.
Expertise/Education Sought: Engineering, computer science, physical science, chemistry, automotive engineering, electrical engineering, industrial engineering, mechanical engineering, materials engineering
Alternatives to Full-Time Employment: Paid internships
Company-Paid Training is available.
Contact: Ms. Barbara James, Senior Personnel Specialist; Phone: 210-522-2223; Fax: 210-522-3990.

SPECIFIC CRUISE SYSTEMS, INC.
1200 WEST RISINGER ROAD
FORT WORTH, TEXAS 76134-5670

Overview: Manufactures automotive cruise controls and security systems. Established: 1979. Ownership: Private.
Key Statistics: Annual Sales: $40 million. Number of Employees: 450.
Hiring History: Number of professional employees hired in 1997: 200. 1996: 100.
Expertise/Education Sought: Mechanical engineering, electrical engineering, accounting, finance, data processing
Alternatives to Full-Time Employment: Paid internships
Company-Paid Training is available.
Contact: Ms. Jane Turner, Director of Human Resources; Phone: 817-293-5313, ext. 263.

SPLIT ROCK SERVICES, INC.
2170 BUCK THORNE PLACE, SUITE 350
THE WOODLANDS, TEXAS 77380-1775

Overview: Provides communications services. Established: 1997. Ownership: Private.
Key Statistics: Annual Sales: $25 million. Number of Employees: 57.
Hiring History: Number of professional employees hired in 1997: 20.
Expertise/Education Sought: Electrical engineering, telecommunications
Alternatives to Full-Time Employment: Paid internships
Contact: For individuals with previous experience: Human Resources Department; Phone: 281-363-9760; Fax: 281-364-6668.
Company URL: http://www.splitrock.net

STERLING SOFTWARE, INC.
300 CRESCENT COURT, SUITE 1200
DALLAS, TEXAS 75201

Overview: Provides business-oriented computer software. Established: 1995. Ownership: Public.
Key Statistics: Annual Sales: $267.8 million. Number of Employees: 1,266.
Expertise/Education Sought: Software engineering/development
International Assignments are available to employees.
Contact: For individuals with previous experience: Mr. Ivan Hughes, Vice President of Organization; Phone: 214-981-1000; Fax: 214-981-1266.
Company URL: http://www.sterling.com

STRUCKMEYER CORPORATION
13737 STEMMONS FREEWAY
DALLAS, TEXAS 75234

Overview: Manufactures medical disposables and provides medical packaging services. Established: 1995. Parent Company: Pacific Device. Ownership: Private.
Key Statistics: Annual Sales: $10 million. Number of Employees: 580.
Expertise/Education Sought: Engineering, manufacturing, quality assurance
Company-Paid Training is available.
Contact: For individuals with previous experience: Mrs. Debbie Whittaker, Human Resources Manager; Phone: 972-247-0407; Fax: 972-620-1262.

SULZER INTERMEDICS, INC.
4000 TECHNOLOGY DRIVE
ANGLETON, TEXAS 77515-2523

Overview: Manufactures surgical supplies, electromedical equipment, and cardiac rhythm instrumentation. Established: 1988. Parent Company: Sulzer Medica. Ownership: Private.
Key Statistics: Annual Sales: $300 million. Number of Employees: 3,020.
Expertise/Education Sought: Accounting, computer programming, marketing, finance, information systems
Alternatives to Full-Time Employment: Part-time jobs, paid internships
Company-Paid Training is available.
International Assignments are available to employees.
Contact: Ms. Terri White, Technical Recruiter; Phone: 409-848-4581; Fax: 409-848-4580. E-mail: terriw@mails.imed.com.
Company URL: http://www.imed.com

SULZER ORTHOPEDICS, INC.
9900 SPECTRUM DRIVE
AUSTIN, TEXAS 78717

Overview: Manufactures surgical and medical instruments and supplies. Established: 1988. Parent Company: Sulzer Medica. Ownership: Private.
Key Statistics: Annual Sales: $100 million. Number of Employees: 500.
Hiring History: Number of professional employees hired in 1997: 50. 1996: 15.
Expertise/Education Sought: Manufacturing engineering, mechanical engineering, quality engineering, marketing, sales, human resources, clinical research, biomedical engineering
Alternatives to Full-Time Employment: Paid internships
Company-Paid Training is available.
International Assignments are available to employees.
Contact: Ms. Dawn Wilkin, Human Resorces Advisor; Phone: 512-432-9900; Fax: 512-432-9200. E-mail: dawn.hitt-wilken@sous.com. Job Hotline: 512-432-9283.

SYNSTRON, INC.
17200 PARK ROW
HOUSTON, TEXAS 77084

Overview: Provides computer systems services. Established: 1996. Parent Company: TechSyn Corporation. Ownership: Public.
Key Statistics: Annual Sales: $113.8 million. Number of Employees: 589.
Expertise/Education Sought: Software engineering/development, hardware engineering, engineering, radio frequency engineering, engineering technology, engineering
Alternatives to Full-Time Employment: Paid internships
Company-Paid Training is available.
International Assignments are available to employees.
Contact: Human Resources; Phone: 281-579-7700; Fax: 281-579-6555.

SYSCO CORPORATION
1390 ENCLAVE PARKWAY
HOUSTON, TEXAS 77077-2025

Overview: Markets and distributes wholesale foods. Established: 1969. Ownership: Public.

Key Statistics: Annual Sales: $14.5 billion. Number of Employees: 30,600.

Hiring History: Number of professional employees hired in 1996: 100.

Expertise/Education Sought: Accounting, finance, management information systems, data processing

Company-Paid Training is available.

Contact: Human Resources Department; Phone: 281-584-1390; Fax: 281-584-2896. Job Hotline: 281-584-2898.

Company URL: http://www.sysco.com

TEJAS GAS CORPORATION
1301 MCKINNEY STREET SUITE 700
HOUSTON, TEXAS 77010

Overview: Operates natural gas pipelines. Established: 1996. Parent Company: Shell Oil Company. Ownership: Public.

Key Statistics: Annual Sales: $204.6 million. Number of Employees: 383.

Hiring History: Number of professional employees hired in 1997: 150.

Expertise/Education Sought: Administration, engineering, technical

Alternatives to Full-Time Employment: Paid internships

Company-Paid Training is available.

Contact: Mr. Wayne Dismukes, Human Resources Administrator. E-mail: waynedismukes@tejasgas.com.

Company URL: http://www.tejasgas.com

TEXAS DEPARTMENT OF CRIMINAL JUSTICE
3009 HIGHWAY 30 WEST
HUNTSVILLE, TEXAS 77340

Overview: Operates correctional facilities. Established: 1995. Ownership: Private.

Key Statistics: Number of Employees: 500.

Expertise/Education Sought: Criminal justice, law, human resources, accounting, administration, maintenance, education, compliance

Company-Paid Training is available.

Contact: For college students and recent graduates: Mr. Jerre Williams, Recruitment Services; Phone: 409-437-4040; Fax: 409-437-4168. For individuals with previous experience: Mr. Art Campos, Recruitment Services; Phone: 409-437-4059; Fax: 409-437-4168.

TEXAS UTILITIES COMPANY
1601 BRYAN STREET
DALLAS, TEXAS 75201-3402

Overview: Provides electric service. Established: 1996. Ownership: Private.

Key Statistics: Annual Sales: $4.4 billion. Number of Employees: 15,461.

Expertise/Education Sought: Information technology, engineering

Alternatives to Full-Time Employment: Paid internships

Company-Paid Training is available.

International Assignments are available to employees.

Contact: For individuals with previous experience: Human Resources; Phone: 214-812-4600; Fax: 214-812-8419.

Company URL: http://www.tu.com

TRANSCONTINENTAL GAS PIPELINE CORPORATION
2800 POST OAK BOULEVARD
HOUSTON, TEXAS 77056-6100

Overview: Provides natural gas pipeline transportation. Established: 1973. Parent Company: Williams Company. Ownership: Public.

Key Statistics: Annual Sales: $2.5 billion.

Hiring History: Number of professional employees hired in 1996: 200.

Expertise/Education Sought: Mechanical engineering, chemical engineering, electrical engineering, industrial engineering, civil engineering, computer programming

Alternatives to Full-Time Employment: Paid internships

Contact: Ms. Glenna Pierpont, Manager of Human Resources; Phone: 713-215-2000, ext. 3663; Fax: 713-215-3195.

TRINITY INDUSTRIES, INC.
2525 NORTH STEMMONS FREEWAY
DALLAS, TEXAS 75207-2401

Overview: Manufactures railroad cars and steel. Established: 1933. Ownership: Public.

Key Statistics: Annual Sales: $2.2 billion. Number of Employees: 12,700.

Expertise/Education Sought: Accounting, finance, management information systems, chemical engineering, mechanical engineering, drafting, computer-aided design

Alternatives to Full-Time Employment: Paid internships

Company-Paid Training is available.

Contact: Ms. Judy Arrington, Director of Human Resources; Phone: 214-631-4420, ext. 814; Fax: 214-589-8810. Job Hotline: 214-589-8943.

TRITON ENERGY, LTD.
6688 NORTH CENTRAL EXPRESSWAY, SUITE 1400
DALLAS, TEXAS 75206-3925

Overview: Produces crude petroleum and natural gas. Established: 1996. Ownership: Public.

Key Statistics: Annual Sales: $134 million. Number of Employees: 270.

Hiring History: Number of professional employees hired in 1997: 50. 1996: 50.

Expertise/Education Sought: Geology, law, geotechnology, treasury management, tax

Alternatives to Full-Time Employment: Paid internships

International Assignments are available to employees.

Contact: For individuals with previous experience: Personnel Services Specialist; Phone: 214-691-5200; Fax: 214-691-0432.

Company URL: http://www.triton energy.com

TRUSSWAY, LTD.
9411 ALCORN STREET
HOUSTON, TEXAS 77093-6753

Mailing Address: PO Box 111577, Houston, TX 77293-1577

Overview: Produces wood trusses for building. Established: 1995. Ownership: Private.

Key Statistics: Annual Sales: $150 million. Number of Employees: 600.

Expertise/Education Sought: Manufacturing, engineering

Alternatives to Full-Time Employment: Paid internships

Company-Paid Training is available.

Contact: For individuals with previous experience: Mr. Ron Chunn, Controller; Phone: 713-691-6900; Fax: 713-699-7330.

UNIDEN HOLDING, INC.
4700 AMON CARTER BOULEVARD
FORT WORTH, TEXAS 76155-2207

Overview: Manufactures electronic parts and equipment for citizens band radios and telephones. Established: 1997. Parent Company: Uniden In Japan. Ownership: Private.

Key Statistics: Annual Sales: $447.8 million. Number of Employees: 550.

Hiring History: Number of professional employees hired in 1997: 130. 1996: 180.

Expertise/Education Sought: Engineering, customer service and support, production

Company-Paid Training is available.

International Assignments are available to employees.
Contact: For individuals with previous experience: Human Resources Department; Phone: 817-858-3372; Fax: 817-858-3605. E-mail: bsuarez@uniden.com.
Company URL: http://www.uniden.com

UNITED TECHNOLOGIES FURUKAWA

6070 GATEWAY BOULEVARD EAST
EL PASO, TEXAS 79905-2019

Overview: Manufactures motor vehicle parts and accessories. Established: 1986. Parent Company: United Technologies Automotive. Ownership: Public.
Key Statistics: Annual Sales: $158.3 million. Number of Employees: 1,700.
Expertise/Education Sought: Mechanical engineering, manufacturing engineering, accounting, computer science
Alternatives to Full-Time Employment: Paid internships
Contact: Personnel Department, 27 Spur Drive, El Paso, Texas, 79906-2019; Phone: 915-775-7900; Fax: 915-775-9631.

WEATHERFORD COMPRESSION

200 NORTH PADRE ISLAND DRIVE
CORPUS CHRISTI, TEXAS 78406-2001

Overview: Manufactures natural gas and air compressors. Established: 1990. Parent Company: Weatherford. Ownership: Public.
Key Statistics: Annual Sales: $160 million. Number of Employees: 570.
Expertise/Education Sought: Mechanical engineering, electrical engineering, design engineering, manufacturing engineering
Alternatives to Full-Time Employment: Paid internships
International Assignments are available to employees.
Contact: Mr. Terry Hatcher, Personnel Director; Phone: 512-693-3119; Fax: 512-289-8393.

WESTERN ATLAS INTERNATIONAL, INC.

10205 WESTHEIMER
HOUSTON, TEXAS 77042

Overview: Provides petroleum exploration, field development, and production services. Established: 1987. Parent Company: Western Atlas, Inc. Ownership: Private.
Key Statistics: Annual Sales: $1.1 billion. Number of Employees: 9,000.
Hiring History: Number of professional employees hired in 1996: 275.
Expertise/Education Sought: Geophysics, geology, mathematics, electrical engineering, mechanical engineering, computer science, petroleum/petrochemical engineering, chemistry, physics
Alternatives to Full-Time Employment: Part-time jobs, internships
International Assignments are available to employees.
Contact: Mr. R. J. Mason, Manager of Industrial Relations, PO Box 1407, Houston, Texas, 77251-1407; Phone: 713-266-5700; Fax: 713-781-0945.
Company URL: http://www.waii.com

For more information, see full corporate profile, pg. 312.

WESTLAKE CA&O CORPORATION

2801 POST OAK BOULEVARD
HOUSTON, TEXAS 77056-6105

Overview: Manufactures alkalis and chlorine. Established: 1997. Ownership: Private.
Key Statistics: Annual Sales: $33.9 million. Number of Employees: 200.
Expertise/Education Sought: Engineering
International Assignments are available to employees.
Contact: For individuals with previous experience: Human Resources Department; Phone: 713-960-9111; Fax: 713-960-1548.

WHOLESALERS, INC.

4733 BALDWIN BOULEVARD
CORPUS CHRISTI, TEXAS 78408-2713

Overview: Provides data processing and preparation services. Established: 1979. Ownership: Private.
Key Statistics: Annual Sales: $6 million. Number of Employees: 68.
Hiring History: Number of professional employees hired in 1997: 55. 1996: 55.
Expertise/Education Sought: Mechanical engineering
Company-Paid Training is available.
Contact: For individuals with previous experience: Human Relations Department.

UTAH

AMERICAN STORES COMPANY

709 EAST SOUTH TEMPLE
SALT LAKE CITY, UTAH 84102-1205

Overview: Operates chain of food and drug stores. Established: 1984. Ownership: Public.
Key Statistics: Annual Sales: $25 million. Number of Employees: 32,000.
Expertise/Education Sought: Accounting, finance, data processing, human resources
Alternatives to Full-Time Employment: Paid internships
Company-Paid Training is available.
International Assignments are available to employees.
Contact: Human Relations Department; Phone: 801-961-5565; Fax: 801-961-5574. Job Hotline: 800-378-9900.

AMERITECH LIBRARY SERVICES

400 DYNIX DRIVE
PROVO, UTAH 84604-5650

Overview: Provides library automation and information access products and services. Established: 1992. Parent Company: Ameritech Corporation. Ownership: Public.
Key Statistics: Annual Sales: $111.5 million. Number of Employees: 1,300.
Expertise/Education Sought: Software engineering/development, system test, systems analysis, program analysis, library/information science
Alternatives to Full-Time Employment: Part-time jobs
International Assignments are available to employees.
Contact: For individuals with previous experience: Human Resources.
Company URL: http://www.amlibs.com

ICON HEALTH & FITNESS, INC.

1500 SOUTH 1000 WEST
LOGAN, UTAH 84321-8206

Overview: Athletic and sporting goods. Established: 1997. Ownership: Private.
Key Statistics: Annual Sales: $750.6 million. Number of Employees: 4,300.
Expertise/Education Sought: Engineering
Alternatives to Full-Time Employment: Paid internships

Icon Health & Fitness, Inc. (continued)

Company-Paid Training is available.

Contact: For individuals with previous experience: Human Resources; Phone: 435-750-5000; Fax: 435-750-5223. Job Hotline: 435-750-3686.

MERIT MEDICAL SYSTEMS, INC.

1600 WEST MERIT PARKWAY
SOUTH JORDAN, UTAH 84095

Overview: Manufactures medical devices. Established: 1987. Ownership: Public.

Key Statistics: Annual Sales: $50.5 million. Number of Employees: 780.

Hiring History: Number of professional employees hired in 1996: 40.

Expertise/Education Sought: Mechanical engineering, electrical engineering, manufacturing engineering, quality engineering

Alternatives to Full-Time Employment: Internships

Company-Paid Training is available.

International Assignments are available to employees.

Contact: Ms. Robin Nielsen, Recruiting Manager; Phone: 801-253-1600; Fax: 801-253-1687. Job Hotline: 801-263-4020.

SINCLAIR OIL CORPORATION

550 EAST SOUTH TEMPLE
SALT LAKE CITY, UTAH 84102-1005

Overview: Refines and markets petroleum products. Established: 1952. Ownership: Private.

Key Statistics: Annual Sales: $1.7 billion. Number of Employees: 5,500.

Expertise/Education Sought: Petroleum/petrochemical engineering, mechanical engineering, management information systems, accounting, transportation, marketing

Contact: Mr. Wendell White, Director of Human Resources; Phone: 801-363-5100.

THIOKOL CORPORATION

2475 WASHINGTON BOULEVARD
OGDEN, UTAH 84401

Overview: Manufactures rocket and missile propulsion systems and related products. Established: 1929. Ownership: Public.

Key Statistics: Annual Sales: $890.1 million. Number of Employees: 5,300.

Hiring History: Number of professional employees hired in 1997: 100. 1996: 15.

Expertise/Education Sought: Accounting, finance, data processing, human resources, marketing, business management, engineering

Contact: Human Relations Department; Phone: 801-629-2316.

VERMONT

CABOT CREAMERY COOPERATIVE

MAIN STREET
CABOT, VERMONT 05647

Mailing Address: PO Box 128, Cabot, VT 05647

Overview: Wholesales dairy products. Established: 1992. Parent Company: Agri-Mark. Ownership: Private.

Key Statistics: Annual Sales: $107.7 million. Number of Employees: 250.

Expertise/Education Sought: Production

Alternatives to Full-Time Employment: Paid internships

Company-Paid Training is available.

Contact: For individuals with previous experience: Mr. Robert Porter, Vice President of Human Resources; Phone: 802-563-2231; Fax: 802-563-2604. Job Hotline: 802-563-2231, ext. 117.

Company URL: http://www.cabotcheese.com

CHAMPLAIN VALLEY TELCOM

ROUTE 100
WAITSFIELD, VERMONT 05673

Mailing Address: PO Box 9, Waitsfield, VT 05673

Overview: Provides local telephone services. Established: 1994. Parent Company: Selectronics, Inc. Ownership: Private.

Key Statistics: Annual Sales: $13.2 million.

Hiring History: Number of professional employees hired in 1996: 15.

Expertise/Education Sought: Telecommunications

Alternatives to Full-Time Employment: Part-time jobs

Company-Paid Training is available.

Contact: For individuals with previous experience: Human Resources Administrator; Phone: 802-496-3391; Fax: 802-496-8342.

Company URL: http://www.wcvt.com

SPECIALTY FILAMENTS

PO BOX 151
MIDDLEBURY, VERMONT 05753

Overview: Manufactures noncellulose organic fibers. Established: 1995. Parent Company: Apex Specialty Materials, Inc. Ownership: Private.

Key Statistics: Annual Sales: $20 million. Number of Employees: 450.

Expertise/Education Sought: Manufacturing

Company-Paid Training is available.

International Assignments are available to employees.

Contact: For individuals with previous experience: Human Resources; Phone: 802-388-4956; Fax: 802-388-3522.

SYMQUEST GROUP, INC.

30 COMMUNITY DRIVE
SOUTH BURLINGTON, VERMONT 05403

Overview: Manufactures electrical equipment and supplies. Established: 1995. Ownership: Private.

Key Statistics: Annual Sales: $11 million. Number of Employees: 100.

Hiring History: Number of professional employees hired in 1997: 12.

Expertise/Education Sought: Education

Alternatives to Full-Time Employment: Paid internships

Company-Paid Training is available.

Contact: For individuals with previous experience: Ms. Marie Grossarth, Administrative Assistant; Phone: 802-658-9880; Fax: 802-658-9885. E-mail: mgrossarth@aol.com.

VIRGINIA

AERIAL MACHINE AND TOOL CORPORATION

US HIGHWAY 58
VESTA, VIRGINIA 24177

Mailing Address: PO Box 222, Vesta, VA 24177

Overview: Manufactures aircraft and marine hardware. Established: 1981. Ownership: Private.

Key Statistics: Number of Employees: 100.

Expertise/Education Sought: Production management, quality engineering

Alternatives to Full-Time Employment: Paid internships

Contact: For individuals with previous experience: Ms. Gail Clark, Sales Director; Phone: 540-952-2006; Fax: 540-952-2231. E-mail: aerialmtcorporation@worldnet.att.net.

AIR-A-PLANE CORPORATION

3040 EAST VIRGINIA BEACH BOULEVARD
NORFOLK, VIRGINIA 23504-4107

Overview: Manufactures air conditioning and heating equipment. Established: 1939. Ownership: Private.

Key Statistics: Annual Sales: $23.2 million. Number of Employees: 280.

Hiring History: Number of professional employees hired in 1997: 50. 1996: 30.

Expertise/Education Sought: Production, electrical engineering, mechanical engineering, HVAC engineering

Company-Paid Training is available.

Contact: For individuals with previous experience: Ms. Kimberly King, Human Resources Manager; Phone: 757-622-5761.

AMERICAN MANAGEMENT SYSTEMS, INC.

4050 LEGATO ROAD
FAIRFAX, VIRGINIA 22033-4003

Overview: International business and information technology consulting firm. Established: 1970. Ownership: Public.

Key Statistics: Annual Sales: $872.3 million. Number of Employees: 7,500.

Hiring History: Number of professional employees hired in 1996: 1,600.

Expertise/Education Sought: Computer science, application developer, computer engineering, management information systems, software engineering/development

Contact: Mr. Allan Jones, Manager of College Recruiting; Phone: 703-267-5084; Fax: 703-267-5894.

Company URL: http://www.amsinc.com

CORPORATE STATEMENT

The Business *AMS helps large organizations achieve their strategic and operational goals through the application of information technology. The company's services range from defining strategies to the full implementation of solutions.*

The Requirements *AMS seeks candidates with innovative problem-solving skills, strong academic performance, entrepreneurial spirit, and a long-term interest in the application of information technologies.*

AMERICAN RED CROSS—BIOMEDICAL SERVICES

1616 NORTH FORT MYER DRIVE
ARLINGTON, VIRGINIA 22209

Overview: Collects and distributes blood, human tissue, and plasma derivatives. Parent Company: American Red Cross.

Expertise/Education Sought: Engineering

Alternatives to Full-Time Employment: Part-time jobs, paid internships

Company-Paid Training is available.

International Assignments are available to employees.

Contact: For individuals with previous experience: Ms. Tracey Hoston, Human Resources Management Specialist; Phone: 703-312-5600; Fax: 703-312-5585. E-mail: hostont@usa.redcross.org. Job Hotline: 800-289-3299.

Company URL: http://www.redcross.org

AMERICAN SAFETY RAZOR COMPANY

HIGHWAY 612
VERONA, VIRGINIA 24482-0979

Mailing Address: PO Box 500, Stanton, VA 24402-0500

Overview: Manufactures razor blades and shaving cream. Established: 1875. Parent Company: Jordan Industries. Ownership: Public.

Key Statistics: Annual Sales: $260.6 million. Number of Employees: 2,100.

Expertise/Education Sought: Engineering, plastics engineering, design engineering, manufacturing engineering, accounting, sales

Alternatives to Full-Time Employment: Paid internships

Company-Paid Training is available.

International Assignments are available to employees.

Contact: For individuals with previous experience: Ms. June Wilson, Personnel Administrator; Phone: 540-248-8000; Fax: 540-248-0522.

AMVEST CORPORATION

1 BOARS HEAD PLACE
CHARLOTTESVILLE, VIRGINIA 22903-4612

Overview: Coal mining. Established: 1960. Ownership: Private.

Key Statistics: Annual Sales: $133.3 million. Number of Employees: 863.

Expertise/Education Sought: Finance, accounting, administration, mining, engineering, geology, drafting

Company-Paid Training is available.

Contact: Mr. Jeff Duncan, Vice President of Human Resources; Phone: 804-972-7722; Fax: 804-977-5902.

ASHWORTH BROTHERS, INC.

450 ARMOUR DALE
WINCHESTER, VIRGINIA 22604

Mailing Address: PO Box 2780, Winchester, VA 22604

Overview: Manufactures conveyer belt systems. Established: 1894. Ownership: Private.

Key Statistics: Annual Sales: $64.3 million. Number of Employees: 600.

Hiring History: Number of professional employees hired in 1997: 6.

Expertise/Education Sought: Manufacturing, human resources, operations

Alternatives to Full-Time Employment: Paid internships

Company-Paid Training is available.

Contact: Mr. Tom Shellhamer, Manager of Human Resources, PO Box 2780, Winchester, Virginia, 22604; Phone: 540-662-3494; Fax: 540-662-3150.

BOEING INFORMATION SERVICES, INC.

7990 BOEING COURT
VIENNA, VIRGINIA 22182-3925

Overview: Provides data processing and systems integration services. Established: 1981. Parent Company: The Boeing Company. Ownership: Public.

Key Statistics: Annual Sales: $165.5 million. Number of Employees: 2,600.

Expertise/Education Sought: Hardware engineering, software engineering/development, systems analysis, computer programming, computer science, telecommunications

Alternatives to Full-Time Employment: Part-time jobs, paid internships

International Assignments are available to employees.

Contact: Ms. Michelle Gill, Employment Manager, PO Box 7000 CV-42, Vienna, Virginia, 22183-7000; Phone: 703-847-1072; Fax: 703-847-1135.

Company URL: http://www.boeing.com

BTG, INC.

3877 FAIRFAX RIDGE ROAD
FAIRFAX, VIRGINIA 22030-7448

Overview: Provides computer systems engineering services. Established: 1982. Ownership: Public.

BTG, Inc. (continued)

Key Statistics: Annual Sales: $400 million. Number of Employees: 1,168.

Hiring History: Number of professional employees hired in 1997: 550. 1996: 400.

Expertise/Education Sought: Hardware technicians, software engineering/development, network management systems

Alternatives to Full-Time Employment: Part-time jobs, paid internships

Company-Paid Training is available.

Contact: Mr. John Lau, Recruiting and Market Manager; Phone: 703-383-8000; Fax: 703-383-4090. E-mail: jlau@btg.com.

Company URL: http://www.btg.com

CENTRAL INTELLIGENCE AGENCY

PO BOX 12727
ARLINGTON, VIRGINIA 22209

Overview: Foreign intelligence agency for United States government.

Expertise/Education Sought: Computer engineering, electrical engineering, computer science

Alternatives to Full-Time Employment: Paid internships

Company-Paid Training is available.

International Assignments are available to employees.

Contact: Recruitment Center; Phone: 800-336-2163; Fax: 703-613-7873. Job Hotline: 800-JOB-SCIA.

Company URL: http://www.odci.gov/cia

CHESAPEAKE CORPORATION

1021 EAST CARY STREET
RICHMOND, VIRGINIA 23219

Overview: Manufactures pulp and paper products. Established: 1918. Ownership: Public.

Key Statistics: Annual Sales: $1.2 billion. Number of Employees: 6,914.

Expertise/Education Sought: Accounting, communications, electrical engineering, mechanical engineering, pulp and paper science, finance, accounting, financial analysis, business (MBA)

Alternatives to Full-Time Employment: Internships

Company-Paid Training is available.

Contact: Ms. Joanne Boroughs, Manager of Employee Services; Phone: 804-697-1147. E-mail: joanne.boroughs.@cskcorp.com.

The CNA Corporation

THE CNA CORPORATION

4401 FORD AVENUE
ALEXANDRIA, VIRGINIA 22302-8268

Overview: Provides operations research and policy analysis. Established: 1942. Ownership: Private.

Key Statistics: Annual Sales: $49.1 million. Number of Employees: 525.

Hiring History: Number of professional employees hired in 1997: 40.

Expertise/Education Sought: Chemical engineering, electrical engineering, aerospace engineering, physics, economics, aeronautical/astronautical engineering

Alternatives to Full-Time Employment: Paid internships

Company-Paid Training is available.

Contact: Ms. Carolyn P. Williams, Manager of Employment; Fax: 703-824-2740. E-mail: resumes@cna.org.

Company URL: http://www.cna.org

For more information, see full corporate profile, pg. 270.

COMARCO SYSTEMS, INC.

1845B FORT MAHONE STREET
PETERSBURG, VIRGINIA 23805-2761

Overview: Provides engineering services. Established: 1997. Parent Company: Comarco, Inc. Ownership: Public.

Key Statistics: Annual Sales: $32 million. Number of Employees: 403.

Expertise/Education Sought: Information technology, engineering, systems engineering

Alternatives to Full-Time Employment: Paid internships

Company-Paid Training is available.

International Assignments are available to employees.

Contact: Human Resources; Phone: 804-861-0294, ext. 221; Fax: 804-861-9840.

Company URL: http://www.comarcosystems.com

COMPREHENSIVE TECHNOLOGIES INTERNATIONAL, INC.

3951 PENDER DRIVE, SUITE 120
FAIRFAX, VIRGINIA 22030

Overview: Provides computer systems and integration design services. Established: 1980. Ownership: Private.

Key Statistics: Annual Sales: $13.9 million. Number of Employees: 250.

Expertise/Education Sought: Languages, information technology

Alternatives to Full-Time Employment: Part-time jobs

International Assignments are available to employees.

Contact: For individuals with previous experience: Ms. Suzanne Berthay, Director of Human Resources; Phone: 703-383-7200; Fax: 703-352-6765. E-mail: hr@cti.com.

Company URL: http://www.cti.com

COMPUTER BASED SYSTEMS, INC.

2750 PROSPERITY AVENUE, SUITE 300
FAIRFAX, VIRGINIA 22031-4312

Overview: Develops computer software. Established: 1978. Ownership: Private.

Key Statistics: Annual Sales: $27.2 million. Number of Employees: 550.

Hiring History: Number of professional employees hired in 1997: 300.

Expertise/Education Sought: Computer programming, program analysis, software engineering/development, networking

Alternatives to Full-Time Employment: Part-time jobs, paid internships

Company-Paid Training is available.

International Assignments are available to employees.

Contact: For individuals with previous experience: Ms. Martie Wood, Employment Manager. E-mail: hr-info@cbsi.com. Job Hotline: 703-849-8080, ext. 4079.

Company URL: http://www.cbsi.com

COMSAT RSI, INC.

1501 MORAN ROAD
STERLING, VIRGINIA 20166-9309

Overview: Manufactures antennae. Established: 1994. Parent Company: Comsat Corporation. Ownership: Public.

Key Statistics: Annual Sales: $102.4 million. Number of Employees: 1,200.

Expertise/Education Sought: Engineering, project management, radio frequency engineering

Alternatives to Full-Time Employment: Paid internships

Company-Paid Training is available.

Contact: Mr. Michael Mackert, Director of Human Resources; Phone: 703-450-5680; Fax: 703-450-2439.

COVANCE LABORATORIES, INC.

9200 LEESBURG PIKE
VIENNA, VIRGINIA 22182-1699

Overview: Conducts research for toxicology and safety assessment of pharmaceuticals and chemicals. Established: 1990. Parent Company: Covance, Inc. Ownership: Public.

Key Statistics: Annual Sales: $126.6 million. Number of Employees: 1,679.

Hiring History: Number of professional employees hired in 1996: 50.

Expertise/Education Sought: Biology, biostatistics, veterinary, chemistry, accounting, finance, business administration, toxicology, laboratory technology

Company-Paid Training is available.

International Assignments are available to employees.

Contact: For college students and recent graduates: Ms. Dolly Bartley, Human Resources—Business Partner; Phone: 703-893-5400; Fax: 703-759-6947. For individuals with previous experience: Ms. Julliete Macelellan, Human Resources—Business Department; Phone: 703-893-5400; Fax: 703-759-6947.

Company URL: http://www.covance.com

DEPARTMENT OF THE NAVY, NAVAL FACILITIES ENGINEERING COMMAND

200 STOVALL STREET, CODE HRM
ALEXANDRIA, VIRGINIA 22332-2300

Overview: Unit of the Department of the Navy responsible for design, construction, and maintenance of all naval shore facilities. Established: 1842.

Hiring History: Number of professional employees hired in 1997: 60. 1996: 40.

Expertise/Education Sought: Civil engineering, environmental engineering, mechanical engineering, electrical engineering, fire protection

Alternatives to Full-Time Employment: Part-time jobs, paid internships

Company-Paid Training is available.

Contact: Ms. Joy A. Bird, Director of Professional Development; Phone: 703-325-8523. E-mail: birdja@hq.navfac.navy.mil.

Company URL: http://www.navfac.navy.mil

DIMON, INC.

512 BRIDGE STREET
DANVILLE, VIRGINIA 24543

Overview: Produces agricultural products and florist supplies. Parent Company: Dimon International, Inc.

Expertise/Education Sought: Finance, management

Alternatives to Full-Time Employment: Paid internships

Company-Paid Training is available.

International Assignments are available to employees.

Contact: Mr. Richard O'Reilly, Senior Vice President of Human Resources; Phone: 804-792-7511; Fax: 804-791-6739.

DYNCORP

2000 EDMUND HALLEY DRIVE
RESTON, VIRGINIA 20191-3400

Overview: Provides construction and engineering services for industry and government. Established: 1946. Ownership: Public.

Key Statistics: Annual Sales: $1 billion. Number of Employees: 14,250.

Hiring History: Number of professional employees hired in 1997: 25.

Expertise/Education Sought: Aerospace science, risk management, business management, law, accounting, human resources, information systems

Alternatives to Full-Time Employment: Part-time jobs, paid internships

Company-Paid Training is available.

International Assignments are available to employees.

Contact: Ms. Deborah Snider, Human Resource Administrator; Phone: 703-264-8702; Fax: 703-264-8714.

Company URL: http://www.DynCorp.com

EG&G SERVICES

3702 PENDER DRIVE, SUITE 400
FAIRFAX, VIRGINIA 22030-6066

Overview: Investment offices. Established: 1990. Parent Company: EG&G. Ownership: Public.

Key Statistics: Annual Sales: $40 million. Number of Employees: 190.

Expertise/Education Sought: Accounting, engineering, administration, auditing, contracts administration/management

Company-Paid Training is available.

International Assignments are available to employees.

Contact: Human Resources Department; Phone: 703-273-3441; Fax: 703-934-9018.

ETHYL CORPORATION

330 SOUTH 4TH STREET
RICHMOND, VIRGINIA 23219

Overview: Produces industrial organic chemicals. Established: 1995. Ownership: Public.

Key Statistics: Annual Sales: $11.5 million. Number of Employees: 56.

Hiring History: Number of professional employees hired in 1997: 8. 1996: 10.

Expertise/Education Sought: Chemical engineering, accounting, administration, engineering, management, marketing, sales, finance, human resources

Alternatives to Full-Time Employment: Part-time jobs, paid internships

Company-Paid Training is available.

International Assignments are available to employees.

Contact: Human Resources Department.

Company URL: http://www.ethyl.com

FORT JAMES CORPORATION

120 TREDEGAR STREET
RICHMOND, VIRGINIA 23218

Overview: Manufactures paper products for consumers, packaging, and industry. Established: 1969. Ownership: Public.

Key Statistics: Annual Sales: $7.7 billion. Number of Employees: 30,000.

Expertise/Education Sought: Mechanical engineering, electrical engineering, management information systems, accounting, finance

Company-Paid Training is available.

International Assignments are available to employees.

Contact: For individuals with previous experience: Human Resources Department, PO Box 2218, Richmond, Virginia, 23217-2218; Phone: 804-649-4286; Fax: 804-649-4340.

Company URL: http://www.fortjames.com

GENERAL DYNAMICS CORPORATION

3190 FAIRVIEW PARK DRIVE
FALLS CHURCH, VIRGINIA 22042-4510

Overview: Manufactures nuclear submarines and land systems. Established: 1899. Ownership: Public.

Key Statistics: Annual Sales: $3.6 billion. Number of Employees: 25,000.

Hiring History: Number of professional employees hired in 1996: 600.

Expertise/Education Sought: Mechanical engineering, electrical engineering, nuclear engineering, design engineering

Company-Paid Training is available.

Contact: Mr. Tommy August, Vice President of Personnel Relations; Phone: 703-876-3400; Fax: 703-876-3550.

GLOBAL ASSOCIATES, LTD.

7600 LEESBURG PIKE, FLOOR 4
FALLS CHURCH, VIRGINIA 22043-2004

Overview: Provides consulting engineering services. Established: 1993. Ownership: Private.

Key Statistics: Annual Sales: $9.4 million. Number of Employees: 130.

Global Associates, Ltd. (continued)

Hiring History: Number of professional employees hired in 1997: 50. 1996: 65.

Expertise/Education Sought: Engineering, computer science

Alternatives to Full-Time Employment: Part-time jobs, paid internships

Company-Paid Training is available.

International Assignments are available to employees.

Contact: For individuals with previous experience: Human Resources; Phone: 703-714-1800; Fax: 703-714-1801.

Company URL: http://www.globalus.com

GLOBAL ONE

12490 SUNRISE VALLEY DRIVE
RESTON, VIRGINIA 20196

Overview: Provides data communications services. Established: 1996. Parent Company: Sprint International. Ownership: Private.

Key Statistics: Annual Sales: $800 million. Number of Employees: 2,500.

Expertise/Education Sought: Information systems, engineering, telecommunications

Alternatives to Full-Time Employment: Part-time jobs, internships

Company-Paid Training is available.

International Assignments are available to employees.

Contact: Mrs. Linda Furr, Human Resources Manager; Phone: 703-689-6000; Fax: 703-318-1969. E-mail: hr.recruiter@globalone.net.

Company URL: http://www.global-one.net

GWALTNEY OF SMITHFIELD, LTD.

601 NORTH CHURCH STREET
SMITHFIELD, VIRGINIA 23430-1221

Overview: Operates meat packing plant. Established: 1981. Parent Company: Smithfield Foods, Inc. Ownership: Public.

Key Statistics: Annual Sales: $592.8 million. Number of Employees: 2,472.

Expertise/Education Sought: Accounting, administration, management, industrial engineering

Contact: Mr. Norman Fisher, Vice President Human Resources; Phone: 757-357-3131, ext. 2744; Fax: 757-357-1568.

HAMILTON BEACH PROCTOR-SILEX, INC.

4421 WATERFRONT DRIVE
GLEN ALLEN, VIRGINIA 23060-3375

Overview: Manufactures household appliances. Established: 1990. Parent Company: Nacco Industries, Inc. Ownership: Public.

Key Statistics: Annual Sales: $395.1 million. Number of Employees: 3,800.

Hiring History: Number of professional employees hired in 1997: 35.

Expertise/Education Sought: Mechanical engineering, electrical engineering, industrial engineering, quality control, marketing

Alternatives to Full-Time Employment: Paid internships

Company-Paid Training is available.

Contact: Mrs. Simone Thomas, Human Resources Administrator; Phone: 804-527-7290. E-mail: s.thomas@hambeach.com.

Company URL: http://www.hambeach.com

IPC TECHNOLOGY

7200 GLEN FOREST DRIVE
RICHMOND, VIRGINIA 23226-3768

Overview: Provides computer maintenance and repair services. Ownership: Private.

Key Statistics: Annual Sales: $13 million. Number of Employees: 160.

Hiring History: Number of professional employees hired in 1996: 77.

Expertise/Education Sought: Software engineering/development, networks

Alternatives to Full-Time Employment: Part-time jobs, paid internships

Contact: For individuals with previous experience: Ms. Stormy Hamlin, Personnel Manager; Phone: 804-285-9300, ext. 3014; Fax: 804-673-9744. E-mail: stormy@ipctech.com.

Company URL: http://www.ipctech.com

ITT DEFENSE AND ELECTRONICS

1650 TYSONS BOULEVARD, SUITE 1700
MCLEAN, VIRGINIA 22102

Overview: Produces electronic defense systems for government and military uses. Established: 1995. Parent Company: ITT Industries. Ownership: Public.

Key Statistics: Annual Sales: $812.9 million. Number of Employees: 14,000.

Expertise/Education Sought: Electrical engineering, mechanical engineering, design engineering, aerospace engineering

Alternatives to Full-Time Employment: Paid internships

Company-Paid Training is available.

International Assignments are available to employees.

Contact: For individuals with previous experience: Ms. Elizabeth Gay, Manager of Human Resources; Phone: 703-790-6300.

LAFARGE CORPORATION

11130 SUNRISE VALLEY DRIVE
RESTON, VIRGINIA 20191-4393

Overview: Manufactures cement, concrete, and construction materials. Established: 1977. Parent Company: Lafarge Coppee. Ownership: Public.

Key Statistics: Annual Sales: $1.6 billion. Number of Employees: 6,800.

Hiring History: Number of professional employees hired in 1997: 50. 1996: 100.

Expertise/Education Sought: Marketing, finance, accounting, human resources, mechanical engineering, industrial engineering, electrical engineering, metallurgical engineering

Alternatives to Full-Time Employment: Paid internships

Company-Paid Training is available.

International Assignments are available to employees.

Contact: For individuals with previous experience: Mr. Thomas Tatum, Senior Vice President of Human Resources; Phone: 703-264-3600; Fax: 703-264-0632.

LITTON PRC

1500 PRC DRIVE
MCLEAN, VIRGINIA 22102

Overview: Provides engineering and information systems, products and services. Established: 1996. Parent Company: Litton Industries, Inc. Ownership: Public.

Key Statistics: Annual Sales: $426.6 million. Number of Employees: 5,500.

Hiring History: Number of professional employees hired in 1997: 1,700. 1996: 1,500.

Expertise/Education Sought: Computer science, information systems, management information systems, electrical engineering

Alternatives to Full-Time Employment: Paid internships

Company-Paid Training is available.

International Assignments are available to employees.

Contact: Staffing Services; Phone: 703-556-1000; Fax: 703-556-2269.

Company URL: http://www.prc.com/ACareer

MEDIA GENERAL CABLE OF FAIRFAX

14650 OLD LEE ROAD
CHANTILLY, VIRGINIA 20151

Overview: Provides cable television services. Established: 1982. Parent Company: Media General, Inc. Ownership: Public.

Key Statistics: Annual Sales: $135.8 million. Number of Employees: 616.

Expertise/Education Sought: Technology, customer service and support, engineering, telecommunications, sales

Contact: For individuals with previous experience: Human Resources Department; Phone: 703-378-8400; Fax: 703-378-3498. Job Hotline: 703-378-3440.

Company URL: http://www.mgcable.com

METRO INFORMATION SERVICES

REFLECTIONICS III, 208 GOLDEN OAK COURT, SUITE 15
VIRGINIA BEACH, VIRGINIA 23452

Overview: Provides computer consulting services. Established: 1973. Ownership: Private.

Key Statistics: Annual Sales: $114 million. Number of Employees: 1,517.

Hiring History: Number of professional employees hired in 1997: 1,300.

Expertise/Education Sought: Program analysis, computer programming

Alternatives to Full-Time Employment: Part-time jobs

Company-Paid Training is available.

International Assignments are available to employees.

Contact: Ms. Kim Gambardella, Head of Data Entry; Phone: 757-486-1900; Fax: 757-306-0257. E-mail: kgambardella@metrois.com.

Company URL: http://www.MetroIS.com

MITRETEK SYSTEMS, INC

7525 COLSHIRE DRIVE
MCLEAN, VIRGINIA 22102-7400

Overview: Provides management consulting services. Established: 1996. Ownership: Private.

Key Statistics: Annual Sales: $75 million. Number of Employees: 650.

Hiring History: Number of professional employees hired in 1997: 30. 1996: 75.

Expertise/Education Sought: Computer science, information systems, electrical engineering, computer engineering, transportation engineering, operations, research, physics, mathematics

Alternatives to Full-Time Employment: Part-time jobs, paid internships

Company-Paid Training is available.

Contact: Ms. Mosetta Blacknon, Director of Human Resources; Phone: 703-610-1950; Fax: 703-610-1952. E-mail: mwb@mitretek.org.

Company URL: http://www.mitretek.org

CORPORATE PROFILE

Mitretek Systems, Inc., is unique among technology companies in that it is a nonprofit, public interest company. The company's mission is to provide scientific engineering services to its clients on technology-based research and development programs. Mitretek is a provider of highly objective, technical services in information, telecommunications, and environmental management technologies. At Mitretek, challenges lie in being able to seize new opportunities. This is done by attracting, retaining, and helping to develop the best professionals possible.

Mitretek is recognized worldwide as a leader in information technologies, telecommunications, computer security, health-care technologies, software engineering, biotechnology, and criminal knowledge. Local, state, federal, and international governments, as well as other nonprofit organizations, have come to depend on Mitretek to deliver recommendations untouched by commercial consideration.

At Mitretek, employees work in partnership with the clients to achieve total solutions by bringing clients the experience, resources, and knowledge base that are embodied in the professional staff of scientists and engineers.

Mitretek is engaged in a variety of capabilities, including biotechnology; criminal justice and public safety; environmental remediation; health information technology; information systems technologies; information security; nuclear and chemical safety; oceanic, atmospheric, and space systems; telecommunications and networking; and transportation systems. These capabilities may be vast and far-reaching, but Mitretek is connected by a common vision of providing integrated technology solutions. Mitretek recognizes the power of technology and is finding ways to deploy that technology to make this a better country in which to live and a healthier, safer world. In every sector, Mitretek is setting the pace for the next century.

Mitretek's mission is to provide innovative technology in all areas of the public interest; therefore, the company supports work in many areas of concern to the public sector. Mitretek has productivity-enhancing, state-of-the-art tools that enable its employees to provide leading-edge solutions to today's and tomorrow's public-sector issues. Using these resources, Mitretek's staff members can streamline processes, evaluate designs, build models, and test single-system components or entire operational concepts, significantly reducing program risk and the time required to realize an operational system.

No matter who the client may be—the U.S. government or a doctor at a hospital—Mitretek uses the same technology-based approach to find solutions to unique problems for the company's stakeholders—the American people.

Numerous staffing opportunities exist for new graduates in the following business units: telecommunications, transportation, law enforcement, information services, biotechnology, energy, environment, and health care. Mitretek Systems, Inc., is looking for those graduates who have a B.S., M.S., or Ph.D. in computer science, information systems, electrical engineering, computer engineering, transportation engineering, engineering economics, operations research, physics, math, or a combination of strong business and technical skills (i.e., MIS or DS). A strong academic record (GPA of at least 3.0 on a 4.0 scale) is a must, and co-op experience is a plus. Interested applicants can visit Mitretek's Web site at http://www.mitretek.org.

MRJ TECHNOLOGY SOLUTIONS

10560 ARROWHEAD DRIVE
FAIRFAX, VIRGINIA 22030-7305

Overview: Provides computer systems and integration design services. Established: 1990. Ownership: Private.

Key Statistics: Annual Sales: $95 million. Number of Employees: 800.

Hiring History: Number of professional employees hired in 1996: 200.

Expertise/Education Sought: Aerospace engineering, networks, data warehousing, systems engineering

Alternatives to Full-Time Employment: Part-time jobs, internships

Company-Paid Training is available.

Contact: For college students and recent graduates: Ms. Carrie Bard, College Recruiter; Fax: 913-432-3318. E-mail: cbard@mrj.com. For individuals with previous experience: Ms. Janet Coooper, Recruiting Manager; Fax: 913-432-3318. E-mail: career@mrj.com.

Company URL: http://www.mrj.com

NATIONAL SCIENCE FOUNDATION

4201 WILSON BOULEVARD
ARLINGTON, VIRGINIA 22230

Overview: Promotes science, engineering, and technical programs.

Hiring History: Number of professional employees hired in 1996: 30.

Expertise/Education Sought: Computer science, contracts administration/management, actuarial, mathematics, engineering, behavioral sciences, education, biology, geology

Alternatives to Full-Time Employment: Part-time jobs, paid internships

Company-Paid Training is available.

Contact: For college students and recent graduates: Miss Madalin Chisley, Personnel Specialist; Phone: 703-306-1185; Fax: 703-306-0260. For individuals with previous experience: Ms. Crystal Matthews, Recruitment Coordinator; Phone: 703-306-1185; Fax: 703-306-0260. Job Hotline: 703-306-0080.

Company URL: http://www.nsf.gov

NAVAL FACILITIES ENGINEERING COMMAND

200 STOVALL STREET
ALEXANDRIA, VIRGINIA 22332

Overview: Responsible for the planning, design, construction, and maintenance of all naval bases world-wide. Division of the Department of the Navy.

Naval Facilities Engineering Command (continued)

KEY STATISTICS: Number of Employees: 21,000.

HIRING HISTORY: Number of professional employees hired in 1996: 80.

EXPERTISE/EDUCATION SOUGHT: Engineering, business

CONTACT: Ms. Joy A. Bird, Director, Professional Development Center; Phone: 703-325-8523; Fax: 703-325-1923. E-mail: birdja@hq.navfac.navy.mil.

ORGANIZATION PROFILE

The Naval Facilities Engineering Command is responsible for the planning, design, construction, and maintenance of all naval shore facilities. Command professionals are also responsible for the acquisition and disposal of real estate, the management of family housing, and the execution of environmental policy throughout the Department of the Navy.

For the military officers of the Civil Engineer Corps and civilians working in the specialized fields of engineering and science, "Service to the Fleet" is both a motto and a purpose. This results in professionals pursuing challenging and satisfying naval careers.

Becoming a civilian engineer with NAVFAC provides excellent opportunities for advancement. Operating within NAVFAC's command is an entry-level engineering intern program—the Professional Development Center. The PDC program provides rapid progression to middle-management and senior grade levels. While how far and how fast one progresses is contingent upon personal drives and motivations, individuals who perform successfully are assured advancement to the middle-management level within two or three years.

Civilian engineers work on a variety of projects. These projects take employees from coast to coast and around the world. Where one goes and what one does is largely self-determined. Career opportunities abound and cover most engineering specializations.

The federal government offers a benefits package that features some of the most attractive benefits available to employees. These benefits include a retirement plan, health insurance, life insurance, vacation, holidays, and sick leave as well as flexible and compressed work schedules.

NAVAL SURFACE WARFARE CENTER, DAHLGREN DIVISION

17320 DAHLGREN ROAD
DAHLGREN, VIRGINIA 22448-5100

OVERVIEW: Navy's principal research, development, test, and evaluation laboratory for surface ship combat systems, ordnance, mines, and strategic systems support. Division of the The Naval Sea Systems Command.

KEY STATISTICS: Number of Employees: 4,400.

EXPERTISE/EDUCATION SOUGHT: Computer engineering, computer science, electrical engineering, electronics engineering, systems engineering, mechanical engineering, aerospace engineering, mathematics, physics

ALTERNATIVES TO FULL-TIME EMPLOYMENT: Internships

CONTACT: College Recruiting Coordinator, Human Resources Department, Code PD, Dahlgren Division; Phone: 800-352-7967; Fax: 540-653-7632. E-mail: recruit@nswc.navy.mil.

COMPANY URL: http://www.nswc.navy.mil/P/RECRUIT/recruit.html

For more information, see full organization profile, pg. 290.

NEWPORT NEWS SHIPBUILDING

4101 WASHINGTON AVENUE
NEWPORT NEWS, VIRGINIA 23607-2704

OVERVIEW: Builds ships and nuclear submarines. Established: 1996. Ownership: Private.

KEY STATISTICS: Annual Sales: $1.9 billion. Number of Employees: 18,000.

HIRING HISTORY: Number of professional employees hired in 1997: 1,200. 1996: 500.

EXPERTISE/EDUCATION SOUGHT: Engineering, information technology, computer programming

ALTERNATIVES TO FULL-TIME EMPLOYMENT: Paid internships

INTERNATIONAL ASSIGNMENTS are available to employees.

CONTACT: For individuals with previous experience: Mr. Jay Dunn, Senior Human Resources Representative; Phone: 757-380-4878; Fax: 757-380-3114. E-mail: employment@nns.com. Job Hotline: 757-380-2142.

COMPANY URL: http://www.nns.com

NEXTEL COMMUNICATIONS

1505 FARM CREDIT DRIVE
MCLEAN, VIRGINIA 22102

OVERVIEW: Provides mobile telephone services. Established: 1987. Ownership: Public.

KEY STATISTICS: Annual Sales: $332.9 million. Number of Employees: 3,600.

HIRING HISTORY: Number of professional employees hired in 1997: 3,000.

EXPERTISE/EDUCATION SOUGHT: Accounting, computer programming, electrical engineering, sales, human resources, telecommunications, marketing

ALTERNATIVES TO FULL-TIME EMPLOYMENT: Part-time jobs, paid internships

COMPANY-PAID TRAINING is available.

INTERNATIONAL ASSIGNMENTS are available to employees.

CONTACT: For individuals with previous experience: Ms. Karen Puleo, Human Resources Manager; Phone: 703-394-3000; Fax: 703-394-3507.

COMPANY URL: http://www.nextel.com

NORFOLK SOUTHERN CORPORATION

3 COMMERCIAL PLACE
NORFOLK, VIRGINIA 23510-9214

OVERVIEW: Provides railway freight transportation services. Established: 1982. Ownership: Public.

KEY STATISTICS: Annual Sales: $4.1 billion. Number of Employees: 23,361.

HIRING HISTORY: Number of professional employees hired in 1997: 90. 1996: 106.

EXPERTISE/EDUCATION SOUGHT: Civil engineering, electrical engineering, mechanical engineering, industrial engineering

ALTERNATIVES TO FULL-TIME EMPLOYMENT: Paid internships

COMPANY-PAID TRAINING is available.

CONTACT: Ms. Lori Leonardos, Employment Manager; Phone: 757-664-5061; Fax: 757-664-5069. E-mail: careers@nscorp.com. Job Hotline: 800-214-3609.

COMPANY URL: http://www.nscorp.com

OVERNITE TRANSPORTATION COMPANY

1000 SEMMES AVENUE
RICHMOND, VIRGINIA 23224-2246

MAILING ADDRESS: PO Box 1216, Richmond, VA 23218-1216

OVERVIEW: Provides trucking and shipping services. Established: 1988. Parent Company: Union Pacific Corporation. Ownership: Public.

KEY STATISTICS: Annual Sales: $961 million. Number of Employees: 13,046.

HIRING HISTORY: Number of professional employees hired in 1997: 3,000. 1996: 3,000.

EXPERTISE/EDUCATION SOUGHT: Computer science, finance, data processing, sales, management information systems, accounting, logistics

ALTERNATIVES TO FULL-TIME EMPLOYMENT: Part-time jobs

COMPANY-PAID TRAINING is available.

INTERNATIONAL ASSIGNMENTS are available to employees.

CONTACT: Ms. Deborah Wickham, Human Resources Manager; Phone: 804-231-8000; Fax: 804-231-8504. Job Hotline: 804-291-5627.

COMPANY URL: http://www.overnite.com

THE PITTSTON COMPANY

1000 VIRGINA CENTER PARKWAY
GLEN ALLEN, VIRGINIA 23060

Mailing Address: PO Box 4229, Glen Allen, VA 23058-4229

Overview: Mines, processes, and markets coal and minerals. Established: 1930. Ownership: Public.

Key Statistics: Annual Sales: $3.1 billion. Number of Employees: 27,000.

Expertise/Education Sought: Marketing, accounting, finance, auditing

Alternatives to Full-Time Employment: Paid internships

Contact: Mr. Joseph Verostic, Director of Human Resources; Phone: 804-553-3600, ext. 3680; Fax: 804-553-3768.

QUALITY SYSTEMS, INC.

8201 GREENSBORO DRIVE, SUITE 1200
MCLEAN, VIRGINIA 22102

Overview: Designs and develops integrated defense systems. Established: 1989. Parent Company: Tracor, Inc. Ownership: Public.

Key Statistics: Annual Sales: $50 million. Number of Employees: 550.

Hiring History: Number of professional employees hired in 1997: 190. 1996: 196.

Expertise/Education Sought: Network management systems, computer programming, database management, systems engineering, software engineering/development, network analysis, training and development

Alternatives to Full-Time Employment: Part-time jobs, paid internships

Company-Paid Training is available.

Contact: Mr. John Sebra, Director Corporate Recruiting; Phone: 703-847-5820; Fax: 703-847-5887. E-mail: recruiting@qsi.com.

Company URL: http://www.qsi.com

REYNOLDS METALS COMPANY

6601 WEST BROAD STREET
RICHMOND, VIRGINIA 23230

Mailing Address: PO Box 27003, Richmond, VA 23261-7003

Overview: Produces aluminum, aluminum products, and plastics. Established: 1928. Ownership: Public.

Key Statistics: Annual Sales: $7 billion. Number of Employees: 29,500.

Expertise/Education Sought: Accounting, finance, chemical engineering, engineering mechanics, information systems

Alternatives to Full-Time Employment: Part-time jobs, paid internships

Company-Paid Training is available.

International Assignments are available to employees.

Contact: For individuals with previous experience: Ms. Maryanne Wince, Headquarters Human Resources; Phone: 804-281-2126; Fax: 804-281-4451.

RICHMOND TIMES DISPATCH

333 EAST GRACE STREET
RICHMOND, VIRGINIA 23219-1000

Overview: Provides newspaper publishing, broadcasting, and cable television services. Established: 1850. Parent Company: Media General, Inc. Ownership: Public.

Key Statistics: Annual Sales: $765.1 million. Number of Employees: 9,849.

Expertise/Education Sought: Data processing, marketing, finance, graphic arts, accounting, journalism, hardware engineering, software engineering/development

Alternatives to Full-Time Employment: Paid internships

Company-Paid Training is available.

Contact: For individuals with previous experience: Ms. Allie Bullock, Employment Supervisor; Phone: 804-649-6000. Job Hotline: 888-562-2864.

ROWE MANUFACTURING, INC.

239 ROWAN STREET
SALEM, VIRGINIA 24153-6932

Overview: Manufactures upholstered household furniture. Established: 1996. Ownership: Public.

Key Statistics: Annual Sales: $142 million. Number of Employees: 1,450.

Expertise/Education Sought: Manufacturing engineering, industrial engineering, mechanical engineering

Alternatives to Full-Time Employment: Paid internships

Company-Paid Training is available.

Contact: Human Resources Department; Phone: 540-389-8671; Fax: 540-375-4908.

Company URL: http://www.rowefurniture.com

RUBATEX CORPORATION

5221 VALLEYPARK DRIVE
ROANOKE, VIRGINIA 24019

Overview: Produces hard rubber and molded rubber products. Established: 1990. Parent Company: RBX Holdings, Inc. Ownership: Private.

Key Statistics: Annual Sales: $109.9 million. Number of Employees: 1,419.

Hiring History: Number of professional employees hired in 1996: 15.

Expertise/Education Sought: Accounting, administration, engineering, marketing, management, sales, chemistry, finance, computer programming, customer service and support

Company-Paid Training is available.

Contact: For individuals with previous experience: Human Resources Department; Phone: 540-561-6000.

Company URL: http://www.thomasregister.com/rubatex

SALLIE MAE, INC.

11600 SALLIE MAE DRIVE
RESTON, VIRGINIA 20193

Overview: Provides federally sponsored credit services. Established: 1996. Ownership: Private.

Key Statistics: Annual Sales: $616.6 million. Number of Employees: 3,727.

Expertise/Education Sought: Technical

Alternatives to Full-Time Employment: Part-time jobs, paid internships

Contact: For individuals with previous experience: Human Resources; Phone: 703-810-3000; Fax: 703-810-5658.

Company URL: http://www.salliemae.com

SHENANDOAH TELEPHONE COMPANY

124 SOUTH MAIN STREET
EDINBURG, VIRGINIA 22824

Mailing Address: PO Box 459, Edinburgh, VA 22824

Overview: Provides local telephone services. Established: 1981. Parent Company: Shenandoah Telecommunication Company. Ownership: Public.

Key Statistics: Annual Sales: $13.8 million. Number of Employees: 111.

Hiring History: Number of professional employees hired in 1997: 25. 1996: 30.

Expertise/Education Sought: Telecommunications, construction, sales, customer service and support

Alternatives to Full-Time Employment: Paid internships

Contact: Ms. Cindy Soltis, Personnel Manager, PO Box 459, Edinburg, Virginia, 22824; Phone: 540-984-4141; Fax: 540-984-8192.

SMA TELECOM, LLC

5 KOGER EXECUTIVE CENTER, SUITE 217
NORFOLK, VIRGINIA 23502

Overview: Provides data processing and preparation services. Established: 1996. Parent Company: SMA. Ownership: Private.

SMA Telecom, LLC (continued)

Hiring History: Number of professional employees hired in 1997: 30. 1996: 30.

Expertise/Education Sought: Accounting, finance, human resources, marketing, management, engineering, sales

Alternatives to Full-Time Employment: Paid internships

Company-Paid Training is available.

International Assignments are available to employees.

Contact: Human Resources; Phone: 757-461-0559; Fax: 757-461-0569.

Company URL: http://www.smatelecom.com

SMITHFIELD PACKING COMPANY, INC.

501 NORTH CHURCH STREET
SMITHFIELD, VIRGINIA 23430-1214

Mailing Address: PO Box 447, Smithfield, VA 23431

Overview: Wholesales meats and meat products. Established: 1981. Parent Company: Smithfield Foods, Inc. Ownership: Public.

Expertise/Education Sought: Manufacturing engineering, administration, accounting, finance, engineering, industrial engineering

International Assignments are available to employees.

Contact: For individuals with previous experience: Mr. Thomas Ross, Vice President of Human Resources; Phone: 757-357-4321, ext. 1380; Fax: 757-357-1366.

SPERRY MARINE, INC.

1070 SEMINOLE TRAIL
CHARLOTTESVILLE, VIRGINIA 22901

Overview: Manufactures navigational instrument systems. Established: 1996. Parent Company: Litton Industries, Inc. Ownership: Public.

Key Statistics: Annual Sales: $52.8 million. Number of Employees: 650.

Hiring History: Number of professional employees hired in 1997: 85. 1996: 80.

Expertise/Education Sought: Electrical engineering, computer science

Alternatives to Full-Time Employment: Part-time jobs, paid internships

Company-Paid Training is available.

Contact: Mr. Charles Lumsden, Manager of Employee Relations; Phone: 804-974-2365; Fax: 804-974-2259.

Company URL: http://www.sperry-marine.com

SPURLOCK INDUSTRIES, INC.

5090 GENERAL MAHONE HIGHWAY
WAVERLY, VIRGINIA 23890-3039

Mailing Address: PO Box 8, Waverly, VA 23890

Overview: Produces industrial organic chemicals. Established: 1996. Ownership: Private.

Key Statistics: Annual Sales: $28.6 million. Number of Employees: 72.

Expertise/Education Sought: Manufacturing

Contact: For individuals with previous experience: Human Resources; Phone: 804-834-3113; Fax: 804-834-2860.

STUDENT LOAN MARKETING ASSOCIATION

11600 SALLIE MAE DRIVE
RESTON, VIRGINIA 20193

Overview: Government student loan association. Established: 1972. Ownership: Private.

Key Statistics: Annual Budget: $3.6 billion. Number of Employees: 4,792.

Hiring History: Number of professional employees hired in 1997: 300.

Expertise/Education Sought: Accounting, law, technical engineering, computer programming, credit/credit analysis

Alternatives to Full-Time Employment: Part-time jobs, paid internships

Company-Paid Training is available.

Contact: Ms. Patricia Scott, Director of Human Resources; Phone: 703-810-3000; Fax: 703-000-0000.

Company URL: http://www.salliemae.com

TARMAC AMERICA

1151 AZALEA GARDEN ROAD
NORFOLK, VIRGINIA 23502-5601

Overview: Manufactures ready-mixed concrete. Established: 1991. Parent Company: PLC. Ownership: Private.

Key Statistics: Annual Sales: $145 million. Number of Employees: 1,092.

Expertise/Education Sought: Accounting, engineering

Alternatives to Full-Time Employment: Paid internships

Company-Paid Training is available.

Contact: For individuals with previous experience: Mr. Ed Pittman, Vice President of Human Resources; Phone: 757-858-6500; Fax: 757-855-7707.

Company URL: http://www.termacamerica.com

TECHMATICS

12450 FAIR LAKES CIRCLE
FAIRFAX, VIRGINIA 22033-3810

Overview: Provides systems engineering services. Established: 1982. Ownership: Private.

Key Statistics: Annual Sales: $47 million. Number of Employees: 650.

Hiring History: Number of professional employees hired in 1997: 300. 1996: 200.

Expertise/Education Sought: Engineering, industrial engineering

Alternatives to Full-Time Employment: Part-time jobs, paid internships

Company-Paid Training is available.

Contact: For college students and recent graduates: Ms. Sophie Ramsay, Career Development Manager, 2231 Crystal Drive, Suite 1000, Arlington, Virginia, 22202; Phone: 703-769-6168; Fax: 703-920-3928. For individuals with previous experience: Ms. Jackie Weisenberger, Technical Recruiter; Phone: 703-802-8300; Fax: 703-802-0412. E-mail: careers@techmatics.com.

Company URL: http://www.techmatics.com

TELECOMMUNICATIONS CONCEPTS, INC.

5554 PORT ROYAL ROAD
SPRINGFIELD, VIRGINIA 22151-2301

Overview: Develops industrial laboratory testing equipment and computer hardware and software. Established: 1989. Ownership: Private.

Key Statistics: Annual Sales: $38.2 million. Number of Employees: 460.

Hiring History: Number of professional employees hired in 1997: 25. 1996: 15.

Expertise/Education Sought: Software engineering/development, computer science, electrical engineering, mechanical engineering, instrumentation

Company-Paid Training is available.

Contact: Ms. Joanne DeAngelis, Human Resources Coordinator; Phone: 703-321-3030; Fax: 703-321-5046. E-mail: tcihr@erols.com. Job Hotline: 703-321-5168.

Company URL: http://www.tcicomm.com

TELOS CORPORATION

19886 ASHBURN ROAD
ASHBURN, VIRGINIA 20147

Overview: Designs integrated computer systems and provides software services. Established: 1993. Ownership: Public.

Key Statistics: Annual Sales: $188.9 million. Number of Employees: 1,374.

Expertise/Education Sought: Software engineering/development, hardware engineering, electrical engineering, data processing, computer programming

Alternatives to Full-Time Employment: Part-time jobs

International Assignments are available to employees.

Contact: Ms. Melissa Levanowitz, Recruiting Assistant; Phone: 703-724-3800; Fax: 703-724-3860.

Company URL: http://www.telos.com

TRW SYSTEMS AND INFORMATION TECHNOLOGY GROUP

12900 FEDERAL SYSTEMS PARK DRIVE, C/O
UNIVERSITY RELATIONS
FAIRFAX, VIRGINIA 22033

Overview: Provides information technology systems and solutions and other technology-based services to government and commercial clients. Established: 1901. Parent Company: TRW. Ownership: Public.

Key Statistics: Annual Sales: $11 billion. Number of Employees: 80,000.

Expertise/Education Sought: Computer science, information systems, electrical engineering, software engineering/development, systems analysis

Alternatives to Full-Time Employment: Paid internships

Company-Paid Training is available.

Contact: For college students and recent graduates: Mr. Seth Feit, Manager of University Relations; Phone: 703-848-6276; Fax: 703-848-6070. E-mail: sfeit@bdm.com. For individuals with previous experience: Ms. Donna Diederich, Manager of Staffing Services; Phone: 703-848-5023; Fax: 888-770-2364. E-mail: hireme@bdm.com.

Company URL: http://www.trw.com

For more information, see full corporate profile, pg. 302.

U.S. DEPARTMENT OF STATE

PO BOX 9317
ARLINGTON, VIRGINIA 22219

Overview: Implements United States foreign policy and maintains diplomatic relations with other countries throughout the world. Established: 1789.

Key Statistics: Number of Employees: 25,000.

Hiring History: Number of professional employees hired in 1997: 300.

Expertise/Education Sought: Computer science, engineering, telecommunications

Alternatives to Full-Time Employment: Part-time jobs, internships

Company-Paid Training is available.

International Assignments are available to employees.

Contact: U.S. State Department, PO Box 9317-P, Arlington, Virginia, 22219. Job Hotline: 202-647-7284.

Company URL: http://www.state.gov

U.S. GEOLOGICAL SURVEY

12201 SUNRISE VALLEY DRIVE
RESTON, VIRGINIA 20192

Overview: Earth and science agency. Division of the Department of Interior.

Expertise/Education Sought: Hydrogeology, hydrology, chemistry

Alternatives to Full-Time Employment: Part-time jobs, paid internships

Company-Paid Training is available.

Contact: For individuals with previous experience: Mr. Richard C. Williams, Chief of Recruiting and Placement; Phone: 703-648-6131.

U.S. PATENT & TRADEMARK OFFICE

2900 CRYSTAL DRIVE
ARLINGTON, VIRGINIA 22202

Overview: Grants patents for the protection of inventions, to register trademarks, and to disseminate information about intellectual property. Parent Company: U.S. Department of Commerce.

Key Statistics: Number of Employees: 5,500.

Expertise/Education Sought: Electrical engineering, biotechnology, organic chemistry, chemical engineering, materials engineering, transportation, construction, agriculture, petroleum/petrochemical engineering, mechanical engineering

Company-Paid Training is available.

Contact: Office of Human Resources, Box 171, Washington, D.C., 20231; Phone: 800-642-5670; Fax: 703-305-9864.

Company URL: http://www.uspto.gov

For more information, see full corporate profile, pg. 310.

VF NETWARE, INC.

WALKER ROAD
MARTINSVILLE, VIRGINIA 24112

Overview: Manufactures textiles. Established: 1984. Parent Company: VF Corporation. Ownership: Public.

Key Statistics: Annual Sales: $250 million. Number of Employees: 6,500.

Expertise/Education Sought: Accounting, finance, mechanical engineering, electrical engineering, textiles

Alternatives to Full-Time Employment: Paid internships

Contact: Mr. Donny Rakes, Employee Development Manager; Phone: 540-634-3635; Fax: 540-634-3752.

WASHINGTON

ECCO WORLDWIDE, INC.

2722 EASTLAKE AVENUE EAST
SEATTLE, WASHINGTON 98102-3143

Overview: Provides telecommunications services. Established: 1995. Ownership: Private.

Key Statistics: Annual Sales: $13 million.

Hiring History: Number of professional employees hired in 1997: 15. 1996: 12.

Expertise/Education Sought: Software engineering/development, web site development, advertising, collection

Alternatives to Full-Time Employment: Paid internships

Company-Paid Training is available.

International Assignments are available to employees.

Contact: Human Resources Department; Fax: 206-505-5505.

Company URL: http://www.usnetwork.com

FLUOR DANIEL HANFORD

PO BOX 1000
RICHLAND, WASHINGTON 99352

Overview: Operates nuclear energy generation facility. Established: 1970. Parent Company: Fluor Daniel, Inc. Ownership: Public.

Key Statistics: Annual Sales: $1.3 billion. Number of Employees: 7,885.

Hiring History: Number of professional employees hired in 1996: 50.

Expertise/Education Sought: Chemical engineering, nuclear engineering, health physics

Alternatives to Full-Time Employment: Paid internships

Company-Paid Training is available.

Contact: Ms. Faye Powers, Human Resources; Phone: 509-376-5141; Fax: 509-373-1010. Job Hotline: 800-438-1305.

GST TELECOMMUNICATIONS, INC.
4001 MAIN STREET
VANCOUVER, WASHINGTON 98663-1896

OVERVIEW: Provides telecommunications services. Established: 1989. Ownership: Public.

KEY STATISTICS: Annual Sales: $41.3 million. Number of Employees: 950.

HIRING HISTORY: Number of professional employees hired in 1996: 800.

EXPERTISE/EDUCATION SOUGHT: Telecommunications

COMPANY-PAID TRAINING is available.

INTERNATIONAL ASSIGNMENTS are available to employees.

CONTACT: Ms. Theresa Osburne, Corporate Recruiter; Phone: 360-906-7200; Fax: 360-906-7132.

COMPANY URL: http://www.gstcorp.com

INTERMEC CORPORATION
6001 36TH AVENUE WEST
EVERETT, WASHINGTON 98203

OVERVIEW: Manufactures bar code readers. Established: 1991. Parent Company: Western Atlas, Inc. Ownership: Public.

KEY STATISTICS: Annual Sales: $350 million. Number of Employees: 1,869.

HIRING HISTORY: Number of professional employees hired in 1996: 100.

EXPERTISE/EDUCATION SOUGHT: Software engineering/development, electrical engineering, quality engineering, mechanical engineering, marketing, public relations, manufacturing engineering

ALTERNATIVES TO FULL-TIME EMPLOYMENT: Internships

COMPANY-PAID TRAINING is available.

INTERNATIONAL ASSIGNMENTS are available to employees.

CONTACT: Mr. Chris Dahl, Manager of Employment; Phone: 425-356-1795; Fax: 425-348-2844. E-mail: cdahl@intermec.com.

COMPANY URL: http://www.intermec.com

PEPSI COLA COMPANY
2300 26TH AVENUE SOUTH
SEATTLE, WASHINGTON 98144-5339

OVERVIEW: Produces soft drinks. Established: 1993. Parent Company: Pepsico. Ownership: Public.

KEY STATISTICS: Annual Sales: $181.9 million. Number of Employees: 1,200.

HIRING HISTORY: Number of professional employees hired in 1997: 62. 1996: 58.

EXPERTISE/EDUCATION SOUGHT: Production, administration, warehousing, sales

ALTERNATIVES TO FULL-TIME EMPLOYMENT: Paid internships

COMPANY-PAID TRAINING is available.

CONTACT: Human Resources Department; Phone: 206-323-2932; Fax: 206-326-7496. Job Hotline: 206-326-7436.

COMPANY URL: http://www.pepsi.com

PHYSIO-CONTROL INTERNATIONAL CORPORATION
11811 WILLOWS ROAD, NE
REDMOND, WASHINGTON 98052-2003

OVERVIEW: Manufactures electromedical equipment. Established: 1994. Ownership: Private.

KEY STATISTICS: Annual Sales: $173.2 million. Number of Employees: 830.

HIRING HISTORY: Number of professional employees hired in 1997: 120.

EXPERTISE/EDUCATION SOUGHT: Electrical engineering, software engineering/development, mechanical engineering

ALTERNATIVES TO FULL-TIME EMPLOYMENT: Paid internships

COMPANY-PAID TRAINING is available.

CONTACT: Human Resources; Phone: 425-867-4000; Fax: 425-867-4227. Job Hotline: 425-867-4130.

COMPANY URL: http://www.physiocontrol.com

PRECISION AIRMOTIVE CORPORATION
3220 100TH STREET, SW, BUILDING E
EVERETT, WASHINGTON 98204-1383

OVERVIEW: Manufactures and overhauls large radial piston engines. Established: 1981. Parent Company: Precision Aerospace Corporation. Ownership: Private.

KEY STATISTICS: Annual Sales: $47.4 million. Number of Employees: 500.

EXPERTISE/EDUCATION SOUGHT: Avionics

ALTERNATIVES TO FULL-TIME EMPLOYMENT: Paid internships

CONTACT: For individuals with previous experience: Mr. Randy Jensen, Engineering Manager; Phone: 425-353-8181; Fax: 425-348-3545.

COMPANY URL: http://www.precisionairmotive.com

R W BECK GROUP, INC.
1001 4TH AVENUE, SUITE 2500
SEATTLE, WASHINGTON 98154-1101

OVERVIEW: Provides engineering services. Established: 1996. Ownership: Private.

KEY STATISTICS: Annual Sales: $54.8 million. Number of Employees: 465.

EXPERTISE/EDUCATION SOUGHT: Consulting engineering

COMPANY-PAID TRAINING is available.

CONTACT: For individuals with previous experience: Mr. Van Finger, Human Resource Director; Phone: 206-695-4700. Job Hotline: 206-695-4524.

SHARP MICROELECTRONICS TECHNOLOGY, INC.
5700 NORTHWEST PACIFIC RIM BOULEVARD
CAMAS, WASHINGTON 98607-9489

OVERVIEW: Researches and develops microelectronics. Established: 1986. Parent Company: Sharp Corporation Japan. Ownership: Private.

KEY STATISTICS: Annual Sales: $100 million. Number of Employees: 250.

HIRING HISTORY: Number of professional employees hired in 1997: 35. 1996: 70.

EXPERTISE/EDUCATION SOUGHT: Electrical engineering, firmware engineering, ECAD engineering, process development, design engineering, test engineering, hardware engineering

COMPANY-PAID TRAINING is available.

CONTACT: For college students and recent graduates: Mr. Edward Vasquez, Human Resources Manager; Phone: 360-834-8720; Fax: 360-834-8611. For individuals with previous experience: Mr. Jim Dwinells, Human Resource Manager. E-mail: jimd@sharpwa.com. Job Hotline: 800-874-2771.

COMPANY URL: http://www.sharpmicroelectronics.com

SIEMENS POWER CORPORATION
PO BOX 130
RICHLAND, WASHINGTON 99352

OVERVIEW: Manufactures energy generation equipment. Established: 1978. Parent Company: Siemens Corporation. Ownership: Private.

KEY STATISTICS: Annual Sales: $500 million. Number of Employees: 1,600.

EXPERTISE/EDUCATION SOUGHT: Mechanical engineering, chemical engineering, nuclear engineering, electrical engineering

INTERNATIONAL ASSIGNMENTS are available to employees.

CONTACT: Ms. Cathy Tuttle, Employee Relations Specialist, 2101 Horn Rapids Road, Richland, Washington, 99352-5901; Phone: 509-375-8100; Fax: 509-375-8224.

TRAMCO, INC.
11323 30TH AVENUE WEST
EVERETT, WASHINGTON 98204-3556

OVERVIEW: Provides commercial aircraft repair and maintenance services. Established: 1988. Parent Company: The BF Goodrich Company. Ownership: Public.

Key Statistics: Annual Sales: $100 million. Number of Employees: 2,000.
Hiring History: Number of professional employees hired in 1997: 900. 1996: 500.
Expertise/Education Sought: Structural engineering
Company-Paid Training is available.
Contact: For individuals with previous experience: Human Resource Department; Phone: 425-423-4040; Fax: 425-514-0155. Job Hotline: 800-347-2341.
Company URL: http://www.bfgoodrich.com

WEYERHAEUSER COMPANY
33663 WEYERHAEUSER WAY SOUTH
FEDERAL WAY, WASHINGTON 98001

Overview: Manufactures and markets pulp, paper and wood products; manages and develops real estate; provides financial services. Established: 1900. Ownership: Public.
Key Statistics: Annual Sales: $11.1 billion. Number of Employees: 39,661.
Hiring History: Number of professional employees hired in 1996: 1,500.
Expertise/Education Sought: Accounting, engineering, human resources, pulp and paper science, forestry, information systems, information technology
Alternatives to Full-Time Employment: Part-time jobs, paid internships
Company-Paid Training is available.
International Assignments are available to employees.
Contact: For college students and recent graduates: Ms. Sally Wagner, College Recruiter; Phone: 253-924-2345. E-mail: resume@wdni.com. For individuals with previous experience: Mr. Don Oars, Recruiting and Staffing; Phone: 253-924-2345. E-mail: resume@wdni.com. Job Hotline: 253-924-5347.
Company URL: http://www.weyerhaeuser.com

WEST VIRGINIA

CENTURY ALUMINUM COMPANY
PO BOX 98
RAVENSWOOD, WEST VIRGINIA 26164

Overview: Produces primary aluminum. Established: 1989. Ownership: Public.
Key Statistics: Annual Sales: $688.9 million. Number of Employees: 2,180.
Hiring History: Number of professional employees hired in 1997: 25. 1996: 24.
Expertise/Education Sought: Engineering, mechanical engineering, electrical engineering, computer programming
Alternatives to Full-Time Employment: Internships
Company-Paid Training is available.
Contact: Mr. Jesse Scott, Senior Human Resources Representative.

CLEARON CORPORATION
95 MACCORKLE AVENUE, SW
SOUTH CHARLESTON, WEST VIRGINIA 25303

Overview: Manufactures alkalies and chlorine. Established: 1995. Parent Company: ICL. Ownership: Private.
Key Statistics: Annual Sales: $76.7 million. Number of Employees: 200.
Hiring History: Number of professional employees hired in 1997: 15.
Expertise/Education Sought: Manufacturing, chemical engineering, engineering, administration, accounting, human resources, sales
Company-Paid Training is available.
International Assignments are available to employees.
Contact: Human Resources; Phone: 304-746-3100; Fax: 304-746-3101.

CUSTOM MANUFACTURING SERVICES
142 BRICK STREET
PRINCETON, WEST VIRGINIA 24740

Overview: Manufactures sheet metalwork. Established: 1994. Ownership: Private.
Key Statistics: Annual Sales: $5 million. Number of Employees: 103.
Hiring History: Number of professional employees hired in 1997: 25. 1996: 40.
Expertise/Education Sought: Automation, sales, engineering, manufacturing
Contact: Mrs. Darla Blankenship, Controller; Phone: 304-431-7170, ext. 216; Fax: 304-431-7176.

EAGLE ENERGY
PO BOX 368
MADISON, WEST VIRGINIA 25130

Overview: Mines bituminous coal. Established: 1989. Parent Company: AT Massey. Ownership: Public.
Key Statistics: Annual Sales: $63.6 million. Number of Employees: 353.
Expertise/Education Sought: Mining
Contact: For individuals with previous experience: Human Resources Department; Phone: 304-245-8236; Fax: 304-245-5467.

PRATT & WHITNEY
BENEDUM INDUSTRIAL PARK
BRIDGEPORT, WEST VIRGINIA 26330

Overview: Manufactures aircraft engines and engine parts. Established: 1970. Parent Company: United Technologies Corporation. Ownership: Public.
Key Statistics: Annual Sales: $120 million. Number of Employees: 405.
Expertise/Education Sought: Military/defense, technical engineering
Contact: For individuals with previous experience: Mr. Denis Parisien, General Manager; Phone: 304-842-5421; Fax: 304-842-7170.

SOUTH ATLANTIC COAL COMPANY
127 NORTH STREET
BLUEFIELD, WEST VIRGINIA 24701-4033

Overview: Wholesales coal. Established: 1974. Parent Company: Mapco Coal. Ownership: Private.
Key Statistics: Annual Sales: $66.5 million. Number of Employees: 174.
Hiring History: Number of professional employees hired in 1997: 3.
Expertise/Education Sought: Accounting, engineering
Contact: For individuals with previous experience: Personnel Department; Phone: 540-326-2600; Fax: 304-325-3500.

WHEELING-PITTSBURGH STEEL CORPORATION
1134 MARKET STREET
WHEELING, WEST VIRGINIA 26003-2971

Overview: Manufactures steel. Established: 1991. Parent Company: WHX Corporation. Ownership: Public.
Key Statistics: Annual Sales: $1 billion. Number of Employees: 5,403.
Hiring History: Number of professional employees hired in 1997: 45. 1996: 75.
Expertise/Education Sought: Marketing, finance, accounting, mechanical engineering, chemical engineering, electrical engineering, metallurgical engineering
Alternatives to Full-Time Employment: Paid internships
Contact: Mr. James Bronchik, Director of Personnel; Phone: 304-234-2523; Fax: 304-234-2213. E-mail: bronchikjg@wpsc.com.

WISCONSIN

AEIC, INC.
202 EAST STREET
BARABOO, WISCONSIN 53913-2515

Overview: Manufactures motors and generators. Established: 1996. Ownership: Private.
Key Statistics: Annual Sales: $10 million. Number of Employees: 200.
Expertise/Education Sought: Manufacturing, sales, engineering
Alternatives to Full-Time Employment: Paid internships
Company-Paid Training is available.
Contact: Human Resources; Phone: 608-356-8437; Fax: 608-356-2478.

APOGEE WAUSAU GROUP
1415 WEST STREET
WAUSAU, WISCONSIN 54402

Overview: Manufactures metal window and door frames and sash. Established: 1968. Parent Company: Apogee Enterprises, Inc. Ownership: Public.
Key Statistics: Annual Sales: $79.4 million. Number of Employees: 775.
Hiring History: Number of professional employees hired in 1997: 15.
Expertise/Education Sought: Engineering, sales, manufacturing engineering
Alternatives to Full-Time Employment: Paid internships
Company-Paid Training is available.
Contact: Mr. Dennis Franecki, Vice President Human Resources; Phone: 715-843-4045; Fax: 715-843-4350.
Company URL: http://www.wausaumetals.com

APPLETON PAPER, INC.
825 EAST WISCONSIN AVENUE
APPLETON, WISCONSIN 54911

Overview: Produces copy paper, fine, and specialty papers. Established: 1994. Parent Company: AWA. Ownership: Private.
Key Statistics: Annual Sales: $636.2 million. Number of Employees: 4,300.
Expertise/Education Sought: Sales, chemistry, finance, pulp and paper science
Alternatives to Full-Time Employment: Paid internships
Company-Paid Training is available.
Contact: For individuals with previous experience: Mr. Jeff Fern, Human Resources Manager; Phone: 920-991-8520.

ASTRONAUTICS CORPORATION OF AMERICA
4115 NORTH TEUTONIA AVENUE
MILWAUKEE, WISCONSIN 53209-6731

Overview: Designs and manufactures aircraft instruments, navigation systems, robotics equipment and computer systems. Established: 1959. Ownership: Private.
Key Statistics: Annual Sales: $186.4 million. Number of Employees: 2,300.
Hiring History: Number of professional employees hired in 1997: 50.
Expertise/Education Sought: Electrical engineering, mechanical engineering, computer science, business, software engineering/development
Alternatives to Full-Time Employment: Paid internships
Contact: For individuals with previous experience: Ms. Donna McSorley, Senior Personnel Administrator; Phone: 414-447-8200, ext. 113; Fax: 414-447-8497.
Company URL: http://www.astronautics.com

BELOIT CORPORATION
1 SAINT LAWRENCE AVENUE
BELOIT, WISCONSIN 53511

Overview: Manufactures paper processing and recycling equipment. Parent Company: Harnischfeger Industries, Inc.
Expertise/Education Sought: Mechanical engineering, electrical engineering, pulp and paper science
Alternatives to Full-Time Employment: Paid internships
Company-Paid Training is available.
International Assignments are available to employees.
Contact: Human Resources; Phone: 608-365-3311.
Company URL: http://www.beloit.com

BEMIS MANUFACTURING COMPANY
300 MILL STREET
SHEBOYGAN FALLS, WISCONSIN 53085-0901

Mailing Address: PO Box 901, Sheboygan Falls, WI 53085-0901
Overview: Manufactures flexible packaging, film, wallpaper, and roll goods. Established: 1935. Ownership: Private.
Key Statistics: Annual Sales: $200 million. Number of Employees: 1,800.
Expertise/Education Sought: Sales, mechanical engineering, plastics engineering, injection molding
Alternatives to Full-Time Employment: Paid internships
International Assignments are available to employees.
Contact: Mr. Ed Collins, Vice President of Human Resources; Phone: 920-467-4621; Fax: 920-467-8573.

BOLDT GROUP, INC.
2525 NORTH ROEMER ROAD
APPLETON, WISCONSIN 54911-0419

Overview: Provides construction services for pulp and paper, engineering, and construction industries. Established: 1984. Ownership: Private.
Key Statistics: Annual Sales: $310 million. Number of Employees: 2,200.
Hiring History: Number of professional employees hired in 1997: 20. 1996: 30.
Expertise/Education Sought: Civil engineering, mechanical engineering, electrical engineering, accounting, finance
Alternatives to Full-Time Employment: Paid internships
Company-Paid Training is available.
Contact: Mr. Gary Tornes, Vice President of Human Resources; Phone: 920-739-6321; Fax: 920-739-4363. E-mail: gtornes@boldt.com.
Company URL: http://www.boldt.com

BRADLEY FIXTURES CORPORATION
W142 N9101 FOUNTAIN BOULEVARD
MENOMONEE FALLS, WISCONSIN 53051-2348

Mailing Address: PO Box 309, Menomonee Falls, WI 53052-0309
Overview: Manufactures plumbing equipment. Established: 1921. Ownership: Private.
Key Statistics: Annual Sales: $70 million. Number of Employees: 380.
Expertise/Education Sought: Design engineering, manufacturing engineering, mechanical engineering
Alternatives to Full-Time Employment: Paid internships
Company-Paid Training is available.
Contact: Dr. John Wandschneider, Vice President of Human Resources; Phone: 414-251-6000; Fax: 414-251-5817.
Company URL: http://www.bradleycorp.com

W.H. BRADY COMPANY
6555 WEST GOOD HOPE ROAD
MILWAUKEE, WISCONSIN 53223

OVERVIEW: Manufactures engraved name plates. Established: 1981. Ownership: Public.

KEY STATISTICS: Annual Sales: $31.7 million. Number of Employees: 500.

HIRING HISTORY: Number of professional employees hired in 1997: 250. 1996: 240.

EXPERTISE/EDUCATION SOUGHT: Sales, management information systems, mechanical engineering, electrical engineering, industrial engineering, chemistry, chemical engineering, marketing, customer service and support

ALTERNATIVES TO FULL-TIME EMPLOYMENT: Paid internships

COMPANY-PAID TRAINING is available.

INTERNATIONAL ASSIGNMENTS are available to employees.

CONTACT: Ms. Grace LoCoco, Recruitment Manager; Phone: 414-358-6768; Fax: 414-358-6798. E-mail: grace_lococo@whbrady.com. Job Hotline: 414-438-6901.

COMPANY URL: http://www.whbrady.com

BRIGGS AND STRATTON CORPORATION
12301 WEST WIRTH STREET
MILWAUKEE, WISCONSIN 53222-2110

OVERVIEW: Manufactures small engines. Established: 1909. Ownership: Public.

KEY STATISTICS: Annual Sales: $1.3 billion. Number of Employees: 7,661.

EXPERTISE/EDUCATION SOUGHT: Marketing, finance, accounting, electrical engineering, mechanical engineering, chemical engineering, design engineering, data processing

ALTERNATIVES TO FULL-TIME EMPLOYMENT: Paid internships

INTERNATIONAL ASSIGNMENTS are available to employees.

CONTACT: For individuals with previous experience: Human Resources; Phone: 414-259-5333; Fax: 414-256-5155.

COMPANY URL: http://www.briggsandstratton.com

CONSOLIDATED PAPERS, INC.
231 1ST AVENUE NORTH
WISCONSIN RAPIDS, WISCONSIN 54495-2774

OVERVIEW: Manufactures quilted paper and corrugated products. Established: 1894. Ownership: Public.

KEY STATISTICS: Annual Sales: $1.5 billion. Number of Employees: 7,000.

EXPERTISE/EDUCATION SOUGHT: Accounting, marketing, finance, management information systems, sales, manufacturing

ALTERNATIVES TO FULL-TIME EMPLOYMENT: Paid internships

COMPANY-PAID TRAINING is available.

CONTACT: Mr. Charles Basting, Professional Recruiting Manager; Phone: 715-422-3895; Fax: 715-422-4000.

COOPER POWER SYSTEMS, INC.
2300 BADGER DRIVE
WAUKESHA, WISCONSIN 53188-5931

OVERVIEW: Manufactures electrical equipment. Established: 1988. Parent Company: Cooper Industries. Ownership: Public.

KEY STATISTICS: Annual Sales: $217.2 million. Number of Employees: 4,000.

EXPERTISE/EDUCATION SOUGHT: Marketing, sales, accounting, management information systems, computer programming, mechanical engineering, technical engineering, human resources

COMPANY-PAID TRAINING is available.

INTERNATIONAL ASSIGNMENTS are available to employees.

CONTACT: For individuals with previous experience: Ms. Ruth Prince, Manager of Benefits and Salary; Phone: 414-896-2400; Fax: 414-896-2386.

COMPANY URL: http://www.cooperpower.com

ELECTRONIC ASSEMBLY CORPORATION
2121 HARRISON STREET
NEENAH, WISCONSIN 54956-3761

MAILING ADDRESS: PO Box 529, Neenah, WI 54957-0529

OVERVIEW: Manufactures printed circuit boards. Established: 1979. Parent Company: Plexus Corporation Electronic Assembly Corporation. Ownership: Public.

KEY STATISTICS: Annual Sales: $300 million. Number of Employees: 2,100.

HIRING HISTORY: Number of professional employees hired in 1997: 377.

EXPERTISE/EDUCATION SOUGHT: Electrical engineering, accounting, computer programming, manufacturing engineering, industrial engineering, finance, purchasing, quality engineering

ALTERNATIVES TO FULL-TIME EMPLOYMENT: Part-time jobs, paid internships

COMPANY-PAID TRAINING is available.

INTERNATIONAL ASSIGNMENTS are available to employees.

CONTACT: For individuals with previous experience: Human Resources; Phone: 920-722-2826.

COMPANY URL: http://www.plexus.com

J.H. FINDORFF AND SON, INC.
601 WEST WILSON STREET
MADISON, WISCONSIN 53703-3628

OVERVIEW: Provides construction services. Established: 1992. Ownership: Private.

KEY STATISTICS: Annual Sales: $117.1 million. Number of Employees: 300.

HIRING HISTORY: Number of professional employees hired in 1997: 20. 1996: 15.

EXPERTISE/EDUCATION SOUGHT: Civil engineering, structural engineering

ALTERNATIVES TO FULL-TIME EMPLOYMENT: Paid internships

COMPANY-PAID TRAINING is available.

CONTACT: Mr. F. C. Hastings, President; Phone: 608-257-5321; Fax: 608-257-5306.

GOOD HUMOR-BRYERS ICE CREAM
909 PACKERLAND DRIVE
GREEN BAY, WISCONSIN 54303-9007

MAILING ADDRESS: PO Box 19007, Green Bay, WI 54303-9007

OVERVIEW: Produces ice cream and other frozen desserts. Established: 1989. Parent Company: Unilever Corporation. Ownership: Private.

KEY STATISTICS: Annual Sales: $198.8 million. Number of Employees: 2,000.

EXPERTISE/EDUCATION SOUGHT: Manufacturing

ALTERNATIVES TO FULL-TIME EMPLOYMENT: Paid internships

COMPANY-PAID TRAINING is available.

CONTACT: For individuals with previous experience: Human Resources; Phone: 920-499-5151; Fax: 920-497-6523.

GREAT LAKES HOSPITALITY CORPORATION
122 WEST WASHINGTON AVENUE, 10TH FLOOR
MADISON, WISCONSIN 53703

OVERVIEW: Provides hotel or motel management services. Established: 1993. Parent Company: NEVAC. Ownership: Private.

KEY STATISTICS: Number of Employees: 500.

HIRING HISTORY: Number of professional employees hired in 1996: 100.

EXPERTISE/EDUCATION SOUGHT: Accounting, sales, marketing, construction, restaurant management

ALTERNATIVES TO FULL-TIME EMPLOYMENT: Paid internships

CONTACT: For individuals with previous experience: Human Resources Department; Phone: 608-251-6400; Fax: 608-251-6800. E-mail: nevac@mail.tds.net.

COMPANY URL: http://www.greatlakeshotels.com

HARNISCHFEGER INDUSTRIES, INC.
3600 SOUTH LAKE DRIVE
SAINT FRANCIS, WISCONSIN 53235

Mailing Address: PO Box 554, Milwaukee, WI 53201-0554

Overview: Manufactures cranes, paper making machinery, and other industrial equipment. Established: 1884. Ownership: Public.

Key Statistics: Annual Sales: $2.9 billion. Number of Employees: 17,500.

Expertise/Education Sought: Electrical engineering, mechanical engineering, production control, systems analysis, hydraulic engineering, design engineering, manufacturing engineering

Company-Paid Training is available.

International Assignments are available to employees.

Contact: For individuals with previous experience: Human Resources Department; Phone: 414-486-6400; Fax: 414-486-6747. Job Hotline: 800-244-9777.

HOOPER CORPORATION
2030 PENNSYLVANIA AVENUE
MADISON, WISCONSIN 53704

Overview: Constructs electric power lines. Established: 1959. Ownership: Private.

Key Statistics: Annual Sales: $43 million. Number of Employees: 390.

Hiring History: Number of professional employees hired in 1996: 3.

Expertise/Education Sought: Mechanical engineering, electrical engineering

Alternatives to Full-Time Employment: Paid internships

Company-Paid Training is available.

International Assignments are available to employees.

Contact: For individuals with previous experience: President.

KIMBERLY-CLARK CORPORATION
2100 WINCHESTER ROAD
NEENAH, WISCONSIN 54956-9317

Overview: Paper mill, including paper toweling, tissue paper, and towel tissue stock. Established: 1995. Ownership: Public.

Key Statistics: Annual Sales: $2.2 billion. Number of Employees: 15,100.

Expertise/Education Sought: Research and development, marketing, engineering, chemical engineering, mechanical engineering, logistics, finance, sales, information systems

Alternatives to Full-Time Employment: Paid internships

Company-Paid Training is available.

International Assignments are available to employees.

Contact: For college students and recent graduates: Ms. Kay Kedberlein, College Recruiting Coordinator; Phone: 920-721-2602; Fax: 920-721-4219. For individuals with previous experience: Corporate Recruiting Department; Phone: 920-721-5768; Fax: 920-721-4219.

Company URL: http://www.kimberly-clark.com

KOHLER COMPANY
444 HIGHLAND DRIVE
KOHLER, WISCONSIN 53044

Overview: Manufactures plumbing fixtures, engines, and generators. Established: 1873. Ownership: Private.

Key Statistics: Annual Sales: $1.1 billion. Number of Employees: 17,000.

Expertise/Education Sought: Accounting, marketing, engineering, sales

Alternatives to Full-Time Employment: Internships

International Assignments are available to employees.

Contact: Mr. Rick Homiston, Director of Personnel; Phone: 920-457-4441, ext. 7367; Fax: 920-459-1818. E-mail: rick.homiston@kohler-co.e-mail.com.

LAMICO, INC.
474 MARION ROAD
OSHKOSH, WISCONSIN 54901

Overview: Manufactures surgical appliances and wooden crutches. Established: 1963. Ownership: Private.

Key Statistics: Annual Sales: $6 million. Number of Employees: 85.

Expertise/Education Sought: Manufacturing engineering

Contact: For individuals with previous experience: Human Relations Department, PO Box 104, Oshkosh, Wisconsin, 54902-0104; Phone: 920-231-1672; Fax: 920-231-3252.

MANITOWOC CRANES, INC.
2401 SOUTH 30TH STREET
MANITOWOC, WISCONSIN 54220-5919

Overview: Designs and manufactures commercial ice machines and cranes. Established: 1997. Parent Company: Manitowoc Company. Ownership: Public.

Key Statistics: Annual Sales: $100 million. Number of Employees: 642.

Expertise/Education Sought: Engineering, manufacturing

Alternatives to Full-Time Employment: Paid internships

Company-Paid Training is available.

International Assignments are available to employees.

Contact: Ms. Trudy Kamps, Director of Human Resources; Phone: 920-683-6224; Fax: 920-683-6320. E-mail: tkamps@manitowoccranes.com.

MARATHON ENGINEERS/ARCHITECTS/PLANNERS, LLC
2323 EAST CAPITOL DRIVE
APPLETON, WISCONSIN 54913-8028

Overview: Provides consulting engineering services. Established: 1995. Parent Company: Jaakko Poyry. Ownership: Private.

Key Statistics: Annual Sales: $30 million. Number of Employees: 135.

Expertise/Education Sought: Chemical engineering, mechanical engineering, electrical engineering, process development, structural engineering, architecture

Alternatives to Full-Time Employment: Part-time jobs

Contact: For individuals with previous experience: Mr. Richard Moon, Controller; Phone: 414-954-2000; Fax: 414-954-2020.

Company URL: http://www.marathon-eap.com

MEGTEC SYSTEMS
830 PROSPER ROAD
DE PERE, WISCONSIN 54115-3104

Overview: Manufactures industrial furnaces and ovens. Established: 1997. Parent Company: Sequa. Ownership: Public.

Key Statistics: Annual Sales: $71.1 million. Number of Employees: 707.

Expertise/Education Sought: Engineering

Alternatives to Full-Time Employment: Paid internships

Company-Paid Training is available.

Contact: For individuals with previous experience: Human Resources Department; Phone: 920-336-5715; Fax: 920-336-3404.

MENARD, INC.
4777 MENARD DRIVE
EAU CLAIRE, WISCONSIN 54703-9604

Overview: Operates chain of home improvement stores. Established: 1960. Ownership: Private.

Key Statistics: Annual Sales: $2.7 billion. Number of Employees: 19,000.

Hiring History: Number of professional employees hired in 1997: 4,000.

Expertise/Education Sought: Business management, finance, computer programming, computer science, manufacturing engineering, accounting, management, management information systems, information systems, construction management

Company-Paid Training is available.

Contact: Mr. Ron Mehr, Operations and Personnel Manager; Phone: 715-876-2460; Fax: 715-876-5010.

MENASHA CORPORATION

1645 BERGSTROM ROAD
NEENAH, WISCONSIN 54957-0367

Mailing Address: PO Box 367, Neenah, WI 54957-0367

Overview: Manufactures corrugated containers and plastics. Established: 1849. Ownership: Private.

Key Statistics: Annual Sales: $915.3 million. Number of Employees: 5,427.

Expertise/Education Sought: Marketing, sales, accounting, business management, engineering, graphic arts, packaging engineering, information technology, management information systems

Alternatives to Full-Time Employment: Paid internships

Company-Paid Training is available.

International Assignments are available to employees.

Contact: For college students and recent graduates: Ms. Diana Friis, Corporate Recruiter; Phone: 920-751-1000; Fax: 920-751-1904. E-mail: recruit@menasha.com. For individuals with previous experience: Mr. Steven Kromholz, Corporate Manager Staffing; Phone: 920-751-1000; Fax: 920-751-1904. E-mail: recruit@menasha.com.

Company URL: http://www.menasha.com

NELSON INDUSTRIES, INC.

PO BOX 600
STOUGHTON, WISCONSIN 53589

Overview: Manufactures filters and mufflers for industrial machinery. Established: 1939. Parent Company: Cummins Engine Company. Ownership: Private.

Key Statistics: Annual Sales: $228 million. Number of Employees: 2,400.

Hiring History: Number of professional employees hired in 1997: 75.

Expertise/Education Sought: Information systems, computer programming, materials science, acoustic engineering, manufacturing engineering, mechanical engineering, drafting, electrical engineering, industrial engineering, quality engineering

Alternatives to Full-Time Employment: Paid internships

Company-Paid Training is available.

International Assignments are available to employees.

Contact: Ms. Julie Norland, Human Resources; Fax: 608-873-1552. E-mail: jnorland@nelsondiv.com.

NICOLET INSTRUMENT CORPORATION

5225 VERONA ROAD
MADISON, WISCONSIN 53711-4418

Overview: Manufactures electronic instrumentation for search navigation and aeronautical systems. Established: 1992. Parent Company: Thermo Optek. Ownership: Public.

Key Statistics: Annual Sales: $91.6 million. Number of Employees: 456.

Expertise/Education Sought: Chemistry, engineering

Alternatives to Full-Time Employment: Part-time jobs, paid internships

Company-Paid Training is available.

International Assignments are available to employees.

Contact: Mr. Cory Erickson, Manager of Human Resources; Phone: 608-273-6881; Fax: 608-273-6880.

Company URL: http://www.nicolet.com

NORLIGHT TELECOMMUNICATIONS, INC.

275 NORTH CORPORATE DRIVE
BROOKFIELD, WISCONSIN 53045-5825

Overview: Provides data communications services. Established: 1991. Parent Company: Journal Communications. Ownership: Private.

Key Statistics: Annual Sales: $48 million. Number of Employees: 180.

Hiring History: Number of professional employees hired in 1997: 71. 1996: 57.

Expertise/Education Sought: Electrical engineering, computer programming

Alternatives to Full-Time Employment: Part-time jobs, paid internships

Contact: Human Resources. E-mail: jobs@norlight.com.

Company URL: http://www.norlight.com

OMNIQUIP INTERNATIONAL, INC.

369 WEST WESTERN AVENUE
PORT WASHINGTON, WISCONSIN 53074-2233

Overview: Manufactures construction equipment. Established: 1995. Ownership: Private.

Key Statistics: Annual Sales: $264.2 million. Number of Employees: 500.

Expertise/Education Sought: Mechanical engineering

Alternatives to Full-Time Employment: Paid internships

Company-Paid Training is available.

Contact: For individuals with previous experience: Human Resources; Phone: 414-284-5571; Fax: 414-268-8988.

Company URL: http://www.omniquip.com

OSCAR MAYER FOODS CORPORATION

910 MAYER AVENUE
MADISON, WISCONSIN 53704-4256

Overview: Produces processed meats and prepared foods. Established: 1985. Parent Company: Kraft Foods. Ownership: Public.

Key Statistics: Annual Sales: $1.9 billion. Number of Employees: 11,000.

Expertise/Education Sought: Marketing, data processing, technical engineering, law, civil engineering, finance, human resources

Alternatives to Full-Time Employment: Paid internships

Company-Paid Training is available.

International Assignments are available to employees.

Contact: Human Resources Department, PO Box 7188, Madison, Wisconsin, 53707-4256; Phone: 608-241-3311; Fax: 608-285-6741.

OSHKOSH B'GOSH, INC.

112 OTTER AVENUE
OSHKOSH, WISCONSIN 54901

Overview: Manufactures and markets children's clothing.

Expertise/Education Sought: Textiles, management, operations

Alternatives to Full-Time Employment: Paid internships

Company-Paid Training is available.

International Assignments are available to employees.

Contact: For individuals with previous experience: Corporate Human Resources Department, PO Box 300, Oshkosh, Wisconsin, 54902; Phone: 920-231-8800; Fax: 920-231-8621. Job Hotline: 920-232-4410.

Company URL: http://www.oshkoshbgosh.com

OSHKOSH TRUCK CORPORATION

2307 OREGON STREET
OSHKOSH, WISCONSIN 54903

Mailing Address: PO Box 2566, Oshkosh, WI 54903

Overview: Manufactures truck and bus bodies. Established: 1917. Parent Company: Pierce Manufacturing. Ownership: Public.

Oshkosh Truck Corporation (continued)

Key Statistics: Annual Sales: $683.2 million. Number of Employees: 2,700.

Hiring History: Number of professional employees hired in 1997: 70. 1996: 20.

Expertise/Education Sought: Accounting, data processing, marketing, engineering

Alternatives to Full-Time Employment: Part-time jobs, paid internships

Company-Paid Training is available.

Contact: Mrs. Tracey Rymer, Corporate Recruiter; Phone: 414-235-9150; Fax: 414-233-9268.

Company URL: http://www.oshkoshtruck.com

PACKERLAND PACKING COMPANY

1330 LIME KILN ROAD
GREEN BAY, WISCONSIN 54311

Overview: Operates meat packing plant and slaughterhouse for beef. Established: 1994. Ownership: Private.

Key Statistics: Annual Sales: $650 million. Number of Employees: 3,000.

Expertise/Education Sought: Accounting, marketing, sales, engineering

Alternatives to Full-Time Employment: Paid internships

Company-Paid Training is available.

Contact: For individuals with previous experience: Mr. Lee Davis, Vice President of Human Resources; Phone: 414-406-2220; Fax: 414-468-7453.

PERRY-JUDD'S, INC.

575 WEST MADISON STREET
WATERLOO, WISCONSIN 53594-1378

Overview: Provides commercial lithographic printing services. Established: 1995. Ownership: Private.

Key Statistics: Annual Sales: $173 million. Number of Employees: 900.

Hiring History: Number of professional employees hired in 1997: 197. 1996: 40.

Expertise/Education Sought: Accounting, finance, administration, marketing, management, engineering, information systems, electrical engineering, design engineering, computer programming

Company-Paid Training is available.

Contact: For individuals with previous experience: Ms. Victoria Richey, Employment Coordinator; Phone: 414-478-1744; Fax: 414-478-1536. E-mail: vrichey@perrygraphic.com.

PIERCE MANUFACTURING, INC.

2600 AMERICAN DRIVE
APPLETON, WISCONSIN 54915

Mailing Address: PO Box 2017, Appleton, WI 54913

Overview: Manufactures fire department vehicles. Established: 1996. Parent Company: Oshkosh Truck Corporation. Ownership: Public.

Key Statistics: Annual Sales: $200 million. Number of Employees: 1,200.

Hiring History: Number of professional employees hired in 1996: 100.

Expertise/Education Sought: Electrical engineering, mechanical engineering, manufacturing engineering

Company-Paid Training is available.

International Assignments are available to employees.

Contact: For individuals with previous experience: Mr. James Staats, Director of Human Resources; Phone: 414-832-3000; Fax: 414-832-3019.

PLASTIC INGENUITY, INC.

1017 PARK STREET
CROSS PLAINS, WISCONSIN 53528-9630

Overview: Manufactures plastic containers. Established: 1972. Ownership: Private.

Key Statistics: Annual Sales: $45 million. Number of Employees: 602.

Hiring History: Number of professional employees hired in 1997: 68. 1996: 70.

Expertise/Education Sought: Design engineering, mechanical engineering, industrial engineering

Alternatives to Full-Time Employment: Paid internships

Company-Paid Training is available.

Contact: Mr. Randy Paul, Director of Human Resources; Phone: 608-798-6169; Fax: 608-798-4452.

Company URL: http://www.plasticingenuity.com

PRENT CORPORATION

2225 KENNEDY ROAD
JANESVILLE, WISCONSIN 53545-0824

Overview: Provides custom plastic thermal forming services. Established: 1967. Ownership: Private.

Key Statistics: Annual Sales: $70.7 million. Number of Employees: 500.

Expertise/Education Sought: Electrical engineering, control engineering, sales, human resources

Contact: For individuals with previous experience: Human Resources Department.

QUAD/GRAPHICS, INC.

W224 N3322 DUPLAINVILLE ROAD
PEWAUKEE, WISCONSIN 53072

Mailing Address: N63 W23075 Highway 74, Sussex, WI 53089

Overview: Provides commercial printing services. Established: 1971. Ownership: Private.

Key Statistics: Annual Sales: $1.2 billion. Number of Employees: 10,000.

Expertise/Education Sought: Marketing, sales, accounting, data processing, information systems, electrical engineering, mechanical engineering

Alternatives to Full-Time Employment: Paid internships

Company-Paid Training is available.

International Assignments are available to employees.

Contact: For individuals with previous experience: Ms. Pam Hinkle, Career Assistance Manager, N63 W23075 Main Street, Pewaukee, Wisconsin, 53089; Phone: 800-682-5627; Fax: 414-691-7814. Job Hotline: 800-682-5627.

REGAL WARE, INC.

1675 REIGLE DRIVE
KEWASKUM, WISCONSIN 53040

Overview: Manufactures cookware, housewares, and household appliances. Established: 1919. Ownership: Private.

Key Statistics: Annual Sales: $250 million. Number of Employees: 1,474.

Hiring History: Number of professional employees hired in 1997: 800.

Expertise/Education Sought: Plastics engineering, mechanical engineering, design engineering

Alternatives to Full-Time Employment: Paid internships

Company-Paid Training is available.

Contact: For individuals with previous experience: Mr. James Moran, Senior Vice President of Corporate Development, PO Box 395, Kewaskum, Wisconsin, 53040-0395; Phone: 414-626-2121; Fax: 414-626-8565.

REXNORD CORPORATION

4701 WEST GREENFIELD AVENUE
MILWAUKEE, WISCONSIN 53214

Mailing Address: PO Box 2022, Milwaukee, WI 53201

Overview: Manufactures chains, gears, idlers, sprockets, and related power transmission components. Established: 1995. Parent Company: BTR, Inc. Ownership: Private.

Key Statistics: Annual Sales: $686.7 million. Number of Employees: 4,718.

Expertise/Education Sought: Engineering

Company-Paid Training is available.

International Assignments are available to employees.

Contact: For individuals with previous experience: Ms. Jean Kurowski, Manager for Headquarters; Phone: 414-643-3000; Fax: 414-643-3078.

Company URL: http://www.rexnord.com

REYNOLDS CONSUMER PRODUCTS

670 NORTH PERKINS STREET
APPLETON, WISCONSIN 54914-2399

Mailing Address: PO Box 2899, Appleton, WI 54913-2399

Overview: Manufactures coated and laminated paper packaging. Established: 1988. Parent Company: Reynolds Metal Company. Ownership: Public.

Key Statistics: Annual Sales: $220 million. Number of Employees: 1,550.

Hiring History: Number of professional employees hired in 1996: 40.

Expertise/Education Sought: Mechanical engineering, chemical engineering

Alternatives to Full-Time Employment: Paid internships

Company-Paid Training is available.

International Assignments are available to employees.

Contact: For individuals with previous experience: Mr. Richard Gile, Employee Relations Manager; Phone: 920-739-9471; Fax: 920-738-1102.

RITE-HITE MANUFACTURING CORPORATION

8900 NORTH ARBON DRIVE
MILWAUKEE, WISCONSIN 53223-2451

Overview: Manufactures miscellaneous metal products. Established: 1994. Parent Company: Rite Hite Holding Corporation. Ownership: Private.

Key Statistics: Annual Sales: $46.7 million. Number of Employees: 60.

Hiring History: Number of professional employees hired in 1997: 100. 1996: 130.

Expertise/Education Sought: Sales, marketing, engineering

Alternatives to Full-Time Employment: Paid internships

Company-Paid Training is available.

International Assignments are available to employees.

Contact: Human Resources; Phone: 414-355-2600.

SCHNEIDER NATIONAL, INC.

3101 SOUTH PACKERLAND DRIVE
GREEN BAY, WISCONSIN 54313-2545

Overview: Provides truck freight and shipping services. Established: 1977. Ownership: Private.

Key Statistics: Annual Sales: $2.2 billion. Number of Employees: 17,150.

Expertise/Education Sought: Operations, logistics, business

Alternatives to Full-Time Employment: Paid internships

Company-Paid Training is available.

International Assignments are available to employees.

Contact: Mr. Tim Fliss, Director of Human Resources; Phone: 920-592-2000, ext. 3261; Fax: 920-592-3252.

Company URL: http://www.schneider.com

A.O. SMITH CORPORATION

11270 WEST PARK PLACE
MILWAUKEE, WISCONSIN 53223-0974

Overview: Manufactures water heaters, car frames, motors, ceramic fibers, and plastic pipe. Established: 1923. Ownership: Public.

Key Statistics: Annual Sales: $781.2 million. Number of Employees: 7,700.

Hiring History: Number of professional employees hired in 1997: 7. 1996: 12.

Expertise/Education Sought: Human resources, electrical engineering, marketing, chemical engineering, finance, financial analysis, industrial engineering, mechanical engineering, accounting

Alternatives to Full-Time Employment: Part-time jobs, paid internships

Company-Paid Training is available.

International Assignments are available to employees.

Contact: Ms. Donna Cook, Manager of Human Resources, PO Box 23974, Milwaukee, Wisconsin, 53223; Phone: 414-359-4131; Fax: 414-359-4064. E-mail: dcook@aosmith.com.

Company URL: http://www.aosmith.com

SNAP-ON TOOLS COMPANY

2801 80TH STREET
KENOSHA, WISCONSIN 53141-1410

Overview: Manufactures hand tools, power tools, tool storage cabinets, and electronic diagnostic and service equipment. Established: 1920. Ownership: Public.

Key Statistics: Annual Sales: $1.5 billion. Number of Employees: 9,000.

Hiring History: Number of professional employees hired in 1996: 80.

Expertise/Education Sought: Accounting, marketing, electrical engineering, finance, mechanical engineering, design engineering, data processing, computer programming, computer analysis

Alternatives to Full-Time Employment: Paid internships

Company-Paid Training is available.

International Assignments are available to employees.

Contact: Mr. Don O'Boyle, Human Resource Administrator; Phone: 414-656-4825; Fax: 414-656-4961. Job Hotline: 888-526-3600.

Company URL: http://www.snapon.com

SWING-N-SLIDE CORPORATION

1212 BARBERRY DRIVE
JANESVILLE, WISCONSIN 53545-0457

Overview: Investment holding company. Established: 1996. Ownership: Private.

Key Statistics: Annual Sales: $30.9 million. Number of Employees: 550.

Hiring History: Number of professional employees hired in 1997: 20. 1996: 10.

Expertise/Education Sought: Engineering, sales, marketing

Company-Paid Training is available.

Contact: Mr. David Hammelman, Human Resources Vice President; Phone: 608-755-4777; Fax: 608-755-4763. E-mail: hammelmand@swing-n-slide.com.

Company URL: http://www.swing-n-slide.com

TIME INSURANCE

501 WEST MICHIGAN STREET
MILWAUKEE, WISCONSIN 53203

Overview: Provides health, life, and disability insurance. Established: 1978. Parent Company: Fortis, Inc. Ownership: Private.

Key Statistics: Annual Sales: $1 billion. Number of Employees: 4,844.

Expertise/Education Sought: Computer programming, data processing, human resources, actuarial, accounting, finance, health care

Alternatives to Full-Time Employment: Paid internships

Company-Paid Training is available.

Contact: For individuals with previous experience: Ms. Mary Girard, Employment Specialist. Job Hotline: 414-299-8800.

TRI-CLOVER, INC.
9201 WILMOT ROAD
KENOSHA, WISCONSIN 53142-7628

OVERVIEW: Manufactures food products machinery, including meat processing machinery and commercial slicers. Established: 1991. Parent Company: ALFA-LAVAL, Inc. Ownership: Private.

KEY STATISTICS: Annual Sales: $158.5 million. Number of Employees: 1,500.

HIRING HISTORY: Number of professional employees hired in 1997: 60.

EXPERTISE/EDUCATION SOUGHT: Engineering

ALTERNATIVES TO FULL-TIME EMPLOYMENT: Paid internships

COMPANY-PAID TRAINING is available.

INTERNATIONAL ASSIGNMENTS are available to employees.

CONTACT: For individuals with previous experience: Human Resources Department; Phone: 414-694-5511; Fax: 414-697-2452.

COMPANY URL: http://www.triclover.com

TYTON HELLERMANN
7930 NORTH FAULKNER ROAD
MILWAUKEE, WISCONSIN 53224-3423

OVERVIEW: Manufactures plastics. Established: 1982. Parent Company: Bowthorpe International. Ownership: Private.

KEY STATISTICS: Annual Sales: $50 million. Number of Employees: 300.

HIRING HISTORY: Number of professional employees hired in 1997: 102. 1996: 90.

EXPERTISE/EDUCATION SOUGHT: Laboratory technology, industrial engineering, sales

ALTERNATIVES TO FULL-TIME EMPLOYMENT: Part-time jobs, paid internships

COMPANY-PAID TRAINING is available.

CONTACT: For individuals with previous experience: Ms. Ellen Seefeldt, Personnel Assistant; Phone: 414-355-1130, ext. 441; Fax: 414-362-8306. Job Hotline: 414-355-1130, ext. 240.

COMPANY URL: http://www.tyton.industry.net

UNITED STARS HOLDINGS, INC.
1546 HENRY AVENUE
BELOIT, WISCONSIN 53511-3668

OVERVIEW: Manufactures steel pipes and tubes. Established: 1994. Ownership: Private.

KEY STATISTICS: Annual Sales: $67.2 million. Number of Employees: 560.

EXPERTISE/EDUCATION SOUGHT: Metal laboratory analysis

CONTACT: For individuals with previous experience: Mr. Arlyn Poppen, Account Manager; Phone: 608-365-8891, ext. 616; Fax: 608-365-1259.

UPC HEALTH NETWORK
3724 WEST WISCONSIN AVENUE
MILWAUKEE, WISCONSIN 53208

OVERVIEW: Provides long-term care, home health-care, pharmacy, purchasing, therapies, and billing services. Established: 1967. Parent Company: Extendicare Health Services. Ownership: Public.

KEY STATISTICS: Annual Sales: $350 million. Number of Employees: 2,000.

HIRING HISTORY: Number of professional employees hired in 1997: 200. 1996: 200.

EXPERTISE/EDUCATION SOUGHT: Information systems

ALTERNATIVES TO FULL-TIME EMPLOYMENT: Part-time jobs, paid internships

COMPANY-PAID TRAINING is available.

CONTACT: Ms. Nancy Monti, Human Resources; Phone: 800-334-4572, ext. 581; Fax: 414-342-6877. E-mail: nmont@upchn.com. Job Hotline: 414-347-4696.

COMPANY URL: http://www.upchn.com

VOLLRATH COMPANY LLC
1236 NORTH 18TH STREET
SHEBOYGAN, WISCONSIN 53081-3201

OVERVIEW: Produces silverware and plated ware. Established: 1996. Parent Company: Windway Capital. Ownership: Private.

KEY STATISTICS: Annual Sales: $130 million. Number of Employees: 690.

HIRING HISTORY: Number of professional employees hired in 1997: 40.

EXPERTISE/EDUCATION SOUGHT: Manufacturing, management, maintenance, industrial engineering

ALTERNATIVES TO FULL-TIME EMPLOYMENT: Paid internships

COMPANY-PAID TRAINING is available.

INTERNATIONAL ASSIGNMENTS are available to employees.

CONTACT: Mr. David Tanger, Labor Relations; Phone: 414-457-4851; Fax: 414-459-5448.

WACKER CORPORATION
N92 W15000 ANTHONY AVENUE
MENOMONEE FALLS, WISCONSIN 53052-1504

MAILING ADDRESS: PO Box 9007, Menomonee Falls, WI 53052

OVERVIEW: Manufactures construction machinery. Established: 1957. Ownership: Private.

KEY STATISTICS: Annual Sales: $60.2 million. Number of Employees: 525.

EXPERTISE/EDUCATION SOUGHT: Construction

COMPANY-PAID TRAINING is available.

INTERNATIONAL ASSIGNMENTS are available to employees.

CONTACT: Human Resources; Phone: 414-255-0500; Fax: 414-255-0550.

WAUPACA FOUNDRY, INC.
311 TOWER ROAD
WAUPACA, WISCONSIN 54981-1664

OVERVIEW: Manufactures gray ductile iron. Established: 1968. Parent Company: The Budd Company. Ownership: Private.

KEY STATISTICS: Annual Sales: $300 million. Number of Employees: 1,900.

EXPERTISE/EDUCATION SOUGHT: Mechanical engineering, electrical engineering, metallurgical engineering, project engineering, accounting, sales, computer programming

COMPANY-PAID TRAINING is available.

CONTACT: Mr. James Larsen, Vice President and Manager of Human Resources, PO Box 249, Waupaca, Wisconsin, 54981-0249; Phone: 715-258-6611; Fax: 715-258-9268.

WICOR INDUSTRIES, INC.
626 EAST WISCONSIN AVENUE
MILWAUKEE, WISCONSIN 53202-4608

Overview: Manufactures pumps and pumping equipment. Established: 1996. Ownership: Public.

Key Statistics: Annual Sales: $409.9 million. Number of Employees: 2,300.

Hiring History: Number of professional employees hired in 1997: 30.

Expertise/Education Sought: Fuel engineering

Alternatives to Full-Time Employment: Paid internships

International Assignments are available to employees.

Contact: For individuals with previous experience: Human Resources; Phone: 414-385-7000; Fax: 414-291-6880. Job Hotline: 414-291-6756.

WISCONSIN ENERGY CORPORATION
231 WEST MICHIGAN STREET
MILWAUKEE, WISCONSIN 53203

Overview: Electric and gas utility. Established: 1981. Ownership: Public.

Key Statistics: Annual Sales: $1.8 billion. Number of Employees: 4,504.

Expertise/Education Sought: Accounting, computer programming, data processing, mechanical engineering, electrical engineering, industrial engineering, civil engineering

Alternatives to Full-Time Employment: Paid internships

Company-Paid Training is available.

Contact: Human Resources; Phone: 414-221-2345; Fax: 414-221-3340. Job Hotline: 414-221-3091.

WISCONSIN ENGINEERING, INC.
2919 NORTH 8TH STREET
SHEBOYGAN, WISCONSIN 53083-4120

Overview: Manufactures lawn and garden equipment. Established: 1993. Parent Company: WIKOV. Ownership: Private.

Key Statistics: Annual Sales: $110.4 million. Number of Employees: 1,104.

Hiring History: Number of professional employees hired in 1997: 200.

Expertise/Education Sought: Mechanical engineering, engineering

Company-Paid Training is available.

International Assignments are available to employees.

Contact: Mr. Robert Olmr, Vice President of Sales; Phone: 920-459-8690; Fax: 920-457-7437. E-mail: rolmr@aol.com.

WYOMING

LOWER VALLEY POWER AND LIGHT
PO BOX 188
AFTON, WYOMING 83110

Overview: Electric utility. Established: 1937. Ownership: Private.

Key Statistics: Annual Sales: $25.5 million. Number of Employees: 80.

Hiring History: Number of professional employees hired in 1997: 6. 1996: 3.

Expertise/Education Sought: Electrical engineering, electrical technology

Company-Paid Training is available.

Contact: For individuals with previous experience: Mr. Winston Allred, Staff Assistant; Phone: 307-886-3175; Fax: 307-886-5787.

OCI WYOMING, LP
LA BARGE ROAD
GREEN RIVER, WYOMING 82935

Overview: Mines potash, soda, and borate minerals. Established: 1995. Parent Company: OCI Chemical Corporation. Ownership: Private.

Key Statistics: Annual Sales: $178 million. Number of Employees: 490.

Hiring History: Number of professional employees hired in 1996: 10.

Expertise/Education Sought: Mining, refining

Company-Paid Training is available.

Contact: Mr. Barry Bundy, Human Resources Manager; Phone: 307-872-7196; Fax: 307-872-7285.

EMPLOYER DESCRIPTIONS

In-depth information on the business activities and employment opportunities at some profiled companies and government organizations

The Charles Stark Draper Laboratory, Inc.

Description of Organization: Private, nonprofit corporation engaged in applied research, engineering development, education, and technology transfer
Number of Employees: 1,300
Headquarters Location: Cambridge, Massachusetts
Academic Fields of Recruitment Interest: Aeronautical/astronautical, electrical, mechanical, ocean, and systems engineering; computer science
Major Entry-Level Opportunities: Instrument development, electronic design, materials, simulation, software, microelectronics process and packaging, quality, and test engineering and guidance, navigation, and control engineering

BACKGROUND AND OPERATIONS

The Charles Stark Draper Laboratory's mission is to serve the national interest as an independent not-for-profit laboratory engaged in applied research, engineering development, education, and technology transfer. Draper creates value for its sponsors by helping them clarify their requirements and conceptualize innovative solutions to their problems by developing and fielding engineering prototypes, by transitioning its products and processes to industry for production, by providing follow-on support throughout the product/process life cycle, and by promoting and supporting advanced technical education.

Much of Draper's work today revolves around measurement, analysis, simulation, and control of complex dynamic systems. Navy Strategic Systems, Space and Missile Systems, Tactical Systems, Ocean Systems and Special Operations, and Information and Automation Systems comprise the Laboratory's five major business units. Although most of its business is with the U.S. government, Draper also applies its technology to automotive, commercial space, and biomedical markets.

Founded in 1930 by Dr. Charles Stark Draper, "the father of inertial navigation," Draper Laboratory began as the Instrumentation Laboratory at the Massachusetts Institute of Technology. In 1973, the Laboratory became a separate, nonprofit corporation. The Laboratory's pioneering work in inertial navigation has led to the development of the highly precise inertial sensors, software and ultrareliable systems that the world depends on today for safe and accurate guidance, navigation, and control of its commercial and military aircraft, submarines, strategic and tactical missiles, and manned and unmanned spacecraft.

In 1988 Draper Laboratory endowed the Charles Stark Draper Prize, a biennial international award for achievement in engineering and technology, which currently consists of a gold medal and $450,000 in prize money. It is administered by the National Academy of Engineering and its purpose is to focus worldwide attention on the critical role that engineering and technology play in improving the quality of everyday life. The 1997 Draper Prize was awarded to Dr. Vladimir Haensel, the developer of platforming, which is a revolutionary chemical engineering process that is essential in converting petroleum into high-level fuels and also generates the raw materials used in the manufacturing of plastics.

EMPLOYMENT OPPORTUNITIES

Draper Laboratory employs engineers and scientists at all degree levels. The following listing of mission areas indicates the nature and scope of opportunities available.

Strategic Guidance Systems. As design agent for the Navy Strategic Systems Program, Draper Laboratory designed and developed the highly accurate guidance system (MK 6) for the Trident submarine's D5 ballistic missile. Draper is the Navy's prime contractor for technical engineering services in support of the Navy's deployed guidance system. Draper also supports the Air Force through its work on the Peacekeeper and Minuteman ICBM guidance system programs.

Ocean Systems and Special Operations. Providing innovative engineering solutions to complex maritime and special operations customers, this directorate is currently managing projects ranging from state-of-the-art research in miniature robotics and sensor systems to the design and development of controls and displays for the Advanced SEAL Delivery System (ASDS). New efforts are being made to develop remote wireless sensor networks for shipboard damage assessment and condition-based maintenance that will lead to reduced crew size on Navy ships. Related programs are underway to develop leading-edge miniaturized, multi-chip module electronics and microelectromechanical systems.

Draper continues to conduct research in autonomous undersea systems, building on a long history of vehicle developments for the Department of Defense, and currently provides technical support to the Navy's Unmanned Undersea Vehicle (UUV) acquisition office, PMS403. A fault-tolerant, multi-channel ship control computer was designed and built by Draper Laboratory for the SSN-21 *Seawolf*-class submarine. Draper provides in-service engineering support for the Navy's Deep Submergence Rescue Vehicles and the USS *Dolphin*, AGSS-555.

Space Systems. Draper Laboratory continues to build its reputation as a nationally recognized space avionics design and development organization for manned and unmanned space programs. This is based on years of experience in guidance, navigation, and control system design as well as supporting sensor, actuator, and data management systems design. Draper continues to explore and develop new designs for advanced inertial systems as well as precision pointing and tracking concepts in support of future space systems. Draper is working directly with NASA and Boeing in development of the system architecture, the control and data handling system, and the guidance, navigation, and control system for the International Space Station. Draper continues to support the Space Transportation System (Shuttle) for specialized missions, such as docking with the Russian MIR spacecraft.

Tactical Systems. In addition to its strategic systems business, Draper is engaged in the development of a variety of measurement, processing, control, and display systems for the tactical arena. Draper is currently pioneering the development of micromechanical sensors for a broad spectrum of tactical system applications. Draper's micromechanical inertial sensors are finding leading-edge, dual-use applications in commercial products such as automobiles as well as in military products such as guided munitions and precision strike weapons. As a contractor to the Air Force Logistics Centers and Naval Sea Systems Command (NAVSEA), Draper performs design studies for upgrading the avionics in fielded systems, develops prototype hardware, and delivers the production data packages that are needed to support subsequent competitive acquisitions. In this way, Draper provides its sponsors with the objective engineering design services and technical support that are required to facilitate effective technology transfer into fielded systems.

Applied Information and Automation Systems. A fast-growing business segment at Draper Laboratory, Applied Information and Automation Systems encompasses the development and application of information technologies across a broad range of problems including command, control, Advanced Warfighting Experiments; planning and decision systems; high-fidelity simulation and modeling, including synthetic environments; automated software engineering; and multimedia. The mission of the Applied Information and Automation Systems Directorate is to identify and develop business opportunities to apply information, software, and automation technologies to the analysis, design, and implementation of solutions to complex problems in decision-support, command and control, and information management and dissemination. The Directorate provides a range of services including technology development, system design, system evaluation, systems engineering and integration, and program management.

Engineering. Draper's engineering focus is on systems engineering; inertial instruments/mechanical design; electronic design and sensor development; guidance, navigation, and control; software engineering; hardware engineering and prototyping; system integration; test and evaluation; quality management; and information technology.

The principal technologies include mechanical design and analysis, including autonomous vehicles; CAD, data management, and prototype development; precision inertial instruments (gyros and accelerometers) using mechanical and microminiature silicon technology; systems integration, test, and evaluation; guidance and navigation analysis; control and decision systems; software and information systems engineering; sensor and communications electronics involving photonic, acoustic, and electromagnetic sensors and devices that include signal processing; digital, computer, and network systems, including multi-chip module technology development; power and radiation-hardened electronics, including electronic packaging design and prototype fabrication; computer and information systems support; and quality management.

Within engineering, Draper's Guidance Technology Center is chartered to foster, promote, and recognize contributions to the field of guidance, navigation, and control. The Center is engaged in technology assessments, forecasting, planning, and sponsored research within the laboratory, in universities, and in Russia; participation in government panels and studies; coordination and technical support for publications, seminars, lectures, and symposia; and supporting professional society activities. The Center director is editor-in-chief of the *Journal of Guidance, Control, and Dynamics*, published by the American Institute of Aeronautics and Astronautics.

Independent Research and Development (IR&D) and Corporate Sponsored Research (CSR). Draper's IR&D and CSR programs are designed to develop innovative system concepts and enabling technologies and the technical capabilities needed to increase productivity of the Laboratory for near- and far-term business needs. These programs include strategic systems concepts and technologies; autonomous air, land, sea, and space systems; precision targeting and weapon systems; information integration; dissemination and control systems; and distributed sensors and networks. Related technology developments include both inertial and noninertial MicroElectro-Mechanical Systems (MEMS) sensors and systems, anti-jam tightly coupled GPS/INS systems, power generation and storage for autonomous vehicles and spacecraft, extremely low power electronics for precision guidance systems and autonomous vehicles, new generation inertial sensors using electro-optics, fault-tolerant processor architectures and reliable software. Draper sponsors on-campus research programs with Massachusetts Institute of Technology (MIT) and other universities to promote strong R&D interaction between Draper's staff members and academia. In addition, Draper's IR&D program funds the research and technical supervision of numerous graduate students seeking advanced science and engineering degrees at MIT and other universities.

TRAINING

All Draper employees are given the opportunity to enhance their skills and maximize their capabilities through a comprehensive professional development training program. Draper provides a generous tuition assistance program and encourages employees to attend conferences and seminars and write technical papers. Televised courses from local universities are available in-house.

EDUCATION

Draper Laboratory's mission to promote and support advanced technical education is manifested in its recruitment of graduate and undergraduate students to work on real-world problems while fulfilling part of their academic requirements at affiliated institutions. The Draper Fellowship Program, cooperative engineering programs, and internships are some of the programs administered by the education department.

THE ORGANIZATION AS EMPLOYER

Draper Laboratory's encouragement of an open and experimental approach to research produces an environment of freedom, excitement, and challenge. Academic interaction between Draper and MIT is exercised through joint educational and research programs. Employees' individual growth is highly encouraged through both on-the-job training and formal course work. Members of the technical staff are encouraged to attend professional seminars and to publish in scientific and engineering journals. Recognition for outstanding work is provided through the Distinguished Performance Awards Program; the Employee Recognition Awards Program; and the Best Patent, Paper, and Task Leader Awards. Draper provides a competitive benefits package. In addition, employees have access to many of the facilities and programs at MIT.

APPLICATION AND INFORMATION

Direct inquiries to:

Professional Employment, MS44
The Charles Stark Draper Laboratory, Inc.
555 Technology Square
Cambridge, Massachusetts 02139-3563
E-mail: hr@draper.com
World Wide Web: http://www.draper.com

The CNA Corporation

Description of Organization: The nation's oldest civilian organization engaged in operations research and policy analysis. CNAC is a nonprofit corporation.
Number of Employees: 525, including 240 research analysts
Headquarters Location: Alexandria, Virginia
Academic Fields of Recruitment Interest: Aeronautical, chemical, electrical, marine, nuclear, and systems engineering; applied mathematics; operations research, physical chemistry, physics; statistics; economics and political science (quantitative backgrounds preferred for both)
Major Entry-Level Opportunities for New Graduates: Empirical research and policy analysis

BACKGROUND AND OPERATIONS

The CNA Corporation (CNAC) traces its origins to World War II, when its forerunner, the Anti-Submarine Warfare Operations Research Group, helped the U.S. Navy defeat the U-boat threat to Allied shipping. CNAC today is a broadly based operations research and policy analysis organization dedicated to conceptually sound, analytically rigorous, imaginative, and objective research for clients in national, state, and local governments and nonprofit institutions.

The corporation serves its clients through two divisions: Center for Naval Analyses (CNA), a federally funded research and development center sponsored by the Department of the Navy; and Institute for Public Research (IPR).

The CNAC division's efforts for the Navy and Marine Corps cover a wide range, from support of training and testing activities in the fleet to evaluation of future technologies, forces, and strategies for top-level decision makers in Washington. CNAC is helping the Navy and Marine Corps make the transition to a post–Cold War security environment. Shrinking defense budgets mean harder choices, the kinds of choices that CNAC's research helps illuminate.

CNAC also applies its expertise on behalf of other Defense agencies and such civilian agencies as the Federal Aviation Administration.

IPR provides analytical and technical support to clients whose needs do not fall within CNAC's traditional areas of inquiry. Dedicated to the same high standards of intellectual rigor, IPR is available to government and nongovernment organizations that need services of the highest quality.

CNAC offers many opportunities to conduct and manage independent, original research. A member of the research staff can:

- Help the Navy and Marine Corps keep their forces ready to respond quickly and effectively in crises around the world through improved selection and training of personnel, more efficient logistics systems, and updated tactics.
- Help the Navy and Marine Corps improve the delivery of health care to their personnel and adjust to such changes as the assignment of women to combat units.
- Assess potential future threats to national security and the implications of those threats for the allocation of defense budgets in the 1990s.
- Influence the selection of new systems that will help the Navy, Marine Corps, and other services cope with the unpredictable world of the twenty-first century.
- Work with civilian agencies on a wide variety of nationally important initiatives, from improvements in the efficiency and safety of commercial air travel to productive uses of excess military facilities.

Analysts work directly with decision makers, consulting frequently with those who have responsibility and authority to establish and implement plans and policies. CNAC analysts must have the ability to reason soundly and to apply scientific techniques imaginatively. Researchers must be able to see beyond surface details to the core of a problem and to devise a logical and, above all, practical solution.

EMPLOYMENT OPPORTUNITIES

ENTRY LEVEL

Aeronautical, Chemical, Electrical, Marine, Nuclear, and Systems Engineers; Applied Mathematicians; Operations Research Specialists; Physical Chemists; Physicists; Statisticians; and Economists and Political Scientists (quantitative backgrounds preferred for both). CNAC hires an average of 20 to 30 analysts a year. Research staff positions are normally limited to candidates who hold advanced degrees or are about to receive them. More than 90 percent of the members of the research staff hold advanced degrees; two thirds have doctorates.

New scientists and engineers work with their more experienced colleagues on critical problems that do not fit into the customary categories of science. Two attributes are critical to most of the work done at CNAC. One is the ability to do relevant scientific literature research, because there is a large fund of knowledge at hand on nearly every subject. The other involves the basic statistics and elementary concepts of probability that underlie almost every experimental scientific study.

As an analyst gains experience, the complexity of assignments will increase. Analysts may also be assigned to several projects. Assignments are made to best match a staff member's expertise and interests to the client's needs. To recommend workable solutions to the problem being investigated, an analyst must fully understand the issues involved and ensure that all possible alternatives are considered. Several stages are involved in the process.

First, the problem and the specific tasks are defined. The appropriate literature is reviewed. Consultation with colleagues, clients, and others in the broader research community provides additional insight. Visits to clients' facilities give analysts the chance to observe their operations firsthand. Preliminary data are collected.

At the same time, the analyst is developing an analytical framework for addressing the problem. Once a study plan with specific tasks is laid out, research progresses until preliminary results are obtained. More data may be collected and, if appropriate, models developed. Progress is discussed in informal meetings and more formal presentations. Finally, when definitive results answer the questions, the analyst presents them, both orally and in a written report, directly to the official who must make the decisions.

Success in analytical assignments almost always leads to the opportunity to direct projects. The successful project director may then follow one of two career routes, sometimes alternating between them. A strong track record in project direction may eventually lead to a position in research management. In contrast, a strong track record could also lead to the management of large, highly visible studies of importance to senior decision makers.

FOR EXPERIENCED PERSONNEL

Seasoned analysts, capable of conducting or directing high-quality analysis, are of special interest to CNAC.

TRAINING

New analysts participate in an orientation and training program at the same time they start working on their first assignments. During one-on-one sessions with senior analysts and in group discussions, new staff members learn about CNAC's mission and philosophy, about clients' systems and operations, about their decision-making processes, and about the applicability of analytical techniques to specific types of problems. This formal orientation lasts about six months. After completing the program, analysts continue to gain knowledge and insight through project work, reading, seminars, and discussions with colleagues.

THE ORGANIZATION AS EMPLOYER

Analysts draw on the rich institutional memory of the oldest organization of its type in the United States. Among the research tools available to them, one of the most valuable is the extensive collection of technical materials, research reports, and data and models. An online reference system facilitates access. The library carries dozens of professional journals and houses a collection of thousands of scientific and historical books.

The staff has access to extensive computing services and facilities, including professional programming, networked mainframes and microcomputers, a library of databases, and modern electronic publishing software. Analysts and secretaries are equipped to produce reports and briefings that are polished as well as substantively valid with support from an in-house staff of professional editors and graphic artists.

Staff members are encouraged to belong to professional societies and to publish in professional journals.

Excellent fringe benefits are offered. These include a retirement program with TIAA/CREF, as well as medical, life, and long-term disability insurance and generous vacation and sick leave. CNAC is committed to equal opportunity employment and affirmative action.

FIELD AND OVERSEAS ASSIGNMENTS

After about three years, most analysts have advanced to the point where they are considered eligible for a field assignment. The field program has about 40 analysts on assignment at Navy and Marine Corps installations throughout the country, including Hawaii, and in the United Kingdom, Italy, and Japan. Assignments are based on experience, ability, and, as much as possible, personal preference. Field representatives enjoy unusual opportunities for independent assignments of practical value with operational naval commands and the opportunity to see their work put to immediate use. Assignments are normally for two years.

APPLICATION AND INFORMATION

CNAC accepts resumes and employment inquiries throughout the year. Ph.D.'s and doctoral candidates who are invited to CNAC for interviews are asked to demonstrate their ability to conduct an independent investigation or analysis and present material clearly by giving a seminar approximately one hour long on a subject of their own choice.

Applicants selected will be subject to a security investigation and must meet eligibility requirements for access to classified information. U.S. citizenship is required.

Direct inquiries to:

Ms. Carolyn P. Williams, Manager of Employment
The CNA Corporation
4401 Ford Avenue
Alexandria, Virginia 22302-8268
Fax: 703-824-2740
E-mail: resumes@cna.org
World Wide Web: http://www.cna.org

Computer Associates International, Inc.

COMPUTER® ASSOCIATES
Software superior by design.

Description of Organization: The world's leading developer of client/server and distributed computing software; develops, markets, and supports enterprise management, database and applications development, and business applications for a broad range of mainframe, midrange, and desktop computers; serves most of the world's major business, government, research, and educational organizations

Number of Employees: 11,000 worldwide

Headquarters Location: Islandia, New York

Academic Fields of Recruitment Interest: Primarily computer science, computer programming, computer operations, technical writing, business, finance, marketing, accounting, and human resources

Major Entry-Level Opportunities for Graduates: Programming, datacenter operations, software documentation, sales, marketing, financial services, and human resources

BACKGROUND AND OPERATIONS

Founded in 1976, Computer Associates International, Inc. (CA) has grown from a small, 4-person software start-up to the world's leading independent software company for mission-critical business computing.

Today, more than 11,000 employees, working in 160 offices in forty-three countries, develop, support, and sell hundreds of products. In calendar year 1997, the company posted a record $4.5 billion in revenue. As a Fortune 500 company, CA is listed on the New York Stock Exchange.

For more than two decades, CA has been at the vanguard of the technology front, offering more than 500 software solutions. The company's install-base of almost 100 percent of the world's Fortune 500 companies continues to be its primary strength.

CA's product portfolio reflects the marketplace's increasing requirements for broad-based, multiplatform solutions to meet the demands of enterprise-wide processing. From network and systems management to database and application development and the Internet/Intranet, CA solutions cover virtually the entire spectrum of business computing.

The company's informal structure not only fosters a nonbureaucratic workplace, but also promotes consistent communication and increased interaction between employees. Headquarters for CA is Islandia, New York, 50 miles east of New York City on Long Island. This site is where the company's research, development, marketing, sales, and administration are located. The six-story building includes an on-site fitness and sports center complete with basketball, volleyball, weight-training, and aerobics facilities. Parents can take advantage of CA's licensed child development center, which cares for children from infancy through kindergarten. The company maintains a smoke-free environment.

CA also recognizes that service and support greatly enhance the value of every software solution, since reliable products and client care create long-lasting relationships.

The foundation of CA is its research and development organization. This engine drives CA's growth and success by providing technology solutions to real business issues.

Enterprise Management Software. CA's enterprise management software delivers the functionality necessary to manage today's increasingly large and complex heterogeneous computing environments.

CA solutions address critical areas of enterprise management for multiplatforms that range from the desktop to the mainframe to the Internet. Voted the best enterprise management solution by top publications, including *LAN Times, CIO, InformationWeek,* and others, Unicenter® TNG™ (The Next Generation) delivers end-to-end management of the entire IT environment. Its rich set of functions are built on an object-oriented and open architecture, making Unicenter® TNG™ highly scalable and able to provide coverage for a broad range of areas, including help desk, performance, security, and event management.

Information Management. CA offers revolutionary object and relational databases, Jasmine™ and OpenIngres®, along with distributed/replicated database technology for a wide variety of platforms and operating systems. A true object-oriented database and multimedia development environment, Jasmine™ is scalable, is distributed, and lets developers build intuitive applications complete with real-world representations of complex activities and data. Applications built with Jasmine™ can be deployed in traditional client/server environments or on the Internet through a Web browser.

The OpenIngres® product suite comprises a variety of modules providing a complete database management and data access environment. OpenIngres is ideally suited for advanced data warehousing and Internet commerce applications. The company's Impact/2000® solution is a powerful impact analysis tool for estimating and planning Year 2000 initiatives. CA application development products integrate object-oriented capabilities with visual programming tools. This combination supports rapid application development of leading-edge technology-based solutions without sacrificing the stability and reusability needed to address today's production business systems. Product applications are compatible with multiple vendors' DBMSs.

Business Application Software. CA's business application software promotes operational efficiency and sharing of information by utilizing high-performance database and transaction processing across multiple platforms. These solutions automate a broad range of functions for horizontal and vertical markets, including financial, manufacturing, accounting, banking, human resources, logistics, and business-decision management.

FUTURE DIRECTIONS

CA's ability to bring dependable products to market is a result of its unique strategy to provide mission-critical software that runs businesses. This single-minded focus enables the company to deliver software solutions for all kinds of businesses. While other software companies develop products that operate on one hardware platform only, CA's policy is vendor neutrality. Since its inception, CA has pursued a four-prong growth strategy. The key points of this market-driven initiative are internal product development, targeted acquisitions of complementary technology, effective integration of new products, and strategic alliances.

CA's research and development budget is nearly $300 million. This money is earmarked specifically for upgrading existing products, integration efforts, and creating new software solutions. CA is the leading business software company because of its distinct ability to provide integrated solutions for today's distributed computing environments.

EMPLOYMENT OPPORTUNITIES

CA is committed to employing high-quality personnel. The company actively pursues entry-level candidates to become productive members of the CA team. Most promotional opportunities come from within the organization. Internal promotion is so successful at CA that it is the proven career path for employees.

Project Managers/Implementation Consultants. Professionals in this position plan, install, implement, and manage Unicenter® TNG™ at multiple client sites. Strong UNIX, Windows NT, and Novell experience and exposure to using and implementing system and network tools are a plus. Knowledge of systems management disciplines such as security, disaster recovery, help desk operations, and software distribution are a plus. Strong project management and communication skills along with the ability to work with multiple operating systems are ideal. Previous experience with Unicenter® TNG™ is preferred. Extensive travel is required.

Systems Engineers. These positions promote leading-edge technology throughout the sales cycle. Pre-Sales Systems Engineers work closely with CA clients to determine their IT needs and strategies. Candidates should be able to present and demonstrate CA's technology to a variety of audiences. CA product knowledge is a plus. Additional responsibilities include implementation and configuration of CA software at client sites. Some travel is required.

Computer Operations Specialists. Lead operators are responsible for nightly production jobs that involve monitoring production and supervising a staff of operators. Experience with MVS, VM, JCL, VTAM, and/or utilities is necessary, and they must be able to work with support personnel to resolve production issues.

Junior Programmers. A Junior Programmer develops, supports, and enhances systems and network management, database and application software. CA provides a dynamic training program, one that immerses candidates in key industry technologies and CA's technology strategy. Candidates should have both an educational and working knowledge of C, C++, UNIX, DOS/Windows, and networking technologies.

Technical Writers. Opportunities exist for technical writers for the Documentation Department. Candidates must have a B.A. in English or journalism (or equivalent), with additional study in computer science and at least two years of experience writing end-user documents for software; must be able to learn quickly and work independently; and must have experience using Microsoft Word; writing Windows NT Help systems is a plus. Writing samples are required.

Kernel Developers. CA is looking for experienced UNIX and NT Kernel Developers to join CA's Unicenter® ENF group. Employees in this position can extend their knowledge of the internals of NT and UNIX Kernels, including file systems, ipc, device drivers, process management, networking, and streams. The ideal candidate must have experience shooting panics on NT, Solaris, HP-UX, AIX, or other UNIX Kernels and must be self-motivated and have the ability to work well with others. Employees may be called upon to consult with other groups at CA to help resolve difficult issues. Travel to client sites may be required.

Sales Professionals. The successful candidate possesses a minimum of three years of software sales success. Responsibilities include promoting CA's systems or database software solutions to the Fortune 500. Excellent earnings potential is available to highly motivated, entrepreneurial candidates with proven track records. Some travel is required.

Client Service Representatives. The ideal candidate is a take-charge person who is capable of handling all areas of account management, including lead creation and direction along with complete client administration, and has a desire to expand his or her sales skills in the information technology industry. The CSRs support the efforts of the marketing specialist to generate revenue. Applicants must be extremely detail oriented, highly responsive, energetic, and ready to do whatever it takes to get the job done.

CA-Starter Program. These temporary positions offer opportunities to explore the company for future employment.

THE ORGANIZATION AS EMPLOYER

Computer Associates recognizes that its success is due to the dedication of its employees. CA trains its employees in all areas of the organization and fosters a flexible and innovative working environment.

CA offers salary and incentive packages that are unsurpassed in the industry, as well as unlimited opportunities for growth. The company's benefits package is continually reviewed and updated to keep it competitive within the industry.

CO-OP AND INTERNSHIP PROGRAMS

Computer Associates actively recruits co-op employees and interns to train for full-time positions.

APPLICATION AND INFORMATION

Computer Associates recruits at and lists openings with universities nationwide. Students should contact their placement office for listings or send a resume and letter describing their interests and abilities.

Direct inquiries to:

Human Resources Department
Computer Associates International, Inc.
One Computer Associates Plaza
Islandia, New York 11788-7000
Phone: 800-454-3788 (toll-free)
Fax: 800-962-9224 (toll-free)
World Wide Web: http://www.cai.com
Resumes may be sent in ASCII text format to: resumes-usa-rl@cai.com

All product names referenced herein are the trademarks or registered trademarks of their respective companies.

Dialogic Corporation

Description of Organization: Leading manufacturer of high-performance, standards-based computer telephony components
Number of Employees: 1,000
Headquarters: Parsippany, New Jersey
Academic Fields of Recruitment Interest: Computer science, software engineering, computer engineering, sales, computer programming

BACKGROUND AND OPERATIONS

Dialogic Corporation, founded in 1983, is the leading manufacturer of high-performance, standards-based computer telephony (CT) components. Dialogic voice and fax processing products enable anytime/anywhere access to information via traditional telecommunications, computer and wireless networks, and the Internet. Applications such as unified messaging, Internet-based voice/fax communication, switching systems, and voice response are all based on Dialogic products.

Computer telephony is the term used to encompass a wide variety of technologies and applications that use the power of a computer (often a server on a local area network) to add intelligence to telephony functions and to mesh these functions with data processing. Today, telephony functions, such as receiving and sending messages, routing telephone calls, and taking action based on caller-entered touchtones, can be controlled by a voice processing server—a PC on a LAN equipped with voice hardware and software. The overlay of computing intelligence on telephony functions makes possible applications that streamline many common business practices. These applications can help to improve customer service, make salespeople more productive, and allow employees using desktop computers to be more efficient.

Headquartered in Parsippany, New Jersey, Dialogic has regional headquarters and sales offices in thirteen countries around the globe. With a worldwide workforce of over 1,000 full-time employees, Dialogic is strategically positioned to offer its global partners a wide range of marketing, technical, training, and other support services. Dialogic headquarters and regional offices include North America (Parsippany, New Jersey); Europe (Brussels); greater China (Beijing); South Asia and Australia/New Zealand (Singapore); Japan and Korea (Tokyo); and Latin America (Buenos Aires).

Dialogic delivered the first high-performance voice processing board to the market, built on open standards. The company is recognized as the world leader in developing and manufacturing high-performance CT components and is a leader in promoting open standards. Dialogic operates globally, with seventeen offices in thirteen countries.

Dialogic is the leading supplier of standards-based voice and fax processing components, with 64 percent market share and 3 million ports deployed worldwide.

Dialogic's next generation DM3 architecture is the foundation for a range of products that set industry benchmarks for performance and density. For example, QuadSpan™ is the first product in its class to support up to 120 ports of combined digital processing (DSP) and network interface on a single card in a single slot.

Dialogic products power a new class of enhanced services being delivered by the telecommunications carriers, such as voice-activated dialing. These servers are deployed in thirty-five wireless and wireline networks in eighty countries.

Dialogic has been a leader in defining and promoting standards. The company has played a leading role in founding standards organizations, such as Enterprise Computer Telephony Forum (ECTF) and Voice Over IP (VOIP) Forum. In addition, Dialogic is an active member and contributor to standards organizations, such as the American National Standards Institute (ANSI), the European Computer Manufacturers Association (ECMA), the European Telecommunications Standards Institute (ETSI), the International Organization for Standardization (ISO), the International Telecommunications Union (ITU), the Signal Computing System Architecture (SCSA), and other concerned organizations.

EMPLOYMENT OPPORTUNITIES

Professionals thrive in a take-it-to-the-edge environment as Dialogic pushes its products to facilitate a wide range of applications. Employees are part of a dynamic team that is dedicated to developing new ways to link people together and provide anywhere, anytime access to information.

On the technical side, Dialogic employees develop device drivers, help clients define new applications, design firmware, or write specifications and documentation. On the business side, they have an influence on product marketing or finance and accounting activities. From all perspectives, a career at Dialogic is a rewarding challenge from day one.

Dialogic is looking for today's top college graduates who offer a technical B.S., M.S., and/or M.B.A and knowledge of Windows/NT, UNIX, C/C++, DSP, firmware, multimedia, voice processing, hardware and software engineering, sales and marketing, computer telephony, or technical writing.

THE ORGANIZATION AS EMPLOYER

Of course, working with an industry leader is not the only benefit employees enjoy as part of a career with Dialogic. The highly competitive benefits package includes medical, vision, prescription, dental, and life insurance coverage. In addition, employees enjoy eighteen days of paid time off each year, eight holidays, profit sharing, and corporate memberships. They also have the option of taking advantage of tuition reimbursement, stock purchase, and retirement plans.

APPLICATION AND INFORMATION

Those interested in becoming a member of Dialogic's team should forward their resume, indicating area of interest, to:

Frank Johnson
Dialogic Corporation
1515 Route 10
Parsippany, New Jersey 07054
E-mail: johnsonf@dialogic.com
Fax: 973-993-5762

Ford Motor Company

Description of Organization: Automotive
Number of Employees: 370,000
Headquarters: Dearborn, Michigan
Academic Fields of Recruitment Interest: All

BACKGROUND AND OPERATIONS

Everyone sees it almost every day, on TV commercials, in traffic, maybe even on the car in your own driveway. It's the Ford Motor Company logo, and it's one of the best-known and most-respected logos in the world. It represents the second-largest company on the planet, a company with thousands of opportunities, successes, and challenges.

Since 1903, Ford Motor Company has blazed new trails in both the automotive industry and the manufacturing world as a whole. It began as a dream to manufacture an automobile everyone could afford. And today, Ford extends well beyond the realm of automobile manufacturing and offers tremendous opportunities in just about every field imaginable.

EMPLOYMENT OPPORTUNITIES

ENTRY-LEVEL

For starters, new graduates can join the Ford team as part of the Ford College Graduate Program. There are plenty of career opportunities and challenges available to those with bachelor's and master's degrees in nearly all disciplines, including engineering, finance, computers, and marketing. Engineers may work on Ford's leading-edge alternative fuel vehicle programs or design the next generation of climate control products. Software developers could get a chance to work on embedded systems for Ford Motorsports. A finance MBA might be the one who coordinates the Chassis five-year business plan. New employees even find themselves in a manufacturing or assembly plant, supervising as many as 35 people on their first day. There are literally thousands of opportunities available at Ford.

Ford Automotive Operations. How successful is Ford? In terms of manufacturing, their record is staggering. In 1997 alone, they sold 6.9 million vehicles worldwide, a global sales record. For that same period, they reported a 3.9 percent return on automotive sales and auto revenues grew by 4.2 percent to $123 billion.

Those are some impressive numbers. But that's just the beginning. With 111 plants in 38 countries, Ford is the world's largest manufacturer of trucks and the world's second-largest manufacturer of cars and trucks combined. That's dominance. In 1997, Ford had four of the ten bestselling vehicles in the United States. Even more impressive, the 1997 J.D. Power Initial Quality Surveys reported a 28 percent improvement in Ford quality.

Of course, as the world's largest manufacturer of trucks, it should come as no surprise that Ford produces the bestselling nameplate in history, the Ford F-Series. Since 1948, more than 26 million F-Series trucks have been produced worldwide.

Right now, there are over 15,000 Ford dealerships in more than 200 countries, and they've just completed their seventh consecutive quarter of increased profits.

Looking ahead to the twenty-first century, Ford Automotive Operations has launched Ford 2000, a company-wide transformation project. And as they prepare to meet the challenges of the new millennium, Ford has already seen dramatic leaps in product quality, efficiency, and profitability. In fact, shareholder value increased 57 percent last year alone.

Billions in sales, millions of vehicles, thousands of dealerships, record revenues—new employees have only to imagine where all this can take them.

Technology First. Graduates who are interested in building a career in technology should consider Ford. Technology is a crucial factor in Ford's approach to automobile manufacturing. State-of-the-art technology is used in all aspects of business, from automobile design to administrative support. For example, by using computer simulation techniques, Ford has created a faster and cheaper method to design final vehicle assembly processes. Also, Ford professionals utilize new software tools that allow them real-time access to product engineering drawings, providing information within a matter of minutes instead of the usual weeks. As technology improves, Ford will continue to pursue faster, better methods of doing business.

Building a Better Tomorrow. Ford is also leading the industry in protecting the environment. That's right, a car company is concerned about the environment. Currently, Ford offers more alternative fuel vehicles than any other manufacturer, including five models to be introduced in Europe in 1999 that use compressed natural gas or liquefied petroleum gas. Recycling also plays a big part in Ford's commitment to the environment. It has used more than 1 million new tires made from recycled materials, and it's the first auto maker in Europe to use new batteries recycled from old ones. In addition, Ford uses 50 million recycled plastic bottles each year for grille reinforcements, luggage racks, and door padding. New employees can contribute their skills to help Ford find even more environmentally friendly solutions and create a better planet for the future.

Setting the Pace in Safety. Safety is a key element of Ford's automotive designs. With more top safety ratings in the United States for model year 1997 than all other auto makers combined, its commitment to safety is unparalleled. In addition to being the leader in dual-air-bag protection, Ford continuously invests in the development of new safety technologies. SecuriLock and SecuriTire are the latest examples of Ford ingenuity at work. SecuriLock, Ford's passive antitheft system, uses a specially encoded key. Without it, the vehicle cannot be started, not even by hot-wiring. And SecuriTire, offered by Lincoln, allows drivers to ride up to fifty miles without tire air pressure. And that's just the beginning. Ford's commitment to safety is ongoing, promising more and better safety devices in the future.

FUTURE DIRECTIONS

Keeping its sights set on tomorrow has kept Ford at the top of the industry since the beginning. So it's no surprise that it has big plans for future growth.

Emerging markets such as Central Europe, Eastern Europe, and India have tremendous potential. So Ford is expanding its dealer network and is pursuing manufacturing opportunities in these areas. In fact, new manufacturing plants in China, Belarus, and Vietnam are all the result of specifically tailored strategies and investments. By starting a career with Ford right out of school, graduates will have an impact on a truly global company.

But excellence in automotive manufacturing is not all Ford Motor Company has to offer. Enterprises such as Ford Credit and Visteon provide expanded services that benefit the company and its customers alike.

FORD CREDIT

Ford Credit, established in 1959, is the world's largest automotive finance company and the second-largest finance company overall, with more than $133 billion in assets. A wholly-owned subsidiary of Ford Motor Company, Ford Credit operates in 35 countries with 15,900 employees at 290 locations. In 1997 alone, Ford Credit provided financing for nearly 4 million new and used vehicles through more than 11,000 dealerships around the globe.

In addition to retail financing for new, used, and leased vehicles, Ford Credit offers wholesale financing, mortgages, and capital loans for dealers, as well as insurance for both dealers and retail customers. The company also serves dealers and customers outside the Ford franchise system through PRIMUS Automotive Financial Services, Inc. PRIMUS provides private-label financing for Mazda, Jaguar, and other manufacturers' dealers and customers in the United States and Canada. Operations in the United Kingdom, Australia, India, Indonesia, Thailand, Spain, and Germany also function under the PRIMUS name. So whether it's with PRIMUS or Ford Credit, graduates can start building a solid career with all kinds of international opportunities.

VISTEON

Visteon, Ford's automotive components business, was organized in 1997 and is the world's second-largest manufacturer of automotive systems and components. With operations in 19 countries, Visteon employs more than 75,000 people in 74 locations.

Continuing Ford's tradition of innovation, Visteon has more than 2,000 patents issued and pending the world over. Its state-of-the-art technology enables Visteon to offer a comprehensive range of products and services, including everything from glass to climate control to powertrain systems. Its manufacturing resources are unique in scale and volume, allowing customers to source everything from a single component to a complete, fully integrated system.

Visteon consists of seven systems divisions, each of which operates as a semi-independent business competing for opportunities in many different industries both in the automotive world and beyond. For example, the Interior Systems Division recently won a contract to produce seats for movie theaters, while various automotive companies have turned to Visteon for solutions to design and manufacturing issues. It's the perfect place to put problem-solving skills to use.

THE ORGANIZATION AS EMPLOYER

Over the years, all of Ford Motor Company's achievements can be attributed to its professionals. Ford is committed to attracting, developing, and retaining employees from all backgrounds. Ford invests hundreds of millions of dollars annually on continuing education programs, with a goal that each employee receive a minimum of 40 hours of training per year.

And this dedication to employees is really paying off. For instance, *Working Mother* magazine named Ford one of the 100 Best Companies for Working Mothers for the third straight year.

Starting a career with Ford gives employees everything they're looking for: responsibilities, challenges, and rewards. They will also enjoy Ford's profit-sharing and savings and stock investment programs as well as extensive benefits. In addition to a generous vacation and holiday schedule, flexible starting times, and part-time work arrangements, Ford also offers child-care resource and referral, diversity and worklife planning, tuition payments, and a participative management approach.

The Ford College Graduate Program offers employment opportunities throughout the United States for U.S. citizens and aliens authorized to work in this country. Overseas employment opportunities are available for individuals with the necessary citizenship, residency, or other requirements of the countries in which facilities are located. Ford is an equal opportunity employer committed to a culturally diverse workforce.

APPLICATION AND INFORMATION

Direct inquiries to:

Human Resources Customer Operations
Salaried Recruitment
Department Code: USAUPTJO
P.O. Box 0520
Allen Park, Michigan 48101-0520

Harris Corporation

HARRIS

Description of Organization: Focuses on four major businesses—electronic systems, semiconductors, communications, and office equipment
Number of Employees: More than 29,000
Headquarters Location: Melbourne, Florida
Academic Fields of Recruitment Interest: Computer, electrical, industrial, mechanical, and chemical engineering; materials science; computer science; solid-state physics; production/operations management; technical sales; business
Major Entry-Level Opportunities for New Graduates: Research and development, design, product assurance, manufacturing, technical sales, business

BACKGROUND AND OPERATIONS

Harris Corporation, founded in 1895 as a printing manufacturer, is a worldwide communications and electronics company focused on four core businesses—electronic systems, semiconductors, communications, and office equipment. A Fortune 500 company with sales of $4 billion, Harris employs more than 29,000 people. The company uses advanced technologies to provide innovative and cost-effective solutions for commercial and government customers, and is committed to providing products, systems, and services of the highest quality. Strategic acquisitions and organizational changes along with a commitment to R&D have positioned Harris for profitable growth for the new millennium.

Electronic Systems The organization provides advanced communication and information processing systems to commercial and government customers. The organization has expanded its strong position in defense communication and avionics markets into a growing role as a supplier of vital infrastructure systems for air traffic control, airport management, energy management, railroad and other transportation systems, and commercial satellite communications. Harris is a leading supplier of large deployable satellite antenna systems and tactical flat panel phased array and single-mission antennas.

Semiconductor The organization provides advanced integrated circuits (ICs) and discrete semiconductors for telecommunication, automotive, multimedia, and power products. The organization is developing breakthrough products for new markets, including wireless communications, wireless LANs, power control for PCs, automotive multiplexing, and satellite communications.

Communications The organization is a leading supplier of telecommunication systems, including microwave radios, digital telephone switching equipment, and telecom test systems and tools; radio and TV broadcast transmitters and studio equipment; and tactical and strategic radios. The organization is providing advanced systems for the emerging markets of high-definition television (HDTV), personal communication services (PCS), and wireless telephony in China, Latin America, and Russia.

Lanier Worldwide The organization is one of the world's largest independent distributors of office products, with more than 1,600 sales and service centers in nearly 100 countries. Lanier provides copying, dictation and continuous recording systems, facsimile systems, multifunctional devices, health-care information management systems, transcription services, facilities management solutions, and software and integration support. In April 1998, Lanier signed an agreement to acquire the Copying Systems Business of Agfa-Gevaert Group, which will double Lanier's presence and market share in the multibillion-dollar European office equipment market.

FUTURE DIRECTIONS

The company is emphasizing continuous improvement in quality and productivity, stressing company-wide customer focus and accelerating the product-to-market cycle. Harris is expanding the systems and technology it has developed for defense business to other government agencies and commercial customers in the United States and overseas. Approximately 30 percent of sales come from international business, and that percentage is expected to grow.

EMPLOYMENT OPPORTUNITIES

ENTRY LEVEL

Harris Corporation offers a challenging environment and early responsibility to bachelor's, master's, and doctoral graduates who can provide, in addition to a record of exceptional academic achievement, evidence of personal initiative and of extra effort in academic and extracurricular pursuits.

Computer Science As Harris continues to expand its advanced software package products to include applications for signal processing, communications network control, text processing, and power control, professionals will be challenged within the company's software development positions. Opportunities also exist in Corporate Information Management.

Electrical & Computer Engineering Within these two fields, career choices include software development/systems engineering; antenna, analog, digital, RF, and control systems design; and product assurance.

Industrial Engineering Harris emphasizes work standards and methods to achieve an efficient and effective workflow. Individuals with strong statistical, analytical, communication, and information-gathering skills are welcome.

Mechanical Engineering Working on the structural aspects of antennas and earth stations and on the development concepts for packaging communications equipment, professionals will assist in creating state-of-the-art solutions to design problems.

Materials Science, Solid-State Physics, Chemistry, and Chemical Engineering Opportunities exist in the semiconductor field for materials science graduates who will work with and develop new materials. Those with chemistry, chemical engineering, and solid-state physics degrees can work in process engineering positions related to semiconductor and integrated circuit production.

Technical Sales/Business Development Individuals in this area create customer awareness of Harris products and capabilities and are an important link between the customer and the technical staff. Formal training is provided.

Finance & Accounting Recent graduates with degrees in accounting, finance, or business administration will find substantial challenge and opportunities. Because each division is largely independent, professionals can readily attain a position with division-wide overview and responsibility. They will prepare financial analyses and budgets, conduct audits, and/or measure financial performance.

Human Resources Professionals ensure that Harris employees contribute to the company goals and benefit from the employment experience. Individuals can gain experience in employment, employee relations, manpower planning, management/organizational development, and compensation and benefits administration.

ELECTRONIC SYSTEMS

Aerospace Systems Division: develops and produces aircraft, spacecraft, and missile communications; and information processing systems for defense, civil, international, and commercial markets.

Communication Systems Division: develops and produces terrestrial and satellite communication systems; electronic warfare simulation and evaluation systems; networks for command, control, communication, and intelligence applications; and high-technology applications for special systems, plus communication systems for metropolitan mass transit modernization and fleet management.

Information Systems Division: designs, develops, manufactures, and supports ground-based custom data processing systems and software for the U.S. government, its agencies, and its prime contractors and provides air traffic control systems for civil and military aviation agencies worldwide and integrated airport communication and management systems for domestic and international airport authorities.

GE Harris Railway Electronics (A joint venture): provider of turnkey planning and control systems to revolutionize transportation service reliability, safety, and efficiency.

GE Harris Energy Systems (A joint venture): develops and produces open-system-based, real-time control systems to monitor, control, automate, and optimize the operations of electric, water, and oil and gas utility companies.

ELECTRONIC SYSTEMS SUBSIDIARIES AND OTHER OPERATIONS

Harris Publishing Systems Corporation: expandable prepress and full-page makeup publishing systems.

Harris Technical Services Corporation: engineering and support services to the Department of Defense (DoD) and other government agencies worldwide.

Melbourne Division Support Operations: central services and support for electronic systems divisions within the Melbourne area.

Sector Technical Operations: manufacturing, material, components, and other support of engineering functions for electronics systems divisions. Contact: Electronic Systems Sector College Recruiting Office, P.O. Box 37 (MS2-1430), Melbourne, FL 32902–0037.

SEMICONDUCTOR

Semiconductor: standard, custom, and semicustom integrated circuits and discrete products for analog, mixed-signal, and power-control applications; custom and standard application-specific ICs for communication, automotive, and military applications. Key technologies include CMOs, DI, GaAs, MOS power, and MOV. Contact: P.O. Box 883, Melbourne, FL 32902.

COMMUNICATIONS

Broadcast Division: radio and TV broadcast equipment, including transmitters, antennas, remote-control equipment and broadcast systems, and studio products. Contact: 3200 Wisman Lane, P.O. Box 4290, Quincy, IL 62305-4290.

Digital Telephone Systems Division: digital voice and data network switches, PBX's and telephone sets. Contact: 300 Bel Marin Keys Boulevard, Novato, CA 94949.

Farinon Division: microwave and lightwave transmission systems for voice, data, and video. Contact: 330 Twin Dolphin Drive, Redwood Shores, CA 94065–1421.

Harris Network Management: designs and develops comprehensive, turnkey software and professional service solutions that reduce the time and effort spent deciphering network alerts and provides the statistical information needed to make effective management decisions. It also improves service levels and reduces the cost associated with managing complex, multifunction networks. Contact: 1025 W. NASA Boulevard, Melbourne, FL 32919.

Network Support Systems Division: testing systems and tools to network service providers, subscriber loop products and systems, in-business paging products and systems. Contact: 809 Calle Plano, Camarillo, CA 93012.

RF Communications Group: firmware-driven, digital signal processing based, HF, VHF and UHF radio equipment and systems for long and short distance communications and turnkey communication systems. Contact: 1680 University Avenue, Rochester, NY 14610.

CORPORATE HEADQUARTERS

Corporate headquarters attracts electrical engineering, computer engineering, and computer science undergraduates with an interest in developing the latest technology information systems and tools in cooperation with the Division MIS staffs. Operations Management interests individuals with a technical undergraduate degree and an MBA with a concentration in operations/production management. Contact: 1025 W. NASA Boulevard, Melbourne, FL 32919.

LANIER WORLDWIDE, INCORPORATED

Copying Systems Division/Voice Products Division: fax machines, copying equipment, and dictation equipment. Contact: 2300 Parklake Drive NE, Atlanta, GA 30345.

THE ORGANIZATION AS EMPLOYER

Professionals are counted on to take initiative and accept early responsibility. As their career progresses, they will meet regularly with their supervisor to discuss their performance and to determine which direction their development might take.

Employees can plan to continue as individual contributors or to accept management responsibilities. Whichever path is chosen, Harris is committed to helping employees grow in skills, responsibility, and knowledge.

Benefits: Harris provides competitive salaries and comprehensive benefits. Group medical and dental, personal life, and disability insurance plans are available. Profit-sharing and savings plans help ensure retirement security. Holiday and vacation provisions are competitive. Relocation expenses are covered in a corporate-wide relocation assistance plan. Harris' Education Assistance Program reimburses 100% of tuition and lab fees for graduate course work.

APPLICATION AND INFORMATION

Applicants interested in a career at Harris should contact their Placement Office to learn when a representative will be on campus, write directly to Harris Corporation at the business unit of their choice, or explore the college recruiting Web site at http://www.harris.com/college.

Lincoln Laboratory

Massachusetts Institute of Technology

Description of Organization: Research and development center operated by the Massachusetts Institute of Technology for research in advanced electronics and related fields

Number of Employees: Approximately 2,200

Headquarters Location: Lexington, Massachusetts

Academic Fields of Recruitment Interest: Electrical engineering, computer science, mathematics, applied physics

Major Entry-Level Opportunities for New Graduates: Immediate working assignments to technical group in area of specialty or interest

BACKGROUND AND OPERATIONS

Lincoln Laboratory was established in 1951 by the Massachusetts Institute of Technology and is operated as a federally funded research and development center. The Laboratory's fundamental mission has remained unchanged since its inception: to apply science, by means of advanced technology, to critical problems of national security. The character of the problems has broadened from the initial emphasis on air defense to include space surveillance, ballistic missile defense, tactical battlefield systems, communications, and air traffic control.

Lincoln Laboratory has pioneered advanced electronics technology. Work in the areas of communications, radar, digital data processing, signal processing, and air traffic control extends from basic research through the development of devices and components to the design, construction, and operation of complex systems. Projects are followed from the concept stage through simulation and analysis to the development of hardware and the ultimate test and demonstration of an integrated system.

Solving the broad problems with which Lincoln Laboratory is concerned requires the cooperative effort of many people who represent a wide range of individual experience and a variety of technical specialties and scientific disciplines. To meet this need, the Laboratory employs approximately 2,200 men and women, including 1,100 professional staff members who hold degrees in electrical engineering, computer science, physics, mathematics, and other scientific and engineering disciplines.

The main Laboratory buildings are located in the town of Lexington, Massachusetts, 14 miles from the center of Boston, with easy access to the mountains and seaside recreational areas of New England. The Laboratory also operates major radar field sites at Westford, Massachusetts, and on Kwajalein Atoll in the Marshall Islands of the central Pacific.

EMPLOYMENT OPPORTUNITIES

ENTRY-LEVEL AND EXPERIENCED PERSONNEL

Lincoln Laboratory is interested in scientists and engineers at all degree levels. The professional staff is currently made up of electrical engineers, 41 percent; physicists, 22 percent; mathematicians, 9 percent; and mechanical engineers, 5 percent. Other areas, such as aeronautical/astronautical engineering, computer science, and materials science, are represented in smaller proportions. Overall, two thirds of the professional staff members have earned advanced degrees in their areas of specialty.

The following descriptions of typical work assignments in areas of continuing interest to the Laboratory indicate the type and variety of opportunities available.

Electrical Engineers, Computer Scientists, Mathematicians, Physicists. Many programs at Lincoln Laboratory involve digital data processing, either for more efficient data reduction and analysis or in the form of improved algorithms for signal processing and analysis or executive systems for real-time computation and control. Computer technology projects include recognition of vehicles through laser and synthetic aperture millimeter-wave radar sensors and recognition of space objects by means of high-resolution millimeter-wave radar images, development of a technology for wafer-scale integration called Restructurable VLSI, and development of robust speech recognition algorithms.

In the radar measurements program, projects involve system analysis, the design and construction of sensors through state-of-the-art technology, and the phenomenology of clutter modeling and target signature characterization. A major focus of this work is on the applications of sensors in systems that are intended to accomplish an operational goal, such as defense against missile attacks.

For more than two decades, Lincoln Laboratory has helped the FAA solve air traffic control problems. The objective of the Laboratory's work in this area is to improve airport capacity, with emphasis on periods of inclement weather. Projects involve developing and testing software algorithms for air traffic management and developing the means for implementing these algorithms in existing FAA computer equipment. The Laboratory is also engaged in meteorological research and the application of advanced radar signal and data processing techniques for weather detection. Current projects include the development of systems to detect wind shear phenomena. Development and operation of weather measurement/air traffic control product demonstration systems at major airports to validate and refine the concepts are also under way.

Communication activities involve robust lightweight communication equipment for space applications and ground terminals. Optical communications components are being developed for very-wide-band networks.

A major portion of Lincoln Laboratory's work over the years has been in the area of surveillance. Radar and electrooptical sensors, either ground-, airborne-, or space-based, have been investigated for their application in systems for the detection, tracking, and identification of objects in space, in the atmosphere, and on the ground.

Programs in space surveillance, namely Satellite Tracking, Space Object Identification, and Advanced Electro-optical Deep Space Surveillance, are involved with the surveillance and identification of satellites from ground-based sensors. Satellite surveillance and tracking techniques for objects located in the deep-space environment have been developed using radar at the Millstone Hill Field Site in Westford, Massachusetts, and at the Kwajalein Missile Range facility in the central Pacific.

In the area of surface surveillance, the Laboratory works on problems of detection and identification of moving and stationary objects. Extensive field measurements programs, coupled with simulations, have established a broad database. Current problems include the imaging of objects under foliage, phenomenology, and measurements involving radar and IR instruments for evaluating air vehicle survivability in threatening environments and air defense.

Lincoln Laboratory conducts an extensive program of research on new materials, devices, and techniques as well as investigations of other phenomena that could have a significant impact on future electronic and optical systems. A broad set of research efforts is focused on speech, acoustic, and radar signal processing applications; machine intelligence technology; and neural network theory and applications.

The Laboratory's solid-state programs include research in submicrometer technology, which concentrates on the development and exploratory applications of advanced lithography, thin-film processing, and microfabrication technologies; quantum electronics; electric materials; high-speed electronics for microwave and high-speed digital operation and submillimeter technology; electrooptical devices; analog signal processing technology; and microelectronics.

Engineering programs at Lincoln Laboratory involve mechanical engineering, including structural analysis and finite-element modeling; aerospace engineering, including data analysis, laboratory and wind-tunnel testing, and flight testing; optical systems engineering that specializes in the design, analysis, testing, and packaging of terrestrial, airborne, and spaceborne optical systems; and control systems engineering that designs and implements state-of-the-art feedback control systems for a variety of demanding applications on the ground, in aircraft, and in space.

TRAINING

To support staff members in their future acquisition of specialized skills, Lincoln Laboratory offers a broad program of in-house technical courses. For those who choose to extend their formal education, there is also a liberal tuition assistance program for courses taken at the Massachusetts Institute of Technology or other schools in the Boston area.

THE ORGANIZATION AS EMPLOYER

Lincoln Laboratory recognizes that its contribution to the nation as a federally funded research center is made possible only by the continuing excellence and accomplishments of its professional staff. The Laboratory actively seeks creative professionals at all levels, with and without previous experience. Men and women are hired to work in the field of their academic training. The new employee is placed directly into one of the Laboratory's technical groups and is expected to assume responsibilities commensurate with his or her academic background and previous experience.

It is the policy of the Laboratory to support and encourage the professional growth of its staff. Staff members are encouraged to publish in professional journals, to attend meetings, and to participate in the activities of their scientific and engineering societies. Organizational flexibility, together with the nature and diversity of the Laboratory's projects, make it possible to maximize professional challenge and match assignments and individual interests. The Laboratory does not formally distinguish management from technical career progressions.

Salaries and benefits are competitive with the industry and include medical, dental, and life insurance; retirement plans; tuition assistance for employees and a college scholarship program for staff members' children; a credit union; child-care services; relocation assistance; a liberal vacation policy; and twelve annual holidays. Staff members are encouraged to participate in the recreational, social, and cultural programs of MIT and are offered full access to the Institute's facilities.

APPLICATION AND INFORMATION

Lincoln Laboratory is an Equal Opportunity Employer, M/F/D/V. U.S. citizenship is required. Candidates should send resumes to the address below for information about employment opportunities.

Direct inquiries to:

Human Resources Department
Lincoln Laboratory
Massachusetts Institute of Technology
Box 730
Lexington, Massachusetts 02173-9108
World Wide Web: http://www.ll.mit.edu

Lucent Technologies

Bell Labs Innovations

Lucent Technologies
Bell Labs Innovations

We make the things that make communications work.

Description of Organization: Worldwide telecommunications leader (formerly the systems and technology business of AT&T plus Bell Labs)

Number of Employees: 125,000

Headquarters: Murray Hill, New Jersey

Academic Fields of Recruitment Interest: Computer science/ engineering, electrical engineering, mathematics, materials science, mechanical engineering, physics, systems engineering

Major Entry-Level Opportunities for New Graduates: Information services, manufacturing and applied engineering, research and development

BACKGROUND AND OPERATIONS

Lucent Technologies is creating the technology that's changing the way the world communicates.

Lucent Technologies is building the next generation of global communications. It is the company that invented the transistor, developed the dial tone, the cell phone, the Telstar satellite, and fiber optics.

Today, Lucent is creating bold, new, innovative communications solutions that integrate sound, images, and data: photonics, microelectronics, Internet security, speech synthesis, wireless communication and network systems, and videoconferencing technologies and data networking systems.

Lucent continues to develop innovations in business communications with advanced networks and switching technologies that allow communications networks to handle a trillion bits of information at once.

Expanding rapidly as a customer-driven, high-performance company, Lucent is focusing its energy on high-growth areas while strengthening its commitment to R&D throughout the business. Today, Lucent Technologies is providing the world's best and most innovative communications systems, products, technologies, and customer support. And it's only the beginning.

Bell Labs

As one of the world's foremost industrial R&D organizations, Bell Labs is the "innovation engine" of Lucent Technologies. It consists of development labs integrated with individual operating units and central labs that serve them all.

Bell Labs has provided Lucent with new products in advanced software, digital sound processing, Internet encryption and multimedia network systems, and wired and wireless networks and multifunctional systems that integrate sound, images, and data in a flash.

Microelectronics Group

The Microelectronics Group develops and produces integrated circuits and optoelectronics products, including the sales and service forces that serve this customer segment.

Business Communications Systems

BCS develops, markets, sells, and services all voice-related products for large business enterprise customers. This division also includes Lucent's newly acquired Octel Messaging Division.

Data Networking Systems

Data Networking Systems provides data networking solutions to enterprise and Internet Service Provider (ISP) customers as well as Internet Protocol and ATM core switching data networking solutions to service provider customers.

Global Service Provider Business

The Global Service Provider Business offers marketing, sales, service, support, and program management for network operators and service providers worldwide. These customers might include local, long-distance, Internet, and wireless service providers as well as global commercial markets.

Wireless Networks Group

This group covers all manufacturing, development, product marketing, and product management for wireless products, software, and support for the service provider markets.

Switching and Access Systems Group

This group develops, manufactures, and provides marketing and management of a broad range of switching/access products and software for the carrier market.

Optical Networking Group

The Optical Networking Group provides manufacturing, development, product marketing, and management for SONET/ SDH, wavelength division multiplexing, and access offers.

Network Products Group

This group develops, manufactures, and provides marketing and management for fiber products, including SYSTIMAX, and power systems products.

Communications Software Group

This group develops software focused in application areas such as intelligent networks, network management and operations, and the Internet.

New Ventures Group

This newly created business division designs innovative ways of getting technology to market faster.

Intellectual Property Division

The Intellectual Property Division of Lucent creates value from Lucent's extensive portfolio of intellectual property.

FUTURE DIRECTIONS

Global communications experts forecast overall growth of almost 15 percent for the next few years. And Lucent Technologies is right at the heart of the fastest-growing areas: wireless systems for network operators, growing at 21 percent per year; communications software, growing at least 14 percent annually; data networking, growing at 20 percent each year—the list continues.

Outstanding customers, the broadest portfolio of technological solutions in the industry, and unparalleled expertise in networking serve to make Lucent a leader in every facet of the telecommunications industry. From silicon chips to the world's most sophisticated switching systems and networking software, Lucent is taking global communications into the twenty-first century.

EMPLOYMENT OPPORTUNITIES

As one of the world's most diverse communications companies, Lucent Technologies offers a wide array of dynamic career opportunities for talented graduates with degrees related to one of the following areas: technical, research and development, sales and marketing, accounting and finance, manufacturing and applied engineering, and administration.

THE ORGANIZATION AS EMPLOYER

Lucent Technologies rewards employee contributions and achievements through one of the most comprehensive compensation and benefits programs in the industry. It is one of the company's strongest commitments.

Lucent provides an attractive salary and comprehensive health and insurance plans from the first day of employment, including medical, dental, vision, life, accidental loss, long-term disability, and long-term care.

Other benefits include an Employee Stock Purchase Plan and excellent savings and pension plans. Lucent also provides an innovative Work and Family Program, such as the Family Resource Programs (child and elder care referral services and educational issues), Adoption Reimbursement, and Family Care Development grants.

Additional Value-Added Programs offered by Lucent Technologies include the Academic Awards Program, the Tuition Assistance Program, the Group Legal Plan, the Employee Mortgage Program, Health Fitness Centers, and Credit Union Membership.

APPLICATION AND INFORMATION

For more information or to apply, interested applicants are encouraged to send a resume to:

Employment Manager, Dept. 169/9631/98
Lucent Technologies
283 King George Road, Room B2C83
Warren, New Jersey 07059
World Wide Web: http://www.lucent.com (to apply on line)

The MITRE Corporation

MITRE

Description of Organization: Operates as a strategic technical partner with the Department of Defense and the Federal Aviation Administration in the areas of communications and information systems

Number of Employees: 4,200

Headquarters Locations: Bedford, Massachusetts, and northern Virginia

Academic Fields of Recruitment Interest: Electrical and computer engineering, computer science, mathematics, physics, and telecommunications

Major Entry-Level Opportunities for New Graduates: Staff positions in planning, analysis, design, and testing of complex information and communications systems

BACKGROUND AND OPERATIONS

The MITRE Corporation is an independent not-for-profit company that provides technical support to the U.S. government, specifically the Department of Defense and Federal Aviation Administration. Working in the public interest, MITRE operates as a strategic partner with sponsoring government agencies. This permits a degree of access and a long-term perspective not available to commercial contractors who compete for government business.

A strong information technology base and an integrated systems approach power all of MITRE's work. MITRE supports the national security and intelligence community with systems engineering and integration work on command, control, communications, computer, intelligence, surveillance, and reconnaissance. MITRE also performs systems research and development work for the Federal Aviation Administration (FAA) and other civil aviation authorities, including designing and implementing innovative solutions to air traffic management systems.

EMPLOYMENT OPPORTUNITIES

ENTRY-LEVEL AND EXPERIENCED PERSONNEL

As electronics technology races ahead toward the twenty-first century and beyond, MITRE's work is in even greater demand, and opportunities are continually arising for both entry-level and experienced candidates from a variety of disciplines. At MITRE, the employees are the most valuable asset. Their technical skills, dedication, and professionalism enable MITRE to exceed its expectations.

Challenge, innovation and cutting-edge thinking are critical for such positions as **Electrical Engineers, Computer Engineers, Computer Scientists, Mathematicians, Telecommunications Specialists,** and **Physicists.** Interested individuals must possess a bachelor's degree or, preferably, an advanced degree in their field of expertise.

Communications Communication system areas of involvement include the analysis, design, and prototyping for a wide variety of next generation communications systems. Hardware areas of involvement include system architecture, modulation coding, antenna and RF design, digital communications, real-time processing architectures, digital signal processing, communications processors, printed circuit board design, networking, and systems test and evaluation. Software areas of involvement include real-time software, C and C++, graphical user interface design, and multimedia, database, World Wide Web applications.

Advanced Information Systems MITRE is involved with advanced system planning, design, research and development in information security, distributed computing, object technology, network and system management, artificial intelligence, human-computer interaction, geospatial information systems, and World Wide Web technologies.

Technical MITRE has technical groups that specialize in workstation systems engineering, economic and decision analysis, reliability and maintainability, and advanced radar systems.

TRAINING

MITRE offers no formal training programs, but each new employee takes part in a variety of orientation programs, beginning on the first day. Through these programs, employees learn how MITRE addresses and solves technical problems, what support facilities are available, and how the corporation and its clients are structured. Most of this orientation is folded into the employee's work assignment. In MITRE's informal atmosphere, technical advice and assistance are readily available.

THE ORGANIZATION AS EMPLOYER

On all levels, management supports and encourages personnel to publish technical articles, present papers, participate in technical meetings and symposia, and serve on technical advisory boards. MITRE's supportive environment fosters individual creativity and innovation.

A firm believer in continuing education, MITRE offers a wide range of educational opportunities. Many employees are enrolled in courses and advanced degree-oriented programs at various colleges and universities in the Boston and northern Virginia areas. Tuition reimbursement is provided by MITRE. In addition, MITRE has an internal educational facility known as the MITRE Institute, which offers full-length courses that bridge the gap between academic training and on-the-job experience. Courses that are offered range from telecommunications engineering to World Wide Web technologies, software engineering, and general systems engineering.

To help employees and clients keep up with technology and products, sharing knowledge is crucial. This is done through an advanced automated information system called the MITRE Information Infrastructure, which uses Internet query tools for access and commercial and MITRE-developed subsystems for input. Staff members also work with a unique office automation environment called Collaborative Virtual Workspace (CVW) that enables them to converse, collaborate, interact, and participate in group discussions regardless of their geographic location anywhere in the world.

MITRE's commitment to its associates is also demonstrated in a highly competitive compensation and benefits package that includes health, dental, life, and disability insurance; tuition reimbursement, and a nationally recognized retirement program.

COOPERATIVE EDUCATION AND SUMMER ASSIGNMENTS

MITRE hires undergraduate and graduate students pursuing electrical and computer engineering, computer science, math, and physics degrees to work on various technical projects throughout the company, thus affording them the opportunity to gain valuable experience and hands-on knowledge in a real-world atmosphere.

OVERSEAS ASSIGNMENTS

In addition to the MITRE Corporation's two major locations in suburban Boston and northern Virginia, numerous international and domestic field sites are in operation. There is always the potential for additional temporary and/or permanent sites, depending on upcoming projects.

CORPORATE SOCIAL RESPONSIBILITY

MITRE understands its role in the community and its responsibility to a society as a whole. Employees are encouraged to participate in meaningful projects that make a difference. Many serve on boards and committees of local, regional, or national organizations that aim to improve social or environmental conditions.

APPLICATION AND INFORMATION

Members of the Corporate Recruitment Office in Bedford or northern Virginia are eager to talk with potential employees. Those who have the technological skills MITRE needs and the desire to pursue a rewarding career with unlimited possibilities for achievement should forward a resume to the address below.

Direct inquiries to:

Corporate Recruitment, Dept. JO98
The MITRE Corporation
202 Burlington Road
Bedford, Massachusetts 01730-1420
E-mail: jobs@mitre.org
Fax: 888-42MITRE
World Wide Web: http://www.mitre.org

National Security Agency

Description of Organization: A federal government agency that coordinates certain domestic and foreign communications challenges, including the collection of foreign secure communications and the protection of U.S. secure communications

Headquarters Location: Fort Meade, Maryland

Academic Fields of Recruitment Interest: Electrical, electronics, and computer engineering; computer science, mathematics

Major Entry-Level Opportunities for New Graduates: Research and development, communications systems engineering, computer engineering, programming and systems analysis, mathematical research, applied mathematics

BACKGROUND AND OPERATIONS

Created by President Truman in 1952, the National Security Agency is charged with the enormous responsibility of helping protect the nation's security. NSA carries out that responsibility in two areas. The first is Signals Intelligence (SIGINT). In this capacity, NSA intercepts, decodes, and analyzes foreign electromagnetic signals to produce vital intelligence reports. SIGINT is particularly important in keeping up with today's rapidly changing world.

The second area is Information Systems Security (INFOSEC). INFOSEC involves the protection of all classified information stored on or sent through government computers, telephones, and other message-sending devices. A vital component in ensuring this protection is the art of cryptography, or the development of codes and ciphers. INFOSEC employees scientifically devise, test, and select cryptographic systems to outwit those who try to intercept these classified communications.

Since the work NSA employees do is so vital to the nation, great care is taken in selecting only America's best and brightest people for positions with NSA. Therefore, positions are limited. An individual chosen to join the NSA family will find it a highly rewarding and exciting career.

FUTURE DIRECTIONS

Individuals intrigued by difficult challenges and eager to apply imagination, initiative, and intellectual curiosity to protecting our nation's security should consider one of the following career paths with NSA.

EMPLOYMENT OPPORTUNITIES

Opportunities exist for full-time, co-op, and summer positions. Depending on NSA's hiring requirements, opportunities may be limited.

ENTRY LEVEL

Applications for the following areas must be received no later than four months prior to the employment date desired.

Mathematicians. NSA mathematicians use their talents within a field that has intrigued man for centuries: cryptology. Possible assignments fall into three categories: consulting with analysts on current communications problems, giving technical assistance to communications systems specialists, and conducting long-range research in communications and computing. The challenge is to use probability, statistics, Fourier analysis, Galois theory, stochastic processes, and other techniques to outwit the world experts in creating or breaking codes and ciphers.

Virtually every mathematical discipline finds some application within NSA, primarily in cryptography and cryptanalysis, but also in engineering, computer security, and communications traffic analysis.

Computer and Electrical/Electronics Engineers. As NSA engineers, employees join a long tradition of excellence. In fact, the Agency's early interest in cryptanalytic research resulted in its being one of the inventors of the modern computer. This pioneering spirit exists to this day and provides NSA engineers with a wide variety of interesting and challenging work.

Depending on their interests, engineers might contribute to assignments like the design of special-purpose computers, antennas, and radar systems; pattern recognition; signal analysis; optics; and the design, development, testing, and evaluation of electronic communication and signal processing systems.

During this process, engineers have access to such highly sophisticated resources as NSA's microelectronics fabrications facility, which features a 20,000-square-foot "Class 10" clean room. Additionally, they have immediate access to some of the finest engineering minds in the country, in the form of senior engineering supervisors.

NSA engineers are also in close working relationships with computer scientists, mathematicians, analysts, contractors, and other people outside of their own discipline, providing the opportunity to explore a variety of fields. After completion of their initial assignment, they may choose from projects in various fields that will increase both technical breadth and responsibility for project management.

Computer Scientists. NSA computer scientists work in one of the world's largest computer and data processing centers. Housed in this center are today's most advanced computers, supercomputers, and a wide array of custom-built machinery, all dedicated to helping NSA employees protect the nation's security and all available to NSA employees well before commercial release.

Some of the challenges to which employees might apply this computer muscle include the design and implementation of software systems; real-time programming; engineering support, from simulation to project management; signals, including analog control; graphical user interfaces; artificial intelligence; neural networks; distributed processing systems; and computer security.

As employees pursue these challenges, they come in contact with an exciting mixture of disciplines, including programming languages, communications, and much more. All told, the efforts of NSA's computer scientists either create new intelligence-gathering tools or devise ways to safeguard these and other valuable tools from espionage.

TRAINING

NSA strongly believes in giving responsibility to new employees right from the start. To enhance this invaluable on-the-job training, employees also receive personalized instruction from some of the most brilliant minds in their field and career counseling from skilled supervisors. In addition, employees are encouraged to learn the intriguing art of cryptology and other work-related subjects by taking classes at NSA's on-site National Cryptologic School.

Furthermore, the Agency fully funds part-time graduate study at nearby universities, including the University of Maryland, Johns Hopkins, and Georgetown. Some employees may even be eligible for full-time study with full tuition and salary.

THE ORGANIZATION AS EMPLOYER

In addition to having an opportunity to use technologies well before they are available for commercial use, NSA employees receive starting salaries competitive with those in private industry. NSA employees are also eligible for an excellent group life insurance plan and may choose from a variety of health benefit plans. Much of the cost is paid by NSA.

The Agency is equally generous with its vacation policy. Entry-level employees receive 13 days of annual leave, 13 sick days, and 10 paid holidays. Employees receive 20 vacation days beginning in their fourth year.

SUMMER AND CO-OP PROGRAMS

At NSA, students have an opportunity to earn while they learn through the Cooperative Education Program and the Summer Employment Program.

Co-op Program: As co-ops, students have the rare opportunity to apply the knowledge acquired in school to a variety of very important real-world problems. The program consists of a series of semester-long work tours. Each tour, co-ops work on a different project and receive an increased amount of responsibility. That way, they get a breadth of experience and a chance to decide what field they like best.

To be eligible for the program, students must be U.S. citizens, be enrolled in their university's cooperative education program, and have a minimum GPA of 3.0 on a 4.0 scale. Additionally, they must be able to work for a total of 12 months prior to graduation, alternating periods of work with periods of full-time study. Appropriate majors are computer and electrical/electronics engineering and computer science.

Co-op salaries are competitive, and students receive full employee benefits. Transportation costs between school and the workplace are paid for by the Agency and tuition assistance is available. Additionally, NSA's Cooperative Education Association provides peer guidance, transportation aid, and housing assistance, as well as general support services.

Students interested in NSA's Co-op Program should apply through their cooperative education office. They may apply as early as their freshman year but no later than six months before they would like to start.

Summer Program: Participants in the NSA Summer Employment Program work the summer following their junior year in assignments similar to those mentioned above. Applicants need a minimum GPA of 3.0. Direct applications should be sent to the NSA Summer Employment Coordinator at the address below by **November 15** for employment the next summer. Individuals who meet with an NSA recruiter on campus have until December 15 to return their forms.

OVERSEAS ASSIGNMENTS

NSA employees may be able to apply for one of a limited number of two- to three-year assignments outside of the country.

APPLICATION AND INFORMATION

The vital nature of NSA's work requires extensive background checks and other procedures before an offer of employment can be made. Therefore, it is important that applicants contact NSA at least four months before graduating from college or before leaving their present position. Students should check with their placement office or write to the address below.

For an elite group of people, NSA is their opportunity to make vital contributions to national security while enjoying early responsibility and continual challenge.

NSA is an equal opportunity employer. U.S. citizenship is required for all applicants and their immediate family members.

Direct inquiries to:

National Security Agency
Attention: S 23 (AYW), Suite 6840
Fort George G. Meade, Maryland 20755-6840
World Wide Web: http://www.nsa.gov:8080

Naval Research Laboratory

Department of the Navy

Description of Organization: The Navy's corporate laboratory for scientific research and advanced technological development directed toward new and improved materials, equipment, techniques, systems, and related operational procedures

Number of Employees: 3,283

Main Site Location: Washington, D.C.

Academic Fields of Recruitment Interest: Aerospace, ceramic, computer, electronics, materials, and mechanical engineering; computer science; chemistry, materials science, metallurgy, physics, geology, geophysics, meteorology, oceanography

Major Entry-Level Opportunities for New Graduates: Various fields in research and development, as listed below

BACKGROUND AND OPERATIONS

At the urging of inventor Thomas Edison, who conceived the idea of establishing a research facility for the Navy, the Naval Research Laboratory (NRL) was established on July 2, 1923. From a modest beginning of five buildings and a few handpicked scientists representing two major research areas—radio and underwater sound—NRL has grown to its present status as the corporate laboratory of the Navy. NRL's main site is in Washington, D.C., and other major sites are located in Stennis, Mississippi and Monterey, California. NRL employs a work force of 3,283.

NRL conducts basic research, translates the results of this research into technology, and assists in transferring this technology to other Navy and industrial laboratories so that it can be incorporated into useful military systems. NRL's efforts encompass the multidisciplinary fields involved in increasing the combat effectiveness of the fleet. Through technology transfer programs, NRL has also helped provide solutions to nondefense problems in such fields as ecology, pollution, transportation, health, and education.

NRL's scientific program has been characterized by continual achievement and flexibility. Certain areas of study have occupied an important place on the laboratory's agenda because of their continuing importance to the Navy. These include basic research related to the Navy's environment of sea, sky, and space. NRL has conducted a wide range of investigations, from monitoring the sun's behavior to analyzing marine atmospheric conditions and measuring the parameters of the deep oceans. Detection and communications capabilities have benefited from research that has explored new portions of the electromagnetic spectrum, extended ranges to outer space, and provided means of transferring information reliably and securely even through massive jamming. Sound in the sea, submarine habitability, ocean and atmosphere forecasting, mapping and charting, lubricants, shipbuilding materials, and fire fighting have also been subjects of continuing interest to NRL.

In addition, the laboratory has been quick to move into new areas opened by the advance of science. It pioneered naval research into space, from atmospheric probes with captured V-2 rockets, through direction of the Vanguard project (America's first satellite program), to involvement in such ongoing projects as the Navy Global Positioning System. It is today, in fact, the Navy's leading laboratory in space research, oceanography, marine meteorology, marine geology, geophysics, fire research, tactical electronic warfare, microelectronic devices, and artificial intelligence. NRL has been quick to recognize and use such technological advances as computers, solid-state electronics, lasers, composite materials, and optical fibers. Finally, it has evaluated serious new threats, such as the effects of intense radiation or new forms of shock and vibration on components or on aircraft, ships, and satellites.

The science program at NRL is carried out by a staff of about 1,889 civilian scientists and engineers. NRL scientists and engineers have developed many improved devices and techniques. For such innovations, the laboratory is awarded patents. Since its founding, NRL has received over 3,765 patents. The practical results of work on projects that are undertaken in response to sponsor requests are made known and available not only to various Navy units but also to the Army, Air Force, Defense Nuclear Agency, and other federal organizations. Moreover, NRL's research is reported to the scientific community in hundreds of presentations, journal articles, and reports every year. In 1985 one of the lab's employees, Dr. Jerome Karle, was awarded the Nobel Prize for chemistry.

In support of its diversified program, the laboratory has an impressive array of modern tools for research, many of which are unique. There are, for instance, a 60-MeV linac; a 5-MV positive-ion and 2-MV electron Van de Graaff; a high-level radiation laboratory; versatile facilities for high-magnetic-field and cryogenics research; light-gas guns for studying hypervelocity impact; a 300 m^3 pressurizable chamber for combustion studies; extensive facilities for studying mechanical properties of materials; a variety of gas and solid-state lasers, including devices of very high power and brightness; one of the world's largest anechoic chambers; a microelectronics facility; a combined wave and wind channel for studying wave interactions; a physical acoustics center; a closed recirculating aquarium system; experimental satellite development and test facilities; advanced scientific computers; a CRAY X-MP/16 supercomputer; a 24-node Intel Touchstone MIMD supercomputer; a 32K-node TMC Connection Machine SIMD supercomputer; a CM5 256-node MIMD computer; CNOC and FNOC computers; and innumerable other instruments needed for present-day investigations. In addition, NRL maintains five aircraft equipped as "flying laboratories."

NRL's diverse activities through the years have been guided by a single aim: to conduct pioneering scientific research and development that will provide improved materials, equipment, techniques, systems, and operations. In the future, as in the past, the laboratory will continue to meet the ever-changing technical needs of the Navy.

EMPLOYMENT OPPORTUNITIES

ENTRY-LEVEL AND EXPERIENCED PERSONNEL

The Naval Research Laboratory offers a wide variety of challenging positions that involve the full range of work from basic and applied research to equipment development.

Ceramic Engineers and Materials Scientists/Engineers. Areas are materials, microstructure characterization, electronic ceramics, solid-state physics, fiber optics, electro-optics, microelectronics, fracture mechanics, vacuum science, laser physics technology, and radio frequency/microwave/millimeter wave/infrared technology.

Electronics Engineers and Computer Scientists. Areas are communications systems, electromagnetic scattering, electronics instrumentation, electronic warfare systems, radio frequency/microwave/millimeter wave/infrared technology, radar systems, laser physics technology, radiowave propagation, electron device technology, spacecraft design, artificial intelligence, information processing, signal processing, plasma physics, vacuum science, microelectronics, electro-optics, fiber optics, solid state, software engineering, computer design/architecture, ocean acoustics, stress analysis, and expert systems.

Mechanical Engineers. Areas are spacecraft design, remote sensing, propulsion, experimental fluid mechanics, experimental structural mechanics, solid mechanics, elastic/plastic fracture mechanics, materials, finite-element methods, nondestructive evaluation, characterization of fracture resistance of structural alloys, combustion, and CAD/CAM.

Chemists. Areas are combustion, polymer science, bioengineering and molecular engineering, surface science, materials, fiber optics, electro-optics, microelectronics, electron device technology, and laser physics.

Physicists. Areas are materials, solid-state physics, fiber optics, electro-optics, microelectronics, vacuum science, plasma physics, fluid mechanics, signal processing, ocean acoustics, information processing, artificial intelligence, electron device technology, radiowave propagation, laser physics, ultraviolet/X-ray/gamma ray technology, electronic warfare, electromagnetic interaction, communications systems, radio frequency/microwave/millimeter wave/infrared technology, and computational physics.

Meteorologists. Area is diagnosing and forecasting worldwide atmospheric conditions that affect operations. Meteorologists specialize in numerical forecasting, atmospheric physics, and forecast applications.

Oceanographers. Areas are ocean dynamics, air-sea interaction, upper ocean dynamics, oceanographic bio-optical modeling, and ocean and atmosphere numerical modeling and prediction.

Geologists and Geophysicists. Areas are marine geology, geophysics, geomorphology, geochemistry, geotechnology, and mapping and charting technology. Emphasis is placed on sedimentary processes, potential fields including gravity and electromagnetics, seismology, geoacoustics, geotechnics, tectonics, and sea-floor acoustic imaging.

SUMMER AND CO-OP PROGRAMS

Summer employment and co-op opportunities are available to college students in engineering, computer science, and the physical sciences. Those who are interested should contact NRL or their university's cooperative education office.

TRAINING

Because the research and development program at NRL is so diversified and technological advances are being made so rapidly, continual education and training are vital to successful job performance and mission accomplishment. As a matter of policy, NRL has encouraged the professional development of its employees at all levels. This development can be achieved in several ways, including individual self-improvement; attendance at professional meetings, conferences, and symposia; interaction with peers; college-level courses; and on-the-job training. Tuition and fees are reimbursable. Short courses in certain program areas are also available during normal working hours. In addition, there are formal graduate programs that enable employees to pursue advanced studies while receiving all or part of their salary, leave, and fringe benefits.

THE ORGANIZATION AS EMPLOYER

NRL places a premium on excellence and achievement. Under a merit-based system, career advancement is based on professional accomplishments. Scientific performance is measured by the quality of papers published and presented to peer groups; patents awarded; honors and awards received; professional society recognition; the number of distinguished American and foreign scientists who visit NRL seeking advice, the exchange of ideas, and the initiation of collaborative efforts; participation in national and international committees; and success in translating research concepts into operational military systems. In addition to basing promotions on accomplishments, NRL has an Incentive Awards Program that acknowledges superior performance through cash awards and special salary increases; special awards are given for beneficial suggestions, patentable inventions, and extraordinary achievements.

NRL employees enjoy all the advantages of federal employment, including a choice of comprehensive health insurance plans, a low-cost life insurance program, a generous retirement plan, and liberal vacation time. NRL has an active Equal Employment Opportunity Program (EEO), including a Federal Women's Program, and is committed to an affirmative employment program for minorities, women, and disabled individuals. The NRL facilities include free parking, a credit union, a cafeteria, a snack bar, and a recreation club with a swimming pool, a basketball court, and a weight and exercise room.

APPLICATION AND INFORMATION

Interested applicants should submit a résumé or an Application for Federal Employment (SF-171) or the Optional Application for Federal Employment (OF-612) to the address given below, or, to find out more about NRL and current career opportunities, applicants may call NRL's Human Resources Office.

Direct inquiries to:

Naval Research Laboratory
Human Resources Office, Code 98 1810PG
Washington, D.C. 20375-5324
202-767-3030

Naval Surface Warfare Center

Dahlgren Division

NSWC
Naval Surface Warfare Center
DAHLGREN DIVISION

Description of Organization: Navy's principal research, development, and test and evaluation laboratory for surface ship combat systems, ordnance, mines, and strategic systems support

Number of Employees: 4,400

Location: Dahlgren, Virginia

Academic Fields of Recruitment Interest: Primary disciplines: computer engineering, computer science, electrical engineering, electronics engineering, systems engineering, mechanical engineering, aerospace engineering, mathematics, and physics. Other disciplines: chemical engineering, chemistry, general engineering, industrial engineering, and operations research analyst.

Major Entry-Level Opportunities for New Graduates: Computer engineering, computer science, electrical engineering, electronics engineering, systems engineering, mechanical engineering, aerospace engineering, mathematics and physics.

BACKGROUND AND OPERATIONS

Under contract to the Defense Department during World War II, Albert Einstein submitted ideas to the Naval Surface Warfare Center (NSWC) for attaching mines to enemy ships. At the Naval Surface Warfare Center, Dahlgren Division (NSWCDD), civilian scientists and engineers are free to explore beyond the known boundaries of technology just as Einstein did. Supplied with the latest equipment and surrounded by experts in advanced scientific fields, new engineers and scientists take on demanding research and development projects that will define surface warfare systems for the next century.

The parent organization of NSWCDD is the Naval Sea Systems Command (NAVSEA), the Navy Department's central activity for designing, engineering, integrating, building, and procuring U.S. Naval ships and shipboard weapons and combat systems. NAVSEA is the largest of the five Navy Systems Commands.

NSWCDD is one of the five Divisions of the Naval Surface Warfare Center. In addition, NSWCDD hosts several tenant activities at the Dahlgren site, the AEGIS Training and Readiness Center, the Joint Warfare Analysis Center, the Naval Space Command, the Panama City site, the Naval Diving and Salvage Training Center, and the Navy Experimental Diving Unit. NSWCDD is involved in warfare analysis, research, design, development, test and evaluation, systems integration, strategic missile systems support, and special and amphibious warfare and fleet services.

Interdisciplinary teams of scientists and engineers, using a systems engineering approach, are engaged in analysis and testing to ensure that Navy and Joint Forces are able to complete their mission in a wartime environment.

The NSWCDD is located in a small town in the tidewater area of Virginia's Northern Neck, with a myriad of outdoor recreational opportunities. The metropolitan areas of Washington, D.C., the nation's capital, and Virginia's capital city, Richmond, are an hour's drive away. These areas offer exciting concerts, clubs, and special events. Also close by are various known amusement parks, famous beaches, landmark sites, and the historic Civil War battlefields of the Fredericksburg area.

FUTURE DIRECTIONS

The NSWCDD's mission is to be the leading expert in the science and engineering of Naval Warfare Systems. Navy, NAVSEA, NSWC, and national interests are best supported by focusing on the following systems: Surface Warfare, Surface Ship Combat, Ordnance, Mines, Amphibious Warfare, Mine Countermeasures, Diving, Special Warfare, and Strategic Systems.

EMPLOYMENT OPPORTUNITIES

NSWCDD is looking for scientists and engineers in many different disciplines. Enlistment in the Navy is not a requirement to work for the organization. NSWCDD consists mainly of civilian employees, although there is a small military representation at the facility. As of September 1997, there were a total of 4,386 civilian employees and 85 military employees at all three locations of NSWCDD.

Positions are available at all degree levels (bachelor's, master's, and doctorate). In some cases, individuals with degrees in related disciplines might qualify for one of the areas for which NSWCDD is in need of employees, depending on course work and experience. For example, a statistician might qualify to be hired as a mathematician. Prospective employees should check with NSWCDD if they are not sure they are qualified for a particular position.

TRAINING

Shortly after starting at NSWCDD, all new scientists and engineers below the Ph.D. level are enrolled in the Junior Professional Development (JPD) program. As part of the JPD program, employees begin a series of training assignments or rotations in other areas of the base. Rotations generally last three to six months and provide new employees with a good opportunity to experience other areas of the base and expand their knowledge into some challenging new fields.

In addition, NSWCDD provides on-the-job training, short courses, and courses via satellite that could lead to an advanced degree. NSWCDD strongly encourages its employees to expand themselves by utilizing all available training opportunities. In support of this, the training budget has been dramatically increased to allow employees a wide spectrum of training choices.

THE ORGANIZATION AS EMPLOYER

NSWCDD offers excellent benefits, travel opportunities, and competitive salaries. Some of these benefits include those listed below.

Job Stability NSWCDD hosts many important Navy programs, and while no military site is completely immune to government cutbacks, NSWCDD is positioned very well for the future.

Flexibility NSWCDD offers its employees flextime, which allows flexible work schedules, accumulation of credit hours (one week employees can work an extra hour Monday to Thursday and leave 4 hours early on Friday), and other benefits to allow for the employees' varying work schedule needs.

Travel Opportunities Some travel is usually required, and employees are often given the opportunity to determine how frequently they travel. This is a great opportunity to see other parts of the country (and sometimes the world) while working with top professionals in other locations.

Potential Promotion Opportunities NSWCDD employees are usually promoted in a timely manner, since they are recognized and appropriately rewarded for good performance with promotions, bonuses, and permanent pay increases. The new personnel demonstration project serves to enhance the ability to be compensated appropriately.

Insurance NSWCDD participates in group life and health insurance programs. This greatly reduces out-of-pocket expenses. Because a large group of people is involved in these programs, the costs are kept low. Employees can choose from a large number of health insurance plans, HMOs, and PPOs to find the insurance that meets their needs.

Leave In addition to ten paid federal holidays, new employees earn thirteen vacation days a year, which increase to twenty days after three years and to twenty-six days a year after fifteen years. Also, employees earn an additional thirteen sick days per year.

Retirement New NSWCDD employees are enrolled in the FERS program (Federal Employees Retirement System). This includes the Thrift Savings Plan, which allows employees to contribute up to 10 percent of their pretax income into their retirement fund, and the government matches the first 5 percent.

SUMMER, CO-OP, AND INTERNSHIP PROGRAMS

Cooperative education is a long-established program at the Naval Surface Warfare Center, Dahlgren Division. The blend of academic study and work experience available through the program has many potential benefits. NSWCDD has openings for outstanding college students majoring in electronics, engineering, electrical engineering, computer engineering, and computer science (math background). In addition, there are a limited number of positions for college students studying aerospace engineering, mechanical engineering, industrial engineering, chemical engineering, physics, operations research analysis studies, and mathematics.

The new Summer Internship Program for college candidates and the Science and Engineering Apprentice Program (SEAP) for high school students are also examples of NSWCDD's commitment to the future generation.

For more information on student programs, interested students should contact NSWCDD via e-mail (recruit@nswc.navy.mil), telephone (540-653-3216), or the Internet (see address below).

CORPORATE SOCIAL RESPONSIBILITY

At NSWCDD, diversity is more than just a buzz word; it is an integral part of the organization. NSWCDD wants its teams to be invigorated by the ideas and energy that come from a good mix of backgrounds and cultures. The organization desires people who are creative, technically superior, and have the ability to contribute as an individual or a team member. NSWCDD is committed to pursing candidates of all background to fill these needs. This diversity enriches the base, its employees, and the community in which the employees live and work.

APPLICATION AND INFORMATION

To apply, e-mail a resume in a standard word processor format (MS Word, Word Perfect, etc.) or as an ASCII text file to ensure that it is legible to recruit@nswc.navy.mil. Applicants also need to mail a transcript. Individuals can also mail a resume and transcript to the address below.

Direct inquiries to:

College Recruiting Coordinator
Human Resources Department, Code PD
Dahlgren Division
Naval Surface Warfare Center
17320 Dahlgren Road
Dahlgren, Virginia 22448-5100
Telephone: 800-352-7967 (toll-free)
Fax: 540-653-7632
World Wide Web: http://www.nswc.navy.mil/P/RECRUIT/recruit.html

Northrop Grumman Corporation

Electronic & Systems Division–Electronic Systems

Description of Organization: A worldwide leader in the research, design, development, and manufacture of radar-jamming countermeasures and combat support systems offering a diverse array of products that are in use by all branches of the U.S. military and international military customers

Number of Employees: 3,000

Location: Rolling Meadows, Illinois

Academic Fields of Recruitment Interest: Electrical and mechanical engineering, computer science, mathematics, physics

Major Entry-Level Opportunities for New Graduates: Software development, radio frequency (RF) and EOIR communications, microwave design, power systems, digital and analog systems design and automatic test systems engineering

BACKGROUND AND OPERATIONS

The Northrop Grumman Corporation is a high-technology company with principal interests in military aircraft, commercial airframe structures, navigation, and electronics. Northrop Grumman's Electronic & Systems Division-ES in Rolling Meadows, Illinois, is a leading producer of electronic countermeasures. All progress and expansion has come about as a result of Northrop Grumman's development and application of new technology in innovative ways to design and build advanced-technology products and provide better services to its customers.

EMPLOYMENT OPPORTUNITIES

ENTRY-LEVEL AND EXPERIENCED PERSONNEL

Systems Engineers. Positions require a minimum of the B.S.E.E. or its equivalent; the M.S.E.E. is desirable.

Threat Analysis/Electronic Warfare (EW) Techniques Development. These engineers need experience in one or more of the following: radar and missile threat analysis, with emphasis on multifunction phased array/pulse Doppler systems; infrared (IR) guided missile threat analysis, with emphasis on advanced reticle-based image processing systems; digital computer modeling and simulation, with emphasis on electronic countermeasures (ECM) and active infrared countermeasures (IRCM) effectiveness; radar, laser, missile, ECM, and IRCM hardware testing and effectiveness evaluation in ground and flight test environments; monopulse countermeasures analysis; and system analysis and vulnerability assessment.

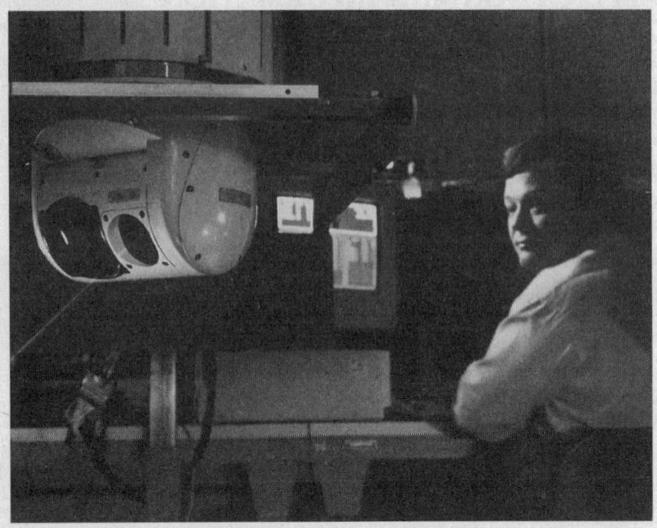

An engineer tests a turret assembly for the IRCM system.

Project Development. Engineers in project development are assigned to a specific development project from concept definition through production start-up. They analyze hardware and software trade-offs, develop simulations and modeling for electromagnetic signal analysis, and solve problems related to electromagnetic/optical/acoustics compatibility, interference, and pulses. A background in defining requirements and functional operations of complex hardware and software systems is required.

RF Systems Design and Analysis. These engineers need experience in angle-of-arrival receiver design, with emphasis on interferometric systems and/or multipath signal processing concept development, simulation, and analysis or in advanced EW systems, RF, intermediate frequency (IF), video, and digital signal processing concept development, simulation, and analysis.

Electro-Optical (EO) and IR Systems Design and Analysis. These engineers need experience in the analysis, specification, and design of EO/IR systems. Particular emphasis is placed upon IR systems in the mid-wave IR, but applications exist in the ultraviolet, visible, and far-infrared special ranges.

Advanced Technology Engineers. Positions require the B.S.E.E. or its equivalent and related experience; the M.S.E.E. or Ph.D. in electrical engineering or physics is desirable.

Microwave. Assignments include active and/or passive microwave and millimeter-wave integrated circuit design, including components such as oscillators, amplifiers, mixers, detectors, filters, and integrated subsystems. Prior computer-aided design experience is desirable.

Receivers. Opportunities exist in conceptual design, fabrication, and test of state-of-the-art receiver systems for ECM/electronic intelligence applications. Familiarity with systems architecture, signal processing, and channelized, set-on, and microscan techniques is desirable.

Analog. These engineers specialize in video amplifiers, filters, and analog-to-digital (A/D) and digital-to-analog (D/A) conversion; they should have a background in feedback theory analysis and computer-aided circuit analysis.

Digital. Areas include design and development of high-speed signal processing and control systems, microprocessor-controlled hardware, D/A and A/D conversion, and digital signal processing utilizing very-high-speed logic, transistor-transistor logic, emitter-coupled logic, complementary MOS, GaAs, gate arrays, and phase alternation lines.

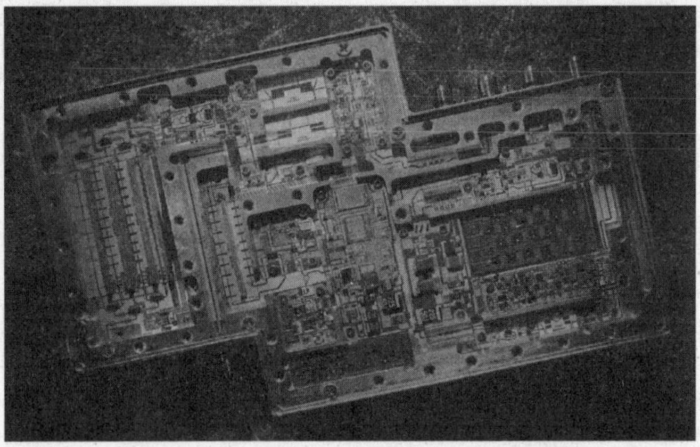

This is an MIC module for the ALQ-135 Advance Repeater Tuning Unit (ARTU). NOTE: Each module contains 99 separate MMIC and ASIC devices, which is the highest level of technology in the world.

Electro-Optical and Infrared. Engineers and scientists are needed for development of a broad range of active and passive systems involving a variety of sensor and source technologies. Areas of particular interest are radiometrics, solid-state infrared lasers, plasma arc-lamps, infrared detectors, LIDAR, and stabilization and control.

Software Engineers. The B.S.E.E., B.S.C.S., or their equivalent is required; the M.S.E.E. or M.S.C.S. is desirable.

Systems Programming. These engineers need a knowledge of software development technologies, real-time operating systems, performance prediction and evaluation, and Ada, assembler, C, JOVIAL, and Pascal.

Software Systems. Opportunities exist for those with experience in software requirements analysis; architecture design; software validation, test specification, and modeling; and interface design.

ECM/EW Software Systems. These engineers need a knowledge of real-time control and embedded computer systems, system- and unit-level diagnostics, object discrimination, algorithm development, and Kalman filtering.

Mechanical. Develop packaging designs for electronic test equipment. Perform thermal, structural, material and tolerance analysis to support design activity. Support design documentation, hardware build and test activities, BSME or equivalent. CAD experience a plus.

Software Architecture. Assignments include initiation of design projects, requirements analysis, algorithm development, and high-level design for large-scale real-time embedded software.

Software Systems Development. Opportunities exist in detailed design, coding, testing, and integration of embedded real-time software using C or Ada in structured design technologies.

Advanced Systems Engineers. The B.S.E.E. or equivalent is required; the M.S.E.E. is desirable.

Electro-Optical and Infrared. Engineers are needed to define advanced electro-optical systems. Applications include IRCM, threat warning systems, automatic test equipment, and others.

Radar Systems Engineering. These engineers participate in the development of advanced ECM system concepts and techniques. The B.S.E.E. or equivalent, including courses on modern signal processing technologies, is required.

Antenna Design Engineers. Positions require the B.S.E.E. or its equivalent; the M.S.E.E. is desirable.

These engineers are responsible for the design and development of airborne ECM antennas. Design experience with phased arrays, monopulse, direction-finder systems, and millimeter waves and the B.S.E.E. or equivalent are required.

Business Applications and Data Processing Analysts. Positions require a minimum of the B.S.C.S. or its equivalent; the M.S.C.S. is desirable.

Systems Analysts–Manufacturing and Purchasing Applications. These analysts participate in new development and support utilizing a multiprocessor IBM 3090 running MVS-XA, COBOL, CICS, and IDMS. New development is on line and batch in a database and VSAM environment. Experience with MVS, COBOL, JCL, CICS, and databases is required. Individuals must have good communication skills and heavy user contact with manufacturing or purchasing as well as large systems design and programming experience. IDMS with ADS/O is desirable.

Northrop Grumman Electronics and Systems Integration Division-ES in Rolling Meadows, Illinois.

THE ORGANIZATION AS EMPLOYER

ESID-ES offers excellent compensation and exciting technical challenges in a modern facility located in Rolling Meadows, Illinois, just 30 miles northwest of Chicago. Benefits include, but are not limited to, health, life, and dental insurance; tuition reimbursement; a 401(k) savings program; paid holidays and vacations; and a retirement plan. Northrop Grumman is a smoke-free workplace.

APPLICATION AND INFORMATION

Resumes should be directed to the address below. U.S. citizenship is required for certain positions.

Direct inquiries to:

Manager, Staffing, T41
Electronic & Systems Division-ES
Northrop Grumman Corporation
600 Hicks Road
Rolling Meadows, Illinois 60008
or send resumes in ASCII format to:
resumes@eiws.northgrum.com

Raytheon Company

Raytheon
EXPECT GREAT THINGS

Description of Organization: Manufactures electronic and aircraft products and appliances and provides energy and environmental services
Number of Employees: 110,000
Headquarters: Lexington, Massachusetts
Academic Fields of Recruitment Interest: Electrical engineering, electronics/electronics engineering, accounting, data processing

BACKGROUND AND OPERATIONS

Throughout its seventy-five-year history, the Raytheon Company has been a leader in developing defense technologies and in converting those technologies for use in commercial markets. From its early days as a maker of radio tubes, its adaptation of World War II radar technology that resulted in the invention of microwave cooking, and its development of the first guided missiles, Raytheon has successfully built upon its pioneering tradition to become a global technology leader. Through strategic acquisitions and mergers, Raytheon has resources of companies with equally distinguished records of innovation. Breakthroughs include the laser, the Surveyor lunar lander, high-resolution satellite imaging of Earth resources, and magnetic anomaly-detection services.

Today, Raytheon is focused on three core business segments: defense and commercial electronics, business aviation and special mission aircraft, and engineering and construction. Raytheon is a top-tier player in each of these segments. Each provides the company with the capabilities it needs to build on its strength as an innovator and to prosper in a highly competitive global economy.

Raytheon's $9.5-billion merger with Hughes defense operations and the creation of Raytheon Systems Company mark an important milestone in the company's history. Built upon the combined forces of Raytheon Electronic Systems, Raytheon E-Systems, and Hughes and Texas Instruments's defense operations, Raytheon Systems Company is now fully equipped to meet the needs of its customers and employees—not just for today or tomorrow but well into the twenty-first century.

DEFENSE ELECTRONICS

Raytheon is a world leader in defense electronics, including missiles; radar; electro-optics, reconnaissance, surveillance, and intelligence; C41; training; simulation and services; and naval and air traffic control systems. Raytheon Systems Company combines the best of Raytheon Electronic Systems, Raytheon E-Systems, Raytheon TI Systems (previously the defense division of Texas Instruments), and the former Hughes Aircraft Company. The newly combined company gives Raytheon the breadth and critical mass necessary to compete as one of the largest defense electronics firms in the world. It also enables Raytheon to achieve significant revenue and cost synergies that benefit the company and its customers and shareholders.

Raytheon is a pioneer in the conversion of defense technologies to meet the needs of commercial markets, successfully transferring sophisticated military applications to make life easier, better, and safer throughout the world. For example, Raytheon is a global leader in designing and building air traffic control (ATC) systems. In 1996 alone, Raytheon Electronic Systems won two major U.S. air traffic control competitions, with an estimated value of $1.6 billion. Additional commercial businesses include image and information management, transportation, wide-area surveillance, marine electronics, and communications and environmental monitoring.

Raytheon Systems Company is organized around five segments, several of which are large enough to be among the top 500 company listings by themselves. These segments are:

Defense Systems: Anti–tactical ballistic missile systems; air defense; air-to-air, surface-to-air, and air-to-ground missiles; naval and maritime systems; ship self-defense systems; torpedoes; strike, interdiction, and cruise missiles; and advanced munitions.

Sensors and Electronic Systems: Ground, shipboard, and airborne fire control and surveillance systems; primary and secondary air traffic control radars; ground, space-based, night vision, and reconnaissance sensors; electronic warfare systems; and Global Positioning Systems.

C3 (Command Control Communications) Systems: Command, control, and communications systems; air traffic control systems; tactical radios; satellite communication ground terminals; wide-area surveillance systems; advanced transportation systems; and simulators and simulation systems.

Intelligence, Information, and Aircraft Integration Systems: Ground-based information processing systems; large-scale information retrieval, processing, and distribution systems; global broadcast systems; airborne surveillance and intelligence systems integration; aircraft modification; and head-of-state aircraft systems.

Training and Services: Training services and integrated training programs; technical services; and logistics and lifetime support.

COMMERCIAL ELECTRONICS

From outer space to ocean depths, Raytheon's commercial electronics business is putting cutting-edge technology to work. Raytheon is developing the Main Mission Antenna transceiver systems for the Motorola IRIDIUM® global satellite communication systems, which is designed to provide wireless voice, paging, data, facsimile, and location services transmissions anywhere on the Earth. The project utilizes Raytheon's advanced gallium arsenide Monolithic Microwave Integrated Circuit (MMIC) technology. Raytheon is also working with Motorola on the development of power amplifiers and receive modules for user handsets.

Other applications include Raytheon's Raylink™ wireless local area network PC card, which can access computer networks without restrictive conventional wire connections. For example, doctors using laptop computers are able to access a patient's medical records from their office, from home, or from a patient's bedside.

Raytheon Marine is the number one supplier of marine electronics in North America and the third-largest supplier in the world. Raytheon products guide sailors and power boaters across oceans and lakes with products such as autopilots, radars, depth finders, and Global Positioning Systems (GPS). Utilizing submarine sonar technology, Raytheon has been an innovator in developing tools to help fishermen. One product, Raytheon's Apelco 365 fishfinder, won the "Innovation in Electronics" award at the International Marine Trade Exhibition and Conference in Chicago. The company also offers ship owners and builders complete solutions to integrated bridge control, communications systems marine radars, and gyro compasses.

BUSINESS AVIATION AND SPECIAL MISSION AIRCRAFT
As the leading provider of business and special mission aircraft, Raytheon delivers a broad line of jet, turboprop, and piston-powered airplanes to corporate and military customers worldwide. The company's activities include aircraft design and manufacture, service, logistics, and support.

All of Raytheon's aircraft are distinguished by superior workmanship, advanced technology, and larger, more commodious cabins. Its commercial aircraft can be adapted for special mission roles. The Hawker 800XP, the Beechjet, and the Beech King Air, for example, have served the U.S. government and its allies, undertaking airborne reconnaissance, search and rescue, and maritime patrol missions.

Key product lines include Hawker 800XP, the world's most popular midsized business jet, with nearly 50 percent of the world market; Hawker Horizon, the new super-midsized transcontinental business jet; Raytheon Premier I, a revolutionary new entry-level jet providing exceptional value; Beech King Air, the most popular turbine business aircraft series ever built; Beech 1900D Airliner, rated among the best regional aircraft by *Professional Pilot* magazine; piston-powered aircraft, high-performance airplanes that include the single-engine Beech Bonanza B36TC and A36 and the twin-engine Beech Baron 58, which carries more payload farther and faster than its piston twin competitors; and Special Mission Aircraft, which are missionized versions of Raytheon commercial aircraft that serve special mission roles for the U.S. military and government and its allies. These include Raytheon's Beech/Pilatus PC-9 MkII aircraft, now renamed the T-6A Texan II, which was selected to be the U.S. Air Force and U.S. Navy twenty-first-century trainer under the multibillion-dollar Joint Primary Aircraft Training System (JPATS) program.

ENGINEERING AND CONSTRUCTION
Raytheon Engineers & Constructors (RE&C) offers full-service engineering and construction capabilities to clients worldwide. It offers a full range of program management capabilities, including project planning, finance, process development, environmental assessment, engineering, procurement, construction, equipment fabrication, start-up, operations, and maintenance services. In forming RE&C, Raytheon acquired more than a dozen well-known engineering and construction firms. Many of these firms trace their roots back to the last century.

RE&C's core business segments include energy, process, metals and mining, pulp and paper, and infrastructure. RE&C's largest current assignment is a $1.1-billion project to design, build, operate, and maintain a 9.5-mile light rail line in northern New Jersey.

FUTURE DIRECTIONS

Raytheon is positioned for growth well into the twenty-first century. Raytheon is a top-tier defense electronics provider with the breadth and critical mass to compete as one of the largest defense electronics firms in the world; well positioned to build upon its proven ability to convert defense technologies to meet the needs of commercial markets; able to leverage its new economies of scale to successfully compete for—and win—additional defense and government procurement programs; and a strong commercial company and leader in engineering arid construction and aircraft manufacturing.

APPLICATION AND INFORMATION

Direct inquiries to:
Raytheon Company
141 Spring Street
Lexington, Massachusetts 02173
Phone: 781-862-6600
Fax: 781-860-2520
World Wide Web: http://www.raytheon.com

Sandia National Laboratories

Description of Organization: Sandia National Laboratories is a U.S. Department of Energy (DOE) multi-program laboratory operated by Sandia Corporation, a wholly-owned subsidiary of Lockheed Martin Corporation.

Number of Employees: 7,500

Headquarters Location: Albuquerque, New Mexico

Academic Fields of Recruitment Interest: Engineering; computer science, information systems/science, systems design/science, mathematics/applied mathematics; chemistry, physics, physical science

Major Entry-Level Opportunities for New Graduates: Engineering, research and development, business applications/systems programming and management information systems, finance and accounting, and technician positions

BACKGROUND AND OPERATIONS

Sandia National Laboratories is one of the largest research and development facilities in the nation. For almost fifty years, Sandia has operated as a multiprogram engineering and science nuclear weapons complex under contract with the U.S. Department of Energy. Today, Sandia applies world-class scientific and engineering creativity and expertise to achieve comprehensive, timely, and cost-effective solutions to the nation's greatest challenges. Sandia applies its problem-solving and innovative technology to energy and environment, manufacturing, transportation, space, the nation's economy, and more.

Sandia employs 7,500 people in its major offices and laboratories in Albuquerque, New Mexico, and Livermore, California. Test ranges are located near Tonopah, Nevada, and on Kauai, Hawaii. The candidate with a strong academic background or applicable work experience can find increasing responsibility and decision-making opportunities in a professional teamwork environment.

Sandia currently employs 4,000 professional and technical staff members. About 35% of the technical staff hold Ph.D. degrees, 41% hold M.S. degrees, and 12% have B.S. degrees. Overall, 65% of the technical staff have engineering degrees and 35% have science degrees. Sandia also employs 1,050 technology program graduates in technician positions. These employees normally have an A.S.T. or B.S.T. degree from an ABET-accredited technology program.

FUTURE DIRECTIONS

Sandia National Laboratories' roots in the nuclear weapons program and nearly half a century of research leave it well positioned to contribute to new initiatives that include nonproliferation treaties and arms control verification for peaceful use of nuclear technology. Sandia is collaborating with universities and medical schools in several areas of biomedical engineering for solutions to cost-effective health care. These advanced technologies and other Sandia-developed data analysis methods continue to enhance quality of living and global and national security.

EMPLOYMENT OPPORTUNITIES

ENTRY-LEVEL AND EXPERIENCED PERSONNEL

Sandia National Laboratories offer unlimited professional potential for top-level people to work with world-class scientists and engineers. Sandia fosters teamwork in an interdisciplinary environment that encourages innovation, improvisation, high quality, and personal satisfaction.

Sandia prefers to hire graduates with technology degrees and bachelor's, master's, and doctoral degrees who were ranked in the top 10% of their class and had a minimum 3.5 GPA. Graduates should have related student work experience, a notable record of extracurricular and leadership activities, and above-average verbal, writing, and quantitative skills. Sandia offers the following opportunities.

Bachelor's and Master's Degrees Sought: Aerospace, ceramic, chemical, civil, computer, electrical, electronics, environmental, geological/geophysical, mechanical, metallurgical, nuclear, safety, software, structural, and systems engineering; chemistry; computer science; engineering mechanics; engineering technology; environmental science/ecology; geology/earth science; geophysics; health physics; information systems; materials science/engineering; mathematics/applied mathematics; optics; plastics engineering/polymer science; and toxicology.

Doctoral Specialties Sought: Aerospace, ceramic, chemical, civil, computer, electrical, electronics, geological/geophysical, mechanical, metallurgical, nuclear, structural, and systems engineering; analytical chemistry; applied mathematics; computational methods and numerical analysis; computer organization and architecture; computer science; electrophysics; engineering mechanics; engineering physics; engineering science; environmental science/ecology; geochemistry; geology/earth science; geophysics; health physics; information science; information systems; materials science/engineering; metallurgy; nuclear physics; operations research; optics; physical chemistry; physics of fluids; plasma physics; plastics engineering/polymer science; polymer chemistry; programming systems and languages; solid-state physics; statistics; systems design/systems science; theoretical physics; and thermal physics.

Technology Degrees Sought (A.S./B.S.): Electronic engineering technology, electrical engineering technology, mechanical engineering technology, mechanical design technology, computer engineering technology, and chemical engineering technology.

TRAINING

Sandia strives to keep its employees as current as the facilities and equipment they operate. Educational programs leading to formal degrees are available at local universities or through support at leading universities throughout the nation. Graduate-level, in-hours courses are provided to give the staff an in-depth understanding of new or evolving techniques that are applicable to Sandia's work.

THE ORGANIZATION AS EMPLOYER

Employees at Sandia receive excellent benefits, including health, vision, and dental insurance; a retirement plan; and paid holidays and vacation. The Laboratories' principal locations, Albuquerque, New Mexico, and Livermore, California, offer a complete range of cultural and recreational activities with the informal living style of the Southwest.

STUDENT INTERNSHIP PROGRAMS

Opportunities are available for undergraduate and graduate students interested in summer and year-round employment and co-op assignments in these disciplines: electrical, mechanical, chemical, and civil engineering; computer science and geosciences; environmental, safety, and health sciences; chemistry; physics; materials sciences; and mathematics.

Selection Criteria: Due to the number of positions available and the level of competitiveness associated with participation, selection is based on academic achievement, prior applicable experience, and the candidate's stated interest.

How and When to Apply: Candidates for summer student internship programs must submit a resume with General, Technical/Science, or Business/MIS specified as an area of interest in the objectives statement and a recent college transcript (including cumulative grade point average) by November 30.

Appointments are available for postdoctoral candidates in a wide variety of engineering and science disciplines. Appointments are normally for a two-year duration and may be extended to three years in some circumstances. Positions are available in New Mexico and California. These positions provide competitive salary, vacation, benefits, and paid relocation.

Direct inquiries to:

Sandia National Laboratories
Staffing Department 3535–0035
P.O. Box 5800, MS1023
Albuquerque, New Mexico 87185-1023
or
World Wide Web: http://www.sandia.gov

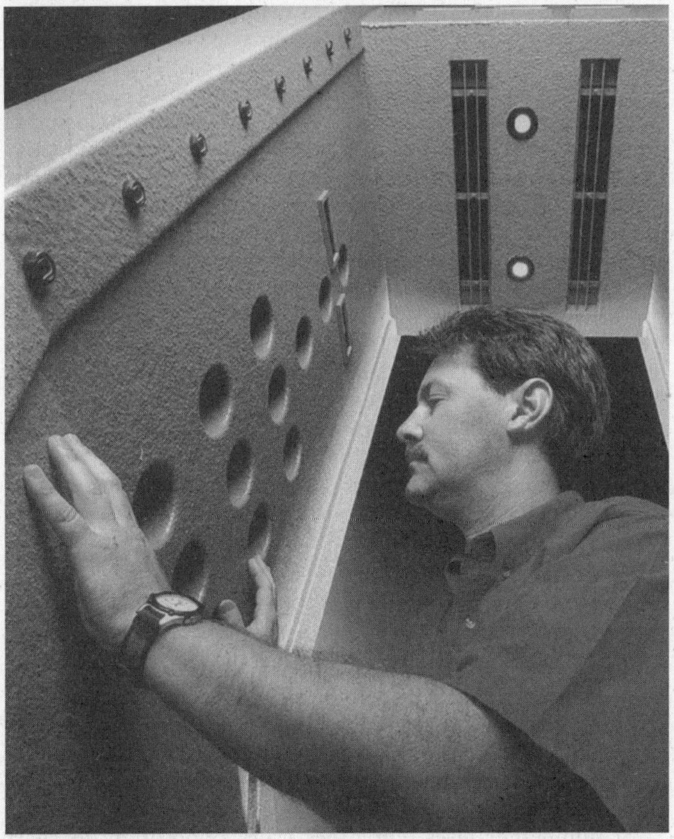

This new walkthrough portal designed at Sandia will improve the safety of the traveling public by detecting minute traces of explosives on airline passengers.

A Sandia engineer checks an automated pit packaging system, an intelligent machine that handles hazardous materials.

Schlumberger

Description of Organization: International enterprise employing advanced technology and services to the energy and measurement and systems industries

Number of Employees: 55,000 worldwide

Headquarters Locations: Houston, Texas; New York, New York; Paris, France; and Tokyo, Japan

Academic Fields of Recruitment Interest: Chemical, civil, computer, electrical, mechanical, and petroleum engineering; computer science; geology, geophysics, physics

Major Entry-Level Opportunities for New Graduates: Field engineering (B.S. and M.S.), research and development (M.S. and Ph.D.)

BACKGROUND AND OPERATIONS

From its very founding in 1927 by the French brothers Conrad and Marcel Schlumberger, the Schlumberger group of companies has built a worldwide reputation for leadership in a variety of technological areas. That first year, they developed a startling new method of evaluating the production potential of oil wells through electrical measurements taken inside the borehole. Schlumberger technical innovation has been continuous ever since.

Today, Schlumberger consists of three business segments operating worldwide: Oilfield Services, Measurement and Systems, and Omnes. Each has had an international technological impact in its field and takes great pride in its role as a leader in applying imaginative engineering solutions to everyday production situations.

Schlumberger Oilfield Services. Schlumberger services the energy industry from top to bottom, start to finish, from seismic acquisition, processing and interpretation, onshore and offshore drilling, and surface and downhole measurements to wire-line and drill-stem formation testing and pumping and stimulation services. Schlumberger Oilfield Services also offers exploration and production companies dependable, highly technical assistance in just about every functional area, in over 75 countries worldwide.

New tools and systems are constantly being employed to optimize oil and gas exploration and production. Schlumberger has taken the lead in development of both the hardware and the software needed to keep the energy industry producing efficiently, safely, and profitably. Services such as well-site computer analysis of real-time drilling, cementing, and imaging of geological data have taken much of the guesswork out of the exploration and production process. Workstation software to integrate this data is also a growing part of the company's business. It is this kind of innovative technology that keeps Schlumberger ahead.

Schlumberger Measurements & Systems. Schlumberger is the world leader in the manufacture of electricity, water, and gas meters. Industry firsts, such as the introduction of solid-state electricity meters and remote reading capabilities, have led to growth now represented by 69 plants in twenty countries. This group also enjoys success in the production of equipment that measures and controls electricity, water, and gas at the primary distribution level.

Automatic meter reading equipment makes the management of natural resources interactive. Schlumberger hardware and software systems read many thousands of meters via radio waves, telephone and cellular lines, and cable television links, helping public utilities to increase efficiency and reduce costs.

Omnes, a joint-venture with Cable and Wireless, provides communications and information technology solutions, wide and local area networks, security, Internet, messaging, and business application solutions for the energy exploration and production sector, allowing information technology systems to communicate anywhere in the world.

As testing and debugging of the fastest electronic components becomes increasingly complex, semiconductor manufacturers are calling on the **ATE Components Division**, a market leader in the design and manufacture of high-speed semiconductor device testers and diagnostic systems. They are in the business of answering the question: "How do you test the world's fastest chips?"

The **Electronic Transactions group** specializes in the rapidly expanding field of automated point-of-sale payment systems and produces products such as the "smart card," a credit card in which microprocessors and memories are imbedded, allowing it to conduct complete financial transactions and store information for future use. The **Retail Petroleum Systems Division** provides solutions for retail operations in service stations of major oil companies and independent fuel retailers. These solutions consist of products such as gasoline dispensers, electronic systems that control dispensers and outdoor peripherals at the station, and perform cash register, Electronic Funds Transfer (EFT), and management information functions.

EMPLOYMENT OPPORTUNITIES

ENTRY LEVEL

Schlumberger has established a history of developing its own leaders through a strong policy of hiring only the best and then promoting from within its ranks. Every group and division offers continuous, advanced training at every operational level and the kind of immediate job responsibility and authority necessary for careers to advance at a self-determined pace.

Schlumberger gives employees every opportunity to chart their own path to success, whether their specialty is software design or computer engineering; mechanical, electrical, petroleum, civil, chemical, manufacturing, or environmental engineering; physics, geology, or geophysics. Positions are open in just about every scientific and engineering discipline for field engineers (at B.S. and M.S. degree levels) and in research and engineering (at M.S. and Ph.D. degree levels).

The company is looking for highly motivated self-starters with career aspirations that go beyond a steady paycheck: men and women who prefer to lead, not follow—people who want to be at the forefront of technological development and who don't shy away from the risks necessary to achieve both corporate and personal goals. Energetic, environmentally conscious, and enthusiastic: These are the qualities that have put Schlumberger in its position as a technological leader, and they are the qualities that will keep it there.

THE ORGANIZATION AS EMPLOYER

Schlumberger believes strongly in attracting and retaining the very best graduates—not only by recognizing and rewarding individual contributions but also by following some very simple philosophies: recruit from colleges and universities, train and develop, promote from within, and challenge and allow flexibility for growth.

Schlumberger also follows a "Statement of Values" put forth by its Chairman and CEO, Euan Baird (who himself started as a field engineer): "Our people, their motivation and dedication to customer service worldwide, in a safe and clean environment, are our main asset. Our commitment to technology and quality is the basis for our competitive advantage. Our determination to produce superior profits is the cornerstone for our future independence of action and growth."

APPLICATION AND INFORMATION

Graduates interested in finding out more about a career with Schlumberger should check their school's placement center library for more comprehensive and specific information on entry-level job opportunities and should meet with Schlumberger representatives who may visit their campus. Or, resumes may be sent to the address below. A cover letter should identify specific technical interests.

Direct inquiries to:

Erwin Muller
Manager of Recruiting Coordination and University Relations
Schlumberger
300 Schlumberger Drive, Dept. PG 98/99
Sugar Land, Texas 77478

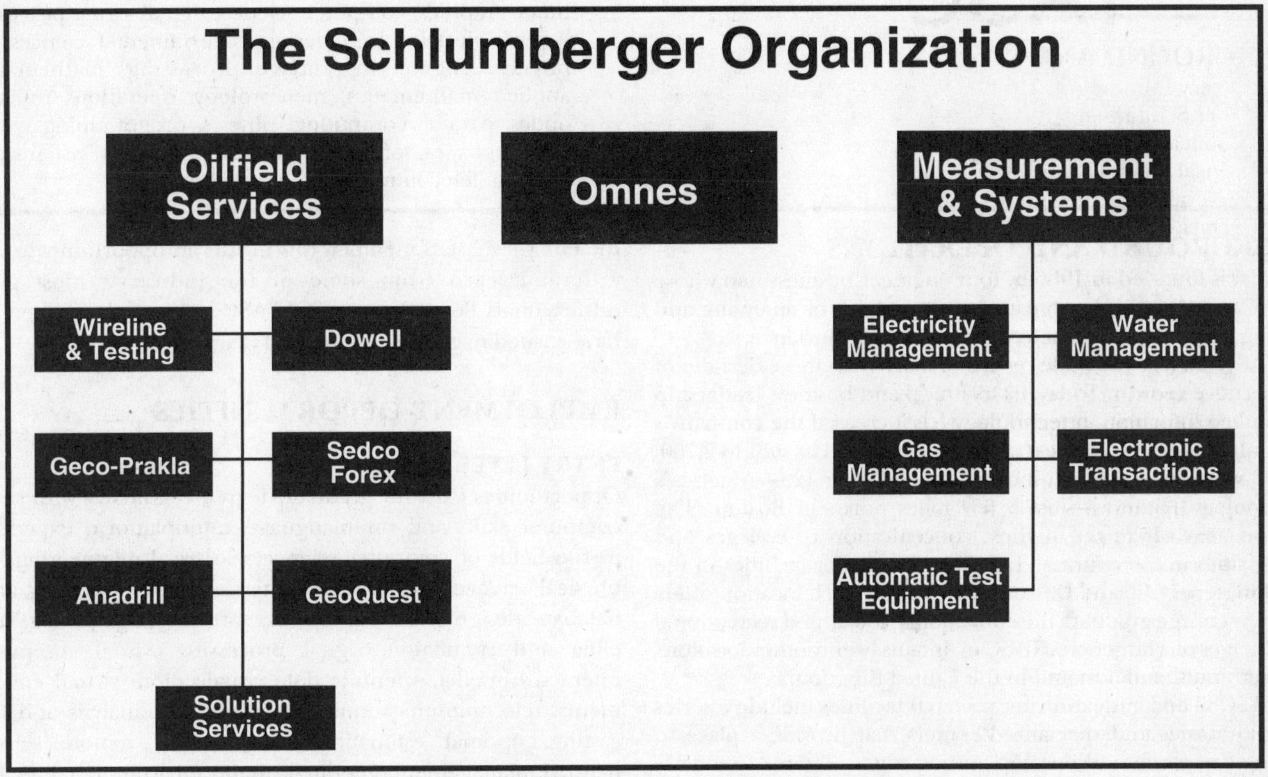

World Wide Web: http://www.slb.com.

TASC

Litton
TASC

Description of Organization: Applied information technology organization (nongovernment); business areas include systems integration, systems engineering, modeling and simulation, document management consulting, geographic information systems, and weather information and products

Number of Employees: 2,700

Headquarters Location: Reading (Boston), Massachusetts

Academic Fields of Recruitment Interest: Computer science, electrical engineering, MIS, physical science

Major Entry-Level Opportunities for New Graduates: Acquisition management, application software development, cartography/photogrammetry, computer engineering, computer graphics, computer science, digital signal processing systems, electrical engineering, environmental sciences, geophysical engineering, image processing, mathematics/applied mathematics, meteorology, operations research, optics, parallel computing, physics, programming systems and languages, software engineering, statistics, systems engineering, telecommunications systems

BACKGROUND AND OPERATIONS

TASC was founded in 1966 by four engineer businessmen whose vision was to establish a creative environment for analyzing and solving technological problems of national importance.

TASC has a remarkable record of more than three decades of consecutive growth. Today, its technical and business leadership in applied information technology has increased the company's annual revenue by in excess of $425 million and its staff to 2,700. TASC's headquarters are located off of Route 128—America's Technology Highway—just a few miles north of Boston. The Boston area offers the highest concentration of colleges and universities in the country. TASC also has major facilities in the Washington, D.C., and Dayton, Ohio, areas. Each location offers the new college graduate the educational, social, and recreational advantages of a larger city. TASC maintains twenty other locations throughout the nation and in the United Kingdom.

TASC's constantly evolving research facilities include a series of laboratories and specialized centers that provide a place to conduct research and develop cutting-edge solutions to today's most complex information challenges. These facilities contain state-of-the-art equipment from leading vendors such as Silicon Graphics, Sun, IBM, DEC, and Hewlett-Packard. They support research in the areas of fusion and visualization, image exploitation and database management, distributed modeling and simulation, wide- and local-area visual information networks, interactive media tools, and collaborative application environments. The facilities also have Internet access to remote supercomputer facilities. TASC has over thirty special research facilities at various locations nationwide to give employees the opportunity to excel at what they do best.

Today, TASC provides Fortune 500 companies and government with a broad range of information technology solutions to give them the efficiencies they need to excel in a dynamic global economy. TASC is looking for that handful of people whose ingenious ideas and unique talents enable them to make a lasting impact on the world. The projects employees work on and the technologies and platforms they work with at TASC literally define the leading edge of technology. From day one, a new college graduate is an active member of a world-class technical team dedicated to bringing solutions to real-world problems. For the entry-level staff member, this means an opportunity to work with and learn from some of the industry's most gifted professionals. Seventy percent of TASC's technical staff members have earned master's and doctoral degrees.

EMPLOYMENT OPPORTUNITIES

ENTRY LEVEL

Opportunities exist for graduate-degree candidates with strong computer skills and multilanguage/multiplatform experience in the fields of computer science, electrical engineering, and physical science who have an interest in the following areas: database design and applications, software engineering, modeling and simulation, signal processing, visual computing, interactive media, scientific data visualization, virtual environments, telecommunications, meteorological analysis and forecasting, optimal estimation and control, remote sensing, network management, geophysical and environmental analysis, human-computer interaction, artificial intelligence, speech processing, or operations research.

TASC's headquarters in Reading, Massachusetts.

TRAINING

TASC is committed to providing every employee with the opportunity to grow and to achieve his or her highest technical and managerial potential. Whether an employee's goal is to move into management or become a key technical contributor, TASC encourages professional development through formal and informal education and training programs. Many of these are provided on-site or are supported by the company's tuition reimbursement program. TASC's career development programs are among the most generous in the industry.

THE ORGANIZATION AS EMPLOYER

TASC offers a very competitive benefits program with a choice of comprehensive medical plans in addition to dental, life, long-term disability, accidental death and dismemberment, and travel insurance. Also offered are an employee computer purchase program, 100 percent pre-approved tuition reimbursement, a lucrative employee-referral program, profit-sharing and stock-ownership plans, and a 401(k) savings plan.

APPLICATION AND INFORMATION

Potential applicants should visit the company's World Wide Web site at: http://www.tasc.com, to learn more about TASC and to see how the company is leading the information revolution.

Individuals interested in a career with TASC should visit their college placement office for TASC's on-campus interview schedule or write to either of the addresses below. U.S. citizenship is required for most positions.

Direct inquiries to:

Manager, College Relations
TASC
55 Walkers Brook Drive
Reading, Massachusetts 01867
Fax: 781-942-2461
E-mail: opportunity@tasc.com

TRW Systems & Information Technology Group

Description of Organization: Provides systems engineering and development, information technology, and technical services to global clients
Annual Sales: $3 billion
Number of Employees: 20,000
Headquarters Location: Fairfax, Virginia
Academic Fields of Recruitment Interest: Computer science, information systems, electrical engineering, software engineering, systems analysis
Major Entry-Level Opportunities for New Graduates: Software engineering, database development, systems integration, business process redesign, information engineering, logistics

BACKGROUND AND OPERATIONS

TRW Systems & Information Technology Group is the new pacesetting organization formed by the fusion of BDM International, Inc., and TRW Systems Integration Group. The acquisition was a strategic move to gain greater mass and synergistic strengths in key market sectors.

Today, 20,000 people worldwide define the spirit and strength of TRW Systems & Information Technology Group. Generating $3 billion per year, it is the tenth-largest global IT enterprise and the fifth-largest federal systems integrator. As part of the $11-billion TRW family, employees have also made vital contributions to the automotive, space, and defense sectors.

At TRW, employees can launch a successful future and enjoy professional opportunities and growth potential backed by world-class resources. Entry-level positions are diverse and span the life cycle of systems and software engineering as well as other technologies for a wide range of customers.

TRW Systems & Information Technology Group serves five high-growth market sectors: defense, intelligence, public sector, commercial, and energy. The company provides a full spectrum of services and solutions, ranging from consulting and analysis to large-scale systems and software development and integration.

Defense

Since the 1950s, TRW has supported all facets of national and international defense. Today, employees can contribute to programs that focus on strategic and tactical systems engineering and integration, logistics and training, test and evaluation, and more.

Intelligence

Supporting the intelligence community is a top priority at TRW. Employee contributions could advance communications and network engineering, signal and image intelligence, reconnaissance and surveillance systems, and other areas.

Public Sector

TRW works with civil government customers—federal, state, and local—to foster important innovations in public safety and integrated justice information systems, intelligent transportation, welfare and human services, and education technologies.

Commercial

TRW employees help meet business goals in industries ranging from automotive and aerospace to pharmaceutical and telecommunications. They can help solve Year-2000 problems, strengthen the supply chain, or capitalize on the Internet.

Energy and Environmental Systems

Contributing to energy systems epitomizes the potential to do work that counts at TRW. Efforts under way address critical challenges such as information management, nonproliferation, nuclear-waste management, and pollution prevention.

EMPLOYMENT OPPORTUNITIES

ENTRY-LEVEL

There are many ways to begin a career at TRW. Development opportunities feature business and technical training opportunities and assignments tailored to an individual's goals and skills. Additional educational programs are available through the company, including mentoring partnerships, software engineering certification, and much more.

Most opportunities require a minimum of a bachelor's or

master's degree in a related technical discipline. Candidates with hands-on experience, including mainframes, client/server, C/C++, OO, Visual Basic, Oracle, PL/SQL, Ada, Java, UNIX, Windows NT, MVS/VMS, VAX, Rational Apex, TCP/IP, UDP, and ISO 9001, find ample opportunities to apply their skills to exciting, real-world challenges.

Beyond the right technical and educational backgrounds, TRW looks for individuals who demonstrate integrity, flexibility, personal initiative, and commitment to the team. With operations from southern California to northern Virginia, employee options are wide open and could include overseas assignments.

THE ORGANIZATION AS EMPLOYER

It's an exciting time to be part of TRW Systems & Information Technology Group. As it capitalizes upon the growth potential in each of the markets it serves, TRW continues to create new career opportunities—at all levels—across a diverse business. Maximizing these opportunities is a shared responsibility between the company and the individual. As professionals gain experience and recognition, TRW strives to provide the training, support, and challenges to realize each individual's potential.

Along with outstanding career challenges, TRW provides a comprehensive benefits package that includes educational assistance, a 401K plan, insurance, and much more.

INTERNSHIPS AND CO-OPS

TRW offers a paid internship/co-op program that is directed toward students majoring in related IT disciplines as well as finance, accounting, and human resources.

APPLICATION AND INFORMATION

It's easy to take the first step to a career at TRW Systems & Information Technology Group. Applicants can interview on campus, apply on line through TRW Web sites (www.trw.com or www.bdm.com) where they'll find specific details about TRW capabilities and career opportunities, or contact the college relations office directly.

College Relations
1501 BDM Way
McLean, Virginia 22102
Fax: 703-848-5006

U.S. Air Force Officer Training School

Description of Organization: Provides the world's most respected air and space force

Number of Employees: Approximately 370,000

Headquarters: Air Force Recruiting Service, Randolph Air Force Base, Texas

Academic Fields of Recruitment Interest: Computer science, aeronautical engineering, software engineering, electrical engineering, meteorology, mathematics, physics, chemistry, biology

BACKGROUND AND OPERATIONS

Leaders have honor, integrity, and vision. An Air Force officer is a leader. Air Force officers are committed to serving their country. Their leadership keeps the Air Force at the forefront of technology and America secure. The Air Force, however, is more than a job. It's a lifestyle that includes rewarding work and leisure activities.

Air Force Officer Training School (OTS) is small and highly specialized. Candidates selected to attend OTS are college graduates and already have an academic foundation. OTS prepares young college graduates for leadership roles in the twenty-first century. The school's motto, "Always with Honor," reflects the ethical and professional standards expected of Air Force officers.

Those eligible for OTS are United States citizens, 18–29, in good health, and able to pass a physical exam. A candidate must also be a graduate of an accredited college or university, have excellent moral character, and score well on the Air Force Officer Qualifying Test. To enter a technical or nontechnical career, candidates must be commissioned before age 35. Pilot or navigator training applicants must pass a flight physical and enter training before age 27½.

Candidates attend OTS at Maxwell Air Force Base in Montgomery, Alabama, in a fast-paced, 13½-week course of study designed to prepare them for the responsibilities of officership. The curriculum includes communication skills, leadership, management, military history, Air Force customs and courtesies, world affairs, air and space power doctrine, and more. Physical conditioning and sports are also part of the curriculum to develop confidence, leadership, and teamwork abilities.

Near the end of OTS training, candidates begin to feel the excitement of graduation—the dining out, a commissioning ceremony, and a graduation parade. Family and friends are encouraged to attend these events. Upon completion of OTS, graduates are commissioned as Air Force second lieutenants.

EMPLOYMENT OPPORTUNITIES

With a bachelor's degree, candidates may apply for a variety of career specialties, including budget, supply management, food service, personnel, aircraft maintenance, and missile operations.

Some career specialties require a technical degree. For example, Information Systems Officer candidates need a degree in computers, meteorology, or mathematics. Many career fields require degrees in physics, chemistry, or biology or in one of the engineering sciences, such as nuclear, aeronautical, astronautical, aerospace, or electrical.

Software Engineers develop and manage communication–computer system software, modify and maintain existing software, and implement software engineering disciplines into Air Force software systems.

Aeronautical Engineers design and direct the development of military aircraft, missiles, and spacecraft and plan and conduct research on aircraft guidance, propulsion, and weapons systems.

Electrical Engineers are involved in projects ranging from developing and maintaining satellite communications systems to working with civil engineers to solve power distribution problems.

Meteorologists manage weather-related activities for everything from a routine training flight to a space launch and assist Air Force operations by analyzing meteorological charts, preparing and issuing forecasts and advisories, and briefing key personnel on weather conditions.

Navigators are an integral part of the flying mission. They master on-board navigation computers, predict the weather, locate ground and aerial targets, operate offensive and defensive radar systems, plot technical maneuvers, and conduct electronic countermeasures. They share responsibility with the pilots for the mission's success.

Pilots receive the best, most intense flying training in the world in aircraft that are second to none. After training, pilots are able to fly an F-15 Eagle challenging an unidentified aircraft or command a C-5 jet transport, flying supplies, equipment, and people wherever they're needed.

TRAINING

Pilot Training Program. Pilot training selectees without a private pilot's license attend the Flight Screening Program before going to OTS. This 5½-week course identifies potential to complete undergraduate pilot training. Training includes aerobatic flight in a single-engine Slingsby Firefly, also known as the T-3A. A solo flight is required to complete the Flight Screening Program.

Navigator Training Program. Navigator trainees attend OTS first, then go to undergraduate navigator training. This course includes air navigation, map reading, celestial navigation, weather, and flight instruments.

THE ORGANIZATION AS EMPLOYER

Air Force standards are high. They must be. Today's young officers are being trained to be the leaders and managers of tomorrow. As they prove themselves, they are given more responsibility. They also enjoy many advantages. An excellent salary, nontaxable allowances for food and housing, and complete medical and dental care are just a few.

Once accepted for OTS, candidates enlist in the Air Force in the rank of staff sergeant. The Air Force pays the cost of OTS, and trainees are paid more than $1000 monthly. They live on base and eat in the OTS dining hall. They pay only for personal items, such as laundry, postage, and telephone calls. A clothing allowance helps defray the cost of uniforms.

Upon graduation, candidates are commissioned as second lieutenants. Nonflying officers incur a service commitment of four years. OTS graduates who attend undergraduate pilot or navigator training incur an additional Air Force commitment.

APPLICATION AND INFORMATION

No matter what you do in life AIM HIGH! If you have an engineering or scientific degree, you may qualify for Air Force Officer Training School. For more information, contact the nearest Air Force recruiter or call 800-423-USAF. (World Wide Web: http://www.airforce.com)

U.S. Army Corps of Engineers

Department of the Army

Description of Organization: The largest public engineering organization in the free world
Number of Employees: 38,000
Headquarters Location: Washington, DC
Academic Fields of Recruitment Interest: Engineering, science, computer science
Major Entry-Level Opportunities for New Graduates: Planning, design, and construction management; operation and maintenance of water resource projects

BACKGROUND AND OPERATIONS

Since the era of George Washington and Thomas Jefferson, the Corps of Engineers has played a vital role in both military engineering and meeting America's civilian infrastructure needs. In the nineteenth century, the Corps explored and mapped the frontier, opened waterways and harbors, built fortifications and public buildings, and undertook protection of the areas that became Yellowstone and other national parks. In the twentieth century, the Corps built the Army and Air Force facilities that saw the nation through five wars and one long cold war, constructed civil water resource projects, and performed work for others, ranging from space shuttle launch sites to wastewater treatment plants.

The Corps operates four major research and development centers that enable it to stay in the forefront of technology. Research at these labs includes coastal engineering, flood control, navigation, environmental quality, water resource planning, surveying and remote sensing, topographic sciences, pollution prevention, and magnetic levitation.

The Corps also serves the nation as one of the leading federal agencies to bring relief to victims of natural disasters such as floods, hurricanes, and earthquakes. The Corps teams provide valuable damage assessment support as well as support for the restoration of water and power to local communities, debris removal, and delivery of potable water and ice. The Corps is also involved in clean-up operations in oil spill disasters in its geographic jurisdictions.

On the international scene, the Corps plays a vital role in supporting the U.S. troops on site during military operations after hostilities to help restore the infrastructure.

The Corps of Engineers is responsible for the stewardship of 11.7 million acres of federal land and water located at more than 460 water resources projects in forty-three states. As the premier engineering organization of the Army and a unique national asset, the Corps applies its capabilities toward ensuring defense preparedness, carefully developing water resources, and preserving the natural environment.

FUTURE DIRECTIONS

The Corps of Engineers has the capability to provide essential planning, engineering, and construction support to meet the nation's needs, both military and civil, of the next century. While some of the Corps's traditional missions, such as military construction, are shrinking, others, such as support for other agencies, are increasing. The Corps has the ability, if called upon to serve, to render assistance to the world's developing nations and transfer essential engineering skills to enhance democratic institutions. The Corps also has the ability to help the United States with its decaying infrastructure problems and help ensure development that is environmentally sustainable for future generations. The Corps, as it has in the past, looks to the future for its engineering challenges and opportunities.

EMPLOYMENT OPPORTUNITIES

ENTRY LEVEL

The Corps of Engineers seeks new members with the broadest possible range of disciplinary backgrounds in support of its diverse missions and operations worldwide. Entry-level opportunities in engineering, science, and computer science are available in most Corps locations. In addition, increased salary rates are offered for certain occupations and geographic locations. Candidates who have completed a full four-year curriculum at an accredited college or university leading to a bachelor's degree are qualified for appointment at the GS-5 grade level or, under certain circumstances, at the GS-7 level. Candidates with a master's degree or a doctorate may be qualified for an entry-level appointment at the GS-9 level or higher.

FOR EXPERIENCED PERSONNEL

The Corps typically seeks individuals who have a combination of education and work-related experience in fields that encompass water resource planning and management; engineering investigations and planning; engineering and architectural design; electronic data processing; information science and technology; research and development; real estate acquisition, management, and disposal; financial management; environmental protection; and natural resource management.

TRAINING

In addition to intensive on-the-job training under senior professionals, the Corps provides tuition reimbursement for job-related courses, programs for professional development, and other advanced study opportunities.

THE ORGANIZATION AS EMPLOYER

Because of the diverse responsibilities of the Corps, few careers can offer such a range of opportunities and involvement in both the national and international arenas. In addition to the challenge, the Corps, as part of the executive branch of the government, offers excellent personal benefits, including paid vacation time in the amount of thirteen to twenty-six days per year, based on length of service; sick leave, which is accrued at thirteen days per year; and ten paid federal holidays a year. Federal employees are also covered by an excellent retirement plan, which is transferable and includes tax-deferred savings alternatives; a life insurance program offering variable options; and a choice of a variety of health insurance plans.

STUDENT EMPLOYMENT PROGRAMS

The Corps of Engineers offers a wide range of employment opportunities for students under the Federal-wide Student Educational Employment Program. Students can be employed either full- or part-time on a yearly or seasonal basis. Cooperative education and other career experience positions related to a student's academic major are available, which can lead to permanent appointments with the Corps upon graduation. Also offered are extensive summer employment opportunities in a variety of occupations.

APPLICATION AND INFORMATION

Interested individuals may obtain more information about career opportunities with the U.S. Army Corps of Engineers by writing to, calling, or visiting the Corps Human Resource Office in the geographic area where they are seeking employment. Prospective employees also visit the Corps' home page on the World Wide Web (http://www. usace.army.mil.). Also, where applicable, the Corps Human Resource Office may be accessed directly through the Internet, or inquiries about job openings may be made through the Job Information Line. After contacting the Corps Human Resource Office, applicants may be directed to contact the Army Civilian Personnel Operations Center that provides service to civilian personnel in that geographic area; they will further assist in the application processing procedures.

NORTH ATLANTIC DIVISION, 90 Church St., New York, NY 10007-2979. 212-264-0200

Baltimore: PO Box 1715, Baltimore, MD 21201-1715. 410-962-2087, E-mail: wrichardson@cenadl.nad.usace.army.mil

Philadelphia: Wanamaker Building, 100 Penn Square East, Philadelphia, PA 19107-3390. 215-656-6870, E-mail: cenap-hr@usace.army.mil

New York: 26 Federal Plaza, Room 1923, New York, NY 10278-6090. 212-264-0200

Norfolk: 803 Front St., Norfolk, VA 23510-1096. 804-441-7636

Army Civilian Personnel Operations Center: Civilian Personnel Operations Center, Aberdeen, MD 21005

NORTH CENTRAL DIVISION, 111 N. Canal St., 14th Floor, Chicago, IL 60606-7205. 312-353-6330

Detroit: 477 Michigan Avenue, Detroit, MI 48231-1027. 313-226-6427

St. Paul: Army Corps of Engineers Center, 190 Fifth St. East, Saint Paul, MN 55101-1638. 612-290-5480

Buffalo: 1776 Niagara St., Buffalo, NY 14207-3199. 716-879-4137

Rock Island: Clock Tower Building, PO Box 2004, Rock Island, IL 61204-2004. 309-794-5152

OHIO RIVER DIVISION, Federal Office Building, 550 Main St., Room 10526, Cincinnati, OH 45202. 513-684-3066, E-mail: ceorl-hrcn-s@smtp.ord.usace.army.mil

Louisville: 600 Martin Luther King Place, Louisville, KY 40201-2230. 502-582-5886

Pittsburgh: William S. Morehead Federal Building, 1000 Liberty Ave., Room 2103, Pittsburgh, PA 15222-4186. 412-644-6839

Huntington: 502 Eighth St., Huntington, WV 25701-2070. 304-529-5662

Nashville: 801 Broadway, Nashville, TN 37202-1070. 615-736-5654

MISSOURI RIVER DIVISION, 12565 West Center Rd., Omaha, NE 68144-3869. 402-697-2408

Omaha: 215 North 17th St., Omaha, NE 68102-9959. 402-221-4071, E-mail: cemro-hr@mro01.usace.army.mil

Kansas City: 601 E. 12th St., 700 Federal Bldg., Room 609, Kansas City, MO 64106-2896. 816-426-3654

SOUTH ATLANTIC DIVISION, 77 Forsyth St., SW, Atlanta, GA 30335-6801. 404-331-6683

Jacksonville: 400 W. Bay St., Room 888, Jacksonville, FL 32232-0019. 904-232-3741

Mobile: 109 St. Joseph St., Mobile, AL 36628-0001. 334-690-2524

Savannah/Charleston: 100 W. Oglethorpe Ave. Savannah, GA 31402-0889. 912-652-5413

Wilmington: PO Box 1890, 69 Darlington Ave. Wilmington, NC 28402-1890. 910-251-4871, 910-251-4876 (job information line)

Army Civilian Personnel Operations Center: Southeast Civilian Personnel Operations Center, Fort Benning, GA 31905

LOWER MISSISSIPPI VALLEY DIVISION, 820 Crawford St., Room 300, Vicksburg, MS 39181-0080. 601-634-5158

New Orleans: PO Box 60267, New Orleans, LA 70160-0026. 504-862-2800, E-mail: smtp:"lopezvan@smtp.lmn.usace.army.mil

Memphis: 167 North Main St., B202, Memphis, TN 38103-1894. 901-544-3105

Vicksburg: 820 Crawford St., Room 300, Vicksburg, MS 39181-0080. 601-634-5158

St. Louis: 1222 Spruce St., Saint Louis, MO 63103-2833. 314-331-8549, E-mail: celms-hr@smtp.lms.usace.army.mil

SOUTHWESTERN DIVISION, Division Human Resource Office, 819 Taylor St., Fort Worth, TX 76102-0300. 817-334-2208, 800-453-8907 (toll-free) (Recruitment for Albuquerque, NM, and Fort Worth and Galveston, TX)

Tulsa: 1645 South 101st East Ave., Tulsa, OK 74128-0061. 918-669-7346

Little Rock: 700 West Capitol Ave., Little Rock, AR 72203-0867. 501-324-5659

NORTH PACIFIC DIVISION, 220 NW 8th Ave., Portland, OR 97209. 503-326-3786

Portland: 333 SW First Ave., Portland, OR 97208-2946. 503-326-6976

Walla Walla: Human Resource Office, 201 North 3rd Ave., Walla Walla, WA 99362-1876. 509-527-7026

Seattle: 4735 East Marginal Way, South, Seattle, WA 98124-2255. 206-764-3416, 206-764-3739 (job information line), Fax: 206-764-6685

Alaska: PO Box 898, Anchorage, AK 99506-0898. 907-753-2824, E-mail: cenpa-hr@npa01.usace.army.mil, World Wide Web: http://www.usace.army.mil/alaska/hr

SOUTH PACIFIC DIVISION, 630 Sansome St., Room 952, San Francisco, CA 94111-2206; 415-705-1627.

Human Resource Division Office, 1325 J St., Suite 1450, Sacramento, CA 95814-2922. 916-557-5310 (job information line) (Recruitment for Sacramento, Los Angeles, and San Francisco, CA)

U.S. ARMY ENGINEERING AND SUPPORT CENTER, HUNTSVILLE, PO Box 1600, Huntsville, AL 35805-1957. 205-895-1230

U.S. ARMY ENGINEER, NEW ENGLAND, 424 Trapelo Rd., Waltham, MA 02254-9149. 617-647-8561

U.S. ARMY ENGINEER, TRANSATLANTIC PROGRAMS CENTER, PO Box 2250, Winchester, VA 22604-1450. 540-665-3731 (Recruitment for Europe, Middle East, and Africa), E-mail: cetac-hr@smtp.tad.usace.army.mil

Army Civilian Personnel Operations Center, Fort Belvoir, 5701 21st St., Suite 10, Fort Belvoir, VA 22060-5919

U.S. ARMY ENGINEER, PACIFIC OCEAN, Building 230, Ft. Shafter, HI 96858-5440. 808-438-1098 (Recruitment for Hawaii, Korea, and Japan)

WASHINGTON, DC METROPOLITAN AREA

U.S. Army Humphrey Engineer Center Support Activity, 20 Massachusetts Ave., NW, Washington, DC 20314-1000. 202-761-1885/1890

Army Civilian Personnel Operations Center, Fort Belvoir, 5701 21st St., Suite 10, Fort Belvoir, VA 22060-5919

CORPS LABORATORIES

U.S. Army Cold Regions Research and Engineering Laboratory, 72 Lyme Rd., Hanover, NH 03755-1290. 603-646-4132

Waterways Experiment Station, 820 Crawford St., Vicksburg, MS 39181-0080. 601-634-5069

U.S. Army Construction Engineering Research Laboratory, 2902 Newmark Drive, Interstate Research Park, Champaign, IL 61821-1076

U.S. Army Topographic Engineering Center, 7701 Telegraph Rd., Building #2592, Fort Belvoir, VA 22310-3864. 703-428-6253/6255, Fax: 703-428-6257

U.S. Nuclear Regulatory Commission

Description of Organization: A federal government agency that regulates the civilian use of nuclear power and nuclear materials

Number of Employees: 3,100

Headquarters Location: Washington, D.C.

Academic Fields of Recruitment Interest: Nuclear, mechanical, chemical, structural, electrical, environmental, materials, and human factors engineering; chemistry; reactor physics; molecular/radiation biology; materials science; health physics; geology; hydrology; seismology; statistics related to nuclear material control and accounting

Major Entry-Level Opportunities for New Graduates: Course work and/or experience with fracture analysis, nondestructive testing, thermal hydraulics, metallurgy, stress corrosion, cracking, computer code development and validation, seismic analysis and effects, geophysics, materials engineering/science, nuclear criticality safety, structural engineering, probabilistic risk assessment, advanced digital microprocessor-based instrumentation and control systems, chemical engineering, physics, and health physics (modeling).

BACKGROUND AND OPERATIONS

The mission of the U.S. Nuclear Regulatory Commission (NRC) is to ensure adequate protection of the public health and safety, the common defense and security, and the environment in the use of nuclear materials in the United States. The NRC's scope of responsibility includes regulation of commercial nuclear power reactors; nonpower research, test, and training reactors; fuel cycle facilities; medical, academic, and industrial uses of nuclear materials; and the transport, storage, and disposal of nuclear materials and waste.

NRC grants licenses for nuclear facilities as well as for the possession, use, and disposal of nuclear materials; establishes the guidelines for the development and implementation of licensed activities; inspects nuclear facilities and materials; and enforces compliance with its strict standards and requirements.

NRC's varied activities are carried out at several program offices located at the Headquarters in Rockville, Maryland. In addition, regional offices are located in Philadelphia, Pennsylvania; Atlanta, Georgia; Chicago, Illinois; and Dallas, Texas.

At the **Office of Nuclear Reactor Regulation** (NRR), NRC employees play a vital role in evaluating all license applications for nuclear reactors, issuing licenses, and regulating site selection, design, construction, operation, maintenance, and decommissioning of facilities.

The charter of the **Office of Nuclear Material Safety and Safeguards** (NMSS) is to protect public health and safety, the national security, and the environment by directing licensing and inspection activities associated with nuclear fuel cycle facilities, uses of nuclear material, safeguarding of nuclear material, management and disposal of low- and high-level radioactive nuclear waste, and decontamination and decommissioning of facilities and sites.

Employees at the **Office of Nuclear Regulatory Research** provide independent expertise and information for making timely regulatory judgements, anticipating problems of potential safety significance, resolving safety issues, and developing technical regulations and standards.

The **Office for Analysis and Evaluation of Operational Data** analyzes and evaluates operational safety data, identifies issues that require NRC or industry action, and provides generic operational safety data and systems to the industry and other agencies.

Regional Offices conduct inspection, enforcement, investigation, licensing, and emergency response programs within regional boundaries that the headquarters offices originate.

EMPLOYMENT OPPORTUNITIES

ENTRY-LEVEL PROGRAMS

Intern Program. Recruiting, training, and retaining high-quality employees is the key to NRC's effectiveness in carrying out its regulatory responsibilities. The Intern Program was established by NRC to attract and develop high-caliber, entry-level technical employees who have the potential to carry out the agency's regulatory health and safety mission. Through on-the-job training, formal classroom training, and individually tailored developmental rotational assignments, recent graduates in engineering, health physics, and other technical disciplines are given a broad exposure to the regulatory work of NRC and gain a broad perspective of the role of the NRC as an independent regulatory agency.

The NRC Intern Program is a two-year program that includes structured technical training, developmental assignments, and optional courses. Since relevant technical knowledge is gained through extensive training and developmental assignments, an education specifically in the nuclear field and experience in the nuclear industry are not prerequisites for the program. Typically, successful applicants are recent graduates with a bachelor's, master's, or doctoral degree in the field of nuclear, mechanical, chemical, structural, electrical, environmental, materials, or human factors engineering; chemistry; reactor physics; molecular/radiation biology; materials science; or health physics.

EXPERIENCED APPLICANTS

Most vacancies are for individuals with engineering and scientific experience as nuclear, mechanical, electrical, computer, environmental, materials, and chemical engineers and health physicists. Of particular interest is inspecting, licensing, operations, and systems experience in the areas of radiation protection, radiation dosimetry, fuel fabrication, and radioactive waste management. Of continuing interest is recent experience with fracture analysis,

"The whole experience—the training, the opportunities, the team approach to problem-solving—my time with the NRC has been the high point of my career. If you're an engineer, this is the place to be."

Sher Bahadur
Chief, Waste Management Branch
NRC since 1984

"I've moved up faster here than I would have anywhere else. NRC sees you not only for what you've done, but for what you're capable of. The career possibilities are varied and exciting."

Deidre Spaulding
Electrical Engineer
NRC since 1990

"NRC offers serious challenge...great people...real rewards. It's an exciting environment to work in because of the range of opportunities and emphasis on world-class quality."

Larry Campbell
Reactor Engineer
NRC since 1990

nondestructive testing, thermal hydraulics, metallurgy, stress corrosion, cracking, computer code development and validation, seismic analysis and effects, geology, hydrology, geophysics, materials engineering and materials science probabilistic risk assessment, and advanced digital microprocessor-based instrumentation and control systems.

NRC anticipates additional staffing needs for criticality safety experts, nuclear process/nuclear engineers, materials/chemical/fire protection engineers, environmental engineers, health physics modelers, physical protection specialists, and nuclear material control and accounting specialists.

TRAINING

NRC provides on-the-job training, supervisory and management training, and specialized training, giving employees unprecedented opportunities to gain expert knowledge, develop their careers, and stay a step ahead of changing technology and industry information.

Courses that provide education in a variety of areas are offered on-site or at local colleges and universities. NRC may pay tuition and fees for job-related courses taken at approved colleges and universities at the undergraduate or graduate level.

THE ORGANIZATION AS EMPLOYER

NRC offers competitive salaries and a three-part retirement program that includes Social Security, a basic annuity, and the Thrift Savings Plan. Also offered are generous leave benefits and the option to participate in the Federal Employees Health Benefits Program and Life Insurance Program.

APPLICATION AND INFORMATION

Individuals interested in a career with NRC should send a résumé that includes their name, address, phone numbers, Social Security number, and citizenship status. The résumé should also include the colleges and universities attended, type and year of degrees received, describe paid and nonpaid work experience related to the job for which they are applying, and include other qualifications such as training courses, skills, honors, and awards. Individuals who graduated within the past five years should also include an unofficial copy of their college transcripts. U.S. citizenship is required.

Direct inquiries to:

U.S. Nuclear Regulatory Commission
Office of Human Resources
Mail Stop O-3E17A (Dept. A-98112)
Washington, D.C. 20555
For the current status of NRC job opportunities, applicants should call: 800-952-9678 (toll-free).

U.S. Patent and Trademark Office

Department of Commerce

Description of Organization: The U.S. Patent and Trademark Office (PTO) is a fully fee-funded agency of the U.S. Department of Commerce. The role of the PTO is to grant patents for the protection of inventions, to register trademarks, and to disseminate information about intellectual property.

Number of Employees: 5,500

Headquarters Location: Arlington, Virginia

Academic Fields of Recruitment Interest: Applicants from all engineering and computer-related disciplines are considered. Career opportunities for B.S., M.S., and Ph.D. degree graduates are available for scientists such as physicists, chemists, biotechnologists, biologists, and microbiologists, and those in related sciences.

BACKGROUND AND OPERATIONS

The Department of Commerce (DOC) promotes job creation, economic growth, sustainable development, and improved living standards for all Americans by working in partnership with businesses, universities, communities, and workers to build for the future and promote U.S. competitiveness in the global marketplace; by strengthening and safeguarding the nation's economic infrastructure; by keeping America competitive with cutting-edge science and technology and an unrivaled information base; and by providing effective management and stewardship of the nation's resources and assets to ensure sustainable economic opportunities.

For more than 200 years, the basic role of the Patent and Trademark Office has remained the same: to promote the progress of science and the useful arts by securing for limited times to inventors the exclusive right to their respective discoveries (Article 1, Section 8 of the United States Constitution). During this time, American industry has flourished. New products have been invented, new uses for old ones discovered, and employment opportunities created for millions of Americans.

The PTO is a noncommercial federal entity and one of fourteen bureaus in the Department of Commerce. The office occupies a combined total of more than 1.4 million square feet in numerous buildings in Arlington, Virginia. The office employs more than 5,000 full-time-equivalent (FTE) staff members to support its major functions—the examination and issuance of patents and the examination and registration of trademarks.

The PTO has evolved into a unique government agency. Since 1991—under the Omnibus Budget Reconciliation Act (OBRA) of 1990—the PTO has operated in much the same way as a private business, providing valued products and services to customers in exchange for fees, which are used to fully fund operations. The primary services provided include processing patents and trademarks and disseminating patent and trademark information.

Through the issuance of patents, technological advancement is encouraged by providing incentives to invent, invest in, and disclose new technology worldwide. Through the registration of trademarks, the PTO assists businesses in protecting their investments, promoting goods and services, and safeguarding consumers against confusion and deception in the marketplace. By disseminating both patent and trademark information, the PTO promotes an understanding of intellectual property protection and facilitates the development and sharing of new technologies worldwide.

The Patent and Trademark Office determines the patentability of discoveries of inventors throughout the world. This determination is made by approximately 2,500 highly trained scientists and engineers (patent examiners). An examiner analyzes the subject matter of a patent application and the pertinent prior art, i.e., patents and other published technical materials, and determines whether the claimed invention is patentable. He or she applies procedural and substantive law and grants or rejects the claims of the application. Generally, examiners specialize in one of the engineering or scientific fields described below.

Intellectual property—the products of the human mind—is the world's only inexhaustible resource and a growing part of America's economy. The Patent and Trademark Office helps protect this vital resource by granting patents and registering trademarks. Nearly 6 million patents have been granted since the patent system was established in 1790. A patent protects the inventor's exclusive right to profit from their invention. Today, this protection lasts for twenty years from the date of application. Intellectual property protection allows firms to maintain a leading edge in global markets and encourages businesses to risk investment for research, development, and marketing.

The mission of the PTO is to promote industrial and technological progress in the United States and strengthen the national economy by administering the laws relating to patents and trademarks; advising the Secretary of Commerce, the President of the United States, and the administration on patent, trademark, and copyright protection; and advising the

Secretary of Commerce, the President of the United States, and the administration on the trade-related aspects of intellectual property.

EMPLOYMENT OPPORTUNITIES

FOR EXPERIENCED PERSONNEL

Immediate patent examiner positions are available in the following specialties: networks, image processing, speech processing, VLSI, artificial intelligence, software, television and display technology, electrical engineering, biotechnology and organic chemistry, chemical and material engineering, transportation, construction and agriculture, power engineering, petroleum engineering, mechanical engineering/products, communications/information process, physics, optics, and system components.

Electrical and Electronics Engineers typically judge the patentability of inventions in all types of computer hardware, telecommunications equipment, photography, illumination, transmission systems, information transmission storage and retrieval, radio, television, data processing systems, electrical switches, semiconductor devices, radiant energy, and optics.

Mechanical Engineers review the patentability of inventions in areas such as conveyors, aeronautics, motor vehicles, tools, surgery, power plants, internal-combustion engines, and heat generation.

Chemical Engineers examine the patentability of inventions in areas such as petroleum chemistry, metallurgy, processing of plastics, fuels, medicine, natural resins, photography, foods, microbiology, and phase separation.

Advancement opportunities are excellent. Promotions are based on individual ability and work performance. Generally, individuals are considered for promotion after a minimum of one year at the preceding level. Patent examiners in the engineering discipline at the GS-5 or GS-7 level are eligible for promotion after six months of service. Promotion to a supervisory position is competitive. This career ladder is one of the best in the federal government, and salaries are competitive with private industry.

TRAINING

The PTO trains employees to apply their technical and analytical skills in the patent examining process. Classroom and on-the-job training enhances skills and provides employees with a knowledge of the legal analysis that is needed. On-the-job training needs are determined with the supervisor and may include individual hands-on experience to facilitate understanding of the complexities of patent examining.

The Patent Academy provides formal training, beginning with an initial two-week program to provide the examiner with the basics. Several additional courses are provided by the Patent Academy during the first year to cover all advanced examining functions and legal concepts.

THE ORGANIZATION AS EMPLOYER

Benefits at the PTO compare to those in top-ranked corporations in the United States. Benefits include thirteen days of annual vacation leave the first three years, twenty days for the next twelve years, and twenty-six days after fifteen years; ten paid holidays each year; thirteen days of sick leave each year, which accrue with no maximum limit; a variety of work schedule options; paid overtime; performance awards and productivity gainsharing; an excellent retirement system/package, including a Thrift Saving Plan (TSP) that allows tax-deferred contributions of up to 10 percent of an employee's salary, with a matching contribution of 5 percent by PTO; Family Friendly Leave Act and Family Medical Leave Act, which allow leave when necessary for family problems; outstanding health benefits, with a large choice of providers; term life insurance options; two weeks of military leave for all armed forces reservists; court leave for jury duty; and coverage for on-the-job injury.

APPLICATION AND INFORMATION

To apply, potential employees should visit the Web site at http://www.uspto.gov or mail a resume and transcript to the address below.

Office of Human Resources
U.S. Patent and Trademark Office
Box 171
Washington, D.C. 20231
Telephone: 800-642-5670
Fax: 703-305-9864

Western Atlas International

WESTERN ATLAS

Description of Organization: World leader in petroleum exploration, field development, and production services
Number of Employees: 9,000 worldwide
Headquarters Location: Houston, Texas
Academic Fields of Recruitment Interest: Petroleum, electrical, and mechanical engineering; computer science; mathematics; physics; chemistry; geophysics; geology
Major Entry-Level Opportunities for New Graduates: Geophysical and geological data acquisition, data processing, analytical services, computer programming, engineering design, research and development

BACKGROUND AND OPERATIONS

Western Atlas International is a leading supplier of information technology services for the oil and gas industry. The services provided by Western Atlas extend from evaluating the hydrocarbon potential of basins, to identifying productive zones and optimizing hydrocarbon recovery through cost-effective reservoir description and management. The three major divisions of Western Atlas International are recognized leaders in the upstream petroleum services industry.

Western Geophysical is a leading geophysical contractor, conducting seismic surveys on land, offshore, and in shallow-water regions and providing data processing services worldwide. Western has helped discover major petroleum fields in frontier provinces and is helping to enhance recovery in mature basins.

Western Atlas Logging Services provides a complete range of downhole petrophysical and geophysical data acquisition, processing, and analysis services for every phase of drilling and production. Logging units, instrumentation, and data analysis systems are marketed through the New Ventures/Export Sales group.

E&P Services provides value-added interpretation and reservoir management services. These services include geological, geophysical, and engineering applications in the evaluation of oil and gas potential of client properties and reservoirs.

Western's research scientists are involved in a combination of basic research and practical problem solving.

EMPLOYMENT OPPORTUNITIES

ENTRY LEVEL

Degreed individuals who are free to travel are sought for positions in field services. Positions are available for both land and marine operations.

Geophysics, Geology, Math, Physics, and Chemistry Majors. Opportunities exist for field engineers, processing technicians, and analysts. These are normally considered entry-level positions for degreed applicants in geophysics, geology, math, physics, or chemistry. Individuals selected for these positions are trained in computer processing, field operations, or laboratory procedures as required.

Electrical and Mechanical Engineers and Technicians. Technicians holding two-year and four-year diplomas and B.S.E.E. or B.S.M.E. graduates are sought for positions in design, manufacturing, and field operations. Many positions require freedom to travel and/or relocation.

Computer Scientists. Opportunities exist in all programming areas to develop software for mainframe systems, workstations, and PCs.

FOR FOREIGN NATIONALS

Because of the worldwide nature of Western Atlas International operations, U.S. citizenship or permanent visa status may be required.

TRAINING

On-the-job training is supported by classroom lectures and seminars in certain technical areas. A wide selection of videotapes is available for employees to increase their knowledge in their chosen fields.

THE ORGANIZATION AS EMPLOYER

Western Atlas offers a comprehensive benefits program that can be structured to meet individual needs. It includes life, dental, and medical insurance; credit union; educational reimbursement; paid holidays and vacations; and a profit-sharing plan.

CORPORATE SOCIAL RESPONSIBILITY

Western Atlas has full-fledged health and safety departments and is actively involved in environmental preservation and recycling programs. The company is a major participant in United Way and Blood Bank programs.

APPLICATION AND INFORMATION

Western Atlas International recruits college students in the United States and many other countries. Qualified candidates are encouraged to contact their campus recruitment office. If this is not possible, applicants are encouraged to submit a resume for review and contact by the company.

Direct inquiries to:

R. J. Mason
Manager, Industrial Relations
Western Atlas International
P.O. Box 1407
Houston, Texas 77251-1407
Telephone: 713-266-5700
Fax: 713-781-0945
World Wide Web: http://www.waii.com

INDEX BY INDUSTRY

Companies are indexed under the following categories in this section.

INDEX BY INDUSTRY

This index classifies companies according to the primary industries in which they operate.

Computer Systems and Peripherals

Conservation Agencies and Organizations

Construction and Contractors

Consumer and Household Products

Medical Products and Services

Mining and Metals

Nursing Homes/Convalescent Care

Office Equipment

Oil, Gas, and Petroleum Products

Optical

Packaging

Paper and Related Products

Stone, Clay, and Glass Products

Telecommunications

Textiles and Clothing

Tobacco Products

Transportation and Shipping

Travel

Utilities

Waste Management

Waste Treatment and Purification

Wholesale Trade and Distribution

ALPHABETICAL INDEX

INDEX OF HIRING NEEDS

An index of profiled companies by areas of expertise sought.

Accounting

Account Management

Acoustic Engineering

Actuarial

Adhesives

Administration

Administration Supervision

Advanced Degreed Mathematicians

Advertising

Aeronautical/Astronautical Engineering

Aeronautical Engineering

Aerospace Engineering

Aerospace Science

Agricultural Economics

Agricultural Engineering

Agricultural Science

Agriculture

Allied Health

Analog Engineering

Analysis

Analytical Chemistry

Application Developer

Applications Engineering

Applied Engineering

Architectural Engineering

Architecture

Chemical Safety

Chemistry

Circuit Design

Civil Engineering

Computer Science

Construction

Construction Engineering

Construction Management

Consulting

Data Warehousing

Defense Systems

Demolition

Dentistry/Dental Hygiene

Design

Design Engineering

Electrical Technology

Electronic Design and Development

Electronics/Electronics Engineering

Engineering Consulting

Engineering Management

Engineering Mechanics

Engineering Science

Engineering Technology

Financial Analysis

Fire Protection

Firmware Engineering

Flight Attendant

Food Chemistry

Food Engineering

Food Science

Food Service Management

Food Services

Forestry

Foundry Engineering

Franchising

Fuel Engineering

General Management

General Nursing

Geography

Geologist

Geology

Geophysics

Geoscience

Liberal Arts

Library/Information Science

Life Sciences

Logistics

Machinery

Maintenance

Management

Management Consulting

Management Information Systems

Manufacturing

Manufacturing Engineering

Market Research

Materials Engineering

Materials Management

Materials Science

Material Testing

Mathematics

Mechanical Engineering

Medical Technology

Merchandising

NOTES